ION IMPLANTATION IN SEMICONDUCTORS
Science and Technology

ION IMPLANTATION IN SEMICONDUCTORS
Science and Technology

Edited by
Susumu Namba
Faculty of Engineering Science
Osaka University
Toyonaka, Osaka, Japan

Springer Science+Business Media, LLC

Library of Congress Cataloging in Publication Data

International Conference on Ion Implantation in Semiconductors and Other Materials, 4th, Osaka, 1974.
   Ion implantation in semiconductors.

   Includes bibliographical references and index.
   1. Ion implantation—Congresses. 2. Semiconductors, Effect of radiation on—Congresses. 3. Metals, Effect of radiation on—Congresses. I. Namba, Susumu, 1928-   II. Title.
TK7871.85.I5762 1974   621.3815'2'028   75-8985
ISBN 978-1-4684-2153-8   ISBN 978-1-4684-2151-4 (eBook)
DOI 10.1007/978-1-4684-2151-4

Proceedings of the Fourth International Conference on Ion Implantation in Semiconductors and Other Materials held at the Osaka Chamber of Commerce and Industry, August, 1974

© Springer Science+Business Media New York 1975
Originally published by Plenum Press, New York 1975
Softcover reprint of the hardcover 1st edition 1975

United Kingdom edition published by Plenum Press, London
A Division of Plenum Publishing Company, Ltd.
Davis House (4th Floor), 8 Scrubs Lane, Harlesden, London, NW10 6SE, England

All rights reserved

No part of this book may be reproduced, stored in a retrieval system, or transmitted, in any form or by any means, electronic, mechanical, photocopying, microfilming, recording, or otherwise, without written permission from the Publisher

Preface

The technique of ion implantation has become a very useful and stable technique in the field of semiconductor device fabrication. This use of ion implantation is being adopted by industry. Another important application is the fundamental study of the physical properties of materials.

The First Conference on Ion Implantation in Semiconductors was held at Thousand Oaks, California in 1970. The second conference in this series was held at Garmish-Partenkirchen, Germany, in 1971. At the third conference, which convened at Yorktown Heights, New York in 1973, the emphasis was broadened to include metals and insulators as well as semiconductors. This scope of the conference was still accepted at the fourth conference which was held at Osaka, Japan, in 1974.

A huge number of papers had been submitted to this conference. All papers which were presented at the Fourth International Conference on Ion Implantation in Semiconductors and Other Materials are included in this proceedings. The success of this conference was due to technical presentations and discussions of 224 participants from 14 countries as well as to financial support from many companies in Japan. On behalf of the committee, I wish to thank the authors for their excellent papers and the sponsors for their financial support.

The International Committee responsible for advising this conference consisted of B.L. Crowder, J.A. Davies, G. Dearnaley, F.H. Eisen, Ph. Glotin, T. Itoh, A.U. MacRae, J.W. Mayer, S. Namba, I. Ruge, and F.L. Vook.

The Committee responsible for organizing this conference consisted of S. Namba (Chairman), F. Fujimoto, M. Fujimoto, S. Furukawa, N. Itoh, T. Itoh, K. Masuda, T. Takagi, Y. Tarui, T. Tokuyama, M. Uenohara, and K. Yata.

On behalf of all participants of this conference, I should like to thank the above persons for their help in making the conference successful.

August 1974
Osaka, Japan

Susumu Namba
Conference Chairman

Contents

III-V COMPOUND SEMICONDUCTORS I

Ion Implantation in III-V Compounds 3
  F.H. Eisen

Anodic Oxidation and Profile Determination of
Ion Implanted Semi-insulating GaAs 19
  H. Müller, J. Gyulai, J.W. Mayer, F.H. Eisen and
  B. Welch

Encapsulation of Ion Implanted GaAs 27
  P.L.F. Hemment, B.J. Sealy and K.G. Stephens

Ion Implantation of Cd and Te in GaAs Crystals 35
  K. Gamo, M. Takai, M.S. Lin, K. Masuda and S. Namba

Implantation of Silicon into Gallium Arsenide 41
  T. Miyazaki and M. Tamura

The Effects of Ion Dose and Implantation Temperature
on Enhanced Diffusion in Selenium Ion-Implanted
Gallium Arsenide 47
  Y. Kushiro and T. Kobayashi

III-V COMPOUND SEMICONDUCTORS II

The Effects of Ion-Implanted Ga, As, and P on the
Subsequent Diffusion of Ion-Implanted Zn in
$GaAs_{0.6}P_{0.4}$ 57
  E.B. Stoneham and J.F. Gibbons

Compensating Layers in GaAs by Ion Implantation:
Application to Integrated Optics 65
  P.N. Favennec and E.V.K. Rao

Electrical Properties of Proton Bombarded N-type GaAs . . 73
   H. Harada and M. Fujimoto

Photoluminescence of Zinc Implanted n-type GaAs 83
   T. Itoh, T. Matsumoto and J. Kasahara

Large Increase of Emission Efficiency in Indirect
GaAsP by N-Ion Implantation 89
   Y. Makita, S. Gonda, H. Tanoue, T. Tsurushima and
   S. Maekawa

Electrical and Photoluminescent Properties of
Zinc Implanted $GaAs_{0.62}P_{0.38}$ 95
   H. Okabayashi

Defect-Free Nitrogen Implantation into GaP 101
   T. Shimada, Y. Shiraki and K.F. Komatsubara

Ion Implantations of Mg and Zn into n-type GaP 107
   T. Inada and Y. Ohnuki

In-Depth Profile Detection Limits of Nitrogen in
GaP, Nitrogen, Oxygen and Fluorine in Si by
SIMS and AES 115
   J.C.C. Tsai and J.M. Morabito

Behaviors of Ga and P Damages Introduced by
Ion-Implantation into GaP 125
   H. Matsumura and S. Furukawa

ESR Studies of Annealing Behavior of Nitrogen-
Implanted GaP 133
   T. Matsumori, K. Miyazaki and S. Shigetomi

PROFILES

Profiles; How Well Can Experimental Results be Explained
by Theories? 143
   S. Furukawa and H. Ishiwara

Measurement of Arsenic Implantation Profiles in
Silicon Using an Electron Spectroscopic Technique 155
   S. Ludvik, L. Scharpen and H.E. Weaver

Atom and Carrier Profiles in As Implanted Si 163
   M. Iwaki, K. Gamo, K. Masuda, S. Namba, S. Ishihara,
   I. Kimura and K. Yokota

CONTENTS

The Influence of Postprocessing on the Electrical
Behavior of Implanted Arsenic Distributions in
Silicon 169
   H. Ryssel, H. Kranz, K. Schmid, I. Ruge and
   H.S. Rupprecht

Anomalous Annealing Behavior of Arsenic Implanted
Silicon as a Function of Dose and Energy 177
   W.K. Chu, H. Müller, J.W. Mayer and T.W. Sigmon

Deviated Gaussian Profiles of Implanted Boron and
Deep Levels in Silicon 183
   Y. Ohmura, K. Koike, H. Kobayashi and K. Murakami

Redistribution of Boron in Silicon Through High
Temperature Proton Irradiation 189
   P. Baruch, J. Monnier, B. Blanchard and C. Castaing

Range Distributions of Boron in Silicon Dioxide and
the Underlying Silicon Substrate 193
   K. Wittmaack, F. Schulz and B. Hietel

Redistribution of Background Impurities in Silicon
Induced by Ion Implantation and Annealing 201
   W.K. Hofker, H.W. Werner, D.P. Oosthoek and
   N.J. Koeman

Implantation of Boron and Lithium in Semiconductors
and Metals 211
   J.P. Biersack and D. Fink

Influence of Annealing and Radiation Damage on
Electrical Carriers Profiles in Phosphorus Implanted
Silicon Along the $|110|$ Axis. 219
   F. Cembali, R. Galloni, R. Lotti and F. Zignani

II-VI COMPOUND SEMICONDUCTORS AND OTHER MATERIALS

Ion Implantation of As in CdTe: Electrical
Characteristics and Radiation Damage 229
   J.C. Bean, J.F. Gibbons, T.J. Magee and J. Peng

Direct Measurement of Impurity and Damage Distribution
in Ion Implanted ZnTe by Cathodo and Photoluminescence . . 235
   D. Demars, M. Quillec, M. Ravetto, J. Marine and
   G. Guernet

Properties of Al and P Ion-Implanted Layers in ZnSe . . . 245
   Y.S. Park, B.K. Shin, D.C. Look and D.L. Downing

N-Ion Implantation into ZnSe 253
   C.H. Chung, H.W. Yoon, H.S. Kang and C.H. Tai

Energy Level Analysis of N^+ Ions Implanted CdS 261
   Y. Machi and S. Adachi

Deep Penetration of Implanted Po in CdS 267
   P.F. Engel and F. Chernow

ZnS:Mn DC Electroluminescent Cells by Ion Implantation
Techniques 275
   T. Takagi, I. Yamada and A. Sasaki

Ion-Bombardment-Induced Surface Expansion of Solids . . . 285
   H. Tanoue and T. Tsurushima

Nitrogen Implantation in SiC: Lattice Disorder and
Foreign-Atom Location Studies 291
   A.B. Campbell, J. Schewchun, D.A. Thompson, J.A. Davies and
   J.B. Mitchell

METALS

Enhancement of the Superconducting Transition
Temperature by Ion Implantation in Aluminium Thin
Films 301
   O. Meyer

The Influence of Heavy Ion Bombardment on the
Superconducting Transition Temperature of Thin Films . . . 309
   G. Linker and O. Meyer

Chemical Aspects of Ion Implantation 317
   G.K. Wolf, W. Fröschen and U. Sahm

Ion Impact Chemistry in Thin Metal Films; Argon, Oxygen and
Nitrogen Bombardment of Tantalum 325
   K.H. Goh, K.G. Stephens and I.H. Wilson

Iron Surface Treatment by Boron Implantation 335
   T. Takagi, I. Yamada and H. Kimura

Ionized-Cluster Beam Deposition 341
   T. Takagi, I. Yamada and K. Yanagawa

CONTENTS

Effects of Ion Bombardment on Metal-Silicon Interface . . . 347
   H. Nishi, T. Sakurai and T. Furuya

Lattice Location of Deuterium Implanted into
W and Cr 355
   S.T. Picraux and F.L. Vook

The Formation of Substitutional Alloys in fcc Metals
by High Dose Implantations 361
   J.M. Poate, W.J. DeBonte, W.M. Augustyniak and
   J.A. Borders

The Effect of Ion Implantation on the Corrosion
Behaviour of Fe 367
   V. Ashworth, D. Baxter, W.A. Grant, R.P.M. Proctor
   and T.C. Wellington

A Rutherford Backscattering and Channeling Study of
Dy Implanted into Single Crystal Ni 375
   G.A. Stephens, E. Robinson and J.S. Williams

RADIATION DAMAGE I

Defect Production in Semiconductors 385
   J.C. Bourgoin

Transport of Ion Deposited Energy by Recoiling
Target Atoms 399
   D.K. Brice

Ion Implantation Through Surface Layers: A Truncated
Gaussian Model 405
   A.V.S. Satya and H.R. Palanki

The Effects of Non-Gaussian Range Statistics on
Energy Deposition Profiles 413
   S.W. Mylroie and J.F. Gibbons

Projected Range Distribution of Implanted Ions in a
Double-Layer Substrate 423
   H. Ishiwara, S. Furukawa, J. Yamada and M. Kawamura

Generation of Knock-Ons in Solids Bombarded with
Energetic Ions and Energy Partition Relations 429
   T. Tsurushima and H. Tanoue

Defects in Ion Implanted SiO_2 Layers on Si 437
   E.P. EerNisse and C.B. Norris

Secondary Defects in Boron Implanted Silicon 439
    G.P. Pelous, D.P. Lecrosnier and P. Henoc

Ternary Defects Resulting from the Implantation of
B, F, BF, and BF_2 Ions into Silicon, Their Formation
and Effect Upon Device Properties 449
    S. Prussin

Techniques for Studying Ion Implantation in Diamond . . . 457
    J.F. Morhange, R. Beserman, J.C. Bourgoin,
    P.R. Brosious, Y.H. Lee, L.J. Cheng and J.W. Corbett

The Characterization of Implanted Layers Using the
Conductivity Modulation Effect 463
    M.J. Howes, D.V. Morgan and P. Ashburn

RADIATION DAMAGE II

Backscattering and ESR Studies in Heavily
Damaged Layer 473
    K. Masuda

On the Determination of Defects Distribution
in Implanted Layers by Means of Backscattering
Technique 485
    N. Matsunami and N. Itoh

The Use of Molecular Ions for Implantation Studies in
Si and Ge 493
    J.B. Mitchell, J.A. Davies, L.M. Howe, G. Foti and
    J.A. Moore

Damage Production and Annealing in Implanted Silicon
as Studied by Optical Reflectivity Profiling 501
    E.T. Yen, B.J. Masters and R. Kastl

Displacement Damage in Ne Implanted Magnesium Oxide . . . 511
    B.D. Evans

An EPR Study on High Energy Ion Implanted Silicon . . . 519
    Y.H. Lee, P.R. Brosious, L.J. Cheng and
    J.W. Corbett

ESR Line Width of Conduction Electrons in P^+ Ion
Implanted Si 525
    T. Shimizu, S. Hasegawa and H. Karimoto

CONTENTS

ESR Studies on Annealing Behavior of Heavily
Damaged Silicon 533
   K. Murakami, K. Masuda, K. Gamo and S. Namba

ESR Studies of Ion Implanted Si-SiO$_2$ Structure 539
   T. Izumi and T. Matsumori

Recovery of Silicon Layers Damaged by Low Energy
Ion Bombardment 547
   R. Prisslinger, S. Kalbitzer, H. Kräutle,
   J.J. Grob and P. Siffert

HIGH DOSE IMPLANTATION

Electrical Behaviour of Heavily-Doped Ion Implanted
Layers in Silicon 555
   J.H. Freeman, D.J. Chivers, G.A. Gard,
   G.W. Hinder, B.J. Smith and J. Stephen

High Dose Phosphorus-Germanium Double Implantation
in Silicon 571
   N. Yoshihiro, M. Tamura and T. Tokuyama

Control of Secondary Defects by Tin Diffusion
in Ion Implanted Silicon Crystals 577
   G. Nakamura, Y. Yukimoto and Y. Hirose

Chemical Composition of High Dose Implants in Silicon
and Germanium 585
   H. Kräutle, A. Feuerstein, H. Grahmann,
   S. Kalbitzer, F. Hasselbach and M. Prager

An Experimental Equipment for Ion Implantation . . . 591
   M. Setvak, J. Kral, Z. Hulek, L. Pina and A. Cako

DEVICES I

Ion Implantation into Polycrystalline Silicon 599
   T. Hirao, T. Ohzone, S. Takayanagi and
   H. Hozumi

Ion Implantation of Impurities into Polycrystalline
Silicon 605
   T. Tsuchimoto, I. Yudasaka and T. Shirasu

The Effect of Proton Bombardment on Porous Silicon
Formation 613
   T. Yashiro, K. Saito and T. Suzuki

P-Type Doping Observed in Silicon Implanted with
High Energy Carbon Ions 619
    J. Stephen, B.J. Smith and P.J. Hammersley

Negative and Anisotropic Magnetoresistance in
Phosphorus Implanted Silicon 627
    T. Itoh, M. Higashiura and H. Sato

Enhanced Residual Disorder in Silicon from Recoil
Implantation of Oxygen and Nitrogen by Arsenic Implants
Through Dielectric Layers 633
    T.W. Sigmon, W.K. Chu, H. Müller and J.W. Mayer

An Analysis of Arsenic Ion Implantation for Use in
Silicon Bipolar Devices 641
    J. Chisholm, J. Stephen, J. Turner, P. Dobson,
    R. Francis and E. Williams

Ion-Implanted Profiles for High-Frequency (>100GHz)
Impatt Diodes 647
    D.H. Lee and R.S. Ying

Noise Characteristics in the Low Frequency Range
of Ion-Implanted-Base-Transistor (NPN TYPE) 655
    T. Koji

DEVICES II

The Measurement of Doping Uniformity in Ion
Implanted Wafers 665
    J. Stephen, B.J. Smith and G.W. Hinder

Stress Adjustment in Si_3N_4 Films by Ion Implantation . . . 673
    G.W. Reutlinger and R.A. Moline

Enhanced Oxidation of Silicon by Ion Implantation and
its Novel Applications 681
    K. Nomura, Y. Hirose, Y. Akasaka, K. Horie and
    S. Kawazu

Ion Implanted Buried Layers Applied for Nuclear
Detector Telescopes 689
    A. Kostka and S. Kalbitzer

Limitations of the C-V Technique for Ion-Implanted
Profiles 695
    C.P. Wu, E.C. Douglas and C.W. Mueller

CONTENTS

Gold Implantation in Silicon: MOS C-V Characterization . . 697
   F.N. Schwettmann, J.M. Herman, III and
   T.M. Mosman

Threshold Voltage Shift of MOS-Transistors by
Ion Implantation of B, Al, Ga, P and As 703
   H. Runge

Noise Characteristics of Ion-Implanted MOS
Transistors 709
   K. Nakamura, O. Kudoh, M. Kamoshida and
   Y. Haneta

Implantation Profile and Buried-Channel Depth in
Ion Implanted MIS Structures 717
   B. Höfflinger and L. Gabler

Electrical Behavior of Boron-Implanted MOS
Transistor 723
   S. Kawazu, N. Kotani and Y. Watakabe

List of Authors 729

Index 733

III-V COMPOUND SEMICONDUCTORS I

ION IMPLANTATION IN III-V COMPOUNDS*

F. H. Eisen

Science Center, Rockwell International

Thousand Oaks, California 91360

ABSTRACT

In this review some of the factors which are important in ion implantation doping of III-V compound semiconductors will be discussed. Emphasis will be placed on the results so far obtained in GaAs. A discussion of the results in other III-V materials will be included as well as some comments on the application of ion implantation to the fabrication of devices in semiconducting III-V compounds.

There are several factors which are important in achieving successful doping of semiconductor materials by ion implantation. Usually, the implanted dopant atoms must occupy substitutional positions in the lattice into which they are implanted. Direct measurements of lattice locations of implanted dopants have been made in only a few cases in III-V compounds. These include cadmium [1] and tellurium [2,3] in GaAs and bismuth in GaP [4]. In other cases, the lattice location must be inferred from the results of the electrical measurements. It is necessary that the number of defects introduced during implantation be minimized by appropriate thermal treatment. This may involve performing the implantations at elevated temperatures as well as post-implantation annealing of the implanted materials. III-V compound materials may tend to disassociate during post-implantation annealing. The easiest method of

---

*Work partially supported by the Advanced Research Projects Agency of the Department of Defense and was monitored by Air Force Cambridge Research Laboratories under Contract No. F19628-74-C-0038.

preventing this disassociation is to cap the implanted sample with
a material which acts as a mask against the diffusion of the
constituents of the compound and the implanted dopant, and which
will adhere to the implanted sample during annealing. The choice
of appropriate materials for this cap will be discussed. The final
consideration which is important in evaluating the effectiveness of
ion implantation doping is the ratio of number of carriers introduced in the semiconductor material by implantation to the number
of implanted dopant atoms. In the case of n-type dopants in GaAs,
a one-to-one relation between implanted dopant atoms and resulting
carriers has generally not been observed.

There are several problems which may be encountered in achieving successful doping of III-V compounds. Changes in the substrate
material may occur during implantation or the subsequent annealing.
One possibility is the redistribution of chromium in Cr-doped semi-
insulating GaAs. An example of this is discussed below. Diffusion
of the implanted dopants may take place resulting in a doping profile which is substantially different from the expected range profile for the implanted dopants. Such diffusion seems to be a common
feature in implantation in GaAs. Diffusion of defects may also
occur. This is thought to be responsible for the intrinsic regions
observed in GaAs implanted with p-type dopants [5]. Finally, doping
of a compound semiconductor by implantation produces a non-
stoichiometry in the material [6]. This non-stoichiometry may have
important effects on the electrical activity resulting from the
implantation. It has been suggested, for example, that in implanting
an n-type dopant into GaAs an equal amount of gallium should also
be implanted [6]. An interesting example of such double implantation
is presented in the paper by Stoneham and Gibbons [7].

Channeling effect [8] data on the annealing of lattice disorder
in various III-V compounds are shown in Fig. 1 [9]. The results
for low-dose implants in GaAs show appreciable annealing in the
temperature range from 200 to 400°K. High-dose implants in which
an amorphous layer is produced, require higher temperatures, but one
might expect that the annealing would be complete at temperatures
of about 800°K (500°C). Some channeling effects measurements have
shown that there is still measurable disorder after annealing at
temperatures near 900°C. The data of Guivarc'h, et al., in Fig. 2
show that after annealing at 880°C there is still a measurable
damage peak in a GaAs sample implanted with 1 MeV zinc ions [10].
Annealing to 950°C results in the channeled backscattering yield
being almost the same as that before implantation. Measurements of
recovery of the electrical properties of damaged samples also indicate that high annealing temperatures are required for ion-implanted
III-V compounds. For example, it is known that in neutron-irradiated
GaAs a recovery stage occurs between 600 and 700°C [11]. It would
be expected that the annealing damage in ion-implanted GaAs would
require temperatures at least as high as those required to anneal

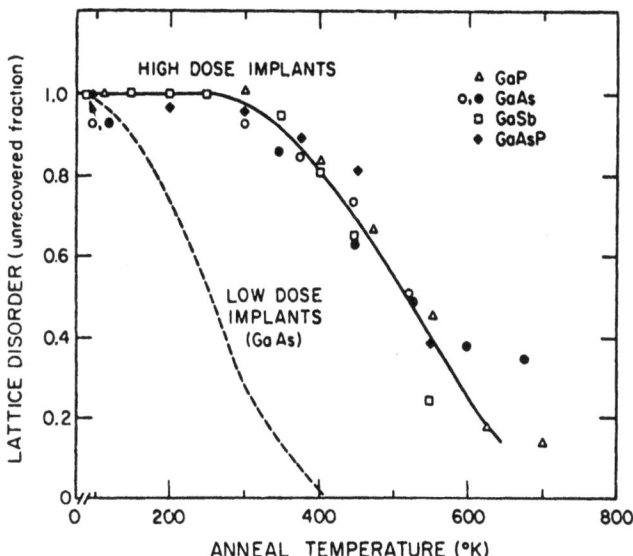

Fig. 1. Channeling effect measurements of the annealing of lattice disorder in several III-V compounds. The solid curve shows the annealing results when an amorphous layer has been formed [9].

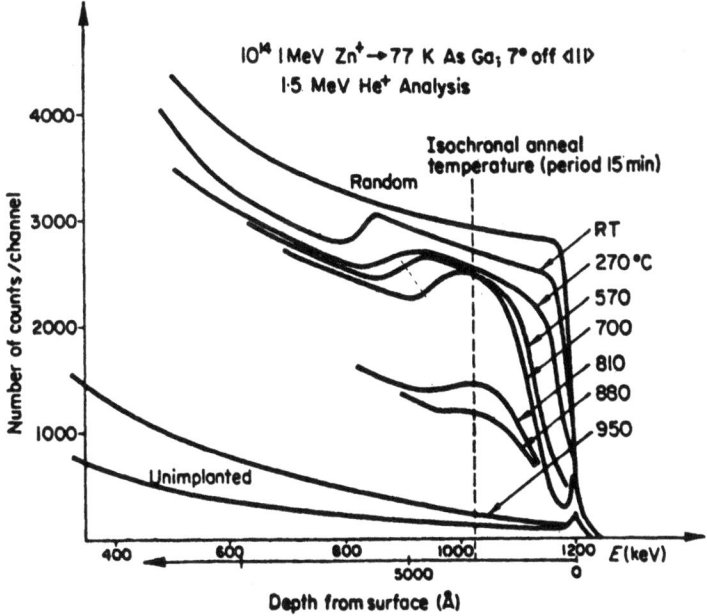

Fig. 2. Aligned and random spectra for a 1 MeV zinc-implanted sample using 1.5 MeV He$^+$ with isochronal annealing (15 min) as a parameter [10].

neutron damage. In fact, in work with p-type implants in GaAs to
produce lasers, it has been found that annealing at 900°C is required
in order to achieve efficient optical emission [12]. These various
results suggest that in GaAs, for example, annealing temperatures
as high as 900°C may be required to remove the defects produced
during implantation.

Figure 3 shows results on the annealing of a GaAs sample
implanted with zinc at 1 MeV [13]. The annealing was carried out
at 600°C with no protective covering on the GaAs surface. The decrease in the number of holes/cm^2 observed at times greater than
about 40 min is due to outdiffusion of the zinc, and/or decomposition of the GaAs substrate. This illustrates the need for protecting
of the surface of the sample during annealing. Release of arsenic
from implanted surfaces has been observed from temperatures as low
as 300°C [14].

SiO_2 has been widely used as an annealing cap on implanted GaAs
and other III-V compounds. As we shall see below, good results have
been obtained for p-type implantation in GaAs using an SiO_2 cap.

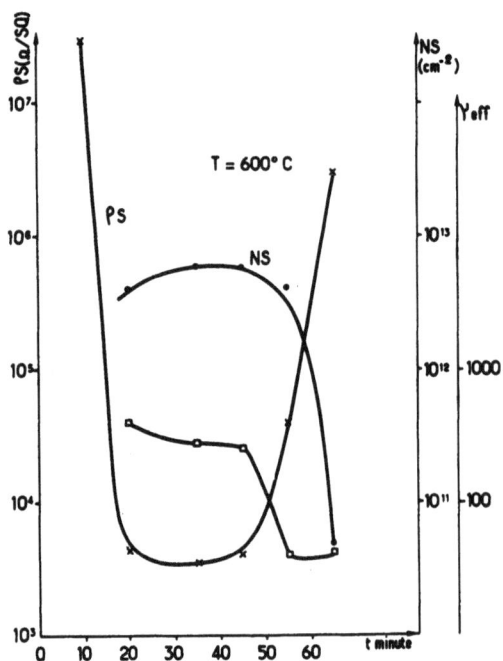

Fig. 3. Dependence of sheet resistance, surface carrier concentration, and mobility on annealing time at 700°C, with no surface protection, for an n-type GaAs sample implanted at room temperature
with 10^{15} 1.0 MeV zinc ions/cm^2 [13].

However, it is known that gallium diffuses readily in SiO_2 [15] suggesting the possibility that gallium vacancies may be formed when a GaAs sample is annealed with an SiO_2 cap. The formation of gallium vacancies in samples implanted with p-type dopants might facilitate the substitutional location of the dopant and thus be advantageous. However, in the case of n-type doping the possibility exists that gallium vacancies may react with the implanted dopant to form gallium vacancy-dopant complexes which act as compensating acceptors in the implanted material. Photoluminescence data obtained by Harris, et al, [16] in tellurium-implanted GaAs samples has been interpreted as indicating that when SiO_2 was used as an annealing cap the number of such gallium vacancy-tellurium complexes formed was much greater than when Si_3N_4 was used as an annealing cap. Their work also showed better electrical activity after annealing at 750°C in the cases where Si_3N_4 was used as compared with SiO_2. Bell, et al., [17] have observed that gallium oxide was formed on the surface of GaAs samples implanted with doses of tellurium of 10^{15} ions/cm^2 or greater, when these samples were annealed with an SiO_2 cap; whereas, the oxide formation did not occur when the nitride cap was used. The results of these workers suggest that Si_3N_4 is to be preferred over SiO_2 as an annealing cap for n-type implants in GaAs and perhaps for n-type doping in other III-V compounds as well.

Sato [18] has observed the formation of an n-type layer on the surface of Cr-doped semi-insulating GaAs annealed to 750°C while covered with an SiO_2 layer made by thermal decomposition of alchoxyl silicate. The data in Fig. 4 show that an n-type layer several thousand Ångstroms deep was formed. Such layers were not found when Si_3N_4 was used to protect the surface of the GaAs during annealing. These results show that the behavior of implanted Cr-doped substrate materials during annealing may depend upon the surface conditions, and that it is possible that in some cases spurious indications of n-type doping by implantation could result.

The effects of variations in substrate material are illustrated in Table I which shows the data of Hunsperger for sulfur implantation into Cr-doped semi-insulating GaAs obtained from various suppliers [19]. The letters A through D indicate different suppliers, whereas the subscripts on those letters indicate different boules form a given supplier. It can be seen that there is an appreciable spread in the doping efficiency and the mobilities obtained by implanting into these different GaAs substrates. Hunsperger believes that the differences are due primarily to a difference in chromium content between the various substrate materials.

Beryllium, magnesium, zinc, and cadmium have all been investigated as p-type dopants in GaAs [20,21]. Figure 5 shows the results of annealing magnesium implants of various doses at different annealing temperatures [21]. It can be seen that in most cases

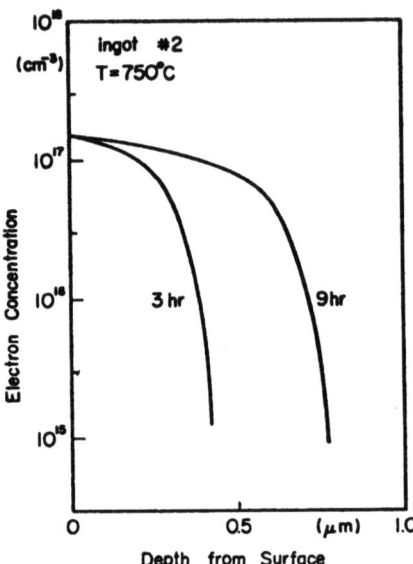

Fig. 4. Profile of the electron concentration in the conductive layer on a Cr-doped GaAs sample annealed with an SiO_2 cap.

Table I

Sample	N_S (cm^{-2})	μ (cm^2/Vsec)	Doping Efficiency (%)
A_1 Bottom	2.1×10^{12}	2660	14
A_1 Top	3.0×10^{12}	2910	20
A_2 Top	5.6×10^{12}	2180	37
A_2 Bottom	3.6×10^{12}	2580	24
A_3	7.0×10^{12}	2630	46
B_1	4.1×10^{12}	2840	27
C_1	4.4×10^{12}	3170	29
C_2 Top	5.0×10^{12}	3620	33
C_2 Bottom	4.4×10^{12}	3410	29
D_1 Top	1.0×10^{12}	2540	6.7
D_1 Bottom	1.2×10^{12}	2410	8

Dose = 5×10^{12} S$^+$/cm^2, 30 keV
1×10^{13} S$^+$/cm^2, 150 keV

Annealed for 20 minutes at 800°C

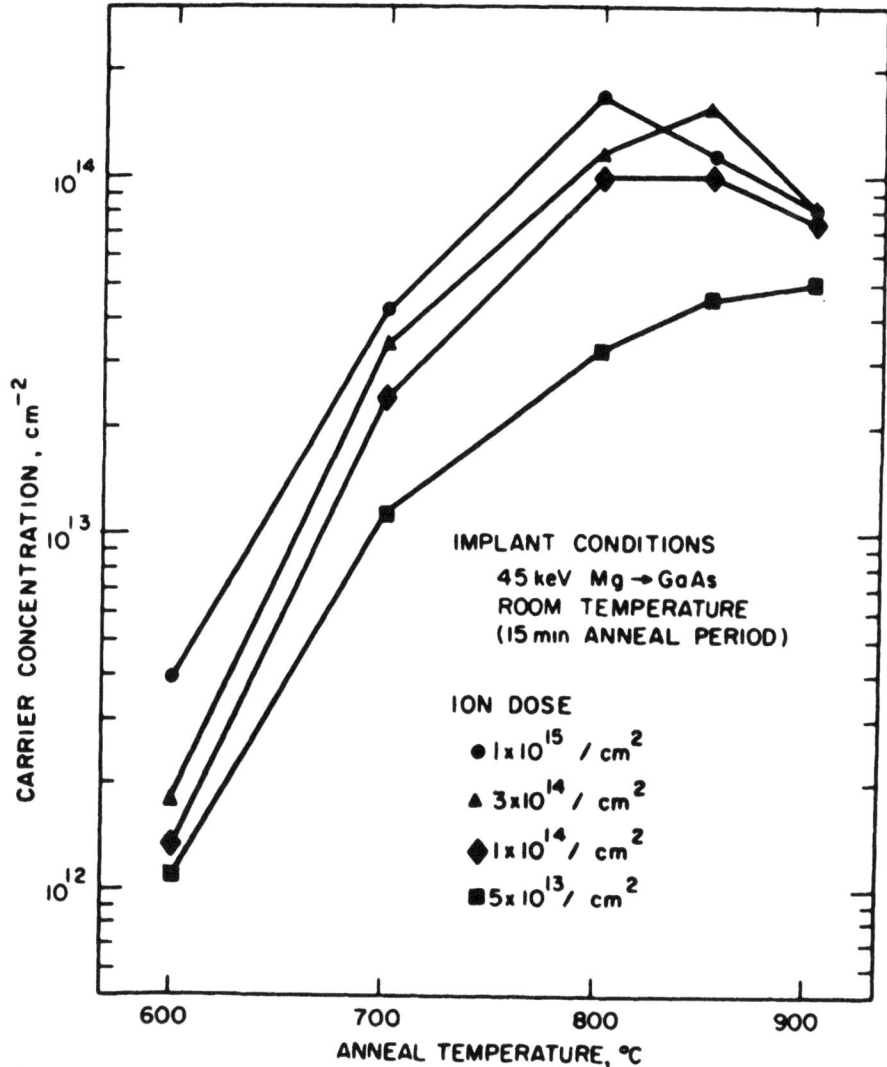

Fig. 5. Sheet electron concentration of Mg-implanted layers in GaAs.

maximum electrical activity was achieved at an annealing temperature of 800°C. This is about the annealing temperature required to achieve activation of zinc and cadmium implants also. In the case of beryllium, only a 600°C anneal is required [21]. This may be due to the lower amount of damage produced during beryllium implantation as compared to that produced during the implantation of heavier ions. For these p-type dopants, at doses less than or equal to about 10^{14} ions/cm^2, there is a one-to-one correspondence between the number of holes and the number of ions implanted. In most cases, the maximum hole concentrations observed seems to be approximately equal to the solid solubility of the implanted dopants.

The formation of an intrinsic layer beneath the p-type layer has often been observed in the implantation of p-type dopants into GaAs [5]. Thicknesses of greater than 100μ have been observed for these layers. Their thickness is minimized by implanting through SiO_2 rather than implanting bare GaAs. This observation and other evidence suggests that the intrinsic regions may be due to the diffusion of arsenic vacancies deep into the material [5]. It has been possible to decrease the thickness of these intrinsic regions to zero by annealing at temperatures of 900°C [5], so that their influence on the performance of devices made by the implantation of p-type dopants can be effectively eliminated.

The results mentioned above for implantations of p-type dopants in GaAs were obtained for implantations carried out at room temperature using SiO_2 as the annealing cap. In the case of the implantation of n-type dopants it seems to be desirable to implant at temperatures above room temperature. Harris, et al. [22], for example, found that lower electrical activities were produced when an amorphous layer was formed during the implantation of silicon into GaAs than when an amorphous layer was not formed. Data on the disorder produced by tellurium implantation into GaAs as a function of implant temperature [3] indicate that there is a rapid decrease in disorder with increasing implant temperature in the vicinity of 150°C. Early results on tellurium implantation, with annealing at 750°C, indicated that better electrical activity was obtained when the implantation was performed at 150°C than when it was performed at room temperature [16]. Davies has recently obtained data for sulfur implantation into GaAs indicating that higher electrical activities are obtained at implantation temperatures between about 150°C and 450°C than when the implantation is carried out at room temperature [23]. Factors which indicate that Si_3N_4 may be a better annealing cap for n-type implants have been discussed above.

The electron concentration profiles resulting from sulfur implantation in GaAs tend to be much deeper than the LSS range distribution for the implanted sulfur, and to show maximum electron concentrations which are only about 1 to 2×10^{17} no matter how high the dose of the implanted sulfur. The results of Sansbury and Gibbons [24] for room temperature sulfur implants annealed at 750°C with an SiO_2 cap are shown in Fig. 6. It can be seen that the depth of the electrical profile increases as the sulfur dose is increased. Recent work on sulfur implantation using an implantation temperature of 350°C and annealing with a Si_3N_4 cap also gave a deep electrical profile. Data for doses of 10^{13} and 10^{14} 100 keV sulfur ions/cm^2 are shown in Fig. 7 for an annealing temperature of 850°C [25]. In

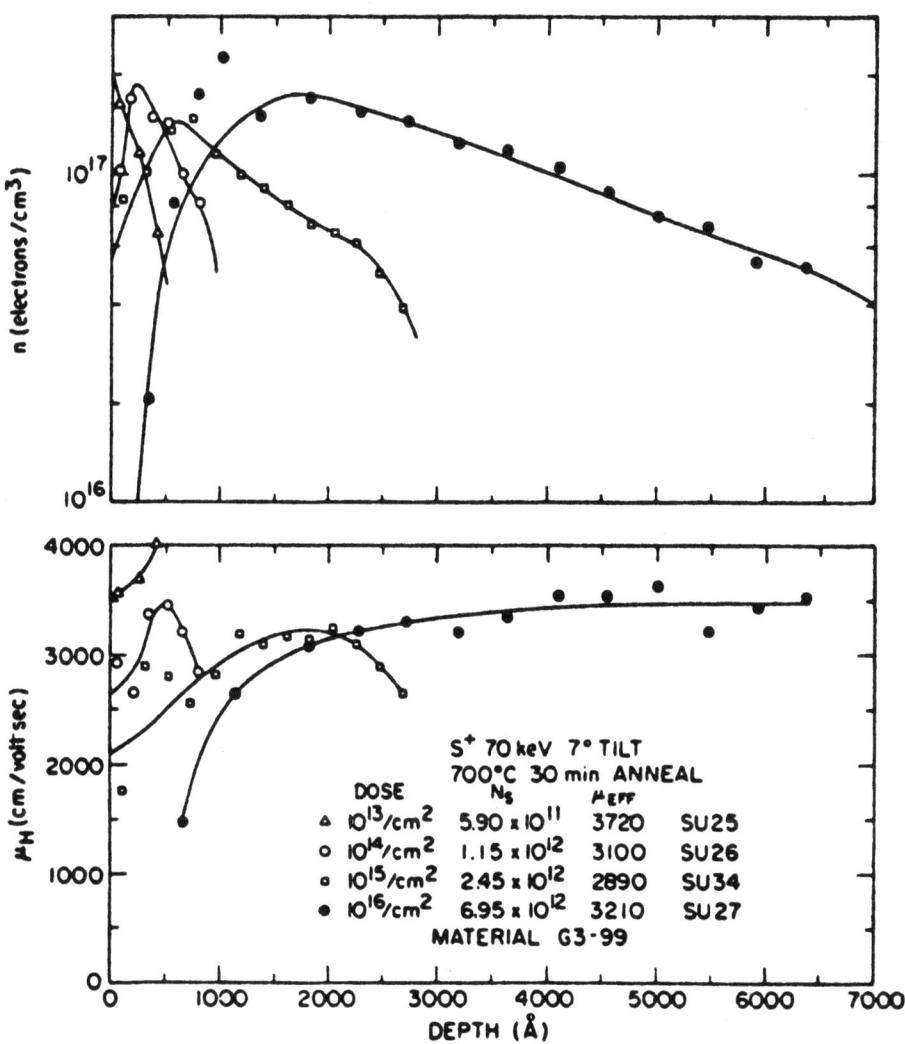

Fig. 6. Electron concentration and mobility profiles for various doses of sulfur implanted in GaAs [24].

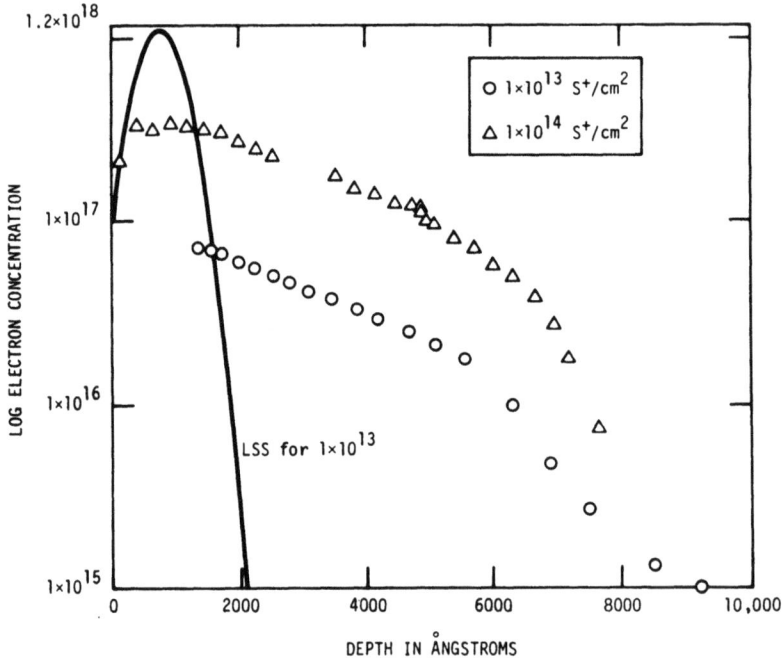

Fig. 7. Electron concentration profiles for 100 keV sulfur implants in GaAs. The Gaussian profile calculated from LSS range parameters for 10^{13} ions/cm^2 is given by the solid line [25].

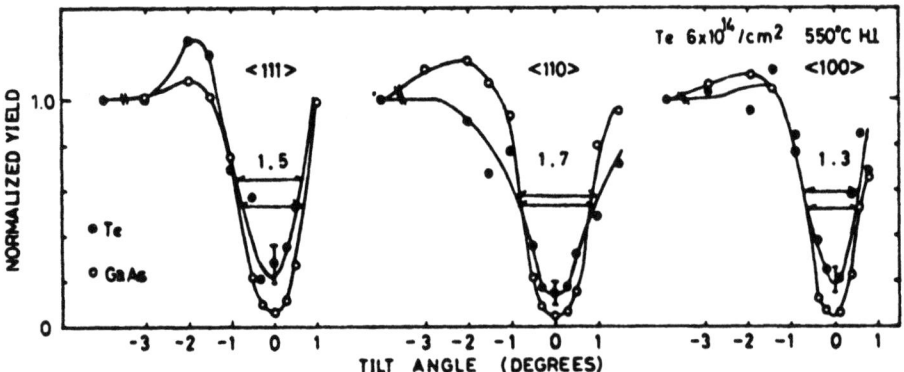

Fig. 8. Channeling angular dips for 1 MeV He$^+$ backscattered from GaAs and Te atoms along each of the major axis. The sample was implanted at 550°C to a dose of 6×10^{14}/cm^2 [2].

this work the percentage electrical activity obtained for the implanted sulfur was appreciably higher than in the work of Sansbury and Gibbons. Profile results show that this is due to a greater depth of the doping rather than having achieved higher peak electron concentration. Sansbury and Gibbons suggested that the large depth of the electrical profile in their work may be due to enhanced diffusion of the sulfur, which occurs during annealing because of the high defect concentration present in the implanted samples.

Lattice location data using the channeling effect technique are available for tellurium implanted into GaAs. Figure 8 shows the angular dependence of the scattering yield of helium ions from the GaAs and from tellurium implanted at 550°C for three principal axes of the GaAs system [2]. The low minimum yield from the implanted tellurium and the agreement between the angular width of the GaAs and the tellurium yield curves for the three axes indicate that the implanted tellurium is located in substitutional positions. The data of Eisen, et al., [3] also indicates that tellurium implanted at 350°C is substitutional following implantation and following annealing at temperatures as high as 900°. The electrical activity obtained as a result of tellurium implantation depends strongly on the annealing temperature in the range between 750 and 900°C. Data for these two temperatures and several doses of implanted tellurium are shown in Fig. 9 [3]. The maximum doping efficiency obtained for a 10^{14} dose was about 30% in this work. There is some indication from profile measurements that the electrical activity of the implanted tellurium was limited by solubility effects for a dose of 10^{14} tellurium ions/cm^2. Considerable scatter in the results is thought to be due to problems with the adherence of the Si_3N_4 annealing cap. In some cases, bubbling and peeling of the nitride cap occurred during annealing.

In summary, in GaAs good electrical activity of implanted p-type dopants has been achieved. For n-type dopants, 100% activity is usually not observed. It is sometimes difficult to obtain good reproducibility in the n-type implantation case and annealing temperatures of at least 800 to 900°C seem to be required.

Work in several laboratories on the implantation of zinc in GaP and GaAsP indicates that p-type doping works well in these materials also [26,27]. Most workers have used room temperature implantation and an SiO_2 annealing cap with annealing temperatures of 800 to 900°C. The results show good doping efficiency and good luminescence efficiency from light-emitting diodes made by the implantation of zinc into n-type substrates. Work on the implantation of nitrogen in GaAsP and GaP, as a radiative recombination center, indicates that in the case of this group V dopant hot implants work better than do room temperature implants [28]--a

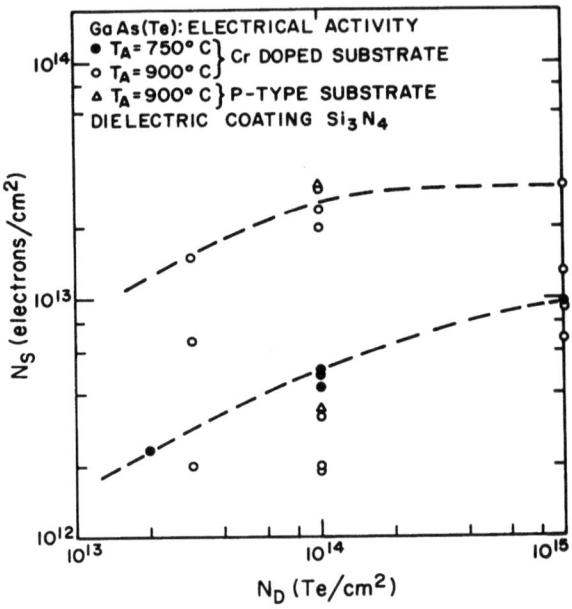

Fig. 9. The measured number of electrons/cm^2 in Te-implanted GaAs after anneal as a function of the number of implanted ions/cm^2 [4].

result similar to that seen for tellurium implantation in GaAs.

The implantation of bismuth in GaP has been fairly extensively investigated by Merz, et al. [4]. It was found that hot implants result in lower disorder following implantation and slightly better luminescence intensity. Channeling effect measurements indicate that the implanted bismuth was substitutional and that the bismuth atoms were located on the phosphorous sublattice. Luminescence intensities obtained using bismuth implantation were disappointingly low. It is thought that strain is partially responsible for these low intensities. However, the authors concluded that competing damage centers were probably also present and accounted for at least partially for the low intensity [4]. In view of the results obtained with n-type doping in GaAs, it might be interesting to repeat this work on bismuth implantation in GaP using a Si_3N_4 annealing cap rather than SiO_2 which was used by Merz, et al.

Doping of InSb by zinc implantation and of InAs by sulfur implantation has been reported [24]. It was possible to make good infrared detectors by ion implantation but no data on doping efficiency, carrier concentrations achieved, or anneal temperature effects has been published.

ION IMPLANTATION IN III-V COMPOUNDS

One application of ion implantation doping in III-V compounds, which has been reported by two laboratories, is the use of implantation to produce the channel regions for a GaAs Schottky barrier FET. Results have been reported by Hunsperger, et al. [30], and by Welch, et al. [31] using sulfur implantation to produce the n-type layer required for the FET, on a semi-insulating GaAs substrate. Such devices require n-type layers a few thousand Angstroms thick with electron concentrations near $10^{17}/cm^3$. These requirements are similar to the results shown for sulfur implantation in Fig. 8. The use of sulfur implantation to produce a modified Read-type structure for an IMPATT diode in GaAs has also been reported [32]. Another possible application of implantation to the FET device would be to produce regions of enhanced electron concentration in the source and drain region so as to achieve better source and drain contacts. The use of zinc implantation to produce p-n junction lasers has been reported, and good optical efficiency has been obtained [11]. Zinc implantation in GaAsP can be used to fabricate light-emitting diodes which equal in brightness those made by diffusion, while offering advantages in uniformity of light output and reproducibility of processing [33].

The achievement of reproducibility in implantation doping is extremely important. In many cases this has not been possible so far in III-V compounds simply because we do not yet have all the variables which may have an effect on the outcome of the implantation under control. Two factors which seem particularly important at the present are the substrate and the annealing cap. In general, we need better substrate materials, and we must compare the results of implantation into substrates of different types for the same semiconductor material, and into substrates obtained from different sources. The annealing cap must be reproducible. It may be desirable, in some cases, to do control experiments in which the changes in unimplanted substrates which are capped and annealed, are determined as a means of evaluating the effectiveness of a given cap layer.

It would be interesting to use Si_3N_4 caps on p-type implants in GaAs. Such experiments might help test some of the ideas which have been put forward to explain the results of n-type implantation in GaAs.

It should be apparent from the few profiles presented in this review that measurements of the sheet electron concentration achieved by implantation are not sufficient to fully evaluate the results of the implantation. In many cases, the electrical profile is far deeper than would be estimated from LSS range statistics. Accurate measurements of profiles will be very important in evaluating the results of implantation experiments in compound semiconductors. Anodization techniques such as that described by Müller, et al., in another paper at this conference should be very important in the determination of such profiles.

It has been possible to enumerate only some of the problems and results observed so far in implantation doping in III-V compounds. Much remains to be done in achieving reproducible doping, and in understanding the results. The progress which has been made to date, and the applications to devices which have already been reported, suggest that the future of implantation in III-V compounds is sufficiently promising to justify the effort involved.

ACKNOWLEDGEMENT

It is a pleasure to acknowledge numerous discussions with H. Müller and J. W. Mayer. Some of the ideas in this review are drawn from "Ion Implantation and Channeling," by F. H. Eisen and J. W. Mayer to be published by Plenum Press in Volume 6 of Treatise on Solid State Chemistry, edited by H. B. Hannay.

REFERENCES

1. G. Ilic, G. T. Ewan, and J. L. Whitton, Rad. Eff. $\underline{18}$, 47 (1973).
2. M. Takai, K. Gamo, K. Masuda, and S. Namba, Japan. J. Appl. Phys. $\underline{12}$, 242 (1973).
3. F. H. Eisen, J. S. Harris, B. Welch, J. D. Haskell, R. D. Pashley, and J. W. Mayer, in Ion Implantation in Semiconductors and Other Materials, ed. by B. L. Crowder (Plenum, New York, 1973), p. 631
4. J. L. Merz, D. L. Mingay, W. M. Augustyniak, and L. C. Feldman, in Ion Implantation in Semiconductors, ed. by I. Ruge and J. Graul (Springer, Berlin, 1971), p. 182.
5. R. G. Hunsperger and O. J. Marsh, Met. Trans. $\underline{1}$, 603 (1970).
6. R. Heckingbottom and T. Ambridge, Rad. Eff. $\underline{17}$, 31 (1973).
7. E. B. Stoneham and J. F. Gibbons, this conference.
8. J. W. Mayer, L. Eriksson, and J. A. Davies, Ion Implantation in Semiconductors, (Academic Press, New York, 1970).
9. S. T. Picraux, Rad. Eff. $\underline{17}$, 261 (1973).
10. A. Guivarc'h, P. N. Favennec, and G. P. Pelous, Radiation Damage and Defects in Solids (The Institute of Physics, London, 1973), p. 429.
11. L. W. Aukerman, P. W. Davis, R. D. Graft, and T. S. Shilliday, J. Appl. Phys. $\underline{23}$, 3590 (1963).
12. M. K. Barnoski, R. G. Hunsperger, and A. Lee, Appl. Phys. Lett. $\underline{24}$, 627 (1974).
13. P. N. Favennec, in Ion Implantation in Semiconductors, ed. by I. Ruge and J. Graul (Springer, Berlin, 1971), p. 174.
14. S. T. Picraux, in Ion Implantation in Semiconductors and Other Materials, ed. by B. L. Crowder (Plenum, New York, 1973), p. 641.
15. A. S. Grove, O. Leistiko, and C. T. Sah, J. Phys. Chem. Solids $\underline{25}$, 985 (1964).

16. J. S. Harris, F. H. Eisen, B. Welch, R. D. Pashley, D. Sigurd, and J. W. Mayer, Appl. Phys. Lett. 21, 601 (1972).
17. E. C. Bell, A. E. Glacum, P. L. F. Hemment, and B. J. Sealy, Rad. Eff. (to be published).
18. Y. Sato, Japan. J. Appl. Phys. 12, 242 (1973).
19. R. G. Hunsperger, Solid State Elect. (to be published).
20. R. G. Hunsperger and O. J. Marsh, J. Electrochem. Soc. 116, 488 (1969).
21. R. G. Hunsperger, R. G. Wilson, and D. M. Jamba, J. Appl. Phys. 43, 1318 (1972).
22. J. S. Harris, in *Ion Implantation in Semiconductors*, ed. by I. Ruge and J. Graul (Springer, Berlin, 1971), p. 157.
23. E. Davies, to be published.
24. J. D. Sansbury and J. F. Gibbons, Rad. Eff. 6, 269 (1970).
25. F. H. Eisen, B. Welch, H. Müller, and J. W. Mayer (unpublished).
26. P. M. Hemenger and B. C. Dobbs, Appl. Phys. Lett. 23, 462 (1973).
27. T. Itoh and Y. Oana, J. Appl. Phys. 44, 4982 (1973).
28. B. G. Streetman, R. E. Anderson, and D. J. Wolford, J. Appl. Phys. 45, 974 (1974).
29. P. J. McNally, Rad. Eff. 6, 149 (1970).
30. R. G. Hunsperger and N. Hirsch, Elect. Lett. 9, 577 (1973).
31. B. Welch, F. H. Eisen, and J. A. Higgins, J. Appl. Phys. 45, 3685 (1974).
32. J. J. Bernez, R. S. Ying, and D. H. Lee, Elect. Lett. 9, 157 (1974).
33. D. L. Dexter (private communication).

ANODIC OXIDATION AND PROFILE DETERMINATION OF

ION IMPLANTED SEMI-INSULATING GaAs*

H. Müller,** J. Gyulai*** and J. W. Mayer

California Institute of Technology
Pasadena, California 91109

and
F. H. Eisen and B. Welch

Science Center, Rockwell International
Thousand Oaks, California 91360

ABSTRACT

A new anodic oxidation process of GaAs, using N-Methylacetamide, adjusted to a pH value of 8.3 and a constant current source, was employed to evaluate Te and Si implanted profiles in semi-insulating GaAs. The samples were implanted at 350°C and encapsulated in Si_3N_4 for anneal at temperatures up to 900°C. Differential Hall measurements indicated strong deviations from predicted Gaussian profiles and maximum carrier concentrations below 10^{18} cm^{-3}.

---

*This research was partially supported by the Defense Advanced Research Projects Agency, the Department of Defense, and was monitored by Air Force Cambridge Research Laboratories under Contract No. F19628-74-C-0038.

**Permanent Address: Lehrstuhl für Integrierte Schaltungen
  Technical University, Munich, Germany.

***Permanent Address: Central Research Institute for Physics
  Budapest, Hungary. Supported in part by
  NSF - International Programs.

INTRODUCTION

Ion implantation of p-type dopants into GaAs has been successfully performed with doping efficiencies up to 100% for Zn, Cd, Mg, and Be.[1,2] For implantation of n-type dopants, which offers interesting high speed device applications, the situation appears different. Si, S, C, Se and Te implants [3,4] were studied and doping efficiencies were found to be a few percent for doses of 10^{14} cm^{-2} and about 15 - 20% for doses around 10^{13} cm^{-2}. Investigations regarding implant temperature indicated that evaluated temperatures ($\geq 200°C$) result in higher electrical activity after a high temperature heat treatment. It was also found that temperatures below 800°C are not sufficient to produce high electrical activity. For higher anneal temperatures SiO_2 is insufficient as a protective cap layer because of Ga outdiffusion and therefore other dielectrics as Si_3N_4 or AlN have to be used to prevent decomposition of the GaAs. Profile determination of ion implanted dopants so far has been carried out by etching techniques and generally showed large deviations from predicted LSS distributions. In this paper we report results on Te and S implants into Cr-doped semi-insulating GaAs. Electrical carrier concentration profiles were determined by differential conductivity and Hall measurements using a recently developed anodic oxidation process, which allowed successive layer removal to be performed in situ. For comparison results obtained by C-V techniques are discussed.

EXPERIMENTAL TECHNIQUES

Implantation of S and Te was generally performed into Cr-doped semi-insulating GaAs with energies of 100 keV and 400 keV at a target temperature of 350°C. Reactively sputtered Si_3N_4 was used as a protective layer during annealing at temperatures between 750°C and 900°C in H_2 atmosphere. For electrical measurements a van der Pauw type pattern was mesa etched to about 3 µm and alloyed Au-Ge-Ni contacts provided ohmic contact to the implanted layers. For differential conductivity and Hall effect measurements an anodic oxidation process was applied which uses N-Methylacetamide adjusted to a pH-value of 8.3. The main difference compared to results reported in the literature on native oxide growth [5,6] is that a constant current source can be employed. The difference between a final cell voltage and the initial voltage drop can be directly used as a reference for the oxide thickness. It was also found that at low pH values simultaneous dissolution of the oxide can occur. This could be avoided by using pH-values of the solution above pH = 8. Results on oxide thickness and amount of GaAs removed as a function of the forming voltage were 20.2 Å/V and 13.4 Å/V respectively. Anodizations of implanted layers typically were carried out at 20 V for each step, which corresponds to 268 Å of layer removal and after each anodization the oxide was dissolved in dilute H_3PO_4 and sheet resistivity and Hall effect were measured. A detailed description of the anodization process is given in Ref. 7.

TABLE 1

	Dose [cm^{-2}]	T_{anneal} [°C]	Time [min]	N_{seff} [cm^{-2}]	ρ_s [Ω/]	μ_{Heff} [cm^2/Vsec]
	5x10^{12}	800	10	1.15x10^{12}	2146	2527
		750	10	7.2 x10^{11}	3634	2364
Te	10^{13}	800	10	3.93x10^{12}	565	2808
		850	10	6.66x10^{12}	269	3480
		900	10	5.71x10^{12}	221	4938
		850	120	7.48x10^{12}	215	3882
S	10^{13}	850	5	1.72x10^{12}	915	3964
	10^{14}	850	10	1.27x10^{13}	141	3466

RESULTS AND DISCUSSION

A. Doping Efficiency for Te and S Implants

Results obtained for Te and S implants for a dose of 10^{13} cm^{-2} and various anneal temperatures are summarized in Table 1.

For Te doses of 10^{13} cm^{-2} and short time anneals (10 min) the electrical activity increases from 7% to 59% for temperatures from 750° to 900°C. After 900°C annealing high mobilities are observed. For compairson a 2 hr anneal at 850°C shows a somewhat higher value for N_{seff}, but a decrease in the mobility.

B. Profile Results

In Fig. 1 carrier concentration profiles are shown for 10^{13} cm^{-2} Te implants, annealed at various temperatures according to Table 1. A steady increase in the maximum carrier concentration is observed with higher anneal temperatures. For 900°C, 10 min anneal and 850°C, 2 hrs. identical profiles are obtained in the higher concentration regions with a maximum concentration of 3.5 x 10^{17} cm^{-3}. This value is only 35% of the expected value from a Gaussian LSS distribution which is included in the figure. A strong deviation is observed. In case of 850°C, 2 hr. anneal a deeply penetrating tail is observed, which reaches a concentration of 2 x 10^{16} cm^{-3} after removal of 0.59 μm. In contrast to that the profile obtained at 900°C, 10 min could be measured only to a concentration of 1.6 x 10^{17} cm^{-3}. The sheet resistivities in both cases for the last layer removal were ≥ 15 kΩ/ which means that there is a considerable difference in the mobilities in the tail region. Obviously a long time anneal causes higher electrical activity in the tail regions. This is shown in Fig. 2 where corresponding differential mobilities are plotted versus depth. For low temperature annealing a cutoff in the mobility occurs at shallower depths than following high temperature. For 850°C, 120 min an almost constant mobility around 4500 cm^2/Vsec is approached. Compared to mobility

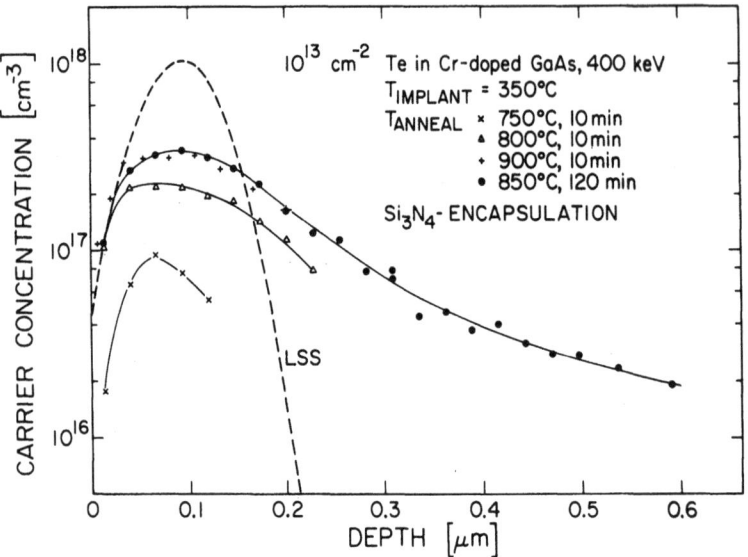

Fig. 1

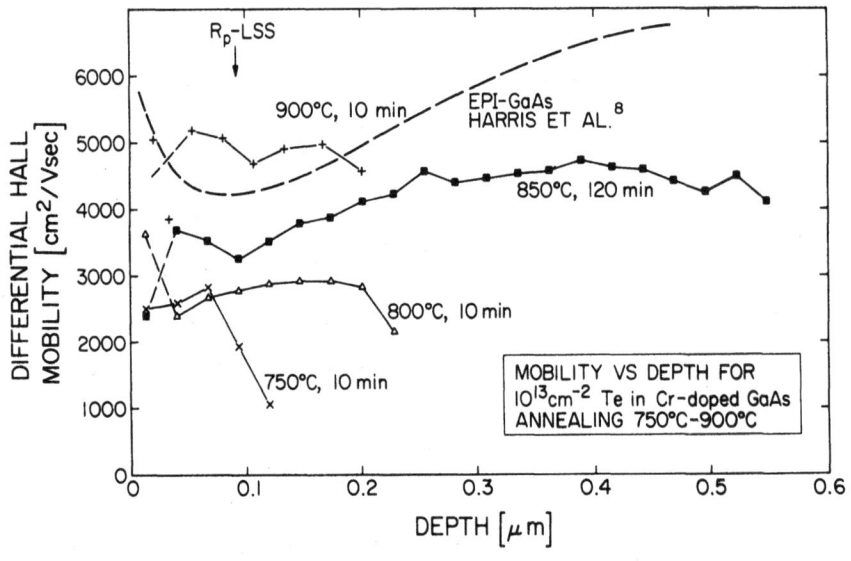

Fig. 2

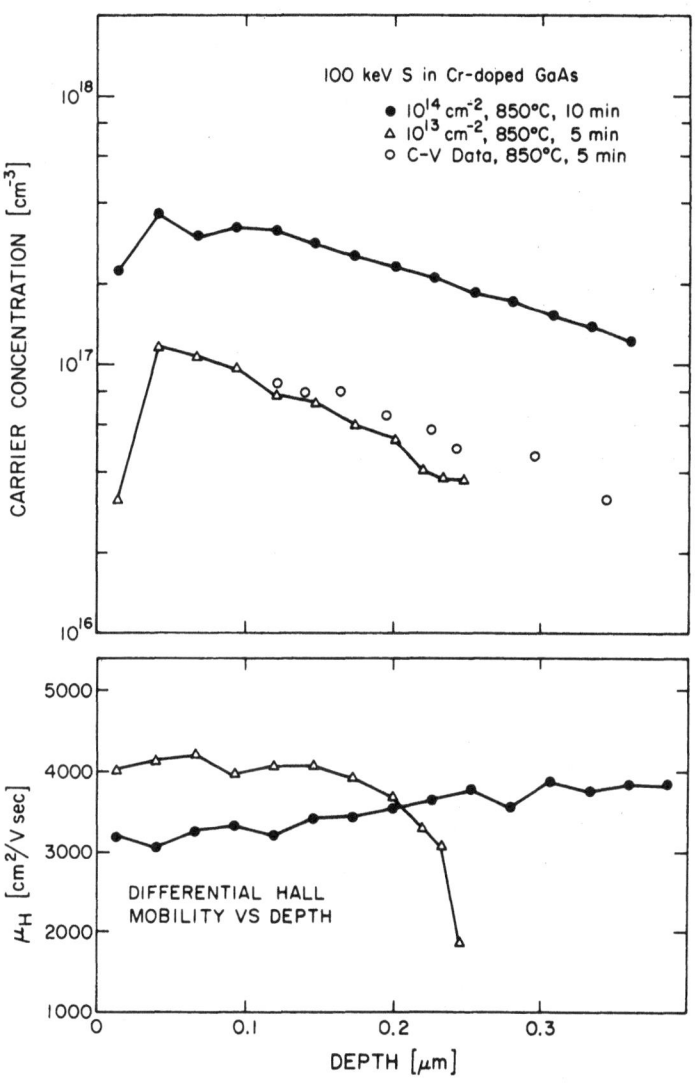

Fig. 3

data obtained from epitaxial n-type layers,* which are shown by the dashed curve in Fig. 2, these values are lower probably due to a scattering limitation in the Cr-compensated material. However fairly high mobilities can be obtained and especially in the higher concentration regions (0.1 - 0.2 µ) non-electrically active Te does not seem to have a strong influence on the mobilities.

In order to gain more information about tail formation S-implants were investigated. The critical angle for channeling at 100 keV is 4° compared to 3.1° for 400 keV Te and the predicted Gaussian distribution is almost identical to a 400 keV Te distribution. Diffusion coefficients for S are reported to be two orders of magnitude higher than for Te. In Fig. 3 profile results for 10^{13} cm^{-2} and 10^{14} cm^{-2} according to annealing in Table 1 are shown. Very flat distributions are obtained and compared to Te the slopes of the exponential tails are different by about a factor of 2. This indicates that diffusion either during the implantation or the subsequent annealing may be the dominating mechanism for tail formation.

C. Comparison with C-V Measurements

In the concentration range reported above, the C-V technique is a very useful tool to investigate effects of incomplete annealing which is difficult to obtain by differential conductivity and Hall measurements because of high sheet resistivities. In order to check on the agreement of both techniques as a typical example a 10^{14} cm^{-2} S implant annealed at 850°C, 10 min was measured first by C-V and then differential conductivity and Hall measurements were performed. The C-V data are plotted as closed circles in Fig. 3 and indicate reasonable agreement between the two techniques. In Fig. 4, S profiles obtained by C-V measurements are shown for different anneal temperatures. Again, in the medium concentration region reasonable electrical activity is observed following low temperature anneals. For higher anneal temperatures the tail region becomes more electrically active.

---

*Conversion of concentration versus mobility data to depth versus mobility was obtained using the experimental data of the 850°C, 120 min anneal.

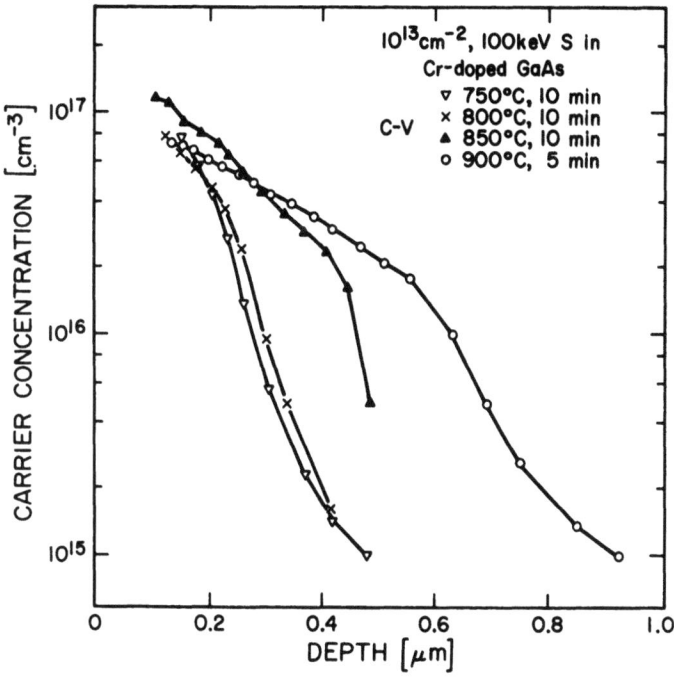

Fig. 4

CONCLUSIONS

With respect to mobility values obtained from epitaxial n-type GaAs, Te and S implants can be successfully used for n-type doping of semi-insulating GaAs in a dose range of 10^{13} cm^2. Annealing at 850°C or higher is required for high electrical activation of the implanted ions. Maximum carrier concentrations are in the range of $< 10^{18}$ cm^{-3}. Elevated target temperatures $\geq 350°C$ have been employed to prevent stable defect formation. However strong deviations from expected shallow Gaussian distributions are found for both S and Te.

REFERENCES

1. R.G. Hunsperger, and O.J. Marsh, Rad. Effects. 6, 263 (1970).
2. R.G. Hunsperger, et al., J. Appl. Phys. 43, 1318 (1972).
3. J.D. Sansbury, Thesis, Stanford University (1970).
4. F.H. Eisen, et al., Proc. 3rd Conf. Ion Implantation, p. 631, (1972).
5. R.A. Logan, B. Schwartz and W. Sundberg, J. Electrochem. Soc. Solid State Sc. Techn. 120, 1385 (1973).
6. C.J. Dell'Oca, G. Yan and I. Young, J. Electrochem. Soc., Electrochem. Sc. 118, 89 (1971).
7. H. Müller, F.H. Eisen and J.W. Mayer, to be published.
8. J.S. Harris, private communication.

ENCAPSULATION OF ION IMPLANTED GaAs

P.L.F. Hemment, B.J. Sealy & K.G. Stephens

University of Surrey

Department of Electronic and Electrical Engineering
Guildford, Surrey

ABSTRACT

Electrical activity from donor ions implanted into GaAs has only been achieved after a post implantation anneal cycle to temperatures in excess of 650°C. SiO_2 coatings have proved unsatisfactory whilst Si_3N_4 layers are good up to at least 750°C. From preliminary measurements Ga_2O_3 films also show promising characteristics.

INTRODUCTION

We have used SiO_2, Si_3N_4 and Ga_2O_3 films as protective layers for GaAs and present our results to date which compare the properties of these coatings on implanted GaAs using Rutherford backscattering (RBS), electron spectroscopy (ESCA), transmission and reflection electron microscopy (TEM & RHEED) and electrical measurements.

EXPERIMENTAL

Samples were prepared from either bulk chromium doped semi-insulating GaAs, Te doped GaAs (n $\sim 10^{17} cm^{-3}$) or epitaxial layers (n $\sim 2-5.10^{15} cm^{-3}$). Implants were carried out at about 8° to the surface normal using Te or Sn ions of energy 50-600 keV and with an average dose rate between 0.05-0.5 $\mu A.cm^2$. After implantation at room temperature or at about 200°C, specimens were encapsulated and then annealed in a nitrogen atmosphere at temperatures between 600-750°C. SiO_2 was formed at about 350°C by the pyrolytic oxidation of silane, Si_3N_4 was sputter deposited at 350-450°C, and Ga_2O_3 was grown by oxidation in an oxygen atmosphere at 500°C and 600°C[1].

After dissolution of the encapsulant the activity profile was measured for some specimens using the Van der Pauw[2] method in conjunction with a chemical stripping technique[3].

RESULTS

After annealing we have found low percentage activity in Sn and Te implanted GaAs coated with SiO_2 (Fig. 1). Most often the peak value was displaced deeper into the material than the atomic profile determined from RBS measurements. (Note the increased activity for 180°C implants compared with room temperature implants, Fig. 1). In agreement with Harris[4] results from ESCA and RBS suggest that gallium diffused through the SiO_2 to the outside surface after annealing at 600°C leaving an arsenic rich surface layer (Fig. 2). At 750°C a build-up of gallium at the GaAs-SiO_2 interface was observed (Fig. 2) and found by transmission electron microscopy to be due to a thin non-uniform layer of β-Ga_2O_3. Stereo TEM and ESCA measurements with depth profiling by argon ion etching showed that the oxide penetrated to a depth comparable to that of the peak in the atomic profile (Fig. 3). This observation explained why the surface layers were of high electrical resistance and why the peak of the activity profile was deeper than the peak of the atomic profile (Fig. 1). There were several exceptions to this result in which gallium oxide did not seem to form in large concentrations due to an improvement of the encapsulation properties of some SiO_2 layers. In these cases the activity peak was very near the theoretical LSS range (Fig. 4) (but the percentage activity was less than 1%, $\hat{n} \simeq 10^{17}$/cc).

In contrast to the SiO_2 results we have found no Ga_2O_3 formation and no outdiffusion of gallium or arsenic through sputtered Si_3N_4 layers when annealed up to 750°C. For example Fig. 5 compares SiO_2 and Si_3N_4 coatings on GaAs implanted with 1.10^{15} Sn^+ ions cm^{-2} and annealed at 700°C. The implants were carried out at 180°C to limit the formation of gross disorder during the implant. It is clear that the Ga/As ratio is close to the expected value of unity for the Si_3N_4 coating but is very different for the SiO_2 layer. After annealing at 600°C or 750°C ESCA showed an increased yield of tellurium at the GaAs/SiO_2 interface, with the Te yield at 600°C being larger than that at 750°C, which did not occur when Si_3N_4 was used (Fig. 3, Fig. 6). A similar specimen implanted at room temperature with 2.10^{14} Te^+ ions cm^{-2} at 150 keV was examined by RHEED after annealing near 600°C without an encapsulant and found to have a polycrystalline phase which was thought to be Ga_2Te_3. Thus the increased Te yield seen by ESCA may correspond to the formation of Ga_2Te_3 at the interface as well as β-Ga_2O_3. A specimen implanted at room temperature with 5.10^{15} Te ions cm^{-2} at 50 keV, followed by annealing at 750°C with a Si_3N_4 coating was found by RHEED to be a perfect single crystal of GaAs and to contain no extra diffraction spots or

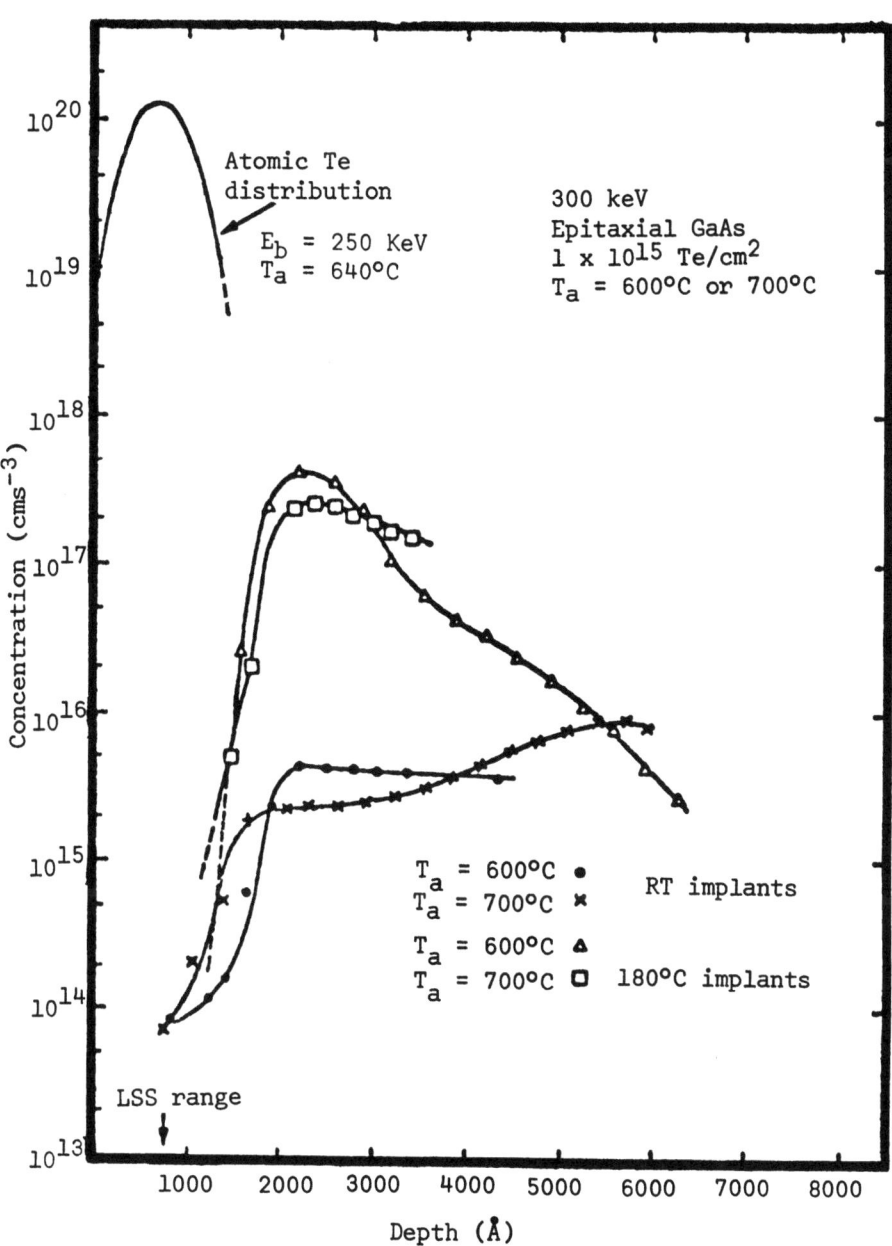

Fig. 1: Activity profile (SiO$_2$)

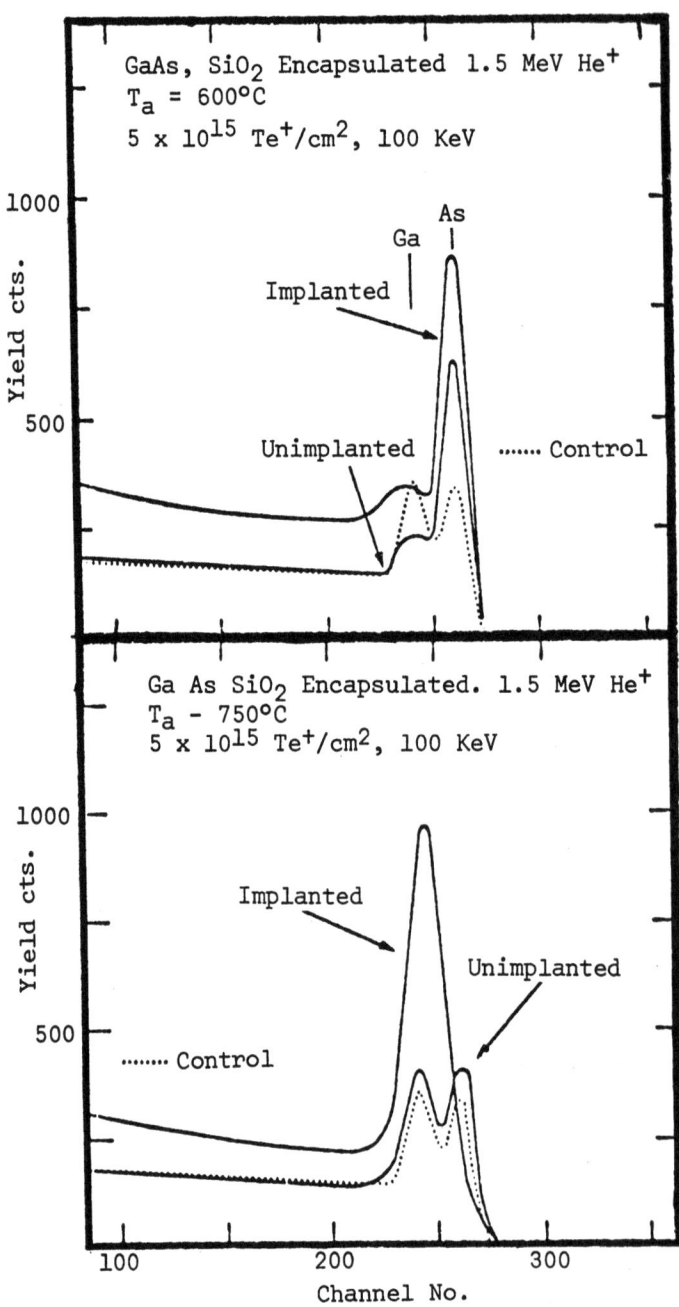

Fig. 2: Surface composition (RBS) (RT). (a) Ta=600°C; (b) Ta=750°C

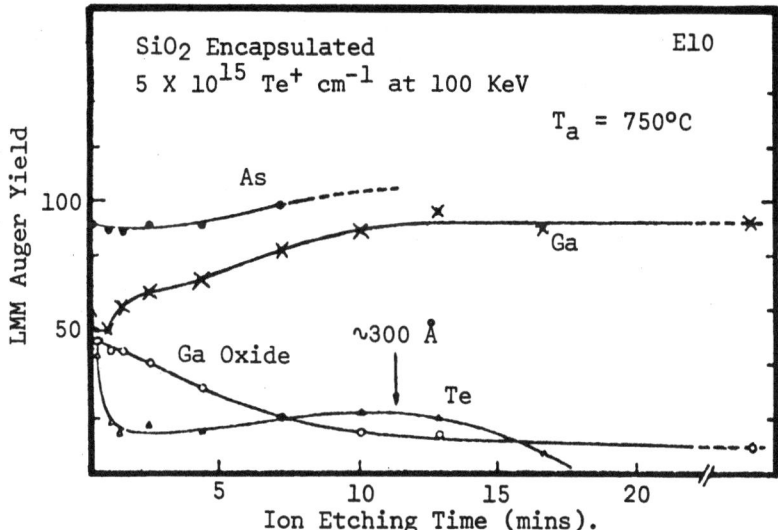

Fig. 3: ESCA profile using SiO_2 (RT).

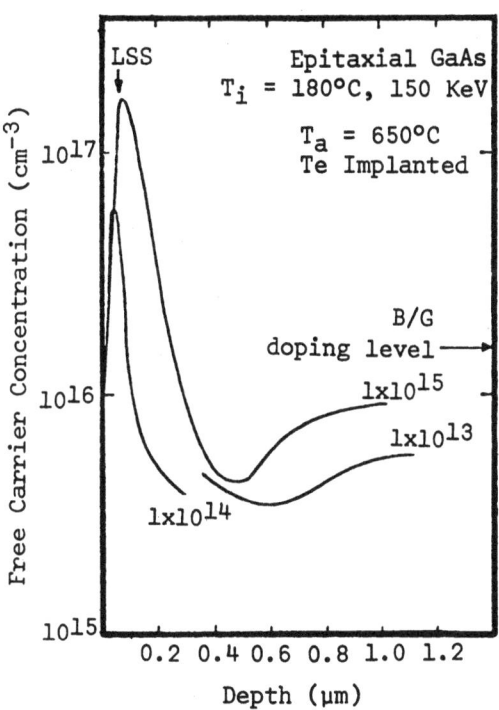

Fig. 4: Activity profile (SiO_2)

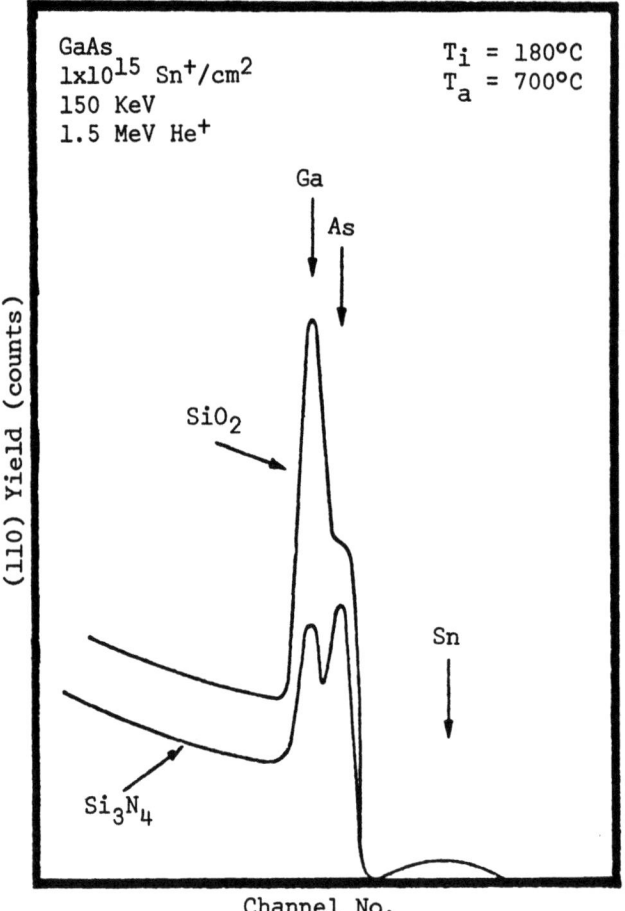

Fig. 5: Surface Composition (RBS)

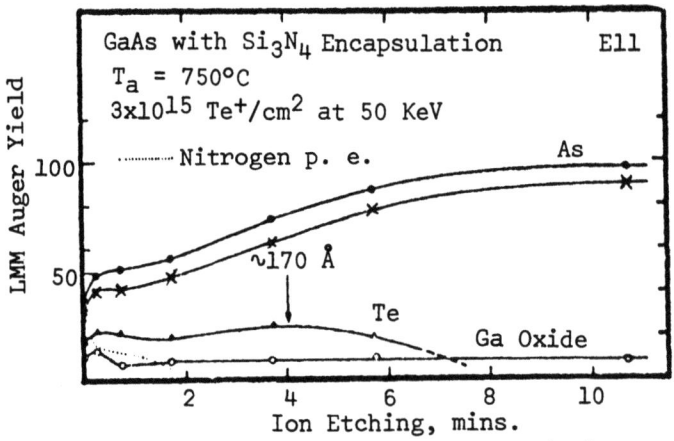

Fig. 6: ESCA profile using Si_3N_4 (RT)

lines implying no outdiffusion of tellurium and no surface Ga_2Te_3 or Ga_2O_3. RBS measurements also indicated that about 60% of the implanted Te resided on or near lattice sites. From the above results and in agreement with several publications[4,5] we conclude that Si_3N_4 is a better encapsulant for GaAs than is SiO_2. However, the fact that Ga_2O_3 forms so readily when SiO_2 coatings are used leads us to suggest the possibility of using Ga_2O_3 as an encapsulant for GaAs.

The thermal oxides we grew were in general very uniform in thickness and seemed to have reproducible properties from specimen to specimen. However, oxidation for one hour at 500°C produced blue films of about 800 Å thick for bulk n-type GaAs but similar treatment on bulk semi-insulating GaAs and epitaxial layers (n ∿ $2-5.10^{15} cm^{-3}$) resulted in a reduced growth rate. ESCA indicated that arsenic was present in an oxide grown at 610°C but after annealing in nitrogen at 680°C for one hour this arsenic concentration became smaller due to outdiffusion. TEM showed that the major phase of both the as grown and annealed specimens was $\beta-Ga_2O_3$. Such films were found to be mechanically stable to temperatures greater than 700°C.

TABLE 1: Structure of ion implanted GaAs surfaces after annealing using SiO_2, Si_3N_4 and Ga_2O_3 coatings.

Technique	T.anneal	Nothing	SiO_2	Si_3N_4	Ga_2O_3
RBS	600	-	As-rich[2], (Ga loss)	No change	No change
	750	-	Ga-rich	No change	-
ESCA	600	Ga_2O_3[1]	Ga_2O_3, Te[3]	-	-
	750	-	Ga_2O_3, Te[3]	Ga_2O_3	-
TEM	600	$\beta-Ga_2O_3$	$\beta-Ga_2O_3$	No oxide	-
	750	$\beta-Ga_2O_3$	$\beta-Ga_2O_3$	No oxide	
RHEED	600	Ga_2Te_3	-	No oxide	No oxide, good crystallinity
	750	-	-	No oxide	-

- Specimens not measured under these conditions
1. Unimplanted specimen, As-rich beneath Ga_2O_3
2. Probable Ga_2O_3 removal by etch
3. Increased Te yield at surface, i.e. outdiffusion

C-V measurements carried out on Au-Ga_2O_3-GaAs structures formed on bulk material indicated carrier concentrations near $10^{17} cm^{-3}$ which was the value measured before oxidation. However some values as low as $3.10^{16} cm^{-3}$ were recorded which may be as a result of compensation by acceptor oxygen centres. Some oxides ($\simeq$ 300-400 Å thick) grown after implantation of 600 keV Te ions were formed by heat treatment for ½ hour at 500°C, at 550°C and at 600°C. RBS showed that beneath these oxides stoichiometry was preserved probably because excess arsenic diffused out during the oxide growth. Subsequent annealing of similar specimens at 700°C in nitrogen gave varying electrical results suggesting that reproducibility depends on growth rate, density and composition of the oxide. However, 16% donor activity has been found for a dose of $2.5.10^{13} Te^{+}$ ions cm^{-2} implanted at 200°C. The mean mobility of this layer was about 2000 cm^2/v.sec which is similar to values found by us and others[4,5] when using Si_3N_4 coatings.

CONCLUSION

We have shown that SiO_2 is a poor encapsulant, compared with Si_3N_4, because of the formation of a non-uniform layer of Ga_2O_3 and the out-diffusion of gallium and tellurium. For anneals up to 600°C, thin uniform layers of Ga_2O_3 are as good as Si_3N_4 coatings. Preliminary experiments using Ga_2O_3 coatings have shown that donor activity after ion implantation can be obtained with good mobilities but results are not yet reproducible. Table 1 summarises our results to date.

ACKNOWLEDGEMENTS

The authors thank Mr. E.C. Bell for valuable comments on the work, Mr. T. Tunkasiri for carrying out the RHEED measurements, Mr. A.E. Glaccum for carrying out some of the RBS experiments, the staff of the accelerator laboratory for assisting with the implants, and Mr. J. Shannon of Mullard Research Laboratories for arranging the deposition of Si_3N_4 films.

REFERENCES

1) B.J. Sealy & P.L.F. Hemment. Thin Solid Films, 22, S39, (1974).
2) L.J. van der Pauw. Philips Res. Rept. 13, 1, (1958).
3) E.C. Bell, A.E. Glaccum, P.L.F. Hemment, K.G. Stephens & J.E. Tansey, to be published.
4) J.S. Harris, F.H. Eisen, B. Welch, J.D. Haskell, R.D. Pashley & J.W. Mayer. Appl. Phys. Lett., 21, 601, (1972); and Yorktown Heights Conf. Ion Implantation in Semiconductors and other Materials, ed. B.L. Crowder, Dec. 1972.
5) J.M. Woodcock, J.M. Shannon & D.J. Clark. to be published.

ION IMPLANTATION OF Cd AND Te IN GaAs CRYSTALS

K. Gamo, M. Takai, M.S. Lin, K. Masuda and S. Namba

Department of Electrical Engineering
Faculty of Engineering Science, Osaka University
Toyonaka, Osaka, Japan

70keV Cd and Te ions were implanted in GaAs crystals at various temperature to investigate lattice location, defects and their annealing characteristics by means of He ion channeling and photoluminescence measurements. It was found by channeling angular scan measurements that high substitutional fraction of Te and Cd was obtained by the implantations between 200 and 300°C. For implantations above 200°C defect peak appears at the depth deeper by 3~4 times than the LSS projected range. It was observed that implantations at 500°C through SiO_2 and Si_3N_4 film on the surface was ineffective for the improvement both in the substitutional fraction and defect density. From photoluminescence measurements it was found that As pre-implantation was effective to suppress the formation of As vacancy.

INTRODUCTION

Many investigations on ion implantation in GaAs have been reported[1~6]. Channeling measurements on ion implanted GaAs have been done by several authors[3-6] and it was observed that lattice location of implanted atoms and defects produced by the implantation strongly depended on implantation temperature. It is necessary to do angular scan measurements in order to know exact lattice location. So far, however, very few have been reported on channeling angular scan measurements for implanted GaAs[3,4]. We investigated the lattice location of Cd and Te implanted in GaAs as a function of implantation temperature by measuring channeling angular distribution. We also investigated the effect of the encapsulation by SiO_2 and Si_3N_4 during hot implantation and dual implantation of As on the lattice location and defects.

EXPERIMENTAL PROCEDURE

The sample preparations, implantation and annealing procedures are written elsewhere in detail[6]. Some hot implantations at 500°C were made with 200keV Cd ions through ~1000Å thick SiO_2 or Si_3N_4 encapsulating films. The As pre-implantations were made at room temperature before Te implantations.

The channeling measurements were carried out by a conventional apparatus with a 0.8 to 1.8MeV He^+ beam. On the discussion of lattice location or percentage attenuation along the lattice rows, the correction due to the dechanneled fraction of the probe beam was not made. Experimental procedures for photoluminescence measurement is written elsewhere[9].

RESULTS AND DISCUSSION

The results of lattice location (percentage attenuation) and defect (minimum yield χm) measurements are shown in Fig. 1 as a function of implantation temperature. The χm was taken at the channel number just behind the damage peak of the He backscattering spectra. For Te implantation, it is clear from Fig. 1 that most Te atoms lie along the atomic rows at 200°C and that the fraction lying along the atomic rows decreased slightly with increasing the implantation temperature above 400°C. For Cd implantation, the decrease of the fraction above 400°C was remarkable.

The χm showed rapid decrease with increasing temperature below 200°C and was almost independent of temperature above 200°C (see also Fig. 4).

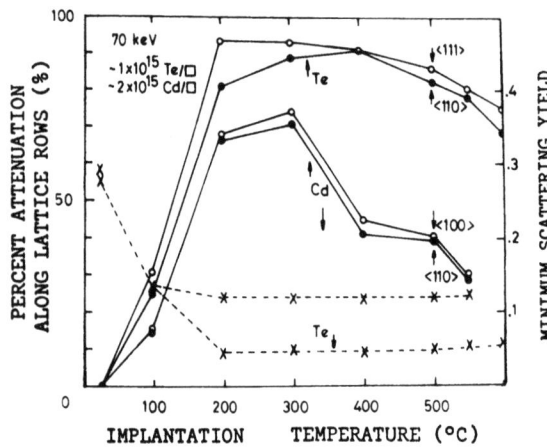

Fig. 1 Dependence of percentage attenuation and χm on implantation temperature. Samples are unannealed.

From the results shown in Fig. 1, the temperature range can be divided in three regions, i.e, below ~100°C, between ~200 and ~300°C and above ~400°C. The angular scan measurements were carried out at typical temperature at these three regions in order to study lattice location in detail. For Te implantations, the results are published elsewhere[6]. It was found that Te implanted above 200°C locate on the substitutional site and those implanted at room tempera-

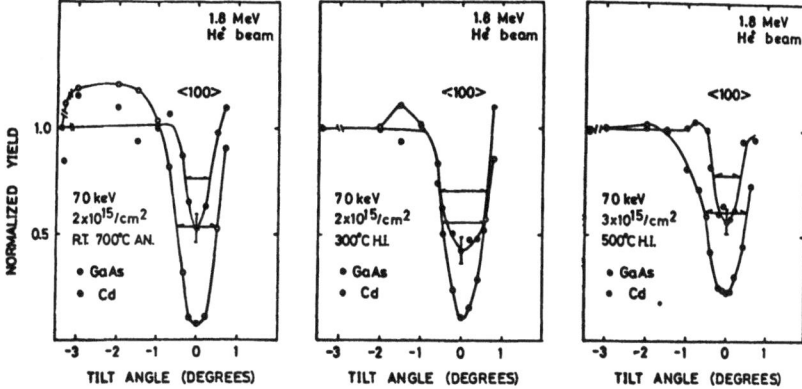

Fig. 2 Channeling angular dips for 1.8MeV He$^+$ backscattering from GaAs and Cd along the <100> axis. Implantation temperature is room temperature (a) (annealed at 700°C), 300°C (b) and 500°C (c).

ture locate on the site displaced by 0.4~0.5Å from the lattice sites even after annealing at 800°C.

For Cd implantation, the results of the angular scan measurements are shown in Fig. 2. The implantation temperature was room temperature (Fig. 2a), 300°C(Fig. 2b) and 500°C(Fig. 2c). For both room temperature and 500°C, the width of the Cd dip was narrower than that of the GaAs and the depth of the Cd dip was shallower than that of the GaAs. These results mean that Cd implanted at room temperature or 500°C locate on the sites displaced from the lattice sites. From Lindhard's channeling model[7], the displacement was estimated to be ~0.5Å from the <100> axes for the two cases. For implantations at 300°C, the width of the Cd dip along the <100> axis was in agreement with that of the GaAs dip but the depth of the Cd dip was shallower than that of the GaAs dip. (Fig. 3b) The angular scan measurements through the <111> and <110> axes also showed the same width of the dip for both Cd and GaAs (not shown). These results mean that about 70% of Cd implanted at 300°C locate on the substitutional sites.

Figure 3 shows the He backscattering spectra for GaAs crystals implanted at 300°C with Cd and Te to a dose of $2 \times 10^{15}/cm^2$. The spectra for unimplanted GaAs is also shown for comparison. The surface peak for the implanted GaAs was almost the same with that for the unimplanted GaAs. This means that the damage near the surface is almost annealed out during the hot implantation. The defect peaks appeared at the depth deeper than the projected range of about 200Å. The peak for Cd implantation (~900Å) was deeper than that for Te implantation (~620Å) and the dechanneling rate for Cd implantation was higher than that for Te implantation.

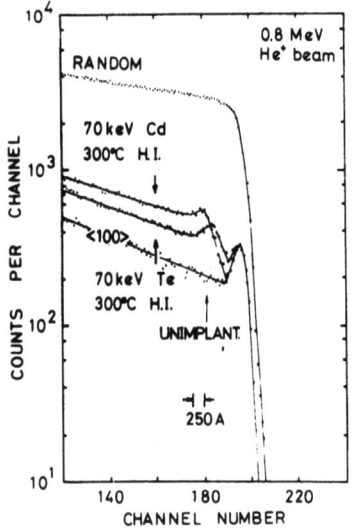

Fig. 3 0.8MeV He$^+$ backscattering spectra.

Figure 4 shows the χ_m along the <100> axis as a function of dose for GaAs crystals implanted at various temperature above 150°C. The χ_m for unimplanted samples was taken at the same channel as implanted samples. The χ_m was almost independent of temperature. For samples implanted to a dose of $2 \times 10^{15}/cm^2$, the χ_m obtained after annealing at 800°C(about 0.07) was still higher than the χ_m for the unimplanted sample. This indicates some lattice disorder still remains after the annealing. The amount of Cd retained after the annealing reduced by about 50% of the initial amount due to the outdiffusion. Samples implanted together with As show almost the same χ_m as those implanted without As (shown by a crosses in Fig. 4). Samples implanted through the encapsulating film shows slightly higher χ_m, especially in case of the SiO_2 film (shown by solid circles). The dose for these encapsulated samples is the amount embedded in the GaAs samples through the films.

Table 1 shows the summary of the lattice location and defect (χ_m) measurements for samples implanted in various conditions. Picraux observed the evaporation of As from a GaAs surface during annealing[8]. This fact suggests that the decrease of the percentage attenuation for implantations above 400°C (see Fig. 1) might be due to the evaporation of As or the formation of As vacancies. The implantations through the encapsulating films and the dual implantation with As was made in order to test this possibility. As shown in the Table, no improvement was observed in the lattice location and defects (χ_m) for both

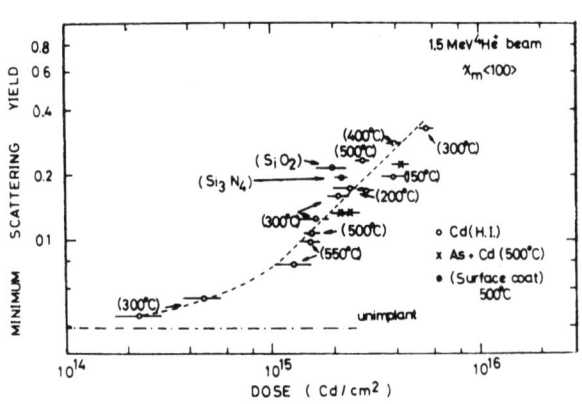

Fig. 4 χ_m as a function of dose.

Table 1 Comparison of the percentage attenuation and the χm for the various implantations.

SAMPLE	%<100>	χm	%<110>	χm	dose(/cm^2)
R.T. IMPLANT.+500°C ANNEAL	40	0.17	40	0.13	2.8×10^{15}
500°C H.I.	40	0.24	30	0.24	2.8×10^{15}
SiO$_2$ COAT 500°C H.I.	30	0.22	--	--	2.0×10^{15}
Si$_3$N$_4$ COAT 500°C H.I.	30	0.19	30	0.17	2.2×10^{15}
As R.T. IMPLANT.+500°C H.I.	50	0.14	40	0.14	2.2×10^{15}

Probing beam: 1.5MeV ^{4}He$^+$

the implantations through the film and the dual implantation. These results suggest that the surface decomposition does not have the prominent effect on the decrease of the percentage attenuation.

The effect of As dual implantation was also studied by photoluminescence measurements. Figure 5 shows the photoluminescence spectra observed for samples implanted at 300°C with Te to a dose of 2×10^{14}/cm^2 and annealed at 800°C. Samples implanted without As exhibited an emission at 9140Å together with an edge emission, while samples implanted with As exhibited only the edge emission. This suggests that the emission at 9140Å is due to the As vacancy or the As vacancy associated complex defect. This was also supported from the annealing characteristics of unimplanted samples[9]. The activation energy for the formation of the emission was found to be 2.0eV which agreed well with the reported values of the activation energy for the formation of the As vacancy[10,11]. Therefore, the present results suggest that the formation of As vacancy can be suppressed by As dual implantation.

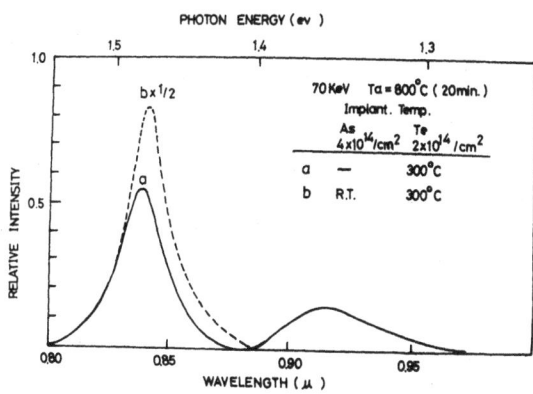

Fig. 5 Effect of As pre-implantation on photoluminescence of Te implanted GaAs.

CONCLUSION

1) The high substitutional fraction of Te and Cd can be obtained by the implantation between 200 and 300°C. Most of Te and Cd locate on sites displaced by ~0.5Å for room temperature implantation.
2) The hot implantation at 500°C through a SiO_2 and a Si_3N_4 film give no improvement in lattice location and defect density.
3) As pre-implantation can suppress the formation of As vacancy.

ACKNOWLEDGEMENTS

We are very grateful to Drs. T. Tsurushima and H. Tanoue of Electrotechnical Laboratories for the 200keV Cd implantation, and to Mr. K. Kawasaki of Osaka University for his help in performing the other implantations.

REFERENCES

1) R.G. Hunsperger and O.J. Marsh: Metal. Trans. **1** 603 (1970)
2) T. Itoh and Y. Kushiro: Proc. 2nd Intern. Conf. Ion Implantation in Semiconductors. ed. I. Ruge and J. Graul (Springer Verlag, 1971) p.168
3) R. Bicknell, P.L.F. Hemment, E.C. Bell and J.E. Tansey: Phys. Status Solidi **12** K9 (1972).
4) J.S. Harris, F.H. Eisen, B. Welch, J.D. Haskell, R.D. Pashley and J.W. Mayer: Appl. Phys. Letters **21** 601 (1972).
5) K. Gamo, M. Takai, K. Masuda and S. Namba: Proc. 4th Conf. Solid State Devices (Tokyo, 1972) Suppl. J. Japan Soc. Appl. Phys. **42** 130 (1973).
6) M. Takai, K. Gamo, K. Masuda and S. Namba: Japan. J. Appl. Phys. **12** 1926 (1973).
7) J. Lindhard: kgl. Danske Videnskab. Selskab. Mat. Fys. Medd. **34** No. 14 (1965).
8) S.T. Picraux: Ion Impalntation in Semiconductors and Other Materials ed. B.L. Crowder (Plenum Press, 1973) p.641.
9) M.S. Lin, K. Gamo, M. Takai, K. Masuda and S. Namba: Proc. 6th Intern Conf. Electron and Ion Beam Science and Technology (San Francisco, May 1974) (to be published).
10) H.R. Potts and G.L. Pearson: J. Appl. Phys. **37** 2098 (1966).
11) E. Munoz, W.L. Snyder and J.L. Moll: Appl. Phys. Letters **16** 262 (1970).

IMPLANTATION OF SILICON INTO GALLIUM ARSENIDE

T. Miyazaki and M. Tamura

Central Research Laboratory, HITACHI Ltd.

Higashi-koigakubo, Kokubunji, Tokyo, Japan

ABSTRACT

The electrical properties of silicon implanted layers in gallium arsenide have been investigated for the purpose of formation of n-type layers with high electron concentration. The dependence of sheet resistivity, surface carrier concentration and mobility on dose, implantation temperature and annealing temperature was determined. Peak doping levels are about $3 \times 10^{18} cm^{-3}$ for 340°C implantation with $10^{15} cm^{-2}$ dose. Diffusion of implanted silicon was also confirmed.

INTRODUCTION

The doping of GaAs by ion implantation has been studied by many investigators to make p- and n-type layers using various ion species. The implantation of p-type dopants reached the maximum hole concentrations which can be achieved by the introduction of the dopants during the crystal growth process. In contrast, the implantation of n-type dopants gave merely the electron concentrations at least an order of magnitude lower than those which can be attained during crystal growth. Sansbury et al[1] and Harris[2] experimentally verified that implanted Si acts as an n-type dopant after annealing at high temperatures. However, the electrical activity of conversion of Si to donor was very low. Recently, Harris et al reported the great usefulness of hot implantation technique to achieve the high electrical activity from implanted Te[3]. In this work, we have examined implantation temperature and annealing conditions in order to obtain better results from the implantation of Si into GaAs.

EXPERIMENTAL

Samples were prepared from boat-grown Cr-doped semi-insulating GaAs with resistivities above 10^8ohm-cm. All wafers were [100] oriented. Si implantations were performed under the following conditions; incident energies: 50 to 170keV, substrate temperatures: r.t. to 650°C and doses: 10^{11} to 10^{16}cm^{-2}.

Following implantations, samples were annealed for 60min at temperatures between 100 and 900°C in N_2 or H_2 atmosphere. During annealing at temperatures above 500°C, samples were coated with an insulator film to prevent decomposition of GaAs substrates and outdiffusion of dopant atoms from the substrate surfaces.

The effect of protective films were made clear by Harris et al [3]. In the present work, several kinds of insulator films such as Si_3N_4, SiO_2, Al_2O_3 and amorphous Si were used. The highest electrical activity was attained in both cases of Si_3N_4 and Si films. As the other oxide films formed deep electron traps near the GaAs surface and, in addition, diffusion of Ga and As through these films was observed even during deposition, oxide films seemed not to be suitable for the purpose intended. Therefore, for annealing above 500°C, Si_3N_4, Si or those combined double layer films were applied to the protective layer.

RESULTS AND DISCUSSIONS

Si ions with 50keV incident energy were implanted in Cr-doped GaAs at r.t.. The value of the r.t. resistivity decreased with increasing dose and annealing up to 400°C increased it again to higher value. For the lightly implanted specimen a further anneal at 500°C increased the surface resistivity still further close to the value of unimplanted surface. These increases were attributed to annealing out of damage effects as shown by Sansbury et al[1].

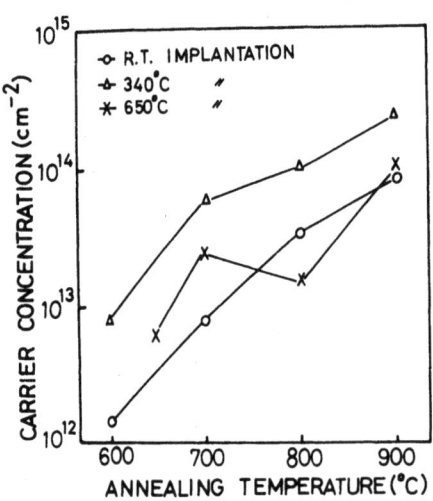

Fig. 1. Electron concentration vs. annealing temperature for 10^{15}cm^{-2} Si implants with implantation temperature as a parameter.

Electron diffraction and backscattering measurements showed the amorphous layer was formed above a dose of 10^{13}cm^{-2}. The amorphous layer was recrystallized by annealing at high temperatures. The recovery of amorphous region depended upon the dose and the annealing temperature of 550°C was required for single crystal formation at the dose of 10^{15}cm^{-2}. In contrast, for implantations at temperatures above 200°C, no formation of the amorphous layer was observed even at a dose of more than 10^{15}cm^{-2}.

Annealing at above 600°C brought out a rapid drop in resistivity and the implanted Si ions were electrically active as donor impurities. For the 10^{15}cm^{-2} implants, Hall carrier concentrations are plotted as a function of annealing temperatures in Fig. 1. Implantation temperatures were varied at r.t., 340°C and 650°C. Surface carrier densities for r.t. and 340°C implantations increased monotonically with annealing temperatures and surface mobilities(not shown here) decreased approximately from 3000 to 1000 cm^2/V·sec. Hot implanted samples showed higher values of the carrier concentration than those of r.t. implanted samples, especially for low annealing temperatures. For the 340°C implantation, the maximum electrical activity of 25% was obtained after the 900°C annealing. For the lightly implanted case(10^{13}cm^{-2} implants), more than 50% activity was attained at the 900°C annealing.

The result of the 650°C implantation was somewhat different from others. In the present case, a 500Å thick Si$_3$N$_4$ film was coated during implantation and the total dose implanted in GaAs was about 5% less than in the case without any protective film. As-implanted layer showed n-type conduction. As the annealing temperatures rose, both conductivity and carrier concentration increased. At the temperature of 800°C, on the contrary, both values decreased slightly. The maximum electrical activity was approximately 10%.

Secondary defects were observed by transmission electron microscopy for the above samples. In the case of r.t. implantation, defects observed at the 600°C annealing were many small black dotted ones, while at the 800°C annealing dislocation loops were observed as a result of the growth of black dotted defects[4]. On the other hand, in the case of hot implantation the black dotted defects and dislocation loops were formed in the 340°C and 650°C as-implanted layer, respectively. As the annealing temperature increased, loop size increased and loop density decreased. Densities of defects or loops were considerably lower in hot implantations than in r.t. implantation. Three types of loops were observed for the 650°C implantation; crystallographic circular loops(about 3000Å in diameter), double loops and faulted loops. Among them faulted loops disappeared after the annealing above 800°C.

In order to obtain a more detailed evaluation of n-type layers, differential Hall effect measurements were performed with layer removal. Fig. 2 shows the result of implanting two different doses into semi-insulating wafers and annealing at 900°C for 60min with Si$_3$N$_4$ coating. In both implantations, the peak electron concentrations appeared near the surface and exceeded 10^{18}cm^{-3}. The maximum concentration of the 10^{15}cm^{-2} implanted sample was 2.8×10^{18}cm^{-3} which is approximately equal to the limit value of electron concentration for Si doped bulk material. The profiles are deeper than the LSS distribution and the distributions of electron concentration are nearly constant at the depth from the projected range to the position of junction(measured by angle-lapping and staining). The result indicates the deepness of the profiles is due to diffusion of implanted Si for the annealing at high temperatures. Another effect of increasing the dose is the lowering of mobility, especially near the surface. Comparing the mobility in implanted layers with the experimental data of Sze et al[5] for bulk GaAs single crystals, the implanted layers followed very favorably the theoretical ionized impurity scattering curve except for high electron concentration near the surface. The lowering of the mobility in the surface layer suggests the presence of unannealed centers in the heavily implanted regions.

Temperature dependences of mobility and surface carrier concentration for the 10^{13}cm^{-2} implanted samples were measured at temperatures from 77 to 300°K. The result was characterized by an almost constant value for the as-annealed sample. After etching the surface layer by 1500Å thickness, however, the temperature dependence became conspicuous. The mobility increased from 3900 to 21000cm^2/V·sec as

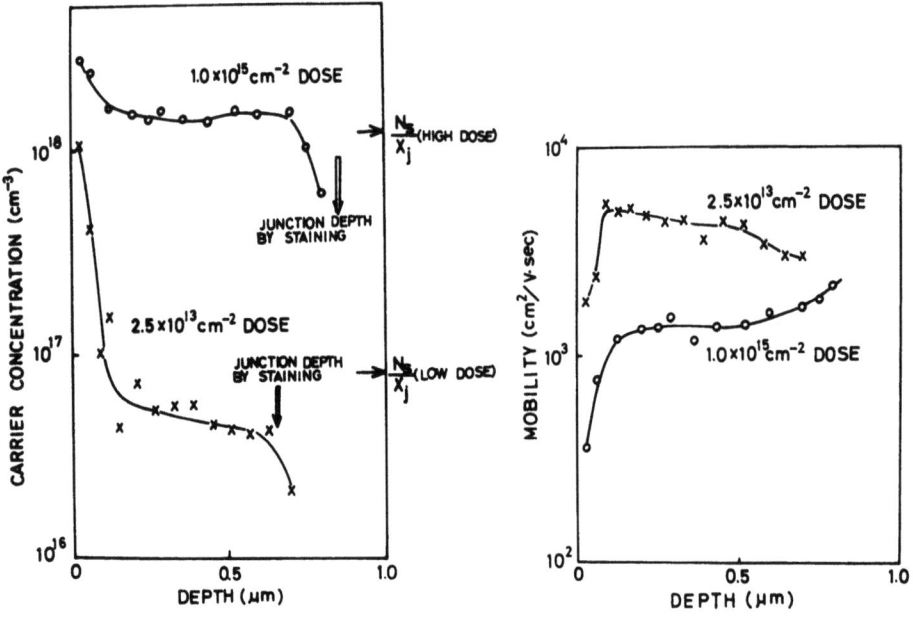

Fig. 2 Electron concentration and mobility profiles of two 170keV Si implants annealed at 900°C for 60min.

the temperature decreased from 300 to 77°K and varied as $T^{-1.2}$, which suggests the contribution from the polar-optical-mode scattering.

The diffusion of implanted ions was confirmed first by Sansbury et al[1]. The junction depth was dependent upon the implanted dose and temperature, in addition to the annealing time and temperature. In the present experiment, the dose was varied from 10^{14} to 10^{16} cm^{-2} and implantation energy and temperature were fixed at 50keV and room temperature, respectively, in order to use the implantation in place of the predeposition in the diffusion process. The junction depth was measured by bevelling and staining. If we assume the diffusion profile of implanted ions(not carrier concentrations) to be given by the Gaussian function

$$C(x_j) = \frac{Q}{\sqrt{\pi Dt}} \exp[-(x_j - R_p)^2/4Dt] \qquad (1)$$

where $C(x_j)$ is the impurity concentration at x_j, x_j the junction depth, D the diffusion coefficient, R_p the LSS projected range and Q the total amount of impurities which is equal to the implanted dose. Now, another assumption is made: the values of $C(x_j)$ are the same for different doses. Then, the diffusion coefficient at different temperatures can be obtained by the measurements of x_j, t and Q. In Fig. 3 the diffusion coefficient is plotted as a function of reciprocal temperature and expressed as

$$D = 7.5 \times 10^{-10} \exp(-1.1/kT). \qquad (2)$$

Compared with the estimated diffusion coefficient of 5×10^{-13} cm^2/sec at 700°C[1], the value of D in Fig. 3 is one order of magnitude lower than it. The discrepancy may be due to the difference of annealing conditions, substrate materials used, implantation energies and the assumptions for diffusion profiles.

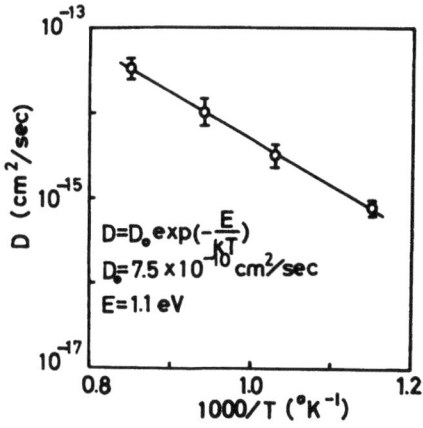

Fig. 3 Diffusion coefficient of electrically active Si implants. Energy: 50keV, doses: 10^{14}, 10^{15} and 10^{16} cm^{-2}.

Implantation of Si was applied to fabrication processes of GaAs devices. An epitaxial p-type layer with carrier concentration of $2.5 \times 10^{17} cm^{-3}$ was implanted with Si at 340°C. Implantation energy was 170keV and the total dose was $10^{15} cm^{-2}$. After the implanted layer was covered by an Si_3N_4 film, the sample was annealed for an hour at 900°C. Capacitance-voltage measurement on a planar n^+-p diode indicated a close relation of $C^{-2} \propto V$ which shows an abrupt junction profile. The breakdown voltage of an as-annealed diode was approximately 10V at 1mA/cm². After etching the implanted layer of 1500Å thickness from the surface, the breakdown voltage increased to 14.5V, which is nearly equal to the theoretically predicted value. The n value for the forward current-voltage curve of the etched diode was estimated to be 2.07, indicating that the recombination current dominates.

CONCLUSIONS

The conclusions derived from the present work are as follows:
1) The formation of amorphous layer during implantation in GaAs is strongly temperature and dose dependent. A dose above $10^{13} cm^{-2}$ converted the implanted region into the amorphous layer for r.t. implantation. In contrast, no formation of the amorphous layer was observed even at the $10^{15} cm^{-2}$ dose for high temperature implantation above 200°C.
2) The electrical activity of implanted Si is also dependent upon implantation temperature. It was found that hot implantation created n-type layers in semi-insulating GaAs which had peak electron concentrations nearly equal to the maximum value attainable in Si doped bulk crystals.
3) Low energy implantation was used as a predeposition for the diffusion process and diffusion coefficient was estimated.

ACKNOWLEDGMENT

The authors wish to thank T. Tokuyama for his encouragement throughout this work and T. Ikeda and N. Nakamura for their help with sample preparations.

REFERENCES

1. J. D. Sansbury and J. F. Gibbons, Rad. Effects, 6, 269(1970).
2. J. S. Harris, Ion Implantation in Semiconductors, edited by I. Ruge and J. Graul(Springer-Verlag, Berlin,1971), p.157.
3. J. S. Harris et al, Appl. Phys. Letters, 21, 601(1972).
4. M. Tamura and T. Miyazaki, Proc. 5th Conf. on Solid State Devices (Tokyo), 12-6(1973).
5. S. M. Sze and J. C. Irvin, Solid State Electronics, 11, 599(1968).

THE EFFECTS OF ION DOSE AND IMPLANTATION TEMPERATURE ON ENHANCED

DIFFUSION IN SELENIUM ION-IMPLANTED GALLIUM ARSENIDE

Y. Kushiro and T. Kobayashi

Kokusai Denshin Denwa Company

Nakameguro, Tokyo 153, Japan

ABSTRACT

Electrical properties and photoluminescence have been studied in GaAs implanted with Se at room temperature, 200 °C and 400 °C. For the low dose case between 10^{12} and 10^{13} ions/cm^2, electrical activities for hot implants were larger by a factor of two than those for the RT implant, whereas at a dose of 10^{14} ions/cm^2 the difference between them was reduced. The profiles of carrier concentration were characterized by deep diffusion of implanted atoms. The diffusion is enhanced with the increase of dose. Implantation temperature had little effect on the enhancement for the low dose case, while at a dose of 10^{14} ions/cm^2 hot implants increased the depth of the diffusion tail. A study of photoluminescence has indicated that two kinds of defects were introduced by implantation; one is dominant at low doses and responsible for a broad band emission centered at 1.31 eV, and the another is Ga vacancy-Se donor complex, which is increased with dose and affects the enhanced diffusion. It is suggested that the broad band at 1.31 eV is attributed to As vacancy.

INTRODUCTION

Anomalous penetration of implanted ions has been frequently observed in GaAs. As for donor impurities, Sansbury and Gibbons (1) have obtained accurate profile of S and Si implanted in GaAs, and they have observed a strong diffusion enhancement attributed to implantation-produced damage. Foyt et al. (2) have studied doping efficiency and carrier concentration profile in Se-implanted GaAs, and they have reported hot implant yielded a high doping efficiency and enhanced diffusion was not observed. This paper reports the effects of dose and implantation temperature on the enhanced diffusion and other electrical properties, and the defect structure which is produced by implantation and affects such properties.

EXPERIMENTAL PROCEDURE

50 keV Se ions analyzed with magnet were implanted into (100) melt-grown GaAs wafers. Implantation was performed at room temperature, 200 °C, and 400 °C, with doses ranging from 10^{12} to 10^{14} ions/cm^2. The substrates were tilted 7° off axis to minimize ion channeling effect. The substrate materials used consisted of O_2-Cr doped semiinsulating GaAs samples which implanted for measurements of Hall effect and photoluminescence, and undoped n-type wafers with carrier concentration of 10^{15}-10^{16} cm^{-3} used for capacitance-voltage measurements. After implantation the samples were coated with a chemical-vapour-deposited layer of SiO_2 about 3000 Å thick. The following anneals were carried out in vacuum evacuated to 10^{-7} Torr at a temperature in the range 600-900 °C for 5 min.

The electrical properties of the implanted layers in semiinsulating GaAs were determined from Hall effect measurements using the van der Paw method. A Keithley 180 digital nanovoltmeter was used for the measurements. The measurements of carrier concentration profile in the implanted layers were done by usual accurate point by point C-V measurements and by an automatic profiler using a feedback method proposed by Miller (3). The agreement of the results between both methods was obtained with good accuracy. Photoluminescence measurements were performed at 68 °K in continuous flow cryostat. The exciting radiation was supplied by a CW Argon ion laser operated at 488 nm. The emitted light from the front surface was analyzed by a Perkin-Elmer E1 grating monochrometer and detected with an RCA 7102 photomultiplier cooled with dry ice. The data shown are corrected for its responce.

RESULTS AND DISCUSSION

Samples were implanted with Se at RT, 200 °C and 400 °C at doses of 10^{12}, 10^{13} and 10^{14} cm^{-2}. The results of Hall effect measurements on these samples are shown in Fig.1 and 2. A marked difference was observed between sheet carrier concentrations for the RT implant and those for hot implants. Electrical activities for hot implants were larger by a factor of two than those for the RT implant, for the low dose case between 10^{12} and 10^{13} cm^{-2}; after 850 °C anneals approximately 50-90 % activities were obtained for the hot implants. The effective Hall mobility seems to be little affected by implantation temperature for the low dose case. Annealing to 800 °C increased the mobility to about 3000 cm^2/V·sec, and further annealing has had little effect on the mobility. This result indicates that the most

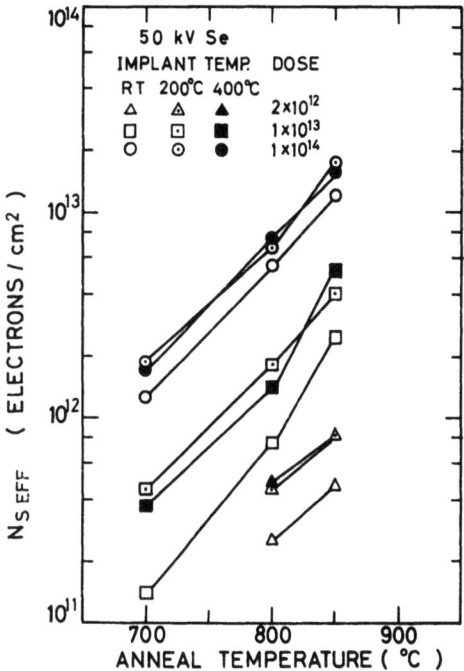

Fig.1. Effective sheet electron concentration vs. anneal temperature for RT, 200, and 400 °C implants with doses as a parameter.

disorder regions formed by implantation are annealed out at 800 °C and the mobility seems to be limited by a large number of deep level impurities contained in the semiinsulating substrate. At a dose of 10^{14} cm^{-2}, on the other hand, electrical activities were only 13-18 % even after 850 °C anneals, and also the mobilities were lower than those for lower doses. The 10^{14} cm^{-2} implant, therefore, seems to form unannealed disorder regions even in the hot implantation. The sheet carrier concentration increased further at higher anneal temperatures; another n-type conduction, however, appeared prominently after an 900 °C anneal. This conduction may be due to outdiffusion of Cr during the anneal from GaAs into a SiO_2 film (4).

The carrier concentration profile has been characterised by a deep diffusion of Se, as shown in Fig.3. It should be noted that a projected range for 50 keV Se in GaAs is only 220 Å (5). In the case of the 200 °C implant the diffusion coefficients at 800 °C are estimated to be $6 \cdot 10^{-14}$, $8 \cdot 10^{-14}$, and $1.3 \cdot 10^{-13}$ cm^2/sec for the $3 \cdot 10^{12}$, $1 \cdot 10^{13}$, and $1 \cdot 10^{14}$ cm^{-2} doses, respectively; while the thermal equilibrium diffusion coefficient of Se at 800 °C is only $1 \cdot 10^{-16}$ cm^2/sec (6). The diffusion of Se during the anneals is apparently enhanced by the presence of implantation-caused defect which increases likely

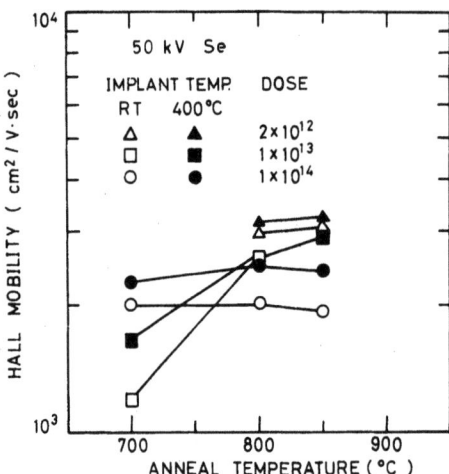

Fig.2. Effective electron mobility vs. anneal temperature for RT, 200, and 400 °C implants with dose as a parameter.

with dose. For the low dose case the hot implants increased carrier concentration near the surface, while at a 10^{14} cm^{-2} dose they increased only the depth of the tail. The effects of hot implant on electrical activity shown in Fig.1 are well explained by the difference in these profiles. Moreover, these profiles suggest that hot

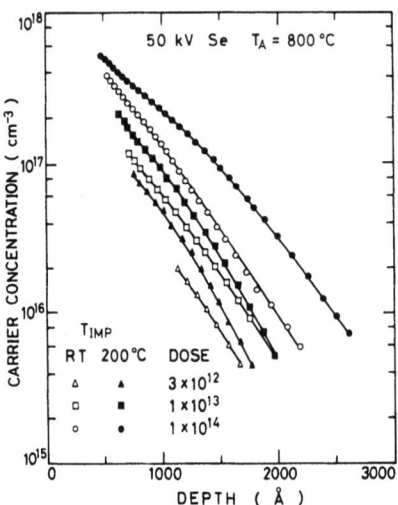

Fig.3. Carrier concentration profile vs. depth for RT and 200 °C implants with dose as a parameter.

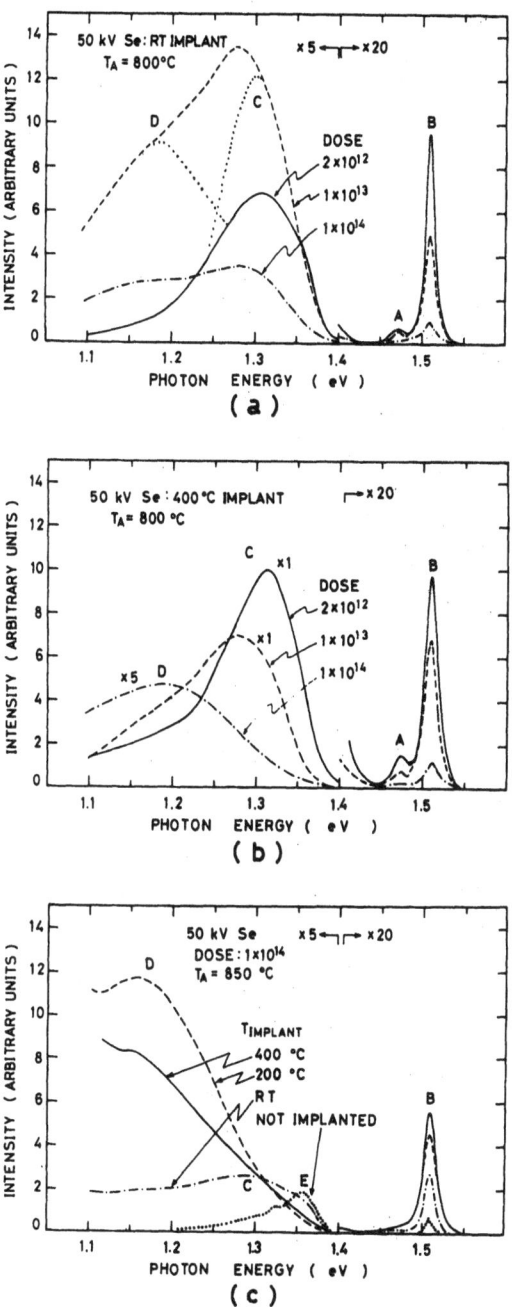

Fig.4. Photoluminescence spectra of n-type layers in semiinsulating GaAs at 68 °K. (a) samples for RT implant and 800 °C anneal, (b) for 400 °C implant and 800 °C anneal, and (c) for 10^{14} cm^{-2} implant and 850 °C anneal.

implants at high dose increase the defect responsible for the enhanced diffusion.

A study of photoluminescence has revealed the structure of the defects, and the results are shown in Fig.4. A near-bandgap emission B is peaked at 1.509 eV which is lower about 6 meV than a bandgap energy at 68 °K. A weak emission peak A at 1.48 eV has a lower energy about 35 meV than a bandgap energy. Since the energies of 6 and 35 meV are approximately the ionization energies of hydrogenic donors and acceptors respectively, the transitions responsible for these emissions are identified as donor-valence band and conduction band-acceptor (7). The donor is presumably Se and the acceptor is thought to be Si in As site, Si_{As}. A broad band emission C peaked at 1.31 eV has been reported for neither melt-grown nor epitaxial grown GaAs, but it has been observed in electroluminescence of Zn-implanted GaAs diode (8). Another broad band emission D peaked at 1.20 eV is usually observed in melt-grown n-type GaAs and identified as so-called self-activated luminescence attributed to Ga vacancy-donor complex (9). A relative weak emission E at 1.36 eV was observed in unimplanted but annealed samples (see Fig.4c), and it is identified as Cu acceptor and its phonon replica (10). Cu may be accidentally introduced during anneals. A near-bandgap emission B is decreased with the increase of dose, and at the highest dose every emission is decreased. The high dose implant seems to form either killer center or deep level defect undetected by the present detector. At any rate unannealed disorder regions reduce the intensity of emissions shown.

It is particularly noteworthy that the peak A was only observed when the peak C was dominant. Since the peak A is thought to be Si_{As}, the peak C may be attributed to As vacancy or As vacancy associated defect. Then, the increase of electrical activities by hot implants for the low dose case is explained as to be due to the increase of As vacancy, as seen in Fig.4 (a) and (b). For the case of 10^{14} cm^{-2} implant, hot implants enhance the formation of Ga vacancy, as shown evidently in Fig.4 (c). It can be also seen in the figures that Ga vacancy-Se donor complex becomes dominant with the increase of dose. Ga vacancy is suspected, therefore, to be responsible for the enhanced diffusion of Se. The relation of Se diffusion with Ga vacancy is not unlikely, since it has been reported that the diffusion coefficient of Se increases with As pressure (6). The mechanism of the defect formation and enhanced diffusion, however, are not evident in the present stage, and the further study should be necessary.

CONCLUSION

The increase of ion dose results in the enhancement of Se diffusion during anneal due to implantation-produced defect, and Ga vacancy probably corresponds to the defect. For the low dose case between 10^{12} and 10^{13} cm^{-2}, hot implants result in the increase of carrier concentration near the surface, and this effect may be attributed to the formation of As vacancy. At higher dose, however, hot implants increase Ga vacancy formation and so enhance the diffusion of Se.

ACKNOWLEDGEMENTS

The authors wish to thank O. Shinbori for measurements of photoluminescence and T. Seimiya for depositing SiO_2 layers.

REFERENCES

1. J. D. Sansbury and J. F. Gibbons, Ion Implantation, edited by F. H. Eisen and L. T. Chadderton (Gordon and Breach, London, 1971) P.253

2. A. G. Foyt, J. P. Donnelly, and W. T. Lindley, Appl. Phys. Letters 14, 372 (1969)

3. G. L. Miller, IEEE Trans. Electron Devices ED-19, 1103 (1972)

4. Y. Sato, Japan. J. Appl. Phys. 12, 242 (1973)

5. W. S. Johnson and J. F. Gibbons, Projected Range Statistics in Semiconductors, dist. by Stanford University Bookstore (1969)

6. D. L. Kendall, Semiconductors and Semimetals, vol.4, edited by R. K. Willardson and A. C. Beer (Academic Press, New York, 1968) p. 163

7. H. B. Bebb and E. W. Williams, Semiconductors and Semimetals, vol.8, edited by R. K. Willardson and A. C. Beer (Academic Press New York, 1972) p.182 and p.321

8. R. G. Hunsperger and O. J. Marsh, Metallugical Trans. 1, 603 (1970)

9. E. W. Williams, Phys. Rev. 168, 922 (1968)

10. E. W. Williams, Brit. J. Appl. Phys. 18, 253 (1967)

III-V COMPOUND SEMICONDUCTORS II

THE EFFECTS OF ION-IMPLANTED Ga, As, and P ON THE SUBSEQUENT
DIFFUSION OF ION-IMPLANTED Zn in GaAs$_{0.6}$P$_{0.4}$

E. B. Stoneham and J. F. Gibbons

Stanford Electronics Laboratories, Stanford University

Stanford, California 94305   USA

INTRODUCTION

Interesting results have been obtained in experiments in which samples of GaAs$_{0.6}$P$_{0.4}$ implanted with Zn were also implanted with other ions such as Ga, As, and P prior to annealing. In particular, the diffusion rate of the implanted Zn during the anneal can be strongly inhibited by the co-implantation of As or P, whereas Ga implantation has only a minor effect.

EXPERIMENTAL

The material used in this study was n-type GaAs$_{0.6}$P$_{0.4}$ grown by vapor phase epitaxy on N+ GaAs substrates. The free electron concentration in the GaAs$_{0.6}$P$_{0.4}$ ranged from 1×10^{17} cm^{-3} to 3×10^{17} cm^{-3}. In each sample, ions were implanted one species at a time at energies of 60 keV for Zn or Ga, 65 keV for As, and 32 keV for P. These energies give a projected range of about 300 Å for each ion. Prior to annealing, the samples were coated with a 1000 Å layer of Si$_3$N$_4$ by plasma deposition at 400°C. After the samples were annealed in sealed, evacuated quartz ampoules, their hole concentration profiles were obtained from van der Pauw measurements in conjunction with an anodic oxidation and stripping technique for removing thin layers of material [1].

Figure 1 shows the results for three samples, each of which was implanted at room temperature with roughly 10^{15} cm^{-2} of Zn. The ones labeled Zn + As and Zn + Ga had additional room temperature implants of 10^{15} cm^{-2} of As and 10^{15} cm^{-2} of Ga respectively. All three samples were annealed at 750°C for one hour. We note that the diffusion depth in the Zn + As sample is significantly smaller

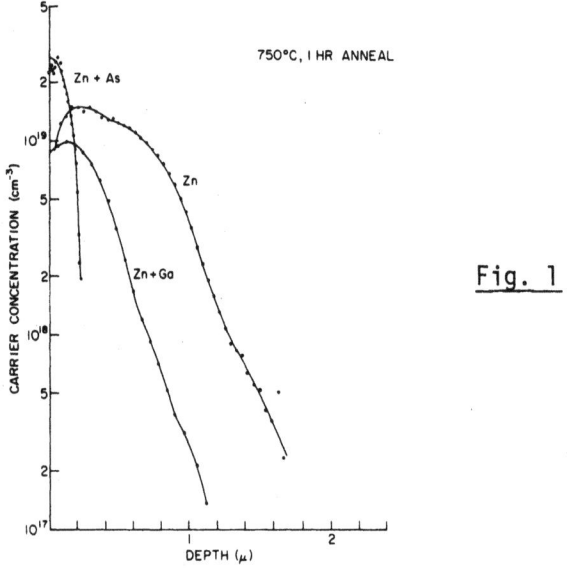

Fig. 1

than that in the Zn sample. The difference between the curves for Zn and Zn+Ga, on the other hand, is much less significant.* Evidently, As inhibits the Zn diffusion while Ga has little effect on it.

Similar experiments in which P was implanted with Zn gave results similar to Zn+As, showing that P, like As, inhibits Zn diffusion. Implants of As and Ga with Zn, or P and Ga with Zn, on the other hand, gave Zn diffusion depths similar to those of Zn and Zn+Ga. It appears, therefore, that Ga cancels the effect of As or P on Zn diffusion.

In general, the implants in this study can be described by the expression $\underline{a}$Zn + $\underline{b}$Ga + $\underline{c}$As + $\underline{d}$P, where $\underline{a}$, $\underline{b}$, $\underline{c}$, and $\underline{d}$ are the respective amounts of each implanted ion. For a constant value of $\underline{a}$, the junction depth x_j after annealing depends primarily on the difference between $(\underline{a}+\underline{b})$ and $(\underline{c}+\underline{d})$ [i.e., the number of implanted gallium-site atoms minus the number of implanted arsenic-site atoms]. This relationship is plotted in Fig. 2 for several experiments. The numbers in parentheses give $n = (\underline{a}+\underline{b}+\underline{c}+\underline{d})/\underline{a}$, indicating the relative total implant dose for each experiment. The junction depth shows a strong dependence on the sign of $(\underline{a}+\underline{b})-(\underline{c}+\underline{d})$, but no major dependence on the value of n.

---

*In fact, more recent results indicate that the difference is primarily due to a difference in Zn doses for the two samples. The Zn sample, having a higher Zn dose, has a greater diffusion depth due to the concentration dependence of the diffusion coefficient of Zn in GaAsP.

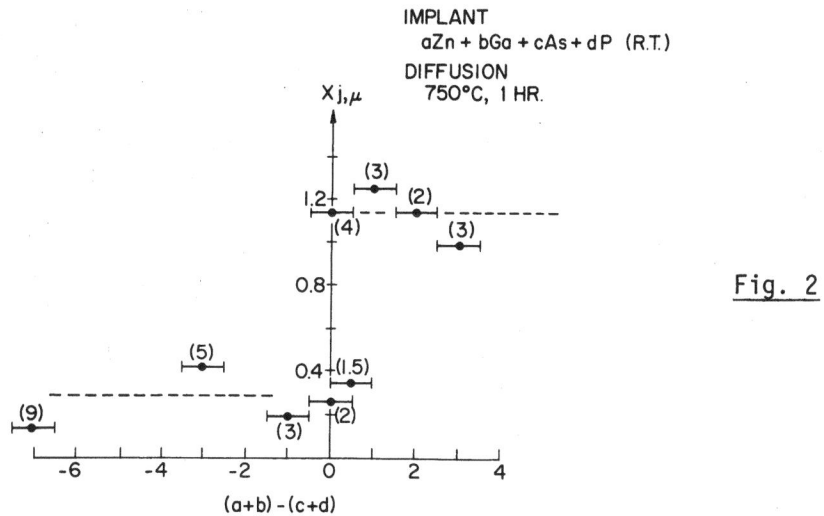

Fig. 2

A more detailed investigation of the diffusion inhibition was conducted for implants of As with Zn (b=d=o). The Zn dose was standardized at 5×10^{14} cm^{-2}. Figure 3 shows the results of an 800°C, 1 hour anneal for samples with various additional implants of As. The diffusion inhibition effect seems to saturate for As doses greater than twice the Zn dose. In light of the interstitial-substitutional model for Zn diffusion in GaAs [2,3] one might surmise that substitutional diffusion becomes the dominant mode when the diffusion inhibition effect saturates. That is, a sufficient excess of As might create enough Ga vacancies in the lattice to insure that the interstitial component of Zn is too small to dominate the Zn diffusion.

A series of experiments was conducted to determine whether Zn behaves as a substitutional diffusant when a large excess of As is implanted with it. A wafer co-implanted with 5×10^{14} cm^{-2} of Zn and 4×10^{15} cm^{-2} of As at room temperature was subdivided into a set of samples. Separate members of the set were then annealed for 1, 4, or 16 hours at 750°, 800°, 850°, or 900°C. Preliminary van der Pauw measurements were used to determine the effective surface concentration of holes in each sample; the samples were then profiled to determine the junction depths. The effective hole surface concentration expressed as a fraction of the Zn dose is plotted against junction depth for these samples in Fig. 4. There appears to be a unique relationship between activity and diffusion depth that does not depend on the annealing temperature. The shape of the curve in Fig. 4 suggests the existence of a stable damage layer near the surface within which the holes introduced by the Zn are compensated. Further evidence for this will be mentioned in connection with transmission electron microscopy data (see discussion below).

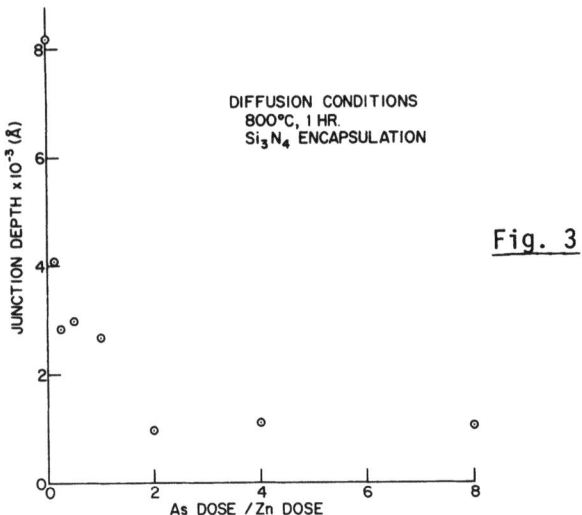

Fig. 3

Hole concentration profiles for the samples annealed at 850°C are shown in Fig. 5.* The junction depth is approximately proportional to the square root of diffusion time as one would expect for purely substitutional diffusion in which the diffusion coefficient is concentration-independent. However, the profiles do not have the Gaussian shape usually associated with concentration-independent diffusion from a δ-function impurity source.

DISCUSSION

An effective diffusion coefficient was computed for each sample from the junction depth, the background electron concentration, the Zn dose, and the annealing time. The average effective diffusion coefficient for each annealing temperature is plotted against reciprocal temperature in Fig. 6. A linear approximation has a slope corresponding to an activation energy of about 3.5 eV, which is considerably higher than the value of 1.39 eV reported for substitutional diffusion of Zn in GaAs [3]. The significance of the data in Fig. 6 is highly questionable, however, since the shapes of the profiles in Fig. 5 indicate that the diffusion coefficient is not simply a function of temperature alone. Furthermore, the mechanism by which the additional implant of As or P acts to inhibit Zn diffusion is not yet clear. The prediction of the interstitial-substitutional model (i.e., that excess As or P forces more Zn into substitutional positions where it diffuses slowly) is probably the best qualitative explanation for the data, although perturbations (such as the effect of Ga vacancies generated at the surface and/

---

* Improved measurements of stripping depths indicate that the diffusion depths in Fig. 5 are too small by a factor of about 1.25.

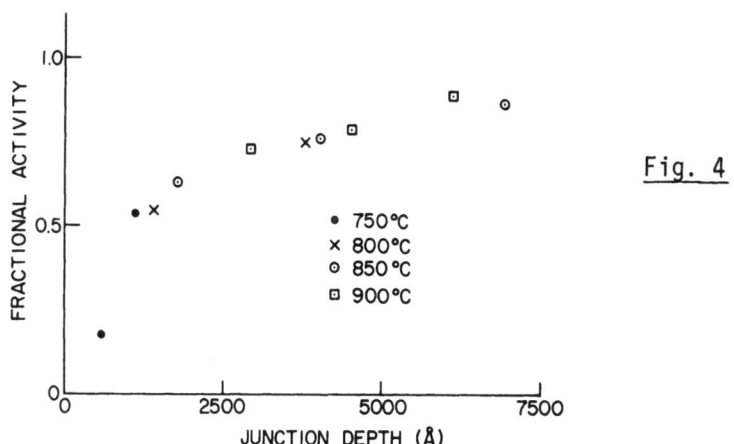

Fig. 4

or damage-enhanced substitutional diffusion) may have to be included to account for the odd shapes of the profiles in Fig. 5. Another possibility is that implanted As or P might precipitate Zn out as a compound or trap it in a second metallurgical phase. Observation of such precipitates and second phases in Zn-implanted GaAs annealed at 600°C for 30 minutes has been reported [4]. To study this possibility, samples of GaAs implanted with Zn or Zn plus eight times as much As were annealed with a Si_3N_4 cap at 850°C for one hour and then examined by transmission electron microscopy. Angle lapping and staining revealed a junction depth of 6.4 µ for the sample implanted with Zn and 0.6 µ for the sample with the additional As implant, indicating that diffusion inhibition had definitely occurred. In both samples, however, the TEM analysis revealed only dislocation loops plus a polycrystalline layer about 500 Å thick at the surface; no precipitates or second phases have been observed in any of the cases so far examined. The dislocation loops are probably the remnants of implantation damage, while the polycrystalline layer is the result of incomplete recrystallization during annealing of the crystal near its interface with the Si_3N_4 encapsulant. As previously mentioned, this observation correlated with the data in Fig. 4 since defects in the polycrystalline layer may compensate the Zn effectively (Zn may also precipitate on dislocations in the polycrystalline layer). An unimplanted control sample showed no dislocations or polycrystalline regions.

Some further experimental results worth noting are as follows. Samples of $GaAs_{0.6}P_{0.4}$ were implanted with Zn and As at 450°C and were annealed at 750°C for one hour. No diffusion inhibition effect was observed regardless of the order in which the two species of ions were implanted. Other samples were implanted with Zn and Kr at room temperature or at 450°C and annealed at 750°C for one hour. It was hoped that the Kr would merely produce damage equivalent to that

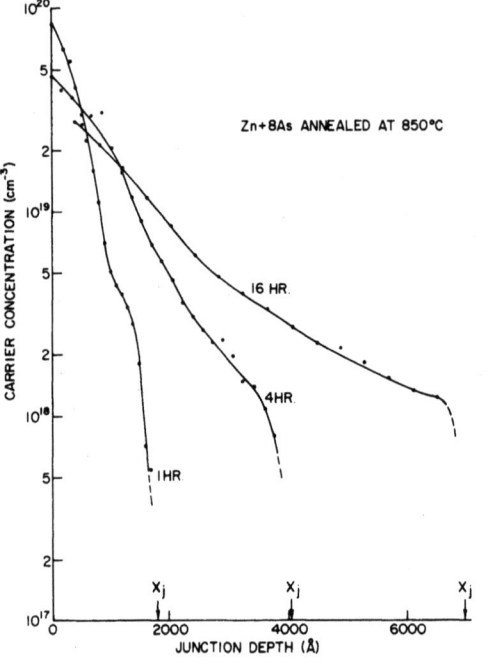

Fig. 5

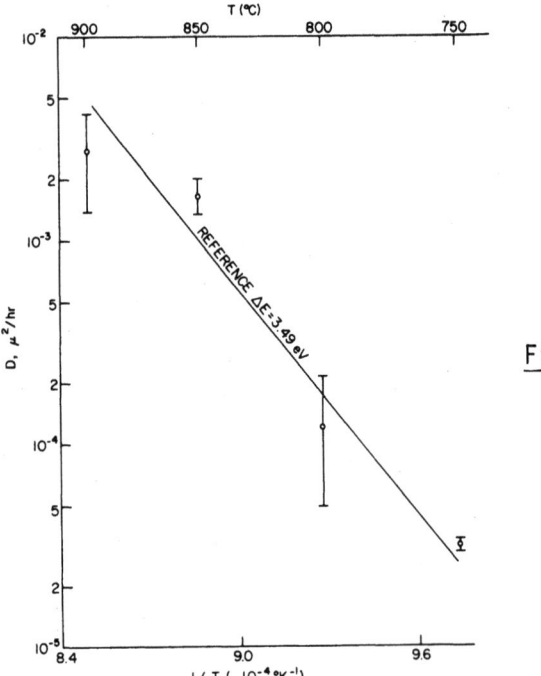

Fig. 6

produced by a similar implant of As, allowing us to separate the effects of the damage from the effects of the As itself. Instead, these samples showed very low Zn activity, indicating either that the Zn is not in electrically active sites or that traps associated with the Kr have reduced the number of free holes.

Further experimentation is now in progress to help clarify the mechanism of the inhibited diffusion observed in this work and to explain why it occurs for room-temperature implants but not for implants carried out at 450°C.

Acknowledgements

The authors wish to thank Drs. T. J. Magee and J. Peng of Stanford Research Institute for their excellent transmission electron microscopy work.

This work was supported jointly by the Army Research Office - Durham, and the National Science Foundation.

References

1. E. B. Stoneham, "A Nonaqueous Electrolyte for Anodizing GaAs and $GaAs_{0.6}P_{0.4}$," to be published in J. Electrochem. Soc., October 1974.

2. L. L. Chang and G. L. Pearson, J. Appl. Phys. $\underline{35}$, 1960 (1964).

3. A. C. Wang, "Zinc Diffusion in Intrinsic and Heavily Doped Gallium Arsenide," Ph.D. Thesis, Stanford University, Stanford, California (1970).

4. M. A. Littlejohn, private communication.

COMPENSATING LAYERS IN GaAs BY ION IMPLANTATION : APPLICATION

TO INTEGRATED OPTICS *

P.N. Favennec and E.V.K. Rao

Centre National d'Etudes des Télécommunications

22301 Lannion, France

ABSTRACT

We show that the compensation in O^+ ions implanted GaAs is of two different origins : implantation induced defects and implanted oxygen atoms. We give a few examples of optical waveguides in near IR region realized by H^+ implantation and we discuss the using O^+ implanted layers to realize light guides as the compensation at high anneal temperatures is stable.

INTRODUCTION

Hitherto, compensating layers in GaAs are obtained by H^+, He^+, B^+, F^+ implantations utilizing the free carrier trapping property of implantation induced defects [1 to 3] but a complete regeneration of free carriers takes place at anneal temperatures in the range of 400°C to 600°C. Favennec et al [4] have obtained high temperature resistant compensation by implanting oxygen in GaAs.

In this paper, we show that compensation in O^+ implanted GaAs can be obtained by chemical doping effect in addition to the usual defect compensation process and the former alone is responsible for compensation at temperatures higher than 600°C. Then we give a few examples of optical waveguides realized by H^+ implantation and we discuss the interest of using O^+ implanted layers for the realization of light guides in near IR region.

* Work supported in part by DGRST contract n° 73-7-1349

EXPERIMENTAL TECHNIQUES

Proton and oxygen ion implantations in a direction 7° off the principal crystallographic axis are carried out in n-type GaAs substrates at room temperature in a Van de Graaff accelerator. Lightly doped <111> substrates (3.10^{16} cm^{-3}) are used for the investigation of electrical properties, whereas highly doped <100> substrates (10^{18} cm^{-3}) are utilized for the realization of optical waveguides. The carrier concentration profiles are obtained from differential capacitance measurements with the help of aluminium schottky barrier contact. The light guiding properties of the layers are investigated by coupling 1,15 μm He Ne laser beam into and out of cleaved faces perpendicular to the implanted surface with the help of an edge focussing coupling system.

RESULTS AND DISCUSSION
Compensation In Unannealed Layers

In figure 1, we have presented the uncompensated fraction of carrier concentration against the depth taken over x_p ; where x_p is the depth at which the minimum in carrier concentration is measured. In table I, we have given the implantation conditions and compensation properties of O^+ implanted layers and compared them with those of H^+ [2] and B^+ [3] implantations. The carrier removal ratio, K, represents the ratio of the integral concentration of removed or compensated carriers to the total dose of implanted ions. R_p [5] is the projected ion range and $\bar{R}_{pd}/R_p$ [6] is the ratio of the projected depth of defects peak to that of ions peak.

Both fig. 1 and Table I demonstrate that a maximum compensation in carrier concentration occurs at a depth that corresponds to $\bar{R}_{pd}$. (See last two columns of Table I). This suggest that the present compensation in O^+ implanted samples is of the same origin as in case

TABLE I

Curve	Substrate (cm^{-3})	Ion	Energy (MeV)	Dose (cm^{-2})	x_p (μm)	K	$(R_p)_{LSS}$[5] (μm)	$\frac{x_p}{(R_p)_{LSS}}$	$\frac{\bar{R}_{pd}}{(R_p)_{LSS}}$[6]
(a)[2]	1.3×10^{16}	H^+	0.15	10^{11}	1.1	4	1.5	0.73	0.8
(b)[3]	5×10^{16}	B^+	1	10^{10}	1.15	200	1.55	0.74	0.8
(c)	3×10^{16}	O^+	1	9×10^9	0.9	200	1.25	0.72	0.8

COMPENSATING LAYERS IN GaAs BY ION IMPLANTATION

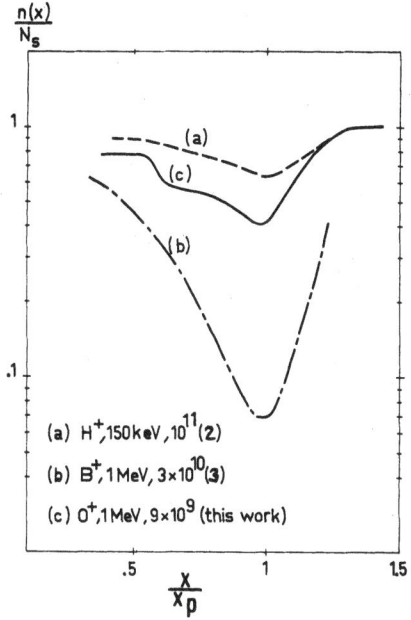

Fig. 1 : Distribution of uncompensated carriers after protons, boron and oxygen implantations.

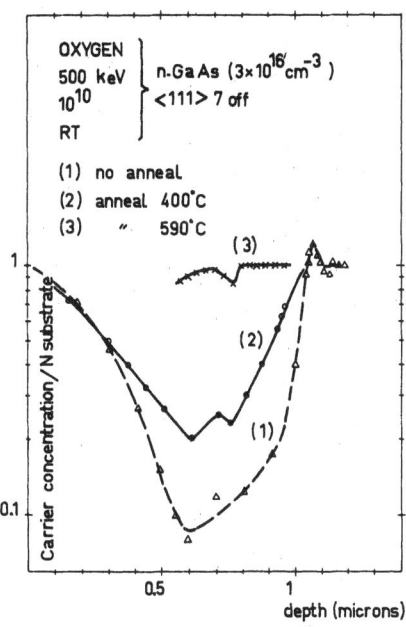

Fig. 2 : Uncompensated carrier distributions after isohronal anneals (15 mn).

H^+ and B^+ implanted layers and that the physical phenomenon responsible is the trapping of carriers by defects. Also the estimated values of K indicate that the compensation efficiency of oxygen ions, being almost the same as that of boron, it is at least 50 times higher then that of H^+ implanted layers. This can be due to heavier masses of B^+ and O^+ resulting in higher concentration of defects.

Compensation In Annealed Samples

In figure 2, we have presented the uncompensated carrier distributions measured after successive isochronal anneals in case of a sample implanted at 0,5 MeV with a dose of 10^{10} O^+/cm^2. Theoretical values of R_p [5] and $\bar{R}_{pd}$ [6] for 0,5 MeV implantation are respectively 0,7 μm and 0,6 μm. The following observations can be made from figure 2 :

i) Carrier distribution in unannealed layer indicates a maximum of compensation at 0,6 μm (Curve 1). This value is in close agreement with the calculated $\bar{R}_{pd}$ value of 0.6 μm.

ii) After 400°C anneal compensation peaked at 0,6 μm decreases and a second maximum in compensation appears at 0,75 μm (Curve 2).

iii) Finally after 600°C anneal, there is almost a complete anneal of compensation peaked at $0.6\,\mu m$, whereas a small but definite maximum in compensation prevails at $0.75\,\mu m$ (Curve 3). This value is in agreement with the theoretical $\bar{R}p$ value of $0.7\,\mu m$.

The above observations suggest that the compensation in O^+ implanted GaAs occurs from two different origins : implantation induced defects with a compensation maximum situated at $\bar{R}pd$, and implanted oxygen ions exhibiting a maximum in compensation at a depth equal to $\bar{R}p$.

Further confirmation of the above results and an estimation of carrier removal ratio K, is carried out by conducting the following experiments : different samples are implanted at 1 MeV with varying doses from 10^{10} to 7×10^{10} O^+/cm^2 and annealed to 590°C in order to be able to measure compensation due to oxygen. In figure 3, we have plotted the fractionnal uncompensated carrier concentrations against the difference in depth taken relative to $(\bar{R}p)_{exp}$. $(\bar{R}p)_{exp}$ represents the depth at which the minimum in carrier concentration is experimentally found. In the inset, we have presented an example of our estimation of K for the particular case of 590°C anneal which we estimate to be about 1.

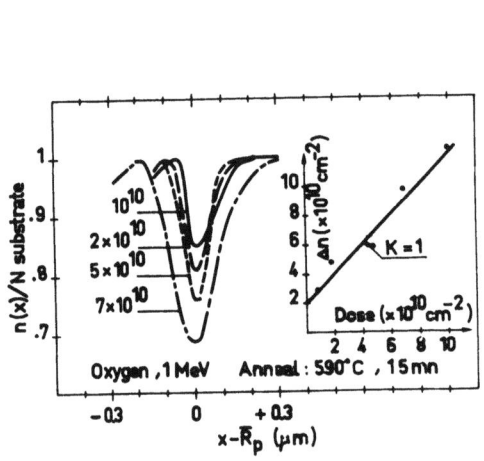 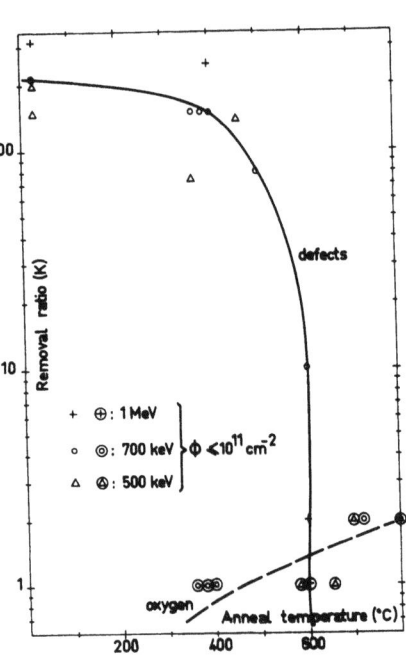

Fig. 3 : Carrier distributions after 590°C anneal. In the inset, estimation of the removal ratio K.

Fig. 4 : Removal ratio K for defects and oxygen (circle dots)

APPLICATION OF COMPENSATING LAYERS
H$^+$ Implanted Layers

Compensating layers in GaAs find an important application in the realization of optical waveguides in the near IR region. Garmire et al [7] have shown the possibility of light guiding in proton bombarded GaAs layers where the compensation is due to defects. Since the layers realized by implantation present a great amount of absorption of guided light, they need annealing to decrease absorption and one has to find a compromise between the amount of compensation needed and the absorption that can be tolerated.

Our study of defect compensated layers for guiding is carried out by conducting room temperature 500 KeV proton implantation and in Table II, we give some typical examples of our guiding experiments where the guiding length in each case is 6 mm. Figures 5 represents the photographs showing the guiding in H$^+$ implanted and annealed layers. Here, we are contented to remark on the qualitative guiding properties of these layers with respect to anneal temperature and it is clear that no guiding is observed above an anneal temperature of 600°C. This can be explained as due to recovery of charge carriers in the implanted layer.

TABLE II

Dose (H$^+$/cm^2) 500 KeV	5×10^{13}				2×10^{14}			2×10^{15}	
Anneal temperature (°C)	25	350	450	600	25	450	600	25	450
Guiding	no	weak	strong	no	no	strong	no	no	no

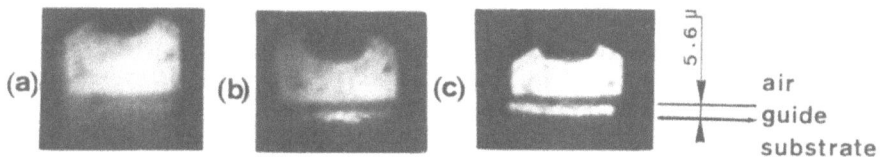

Fig. 5 : These photographs show the guiding in H$^+$ implanted and annealed samples : a) 5×10^{13} H$^+$/cm^2, 500 keV, TA = 350°C
b) 5×10^{13} H$^+$/cm^2, 500 keV, TA = 450°C
c) 2×10^{14} H$^+$/cm^2, 500 keV, TA = 450°C

O^+ IMPLANTED LAYERS

O^+ implanted GaAs layers exhibit compensation due to chemical doping effect depending on the anneal temperature. We have conducted low and high doses oxygen implantations in order to investigate independently the guiding properties of defect compensated and oxygen compensated layers. The implantation energy is chosen to obtain compensated layers of at least 3 m thick.

In case of layers implanted with low doses, 10^{10} to 10^{12} O^+/cm^2, and annealed to temperatures of about 450°C, we did not observe any guiding. This result is in a way not surprising when we consider the behavior of K with respect to anneal temperature (figure 4) which indicates the presence of unannealed defects that can absorb the guided light. In proton bombarded layers, the absorption of guided light is find to decrease by annealing to temperatures less than 450°C. This is perhaps due to the nature and/or anneal behavior of defects participating in compensation and can be well obvious when we compare 10^{12} O^+/cm^2 and 5×10^{13} H^+/cm^2 implanted layers, that are supposed to have the same number of compensating defects.

Investigation of the guiding properties of chemically compensated layers is carried out by conducting different high integral dose (10^{14}, 2.5×10^{14}, 5×10^{14} O^+/cm^2) multi implantations of oxygen ions. Very weak guiding in these layers is observed only when they are annealed to a temperature of about 500°C. In case of few samples annealed to temperatures above 500°C, the film deposited protective SiO_x layer is found to create a number of diffusing centers for the guided light and this did not facilitate our observations. Presently our efforts are concentrated in finding non reactive protective layers on GaAs surface, and then we hope to obtain low loss guides since the compensation is due to a chemical doping effect unlike proton bombarded layers.

CONCLUSIONS

In this paper we have shown that the compensation in oxygen implanted GaAs layers is of two different origins : implantation induced defects and implanted oxygen atoms. The later is alone the stable compensating process at high anneal temperatures, $> 600°C$, and this can only be due to chemical doping effect of oxygen in GaAs lattice. We have also shown that, in general, compensating layers can be utilized as efficient waveguides in the near IR region, and put forward the idea that oxygen implanted layers can find an interesting application to realize optical waveguides as the compensation at high anneal temperatures is stable.

REFERENCES

[1] A.G. Foyt, W.T. Lindley, C.M. Wolfe and J.P. Donnelly
Solid State Electronics, 1969, 12, 209

[2] B.R. Pruniaux, J.C. North and G.L. Miller
Ion Implantation in semiconductors, 1971 (Springer-Verlag) p.212

[3] D. Eirug Davies, J.K. Kennedy and A.C. Yang, Appl. Phys. Letts.
1973, 23, 11

[4] P.N. Favennec, G.P. Pelous, M. Binet and P. Baudet
Ion Implantation in Semiconductors and Other Materials, 1973,
p. 621

[5] W.S. Johnson and J.F. Gibbons
Projected Range Statistics in Semiconductors, 1969

[6] P. Sigmund and J.B. Sanders
Int. Conf. on Applications of Ion Beams, Grenoble 1967, p. 215

[7] E. Garmire, H. Stoll, A. Yariv and R.G. Hunsperger
Applied Physics Letters, 1972, 21, 3

H. Harada and M. Fujimoto

Musashino Electrical Communication Laboratory

Musashinoshi Tokyo Japan

ABSTRACT

The effect of 60-380 KeV proton bombardments on the electrical properties of N-type GaAs has been studied by C-V and Hall-effect measurements. The thickness of the semi-insulating layer produced by the bombardments is nearly equal to the projected proton range. From Hall-effect measurements, it is found that the donor level of 0.15 eV is created by proton bombardments.

INTRODUCTION

It is known that the high resistivity layer in GaAs is produced by proton bombardments. The proton bombardments are known to be useful in fabricating GaAs devices, such as junction isolation diodes, Schottky barrier diodes, semi-insulated gate FET, stripe geometry laser diodes and GaAs IMPATT diodes.
This paper reports the electrical properties of the proton bombarded GaAs layers studied to determine compensation phenomena.

EXPERIMENTAL

(100) plane wafers from non-doped GaAs ingots were used for C-V measurements. Schottky barriers were formed in vacuum by vapor deposition of Ti and Au. Hall-effect measurements were carried out by use of the vapor phase epitaxial GaAs layers grown on the semi-insulating Cr doped substrates.

RESULTS AND DISCUSSIONS

1. Depth Distribution of Free Carriers

Figs. 1 and 2 show the dependence of the free carrier concentration profiles on the bombarded proton energy. Rp and Rp' in the figures indicate the projected range of proton reported by Furukawa et al (1) and Tsurushima et al (2), respectively. As the LSS theory is not applicable for a light ion, such as proton, Tsurushima et al calculated the projected range by use of the some extended electronic stopping power derived by both the LSS and Bethe-Bloch theory. Fig 3 shows the carrier concentration profiles created by 60 KeV proton for several doses. The carrier concentration peaks near the end of the profiles increase with increasing the proton dose. Foyt et al (3) reported a similar peak in the proton bombarded GaAs measured by C-V method. However, there is no such peak in the carrier profiles reported by Pruniaux et al (4). They have measured the carrier concentration profiles by AC method similar to that used by Copeland. The carrier concentration profiles determined by C-V and Copeland method were compared, as shown in Fig. 4. The peak in the carrier profiles measured by C-V method is due to the electron emission from the deep donor or compensating acceptor produced by the proton bombardments.

Annealing behaviors are shown in Fig. 5. The compensated layer is affected by 300°C annealing for 10 min. and annealed out at 500°C for 10 min. Pruniaux et al reported that the semi-insulating properties of the bombarded layer disappeared at annealing temperature in excess of 350°C.

Fig. 6 shows the dependence of the carrier removal rate, Δn, of proton doses for 60 and 150 KeV. The removal rate is obtained from the carrier removal peak in the profiles. For 60 KeV proton, $\Delta n \simeq 8 \times 10^4 \cdot D$, and, for 150 KeV, $\Delta n \simeq 4 \times 10^4 \cdot D$ are obtained, respectively, where D is the proton dose per square cm. Since the projected range of 150 KeV proton is about two times larger than that of 60 KeV, the removal rate per one proton does not depend on the bombarded energy. Following Kinchin and Pease, the total number (N_t) of the displaced atoms produced by a light moving atom energy E_o is given by

$$N_t = \{P(E_o - L_c) + bL_c\}/E_d \qquad (1)$$

where P is the order of 10^{-3}–10^{-4}, L_c is the critical energy of

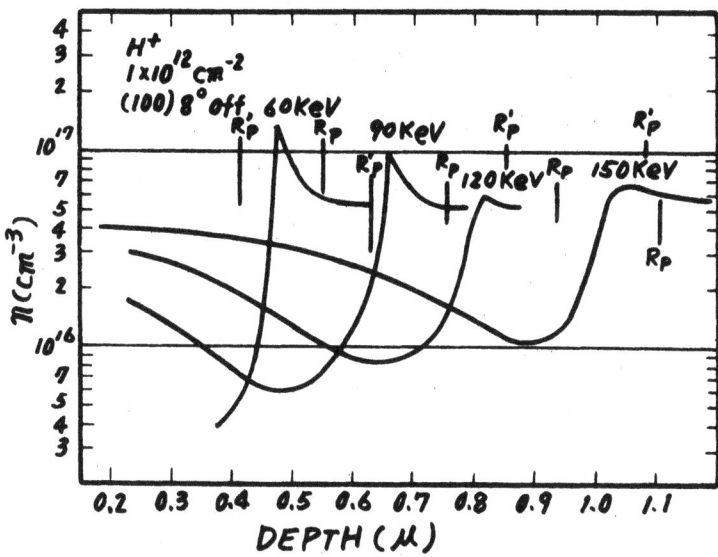

Fig. 1. Carrier concentration profiles created by 60-150 KeV proton bombardment

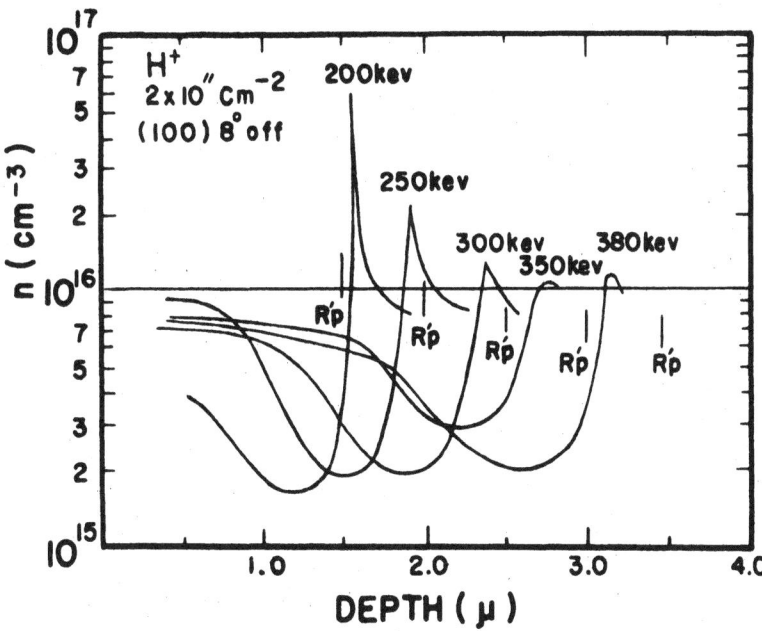

Fig. 2. Carrier concentration profiles created by 200-380 KeV proton bombardment

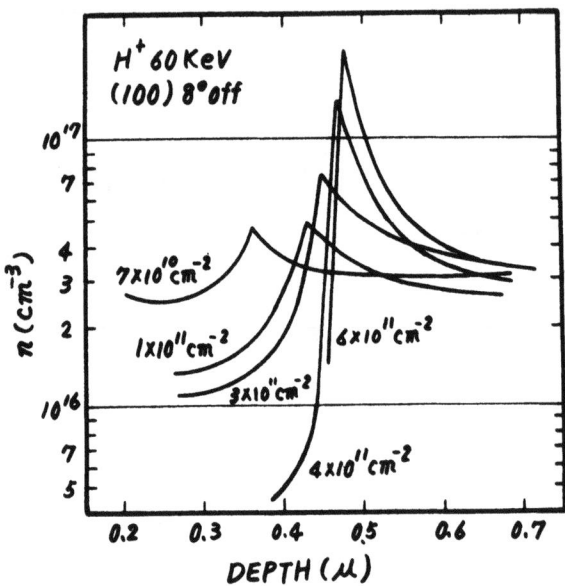

Fig. 3. Carrier concentration profiles created by proton bombardment with 60 KeV for several doses

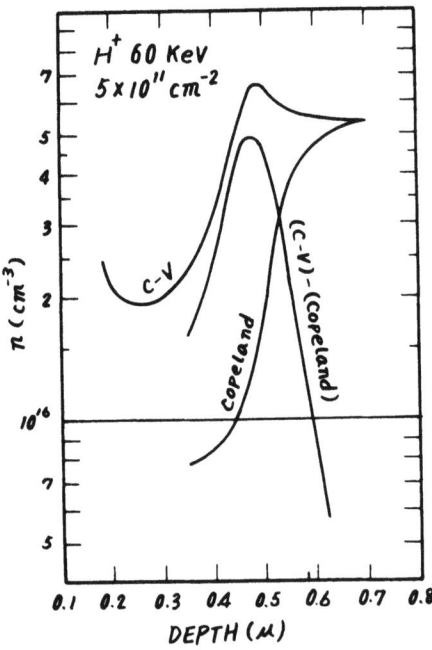

Fig. 4. Carrier concentration profiles determined by Copeland and C-V methods

ELECTRICAL PROPERTIES OF PROTON BOMBARDED N-TYPE GaAs

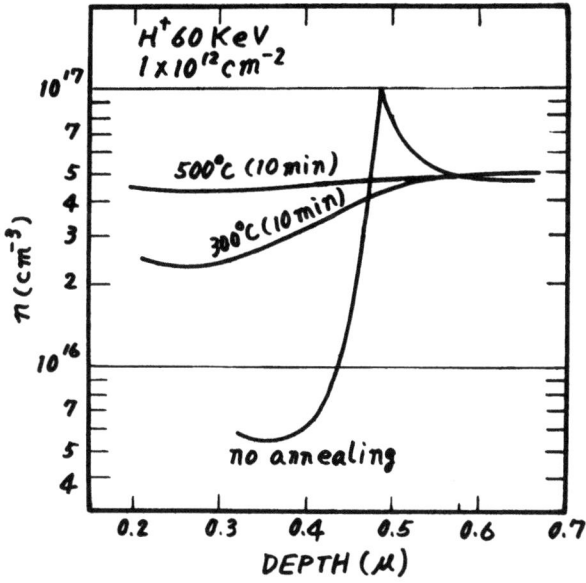

Fig. 5. Annealing characteristics of carrier concentration profiles

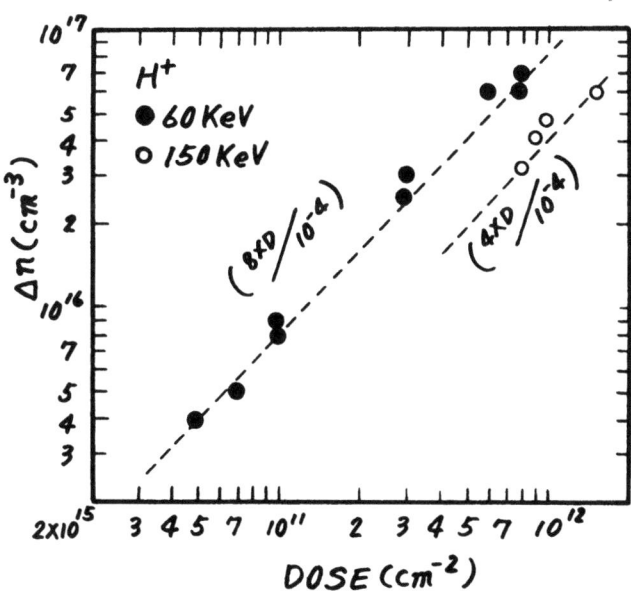

Fig. 6. Carrier removal rate, Δn, as a function of proton dose for 60 and 150 keV

electronic excitation, b is ½ in this case and E_d' is the deformation potential. Since L_C for proton is about 321 eV, N_t weakly depends on E_o in the energy range of this experiment. Eq.(1) indicates that L_C is the most important term for obtaining N_t. L_C for He ion is 1.28 KeV.

Fig. 7 shows the carrier concentration profiles produced by He ion bombardments. $\Delta n = 5 \times 10^5$ D has been obtained for 150 KeV He ion. This value is in good agreement with the one expected from Eq.(1).

2 Hall Effect Measurement

Fig. 8 shows the carrier concentration temperature dependence of proton bombarded GaAs. It is found that proton bombardment induces a deep donor. The temperature dependence of the bombardment-induced deep donor density seems to increase with increasing the proton dose. This behavior is due to the non-uniform depth distribution of free carriers in the proton bombarded layers as shown in Fig. 1. As the complex center of the bombardment-induced defects is not produced in this experimental range of proton dose, the defect created is thought to be the isolated point defect.

The average electrical conductivity, $\tilde{\sigma}$, in the layer with the non-uniform carriers along the z direction is given by

$$\tilde{\sigma} = \frac{1}{t}\int_0^t \sigma(z)\, dz \qquad (2)$$

where t is the thickness of the layer and $\sigma(z)$ is the conductivity at z. Assuming that the proton bombarded layer can be divided into binary layers, where one has weak proton bombardments effects (suffix 1) and the other has strong effects (suffix 2), the average conductivity is given by

$$\tilde{\sigma} = \frac{t_1}{t}\sigma_1 + \frac{t_2}{t}\sigma_2, \qquad t = t_1 + t_2 \qquad (3)$$

For simplicity, assuming the difference of mobility between layers 1 and 2 to be negligible, measured carrier concentration, $\tilde{n}$, is given by

$$\tilde{n} = \frac{t_1}{t}n_1 + \frac{t_2}{t}n_2 \qquad (4)$$

where the relation $\sigma = ne\mu$ is used.

Since proton bombardment produces deep centers, the term $(t_1/t)n_1$ can be estimated by $\tilde{n}$ at low temperature. Then, $(t_2/t)n_2$ is obtained by extrapolation of $(t_1/t)n_1$ into the high temperature region, as shown in Fig. 9. From Fig. 9, the 150 meV energy is obtained for both samples of 3×10^{11} and 5×10^{11}/cm^2 doses. Pruniaux et al have reported two ionized levels of 0.8 and 0.4 eV from

ELECTRICAL PROPERTIES OF PROTON BOMBARDED N-TYPE GaAs

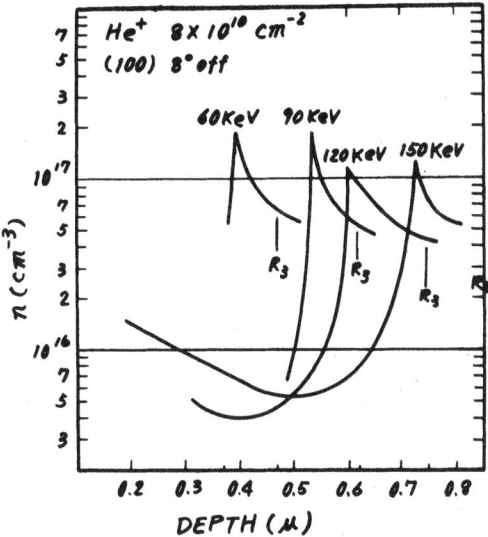

Fig. 7. Carrier concentration profiles created by 60-150 KeV He ion bombardment

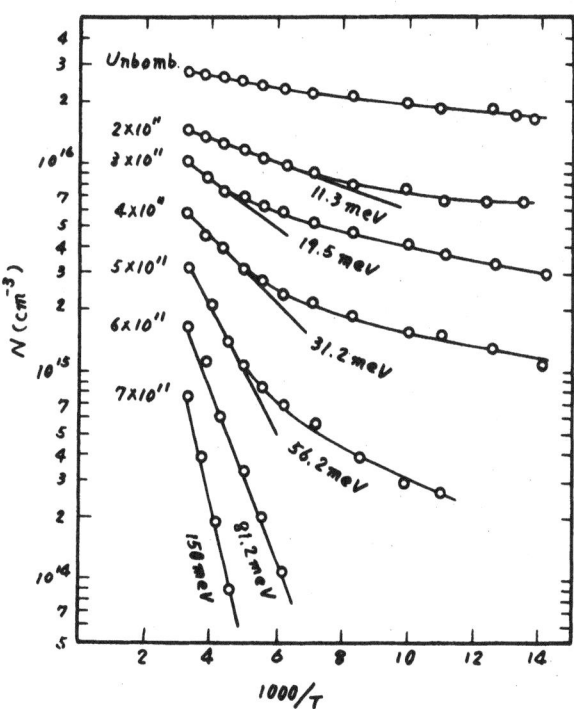

Fig. 8. Carrier concentration as a function of temperature for a series of proton doses

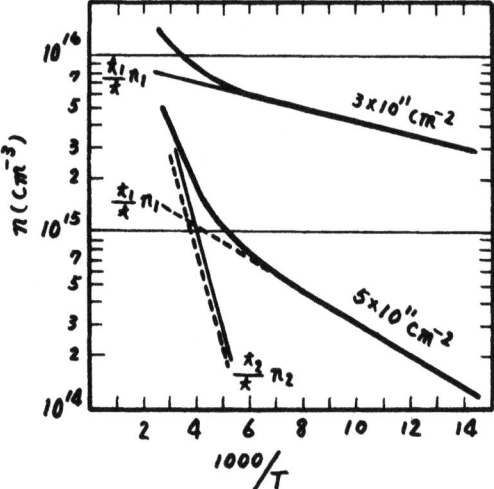

Fig. 9. Carrier concentration obtained by the binary layer model

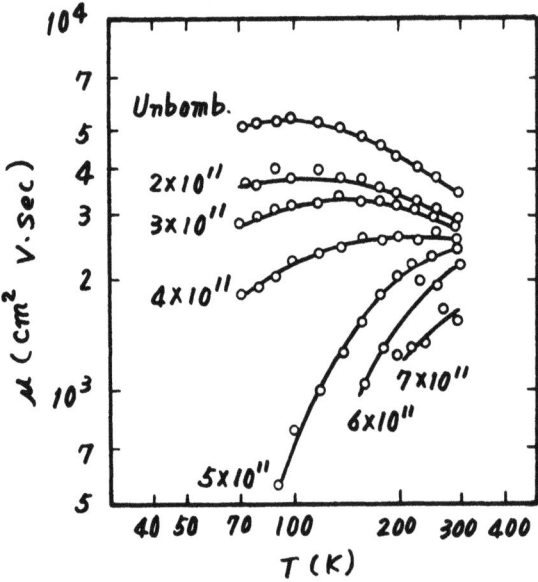

Fig. 10. Hall mobility as a function of temperature for a series of proton doses

resistivity temperature dependence between 77 and 500 K. Brehm and Pearson (5) have reported donor energy levels in γ-irradiated n-GaAs, 0.13, 0.16 and 0.3 eV from Hall-effect and 0.7 and 0.6 eV from photo-luminescence measurements.

Hall mobility temperature dependence for various doses is shown in Fig. 10. The decrease in Hall mobility indicates that the total number of ionized impurity centers increases with increasing the proton dose.

CONCLUSION

The depth distribution of free carriers by proton and He ion bombardments in GaAs has been studied by C-V measurements. The shape of the carrier profiles created by proton was similar to that by He. The carrier concentration peaks appear near the end of the proton and He ion travels. The peaks are found to be due to the electron emission from deep centers during C-V measurements. The carrier removal rate can be explained qualitatively by Kinchin and Pease's equation.

The 150 meV deep donor is obtained by Hall-effect measurement. However, more work is necessary for the detection of deeper level reported by Pruniaux et al. The compensation of carriers may be due to the deep donor and acceptor created at the same rate by the bombardments. The acceptor compensates the shallow donor and the electrical properties of bombarded GaAs are determined by the deep donor, which does not ionize under usual conditions.

ACKNOWLEDGEMENT

The authors are grateful Drs. H. Toyota and Y. Sato for their useful discussions and encouragement. Thanks are also due to Mr. T. Honda for his cooperation and technical assistance.

REFERENCES

(1) S. Furukawa et al, J. J. A. P. 11, 134 (1972)

(2) T. Tsurushima, J. Phys. Soc. Japan 31, 1695 (1971)

(3) A. G. Foyt et al, Solid State Elec., 12, 209 (1969)

(4) B. R. Pruniaux et al, 2nd Int. Conf. Ion Impl. in Semi. p.212

(5) G. H. Kinchin and R. S. Pease, Rep. Prog. Phys., 18, 1 (1955)

(6) G. E. Brehm and G. L. Pearson, J. Appl. Phys., 43, 568 (1972)

PHOTOLUMINESCENCE OF ZINC IMPLANTED n-TYPE GaAs

T. Itoh, T. Matsumoto and J. Kasahara*

School of Science and Engineering, Waseda University

Shinjuku-ku, Tokyo, Japan

ABSTRACT

Photoluminescence measurements of ion-implanted GaAs were performed to investigate recovery of induced damages. Silicon-doped n-type GaAs samples with (100) orientation were implanted with Zn in a dose range of 5×10^{13} to 5×10^{15} ions/cm^2 and were subsequently annealed for 20 min at temperatures between 600°C and 900°C. A well-defined single emission peak due to Zn acceptor level was obtained in samples implanted with a dose of 1×10^{15} ions/cm^2 and annealed at 800°C. It was found that the implanted-and-annealed samples contained fewer defects than unimplanted ones which were given the same heat treatment. Compensation of thermally induced Ga-vacancies by implanted Zn atoms plays a dominant role in the recovery process. In-depth variation of photoluminescence spectrum was also studied by succesively etching the implanted surface. This technique was found very useful for profiling implanted atoms and vacancies.

INTRODUCTION

The annealing of induced lattice defects is an established technique in the case of ion implantation into silicon. However, annealing of ion-implanted GaAs is not so simple, since thermal dissociation during the annealing process may bring in additional imperfec-

*Present address: Sony Corporation Research Center, Hodogaya, Yokohama, Japan

tions. Although a number of studies have been made, the nature of the induced defects and the annealing mechanism are not well understood yet. In the present work, the annealing process was investigated by means of photoluminescence measurements. Photoluminescence is very useful means for this purpose because the species of defects and impurities can be identified. Another advantage is that the analyzed region can be restricted to a very shallow surface layer. In order to discriminate the defects due to ion implantation from those associated with thermal dissociation, both implanted and unimplanted samples were annealed under the same conditions and their photoluminescence spectra were compared.

Preliminary results on unimplanted samples, which will be published in a seperate paper [1], are summarized as follows:
(1) Annealing above 600°C causes monotonic decrease of the total emission intensity with increasing the annealing temperature. This is ascribed to thermal dissociation of GaAs.
(2) Annealing at 600°C gives rise to a considerable amount of As-vacancies. Resultant migration of Si atoms from Ga-site to As-site was observed.
(3) Annealing above 800°C assists the formation of Ga-vacancies. This is explained by the outdiffusion of Ga through the SiO_2 film [2,3].
The results on implanted samples with different doses and annealing at different temperatures are presented in this report. These results together with above-mentioned observations on unimplanted samples will elucidate the annealing mechanism. The in-depth variation of impurity and/or defect concentrations were obtained, for the first time so far as we know, from photoluminescence spectra measured with successive etching of the surface layer.

EXPERIMENTAL

Si-doped n-type (100) GaAs substrates were used in this study. Substrates had a carrier concentration of 1.27×10^{18} cm^{-3} and an electron mobility of 2350 cm^2/V sec. 20 KeV Zn ions were implanted in the range of 5×10^{13} to 5×10^{15} ions/cm^2 at room temperature with a beam flux density of 400 nA/cm^2. Both surfaces of the samples were covered with 3000 to 4000 Å RF-sputtered SiO_2 films before annealing. Annealing was made at temperatures between 600°C and 900°C for 20 min in vacuum. Photoluminescence measurements were made at 77 K after etching the SiO_2 films. A focused Ar-laser beam (60 mW, 5145 Å) was used as the excitation source. Photoluminescence spectra were measured by Nikon P-250 monochromator equipped with RCA-7102 photomultiplier. Successive etching for in-depth measurements was made using NaOH etchant (1N-NaOH/H_2O_2 = 20/1) at an etching rate of 0.15 μm/min.

RESULTS AND DISCUSSION

The dependence of the total emission intensity (I_t) on the annealing temperature is shown in Fig. 1 for both implanted and unimplanted samples. After annealing, I_t of most of the samples exceeded that of unimplanted samples. Though not shown in this figure, this relation was reversed for annealing below 400°C. Since I_t can be regarded as a measure of the lattice perfection, it is concluded that the induced disorder is considerably recovered by annealing above 600°C. Figure 2 shows the intensity of the emission peak due to Zn acceptor (1.473 eV) as a function of the dose. Typical line shapes of this emission peak are illustrated in Fig. 3. The ratio of the emission intensity for V_{As}-complex (1.27 eV) to that for Zn acceptor is plotted against the dose and annealing temperature in Fig. 4.

At the lowest dose of 5×10^{13} ions/cm^2, the annealing behavior was not much different from that of the unimplanted sample, and the emission peak for Zn acceptor was not clearly separated from nearest emission peaks due to other origin. The peak due to Zn acceptor was the dominant one in the photoluminescence spectra for samples with dose of more than 2×10^{14} ions/cm^2 where the Zn concentration is higher than that of Si. The emission intensity of this level, and of I_t, increased as the increase of dose up to 1×10^{15} ions/cm^2 and then decreased for further increase of dose. The decrease in the high dose region can be explained by the segregation of Zn atoms; its concentration in the implanted region exceeds the solid solubility limit.

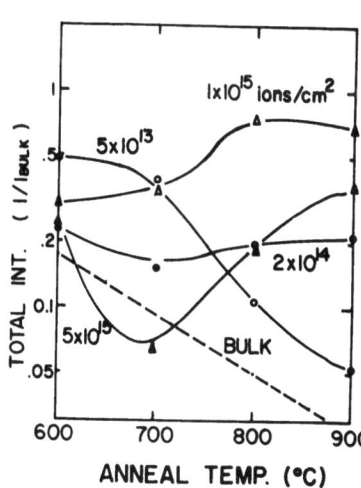
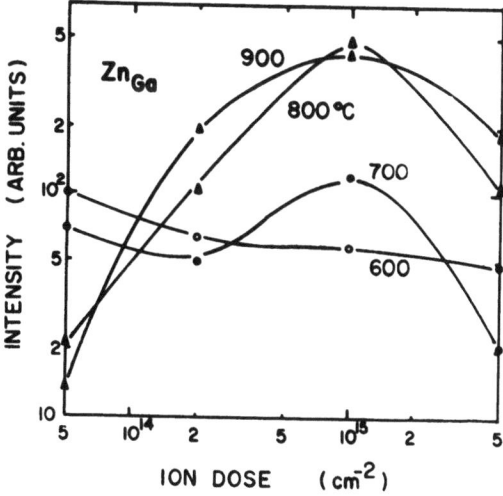

Fig. 1. Total emission intensity vs. annealing temperature.

Fig. 2. Emission intensity from Zn acceptor vs. Zn ion dose.

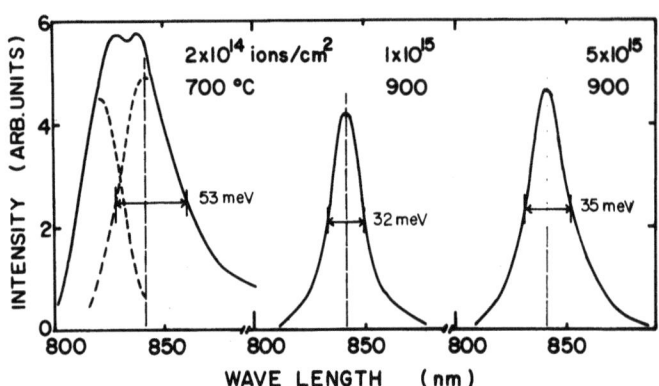

Fig. 3. Difference of line shapes at about 1.47 eV.
(reduced scale; left x1, center x1/5, right x 1/2)

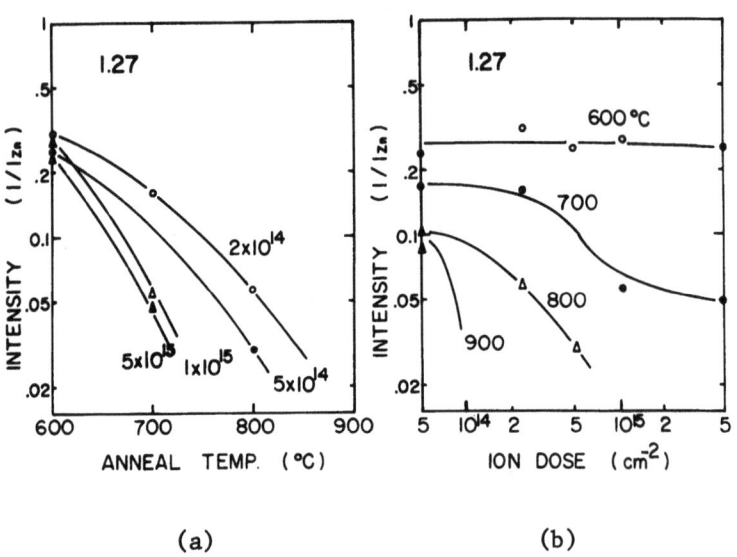

(a) (b)

Fig. 4. Intensity of the emission peak at 1.27 eV as a function of (a) annealing temperature and (b) ion dose.

When Zn dose exceeded 1×10^{15} ions/cm^2, a single peak for Zn-acceptor was obtained after annealing at 900°C, as shown in Fig.3. The shape of the peak was not simple but extended to the lower energy side. This complex nature of the peak was well observed in samples with lower Zn dose and annealing at lower temperatures, showing the presence of emission due to Si-acceptors.

The ion dose range can be devided into three groups from Fig.4; less than 1×10^{14}, 2 to 5×10^{14} and above 1×10^{15} ions/cm^2. The emission peak related to V_{As} was not observed for samples with high dose and high temperature annealing. This result can be explained in the following way. As-vacancies are partly compensated by the migration of Si atoms from Ga site to As site, but compensation is not sufficient. There are many V_{As} after annealing at 600°C even in highly dosed samples. After annealing at higher temperature, V_{Ga} formation is dominant. The emission from V_{As} related defect levels is then reduced.

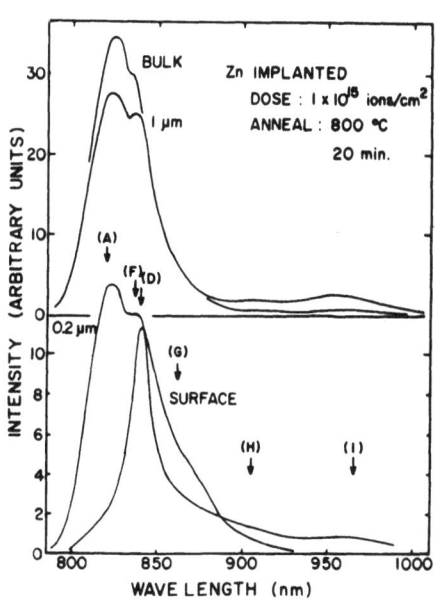

Fig. 5. In-depth variation of the spectrum

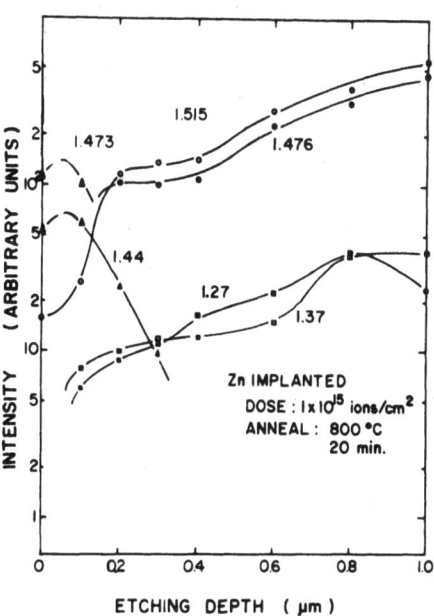

Fig. 6. Emission intensity vs. depth

When a large amount of Zn ions are implanted, most of them occupy Ga substitutional site, and Si atoms in Ga site are swept out to vacant As or interstitial site. It is found that Zn ion implantation assists compensation of V_{As}, though there are few V_{Ga} in the case of high ion dose. The emission spectrum of samples with dose of 5×10^{15} ions/cm^2 contained 1.476 eV peak, but that of samples with dose of 1×10^{15} ions/cm^2 did not show any emission related to defect levels. Such a well defined peak can be obtained only when implanted Zn concentration exceeds the Si concentration and a substrate is annealed at an appropriate temperature; i.e. dose of 1×10^{15} ions/cm^2 and annealing at 800°C.

Figures 5 and 6 show in-depth variation of photoluminescence spectrum for a sample with 1×10^{15} ions/cm^2 and 800°C annealing. The peak due to Zn acceptor disappeared at 0.2 μm from the surface. This value is consistent with the electrically measured p-region depth. In the region deeper than 0.2 μm, the band edge emission at 1.515 eV, two peaks due to V_{As}-complex at 1.27 and 1.37 eV, and a peak due to V_{Ga}-Si complex at 1.476 eV were observed. The intensity of these emissions appeared to saturate at around 1 μm depth. This may indicate that any influence of ion implantation did not extend further. No definite conclusion, however, can be derived until measurement are carried out in much deeper layer and also in unimplanted samples. The total emission intensity, though not shown in Fig. 6, also showed monotonic increase with depth from the surface. This fact suggests that the surface layer was of inferior quality as compared with the inside bulk even though the best annealing treatment was made. It should be noted that the technique described here is a very useful means for obtaining in-depth profiles of impurities and/or defects with relatively high resolution.

In summary, it was verified from photoluminescence measurements that As- and Ga-vacancies produced during implantation and annealing were compensated by dopant Si and implanted Zn atoms, respectively. Thus, the implanted-and-annealed samples showed fewer defects than unimplanted-and-annealed ones. A new technique using photoluminescence was introduced for profiling concentration of impurities and/or defects.

REFERENCES

1. T. Itoh and J. Kasahara, to be published.
2. J. Gyulai, J.W. Mayer, I.V. Mitchell and V. Rodriguez, Appl. Phys. Letters, 17, 332 (1970)
3. L.L. Chang, L. Esaki and R. Tsu, Appl. Phys. Letters, 19, 143 (1971)

LARGE INCREASE OF EMISSION EFFICIENCY IN INDIRECT GaAsP BY N-ION IMPLANTATION

Y. Makita, S.Gonda, H.Tanoue, T.Tsurushima and
S. Maekawa
Electrotechnical Laboratory, Tanashi Branch

5-4-1 Mukodai-machi, Tanashi-shi, Tokyo, Japan

ABSTRACT

Photoluminescence from N-ion-implanted, indirect $GaAs_{1-x}P_x$ (x=0.52) samples was measured as a function of annealing temperature. Samples were implanted at 350°C with N ions to concentration $10^{17} \sim 10^{20}$ cm^{-3}. Two conspicuous bands, in the general locations of 599.5 nm and 618 nm at 2K, appear as a result of annealing. They are due to isoelectronic impurities and the narrow band (599.5 nm) is attributed to 'A' line and the broad band (618 nm) is to NN line. In the measurement at 77K integrated intensity of lightly implanted samples exceeds 1000 times as large as that of unimplanted one. Emission intensity becomes nearly the same order as that of direct GaAsP. This is attributed to the emission due to the NN line.

INTRODUCTION

Nitrogen atoms in GaAsP are known to be very effective radiation recombination centers. The introduction of N atoms by ion implantation is a useful method to obtain a desired concentration and a well-controlled spatial distribution of the N atoms. The GaAsP wafers implanted with N atoms at several hundred°C become luminescent after appropriate annealing treatment [1-4]. Our previous measurements show that the photoluminescence intensity of N-implanted $GaAs_{1-x}P_x$ (x=0.52) at 2K increases by several times the intensity of unimplanted ones [4]. In this paper we present a very large photoluminescence intensity increase, of the order of 1000 times, observed at 77K.

EXPERIMENTAL

The samples used in this experiment are $GaAs_{1-x}P_x$ (x=0.52) single crystals grown on GaAs substrates by vapor-phase epitaxial method. Donor or acceptor impurity is not intentionally doped. The (100) surface was polished, optically finished, etched and rinsed in 0.5% Br methanol solution.

Implantation was accomplished at 350°C with a mass separated beam of N ions. Six samples were prepared with concentration of 10^{17}, 3×10^{17}, 10^{18}, 3×10^{18}, 10^{19}, and 10^{20} cm^{-3}. In case of $10^{20} cm^{-3}$, for instance, the energies were 50, 60, 100 and 220keV with fluence of 3.5×10^{14}, 2.0×10^{14}, 1.1×10^{15} and 2.9×10^{15} cm^{-2} respectively. In this four stage implantation a theoretical calculation predicts that the concentration profile of the implanted nitrogen has a plateau in the range between 0.12 μm and 0.50 μm below the surface. The implanted wafer was coated with SiO_x by Ar^+ sputtering to prevent the decomposition of GaAsP during annealing at high temperatures. Annealing was made in vacuum of 10^{-6} Torr for 1 hour at 500°C, 600°C, 700°C and 800°C and for 30 minutes at 850°C.

The photoluminescence was measured at 2K and 77K. The excitation was done by a cw Ar laser emitting maximum power of 1 W at 514.5 nm.

RESULTS AND DISCUSSIONS

Figures 1a and 1b show the dependence of photoluminescence spectra on the annealing temperature at 2K and 77K respectively. The excitation power density is about 10W cm^{-2}. In the figures the spectrum of an unimplanted and noncoated wafer is also shown, in which the emission due to free excitons at X point, Egx, is seen at 591.5 nm (2K) and 588 nm (77K). The LA phonon replicas (Egx-LA and Egx-2LA) are also obtained, which are centered at 602 nm and 611 nm (2K) or 596 nm and 606 nm (77K). At 2K in the implanted wafer, the phonon replicas disappear and there are obtained two peaks centered at 599.5 nm and 618 nm. The former is distinctly observed in the sample annealed at lower temperatures, but becomes very weak at high annealing temperature. In contrast, the latter appears only after the former grows sufficiently, and becomes predominant at high annealing temperature. The spectrum coincides with that of conventionally N doped GaAsP [5] and diode [6]. The peaks at 599.5nm and 618 nm are, therefore, the emissions due to the radiative recombinations of excitons bound to substitutional, isolated N atoms (A-line) and their pairs (NN line) respectively. At 2K the A line appears in the sample before annealing. This is

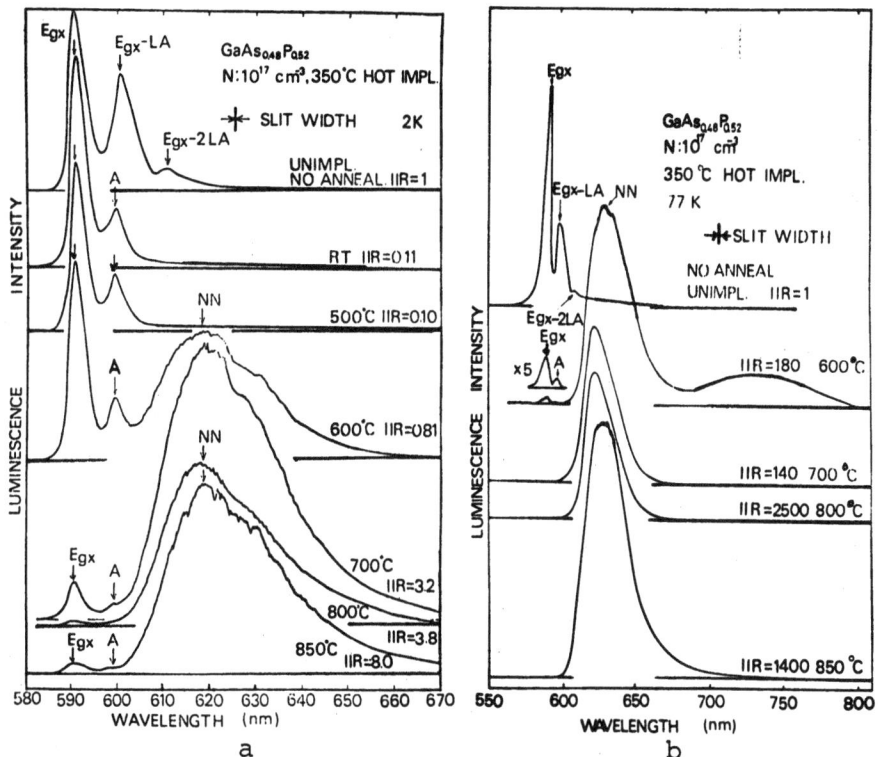

Fig. 1 Photoluminescence spectra at 2K (a) and 77K (b) for GaAs$_{0.48}$P$_{0.52}$ implanted with N ions to 10^{17}cm^{-3} and annealed at various temperatures. IIR stands for Integrated Intensity Ratio normalized by the intensity of the original sample.

different from the case of direct GaAsP implanted with N, where A line is not observed in principle unless the line is resonantly enhanced [5]. The fact that only A line appears and no NN line appears suggests that substitutional fraction of N atoms is still very small at this stage.

In the measurement at 77K before annealing no appreciable signals are observed, probably due to the large damages and strong local strains. After 600°C annealing, Egx (588 nm), A line (596 nm), NN line (625 nm) and another peak (720 nm) are observed. It should be noted that at 77K the contributions of Egx and A line are very small compared with the spectra at 2K, which is coincident with the case of GaAsP wafers with N incorporated during growth.

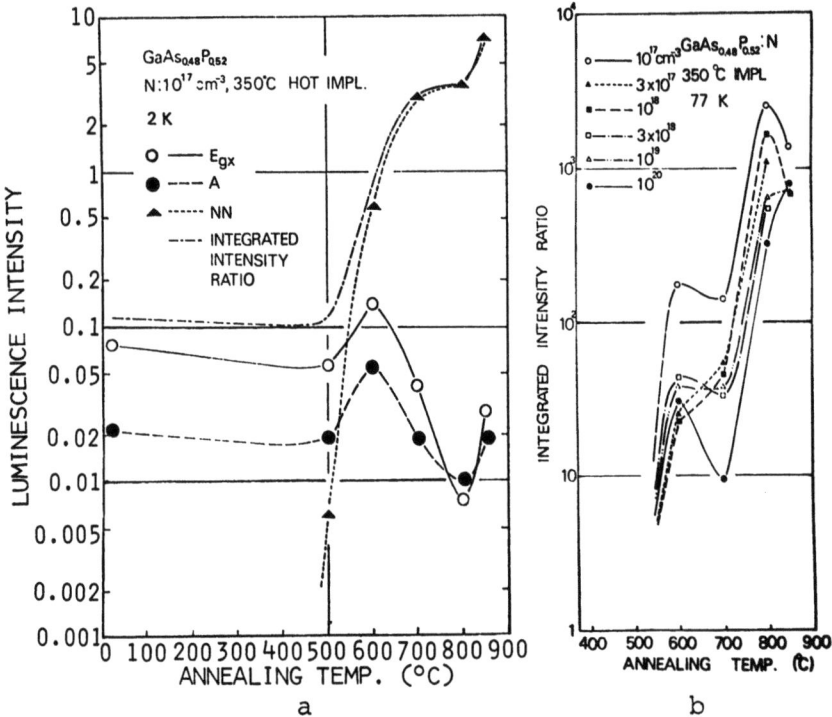

Fig. 2 Annealing temperature dependence of the integrated intensity ratio (IIR) at 2K (a) and 77k (b) for $GaAs_{0.48}P_{0.52}$ implanted with 10^{17} N ions cm^{-3}. The luminescence intensity, 1, in the ordinate corresponds to that of the unimplanted, noncoated crystal.

In Fig. 2 the dependence of total integrated intensity ratio (IIR) on annealing temperature is shown, together with IIR of each peak. In Fig. 2b only the behavior of the total IIR is plotted, considering that total IIR is equivalent to their integrated intensity of the NN line, because the contributions of the Egx and A line to the total intensity are negligibly small. At 77K a 800°C annealing brings about large increase of IIR under the condition of $10^{17} cm^{-3}$. According to Fig. 2b, the IIR decreases with increasing the implanted number of atoms. It is apparent that the induced damage is the principal factor to determine the emission efficiency of the implanted specimen. The knicks at 800°C in Fig. 2a is due to the annealing effect of SiO_x coated onto GaAsP wafer [4]. The disappearence of the NN lines at higher annealing temperature, observed in direct $GaAs_{1-x}P_x$ implanted with N [2], is not observed in this case.

On inspecting the figures at 2K, we may describe the annealing process as follows. Before annealing the strong peak is the one due to the radiative recombination of free excitons and no distinct phonon replicas are observed appreciably. Therefore it is considered that the crystal is highly damaged and phonon spectrum is much disturbed. The emission due to radiative recombinations at damages in indirect GaAsP is extremely weak compared with the direct case [1,2]. After annealing at 500°C, the A line is still weak, showing that the substitution process occurs insufficiently. The next step of annealing at 600°C shows the increase of the Egx line, indicating that the damages decrease to a large extent. The A line is also clearly observed and furthermore the broad NN line grows rapidly, indicating that the substitution occurs rapidly and the numbers of substituted N atoms become large enough to form NN pairs. The intensity of the A and Egx lines is transferred to that of the NN line at about 700°C. This is probably because the numbers of NN pairs become far larger than that of the isolated N atoms.

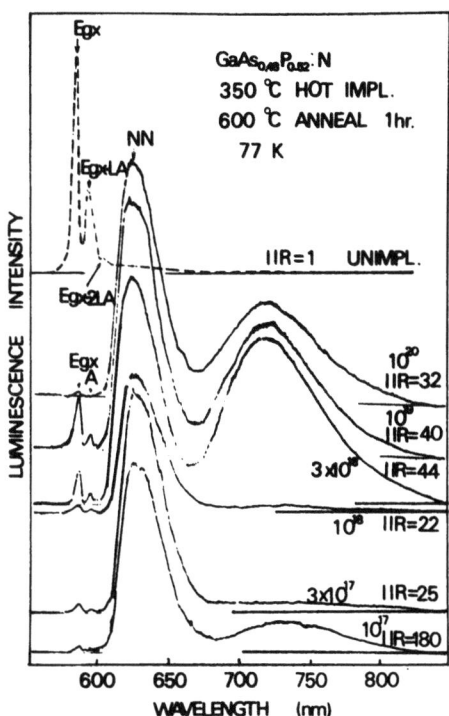

Fig. 3 Photoluminescence spectra at 77K for $GaAs_{0.48}P_{0.52}$ implanted with N ions to 10^{17}, 3×10^{17}, 10^{18}, 3×10^{18}, 10^{19}, and 10^{20} cm^{-3} and annealed at 600°C for 1 hr.

It should be remarked that new emission peak is observed under the appropriate condition of annealing temperature and N concentration. This is the 720 nm peak as seen in Fig. 1b. As is shown in Fig. 3 the intensity of this broad peak is strongly dependent on implanted numbers of N atoms. This peak is observable only at relatively low annealing temperatures, and scarcely discernible before the appearance of A line and disappear at higher annealing temperatures. These facts suggest that the energy level corresponding to this peak is metastable and presumably due to the complexes between N atom at P site and interstitial N atom.

In conclusion, the emission intensity of indirect GaAsP implanted with N atoms increases by about 1000 times at 77K compared with that of the unimplanted one. This value tells us that the total emission intensity of the indirect band gap GaAsP is in the same order of magnitude compared with that of direct band gap GaAsP. This is attributed to the emission due to the isoelectronic traps produced by the implanted N atoms.

ACKNOWLEDGEMENT

Thanks are due to S.Mukai for stimulating discussion and H.Fujita of Mitsubishi Monsanto Co. for supplying crystals. The authors are grateful to Dr. K.Sakurai and Dr. J.Shimada for their continual interest.

REFERENCES

[1] Y.Makita, S.Gonda, H.Tanoue, T.Tsurushima and S.Maekawa : Japan. J. appl. Phys. 13 (1974) 563.

[2] Y.Makita and S.Gonda : Japan. J. appl. Phys.13 (1974) 565.

[3] B.G.Streetman R.E.Anderson and D.J.Wolford : J. appl. Phys. 45 (1974) 974.

[4] S.Gonda, Y.Makita, S.Maekawa, H.Tanoue and T.Tsurushima : to be published in Japan. J. appl. Phys. 13 (1974) No. 9.

[5] N.Holonyak Jr., R.D.Dupuis, H.M.Macksey, M.G.Craford and W.O.Groves : J. appl. Phys. 43 (1972) 4148.

[6] M.G.Craford, R.W.Shaw, A.H.Herzog and W.O.Groves : J. appl. Phys. 43 (1972) 4075.

ELECTRICAL AND PHOTOLUMINESCENT PROPERTIES

OF ZINC IMPLANTED $GaAs_{0.62}P_{0.38}$

H. Okabayashi

Central Research Laboratories
Nippon Electric Company, Ltd.
Kawasaki, Japan

ABSTRACT

High temperature annealing characteristics of electrical and photoluminescent properties of $GaAs_{0.62}P_{0.38}$ implanted with 50 keV Zn at room temperature have been studied. Electrical activities, approximately independent of Zn dose, increased with annealing temperature, and reached about 100% at 950°C. Previously reported abnormally high electrical activities (~300% at 800°C) were not observed. A close correlation between photoluminescence intensity and effective surface carrier concentration was found. Qualitative explanation of annealing behaviors is presented.

INTRODUCTION

Several investigations have been reported on electrical and optoelectronic properties of Zn implanted $GaAs_{0.6}P_{0.4}$ [1] - [3]. Annealing characteristics of electrical properties of 400°C-implanted samples have been studied in fair detail [2]. However, large differences in electrical properties between room temperature- and hot-implantations have been reported [3].

This paper mainly reports annealing characteristics of electrical and photoluminescent properties of $GaAs_{0.62}P_{0.38}$ implanted with 50 keV Zn at room temperature. Also described is influence of 400°C-implantation on electrical properties. Qualitative explanation on annealing behaviors are also given.

EXPERIMENTAL

$GaAs_{0.62}P_{0.38}$ crystals (from Monsant Corp.) were Te-doped ($5 \sim 10\times10^{16}$ cm^{-3}) n-type, epitaxially grown on GaAs. Implantation was carried out with mass-separated 50 keV Zn. Implantation dose rate was $\sim 4\times10^{15}$/cm^2/hour. Implanted samples were coated with sputtered SiO_2 and then annealed in N_2 flow. Effective surface carrier concentration N_S and effective Hall mobility μ_H were determined from Hall effect measurements. Carrier profiles were obtained from differntial Hall effect measurements. Photoluminescence (PL) due to near-band-edge transitions was excited with a 5145Å Ar laser beam at room temperature (R.T). PL intensity was defined as (peak hight of a spectral band) × (half width of a spectral band).

RESULTS

Electrical Properties Figure 1(a) shows the results of 40 minute-isochronal annealing on electrical properties of R.T-implanted samples. The effective surface carrier concentration N_S increases with annealing temperature. The activation energy of N_S is ~ 1.5 eV at lower annealing temperatures, and decreases by a factor of 2 or more at higher temperatures. The temperature where activation energy changes decreases for the highest dose. It should be noted that μ_H shows nearly constant value at above 750°C, independent of dose and, therefore, of N_S. Electrical activities (N_S/dose) derived from these data show little dose dependence and were $\sim 10\%$ at 750°C, $\sim 60\%$ at 850°C and $\sim 100\%$ at 950°C.

Influences of implantation temperature (R.T and 400°C) on N_S are shown in Fig. 1(b). Also shown, for comparison, are results published by Itoh and Oana [2], [3]. The present results show little differences between R.T- and 400°C-implantations, except for the highest dose ($\sim 10^{16}$ cm^{-2}). Abnormally high electrical activities ($\sim 300\%$) were not observed.

Figure 2(a) shows carrier profiles of samples implanted with 4×10^{15} Zn/cm^2 at R.T and annealed at 850°C. Since the Zn ion range R_P and the standard deviation $\triangle R_P$ are expected to be comparable to those in GaAs and GaP, where R_P and $\triangle R_P$ are about 250Å and 100Å, respectively [4], the very deep carrier distribution is due to thermal diffusion of Zn in annealing process. The shapes of carrier profiles are similar to those reported [3].

Hole mobilities, obtained from differential Hall effect measurements, are shown in Fig. 2(b). Hole mobilities consistent with diffused values are obtained, at least, for annealing above 850°C.

Photoluminescent Properties Figure 3(a) shows results of 40 minutes-isochronal annealing in PL intensity and half width of spectral bands for R.T-implanted samples. The PL intensities show similar annealing behaviors to those of N_S in Fig. 1(a). The tendency, wherein the annealing temperature where the change of the

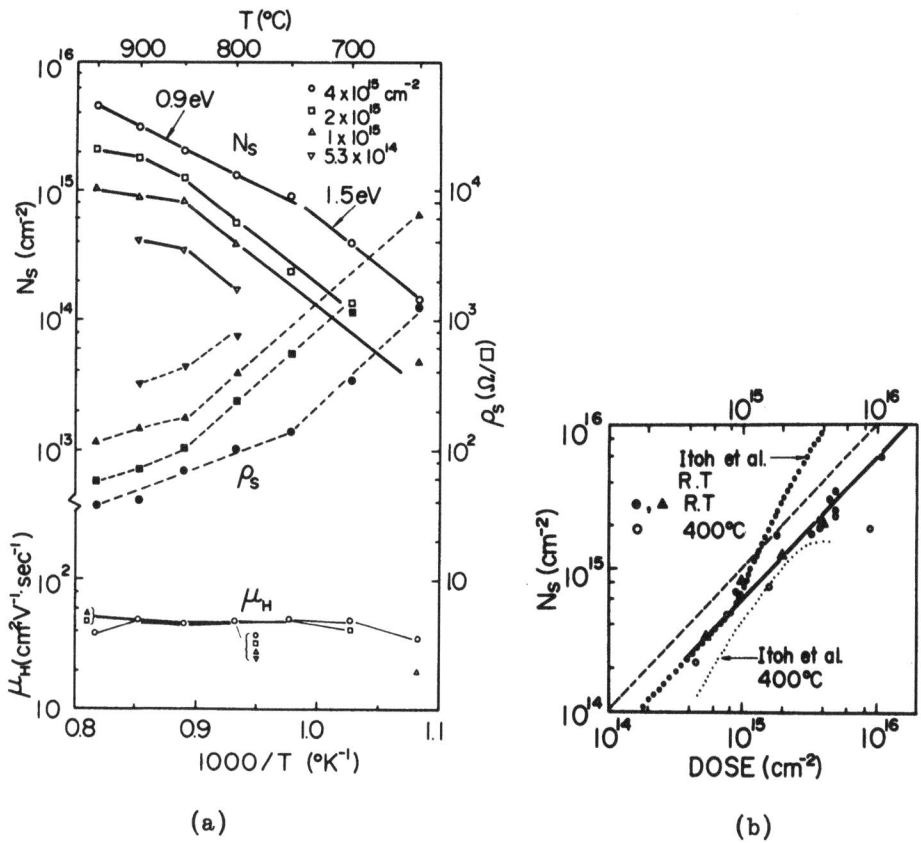

Fig. 1 Results of Hall effect measurements.
(a) 40 minute-isochronal annealing characteristics of electrical properties of samples implanted at R.T.
(b) Influence of implantation temperature on N_S. ● and ▲ (two different implantation machines) (R.T-implant and 850 °C-anneal), and ○ (400°C-implant and 850°C-anneal), this work; ••• (R.T-implant and 800°C-anneal) and ... (400°C-implant and 850°C-anneal), Ref. (2) and (3).

activation energy occurs decreases for higher doses, can be seen a little clearer in this case than in the case of N_S. The half width increases with the annealing temperature and saturates at a value 2.5~3 times as large as that of unimplanted samples. The half width saturation temperature decreases as the dose increases.

As is expected from the similarity of annealing behaviors, a close correlation between N_S and the PL intensity was found, which is represented by the curve in Fig. 3(b). The PL intensity depends only on N_S and is independent of the dose and annealing temperature.

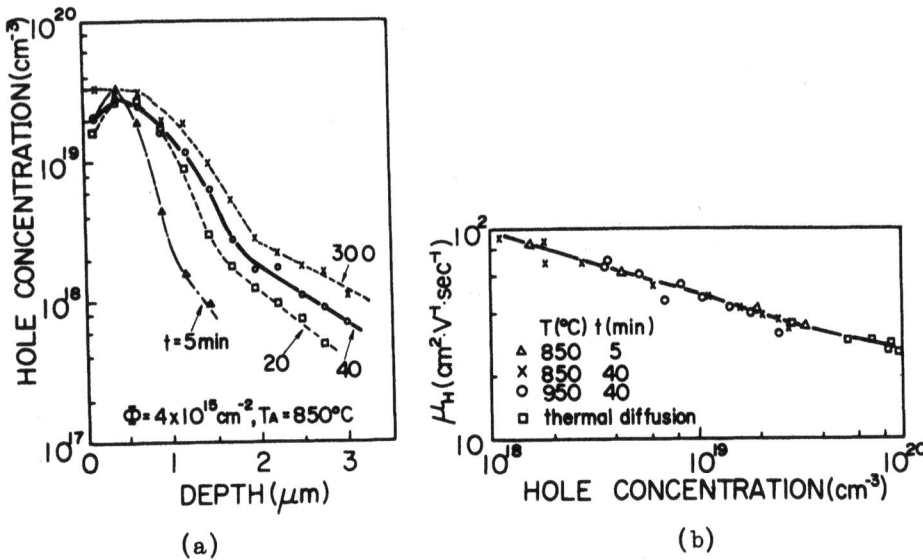

Fig. 2 Results of differential Hall effect measurements.
(a) Carrier (hole) profiles of samples implanted with 4×10^{16} Zn/cm^2 and annealed at 850°C.
(b) Hole mobility as a function of hole concentration.

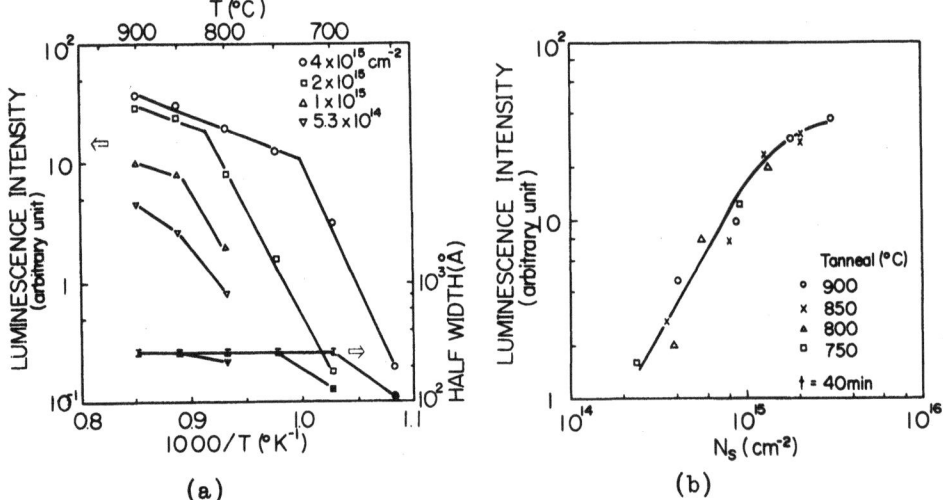

Fig. 3 Results of photoluminescence measurements.
(a) 40 minute-isochronal annealing characteristics of PL intensity and half width of spectral bands. Samples were implanted at R.T.
(b) PL intensity as a function of N_s.

DISCUSSIONS

Annealing behaviors of the Zn-implanted GaAs-GaP system are considered to be mainly influenced by three factors: 1. annealing of implantation-induced defects, 2. Zn solubility 3. Zn diffusibility.

A 100% activation at 700°C can be obtained for Zn in GaAs [5], and Zn-implantation-induced defects show no chemical component dependence on annealing behaviors [6]. In the present results, mobilities and PL intensities at above 750°C can be considered to be little influenced by residual defects, because they are independent of both annealing temperature and dose. Therefore, influence of residual defects on N_S can also be expected to be small.

As carrier profiles depend on Zn diffusion in annealing process, N_S then can be approximated as $N_S \simeq C_S \cdot x_S$, assuming that the Zn atom concentration near depth x_S decreases sharply with the depth, and that the peak Zn atom concentration far exceeds the Zn solubility C_S, where x_S is the depth where the Zn concentration becomes equal to C_S. These assumptions seem to be plausible, at least below 800°C, where the electrical activity is small. Zinc diffusion coefficient D in the GaAs-GaP system depends on both Zn concentration and temperature. D can be expressed as $D \propto C^n \cdot \exp(-E_D/kT)$, where n is a constant and E_D is the activation energy [7]. Since the temperature dependence of C_S can be given by $C_S \propto \exp(-E_S/kT)$, ($E_S \simeq 0.8$ eV) [8], and x_S can be expected to be proportional to the square root of D [9], N_S then can be expressed as $N_S \propto C^{n/2} \cdot \exp[-(E_D + 2E_S)/2kT]$. For the purpose of examining the dose dependence of the electrical activity, C can be replaced by the dose Φ. Then, electrical activity $\alpha (\alpha = N_S/\Phi)$ can be expressed as $\alpha \propto \Phi^m$ (m=0~0.5), because n can be considered to be 2~3 [7]. This conclusion was found to be approximately satisfied by the present results derived from the data in Fig. 1(a) and the previously reported results for 400°C-implantation [2]. From the observed activation energy of ~1.5 eV for N_S at lower annealing temperatures in Fig. 1(a), and from E_S=0.8 eV, E_D is derived to be ~1.4 eV. This small value suggests the interstitial diffusion mechanism, which is the case assumed above in the concentration dependence of D.

The activation energy of N_S for the highest dose and higher temperature annealing in Fig. 1(a) is nearly equal to that of C_S. This can be attributed to the fact that the relative change of the p-layer thickness in this case may be smaller than in the case discussed above, because the p-layer is fairly thicker in this case.

The photoluminescence at above half width saturation temperature, where the half width is 2.5~3 times larger than that for unimplanted n-GaAsP, can be attributed to that from p-layers [10]. The value of N_S at the half width saturation temperature is nearly constant (~4×10^{14}cm^{-2}). This is considered to be due to formation of heavily doped p-layers with a thickness comparable to the absorption length of the exitation beam (~0.2μm [11]). The p-layer thickness increases with the dose, so, the half width saturation temperature decreases as the dose increases.

CONCLUSIONS

Annealing characteristics of electrical and photoluminescent properties of Zn implanted $GaAs_{0.62}P_{0.38}$ have been studied.

Electrical activity α was found to depend little on dose and to vary with annealing temperature as $\alpha \propto \exp(-E/kT)$. These annealing behaviors of α were qualitatively explained from both the temperature and concentration dependence of the Zn diffusibility and the temperature dependence of the Zn solubility.

No abnormally high electrical activities were observed.

Intensities of photoluminescence from p-layers increased with effective surface carrier concentration, independent of annearling temperature. This is mainly due to increase in the p-layer thickness.

ACKNOWLEDGEMENTS

The author thanks S. Asanabe, N. Kawamura and D. Shinoda for their encouragement. He also thanks T. Kawano, K. Mori, T. Suzuki and H. Muta for use of their apparatus and for their comments, T. Koshimura for supplying diffused samples and T. Nozaki and K. Ueki for their help in the experiment.

REFERENCES

1. K. R. Faulker and A. Todkill, Ion Implantation in Semiconductors, editer by I. Ruge and J. Graul (Springer-Verlag, Berlin, 1971), p. 222.
2. T. Itoh and Y. Oana, J. Appl. Phys. 44, 4982 (1973).
3. T. Itoh and Y. Oana, Appl. Phys. Lett. 24, 320 (1974).
4. W.S. Johnson and J.F. Gibbons, Projected Range Statistics in Semiconductors (Standord University Bookstore, Standord, 1970).
5. Y. Yuba, K. Gamo, K. Masuda and S. Namba, Japan. J. Appl. Phys. 13, 641 (1974).
6. S.T. Picraux, Rad. Effects 17, 261 (1973).
7. D.L. Kendall, Semiconductors and Semimetals, edited by R. K. Willardson and A. C. Beer(Academic Press, New York, 1968), p. 163.
8. L. L. Chang and G. L. Pearson, J. Phys. Chem. Solids 25, 23 (1964).
9. L. L. Chang, Solid-State Electron. 7, 853 (1964).
10. M. R. Lorenz and A. E. Blakeslee, Gallium Arsenide and Related Compounds (The Institute of Physics, London, 1972), p. 106.
11. V. K. Subashiev and G. A. Chalikyan, Soviet Phys. Semicond. 3, 1216 (1970).

DEFECT-FREE NITROGEN IMPLANTATION INTO GaP

T. Shimada, Y. Shiraki and K. F. Komatsubara

Central Research Laboratory, Hitachi Ltd.

Kokubunji, Tokyo, Japan

ABSTRACT

Backscattering yields and photoluminescence (PL) spectra were measured in GaP crystals implanted with 200keV-N^+ ions. Backscattering results indicate that implantation at 500°C greatly reduces the radiation damage. The PL intensities of NN lines were maximum in the sample implanted with N^+ ions of 3×10^{14} cm^{-2} at 500°C, and annealed at 1000°C for one hour. The PL intensity is comparable to that of the nitrogen-doped sample during the liquid phase epitaxy which is widely accepted as the best method of introducing nitrogen into GaP crystals.

INTRODUCTION

Nitrogen implantation into GaP crystals is a representative process which provides not only a useful application, but also information concerning the implantation process itself. Implanted nitrogen atoms with subsequent annealing can be substituted for phosphorus sites and formed the isoelectronic luminescent centers[1]. However, the reported photoluminescence (PL) efficiency of implanted samples compared to that of the nitrogen-doped samples by the conventional liquid phase epitaxy (LPE) method is at most 10^{-2} due to the residual defects produced by implantation.

In this paper, we demonstrate the implantation and annealing procedures necessary to introduce the isoelectronic nitrogen-nitrogen pair centers of which the PL efficiency is comparable to that of the centers doped during the LPE growth.

EXPERIMENTALS

The samples used in this study were sulfur-doped GaP crystals grown by the LPE method. The epitaxial layer contains sulfur

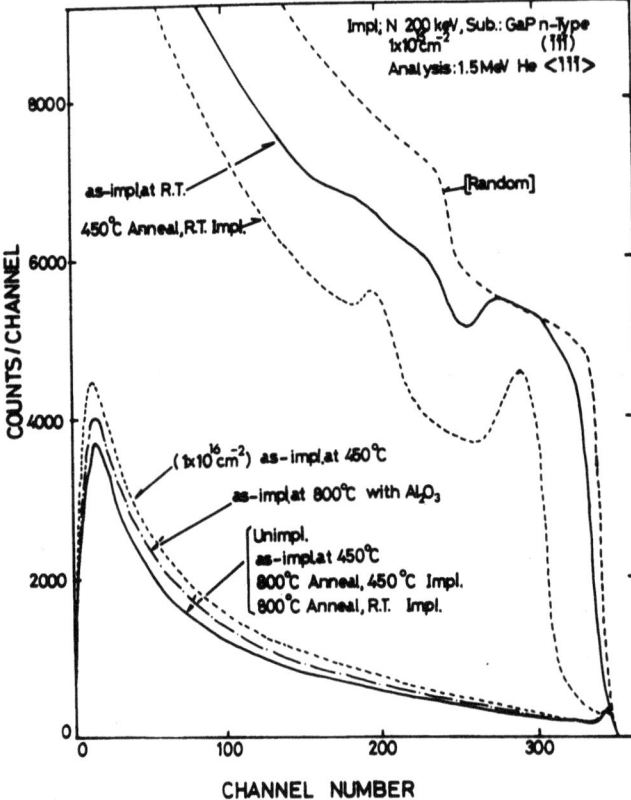

Fig. 1. Backscattering spectra of the implanted GaP for various implantation and annealing conditions.

impurities of $3\sim7\times10^{17}$cm^{-3} and nitrogen atoms of "less than a few" $\times 10^{16}$cm^{-3}.

The samples were implanted with 200keV-N$^+$ ions ($R_p \cong 0.39\mu$m) at doses ranging from 5×10^{13} to 1×10^{16}cm^{-2} at from room temperature 800°C. Samples were coated with aluminum glass only for the 800°C implantation. After implantation, all samples were coated with aluminum glass and PSG (phospho-silicate glass), and annealed at a temperature of up to 1100°C in evacuated silica ampoules.

The lattice disorder produced during implantation was investigated by backscattering experiments using 1.5MeV-He$^+$ ions. The PL spectra were taken by a Perkin-Elmer E-1 type spectrometer at 4.2°K. The samples were excited by a Kr UV laser (350.7/356.4mμ) to measure the PL of the thin implanted layer only.

EXPERIMENTAL RESULTS

Figure 1 shows the backscattering spectra for various implantation and annealing conditions. These data show that hot implantation at a temperature of 450°C produces a lattice disorder less than the backscattering sensitivity. This result is consistent with the result of Bi implantation[2]. In the sample implanted at 800°C, a few percent increase in aligned yield was observed. However, in the samples implanted at room temperature and at 450°C, respectively, and subsequently annealed at 800°C, there were no lattice disorders.

The PL spectra of the nitrogen-undoped and unimplanted crystals, are shown in Fig. 2-A. Only the C-S donor-acceptor pair

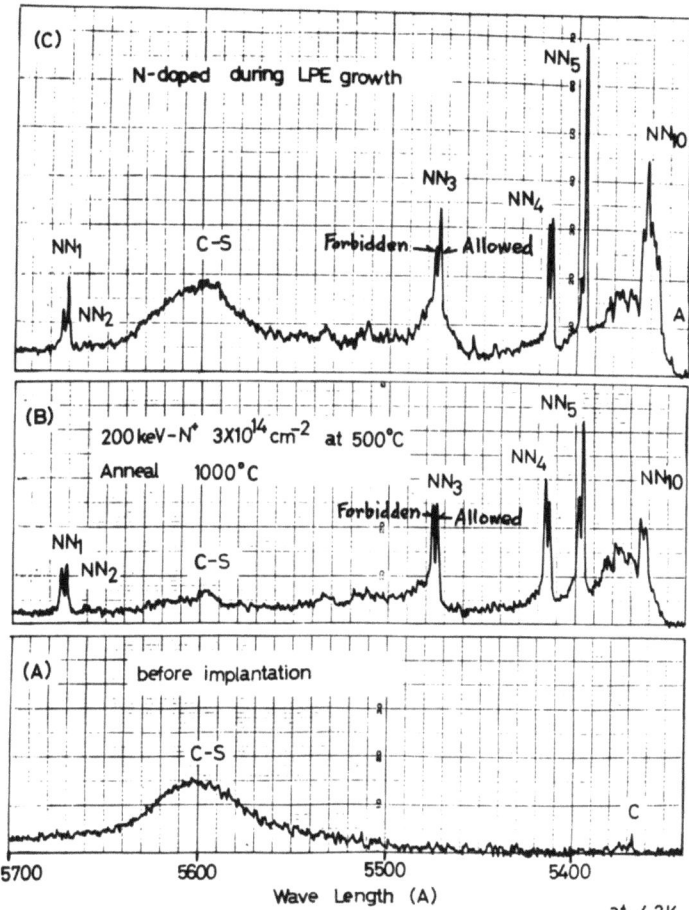

Fig. 2. The PL spectra of the N-undoped and unimplanted GaP (A), of the GaP implanted with 3x10 N ions cm at 500°C followed by the 1000°C anneal (B) and of the N-doped GaP during LPE growth (C).

recombination emission (C-S pair) can be seen clearly.

Figure 2-B shows the PL spectrum of the sample implanted with 3×10^{14} N$^+$ ions cm^{-2} at 500°C, and annealed at 1000°C for one hour. The sharp pair lines in the spectrum are the recombination emissions of excitons trapped at nitrogen-nitrogen pairs (NN line).

Figure 2-C shows, in comparison, the PL spectrum of the sample doped with 4.8×10^{18} cm^{-3} nitrogen atoms during LPE growth.

As the dose of the implanted nitrogen was increased, the NN line intensities increased, whereas the C-S pair intensities decreased. The total intensity of NN lines reached a maximum at a dose level of around 3×10^{14}cm^{-2}.

The PL efficiency vs temperature during implantation is shown in Fig. 3. The highest efficiency was obtained in the sample implanted at 500°C and annealed at 1000°C. It is about 10^3 times as large as that of the sample implanted at room temperature, and annealed at 1000°C.

The annealing temperature dependences of PL intensities of the

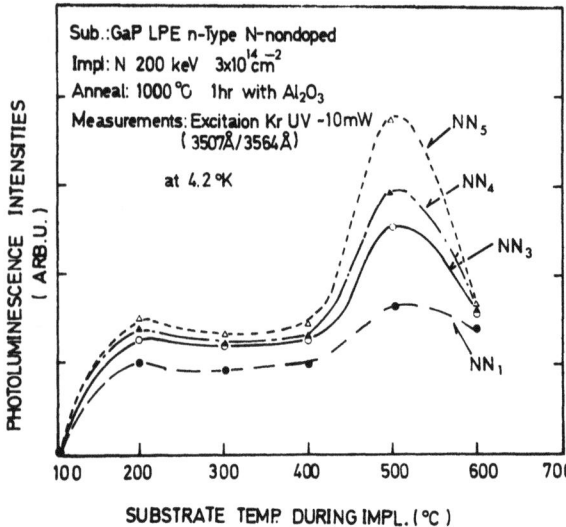

Fig. 3. The PL intensities of NN lines vs temperature during implantation.

NN lines and the C-S pair are shown in Fig. 4. After annealing at relatively low temperatures around 700°C, the C-S pair emission efficiency was restored and reached a maximum after annealing at around 1000°C. On the other hand, the NN lines which were not observed after annealing below 700°C, slightly appear after annealing at around 900°C. After annealing at 1000°C the total intensity of NN lines became about ten times stronger than the C-S pair emission.

DISCUSSIONS

Substrate temperature during implantation is a most important factor in producing NN centers having the highest radiative recombination efficiency. Results of backscattering and PL experiments (Figs. 1 and 3) indicate that the 500°C implantation followed by the 1000°C anneal is a defect-free nitrogen implantation into GaP. The reason why the PL efficiency for the room temperature implantation is very low, is that room temperature implantation produces a non-crystalline state and the crystalline order does not recover even after annealing at elevated temperatures[3]. In the case of high temperature implantation above 600°C, the PL efficiency is low, presumably due to the decomposition of the GaP crystals.

The backscattering spectra show that hot implantation at 450°C greatly reduces the lattice disorder compared with room temperature implantation. However the PL is not observed. These facts indicate that non-radiative defects which cannot be detected by the backscattering measurement exist. Therefore, annealing is necessary even for hot-implantation.

As shown in Fig. 4, only the C-S pair emission is observed after annealing at around 700°C. The reappearance of the C-S pair suggests that the defects produced during implantation are annealed out. However, the NN lines are not observed even after annealing at a temperature of 700∼800°C, suggesting that implanted nitrogen atoms do not substitute for phosphorus sites below 800°C. According to the EPR experiments[4], the implanted nitrogen is on the interstitial site after annealing at a temperature of 700∼800°C.

DEFECT-FREE NITROGEN IMPLANTATION INTO GaP

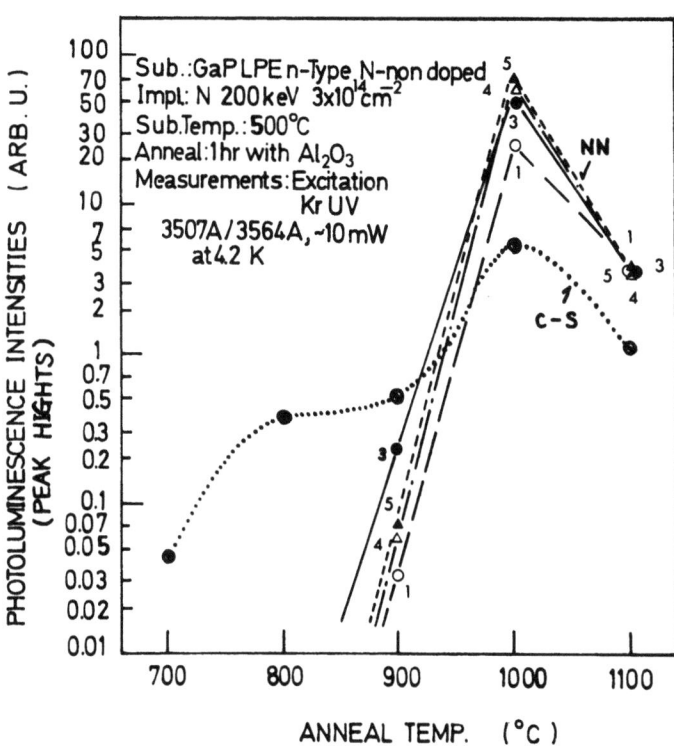

Fig. 4. The PL intensities of the NN lines and the C-S pair vs annealing temperature after implantation.

The rapid increase of the NN line intensities after annealing at a temperature of 900~1000°C, evidences the increase of substitutional nitrogen atoms.

After annealing at 1100°C, both the intensities of the NN lines and C-S pair decrease. This may be caused by out diffusion of the implanted nitrogen and/or decomposition of the GaP crystal.

The intensity ratios of the NN lines of the implanted sample to those of the sample doped by the LPE method are 0.5~0.8. These values might be underestimated because the escape of photocarrier from the implanted layer is not taken into account.

Each NN pair line consists of allowed and forbidden lines as seen in Figs. 2-B and 2-C. The relative intensity of the forbidden line to the allowed line in the implanted sample is remarkably higher than that in the doped sample during LPE growth. This fact suggests that some kinds of defects or strains remain around the NN centers even in the implanted samples with a maximum PL efficiency. These defects or strains seem not to reduce the PL efficiency.

CONCLUSIONS

The backscattering and PL spectra of 200keV-N^+ ion implanted GaP were studied as functions of ion dose, temperature during implantation and anneal temperature after implantation. The results obtained here are summarized as follows. (1) The maximum PL efficiency, which was observed in the sample implanted at a dose of 3×10^{14}cm^{-2} at 500°C, and annealed at 1000°C for one hour, is comparable to that of nitrogen centers incorporated during LPE growth. (2) Hot implantation at around 500°C greatly reduces the lattice disorder compared with room temperature implantation. However, no PL was observed in as-implanted samples. (3) By annealing at 700~800°C, non-radiative defects are annealed out. (4) By annealing at 900~1000°C, the implanted nitrogen substitutes for the phosphorus sites. (5) Some kinds of defects or strains remain around the NN centers even in the implanted samples with a maximum PL efficiency. These defects or strains seem not to reduce the PL efficiency.

A first step toward defect-free nitrogen implantation has been realized.

ACKNOWLEDGMENTS

The authors wish to thank Professor S. Furukawa, Tokyo Institute of Technology for allowing them to use the backscattering equipment and for his advice, Professor T. Matsumori, Tokai University for information about EPR results before publication and Dr. M. Ogirima, Hitachi Central Research Laboratory, for GaP substrates.

REFERENCES

(1) J. L. Merz, E. A. Sadowski and J. W. Rodgers: Solid State Comm., 9 1037 (1971).
(2) J. L. Merz, D. W. Mingay, W. M. Augustyniak and L. C. Feldman: in Proceedings of the Int. Conf. on Ion Implantation in Semiconductors 182 (1971).
(3) T. Shimada, Y. Shiraki, Y. Kato and K. F. Komatsubara: to be published in Proceedings of the Int. Conf. on Lattice Defects in Semiconductors (1974).
(4) Private communication from Prof. T. Matsumori, Tokai Univ.

ION IMPLANTATIONS OF Mg AND Zn INTO n-TYPE GaP

T. Inada and Y. Ohnuki

College of Engineering, Hosei University

Koganei, Tokyo 184, Japan

ABSTRACT

Measurements of Hall-effect and sheet-resistivity and of depth of pn junction were carried out to make an electrical evaluation of p-type layer formed in GaP by either Mg or Zn ion implantation. Photoluminescence spectra were also measured for the implanted GaP both at room temperature and at 100 °K.

INTRODUCTION

Magnesium is known as a p-type dopant species for III-V compound semiconductors. Since an Mg acceptor is small in size and less electronegative than Zn, it can be expected that the bound exciton emission from an Mg-O complex would be shifted towards a higher energy than that from Zn-O complex. [1] However, Lorimor and Weiner [2] pointed out that it was difficult to use Mg as a dopant for GaP during crystal growth because of its strong affinity for oxygen. Ion implantation may provide a nearly oxygen-free system for doping of GaP.

Preliminary results on electrical transport properties for GaP implanted with either 15-keV Mg or Zn ions were reported previously by the present authors. [3] In this study, implantations were carried out at a higher energy up to 150 keV and the dopant behaviors of Mg and Zn introduced into GaP were investigated in some detail. Photoluminescence (PL) spectra were also measured for these implanted samples and an evaluation of the effect of an additional implantation with either oxygen or nitrogen on its emission spectrum was attempted.

EXPERIMENTAL

The substrate used in this study was a < 111 > oriented, 0.7 Ω-cm n-type (S-doped) liquid encapsuled pulled GaP crystal with a background doping of approximately 3×10^{17} cm^{-3}. The implanted ions were mass separated by a 30° analyzing magnet and the beams were electrostatically swept in two perpendicular directions over the sample to achieve a uniform implant. These ions were implanted into the GaP substrate titled about 7° away from the <111> axis to minimize channeling effects. After the ion implantation, all samples were annealed in a flowing, dry nitrogen atmosphere. Prior to anneal, both the front and rear surfaces of implanted substrate were coated with sputtered layers of SiO$_x$ film (0.6 µm in thickness) in order to avoid out diffusion of implanted atoms and decomposition of GaP during the annealing at higher temperature. As previously reported by the present authors [3], a higher temperature anneal is required to achieve electrical conduction type conversion of GaP due to implanted Mg atoms. And, in this study, all samples were annealed at temperatures above 900 °C and an annealing time was chosen for 30 min for all samples reported here.

Four holes were made to the SiO$_x$ film covering the implanted surface by a masking and etching procedure, one at each corner of the squared substrate. Then Ohmic contacts were made to the "stripped areas". For the fabrication of Ohmic contacts to the implanted surface three different processes were employed, that is,

(A) a 15-keV Zn-ion implantation was done at a dose of 2×10^{15} cm^{-2} before annealing and vacuum evaporation of an alloy (80%Au-20%Zn in wt%) was made at a substrate temperature of 550 °C after annealing

(B) vacuum evaporation of the same alloy as mentioned above was made at 550 °C after annealing

(C) vacuum evaporation of metallic gold was done at room temperature after annealing.

In this paper, these three different processes are designated as process A, process B, and process C, respectively.

The carrier concentration per square cm, N_S, and sheet resistivity, ρ_S, in implanted layers were measured by a means of van der Pauw technique. [4] The effective mobility of carrier, μ_H, was calculated from the data of sheet Hall coefficient and sheet resistivity. A method of angle-section and staining was used in measurements of depth of the pn junction formed. Photoluminescence measurements were carried out for the implanted and annealed sample. Photoexcitation was performed with the 4880 Å emission line of an argon laser. After correcting for the wavelength response of photomultiplier, the spectra were plotted as a function of photon energy.

RESULTS AND DISCUSSION

Experimental conditions and room temperature electrical properties of the p-type layer formed by either Mg- or Zn-ion implantation are summarized in Table 1. As shown in Table 1, it was found that there was a large difference in value of N_s between the Mg-implanted samples whose Ohmic contacts were fabricated through processes A and C, even if these samples were implanted at a similar dose. For example, the measured values of N_S were 2.0×10^{14} and 2.7×10^{13} cm^{-2} for samples M-13 and M-24, respectively. These two samples were prepared under similar experimental conditions except the fabrication procedures of their Ohmic contacts (see Table 1). In order to investigate the origin which caused such a difference in measured value of N_s, thicknesses of the p-type layers built up in these samples were measured. And it was revealed that the thicknesses of the p-type layers were approximately 0.6 μm and 0.3 μm for samples M-13 and M-24, respectively. For sample M-13 it was also shown that the thickness of the p-type layer varied with distance in the vicinity of the Ohmic contact area. On the other hand, such a variation of the thickness of the p-type layer could not be observed in sample M-24. The pn junction depth (0.3 μm) obtained from sample M-24 was considerably agreed with the one expected from LSS range-energy theory. [5] For the sample which was prepared with a higher implant dose, 2.0×10^{16} cm^{-2}, the thickness of the p-type layer ranged from 2.3 μm to 13.5 μm when its Ohmic contact was made through process A. Fig.1 is a photograph of angle-lapped and stained sample showing the pn junction formed in sample M-14. However, such a variation of the thickness of the p-type layer with distance could not be observed for sample M-25 whose contacts were fabricated through process C.

These results mentioned above could be explained as follows. In the sample whose contacts were made through process A, Zn atoms existing in the contact regions would diffuse rapidly during annealing not only into the interior of substrate but into the Mg-implanted layer. A diffusion of Zn atoms such as being directed parallel to the implanted surface, would give rise to increases not only in the value of the thickness of the p-type layer but in the value of N_s in the sample whose contacts were fabricated through process A. In the sample whose contacts were made through process C, such a diffusion never occurred and its p-type layer was formed only by the implanted Mg atoms. And it could be concluded that a net value of N_s actually due to implanted Mg atoms could be measured in the sample whose contacts were fabricated through process C. The net value of N_s seemed to be saturated to a value of around 4×10^{13} cm^{-2} for the Mg-implanted GaP prepared in this experiment. For Zn-implanted samples, no large difference in values both of N_s and x_j were found between the samples whose contacts were made through processes B and C. For the sample implanted at a dose of 2.0×10^{15} cm^{-2}, measured thickness of the

TABLE 1

EXPERIMENTAL CONDITIONS AND ELECTRICAL PROPERTIES OF ION-IMPLANTED GaP

SAMPLE NO.	ION	DOSE (cm^{-2})	T_a^* (°C)	O.C.**	N_s (cm^{-2})	ρ_s $(\Omega\text{-}\square^{-1})$	μ_H $(cm^2 V^{-1} s^{-1})$	x_j (μm)
M-12	150-keV Mg	2.0×10^{15}	900	A	2.3×10^{14}	1.2×10^3	22	0.3
M-13	150-keV Mg	1.0×10^{15}	1000	A	2.0×10^{14}	1.6×10^3	20	0.6
M-14	150-keV Mg	2.0×10^{16}	1000	A	3.1×10^{15}	60	34	2.3
M-23	150-keV Mg	2.0×10^{15}	900	C	3.3×10^{13}	5.2×10^3	35	0.3
M-24	150-keV Mg	2.0×10^{15}	1000	C	2.7×10^{13}	6.4×10^3	37	0.3
M-25	150-keV Mg	2.0×10^{16}	1000	C	4.1×10^{13}	4.4×10^3	34	0.3
M-29	15-keV Mg	2.0×10^{15}	1000	C	2.6×10^{13}	5.9×10^3	41	0.1
Z-15	150-keV Zn	2.0×10^{15}	900	B	8.3×10^{14}	1.2×10^3	6.5	0.6
Z-16	150-keV Zn	2.0×10^{15}	1000	B	1.4×10^{15}	1.7×10^2	26	1.5
Z-17	150-keV Zn	2.0×10^{16}	1000	B	8.7×10^{15}	21	34	13.5
Z-25	150-keV Zn	2.0×10^{15}	900	C	1.2×10^{14}	1.5×10^3	34	0.6
Z-26	150-keV Zn	2.0×10^{15}	1000	C	6.5×10^{14}	2.3×10^2	43	2.0
Z-27	150-keV Zn	2.0×10^{16}	1000	C	5.0×10^{15}	33	38	13.5

* ANNEAL TEMP.   ** FABRICATION PROCEDURE FOR OHMIC CONTACTS

Fig. 1. Photograph showing the pn junction formed in sample M-14

p-type layer was approximately 0.6 μm when the sample was annealed at 900 °C, and its thickness increased to the value of ~1.5 μm when the crystal was annealed at 1000 °C. For the sample implanted at a dose of 2.0×10^{16} cm^{-2}, it was found that the pn junction located around 13.5 μm beneath the implanted surface. These results indicated that a diffusion of Zn into the interior of the substrate played a significant role in the final atom location of Zn implanted GaP crystal. It was attempted to calculate an effective diffusion coefficient of Zn implanted GaP from the measured data by simply assuming that the depth of the pn junction would be represented in terms of a diffusion length, i.e., $x_j = \sqrt{D_{eff} \cdot t}$, where, D_{eff} is an effective diffusion coefficient and t is an annealing time. As a result, it was found that the effective diffusion coefficient of Zn equaled to approximately 1×10^{-9} cm$^2 \cdot$sec^{-2} for sample Z-27. It is well known that a diffusion coefficient of Zn in GaP varies with Zn concentration. [6] The effective diffusion coefficient obtained from the calculation mentioned above was found to agree in value with the diffusion coefficient published in Ref. [6] for a corresponding Zn concentration. However, it is difficult to give a more definite conclusion on the diffusion mechanism, at this time, because of lack of knowledge of the concentration-vs-depth profiles as functions of annealing temperature and time and others.

Hall mobility of holes due to implanted Mg was found to be between 30 and 40 cm$^2 \cdot$V$^{-1} \cdot$sec^{-1} at room temperature for samples annealed at 1000 °C. These measured mobilities agree in value with mobilities of holes for Be-diffused GaP (reported by Ilegems and O'Mara [7]) and Zn-diffused GaP (reported by Casey et al [8]). A reversal of sign of the Hall coefficient from positive to negative at measurement temperature around 100 °K. It is interesting to note that open-circuit photo-voltage measured by illuminating the implanted surface with a lamp was reduced in value for the sample in which the reversal of conduction type was observed. The mobility for Zn-implanted samples increased as the temperature lowered, reached a maximum at around 200 °K and then lowered as the temperature lowered. The qualitative characteristics of temperature-vs-mobility curve are similar to the one reported in Ref. [7] for Zn-doped GaP with a corresponding hole concentration.

Measurements of photoluminescence (PL) spectra were done at room temperature and at 100 °K. Implantation conditions and measured peaks in the PL spectra are summarized in Table 2. For some samples an additional implant was done. For Mg-implanted samples the additional implant of oxygen ions was made at 100 keV in order to fit the as-implanted profile of oxygen ions on that of Mg ions implanted at 150 kev. For Zn-implanted sample the additional implant was done at 150 keV.

For Mg-implanted samples M-26 and M-27, PL spectra were found to consist of two separated peaks when the crystal were cooled to

TABLE 2

EXPERIMENTAL CONDITIONS AND PEAKS IN EMISSION SPECTRA FROM ION-IMPLANTED GaP

SAMPLE NO.	ION	DOSE (cm^{-2})	PEAKS AT R.T. (eV)	PEAKS AT $100°K$ (eV)	
M-26	150-keV Mg	2.0×10^{16}	$P_1=1.72$	$P_1=1.70$	$P_2=1.97$
M-27	150-keV Mg 100-keV O	2.0×10^{16} 2.0×10^{15}	$P_1=1.72$	$P_1=1.70$	$P_2=1.98$
Z-22	150-keV Zn	2.0×10^{16}	$P_1=1.79$	$P_1=1.82$	$P_2=2.05$
Z-23	150-keV Zn 150-keV O	2.0×10^{16} 2.0×10^{15}	$P_1=1.79$	$P_1=1.82$	$P_2=2.03$
Z-24	150-keV Zn 150-keV N	2.0×10^{16} 2.0×10^{15}	$P_1=1.73$	$P_1=1.71$	$P_2=1.95$

All samples reported here were annealed at 950 °C for 30 min.

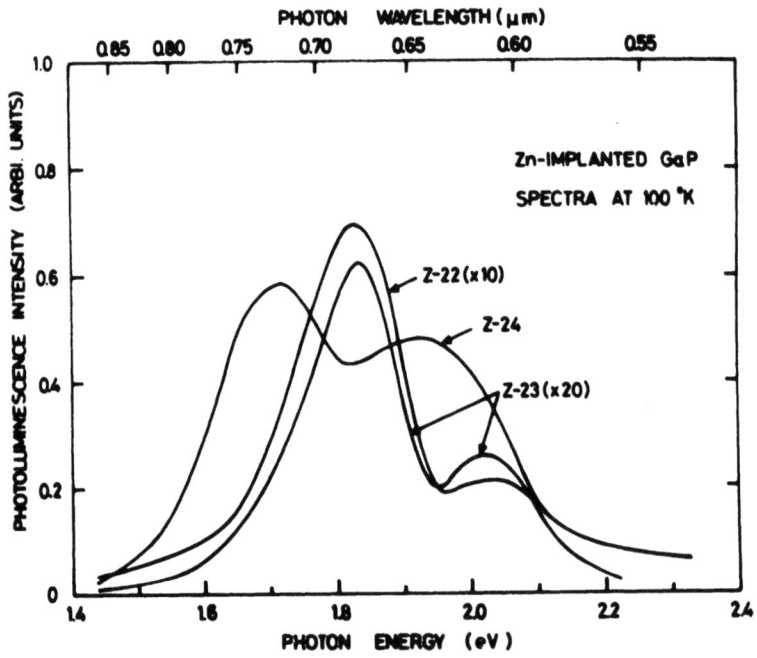

Fig. 2. PL emission intensity vs photon energy

100 °K. The peak in lower energy region is designated as P_1 and the other as P_2 in this text. The peak designated as P_2 could not be detected at room temperature. A spectral shift with the additional implant of oxygen ions was not measurable magnitude. However. the additional implant gave rise to increases in emission intensities at both P_1 and P_2. It was attempted to calculate the ratio of the emision intensity obtained from the sample implanted with both Mg and O ions (M-27) to that from the sample implanted with Mg ions only (M-26). And it was shown that the ratios equaled ~1.3 and ~2.7 for the peaks P_1 and P_2, respectively. This suggested that the additional implant with oxygen ions had a larger influence on the formation of the emission peaked at P_2 than that peaked at P_1.

For Zn-implanted sample (Z-22), a major peak (P_1) was observed at a photon energy of 1.79 eV at room temperature. This peak shifted to 1.82 eV when the sample was cooled to 100 °K. At this temperature, an additional spectral peak (P_2) was appeared. Fig. 2 shows the PL emission intensity-vs-photon energy curves obtained from Zn-implanted samples. As shown in Fig. 2, the additional implant of oxygen ions also yielded increases in emission intensities at both P_1 and P_2. The emission peaked at P_1 (1.79 eV at room temperature) was believed to be due to the recombination between a Zn-O donor-acceptor pair. [9] [10] On the other hand, an additional implant of nitrogen ions reduced the emission peaked at around 1.8 eV. And two different peaks were observed for this sample as shown in Fig. 2. The emission intensities at these two peaks were found to be lower by a factor of approximately 10 comparing with the emission intensity measured at P_1 for sample Z-22.

ACKNOWLEDGEMENTS

The authors thank H. Nakada and S. Iwamoto, students of Graduate School of Hosei University, for their assistances in sample preparation and measurements. The authors would like to thank Prof. Dr. T. Itoh, Waseda University, for affording us facility for photoluminescence measurements.

REFERENCES

1. R. N. Bhargava et al, Appl. Phys. Lett. 20, 227 (1972)

2. O. G. Lorimor and M. E. Weiner, J. Electrochem. Soc. 119, 1576 (1972)

3. T. Inada and Y. Ohnuki, Appl. Phys. Lett. 25, 228 (1974)

4. L. van der Pauw, Philips Res. Repts. 13 1 (1958)

5. J. Lindhard et al, Mat. Fys. Medd. Dan. Vid. Selsk. 33, 1 (1963)

6. H. C. Casey Jr., <u>Atomic Diffusion In Semiconductors</u> Chap. 4, edited by D. Shaw, Plenum Press, London and New York (1973)

7. M. Ilegems and W, C. O'Mara, J. Appl. Phys. 43, 1190 (1972)

8. H. C. Casey Jr. et al, J. Apll. Phys. 40, 2945 (1969)

9. T. N. Morgan et al, Phys. Rev. 166, 751 (1968)

10. C. H. Henry et al, Phys. Rev. 166, 754 (1968)

IN-DEPTH PROFILE DETECTION LIMITS OF NITROGEN IN GaP,
NITROGEN, OXYGEN AND FLUORINE IN Si BY SIMS AND AES

J. C. C. Tsai, Bell Telephone Laboratories,

Incorporated, Reading, Pennsylvania

J. M. Morabito, Bell Telephone Laboratories,

Incorporated, Allentown, Pennsylvania

ABSTRACT

Secondary Ion Mass Spectrometry (SIMS) and Auger Electron Spectroscopy (AES) have been used to determine the distribution of nitrogen in GaP ion implanted with nitrogen, and of nitrogen, oxygen, and fluorine ion implanted into silicon. The in-depth profile detectabilities for nitrogen in GaP are in the low 10^{18} atoms/cm^3 range and upper 10^{19} atoms/cm^3 range for SIMS and AES analyses, respectively. The in-depth profile detectability of SIMS for fluorine and oxygen is $\sim 5 \times 10^{16}$ and 10^{19} atoms/c.c.; AES in-depth detectability limits for oxygen and fluorine are in the 10^{20} atoms/c.c. range.

Both nitrogen and oxygen in silicon have gaussian distributions with the projected range statistics agreeing with the LSS theoretical calculations. For nitrogen in GaP, a profile tail is observed by the SIMS technique. The profile tail could be due to channelled ions during ion implantation and a recoil process during SIMS measurements. For fluorine in silicon, a more pronounced tail was observed which can be attributed mostly to channelling and recoiling during ion implantation. Complicated profiles of fluorine were observed after annealing at temperatures above 500°C for the sample dose of 6×10^{15} ions/cm^2.

INTRODUCTION

Both Secondary Ion Mass Spectrometry (SIMS) and Auger Electron Spectroscopy (AES) have been successfully applied to in-depth profile measurements.[1-15] Quantitative SIMS analyses via calibration with ion implanted phosphorus, boron, and arsenic standards in silicon have been obtained and the results agreed with electrical measurements when such comparisons are made.[10,11] Recently, quantitative AES analyses via calibration by ion implantation[15] or by reactive sputtering[10] have also been demonstrated. This paper will compare the <u>in-depth detectability limits</u> of SIMS and AES for nitrogen in GaP and for electrically inactive or low activity elements such as nitrogen, oxygen, and fluorine in silicon. The detection of ion species by the SIMS technique requires volatilizing a given volume of material by sputtering; while the AES measures the surface concentration of the species. In the latter method, concentration profiles are determined by simultaneous AES analysis and ion sputtering. The sensitivity for in-depth profile measurement for both methods can be readily compared at various depths with a given concentration distribution. This was accomplished by the use of ion implantation by introducing a known quantity of the elements of interest within a specified volume of the sample.

EXPERIMENTAL

GaP and Si Samples

Wafers from gallium phosphide single crystals which had been doped with nitrogen and selenium at 10^{17} and 2×10^{17} atoms/c.c., respectively, and chemically polished were used in this study. Nitrogen implantation was performed at 50KeV to ion doses 1×10^{13}, 1×10^{14}, 1×10^{15} and 5×10^{15} ions/cm^2. Silicon wafers of <100> orientation were implanted with various ions (N^+, O^+ and F^+) at 50KeV to a dose of 10^{15} to 10^{16} ions/cm^2. The wafers were misorientated during ion implantation to avoid channel implantation. Heat treatment or annealing was not performed on these wafers prior to SIMS or AES analysis, except the sample implanted with fluorine. The sputtering depth for every crater from the SIMS measurement on both the GaP and silicon samples was measured with a Proficorder.* Since the GaP sample surfaces were not smooth and flat, difficulties in determining the sputtering rate were observed.

Since the ion doses and the sputtering depths were used for calibrating the SIMS and AES data, the accuracy of ion dose and depth measurements are very important.

*PROFICORDER Pilotor Model 5, The Bendix Corporation.

Due to a lack of absolute determination of ion dose, an indirect estimation of ion dose is used as following. From the electrical results of boron implanted silicon after annealing, the sheet resistance reproducibility is approximately 2.6%. When comparing similar implantation from different machines, the sheet resistance showed a maximum difference of 6%. From these results, the accuracy and reproducibility of the ion doses is estimated to be better than 5%. The accuracy of the depth measurement is approximately 10%. Hence, the overall accuracy for the measured profiles would be $\cong \pm 15\%$.

SIMS and Auger Analyses

The principles of the instrumentation used in SIMS and AES analyses have been discussed in the literature.[9,10,12] The AES profiles were obtained by thin layer removal using argon ion sputtering at 2KeV.[9,10] For SIMS analysis, positive oxygen primary ion beams of 500-1000 nanoamps and 14.5KeV were used except for the measurement of oxygen where an argon primary ion beam was used. Signals from a 70μ diameter area inside the crater were recorded in the multichannel analyzer.

Since both nitrogen ($^{14}N^+$) and the doubly charged silicon ion ($^{28}Si^{++}$) have the same equivalent mass of 14, nitrogen or their oxygen complex ions are difficult to detect in Si without the use of high mass resolution (M/Δm). It is for this reason that nitrogen was ion implanted into GaP wafers.

The negative or positive secondary ions were mass analyzed, counted, and converted into concentration by the following equations:[6]

$$Q = C_{max} \sum_{i=1}^{n} \left[\frac{\text{(ion counts at } x_i)}{\text{peak counts at } x=R_p} (\Delta x) \right] \quad (1)$$

where Q is known from the ion implantation dose, C_{max} is the peak concentration at $x=R_p$, where R_p is the projected ion range, $C(x_i)$ is the concentration at $x=x_i$ and Δx is the sputtering depth increment. When the sputtering depth increment, Δx, and Q are known, C_{max} can be calculated from Eq. (1) and C(x) is obtained from Eq. (2).

$$C(x) = C_{max} \frac{\text{ion counts at x}}{\text{peak count}} \quad (2)$$

The nitrogen (380eV), oxygen (510eV) and fluorine (650eV) Auger peak heights after normalizing to the matrix

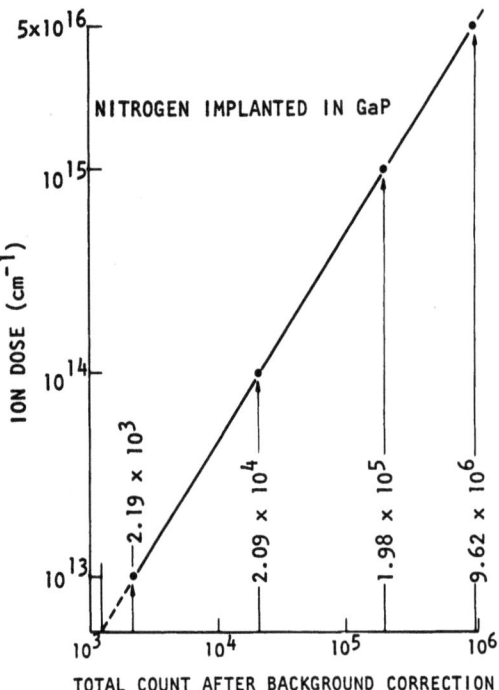

Fig. 1. Nitrogen Implant Dose vs. Total No⁻Ion Counts

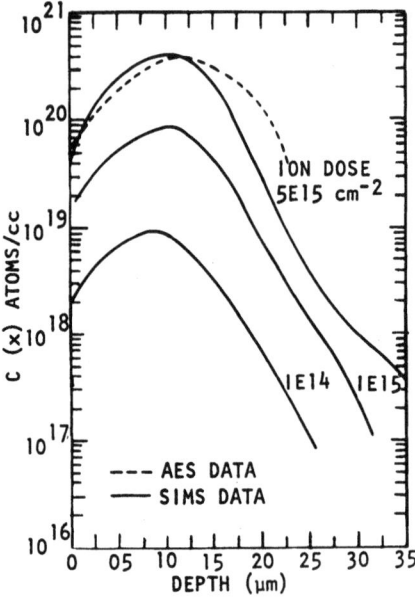

Fig. 2. Nitrogen Profile in GaP from SIMS and AES Data

(Si or P) were converted to concentration using the ion dose introduced into the sample. The depth resolution possible with AES and SIMS was 5-10% of the thickness removed by the sputtering in the absence of sputtering artifacts.

RESULTS

Nitrogen in GaP

The linear relationship observed between the ion doses and the total ion counts by the SIMS technique is shown in Figure 1 after correction has been made for background count. This linear relationship indirectly confirms the validity of the assumption that the secondary ion signal is proportional to the concentration of the element in the sample. A linear relationship between ion dose and concentration has also been reported for boron in silicon.[3,4] Figure 2 shows the concentration profiles of nitrogen calculated from Eq. (2).

The nitrogen concentration from AES data is also shown in Figure 2 where the implantation dose was used to convert the nitrogen (380eV) to phosphorus (120eV) Auger peak height ratios into concentration. For the AES data the sputtering rate was estimated to be $\simeq 35\text{Å/min}$. The peak concentrations from the SIMS and AES measurement agreed with each other; however, the concentration profile from the AES data is broader than that of the SIMS data. The AES method was found to be insensitive to nitrogen ion doses below 1×10^{15} atoms/cm^2.

The profiles shown in Figure 2 showed profile tails. The extent of these profile tails varied from measurement to measurement, indicating that a recoil process was present during the SIMS measurement.[15] The projected range and the standard deviations from Figure 2 are 1160Å and 560Å (ΔR_{p1}) and 470Å (ΔR_{p2}) which agree reasonably well with the theoretical values of 1040Å and 455Å. The theoretical range statistics were estimated from the interpolations of R_p and ΔR_p of carbon and oxygen in GaP by Johnson and Gibbons.[13]

Nitrogen, Oxygen and Fluorine in Silicon

The AES nitrogen concentration profile for a silicon sample which has been nitrogen implanted at 50KeV to a dose of 10^{16} cm^{-2} is shown in Figure 3.

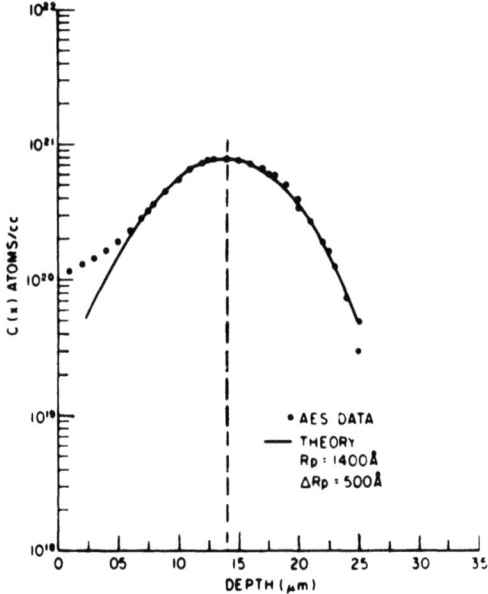

Fig. 3. Nitrogen Profile in Silicon

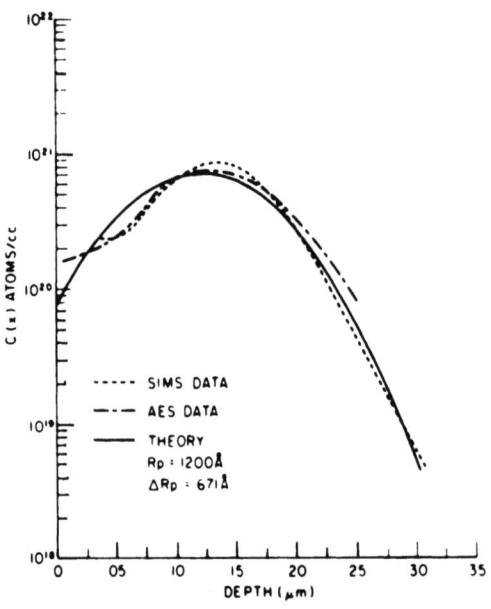

Fig. 4. Oxygen Profile in Silicon

IN-DEPTH PROFILE DETECTION LIMITS OF NITROGEN

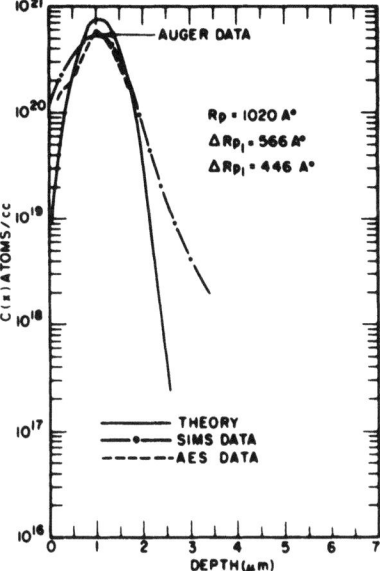

Fig. 5. Fluorine Profile in Silicon

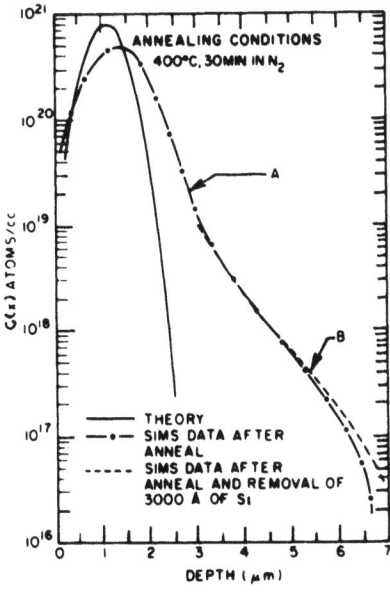

Fig. 6. Fluorine Profile in Silicon, After Annealing

The measured location (R_p) of the peak and the standard deviations (ΔR_p) are 1400Å and 500Å, respectively, which agree with the theoretical calculations of Johnson and Gibbons[13] (R_p=1410Å and ΔR_p=469Å).

The profile of oxygen from implantation at 50KeV to a dose of 10^{16} ions/cm^2 into silicon was measured by both the SIMS and the AES techniques and shown in Figure 4. High background counts were observed from SIMS data and they could be due to residual oxygen in the primary ion source and in the vacuum system. The calculated profile assumed a gaussian distribution with R_p=1200Å and ΔR_p=571Å. The location of the peak from the SIMS data is slightly deeper than the theoretical value which can be attributed to errors in the crater depth measurements.

A fluorine implanted silicon sample (fluorine dose of 6.7x10^{15} ions/cm^2 at 50KeV) was used to establish the detection limits for fluorine in silicon. Figure 5 shows the measured profiles where the dash line represents the Auger data and the dash-dot line represents the SIMS data. The location of the peak is approximately 1020Å which agrees with the theoretical value of 1060Å.[13] However, the profile is asymmetrically shifted towards the bulk and is broader than the theoretical gaussian distribution.

In order to ascertain that the profile tail is not due to a recoil process (the knock-on effect)[15] during SIMS measurement, two samples (A and B) from the same fluorine implantation were annealed at 400°C in nitrogen for 30 minutes. Sample A was used as a control and approximately 3000Å of silicon was removed from Sample B by the anodization and stripping technique. Concentration profile were obtained by the SIMS measurement and they are shown in Figure 7 (-----for Sample A and - . - . for Sample B). The profile tails from both samples agree, hence, they are not due to recoiling atoms during SIMS measurement.

When the samples were annealed at temperatures greater than 500°C, the fluorine profiles showed complicated multiple peaks while the silicon profiles were constant indicating that these profiles were not due to artifacts of SIMS measurements. However, due to space limitations they are not presented in this paper.

TABLE I

DETECTION LIMIT OF IN-DEPTH PROFILE FOR N IN GaP, N, O, F IN Si BY SIMS AND AES TECHNIQUES

Substrate	GaP	Si		
Elements	N	N	O	F
SIMS (atoms/c.c.)	$>10^{18}$	---	$\sim 10^{19}$	5×10^{16}
AES (atoms/c.c.)	$>5 \times 10^{19}$	$>5 \times 10^{19}$	10^{20}	10^{20}

DISCUSSION AND CONCLUSIONS

Quantitative data have been obtained from both SIMS and AES analysis by the use of calibration standards prepared by ion implantation. In general, good agreement between SIMS and AES measurements has been observed, and the in-depth detectability limits of both techniques for N in GaP and for N, O, and F in silicon have been established. These results are summarized in Table I.

It is often of interest to detect impurities in a localized region of 1-5 microns in diameter. The sensitivities of both SIMS and AES techniques for the detection of impurity atoms in such small areas can be estimated from Table I by a simple scaling of the volume measured.

The SIMS data showed deviations from the gaussian distribution expected from LSS theory[14] for nitrogen in GaP and fluorine in Si. Similar deviations (e.g. profile tails) from the theoretical profile have also been observed by SIMS for boron and arsenic ion implanted into silicon.[6,15] Three possible explanations have been considered in the literature.[7,15,16] For the present study we have considered three possible mechnisms, namely, implantation in channeling directions, recoil of the implanted ions during implantation or during SIMS analysis. For nitrogen in GaP the profile tails are likely to be due to channeled ions and a slight recoil process during SIMS measurement. For fluorine in silicon, the profile tails are most likely due to channeled or recoiled ions during the ion implantation.

REFERENCES

1. D. P. Lecrosnier & G. P. Pelous, European Conf. on Ion Implantation, p. 102, (P. Peregrinus Ltd., Stevenage) 1970.
2. V. M. Pistryak, A. K. Gnap, V. F. Kozlov, R. I. Garber, A. I. Fedorenko and Ya. M. Fogel, Sov. Phys. Sol. State 12, 1005, 1970.
3. M. Croset, Revue Techn. Thompson CSF3, 19, 1971.
4. B. Blanchard, N. Hilleret and J. B. Quoirin, Int. Meeting Chemical Analysis by Charged Particle Bombardment, L.A.R.N. Namur, 1971.
5. J. L. Assemat, Proc. II. Conf. Ion Implantation, p. 351, Garmisch-Partenkirchen, 1971.
6. J. C. C. Tsai, J. M. Morabito and R. K. Lewis, Ion Implantation in Semiconductors and Other Materials, p. 87, Plenum Press, New York, 1973.
7. G. Schwarz, M. Trapp, R. Schimko, G. Butzke and K. Rogge, Phys. Stat. Sol. (a), 17, pp. 653-658, 1973.
8. W. K. Hofker, H. W. Werner, D. P. Oosthoek and H. A. M. deGrette, Ion Implantation in Semiconductors and Other Materials, p. 133, Plenum Press, New York, 1973.
9. J. M. Morabito, Thin Solid Films, Vol. 19, p. 21, 1973.
10. J. M. Morabito, Anal Chem. Vol. 46, pp. 189-196, 1974.
11. J. M. Morabito and J. C. C. Tsai, Surface Science, Vol. 33, pp. 422-426, 1972.
12. J. M. Morabito & R. K. Lewis, Anal. Chem., Vol. 45, p. 869, 1973.
13. W. S. Johnson and J. F. Gibbons, Projected Range Statistics in Semiconductors, Stanford University Book Store, 1969.
14. L. J. Lindhard, M. Scharff and H. E. Schiott, KgI. Danske Videnskab Selskab, Mat.-Fys. Model $\underline{33}$, 14, 1963.
15. F. Schulz, K. Wittmaack and J. Maul, Radiation Effects, Vol. 18, p. 211, 1973.
16. R. P. Gittins, D. V. Morgan and G. Dearnaley, J. Phys. D: Appl. Phys., Vol. 5, pp. 1654-1663, 1972.

BEHAVIORS OF Ga AND P DAMAGES INTRODUCED BY ION-IMPLANTATION
INTO GaP

Hideki MATSUMURA and Seijiro FURUKAWA*

Dept. of Phys. Electronics
*Res. Lab. of Precision Machinery and Electronics
Tokyo Institute of Technology, Tokyo 152, JAPAN

ABSTRACT

Individual annealing behaviors of Ga and P damages introduced in GaP by an ion-implantation were measured by a back-scattering method. ^{12}C ions were implanted at room temperature, in order to know effects of damage difference between Ga and P on electrical properties of amphoteric impurity in GaP. As results, it was found that total amount of P damage as implanted state tended to become more than that of Ga damage with decrease of implanted ion doses. Moreover, ^{12}C implanted GaP showed p-type conductivity and carrier concentration also tended to become higher with decrease of doses. And these results supported our expectation that electrical properties of amphoteric impurities in GaP could be controlled, using damage difference between Ga and P caused by ion-implantation.

INTRODUCTION

Ga damage introduced in GaP by the ion-implantation might be different from P damage, since Ga and P atoms have different cross-sections to incident ions. Considering this effect, we expected that we could control the electrical properties of GaP in which amphoteric impurities were introduced. That is, if the amount of Ga damage is more than that of P damage, amphoteric impurities might tend to substitute in Ga lattice sites, and otherwise, those might tend to do in P sites.

G.Carter et al.[1] reported the results of back-scattering mesurements on the individual damages of Ga and P. However, there is no report about the influence of damage difference between Ga and P on electrical properties.

So, at first, we measured 80 keV- and 150 keV-^{12}C implanted GaP samples by the back-scattering method with 1.5 MeV-^{4}He$^+$. As a result, it was found that the total amount of P damage as implanted state tended to become more than that of Ga damage with decrease of implanted ion doses.

Next, we tried to measure the electrical properties of GaP in which ^{12}C ions were implanted. It was found that ^{12}C implanted GaP showed a p-type conductivity at ~ 900°C annealing and that carrier concentration had also a tendency to become higher with decrease of doses. These results supported our expectations; that is, the damage difference between Ga and P affected on the electrical properties of amphoteric impurities in GaP, and the electrical properties of amphoteric impurities could be controlled using this damage difference caused by ion-implantation.

EXPERIMENTS BY BACKSCATTERING METHOD

The damage in GaP, in which 150 keV ^{12}C ions were implanted at room temperature along 7° off <111> axis to doses of 6×10^{14} and 1.8×10^{15} ions/cm^2, were measured by the back-scattering method using 1.5 MeV-^{4}He$^+$. Experimental apparatus was the same as previously reported one [2]. Each implanted sample was annealed for 15 minutes in Ar atmosphere. For an example, the measured raw data of samples with dose of 1.8×10^{15} ions/cm^2 are shown in Fig.1. Ga and P damages were divided from these measured raw data, assuming that Ga dechanneling yeilds at P damage peak were nearly straight lines. And divided Ga

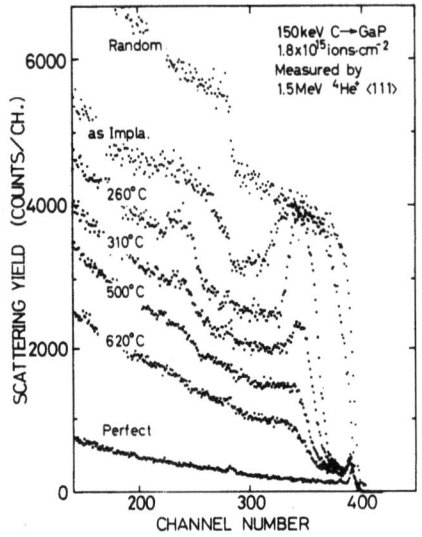

Fig.1 Measured raw data.

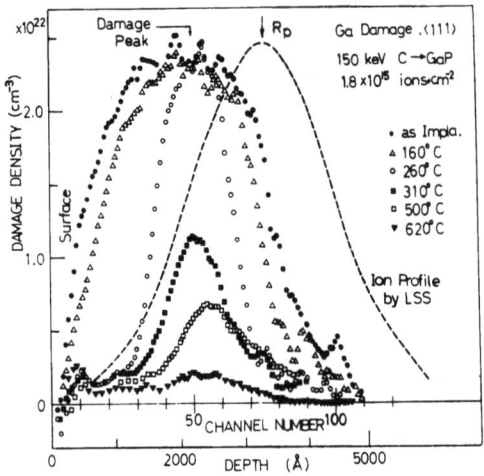

Fig.2 Evaluated Ga damages from raw data.

and P damages were processed by use of the multiple scattering formula by Feldman et al.[3]. But, in this case, dechanneling rates were varied not to be shown negative damages.

Ga damage distributions evaluated from Fig.1 are shown in Fig. 2. It is found, from this figure, that 1) damage peak is 1/3 shallower than the peak of theoretically estimated ion profile. 2) damages begin to recover at annealing from both sides of the peak, especially from the surface side.

Figure 3 shows the annealing behaviors of total amount of Ga and P damages. It is shown that the annealing behavior of P damage is not qualitatively different from that of Ga damage. However, it is found that the total amount of P damage as implanted state has a tendency to become more than that of Ga damage with decrease of implanted ion doses.

In order to decrease the error due to separation of P damage from Ga damage and ensure the above tendency, GaP samples in which 80 keV-^{12}C ions were implanted to doses of 6×10^{14} and 1.5×10^{15} ions/cm^2 were also measured. The influence of damage caused by ^{4}He$^+$ beam was minimized by sliding the sample position bombarded by ^{4}He$^+$ for every 3 µC of measuring charge. And the back-scattered ions were counted, until the statistical error per a channel of P damage became less than several %.

Measured and processed results are shown in Fig.4. For comparison, Ga and P damages are plotted together. From this figure, it can be said that the above stated tendency is confirmed by this experiment. Moreover, this tendency can be found in Fig.2 of Ref.[4], in the case of Bi implanted into GaP, though the authors of Ref.[4] said nothing about the damage difference between Ga and P.

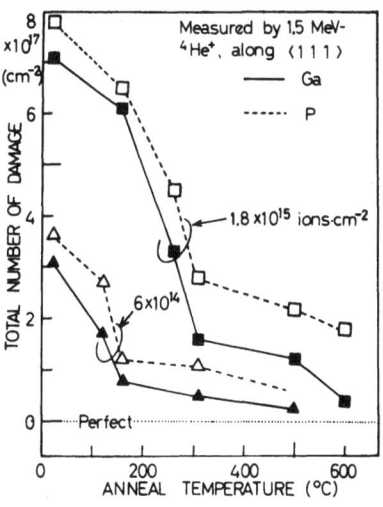

Fig.3 Annealing behaviors of P and Ga damages.

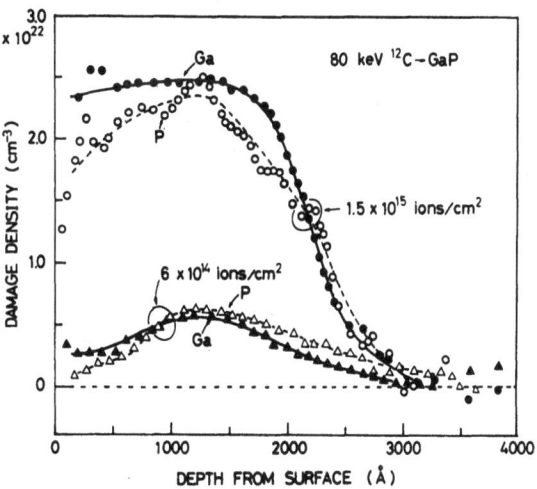

Fig.4 Processed data.

MEASUREMENT OF ELECTRICAL PROPERTIES

Samples were fabricated by ion-implantation into semi-insulating GaP under the same conditions as the samples stated in previous paragraph. The electrical properties of the samples were measured by the van der PAUW method. The surfaces of all samples were coated with sputtered $\sim$1000 Å SiO_2 during annealing and, after 20 minutes annealing in Ar atmosphere, SiO_2 films were removed by dipping the samples in $\sim$20% HF for 1 minutes. Alloys of 5% Zn in In were attached at the four corners of a sample and annealed at $\sim$500 °C in Ar atmosphere for 10 minutes to take ohmic contacts.

^{12}C implanted GaP shows p-type conductivity at $\sim$900 °C annealing and measured results are shown in Fig.5. In this case, since SiO_2 coating films were unstable above $\sim$950 °C annealing, we discuss on the results of the samples annealed at 910 °C. It is found from this figure that carrier concentrations do not increase but decrease with increase of implanted ion doses, though hole mobilities of all samples annealed at 910 °C are equal to unimplanted p-type GaP.

You may say that these electrical properties are caused by a) the damage introduced during ion-implantation, or by b) the effect of ^{12}C ions beyond solubility limit. However, Hemenger et al.[5] reported the results of electrical measurements in the case of 250 keV-Zn implanted GaP with dose of 10^{16} ions/cm^2. Even though the damage in their case is much heavier than that in our sample, the substitutional fraction of implanted ions becomes $\sim$40% at 900 °C annealing according to their report. Considering this fact and our mobility data, it can be said that our electrical results are not caused by a).

Solubility limit of ^{12}C ions in GaP is not well known at present time, but, ^{12}C ions are reported to have a solubility in P lattice site as high as $\sim 10^{18}$ ions/cm^3[6] when other doping methods are used. Moreover, it can be said that, even if implanted ^{12}C concentration exceeded the solubility limit, we could not observed the decrease of the carrier concentration but the saturation of that with increase of implanted ion doses.

Considering above discussions, it can be concluded that, though the total number of ^{12}C ions substituting in both Ga and P lattice sites scarcely changes due to doses, the ratio between ^{12}C ions in Ga site and P site changes.

DISCUSSIONS AND CONCLUSIONS

The ratio between Ga and P damages as implanted state and the surface carrier concentration of 910 °C annealed samples versus implanted ion doses are shown in Fig.6. It can be said from this figure that the tendency shown in the electrical properties is in agreement with the results expected from the tendency of damage properties.

Trumbore et al.[7] reported about the behaviors of amphoteric impurities (Sn, Ge or Si) in GaP, which were introduced by the thermal

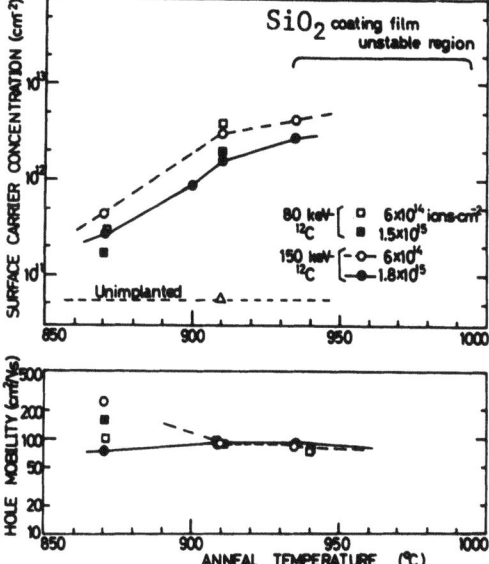

Fig. 5. Electrical properties

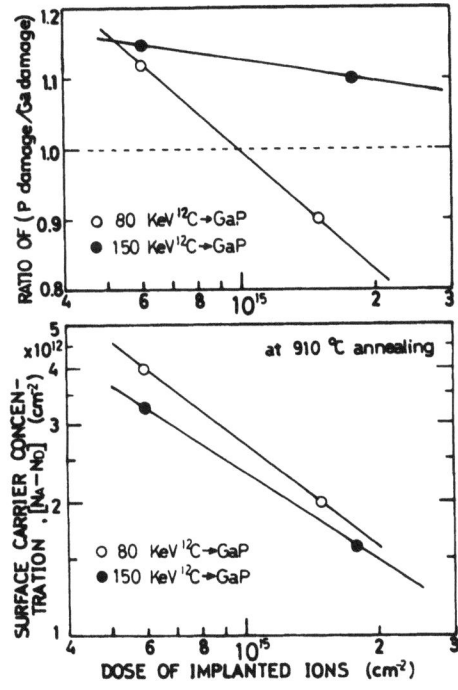

Fig. 6. Relation between damages and electrical properties

gradient technique. It is shown in their report that the carrier concentration due to the amphoteric impurity decreases with decrease of total dose of the amphoteric impurity, in both case of p-type GaP and n-type GaP. Moreover, Dean et al.[6] reported that GaP crystals grown in contact with CO or CO_2 gas were strongly p-type and that the hole concentration increased with pressure of CO or CO_2 (the pressure of CO or CO_2 is considered to correspond to doses of ^{12}C in GaP.). However, the fact shown in Fig.6 is entirely different from the results expected from Refs.[6] and [7]. Therefore, it can be said that the properties of amphoteric impurity shown in Fig.6 are caused by ion-implantation.

Moreover, the correlation between the electrical properties and damage difference supported our expectation that the electrical properties of amphoteric impurities in GaP could be controlled using damage difference between Ga and P caused by ion-implantation.

Above stated discussions are considered to be suggestive in the case of implantation of any other impurities in GaP; that is, above stated discussions suggested that lower implanted dose is desirable for substitution of implanted ions into P lattice sites. For instance, compairing two cases; a) one time implantation with dose of 1×10^{15} ions/cm^2, and b) 10 times cycle implantation (first implantation with dose of 1×10^{14} ions/cm^2 →annealing→second implantation with dose of 1×10^{14} ions/cm^2→ .....repeated 10 times); the case b) is considered to be desirable for the substitution of implanted ions into P sites.

Now, we concluded above stated discussions as following;
1) The damage peak is 1/3 shallower than the theoretically estimated ion profile.
2) The damages begin to recover at annealing from both sides of the peak, especially from surface side.
3) The annealing behavior of P damage is not different qualitatively from that of Ga damage.
4) ^{12}C implanted GaP shows p-type conductivity at ∼900 °C annealing.
5) The total amount of P damage tends to become more than that of Ga damage, with decrease of implanted ion doses.
6) The carrier concentration due to ^{12}C ions tends to become higher with decrease of doses.

And finally
7) The correlation between the electrical properties and damage difference supported our expectation that the electrical properties of amphoteric impurities in GaP could be controlled using damage difference between Ga and P caused by ion-implantation.

ACKNOWLEDGMENTS

We appreciate the members of the steering committee and staffs of a van de Graaff accelerator at Tokyo Institute of Technology at Ookayama. We express our gratitude to the members of Central Research Laboratory of Hitachi Ltd. for the preparation of samples, and to Dr. Ishiwara for his useful advices.

This work is partially supported by a Grant-in-Aid for Scientific Research (A) from the Ministry of Education.

REFERENCES

1) G.Carter and J.L.Whitton, Rad. Effects, 15, pp143-148 (Aug.'72)

2) S.Furukawa, H.Matsumura and H.Ishiwara, Thin Solid Films, 19, pp 399-406 ('73)

3) L.C.Feldman and J.W.Rodgers, J. Appl. Phys., 41, pp3776-3782 (Aug.'70)

4) L.C.Feldman, W.A.Augustyniak and J.L.Merz, Proc. of 1st Int. Conf. on Ion-Implantation, edited by F.H.Eisen and L.T.Chadderton, pp 277-283

5) R.M.Hemenger and B.C.Dobbs, Appl. Phys. Letters, 23, pp462-464 (15 Oct.'73)

6) P.J.Dean C.L.Frosch and C.H.Henry, J. Appl. Phys., 39, pp 5631-5646 (Nov.'68)

7) F.A.Trumbore, H.G.White, M.Kowalchik, C.L.Luke and L.Nash, J. Electrochem. Soc., 112, pp1208-1211 (Dec.'65)

ESR STUDIES OF ANNEALING BEHAVIOR OF NITROGEN-IMPLANTED GaP

T. Matsumori, K. Miyazaki and S. Shigetomi

Department of Electronic Engineering, Faculty of Engineering, Tokai University, Shibuya, Tokyo, Japan

ABSTRACT

Annealing behavior (up to 1000°C) of nitrogen-implanted GaP has been studied using an ESR technique. The spectrum for samples as implanted is symmetrical, Lorentzian and isotropic, and its g-value = 2.0032 ± 0.0004, the line width $\Delta Hms1 \sim 6$ gauss, which is due to the disordered state. Two recovery stages have been observed in the isochronal annealing curve. The activation energy for the first stage (up to 150°C) is 0.22 - 0.23 eV, and for the second (300 - 500°C) is 1.40 - 1.50 eV. The spectrum anneals out at $\sim$500°C. On the other hand, some weak lines due to the anisotropic damage centers have appeared after the $\sim$100°C anneal. Moreover, the other structural line which is enhanced by illumination at 4920 or 7050 Å has been detected. After the 700 or 750 °C anneal, the hyperfine spectrum of the implanted N^{14} (with illumination at 5180 or 7050 Å at 4.2 K) which is evidenced 3 lines centered about a g-value of 2.017, and the spectrum due to the secondary defects of which g-value is 2.0030 ± 0.0004 and $\Delta Hms1 \sim 6$ gauss have been detected.

INTRODUCTION

The ESR of impurity-ion implanted silicon has been extensively studied by many workers and various kinds of centers have been identified. These centers are classified into two groups: (1) damage centers produced in the crystal lattice by ion implantation, and (2) implanted substitutional donor (or acceptor) centers. Moreover, the spectrum due to conduction electrons in implanted silicon has also been observed. However, the ESR studies of ion-implanted compound semiconductors are extremely scarce as compared with silicon, since it may be difficult to detect and analyze the spectra.

In this paper we present new ESR centers generated by nitrogen ion implant in GaP crystals and study the annealing behavior between room temperature and 1000°C.

EXPERIMENTAL PROCEDURE

The substrates for implantation were p-type (zinc doped with $3 \times 10^{17}/cm^3$) GaP crystals grown by liquid-encapsulated-Czochralski method. After mechanical polishing and chemical etching, N^+ ions were implanted in the (111) oriented samples with 200 keV at doses of $1 \times 10^{14} - 2 \times 10^{16}$ ions/cm^2 at room temperature, respectively. The ESR measurements were done using X-band spectrometer with 100 kHz modulation at 77 K and with 80 Hz modulation at 4.2 K.

Isochronal (10 min.) annealing of the implanted samples was performed in dry nitrogen atmosphere for temperatures up to 1000°C, where preliminary PSG coatings were done on the surfaces of the samples to be annealed above 500°C in order to prevent the dissociation of phosphorous atoms from GaP.

Upon observing photosensitive ESR effects, the samples were illuminated by 4180 - 7540 Å radiation from a 300 W halogen lamp.

RESULTS AND DISCUSSIONS

In GaP crystals after nitrogen-ion implant with high doses (higher than 5×10^{15} ions/cm^2) at 200 keV at room temperature, a new paramagnetic center appeared. Fig. 1 shows the derivative of the ESR spectrum observed in samples as implanted with 2×10^{16} ions/cm^2. The intensity of the spectrum increased with increasing ion dose.

Crowder et al. previously observed the ESR signals for Si^+, P^+ and As^+ ion-implanted silicon, and suggested from the g-value (= 2.0059 $\pm$ 0.0005), line shape and line width $\Delta Hmsl$ (=5.2 $\pm$ 0.4 gauss) that are due to the complete disordered, amorphous state produced by the ion implantation [1]. The spectrum we have observed is also symmetrical, Lorentzian, and isotropic with g= 2.0032 $\pm$ 0.0004 and $\Delta Hmsl \sim 6$ gauss and appears only in the case of heavy implants. We therefore conclude that the spectrum arises from the complete disordered state created by nitrogen-ion implants, where there is no crystalline order but there are a lot of dangling bonds.

The annealing of the center was next investigated. Samples implanted with 1×10^{16} and 2×10^{16} ions/cm^2, respectively, were subjected 10 minutes isochronal anneals in nitrogen atomosphere and the relative spin density of the center as a function of annealing temperature as shown in Fig. 2. Two decreasing stages and one reverse annealing stage of the spin density are seen for both doses of 1×10^{16} and 2×10^{16} ions/cm^2. Haskell et al. have

ERS STUDIES OF ANNEALING BEHAVIOR

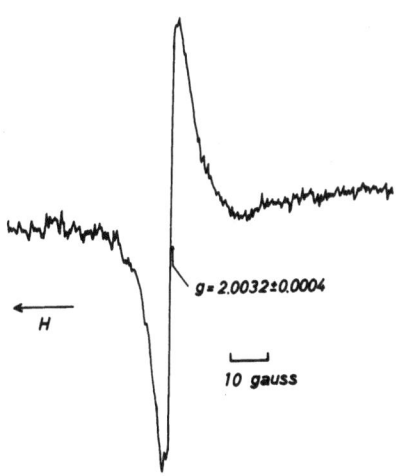

Fig. 1 The derivative of the ESR spectrum due to the disordered state produced in heavily implanted GaP.

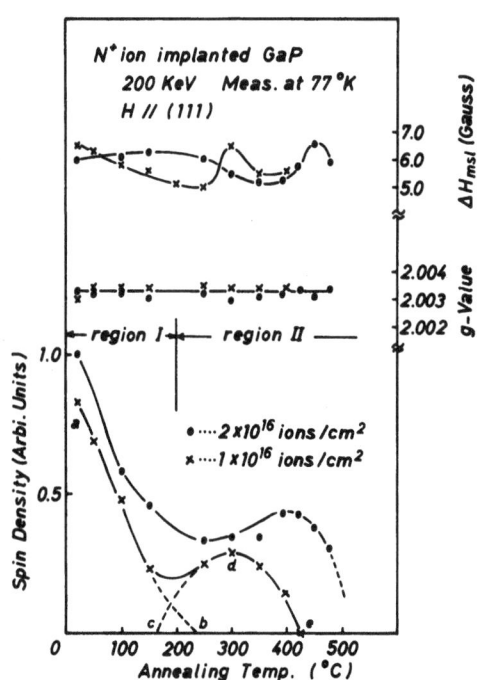

Fig. 2 The isochronal annealing curves of 10 min. of the paramagnetic damage centers produced by the N^+ ion-implant in GaP.

obtained a similar annealing curve for the p-type (Zn-doped) GaP samples implanted with 4×10^{14} Te ions/cm^2, using the backscattering technique [2]. The activation energy for recovery E_{a1} for the stage in the lower temperature range I (room trmperature $-150°C$) is 0.22 - 0.23 eV for doses of 2×10^{16} and 1×10^{16} ions/cm^2, while in the higher temperature range II (300 - 420°C for a dose of 1×10^{16} ions/cm^2; 400 -520°C for a dose of 2×10^{16} ions/cm^2) the activation energy for recovery E_{a2} is 1.40 -1.50 eV. E_{a1} was also confirmed by the results from the isothermal anneals and was estimated 0.23 ± 0.02 eV, which is consistent with the results from the isochronal anneals.

Each annealing curve can be regarded as the superposition of two annealing curves, such as ab and cde, where ab is considered to represent the annealing process of dangling bonds in the completely disordered state and cde the process in which these dangling bonds associate with some impurities already included in the samples before implantation (such as Zn, O, C or S) or with the implanted N atoms halfway to the disappearance, the other disordered state (such as dangling bond-impurity complexes) are resulted, and as increasing the annealing temperature the amount of these centers decreases again. Our consideration of the latter process is supported by the experimintal results that an increase in ΔH_{ms1} is seen around the peak position d as shown Fig. 2, which suggests that some changes in the structure of these damage center actually take place.

The activation energy of 1.40 - 1.50 eV for E_{a2} represents comparatively abrupt changes from the above mentioned disordered state to the perfect amorphous state which preserves the local order of the zincblende lattice. These behaviors correspond to the results from the optical absorption measurements of GaP samples heavily implanted (higher than 2×10^{15} ions/cm^2) with 200 keV nitrogen ions by Shimada et al. [3], where the absorption coefficient α abruptly falls off at around 400°C with increasing the annealing temperature.

In addition to the spectrum due to the disordered states, some sharp lines due to anisotropic damage centers have appeared after annealing above $\sim 100°C$ as shown in Fig. 3 and disappeared below the vanishing temperature of the above mentioned spectrum. These lines are usually too weak to resolve. Numbers of the lines tend to decrease with increasing the annealing temperature. It may indicate that structure of these anisotropic centers successively changes into more simple structure. Some photosensitive lines are also included in these anisotropic lines, which we cannot sufficiently separate from the other lines. These ESR lines are significantly enchanced for the illumination at 7050 Å (1.76 eV) and 4920 Å (2.5 eV), as shown in Fig. 4. It may be interpreted from this result that some defect energy level is present at 1.76 eV above the valence band within the bandgap of the implanted GaP.

Further ESR measurements have been done for samples after

ERS STUDIES OF ANNEALING BEHAVIOR

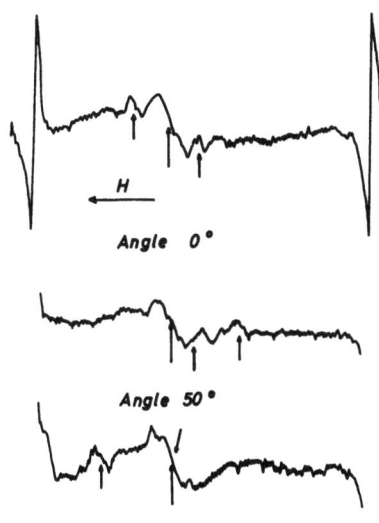

Fig. 3 Angle dependence of the anisotropic damage centers.

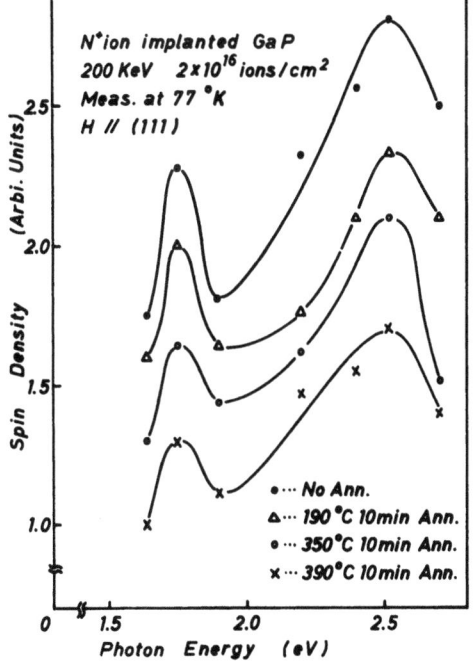

Fig. 4 Photoresponce of the ESR spectrum of the anisotropic damage centers.

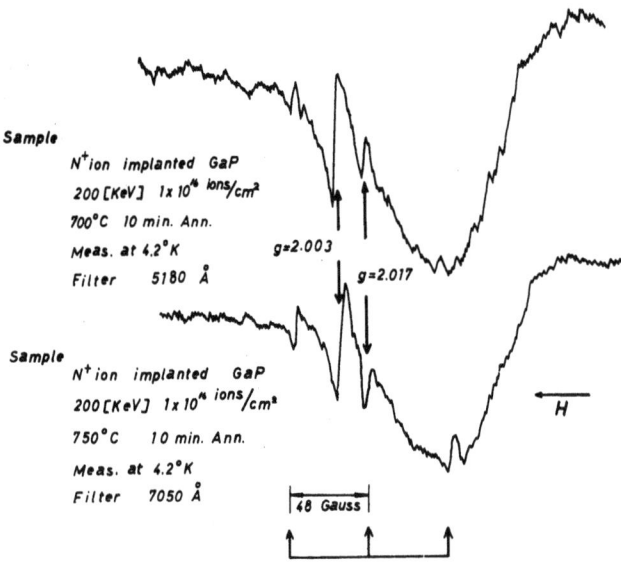

Fig. 5 The hyperfine spectrum of N^{14} implanted in GaP and the spectrum of the secondary defects produced by the high temperature anneals (700 and 750°C).

annealing above 700°C, where the defect centers above mentioned have disappeared. After the 700 or 750°C anneal, the hyperfine spectrum of N^{14} which is evidenced by $(2I + 1) = 3$ lines centered about a g-value of 2.017 and equally spaced with 48 gauss, where $I = 1$ is the nuclear spin of N^{14} has been detected with illumination at 5180 or 7050 Å at 4.2 K as shown in Fig. 5. This spectrum assures the existence of nitrogen centers in the implanted GaP crystals. However, no resonance was observed for the samples annealed at 1000°C after the room temperature implant or the hot implant (at 500°C) were performed. It may suggests that, after annealings at 700 and 750°C, the implanted N atoms can not occupy substitutional P sites and the origin of this hyperfine line therefore appears to be N atoms which occupy interstitial sites. It may be considered that the illumination at 5180 Å signifies the excitation of electrons to the conduction band by band-to-band transition, while the illumination at 7050 Å sigifies that the above mentioned photosensitive defect level (1.76 eV) is responsible for the appearance of the hyperfine line.

On the other hand, the spectrum due to the secondary defects has also been observed in Fig. 5, where the g-value is 2.0030 ± 0.0004 and $\Delta Hmsl$ is ~6 gauss. Apparently this spectrum is similar to the spectrum from the disordered state, however, we have not identified what is the origin of this spectrum.

SUMMARY

The major results of this study are:

(1) A new ESR spectrum has been observed for the GaP crystals after nitrogen implant with high doses at 200 keV at room temperature. The spectrum arises from the complete disordered, amorphous state. The activation energy for recovery is o.22 - 0.23 eV.

(2) The other disorderded state originates after annealing above 150°C, which is presumably due to the dangling bond-impurity complexes and the activation energy for recovery is 1.40 - 1.50 eV.

(3) In addition to these centers, some anisotropic damage centers have been produced in the samples after annealing above ~100°C. The centers are classified in two: non-photosensitive and photosensitive centers. The level of the photosensitive center is present at 1.76 eV above the valence band within the bandgap of GaP.

(4) After the 700 or 750°C anneal, the hyperfine spectrum of N^{14} has been detected with illumination at 5180 or 7050 Å at 4.2 K and the origin of the line appears to be N atoms which occupy interstitial sites.

ACKNOWLEDGEMENTS

The authors are grateful to Dr. K. F. Komatsubara, T. Shimada and Y. Shiraki of Hitachi Central Research Laboratory, Tokyo for supplying the implanted samples and for valuable discussions and Prof. K. Hiraoka and Y. Kimura for their helpful advice.

REFERENCES

[1] B. L. Crowder, R. S. Title, M. H. Brodsky and G. D. Pettit, Appl. Phys. Letters 16, 205 (1970).

[2] J. L. Haskell, W. A. Grant, G. A. Stevens and J. L. Whitton, Proc. 2nd Int. Conf. on Ion Implantation, edited I. Ruge and J. Graul, III - 8, p. 193 (1971).

[3] T. Shimada, Y. Shiraki and K. F. Komatsubara, Proc. 4th Int. Conf. on Ion Implantation.

PROFILES

PROFILES ; HOW WELL CAN EXPERIMENTAL RESULTS BE EXPLAINED BY

THEORIES ?

S. Furukawa and H. Ishiwara*

Res. Lab. of Precision Machinery and Electronics
* Faculty of Engineering
Tokyo Institute of Technology, Tokyo 152 JAPAN

ABSTRACT

This paper reviews the current status of theoretical and experimental investigations on the spatial distribution of implanted ions. The theoretical predictions show reasonably good agreement with the experimental results in both cases of uniform and double-layer substrates. Deformation of distributions due to secondary effects such as sputtering or expansion of the substrate is also discussed.

INTRODUCTION

The recent developement of ion implantation technique as a useful tool in the fabrication of semiconductor devices has stimulated a great deal of experimental and theoretical study, both from basic and applied points of view. Experimentally, many physical methods to measure the spatial distributions of implanted ions have been developed as well as the electrical methods. These physical methods, that is Rutherford backscattering, nuclear reaction, induced X-ray detection, ion micro-analyser, auger electron microscopy, and so on, have enabled us to compare the theories with the experimental distributions with no ambiguity due to the annealing effects.
Simultaneously, some important theoretical studies to explain or to predict experimental results have been also achieved in the recent years, which include problems concerned with lateral spread of ions, asymmetry of distributions, multiple-layer substrates, sputtering or expansion of surfaces during implantation, and so on. Classification of the typical theories on the spatial distributions of ions and damages is shown in Table 1, in which the theories on deformation of the distributions due to secondary effects such as

TABLE I

CLASSIFICATION OF THEORIES

Method			Range					Damage	Authors (Ref.)
			R_p	ΔR_p	Higher Moments	Lateral Spread	Multi-Layer		
Transport Eq. (LSS Theory)	Diff. Eq.	1st Order	○	×	×	×	×	×	1) Johnson, Gibbons
		2nd Order	○	○	×	○	×	×	2) Brice 3) Furukawa Matsumura, Ishiwara
		3rd Order	○	○	○	△	×	×	4) Mylroie, Gibbons
	Integral Eq.		○	○	○	○	×	○	5) Sigmund, Sanders 6) Winterbon, Sigmund, Sanders 7) Winterbon
Intermediate Method	Two Step Method		○	○	×	△	△	○	8) Brice 9) Tsurushima, Tanoue
	Semi-Monte Carlo Method		○	○	×	×	○	○	10) Furukawa, Ishiwara
Monte Carlo Method			○	○	△	△	△	○	

○; calculated, △; possible, ×; impossible or inaccurate

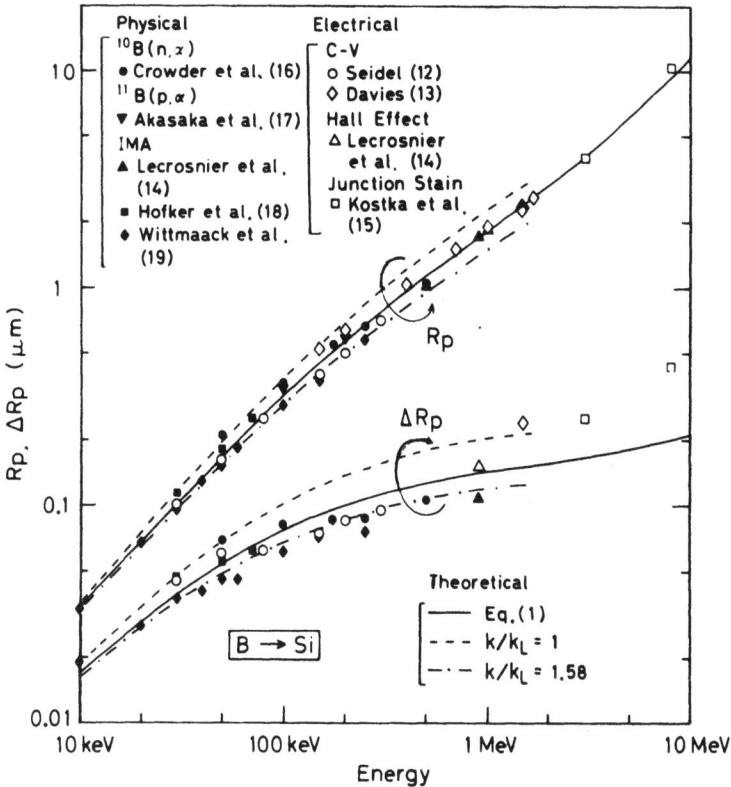

Fig. 1. Comparison of theoretical values with experimental ones in R_p and ΔR_p.

sputtering phenomena are excluded.

In this review, the current status and future problems on the spatial distribution of implanted ions are discussed from the point of view of comparison between the theories and the experiments.

DISTRIBUTIONS IN A UNIFORM SUBSTRATE

a. R_p and ΔR_p

The average projected range R_p and the projected standard deviation ΔR_p are calculated by many researchers as shown in Table 1. It is shown [3] that these values are well evaluated by using the second order differential equations derived from the LSS theory [11]. Comparison with the experimental results is shown in Fig. 1 in case of B ions in Si, in which the middle curves are calculated assuming the following electronic stopping power Se (E).

$$S_e(E)=1.4k_L E^{1/2}/\{1+(E/E_0)^m\} \quad (1)$$

where k_L is the constant defined in the LSS theory. E_0 and m are adjustable parameters and chosen equal to 4.57 MeV and 1.2, respectively, in order to best fit Eq.(1) on the reported values [20]. At present, it appears useful to use Eq.(1), since too few data are reported to determine the parameters in the Brice's electronic stopping power formula [21].

In Fig.1, other curves calculated by assuming the velocity proportional electronic stopping power are also shown in cases of $k/k_L=1$ and $k/k_L=1.58$. The latter value is based on the experimental results by Eisen et al. [22]. The experimental values can be well explained by the lower two curves (Eq.(1) and $k/k=1.58$), though physical measurements in an energy range above 500 keV is needed in order to determine the k value more precisely.

b. Lateral Spread of Ions

Theoretical consideration on the lateral spread of implanted ions has been first done by Sigmund et al., using the power cross sections [5]. Calculations assuming the Thomas-Fermi potential and taking the electronic stopping power into account have been done by Furukawa et al. [3] and Winterbon [7], which show that the standard deviation of the lateral spread is larger than ΔR_p in case when the electronic stopping is dominant. Experimental results measured by the junction stain [23], the Rutherford backscattering [24], and the C-V method [25] support the theoretical predictions.

c. Asymmetry of Distribution

Ion distributions are deviated from the gaussian, especially in an energy range where the electronic stopping is dominant. The deviated distribution can be constructed by using the higher moments of range [6]. Comparison of the experimental distributions with the theoretical ones constructed by the Edgeworth formula is shown in Fig.2. The values of the skewness (n=3 in $<(z-<z>)^n>/<(z-<z>)^2>^{n/2}$) are taken equal to -0.88 [18] and -1.68 [4] for 70 keV and 175 keV, respectively, and the kurtosis (n=4) equal to 3.9 for 70 keV. Agreement with the experimental values is excellent for 70 keV, however the calculated curve for 175 keV is not smooth and agreement is rather poor. It can be said from this figure that in order to discuss such asymmetric distributions as the absolute value of the skewness exceeds unity, it is necessary to take the kurtosis into account as well as the skewness.

d. Reflection of Ions from Substrate Surfaces

Reflection of ions from the substrate surfaces has been investigated by Bottiger et al. [26]. In the simple estimate, the reflection coefficient R (the number of reflected particles divided by the number of incident ions) is expressed as follows,

$$R = \int_{-\infty}^{0} F(z) \, dz \qquad (2)$$

where $F(z)$ is the probability function in an infinite target. However, Eq. (2) has to be corrected somewhat, since it doesn't include the effect that ions backscattered through the surface never return to the substrate. This correction is known as the "surface correction". The values of R calculated by Eq.(2) are the order of 0.1~0.2 in case of K implanted Au, whereas the corrected values are larger by about 50% and they well agree with the experimental results [26].

This reflection effect appears to give rise to two new problems concerned with the prediction of ion distributions; the first is that the ion distributions near the surface will deviate from those predicted by assuming an infinite substrate, and the second that the total dose measured electrically will change whether the reflected particles are ionized or neutral. If all particles are ionized, the total dose is equal to the number of ions remaining in the substrate, whereas if all neutral, it includes the reflected ones. Then, the ion concentration has ambiguity of the order of R.

e. Effects of Sputtering or Expansion of Substrate

Deformation of ion distribution due to sputtering of the substrate surfaces is considered important in case of high dose implantation such as emitter doping of bipolar transistors. It has been theoretically and experimentally discussed by several researchers [27]-[29]. The most remarkable feature of this phenomenon is that the total number of ions N_r remaining in the substrate saturates above a certain dose level. Under the saturation condition, the ion distribution measured from the instantaneous surface does not change any more, and it is expressed by the complementary error function in case when the probability function for each ion is assumed gaussian. Though the saturation levels N_r shown in Fig.3 can be

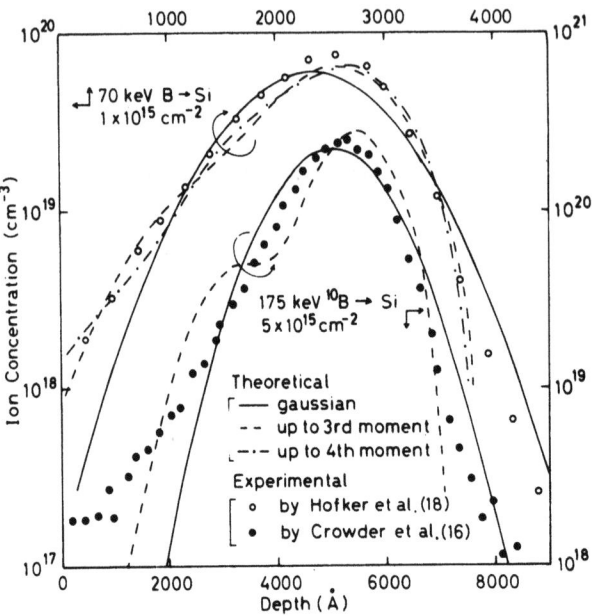

Fig. 2. Comparison in asymmetry of the distributions

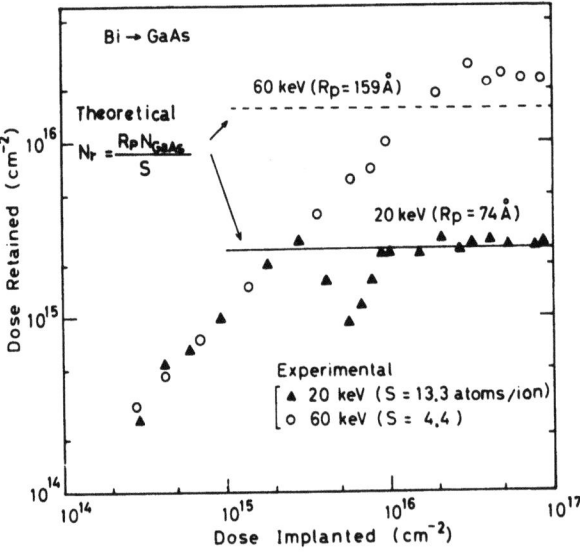

Fig. 3. Comparison of theoretical with experimental saturation levels

well predicted by using the independently measured sputtering coefficient [28], the ion distribution under the saturation condition [29] are experimentally found to penetrate about five times deeper than the theoretical prediction as shown in Fig.4. In order to explain this discrepancy, it may be necessary to take into account further effects such as knock-on effects.

Theoretical predictions on deformation of the ion distributions due to expansion or condensation of the implanted layer have not been reported. Recently, Tanoue et al. [30] and Eernisse [31] have proposed the phenomelogical models explaining the expansion or condensation of the substrate. Then, the deformed distributions may be calculated using these models, in which calculations for a substrate whose atomic density continuously varies with depth are needed.

DISTRIBUTIONS IN A MULTIPLE-LAYER SUBSTRATE

a. Ion Distributions

Theoretical considerations on this problem have been first reported by Furukawa et al. [10] concerned with the total range, in which conservation of the energy distribution of ions is assumed at the interface of layers. One of the conclusions by this analysis

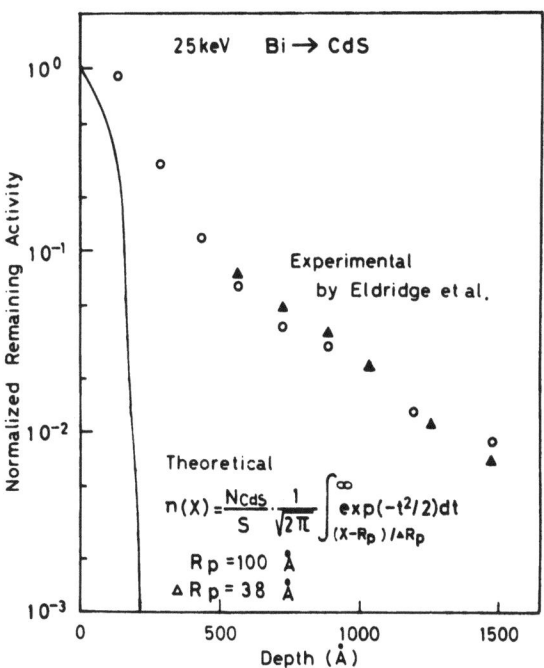

Fig. 4. Saturated ion distributions

is that a discontinuity in the ion concentration generally exists at the interface of layers. Further analyses concerned with the projected range show that the ion distributions in some double-layer substrates including SiO_2-Si can be approximately constructed by combining two distributions in uniform substrates. Detailed discussions will be presented in this proceeding [32].

Figure 5 shows comparison with the experimental results by Combasson et al. [33] in case of 60 keV B in a SiO_2 on Si substrate. The theoretical values are calculated assuming gaussian in each layer, and assuming the density of thermal SiO_2 equal to 2.0. Agreement of the both results is fairly well, especially near the interface. The deviation at the tails will be reduced by considering the higher moments in the range. However, the deviation in the Si layer may remain somewhat, since the theoretical prediction is not so accurate as to discuss the tail in the second layer.

Developement of more general and accurate theories is necessary for more precise discussions.

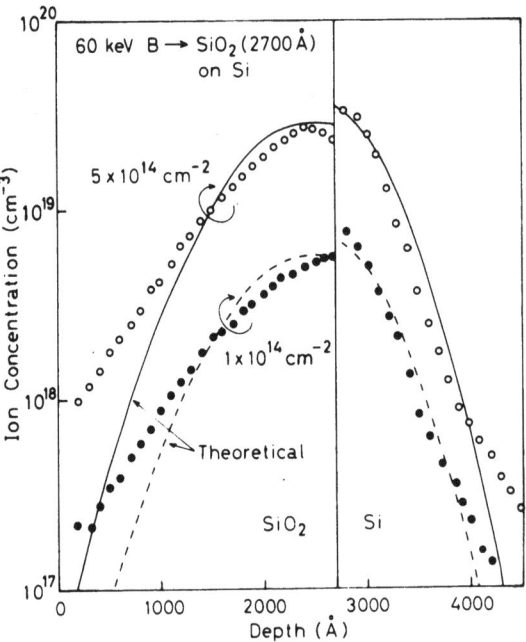

Fig. 5. Theoretical and experimental ion distributions in a double-layer substrate.

b. Knock-on Effect of the First Layer Atoms into the Second

Theoretical and experimental considerations on this problem have been scarecely reported, except a few recent works by Nishi et al. [34] and Moline et al. [35], though they appear important for fabrication of semiconductor devices.

The experimental results of Ar implanted into Mo-Si substrates are shown in Fig.6, in which Mo atoms knocked-on into Si layer are plotted with thickness of Mo layer. The theoretical values in the figure are the ion distribution and the energy deposition curves calculated by Winterbon [36] in which Rp is assumed equal to 700 Å. Though the experimental ion distribution is much sharper than the theoretical one, the energy deposition curve shows reasonably good agreement with the experimental results. And this agreement is considered to suggest that Mo atoms near the interface are knocked-on into the Si layer by Ar ions. Further investigations including the distribution of the knocked-on atoms will be the future problems.

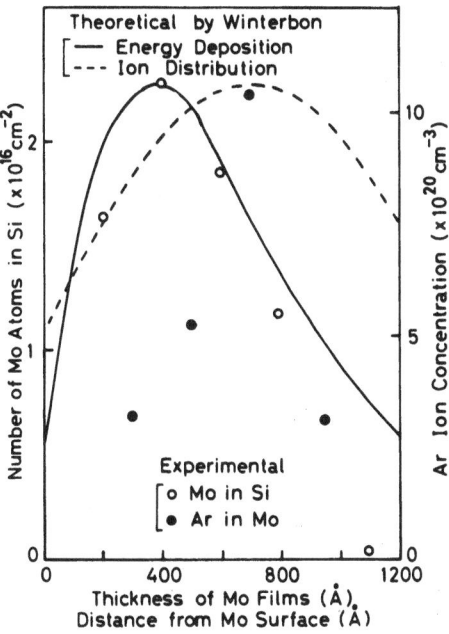

Fig. 6. 150 keV Ar in Mo-Si substrate. Theoretical; $M_2/M_1 = 2.5$, $m = 1/3$.

CONCLUSIONS

In this review, we discussed the current status and the future problems concerned with the ion distributions and the related effects. Conclusions are summarized as follows;

(1) Ion distributions in a uniform substrate can be essentially predicted using the current theories. For more accurate comparison, basic quantities such as the electronic stopping power have to be determined more accurately.

(2) Asymmetry of distributions is enhanced in an energy range where the electronic stopping is dominant. To discuss the asymmetry in this range, it is necessary to take the fourth moment of the range into account.

(3) Deformation of ion distributions by secondary effects such as the sputtering, expansion, or reflection effects can not be neglected under some implant conditions. Detailed investigations are left in the future.

(4) Many data concerned with the ion distribution in a double-layer substrate are now being accumulated. Theoretical analyses applicable to some industrially important substrates have been presented, but not yet to any substrates.

(5) Knock-on effects of the first layer atoms into the second have been scarecely investigated.

ACKNOWLEDGEMENTS

We wish to thank H.Matsumura and M.Nagatomo for their useful discussions and computer calculations.

This work is partially supported by a Grant-in -Aid for Scientific Research (1974-A) from the Ministry of Education.

REFERENCES

1. W.S.Johnson and J.F.Gibbons, "Projected Range Statistics in Semiconductors", (Stanford University Bookstore, 1969)

2. D.K.Brice, Proc. 1st Intern. Conf. on Ion Implantation in Semiconductors, edited by F.H.Eisen and L.T.Chadderton (Gordon and Breach, New York, 1971) p.101

3. S.Furukawa, H.Matsumura, and H.Ishiwara, Japan. J. Appl. Phys. $\underline{11}$, 134 (1972)

4. S.Mylroie and J.F.Gibbons, Proc. 3rd Intern. Conf. on Ion Implantation in Semiconductors and Other Materials, edited by B.L.Crowder (Plenum Press, New York, 1973) p.243

5. P.Sigmund and J.B.Sanders, Proc. Intern. Conf. on Application of Ion Beams to Semiconductor Technology, Grenoble, edited by P.Glotin (1967) p.215

6. K.B.Winterbon, P.Sigmund, and J.B.Sanders, K. Danske Vidensk. Selsk. mat.-fys. Medd., $\underline{37}$, No.14 (1970)

7. K.B.Winterbon, Rad. Effects, $\underline{13}$, 215 (1972)

8. D.K.Brice, Appl. Phys. Letters, $\underline{16}$, 103 (1970)

9. T.Tsurushima and H.Tanoue, J. Phys. Soc. Japan, $\underline{31}$, 1695 (1971)

10. S.Furukawa and H.Ishiwara, J. Appl. Phys., $\underline{43}$, 1268 (1972)

11. J.Lindhard, M.Scharff, and H.E.Schiott, K. Danske Vidensk. Selsk. mat.-fys. Medd., $\underline{33}$, No.14 (1963)

12. T.E.Seidel, Proc. 2nd Intern. Conf. on Ion Implantation in Semiconductors, edited by I.Ruge and J.Graul (Springer-Verlag, Berlin, 1971) p.47

13. D.E.Davies, Can. J. Phys., $\underline{47}$, 1750 (1969)

14. D.P.Lecrosnier and G.P.Pelous, IEEE Trans. $\underline{ED-21}$, 113 (1974) and Proc. European Conf. on Ion Implantation, Reading (Peter Peregrinus, 1970) p.203

15. A.Kostka and S.Kalbitzer, Rad. Effects, $\underline{19}$, 77 (1973)

16. B.L.Crowder, J.F.Ziegler, and G.W.Cole, in Ref. 4, p.257 and J.F.Ziegler, B.L.Crowder, G.W.Cole, J.E.E.Bablin, and B.J. Masters, Appl. Phys. Letters, $\underline{21}$, 16 (1972)

17. Y.Akasaka and K.Horie, in Ref. 4, p.147

18. W.K.Hofker, H.W.Werner, D.P.Oosthoek, and H.A.M.de Grefte, in Ref. 4, p.133

19. K.Wittmaack, J.Maul, and F.Schulz, in Ref. 4, p.119

20. L.C.Northcliffe and R.F.Schilling, Nucl. Data, $\underline{A7}$, 233 (1970)

21. D.K.Brice, Rad. Effects, 18, 13 (1973)
22. F.H.Eisen, B.Welch, J.E.Westmoreland, and J.W.Mayer, Proc. Intern. Conf. on Atomic Collision Phenomena in Solids, Sussex, (1969) p.111
23. Y.Akasaka, K.Horie, and S.Kawazu, Appl. Phys. Letters, 21, 128 (1972)
24. S.Furukawa and H.Matsumura, Appl. Phys. Letters, 22, 97 (1973)
25. H.Okabayashi and D.Shinoda, J. Appl. Phys., 44, 4220 (1973)
26. J.Bøttiger, J.A.Davies, P.Sigmund, and K.B.Winterbon, Rad. Effects, 11, 69 (1971), J.Bøttiger, H.W.Jørgensen, and K.B. Winterbon, Rad. Effects, 11, 133 (1971), and J.Bøttiger and K.B.Winterbon, Rad. Effects, 20, 65 (1973)
27. G.Carter, J.N.Baruah, and W.A.Grant, Rad. Effects, 16, 107 (1972)
28. A.W.Tinsley, W.A.Grant, G.Carter, and M.J.Nobes, in Ref. 12, p.199
29. G.Eldridge, F.Chernow, and G.Ruse, J.Appl. Phys. 44, 3858 (1973)
30. H.Tanoue and T.Tsurushima, this conference
31. E.P.Eernisse, J.Appl. Phys., 45, 167 (1974)
32. H.Ishiwara, S.Furukawa, J.Yamada, and M.Kawamura, this conference
33. J.L.Combasson, J.Bernard, G.Guernet, and M.Bruel, in Ref. 4, p.285
34. H.Nishi, T.Sakurai, T.Akamatsu, and T.Furuta, Appl. Phys. Letters, in press
35. R.A.Moline, private communication
36. K.B.Winterbon, private communication

MEASUREMENT OF ARSENIC IMPLANTATION PROFILES

IN SILICON USING AN ELECTRON SPECTROSCOPIC TECHNIQUE

S. Ludvik, L. Scharpen and H. E. Weaver

Hewlett-Packard Co.

Palo Alto, California 94304

ABSTRACT

A method is described for analyzing implanted semiconductors in which the chemical state of the substrate and impurity atoms is determined from the X-ray photoelectronic spectrum of a narrow surface layer. For the As/Si system, measurements on the As(3d) and Si(2p) lines have established the impurity profile and indicated residual damage after anneal of high dose implants.

INTRODUCTION

With the increasing role of ion implantation in all aspects of semiconductor device fabrication, it has become important to isolate diagnostic techniques which bring out the particular characteristics of the implantation process. The primary goal of such effort is to improve the understanding of the basic implantation process and, at the same time, anticipate its effect on device behavior. Over the years many diagnostic techniques have evolved, and these are summarized in Fig. 1. In most cases where electrical measurements are required, it is necessary to anneal the sample and this step loses some of the information on the initial implant. During the anneal cycle, the impurities and substrate atoms redistribute and at high dose levels where the amount of diffusion, even at low temperatures ($<900^\circ C$), is not a simple function of time, significant changes can occur in the impurity profile. The transition of the substrate to the annealed state raises a number of fundamental questions on the mechanism, and we can list them as follows:
1. Atom redistribution
2. Chemical state of species
3. Residual damage

METHOD	N_T Total Impurities	N_S Carriers	N_d Defects	μ Mobility	COMMENT
HALL MEASUREMENTS —Conductivity —Sheet Resistivity	-	X	-	X	
C-V MEASUREMENTS —Static —Transient	- -	X -	- X	- -	Low Concentrators ($\leq 10^{18}$ cm^{-3}) Defects in junction region only
JUNCTION STAINING	-	X	-	-	Not suitable for shallow junctions
RADIOTRACER SECTIONING	X	-	-	-	
RUTHERFORD SCATTERING (Photon, He)	X	-	X	-	High doses required $>10^{15}$ cm^{-2}
TRANSMISSION ELECTRON MICROSCOPY	-	-	X	-	
ESCA	X	X	X	-	
AUGER SPECTROSCOPY	X	X	X	-	
INFRARED SPECTROSCOPY	-	-	X	-	

Fig.1 Summary of diagnostic methods for analyzing implanted layers

In the present work, our purpose is to describe a method for measuring the implanted profile which does not depend on electrical activation of the impurities. The method is based on the ESCA system (Electron Spectroscopy for Chemical Analysis) developed originally by Siegbahn[1] and his co-workers. Identification of the implanted species and its concentration are determined from the characteristic photoelectrons emitted when the sample is illuminated by a monochromatic X-ray beam. The key characteristic of this technique lies in the fact that, while the incident X-rays are relatively unattenuated as they pass into the sample, the escaping photoelectrons are confined to a depth of 10-50 A° so that the method has an intrinsically high depth resolution. By combining this surface analysis technique with a well controlled ion mill, the implanted sample can be sectioned to determine an accurate profile of the total impurity distribution. In addition, from the energy spectrum of the photoelectrons it is possible to measure energy shifts and broadening of the atomic line and, in this way, identify the chemical state of the impurity ions. Several other techniques that lend themselves to study of the unannealed state should be mentioned, and among these are radiotracer anaylsis[2], Rutherford backscattering[3] and Auger spectroscopy[4]. The principal disadvantages of the radiotracer method are the problem of locating suitable isotopes for the implanted species and that one cannot obtain information on the chemical state of the species. With backscattering, useful information can be obtained on damage and implanted profiles,

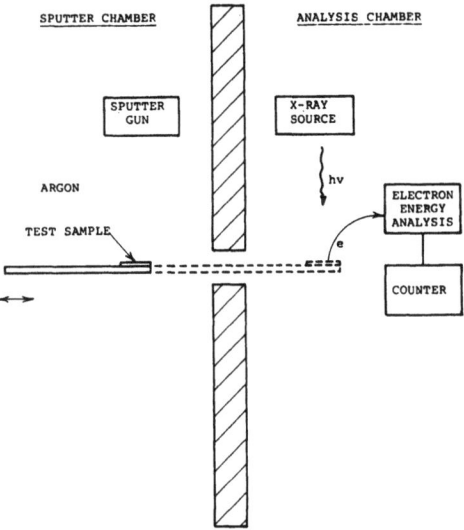

Fig. 2 Basic schematic showing operation of ESCA system

but it depends on the impurity ions being heavier than the target atom. Most closely related to the method to be described here is that of Auger spectroscopy, where illumination of the surface is done by means of electrons rather than X-rays. In that case, the escaping electrons which correspond to the atomic lines are superimposed on a background of secondary electrons, and this limits the resolution of the electronic spectrum.

EXPERIMENTAL TECHNIQUE AND RESULTS

The present study has been carried out using arsenic ($^{75}As^+$) implanted into p-type silicon with the $\langle 111 \rangle$ axis inclined at 7° to the incident beam. Implant energies have been investigated in the range 15-100keV using doses from 3 to 6×10^{15} cm^{-2}. A schematic of the ESCA analysis system is illustrated in Fig. 2. The unit consists of an ESCA spectrometer (Hewlett-Packard 5950A) combined with a focussed-beam argon ion gun in an adjacent chamber. During the sectioning process, the sample is moved alternately between the two chambers. The spectrometer contains an X-ray source operating around 0.8kW, which is then used to produce a highly monochromatic photon beam at an energy of 1486.1eV. The X-ray beam is centered mechanically on the sample, and the energy of photoelectrons leaving an area 1mm x 2mm, are analyzed by a fully automated electron spectrometer which has relative energy resolution within $\pm$0.1eV. Typical usage of the instrument involves doing a preliminary survey scan extending from 0-1000eV to determine suitable atomic lines. The scanning range of the spectrometer is then cut down to a width of 20eV about the chosen lines, and the spectrum around each line is resolved by 256 channels. At the doping concentrations used in these experiments, a dwell time of 13.5 min. for each line was found to be satisfactory, and shorter times could readily be used but at the expense of energy resolution of the line. The ion milling was done be means of an argon ion gun operating at 900V at an ambient pressure of $5-7 \times 10^{-5}$ Torr. The etch rate of this system could be varied from 4-5A° per minute and a mean value of 4.14A° per min. was used for these experiments. The etch rate was determined by a mechanical measurement of the step height after milling the sample for an extended period, typically aroung 120 min. using a Bendix proficorder. The step height by this method correlated well with measurements by a calibrated scanning electron microscope. One disadvantage of the milling system employed here was that the uniformity of the etch rate was only 20% across the sample area, due to the gaussian shape of the beam. This difficulty should be readily overcome, however, by either moving the sample or scanning the ion beam during the etching step.

Examples of the photoelectron spectra obtained at various depths are shown in Fig. 3 for a sample implanted with arsenic at 10^{16} cm^{-2} at 80keV. The species which could be identified were argon (Ar 3s line at 23ev), gold (Au 5p's and 4f's at 59, 75 and 86, 89eV respectively), arsenic (3^d line at 41eV) and silicon (2p's at 100eV and 2s at 151eV). The presence of argon could clearly be attributed to low energy implant of argon during the milling process. The gold lines observed were probably due to

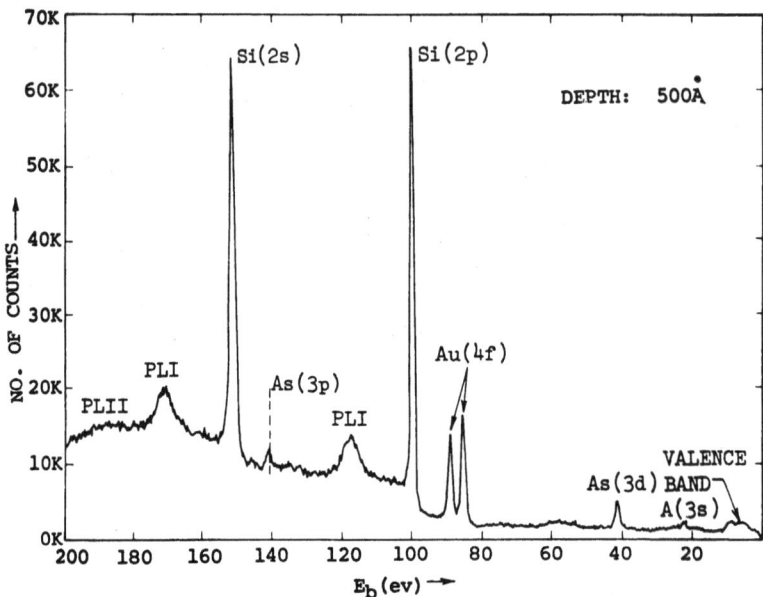

Fig. 3. Typical electronic spectrum for ^{75}As$^+$ implanted silicon
Conditions: 80keV, 10^{16} cm^{-2}

contamination on the sample surface from the holder, and this was
verified by analyses of subsequent samples. The broader peaks
adjacent to the main silicon lines correspond to plasmon modes
which are excited by energy losses of the inner shell electrons to
collective modes of the valence electrons. The plasmon modes are
characteristically observed when an electron beam is passed through
a thin film and their spacing is at integral multiples of $\hbar w_p$ where
w_p is the plasma frequency.

In order to interpret the line intensities and relate them to
the actual impurity concentration, a number of assumptions were
required. Firstly, it was assumed that all the atoms within the
emitting volume act as independent centers and no re-absorption
occurs. In effect, this amounts to a constant depth for the emitting volume independent of the impurity concentration within the
volume. Estimates of the depth of the active region from the work
of Steinhardt, et al[5], suggest a value from 10-50A°. Within these
simplified approximations the impurity concentration is proportional
to the integrated line shape. For the As/Si system, we have used as
the principal line the 3d line for arsenic and the 2p line for silicon.
The arsenic impurity profiles obtained by integrating the 3d line
are shown in Fig. 4 (a) for two implant conditions. The position of
the distributions are clearly resolved and their separation of about
100A° is in good agreement with the theoretical calculations of
Johnson and Gibbons[6]. To this extent the basic measurement technique
proposed here has been validated. The situations where striking
differences are observed with the ESCA technique become apparent
when we compare the line intensities at various stages of annealing.
Fig 4b shows the redistribution of a 40keV arsenic implant at a dose
level of $3 \times 10^{15} cm^{-2}$ after an anneal of 850°C for 30 min. The
integrated As(3d) line intensity provides a relative concentration
scale and these measurements indicate a piling up of arsenic at the
silicon surface together with an overall broadening of the distribution.
The peak of the distribution, however, remains at approximately the
same position which is verified on annealed samples that have been
profiled[7] by Hall measurements and anodic sectioning. Since, at these
high dose levels, the implant forms an amorphous region, the redistribution of arsenic is not surprising and similar effects have
been noted in backscattering studies by Mayer, et al[8]. Additional
effects occurring during the changes of the amorphous region are
found by examining the Si(2p) line intensity which is shown in
Fig 5 as a function of distance into the sample. For the unimplanted sample (C), the line intensity remains constant to within
10-15% into the crystal. Some variation is apparent at the surface
due possibly to oxide formation. This feature is exactly what would
be expected and confirms also that the argon ion milling process does
not introduce cumulative damage into the crystal. By contrast, the
sample (A) immediately after implant demonstrates pronounced
differences and indicates an enhanced silicon concentration whose
peak is displaced about 250A° beyond the arsenic maximum. The

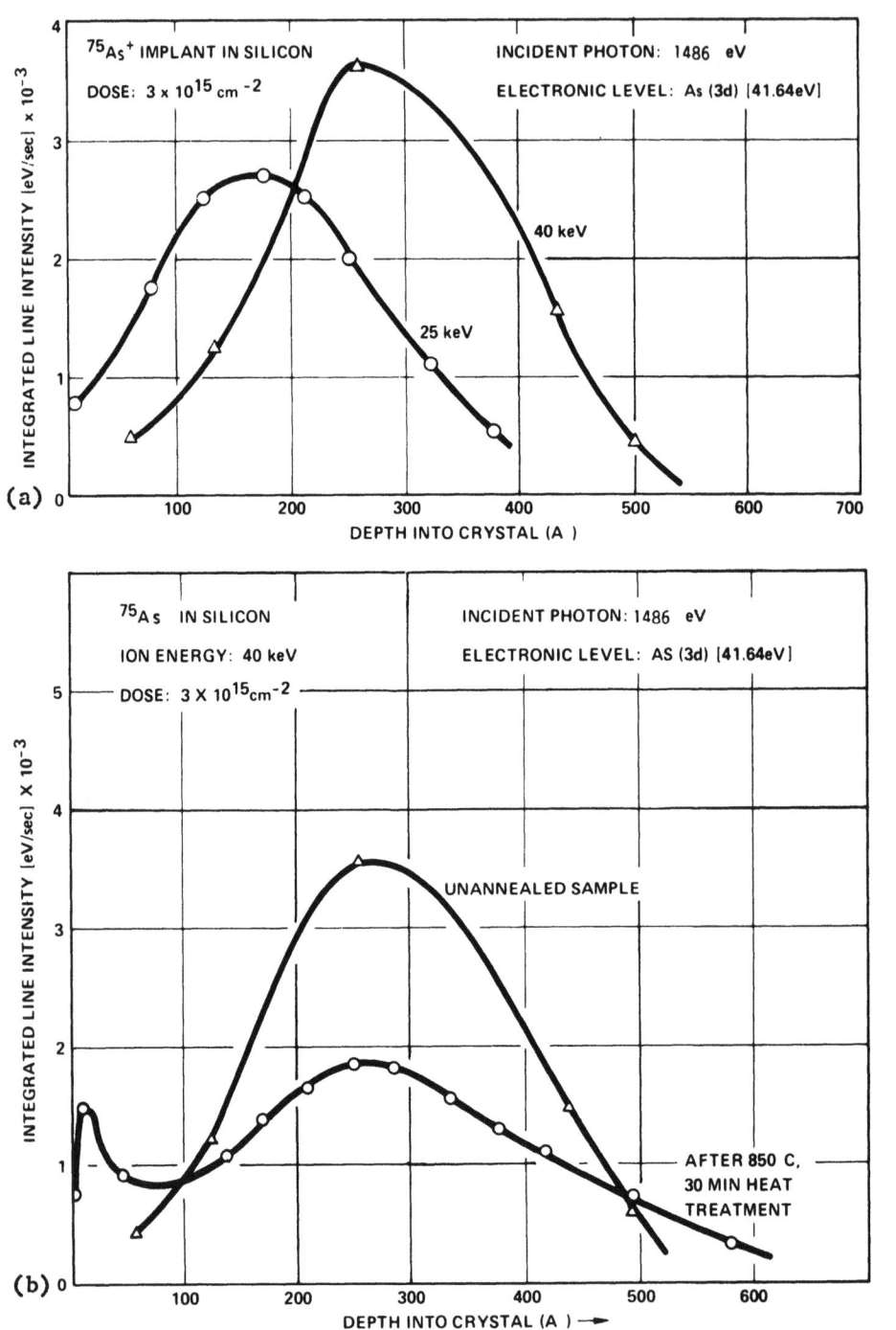

Fig 4 (a) Profiles at two implant energies for unannealed samples
(b) Changes in implant profile during an anneal 850°C, 30 min.

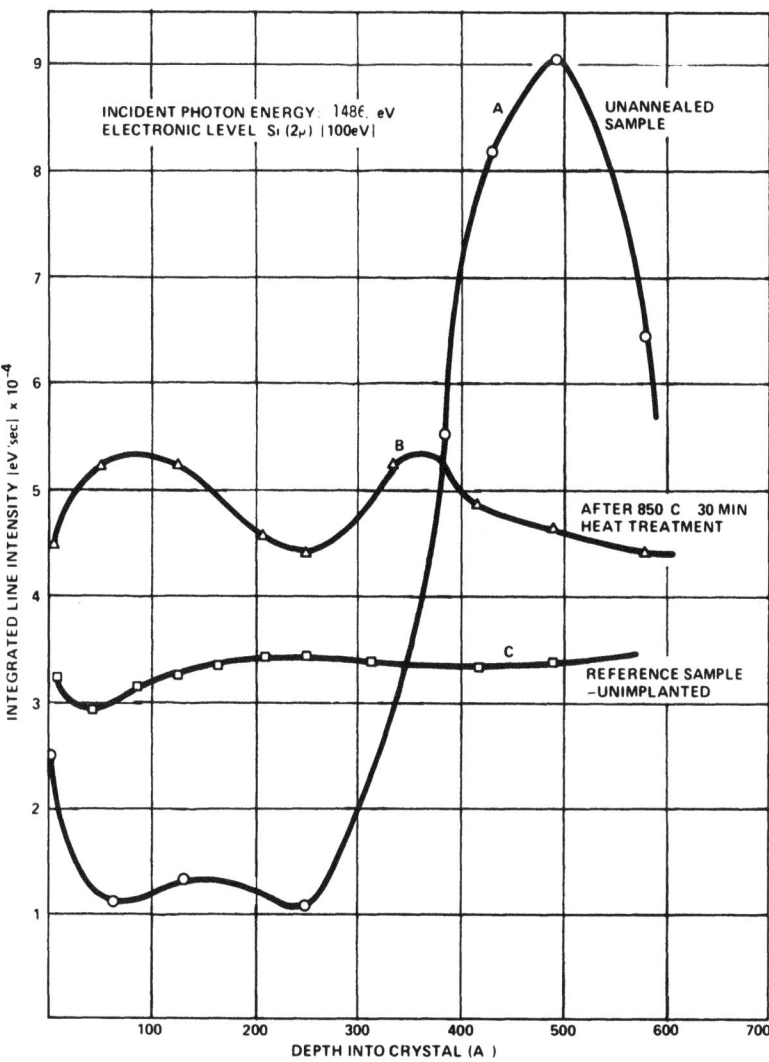

Fig 5. Dependence of Si (2p) line intensity on different sample conditions showing pronounced effect of implant damage.

position of the peak is approximately the theoretical projected range calculated for a silicon ion of energy 30-40keV incident on a randomly oriented silicon crystal. For the substrate atoms the total line intensity is probably not directly proportional to the concentration since re-absorption may occur; however the line intensity variation suggests that the peak is produced by self-implantation of the silicon by surface atoms knocked-on by the incident arsenic ions. After the anneal cycle of 850°C for 30 min. it can be seen in the case of sample (B) that the variations in the silicon line intensity are considerably reduced although some residual damage still appears.

The use of the ESCA method for evaluating implanted layers provides a great deal of information on both the impurity and substrate atoms. The particular advantage of the technique lies in that the sample does not require annealing, and it is possible to make direct measurements immediately after the implantation. The minimum detectable concentration depends on the cross section of the species, and there is a compromise between the line resolution and the sensitivity. In the case of arsenic by reducing the number of channels and increasing the photon intensity, a sensitivity limit around $2 \times 10^{17} cm^{-3}$ appears practicable, although the method is better suited to higher concentrations. Since damage and anneal behavior are of greater concern at these higher doping levels, the system sensitivity does not appear to be a severe problem.

REFERENCES

1. K. Siegbahn et al, "ESCA-Atomic, Molecular and Solid State Structure by Means of Electron Spectroscopy", Almquist and Wiksells, Uppsala (1967)

2. R. Kelly and E. Ruedl, Phys. Status Solidi $\underline{13}$, 15 (1966)

3. D. Powers and W. Whaling, Phys. Rev. $\underline{126}$, 61 (1962)

4. J.C. Riviere, Contemp. Phys. $\underline{14}$, 513-539 (1973)

5. R.G. Steinhardt, J. Hudis and M.L. Perlman, Phys. Rev. $\underline{58}$, 1016-1020 (1972)

6. W.S. Johnson and J.F. Gibbons, "Projected Range Statistics in Semiconductors", (Stanford University Press, 1969)

7. T. Sigmon (private communication)

8. J.W. Mayer, L. Ericksson, S.T. Picraux and J.A. Davies, Can. J. Phys. $\underline{45}$, 4053 (1967)

ATOM AND CARRIER PROFILES IN As IMPLANTED Si

M. Iwaki, K. Gamo, K. Masuda and S. Namba

Faculty of Engineering Science, Osaka University

Toyonaka, Osaka, Japan

S. Ishihara and I. Kimura

Reseach Reactor Institute, Kyoto University

Kumatori, Sennan-Gun, Osaka, Japan

and

K. Yokota

Faculty of Engineering, Kansai University

Suita, Osaka, Japan

ABSTRACT

The annealing behavior of As implanted Si-layers is investigated by measuring the atom concentration profile with radioactive tracer technique and the carrier concentration profile with Hall effect measurements in the same sample. Silicon substrates have been implanted at room temperature with a low (4×10^{13}/cm^2) and high ($5 \sim 6\times10^{14}$/cm^2) dose of ^{75}As ions together with radioactive ^{76}As ions at 45keV and subjected to annealing at 500°C and 600°C for 20min. After annealing at 600°C, the carrier concentration profile corresponds closely to the implanted As ion profile. The carrier mobility observed for 600°C annealing is almost the same as that expected for comparably doped bulk silicon, except near the peak region for the high dose implantation. The results suggest that the recovery of the carrier and mobility takes place from inside toward the surface.

INTRODUCTION

The annealing behavior of electrical properties in arsenic implanted silicon have been studied as a function of the isochronal annealing temperature by means of sheet resistivity measurements and most of the implanted As ions were found to be electrically active after a 550°C anneal for doses of $3 \times 10^{14}/cm^2$ and above [1].

The annealing process, however, seems to vary with the depth because the atom density and defect density depend on the depth. Therefore, it is important to investigate annealing characteristics as a function of the depth. So far, both atom and carrier concentration profiles have been measured independently. Crowder [2] measured the carrier concentration profiles of arsenic implanted in silicon by means of Hall effect measurements combined with layer removal techniques and observed that, if a continuous amorphous layer is present, ions within this region become electrically active for annealing at a temperature range between 550°C and 600°C. Atom concentration profiles of arsenic ions implanted in silicon have been measured by means of neutron activation analysis [3,4] and are in agreement with LSS prediction [5,6] for amorphous targets.

Layer-removal techniques generally involve anodic oxidation and oxide-layer removal by hydrofluoric acid. Anodic oxidation is a good technique for layer removal but, if atom and carrier concentration profiles are measured by using two different samples, the scatter in removed depth between both profiles is contained. Therefore, it is important to measure both atom and carrier concentration profiles in the same sample. Such measurements have not been done.

The purpose of the present work is to investigate the annealing characteristics of As implanted layer in silicon. ^{75}As ions together with radioactive ^{76}As ions have been implanted in silicon and both atom and carrier concentration profiles have been measured in the same sample by means of radiotracer method and Hall effect measurements, respectively, combined with layer removal techniques.

EXPERIMENTAL

Silicon substrates used were 2~6 ohm·cm, B-doped (111) wafers cut from Czochralski grown crystals. All samples were implanted with 45keV As ions together with radioactive ^{76}As ions at room temperature in a direction tilted 8 degree off <111> crystal axis. The radio-isotope ^{76}As was obtained from arsenic by exposing to a neutron flux of 8.15 neutrons/cm^2/sec for 10 hours. The ion beam was magnetically mass separated, but the arsenic ion beam was a mixture of ^{75}As and ^{76}As due to poor mass resolution of the ExB filter. The doses used were $4 \times 10^{13}/cm^2$ (low dose) and $5 \sim 6 \times 10^{14}/cm^2$ (high dose) respectively. After ion implantation, samples were annealed at 500°C and 600°C for 20 min in vacuum.

Carrier concentration profiles were obtained by means of Hall effect and sheet resistivity measurements on van der Pauw method in conjunction with anodic oxidation and HF stripping techniques. The thichness of the removed silicon per one step was about 80 to 100 Å. Before implantation of ^{75}As and ^{76}As ions, the contact regions of the van der Pauw pattern have been produced by diffusing As at 1100°C, which was pre-deposited by ion implantation, in order to facilitate good electrical contacts to the radioactive ion implanted region. Electrical contacts were made by fastening gold wires to the contact regions by means of conducting silver paste.

Atom concentration profiles were obtained by measuring the total number of counts of γ-rays emitted from ^{76}As contained in the HF solvent of each section. Half-life determination and γ-ray spectral analysis were used for identification of the isotopes. The area of the radioactive ion implanted region is 0.503 cm^2, in order to maintain sufficient sensitivity in the radiochemical assay.

RESULTS AND DISCUSSION

The comparison of the As atom profile with the carrier profile for the high dose is presented in Fig.1 (annealing at 500°C) and Fig.2 (annealing at 600°C). For the high dose, the implanted layer was amorphized, which was observed by means of electron spin resonance (ESR) measurement. No difference is observed between the atom profiles before (not shown in the figure) and after annealing except near the surface. A scatter of the atom concentration at the surface may be due to a contamination by ^{76}As (probably neutral fraction of the ion beam). The atom profile is composed of two parts ; in the high concentration region, the profile is nearly gaussian, whereas at the low concentration region a small tail is visible. The projected range and the projected standard deviation of this profile were found to be almost the same as those predicted by the LSS theory [5,6]. For the formation of the tail, a rapid diffusion process has been proposed by Iwaki et al. [4] and recently by Schwettmann [7].

For annealing at 500°C, the carrier recovery is very low and the number of carriers is found to be about 4% of the atom number in the peak region where the amorphous phase has been produced. In the tail region, however, most of the implanted arsenic ions become electrically active. For annealing at 600°C, the carrier concentration profile almost corresponds to the As atom profile. This result agrees with the fact that, if a continuous amorphous layer is present, implanted ions become electrically active during the epitaxial recrystallization of the amorphous layer [2].

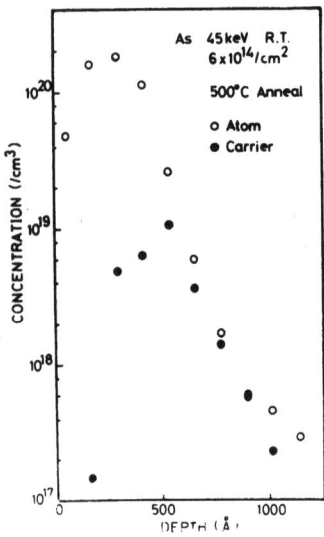

Fig. 1. Comparison between atom and carrier profiles in As implanted Si-layers with the high dose for annealing at 500°C.

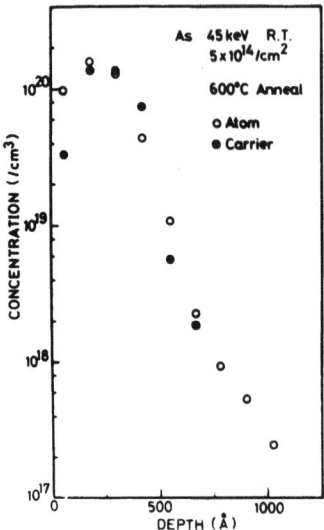

Fig. 2. Comparison between atom and carrier profiles in As implanted Si-layers with the high dose for annealing at 600°C.

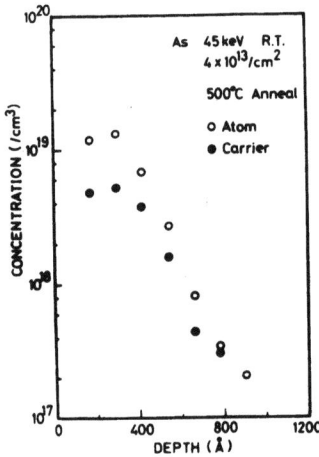

Fig. 3. Comparison between atom and carrier profiles in As implanted Si-layers with the low dose for annealing at 500°C.

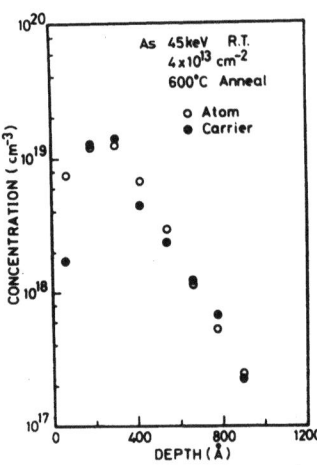

Fig. 4. Comparison between atom and carrier profiles in As implanted Si-layers with the low dose for annealing at 600°C.

The comparison of the As atom profile with the carrier profile for the low dose is presented in Fig.3 (annealing at 500°C) and Fig.4 (annealing at 600°C). For the low dose, the lattice disorder density was about 40% as has been measured by the ESR technique. A significant recovery was observed even near the peak region for annealing at 500°C. The doping efficiency was found to be 20%, and this is an improvement of a factor 5 compared to the high dose implantation. At the tail region, most of the implanted arsenic ions became electrically active, as has been also observed in the high dose implantation. After annealing at 600°C, the carrier concentration profile agrees with the implanted As profile.

By comparing the results in the high and low dose implantations after annealing at 500°C and 600°C, it is clear that the carrier recovery takes place from inside toward the surface.

The carrier mobility is presented in Fig.5 as a function of the carrier concentration. Points adjacent in depth are connected by the arrows and it indicates the direction toward inside. The solid line shows the mobility for n-type bulk silicon as given by Irvin [8]. After annealing at 500°C, it has been found for both doses that the mobility approaches the bulk mobility as the depth from the surface increases. In the peak region, the carrier mobility is lower than the bulk mobility, but in the tail region the carrier mobility is almost the same with the bulk mobility. After annealing at 600°C, the carrier mobility for the low dose implantation is the same with the bulk mobility in the whole region of the depth. For the high dose implantation, however, the carrier mobility between the surface and the peak region is found to be lower than the bulk mobility. This suggests that a certain fraction of the lattice damage centers produced by ion implantation may remain.

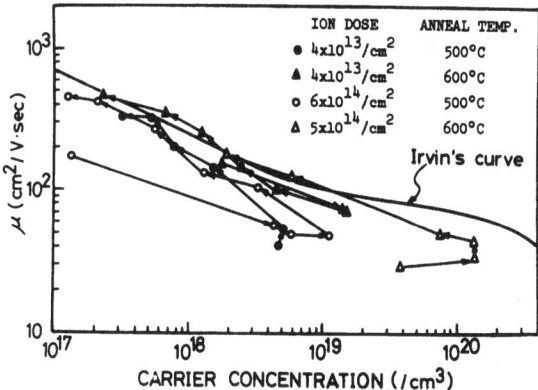

Fig. 5. Hall mobility of As implanted Si-layers as a function of observed carrier concentration of the layer.

By comparing the results of annealing at 500°C and 600°C, it is clear that the recovery of carrier mobility also takes place from inside toward the surface

SUMMARY AND CONCLUSION

45 keV ^{75}As ions together with radioactive ^{76}As ions have been implanted at room temperature in silicon with doses of 4×10^{13}/cm^2 and $5\sim6\times10^{14}$/cm^2 respectively and subjected to annealing at 500°C and 600°C. Atom and carrier concentration profiles have been determined in the same sample. Measured values for the carrier mobility is compared with that expected for comparably doped bulk silicon. From the results we conclude:
1) After annealing at 600°C for 20 min, all implanted arsenic ions become electrically active, but the carrier mobility in the peak region is still low for the high dose implantation.
2) The recovery of the carrier and mobility takes place first in the tail region and proceeds toward the surface.

ACKNOWLEDGEMENT

We would like to thank Mr. K. Kawasaki for his help in performong the experiments.

REFERENCE

1) B.L. Crowder and F.F. Morehead, Jr.: Appl. Phys. Letters 14 (1969) 313
2) B.L. Crowder: J. Electrochem. Soc. 117 (1970) 671
3) J M. Fairfield and B.L. Crowder: Trans. Met. Soc. AIME 245 (1969) 469
4) M. Iwaki, K. Gamo, K. Masuda, S. Namba, S. Ishihara and I. Kimura: Ion Implantation in Semiconductors and other Materials, ed. by. B.L. Crowder, Prenum Press, New York (1973) p.111
5) J. Lindhard, M. Scharff and H.E. Schiott: kgl. Danske Videnskab. Selskab. Mat. Fys. Medd. 33 (1963) 14
6) D.K. Brice: Radiation Effects 6 (1970) 77 and private communication
7) F.N. Schwettmann: Appl. Phys. Letters 22 (1973) 570
8) I. Irvin: Bell System Tech. J. 41 (1962) 387

THE INFLUENCE OF POSTPROCESSING ON THE ELECTRICAL

BEHAVIOR OF IMPLANTED ARSENIC DISTRIBUTIONS IN SILICON*

H. Ryssel, H. Kranz, K. Schmid and I. Ruge

Institut für Festkörpertechnologie

Paul-Gerhardt-Allee 42, 8 München 60, Germany

H.S. Rupprecht

IBM, Systems Product Div., East Fishkill, N.Y., USA

ABSTRACT

In this paper the annealing and diffusion behavior of arsenic distributions are investigated. For the annealing of implanted layers two distinct activation energies (0.943 eV and 3.74 eV) have been measured at low implantation doses. At doses which result in the formation of amorphous layers, a single activation energy of 2.75 eV has been measured. Quenching experiments indicate that slow cooling after annealing can increase the electrical activation up to a factor of two. Profiles of arsenic implants after drive-in diffusions and after radiation enhanced diffusion are shown. Whereas the first ones show a large impurity gradient at the leading edge of about $5 \times 10^{24} cm^{-4}$, a plateau is seen in the second case.

INTRODUCTION

High impurity concentrations are required for emitter and subcollectors in bipolar transistors. Arsenic has turned out to be a very suitable dopant for those applications. The main advantages using arsenic as a subcollector dopant in comparison to antimony are a higher solubility limit resulting in extremely low sheet resistances. In the case of emitter structures arsenic was found to give high gradient profiles which lead to improved device characteristics.

* This work was sponsored by the "Deutsche Forschungsgemeinschaft", Bonn, Germany

These advantages are partially compensated by difficulties in developing a viable diffusion process from a manufacturing point of view. Ion implantation techniques offer an attractive alternative to circumvent those technological shortcomings. Investigations have shown that complex formation occurs at high arsenic concentrations /1/ and that the electrical activation is strongly temperature and dose dependent. In order to elucidate some of the chemical and electrical properties of arsenic doped silicon the following investigations have been carried out.

EXPERIMENTAL TECHNIQUES

The samples used for the measurements were (100)-oriented, 8 Ωcm boron doped silicon wafers from Monsanto. All wafers were tilted 7° ± 2° before implantation to avoid channeling. Implantations were done at 300 K. Annealing was carried out in dry nitrogen atmosphere. Measurements were made on structures using a Van der Pauw geometry published earlier /2/. The radiation enhanced diffusion was carried out with a hydrogen beam. For heating the samples a special set up /3/ was used which allowed the heating of the wafers by the beam itself.

ANNEALING BEHAVIOR

For the annealing experiments amorphous and nonamorphous implants were used. The amorphous dose arsenic in silicon at 300 K

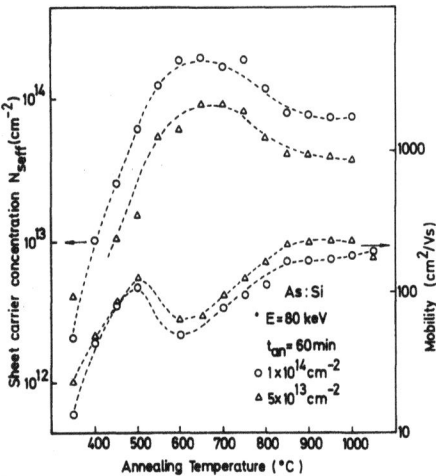
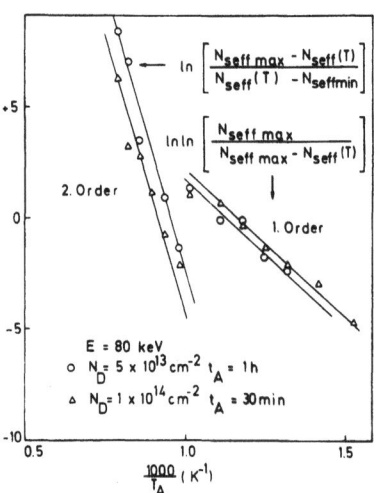

Fig. 1 Isochronal annealing of a 80 keV arsenic implant in silicon. Doses are $5 \times 10^{13} \text{cm}^{-2}$ and 10^{14}cm^{-2}, annealing time is 60 min.

Fig. 2 Arrhenius plots derived from 30 min and 60 min annealing. Energy and doses are 80 keV and $5 \times 10^{13} \text{cm}^{-2}$, 10^{14}cm^{-2} respectively.

is $2 \times 10^{14} \text{cm}^{-2}$ /4/. The doses used were $5 \times 10^{13} \text{cm}^{-2}$, 10^{14}cm^{-2}, $5 \times 10^{14} \text{cm}^{-2}$, 10^{15}cm^{-2}. In Fig. 1 the isochronal annealing behavior of the non-amorphous doses is shown for an annealing time of 60 min From 350°C to 650°C the sheet carrier concentration N_{Seff} increases steeply to a value of 2 times the implanted dose $\emptyset$. For temperatures above 650°C N_{Seff} decreases slowly to a value of approximately $\emptyset$. If reaction kinetics are used to interpret these results, it is seen that the first process is of first order and the second one is of second order. For first order processes ln ln Nseffmax/ (Nseffmax - Nseff (T)) plotted vs 1/T gives a straight line with a slope of Ea1/kT where Ea is the activation energy. In Fig. 2 this is done for samples which were annealed for 30 min and 1 h at each temperature. The mean activation energy from measurements made with annealing times ranging from 30 min to 2 hours was evaluated by the method of least mean squares and found to be Ea1 = (0.943 ± 0.030) eV. For processes of second order ln (Nseffmax - Nseff (T))/ (Nseff(T) - Nseffmin) plotted vs 1/T gives a straight line with slope of Ea2/kT. If this is done for the temperatures above 650°C an activation energy of 3.74 eV is deduced. The first activation process is obviously due to a certain doubly charged complex formation. The second process then must represent the dissociation of this complex and incorporation of arsenic on regular sites.

In Fig. 3 isochronal annealing curves for the amorphous doses are shown. There is a very steep increase of Nseff between 350°C and 500°C. No increase of Nseff to a twofold of the dose can be seen as in the case of non-amorphous implants. These curves cannot

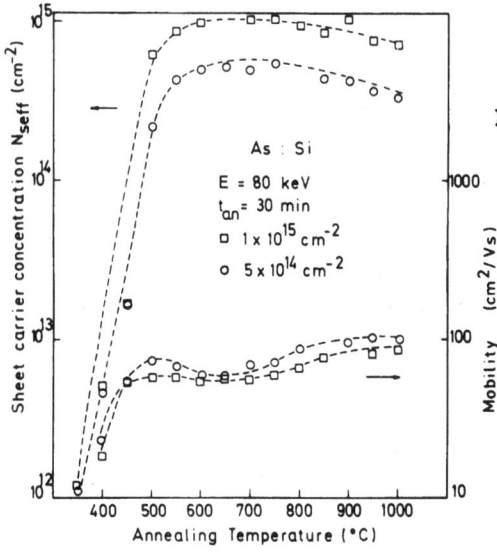

Fig. 3 Isochronal annealing of a 80keV arsenic implant in silicon at doses of $5 \times 10^{14} \text{cm}^{-2}$, and 10^{15}cm^{-2}, annealing time is 30 min.

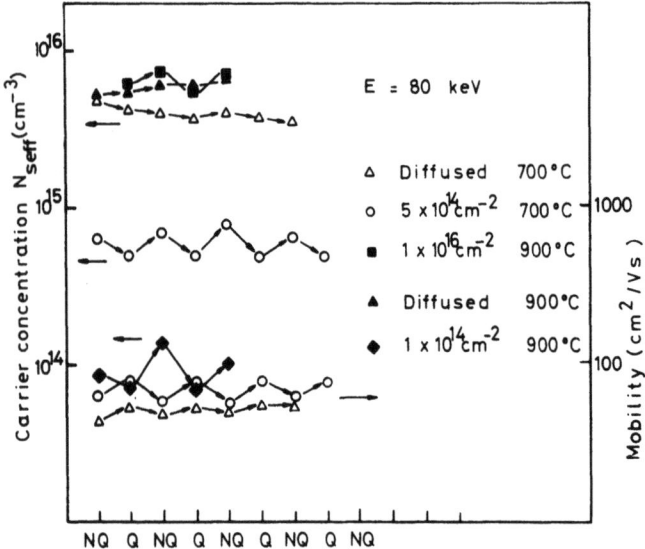

Fig. 4 Nseff and μseff for various doses and annealing temperatures vs successive cycles of quenching (Q) and slow cooling (NQ). For clarity only two mobility curves are shown.

be interpreted by reaction kinetics because of the steep gradient. By means of isothermal measurements, however, an activation energy of 2.75 eV was obtained. In order to obtain those results particular care has to be taken regarding the cooling procedure after annealing. In Fig.4 the effect of the cooling rate on Nseff is demonstrated for diffused and implanted samples. Quenching of the samples results in a markedly lower value for Nseff than for the slowly cooled samples. The magnitude of the effect is strongly dose dependent. Diffused samples seem to approach asymptotically a new equilibrium value after prolonged cycling.

PROFILES

In Fig. 5 the main features of diffused, implanted and distributions obtained by RED are depicted. The largest gradient is obtained by diffusion (Fig. 5a). Implanted profiles (Fig. 5b) always show a characteristic tail at the leading edge as reported earlier /5/. For practical applications it has turned out to be necessary to position the p-n junction away from the radiation damaged region by a drive-in diffusion /6/. In Fig. 5c a typical profile obtained by RED is shown. In this case large gradients and usually a plateau are present.

Drive-in Profiles

For typical emitter application doses are between $5 \times 10^{15} cm^{-2}$

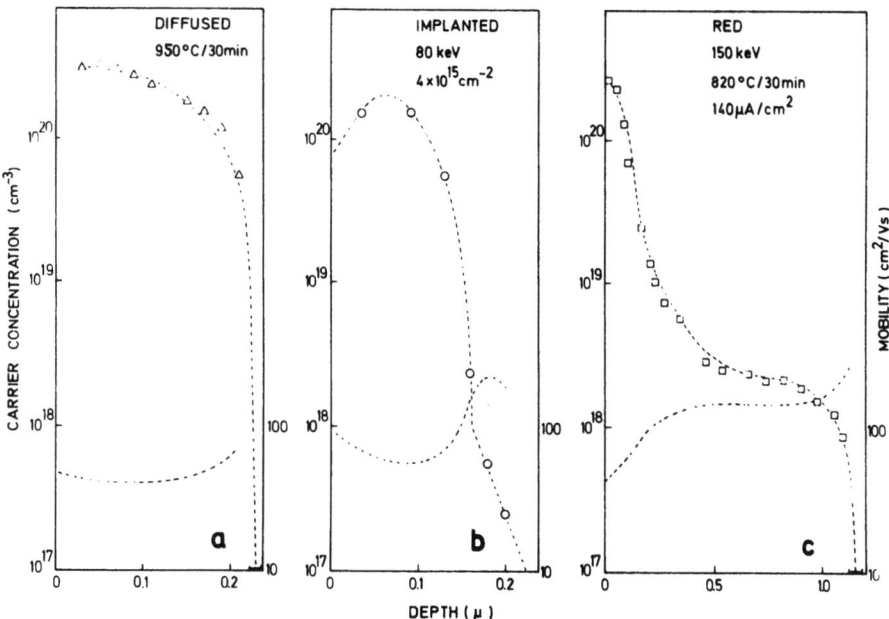

Fig. 5 Comparison of typical implanted, diffused and RED profiles of arsenic in silicon.

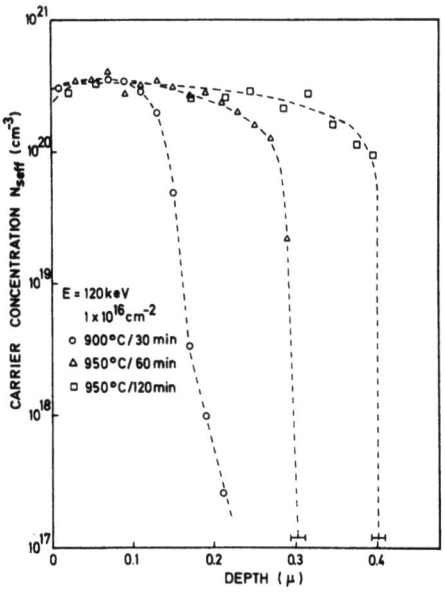

Fig. 6 Implanted arsenic profiles after different drive-in diffusions.

and $10^{16} cm^{-2}$. In Fig. 6 typical profiles obtained by drive-in diffusions are shown along with the original implanted profile. The gradients of the drive-in profiles are about $5 \times 10^{24} cm^{-4}$ as compared to a gradient of $10^{23} cm^{-4}$ in the case of the original implanted profile at concentrations of $10^{17} cm^{-3}$. Similar gradients as for these drive-in profiles were measured on diffused structures (capsule diffusion at 950°C for 120 min, Fig. 5a).

Radiation Enhanced Diffusion

Radiation enhanced diffusion is a means of diffusing selectively at relatively low temperatures. We used temperatures from 500°C to 950°C and current densities ranging from 70 μA/cm² to 250 μA/cm². These current densities are very high compared to those used for boron, phosphorus and antimony /7,8/. They are necessary, however, in the case of arsenic to enhance the diffusion measurable. From backscattering measurements it can be concluded that temperatures higher than 700°C are required during the bombardment in order to avoid radiation damage. After RED the samples have to be annealed in order to restore the carrier concentration and mobility. In Fig. 7 profiles obtained after RED of arsenic implanted layers (80 keV, $10^{16} cm^{-2}$) by hydrogen bombardment at different energies are shown. The arrows indicate the position of the projected range R_p of hydrogen. Fig. 7a depicts the original profile before RED. If the

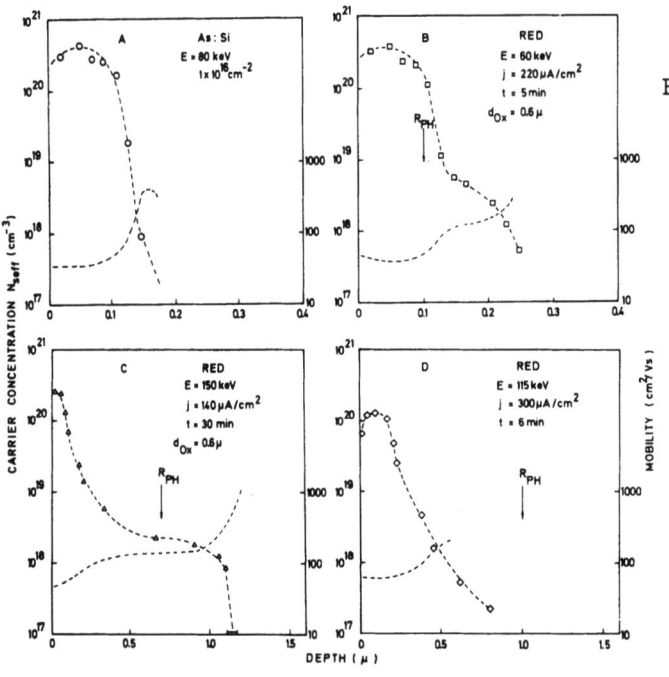

Fig. 7 RED profiles of arsenic implanted layers at different hydrogen energy. A depicts the profile before diffusion. For B and C a SiO_2 layer was used to reduce the hydrogen energy.

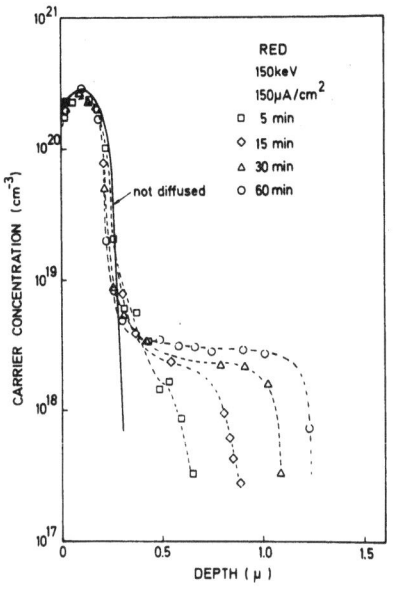

Fig. 8 RED profiles of arsenic implanted layers. Hydrogen energy 150 keV, current density 150 µA/cm^2, diffusion time 5 min, 15 min, 30 min and 60 min.

generation of vacancies which enhance the diffusion takes place within the arsenic distribution or close to it (Fig. 7b and c) a plateau with a steep gradient at the leading edge develops. If the depths of the arsenic and the vacancy generation distributions differ considerably, approximately exponentially shaped tails are formed (Fig. 7c). In Fig. 8 a set of RED profiles with different diffusion times is shown. The hydrogen energy was 150 keV, because of a SiO_2 layer, however, the effective energy was about 70 keV. The reason for the formation of the plateau is a relatively small diffusion length of the generated vacancies. The plateau has a doping concentration of about $(1-2) \times 10^{18} cm^{-3}$. We made most experiments with implanted layers, because of ease to measure doping and mobility profiles by anodic oxidation. Special subcollector structures, however, were investigated by backscattering measurements and showed similar behavior. The improvements of microwave transistors by use of a collector pedestal have been discussed by Ziegler /9/ and Abe /10/. The base-to-collector capacitance can be reduced markedly by providing a wide separation between base and subcollector except in the active transistor region. A detailed discussion of those problems, however, would be beyond the scope of this paper.

CONCLUSION

It was shown that different mechanisms take place during the annealing of implanted layers. In the non-amorphous case the annealing takes place in two steps, two activation energies (0.943 eV and 3.74 eV) are measured. The reason might be the formation of a doubly charged complex at low temperatures and the dissociation of this complex at higher temperatures. Amorphous implants show a single activation energy of 2.75 eV. The cooling rate after annealing is very important in arsenic implanted layers. The difference in Nseff between quenched and unquenched samples may be as high as 100%.

If arsenic implantation is used as predepostion gradients of $\geq 5 \times 10^{24} \text{cm}^{-4}$ can be obtained at the leading edge of the profile. It could be shown that RED can be used effectively to diffuse arsenic. High temperatures and doses are required to achieve this in a reasonable time.

ACKNOWLEDGEMENT

We would like to thank R. Zölch and M. Bleier for performing measurements and M. Schenk, L. Fiesel and E. Traumüller for technical assistance. The assistance of Siemens Laboratories for performing some diffusions and IBM Sindelfingen for supplying subcollector structures is appreciated.

REFERENCES

/1/ R.B. Fair and G.R. Weber, Appl. Phys. Lett. 44, 273 (1973)
/2/ H. Ryssel, K. Schmid and H. Müller, J.Phys. E.: Sci. Instr. 6, 492 (1973)
/3/ H. Ryssel, H. Rupprecht, and H. Kranz, to be published
/4/ J.W. Mayer, L. Eriksson, and J.A. Davies, Ion Implantation in Semiconductors, Academic Press, New York (1970)
/5/ F.N. Schwettmann, Appl. Phys. Lett. 22, 570 (1973)
/6/ F.J. Scavucco, R.S. Payne, K.H. Olson, J.M. Nacci, and R.A. Moline IEEE IEDM, Dec. 1972, Washington
/7/ D.G. Nelson, J.F. Gibbons, and W.S. Johnson, Appl. Phys. Lett. 15, 246 (1969)
/8/ Y. Ohmura, S. Mimura, M. Kanazawa, T. Abe and M. Konaka, Rad. Effects 15, 167 (1972)
/9/ J.F. Ziegler, B.L. Crowder and W.J. Kleinfelder, IBM J. Res. Dev. 15, 452 (1971)
/10/ T. Abe, S. Yamamotu, T. Sakamotu and T. Zhota, Ion Implantation in Semiconductors, Ed. S. Namba, p.121, Jap. Soc. Prom. Sci. 1972

Anomalous Annealing Behavior of Arsenic

Implanted Silicon as a Function of Dose and Energy*

   W. K. Chu, H. Müller and J. W. Mayer
   California Institute of Technology
   Pasadena, California 91109
   and
   T. W. Sigmon
   Hewlett-Packard Laboratories
   Palo Alto, California 94304

ABSTRACT

Anomalous annealing behavior of damage in As implanted Si has been investigated by channeling effect measurements. Room temperature As implants with doses from 10^{14} to $2 \times 10^{16} cm^{-2}$ at energies ranging from 50 to 250 keV into <111> Si were annealed over the range of 600°C to 1100°C. Anneal temperatures $\geq$ 950°C reduce the disorder for ion doses of 0.5 to $1 \times 10^{16} cm^{-2}$ at energies between 50 and 250 keV. Anomalously high residual disorder was found for the intermediate doses of 2×10^{14} to $2 \times 10^{15}/cm^2$ implanted above 100 keV. For this intermediate dose, annealing of implants performed at liquid nitrogen temperatures did not show significant reduction in this damage. However implants at 200° to 300°C indicate a lower amount of damage compared to the room temperature implants.

INTRODUCTION

The high solubility and concentration dependent diffusion coefficient of As in Si suggest use of implanted As for emitter in npn transistors. It has been found that for certain transistor structures high temperature anneals must be performed, often in an oxidizing atmosphere to obtain low leakage emitter-base junctions [1]. In this paper channeling effect measurements are utilized to investigate the disorder remaining in the silicon following implantation and anneal. Parameters varied included implantation energy, dose and dose rate along with the substrate temperature.

---

*Caltech work supported in part by O.N.R. (L. Cooper).

†Permanent address: Lehrstuhl für Integrierte Schaltungen, Munich, W-Germany.

EXPERIMENTAL TECHNIQUES

Silicon wafers of <111> orientation with a resistivity of 2-4 ohm-cm were used throughout the experiments. Arsenic implants were performed at energies ranging from 50 to 250 keV over a dose of 10^{14} to $2 \times 10^{16}/cm^2$ in a random direction. Dose rates of 0.2 to $2\mu A/cm^2$ were employed. Substrate temperatures were varied between liquid nitrogen temperature and 300°C with the majority of the implants performed at room temperature.

Channeling analysis for the implanted layers was carried out with 2.0 MeV He^+ ions. The silicon portion of the aligned spectra was used as a measure of the residual disorder that remained in the samples after the anneal.

RESULTS

A. Annealing Behavior - $10^{16} As/cm^2$, 200 keV

The aligned spectra in Fig. 1 show the relative damage of the Si crystal caused by $10^{16}/cm^2$ arsenic ion implantation and the residual disorder remaining after various heat treatments. The small dip in the random spectrum of the non-annealed sample in Fig. 1 is due to the dilution of Si caused by the high concentration of the As ($\sim$ 3 atomic %) around the projected range R_p. The aligned spectrum of this non-annealed sample has a sharp dip near R_p indicating the crystal is not completely amorphous in the region around R_p. This is believed to be due to some annealing which occurs during the arsenic implantation. Since the Si crystal was not heat-sinked and the implantation current density was typically $2\mu A/cm^2$, the wafer temperature could rise significantly above ambient during implantation. Infrared pyrometer measurements have indicated that wafer temperatures as high as 250°C can occur.

Between 600 and 700°C the surface portion of the damage anneals rapidly; (Fig. 1) as reported earlier [2]. However, the deeper damage region located at $R_p + \Delta R_p$ requires annealing at temperatures above 950°C. At 950°C or above, the aligned spectrum is identical to that of an un-implanted Si crystal and indicates that the As atoms are highly substitutional (> 90%).

B. Dose and Energy Dependence, T_A = 950°C

We investigated the dose dependence of samples that had been annealed at 950°C for 15 min. Figure 2 shows the aligned spectra for samples implanted at 250 keV for arsenic doses between 10^{15} to $10^{16}/cm^2$. The dashed line represents the random spectrum for these targets. The data in Fig. 2 indicates that the samples with low or high dose have very low residual dis-

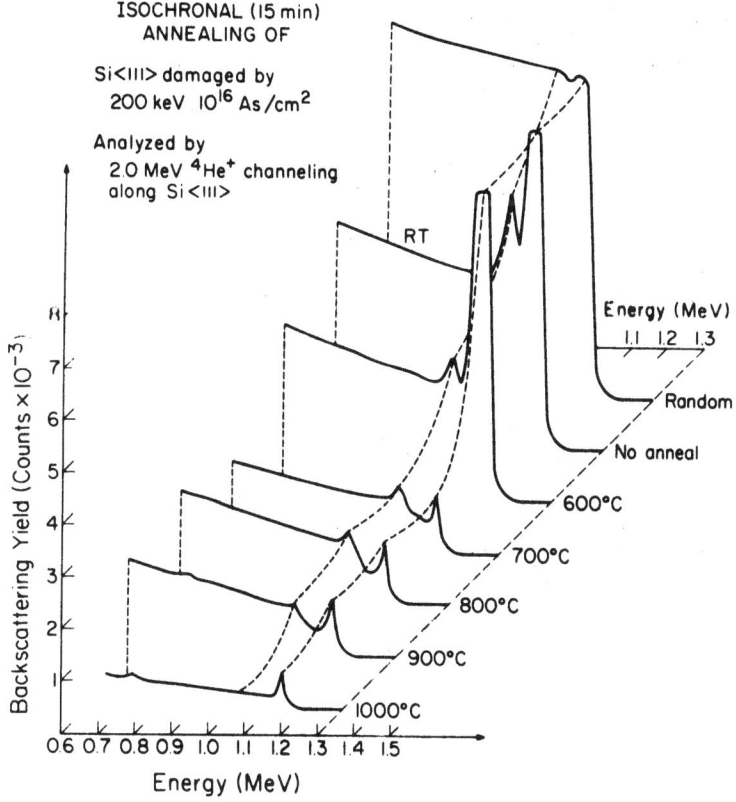

Fig. 1

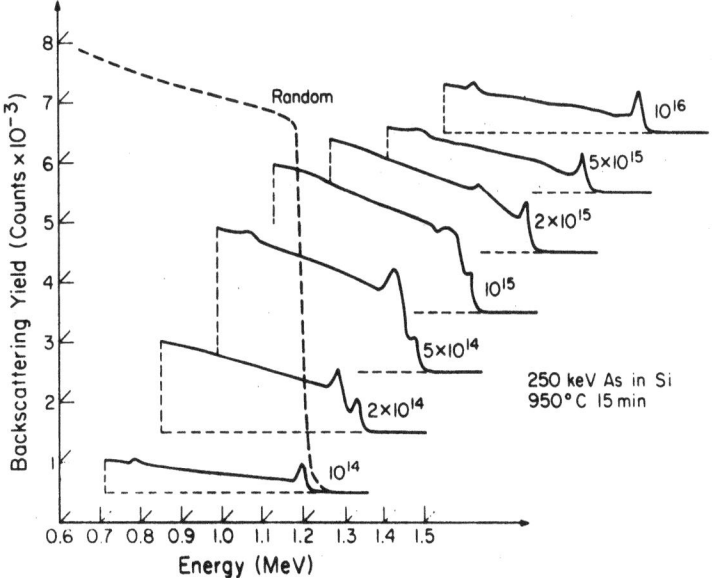

Fig. 2

order. However, it is seen that in the dose region between 2×10^{14} to $2 \times 10^{15}/cm^2$, a significant amount of disorder exists. Again, there appear two damaged regions; one located at $\simeq R_p$ ($\simeq$ 1400Å for 250 keV As in Si) and the other at $\simeq R_p + \Delta R_p$. The amount of damage observed in each depends upon the implanted ion dose.

The dose dependence of disorder has been investigated at implantation energies of 50, 100, 150, 200 and 250 keV. The most striking result is the absence of the residual disorder for the 50 keV arsenic implants after a 950°C anneal. This occurred over the entire dose range studied. For the 100 keV implants the maximum amount of residual damage is found at a dose of $2 \times 10^{14} cm^2$. For 200 and 250 keV implants the maximum amount of residual damage was found at a dose of about $5 \times 10^{14}/cm^2$ equivalent to an As concentration of about $5 \times 10^{19}/cm^3$.

The energy dependence of the residual damage for $5 \times 10^{14}/cm^2$ implants is given in Fig. 3. For the 50 keV implant, no damage is observed. The surface peak is due to a thin layer of oxide grown during annealing. For 200 keV implants the aligned yield at the peak of the damage approaches half of the random yield. For 250 keV implant, the peak of the damage is lower ($\sim$ 1/3 of the random yield) and appears at a larger depth.

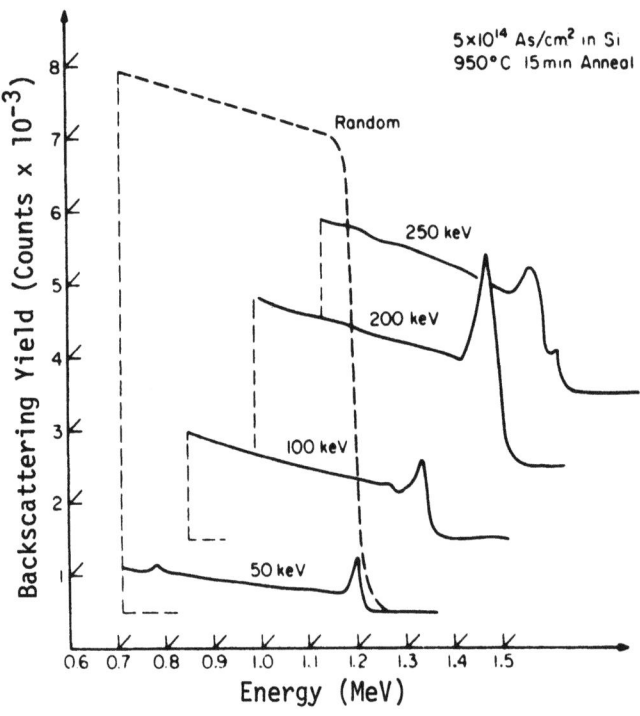

Fig. 3

C. Implantation and Anneal Temperature Dependence

In the intermediate dose range, the residual damage appears to be relatively stable for anneal temperatures between 950 and 1100°C. Figure 4 shows aligned spectra for a 5×10^{14} As/cm^2, 200 keV R.T. implant before anneal and after anneal at 1000 and 1100°C for 30 min. For the non-annealed sample the aligned spectra in the implanted region coincides with the random spectra. The same results were found for 200 keV implants at doses between 2×10^{14} and 5×10^{15} As/cm^2. As shown in Fig. 4, appreciable disorder remains even after anneal at 1100°C.

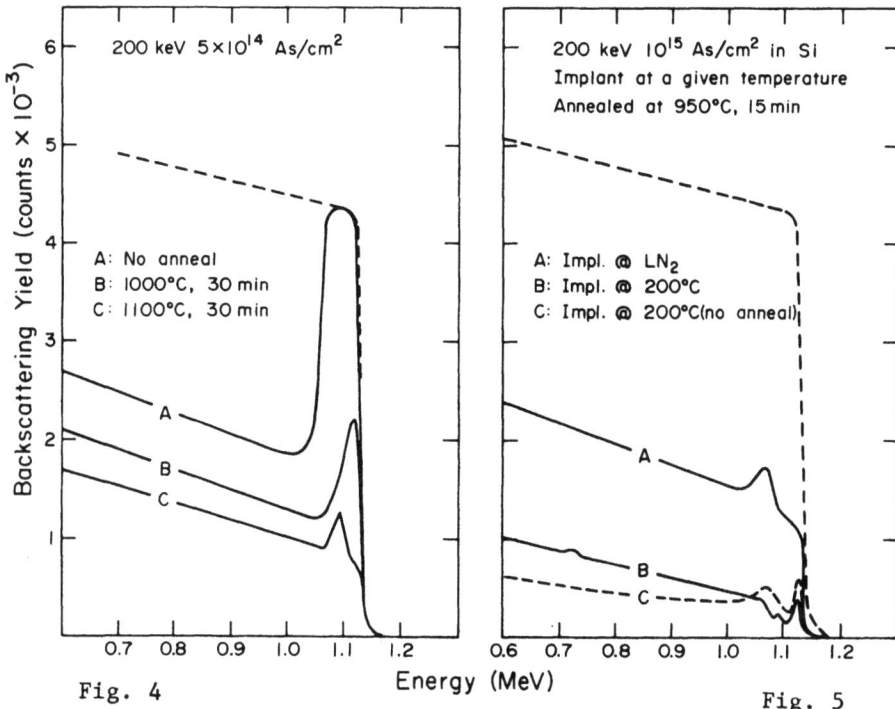

Fig. 4   Fig. 5

Residual disorder is found in samples implanted at LN$_2$ temperatures in the intermediate dose region. Figure 5 shows the aligned spectra for a sample implanted with 5×10^{14} As/cm^2 at 200 keV. The high residual disorder shown in Fig. 5 was also found in comparison with samples implanted at Hughes Research Laboratories at temperatures of 150 to 160°K (5×10^{14} As/cm^2, dose rate 0.6 and 0.06 µA/cm^2).

The amount of disorder can be reduced significantly by implantation at elevated substrate temperatures. The lower curves in Fig. 5 show the aligned spectra for a 200 keV, 10^{15} As/cm^2 sample implanted at 200°C. The spectra before anneal shows some residual disorder at $R_p + \Delta R_p$ that remains nearly unchanged for an anneal at 900°C.

D. Low Temperature Oxidation

Growth of oxide layers at 850°C in steam is sufficient to remove residual disorder if the amount of Si consumed is greater than $R_p + \Delta R_p$. The disorder can be reduced by growth of thinner oxide layers. Figure 6 shows aligned spectra for a 10^{15} As/cm^2 implant following anneal at 950°C and after 600 and 800Å of Si have been removed by oxide growth and HF stripping. The data show that the disorder is reduced but not completely eliminated.

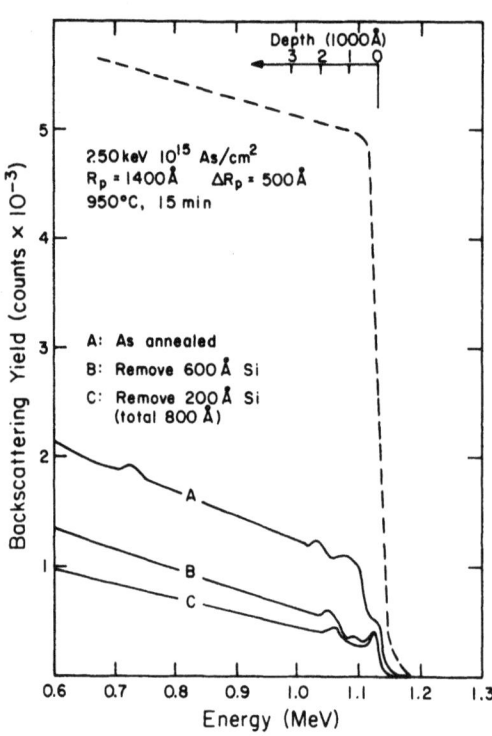

Fig. 6

CONCLUSION

The results obtained for the intermediate dose range of 2×10^{14} to 5×10^{15} cm^{-2} at energies between 100 keV and 250 keV suggest that special care has to be taken when applying high energy implants to devices. A possible alternative to circumvent these problems is to use low energy implantation combined with a following drive in diffusion. Release of residual disorder can be enhanced by carrying out the annealing in an oxidizing atmosphere.

REFERENCES

1. R. S. Payne, R. J. Scavuzzo, K. H. Olson, J. M. Nacci, R. A. Moline, IEEE Trans. E.D. **21** (No. 4) 273 (1974).

2. F. F. Morehead, Jr. and B. L. Crowder, Proc. First Intern. Conf. on Ion Implant. (Gordon and Breach, 1971).

DEVIATED GAUSSIAN PROFILES OF IMPLANTED BORON

AND DEEP LEVELS IN SILICON

Y. Ohmura, K. Koike and H. Kobayashi

Toshiba R&D Center, Komukai, Saiwai-ku, Kawasaki 210

K. Murakami

Toshiba Oita Works, Matsuoka, Oita

ABSTRACT

Deviation from Gaussian profiles for both ^{11}B and ^{10}B implanted in single crystal Si has been investigated for from 60 to 200 keV using the C-V method. Projected ranges for lighter ^{10}B have been found to be slightly smaller than for ^{11}B, which is explained as being due to the larger electronic stopping. Near the concentration peak (from the top to $\sim$ 1/e), profiles are fairly symmetric, which leads to the smaller central third moment compared with semi-empirical LSS calculation by Mylroie and Gibbons.
Deep hole trap distributions due to ^{11}B and ^{10}B have also been obtained with C-V measurement.

INTRODUCTION

Calculations of the first, second and higher order moments for concentration profiles of both implanted species and produced defects have already been published by several authors [1,2]. Experimental deviated Gaussian profiles have been determined independently for ^{11}B and ^{10}B implanted at various acceleration energies in Si [3]. Implantation induced defect profiles have, so far, been determined on high dose implanted samples by back scattering technique [4].

The purpose of this paper is to report on systematic investigation and directly compare the deviated Gaussian profiles for both ^{11}B and ^{10}B implanted in single crystal Si by the C-V method.

For defect profiles, deep hole trap profiles produced by ^{11}B and ^{10}B implantation with doses of as low as 1-10 x 10^{10}/cm^2 will be given.

EXPERIMENTAL

Samples used were chemi-mechanically polished, high resistivity (more than 40 Ωcm), (111) plane oriented p-type Si wafers. In comparison between ^{11}B and ^{10}B ranges, 10-20 Ωcm (100) plane p-type Si wafers with ∿1000Å thermally grown SiO$_2$ films were also employed, to exclude the channeling effect. The ^{11}B and ^{10}B implantations were performed with doses of 2 ∿ 3.5 x 10^{12}/cm^2 using three different, magnetically mass-separated implantation machines, in order to confirm that the depth profile is independent of machines. All the Si wafers were implanted with <111> or <100> axes 7° off the ion beam direction, respectively. After implantation, the samples were annealed for 20 minutes in pure N$_2$ at 800°C. The above-mentioned SiO$_2$ films, grown on Si wafers, were removed with NH$_4$F solution prior to annealing. As will be seen later, integrated carrier concentrations per unit area are the same as implanted doses, which implies that the 800°C annealing temperature is high enough to activate the implanted B. After annealing, Al was evaporated onto the back surface of the wafer and sintered at 500°C. Ti was then evaporated onto the front surface of the wafer, through a metal shadow mask, to form a Schottky barrier electrode. Electrode diameter was about 500 μm. The diameter was constant to within about 2 % throughout the whole measurement. The carrier concentration profile was determined using a Copeland-type automatic C-V profile plotter [5] on the above-mentioned Schottky barrier diode. All the data were obtained within the central circular region (∿10 mm in diameter) of the 50 mm wafer to compare the profile under the same conditions in any respect.

The 0.2—0.3 Ωcm low-resistivity (111) plane p-type Si wafers were employed for deep hole trap profile measurement. Al was also evaporated and sintered on the back surface of the wafer prior to implantation. The implanted dose ranged from 1 to 10 x 10^{10}/cm^2.

RESULTS AND DISCUSSIONS

a. Boron Profiles

Figure 1 shows typical concentration profiles for ^{11}B (solid line) and ^{10}B (dotted line) implanted at 80, 160 and 200 keV with three different implanters, respectively. First, the depth difference should be noted. Profiles for both kinds of ions are separated and those for lighter ^{10}B are shallower, although the differences are very small. The (100) plane Si wafers with 1000Å SiO$_2$ film were employed to completely eliminate the possibility that the depth difference is caused by some channeling effect. These Si wafers with SiO_2 film were cut into two halves. One half was implanted with ^{11}B and the other half with ^{10}B at 190 and 200 keV, respectively. The shallower profiles for ^{10}B were also observed on these wafers. This projectile ion isotope effect can be interpreted qualitatively, within the framework of LSS theory [6], as follows:

The lighter isotope projectile has the longer range, when only the nuclear stopping is taken into account. On the other hand, when both nuclear and electronic stoppings exist, the range for the lighter isotope can be smaller.

The depth relationship for ^{11}B and ^{10}B in this experiment gives strong evidence that the electronic stopping is the dominant mechanism for relative ranges between ^{11}B and ^{10}B. Details on the depth difference will be published elsewhere [7].

To obtain the third moment from the experimental data, a simple two half-Gaussian method by Gibbons and Mylroie [8] has been adopted. Because of the limitation in the Schottky barrier C-V method, the carrier concentration can be determined for only the small range from the peak. As is seen in Fig. 1, the profiles are fairly symmetric from the top to $\sim 1/e$ of the peak for both ^{11}B and ^{10}B in the acceleration energy range used. Figures 2a and 2b show the projected range, the standard deviation and the absolute value of CM_{3p}, the cubic root of the central third moment, for ^{11}B and ^{10}B, respectively. The sign of CM_{3p} is negative. Three lines in Fig. 2a are calculated ranges by Mylroie and Gibbons [2]

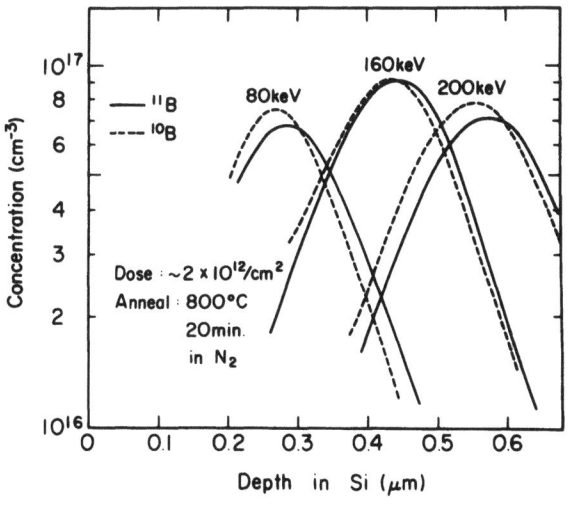

Fig. 1 Concentration profiles for ^{11}B (solid line) and ^{10}B (dotted line) implanted in Si at 80, 160 and 200 keV with three different implanters, respectively.

using the empirical electronic stopping suggested by Eisen [4]. Because of the logarithmic plot, the small difference between ^{11}B and ^{10}B ranges could not be shown in the two figures. It can be pointed out that one could not obtain the correct relationship between ^{11}B and ^{10}B ranges, if one uses the same empirical electronic stopping for ^{11}B and ^{10}B that has no correct dependence on projectile ion and target atom mass number, as in LSS's theory [6]. The projected ranges, the standard deviations and CM_{3p} in the figures are obtained at $1/\sqrt{e}$ of the peak. The magnitudes of projected ranges and standard deviations for ^{11}B and ^{10}B are in fair agreement with those of independently reported depth profiles [9,10], except for the small difference between ^{11}B and ^{10}B projected ranges. The central third moments obtained from around the peak are small compared with the calculation [2] and the high dose B profiles implanted in amorphous Si [3].

b. Deep Level Profiles

Figure 3a shows the carrier concentration profiles for unimplanted and implanted regions. The reduction of carrier concentration in the implanted region is caused by deep hole traps produced by B implantation. About 50-70 carriers per ion were reduced for 200 keV ^{10}B implantation. Figure 3b shows deep hole

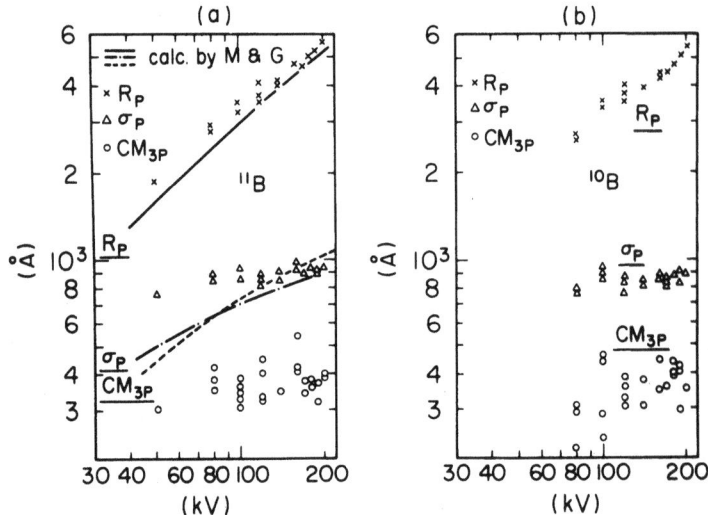

Fig. 2 Plots of R_p, σ_p and the absolute value of CM_{3p} ($CM_{3p} < 0$) for ^{11}B (a) and ^{10}B (b) as a function of acceleration energy. Three lines in (a) show calculations by Mylroie and Gibbons [2].

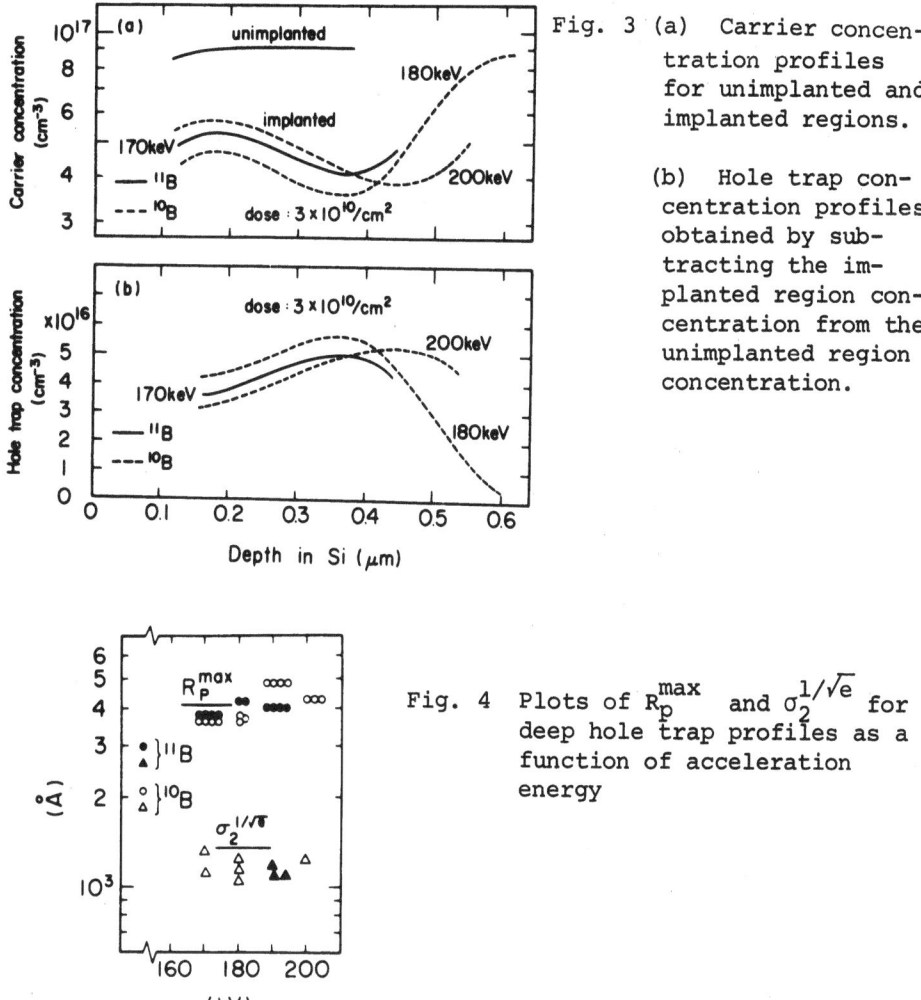

Fig. 3 (a) Carrier concentration profiles for unimplanted and implanted regions.

(b) Hole trap concentration profiles obtained by subtracting the implanted region concentration from the unimplanted region concentration.

Fig. 4 Plots of R_p^{max} and $\sigma_2^{1/\sqrt{e}}$ for deep hole trap profiles as a function of acceleration energy

trap depth profiles obtained by subtraction of implanted region concentration from unimplanted region concentration. The hole trap profiles are in qualitative agreement with theory [11] and defect profiles by other techniques [4]. The profiles are too much skewed to express them with only three moments. In Fig. 4, R_p^{max} and $\sigma_2^{1/\sqrt{e}}$, the standard deviation at $1/\sqrt{e}$ of the peak on the deeper side, are plotted as a function of acceleration energy. There may be some difference between R_p^{max} of hole traps for ^{11}B and ^{10}B. However, R_p^{max} of these flat-topped profiles are so scattered that the difference cannot be confirmed.

SUMMARY

Direct comparison and systematic investigation have been made for implanted ^{11}B and ^{10}B profiles in Si by the C-V method. It was found that the projected range for the lighter ^{10}B is slightly smaller than for ^{11}B. This projectile ion isotope effect can be interpreted as being due to the larger electronic stopping. Profiles for both ^{11}B and ^{10}B are fairly symmetric, which results in smaller central third moments compared with semi-empirical LSS calculations. Implantation induced deep hole trap profiles are extremely skewed.

REFERENCES

[1] K. B. Winterbon, Rad. Effects 13, 215 (1972).
[2] S. Mylroie and J. F. Gibbons, Proc. of International Conference on Ion Implantation in Semiconductors and Other Materials, Ed. B. L. Crowder, Plenum press N.Y. (1973), p.243.
[3] For example, B. L. Crowder, J. Electrochem. Soc. 18, 943 (1971).
[4] For example, F. H. Eisen, B. Welch, J. E. Westmoreland and J. W. Mayer, Proc. International Conference on Atomic Collision Phenomena in Solids, Univ. of Sussex (1969), p 111.
[5] J. A. Copeland, IEEE Trans. Electron Dev. ED-16, 445 (1969).
[6] J. Lindhard, M. Scharff and H. E. Schiott, Mat. Fys. Medd. Dan. Vid. Selsk. 33 (No. 14), 1 (1963).
[7] Y. Ohmura and K. Koike, to be published.
[8] J. F. Gibbons and S. Mylroie, Appl. Phys. Lett. 22, 568 (1973).
[9] T. E. Seidel, Proc. II. International Conference on Ion Implantation in Semiconductors ed. I. Ruge and J. Graul (Springer-Verlag, N.Y., 1971), p 47.
[10] J. F. Ziegler, B. L. Crowder, G. W. Cole, J. E. E. Baglin and B. J. Masters, Appl. Phys. Lett. 21, 16 (1972).
[11] D. K. Brice, Rad. Effects, 6 77 (1970).

REDISTRIBUTION OF BORON IN SILICON THROUGH HIGH TEMPERATURE

PROTON IRRADIATION

P.Baruch[*], J.Monnier[**], B.Blanchard[***], C.Castaing[*]

[*] Groupe de Physique des Solides de l'E.N.S., Tour 23, 2 place Jussieu, 75221 Paris Cedex 05
[**] L.E.T.I., Centre d'Etudes Nucléaires de Grenoble, B.P. 85, Centre de Tri 38041 Grenoble Cedex
[***] S.L.A.G.-E.A.P.C., Centre d'Etudes Nucléaires de Grenoble, B.P. 85, Centre de Tri, 38041 Grenoble Cedex

The irradiation defects are usually assumed to act on impurity diffusion processes through an increase of the effective jump frequency, in a vacancy diffusion mechanism. We have shown that, at least for boron in silicon, the usual enhanced diffusion mechanism is swamped by another interaction mechanism, which causes depletion of impurity in some parts of the crystal, and accumulation elsewhere (1).

The experiment is identical with a radiation enhanced diffusion experiment : high temperature (850 C) irradiation, with protons of energy 150 to 400 keV, at fluences 1 to 6×10^{17} p/cm^2. The boron profile has been measured with secondary ion mass spectrometry (SIMS) and nuclear reactions.

Fig.1 shows the effect of 880 C irradiations with a fluence of 4×10^{17} p/cm^2 on a prediffused sample. The initial diffusion had yielded a surface concentration of 2×10^{20} B/cm^3, a layer depth near 4 µm as shown on curve 1, fig.1. The irradiations were performed at two different energies 250 keV (curves 2 and 2') and 400 keV (curve 3), and the profile was analyzed through SIMS (curves 2 and 3) or through the nuclear reaction $B^{11}(p,\alpha) Be^8$ (curve 2'). It is seen that the post-irradiation profiles show slope reversals near 2 and 4 µm, depth which are very close to the projected ranges - 2.4 µm and 4.3 µm of 250 and 400 keV protons. A peak in boron concentration appears, and then the concentration resumes its normal fall. The good agreement between SIMS and nuclear reaction profiles excludes the possibility of an artefact due, for example, to damage enhancement of ion yield in SIMS measurement.

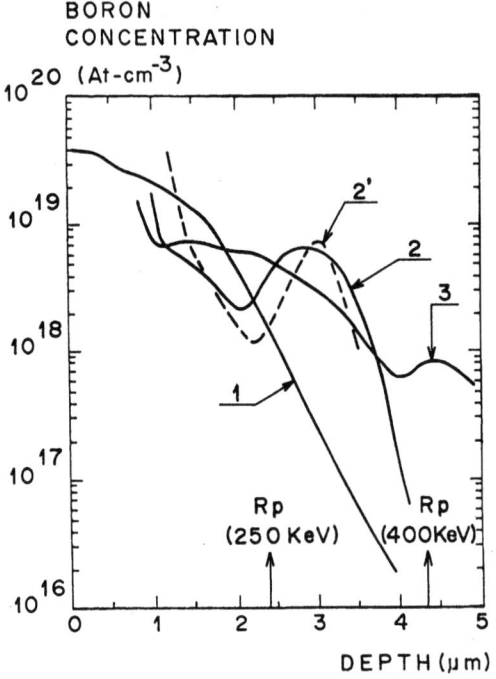

Fig.1. Boron concentration profiles, obtained through SIMS (curves 1,2,3) or through the $B^{11}(p,\alpha)Be^8$ reaction (curve 2'). 1 : initial profile (prediffused). 2-2' : 880 C proton irradiation ; energy 250 keV ; fluence 4×10^{17}p/cm^2. 3 : 880 C proton irradiation ; energy 400 keV ; fluence 4×10^{17}p/cm^2.

The profile anomaly appears superimposed on the normal widening of the initial profile, as expected from a "normal" enhanced diffusion effect, but, near R_p, the boron atoms flow against the concentration gradient showing "up-hill" diffusion.

Since this effect does not depend, apparently, on the initial concentration gradient, it could be expected that, for an initially uniform sample, the irradiation could also produce a change in the impurity concentration – which should not be expected at all from a normal diffusion process.

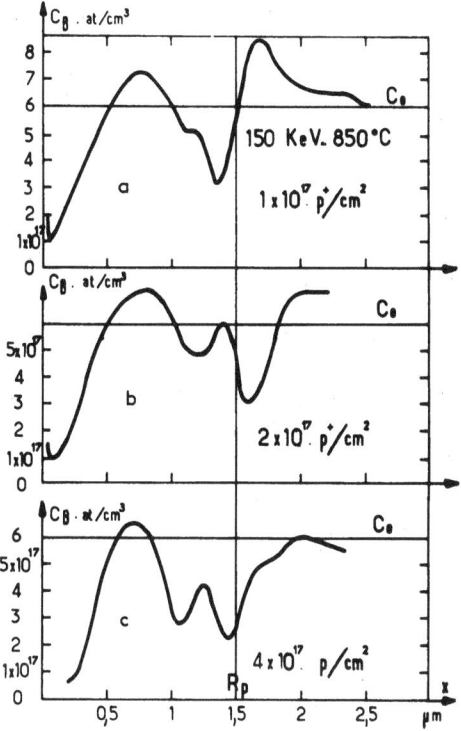

Fig.2. Boron concentration profile (SIMS) of an initially uniform sample ($C_o = 6\times10^{17}$ B/cm^3) irradiated at 850 C by 150 keV protons. a) fluence 1×10^{17} p/cm^2. b) fluence 2×10^{17} p/cm^2. c) fluence 4×10^{17} p/cm^2.

Indeed, violent oscillations show up (Fig.2) in the boron concentration of samples melt-doped with a boron concentration of 6×10^{17}cm^{-3}, irradiated at 850 C by 150 keV protons, with fluences 1×10^{17} (fig.2a), 2×10^{17} (fig.2b), 4×10^{17} protons/cm^2 (fig.2c). At the lowest fluence (1×10^{17} p/cm^2), a dip appears at R_p, with shoulders (fig.2a). Then, as the fluence increases (fig.2b, 2c), a smaller peak grows from the bottom of the valley. There are indications that the electrical activity profile follows the total boron profile.

Since this anomaly shows up near R_p, where the displacement rate is highest, it is connected with the high density of defects during irradiation ; however, since the irradiation is performed at high temperature, there is almost no residual damage.

We propose that the effect can be attributed to the transformation of substitutional boron in other boron-defect associations, with different diffusion coefficients :

a) formation of boron interstitials near R_p, either through direct displacements or through the Watkins interchange mechanism (2) They diffuse out from R_p with a high diffusivity, while substitutional boron flows towards the sink at R_p. A crude calculation shows this model is compatible with the profile 2a (including outdiffusion to the surface).

b) formation of immobile complexes (precipitates, dislocation trapping, etc...?) at higher fluences, giving a boron pile-up in the center, and adding to the depletion on the sides (2b, 2c), in competition with the first mechanism.

Similar effects have also been observed (3,4) in the case of implanted boron, after annealing, and can probably be ascribed to the same mechanisms.

A more detailed description is being published elsewhere (1,5).

REFERENCES

(1) P.Baruch, J.Monnier, B.Blanchard, C.Castaing, submitted to Applied Physics Letters.
(2) G.D.Watkins, in Radiation Effects in Semiconductors, p.97 (Dunod, Paris, 1965).
(3) D.Dieumegard, L.M.Mercandalli, M.Croset, SIMS Meeting of A.N.R.T. (Orsay, 1973 ; unpublished).
(4) W.K.Hofker, H.W.Werner, D.P.Oosthoek, M.A.M. de Greft, Appl. Phys. 2, 265 (1973).
(5) P.Baruch, J.Monnier, B.Blanchard, C.Castaing, communication at the International Conference on Defects in Semiconductors (Freiburg, July 1974), to be published by the Institute of Physics (London).

RANGE DISTRIBUTIONS OF BORON IN SILICON DIOXIDE AND THE UNDERLYING SILICON SUBSTRATE

K. Wittmaack, F. Schulz and B. Hietel

Gesellschaft für Strahlen- u. Umweltforschung mbH
Physikalisch-Technische Abteilung
D-8o42 Neuherberg, Germany

ABSTRACT

Range distributions of 5 - 15o keV boron ions in silicon dioxide have been measured by means of secondary ion mass spectrometry. Mean projected ranges are in good agreement with predictions of the LSS theory (maximum deviation 1o%). Experimental range straggling values are more than 2o% larger than predicted. Moreover the distributions show pronounced deviations from a Gaussian. Comparison with boron range in amorphous silicon indicates 1o to 15% larger mean projected ranges in silicon dioxide. Amorphous oxide layers on single crystal silicon do not prevent the occurrence of channelling tails in the substrate.

INTRODUCTION

The preceding conference on ion implantation in semiconductors and other materials /1/ demonstrated that more experimental data on range distributions of energetic ions in solids are desirable. Compared to monatomic targets very little interest has been devoted to a study of the penetration of ions in compound semiconductors or insulators. The latter are of particular importance in device fabrication where they are used as masking materials or protective surface coatings during implantation doping.

Chu et al. /2,3/ have measured projected range and range straggling of heavy ions in SiO_2, Si_3N_4 and Al_2O_3 at energies between 15o and

3oo keV by the backscattering technique. The projected range values were systematically greater than LSS calculations by a factor of 1.2 to 1.5. The deviations were even more pronounced for the straggling data. Very few investigations have been carried out on range distributions of light ions such as boron in insulators. Volod'ko et al. /4/ deduced range profiles of 3o to 1oo keV boron ions in SiO_2 from measurements of the electrical activity introduced in silicon by implantation through oxide layers of various thicknesses. Combasson et al. /5/ applied in-depth profiling by secondary ion analysis to study boron range in SiO_2 and Si_3N_4, the profile sometimes extending into the silicon substrate.

The purpose of this investigation is to study the energy dependence of boron range distributions in silicon dioxide and to compare the results with theoretical estimates and recent experimental data on boron in silicon. In addition range distributions obtained when implanting through oxide layers are investigated.

EXPERIMENTAL

Range distributions were measured using the ultrahigh vacuum version of our high sensitivity secondary ion in-depth analyzer /6,7/. Either Ar^+ or a mixture of N_2^+ + NO^+ /8/ with energies between 3 and 1o keV were used as primary ions. The sputtering rate varied between o.1 and 1 Å/s. Charge built-up on the oxide specimens was avoided by additionally bombarding the target with a low energy electron beam (3oo eV).

Silicon dioxide films on (111) silicon crystals (n-type, 1 Ω cm, 1o mm diameter) were prepared by two different oxidation techniques. Layers with a thickness of up to 15oo Å could produced by anodic oxidation /9/, whereas thicker layers were obtained by thermal oxidation /1o/. Thicknesses of the films were routinely determined by the talysurf method. Some calibration measurements were also carried out by ellipsometry. The target density of films prepared by anodic oxidation was determined by the backscattering technique.

To allow an accurate determination of the sputtering rate during secondary ion analysis the specimen configuration shown in Fig. 1 was used. Except for the 3 mm diameter central part all the surface oxide was etched away. The implantation area covered a concentric 2 mm diameter area. The constant current density sputtering beam was 4 mm in diameter. From the changes in surface relief produced during in-depth profiling the sputtering rates can be determined for silicon dioxide as well as for silicon. Therefore it is possible to calibrate the depth accurately even in case that profiles extend through the oxide.

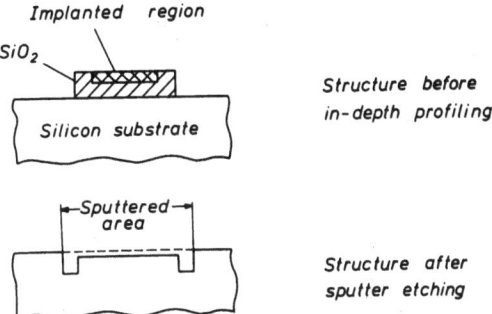

Fig. 1. Specimen configuration before and after in-depth analysis.

The pre-equilibrium variation of the secondary ion yield could be accounted for by subtracting the yield obtained with non-implanted dummies. The boron concentration was calibrated by use of an oxidized boron doped specimen (silicon substrate o.oo4 Ω cm, p-type).

RESULTS AND DISCUSSION

Range profiles of boron in silicon dioxide (layer thickness o.8μm) are presented in Fig. 2. As in the case of boron in amorphous silicon the profiles are skewed towards the surface. The tails on either side of the distribution drop exponentially, i.e. proportional to $\exp(\pm x/\lambda)$. The energy dependence of the slope constant λ as determined from Fig. 2 is shown in Fig. 3. The data for 5 and 1o keV are also included for the long range tail. The short range tailing could not be determined in that case because the tail is outside the surface.

As one can see from Figs. 2 and 3 the boron concentration decreases more rapidly on the long range side of the distribution, the ratio of the corresponding λ-values amounting more than a factor of three at 15o keV. Towards lower energies the difference becomes less pronounced. Extrapolation in Fig. 3 indicates that the boron range distribution might become symmetrical around 5 keV.

From the measured range distributions not only peak position and profile width but also mean projected range <x> and range straggling $\sigma = <\Delta x^2>^{1/2}$ could be evaluated. This was done by fitting Type IV Pearson distributions /11/ to the measured data. Thus moments up to the fourth were included. Fig. 4 gives an example which demonstrates that this type of distribution fits the experimental points very well. The agreement to be obtained with fourth order Gram-Charlier expansions is much worse. The reduced third

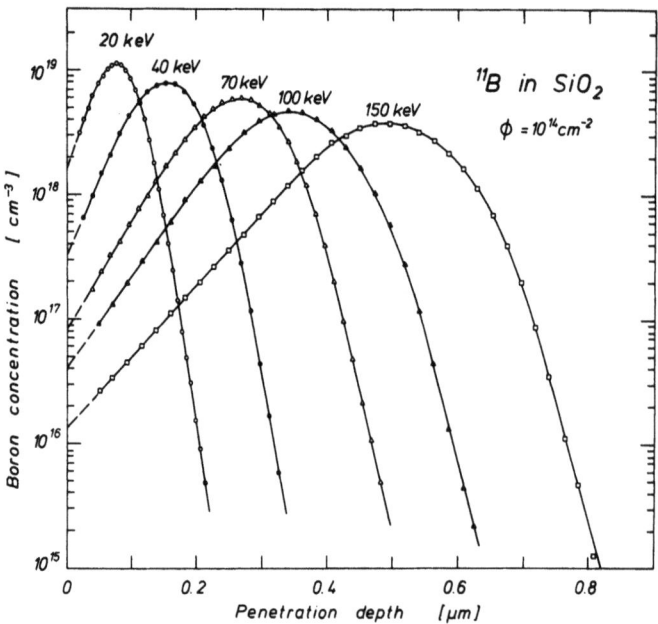

Fig.2. Boron range distributions in thermally grown silicon dioxide.

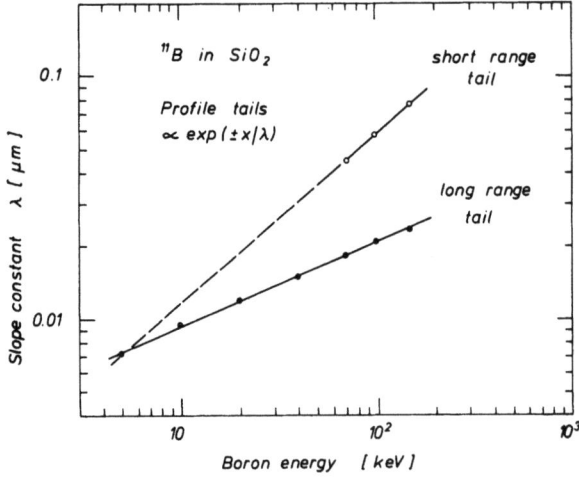

Fig.3. Energy dependence of the slope constant for the exponential tails of the range distributions of boron in silicon dioxide.

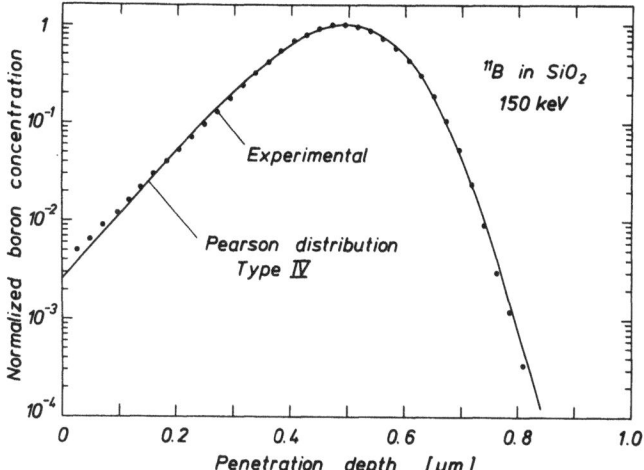

Fig.4. Comparison between experimental range distribution of boron in silicon dioxide and a fitted Pearson distribution.

and fourth moment were found to be $<\Delta x^3>/<\Delta x^2>^{3/2} = -0.7$ and $<\Delta x^4>/<\Delta x^2>^2 = 4.4$ at 150 keV. More details concerning the energy dependence of higher order moments will be reported elsewhere /12/.

The energy dependence of mean projected range and range straggling of boron in silicon dioxide is compared in Fig. 5 with data for an amorphous silicon substrate /13,14/[x]. As one can see range and straggling are always somewhat (10-15%) larger in silicon dioxide than in amorphous silicon. The general trend of the energy dependence, however, is very similiar for either target.

Our results on boron range distributions in silicon dioxide may be compared with some recently reported data /4,5/. Very good agreement is found with respect to either most probable or mean projected range. The range straggling reported by Volod'ko et al./4/ is much (~50%) larger than in our case, the broadening most likely being due to insufficient depth resolution ($\geq 0.02\mu m$). The range straggling reported by Combasson et al. /5/ for 60 keV is about 10% smaller than expected from Fig. 5. However, the Edgeworth expansion

---

[x] The range data in silicon were taken from measurements published recently /13,14/. Unfortunately it come out somewhat later that the implantation energy was 8% smaller than the indicated energy, e.g. 100 keV in ref. /13,14/ should read 92 keV. Fig. 5 shows energy corrected results.

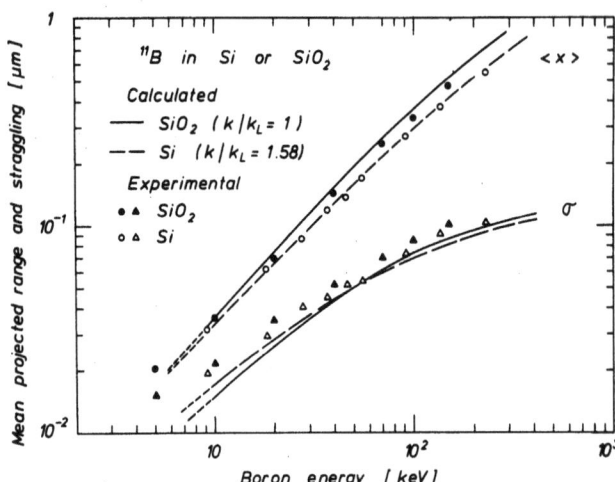

Fig.5. Mean projected range and straggling for boron in amorphous silicon and silicon dioxide. Theoretical curves according to ref. /15/ (silicon dioxide) and ref. /17-2o/ (silicon).

used in ref. /5/ does not seem to fit the experimental data very well. Moreover there were shoulders on the profile which have not been found in our investigations.

For comparison between experimental and theoretical results tabulated data /15/ were used after normalizing to the density of thermally grown silicon dioxide /2,3/. The calculations /15/ are based on the LSS theory /16/. The theoretical curve is very close to the experimental points, the maximum deviation amounting only 1o%. Boron range in silicon can be described very well /17-2o/ by using an electronic stopping constant k which is 1.58 times larger than the Lindhard values k_L /16/.

The agreement between experimental and theoretical results is not as good for the range straggling. For silicon dioxide the straggling is found to be at least 2o% larger than calculated. This deviation may support recent critism /3/ concerning the accuracy of the tabulated straggling values.

In addition to the above studies we have measured range distributions after implantation through the oxide. An example is shown in Fig. 6. In the region of random range the profile looks as expected. However, irrespective the fact that the beam had passed through a 0.04 μm silicon dioxide layer, there is a channelling tail if the implantation is carried out normal to the (111) silicon substrate.

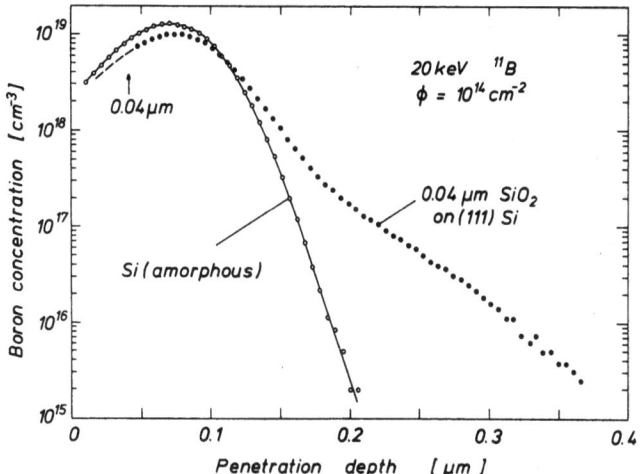

Fig.6. Comparison of boron range in amorphous silicon and in oxide covered single crystal silicon.

According to the results of Fig. 6 one has to be aware of a channelling contribution even in case that multiple scattering /21/ has produced a broad angular distribution of the ions entering the single crystal target. A similar result has been observed for the range distribution of 4o keV xenon ions in silicon covered with a 0.018 μm layer of silicon dioxide /22/.

Acknowledgements

We thank Miss E. Schneider for carefully preparing the specimens, P. Blank for carrying out the backscattering measurements, and J. Meier for operating the accelerator IB 3oo. We are indebted to H. Boroffka, Siemens, Munich, for supplying ellipsometry facilities.

REFERENCES

/1/ Proc. Third Int. Conf. Ion Implantation in Semiconductors and Other Materials, ed. by B.L. Crowder, Plenum Press, New York (1973).
/2/ W.K. Chu, B.L. Crowder, J.W. Mayer, and J.F. Ziegler, in ref. /1/, p. 225.
/3/ W.K. Chu, B.L. Crowder, J.W. Mayer, and J.F. Ziegler, Appl. Phys. Lett. 22 (1973) 49o.
/4/ V.G. Volod'ko, E.I. Zorin, P.V. Pavlov, and D.I. Tetel'baum, Soviet Physics-Solid State lo (1969) 828.

/5/ J.L. Combasson, J. Bernard, G. Guernet, N. Hilleret, and M. Bruel, in ref. /1/, p. 285.
/6/ J. Maul, F. Schulz, and K. Wittmaack, Adv. Mass Spectrometry VI (1974) 493.
/7/ K. Wittmaack, J. Maul, and F. Schulz, Proc. Sixth Int. Conf. Electron and Ion Beam Science and Technology, in press.
/8/ K. Wittmaack, submitted to Int. J. Mass Spectrom. Ion Phys.
/9/ W. Przyborski, J. Roed, J. Lippert, and L. Sarhold-Kristensen, Rad. Effects 1 (1969) 33.
/1o/ A.S. Grove, Physics and Technology of Semiconductor Devices, J. Wiley, New York (1967).
/11/ M.G. Kendall and A. Stuart, The Advanced Theory of Statistics, Vol. 1, C. Griffin & Co., London (1958).
/12/ K. Wittmaack and F. Schulz, to be published.
/13/ K. Wittmaack, J. Maul, and F. Schulz, in ref. /1/, p. 119.
/14/ K. Wittmaack, F. Schulz, and J. Maul, Phys. Letters 43A (1973) 477.
/15/ W.S. Johnson and J.F. Gibbons, Projected Range Statistics in Semiconductors, distr. by Stanford University Bookstore (1969).
/16/ J. Lindhard, M. Scharff, and H.E. Schiott, Mat. Fys. Medd. Dan. Vid. Selsk. 33, No. 14 (1963).
/17/ S. Furukawa, H. Matsumura, and H. Ishiwara, Proc. US-Japan Seminar on Ion Implantation in Semiconductors, ed. by S. Namba, Jap. Soc. Promotion Sci. (1972) p. 73.
/18/ D.K. Brice, Rad. Effects 11 (1971) 227.
/19/ S. Mylroie and J.F. Gibbons, in ref. /1/, p. 243.
/2o/ K.B. Winterbon, private communication.
/21/ S. Schwabe and R. Stolle, Phys. Stat. Sol.(b)47 (1971) 111.
/22/ J.A. Davies, G.C. Ball, F. Brown, and B. Dorney, Canad. J. Phys. 42 (1964) 1o7o.

REDISTRIBUTION OF BACKGROUND IMPURITIES IN SILICON INDUCED BY ION IMPLANTATION AND ANNEALING

W. K. Hofker[†], H. W. Werner, D. P. Oosthoek[†], and N. J. Koeman[†]

Philips' Research Laboratories, Eindhoven, Netherlands

ABSTRACT

In previous investigations it was found that ion implantation and annealing caused an initial homogeneous boron dope of silicon to redistribute. The purpose of this study is to investigate these redistribution effects experimentally in more detail.

For that purpose boron redistributions obtained by different experimental conditions are measured. The implanted ion species, the annealing temperature, the crystal direction of the implanted substrate and the background dope concentration are varied. The redistribution of a phosphorus background was investigated, too.

The redistribution effects observed in the boron and phosphorus background dopes are explained by the an-isotropic properties of a crystal lattice containing implanted ions and defect structures.

INTRODUCTION

Redistribution effects of impurities induced by ion bombardment or ion implantation were often investigated with the substrate kept at an elevated temperature during the irradiation process [1, 2, 3]. In these investigations the bombardment or implantation was by means of light ions, such as protons. The observed effects were explained by thermal diffusion, which is enhanced by generated excess vacancies.

[†] Philips' Research Laboratories, Amsterdam Department, Institute of Nuclear Physics, Oosterringdijk 18, Amsterdam, Netherlands

In our investigations the experimental situation is different in the sense that during the implantation the substrate is kept at room temperature. A further difference is that heavy ions such as Ar^+, Sb^+, As^+ are implanted preferably. After the implantation the sample is annealed and the distribution of the background is measured. By comparing this distribution with the original one the redistribution is found.

A study like this is of interest for the understanding of problems in electronic device technology such as those met with in the case of double doping.

Some results of similar experiments were reported earlier by us. In ref. 4 we mentioned the redistribution of boron, observed after a phosphorus implantation in silicon and subsequent annealing at 600°C. In an other paper [5] we described the redistribution of the ^{10}B isotope of a boron background, present in natural isotopic abundance, by a ^{11}B ion implantation. Peculiarly shaped boron distributions were found. They were explained by assuming enhanced diffusion and the occurrence of precipitation. The experimental results of Monfret [6] are also relevant to these results. He found a peculiar deviation in an implanted and annealed boron distribution after arsenic was implanted and the silicon had subsequently been annealed at 800°C.

The purpose of the present study is to investigate experimentally the parameters which are of interest in these redistribution effects. Those included the dependence on the implanted ion species, the annealing temperature, and the background dope concentration.

The investigations were performed with silicon containing a boron background. In order to see whether redistribution effects can be expected at other background impurities as well, the redistribution of a phosphorus background was investigated too.

THE EXPERIMENTAL METHOD

The boron and phosphorus concentrations were measured as a function of depth with Secondary Ion Mass Spectrometry (SIMS). A Cameca secondary ion mass spectrometer, type IMS 300, was used.

The reliability of the boron measurements was discussed in a previous paper [4]. The results of the boron measurements should be considered with some caution due to the fact that precipitation enhances the ion yield [5]. The results in this paper have not been corrected for this effect.

In case of the measurement of phosphorus concentration profiles sputtering occurred with a primary beam of O_2^+ ions and the secondary P^- ion current was measured. The energy of the primary ions was 14.5 keV. This energy is higher than the energy which was used at the boron measurements. In spite of this higher energy the sputtering yield is lower due to an unfavourable electrical field configuration giving a wider angle between the primary ion beam and the substrate surface. In this situation some distortion of the concentration profile may occur due to recoiling effects. The measurements in this paper however are not so critical that this will lead to misinterpretations.

EXPERIMENTAL RESULTS

In a first experiment the influence of an implantation of H^+ ions and of the inert gas ions Ar^+ and Ne^+ on a boron background was investigated. Implantation of inert gas ions is of interest because chemical interactions between the implanted ions and the substrate or the background impurities are unlikely. The ions were implanted at an energy of 70 keV and a dose of 10^{16} ions cm^{-2} in a random direction in Czochralsky-grown silicon, doped with boron of which the concentration of the ^{11}B isotope was 1.3×10^{19} ions cm^{-3}. Annealing was performed at temperatures up to 1000°C in a nitrogen atmosphere during a period of 40 minutes.
For the Ar^+ ion implantation the results obtained after annealing at temperatures in the range 600-900°C are shown in fig. 1.

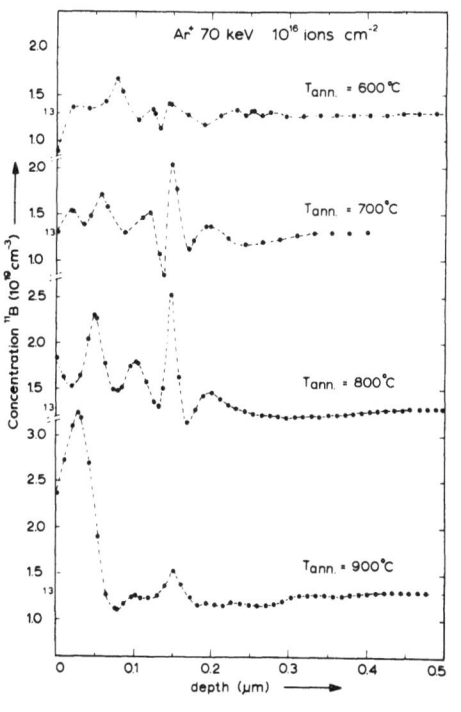

Fig. 1 ^{11}B concentration versus depth after Ar^+ ion implantation (energy 70 keV, dose 10^{16} ions cm^{-2}) and annealing at different temperatures T_{ann}.

At lower annealing temperatures no redistribution effects were found. For the Ne^+ ion implantation only after annealing at 800°C a similar redistribution effect was observed.
For the H^+ ion implantation no effects at all were found.
For the argon implantation a peculiar shaped transient in the boron profile is found at a depth of about 1500 Å (fig. 1).
For the implanted Ar^+ ions the projected range is 700 Å and its standard deviation is 200 Å [7]. Roughly it can be concluded that the transient is situated at the boundary of the implanted and not-implanted region. On both sides of the transient a regular pattern is observed. Often a less pronounced pattern is found. However, from spot to spot on the same slice the boron distributions reproduce in detail.

The dependence on the crystallographic direction of the implanted surface was studied by implanting Ar^+ ions at an energy of 70 keV and a dose of 10^{16} ions cm^{-2} in silicon slices, which were cut perpendicularly (within 0.2°) to a <1 1 0>, <1 1 1> and <9 11 5> direction respectively. Floating zone silicon was used doped by implanting boron (at energies 30, 70 and 150 keV with doses 2×10^{14}, 5×10^{14} and 5×10^{14} ions cm^{-2}, respectively) and annealed at 1000°C.

In these cases identical profiles were obtained, showing that the crystallographic direction of the surface has no influence on the redistribution (fig. 2). The profiles obtained differ from those given in fig. 1. This may be due to different ingot properties.

The influence of the concentration of the boron background on the redistribution effects was investigated by implanting Ar^+ ions in silicon, having a 10 times lower boron concentration than that in the foregoing experiments. Similar redistribution effects were obtained, proving that the level of the boron concentration is not important.

In order to see whether redistribution of background impurities, other than boron, can be expected as well, the influence of an Ar^+ ion implantation on a phosphorus background was also investigated. The phosphorus background was obtained by implantation (energy 70 keV, dose 5×10^{15} ions cm^{-2}).

Fig. 3 shows the phosphorus distribution before and after an Ar^+ ion implantation (energy 70 keV, dose 10^{15} ions cm^{-2}) and subsequent annealing at 700°C. It is observed that in this case too, redistribution of the background occurs, however, the profile is smoother and shows less substructure than is found in the case of a boron background.

REDISTRIBUTION OF BACKGROUND IMPURITIES IN SILICON

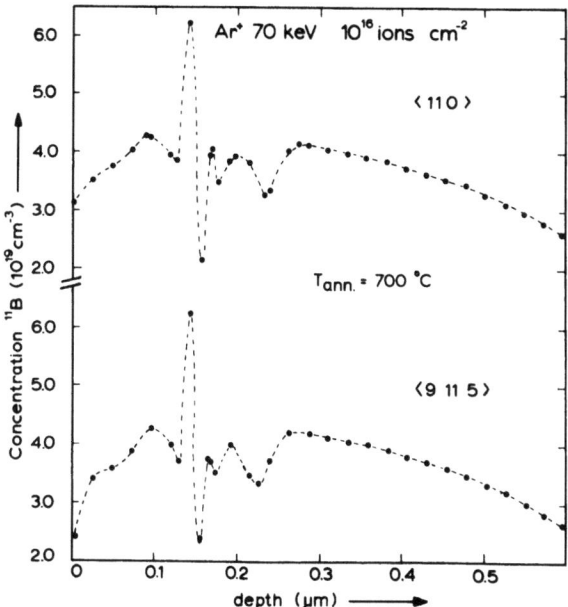

Fig. 2 ^{11}B concentration versus depth after Ar$^+$ ion implantation (energy 70 keV, dose 10^{16} ions cm^{-2}) in silicon with the surface plane perpendicular to a <1 1 0> and <9 11 5> direction, respectively. Annealing was at 700°C. The boron background was implanted.

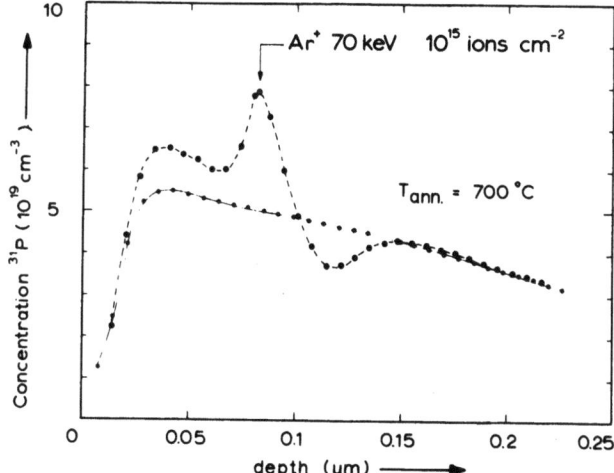

Fig. 3 ^{31}P concentration versus depth before (•——•) and after (•----•) ion implantation (energy 70 keV, dose 10^{15} ions cm^{-2}) and annealing at 700°C.

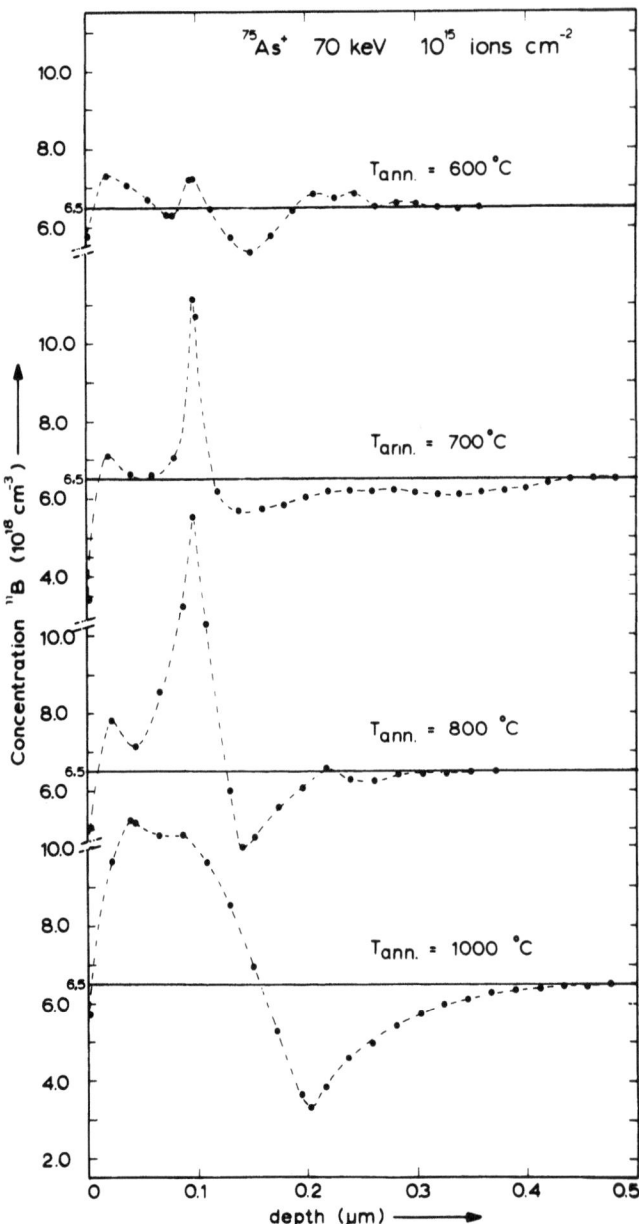

Fig. 4 ^{11}B concentration versus depth after ^{75}As$^+$ ion implantation (energy 70 keV, dose 10^{15} ions cm^{-2}) and annealing at different temperatures T_{ann}.

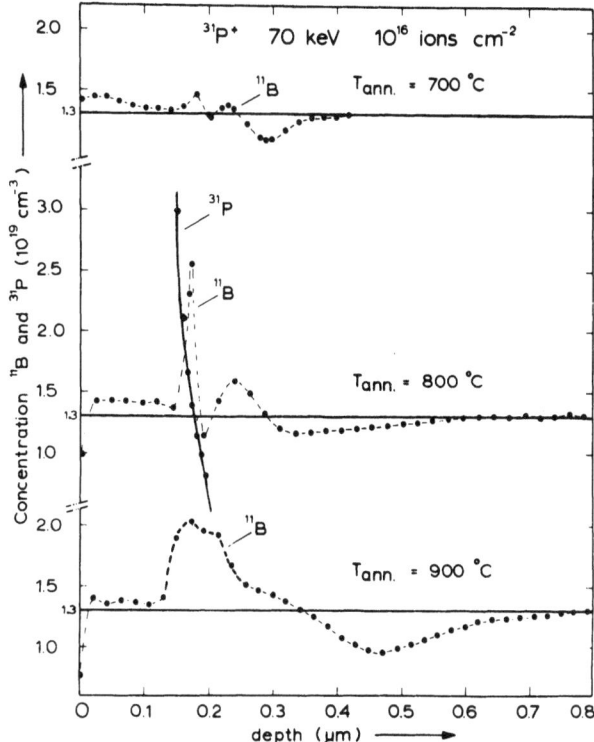

Fig. 5 ^{11}B concentration versus depth after ^{31}P$^+$ ion implantation (energy 70 keV, dose 10^{16} ions cm^{-2}) and annealing at different temperatures T_{ann}.

In the figure the concentration profile of ^{31}P is indicated, too.

In further experiments the influence of As^+ and P^+ ion implantations on a boron background distribution were investigated, because of the special importance of these dopants in electronic device technology.
The results are given in figs. 4 and 5, respectivily.
In the case of the P^+ ion implantation the phosphorus concentration profile was measured too. From the results given in fig. 5 it is observed that the peak in the boron distribution is situated at the boundary of the implanted and not implanted regions.

DISCUSSION AND CONCLUSIONS

From the observation that the redistribution effects increase with the mass of the implanted ions and are independent of the concentration of the background dope, we conclude that the redistribution effects are mainly due to mechanical effects induced by the implanted ions.

With the ions and doses used in this work an amorphous zone is created. Such a zone recrystallizes epitaxially on annealing at about $600°C$. It might be supposed that during this recrystallization process redistribution of the boron occurs by selective segregation. However, only slight redistribution effects are found after annealing at $600°C$. Therefore such an effect is of no interest in this case. However after annealing at temperatures in the range $600-1000°C$ there are still many lattice defects, such as loops, stacking faults, etc. caused by the ion implantation. The dimensions and configurations of these defect structures are strongly dependent on the temperature of annealing. Therefore we are dealing in this temperature range with a strained unstable lattice. One may expect that the elastic stress and strain in such a lattice enhance the thermal diffusion of impurities to energetically more favourable positions.
A similar effect is observed in plastically deformed silicon[8].
This explains the irregular background distributions that are obtained. The transient in the background situated at the boundary between the implanted and not imlanted region (see e.g. the profile at $T_{ann} = 800°C$ in fig. 5) can in this way be explained by the elastic stress and strain in this boundary region.

The observed differences between the boron distributions obtained by Ar^+, As^+ or P^+ ion implantations (figs. 1, 4 and 5) is explained by the different behaviour of these ions in a silicon lattice during annealing. The solubility of argon in silicon is very low, whereas on the other hand the implanted arsenic and phosphorus ions will readily occupy substitutional lattice positions. Therefore the defect structures in the latter cases differ considerably from those of the Ar^+ implantation which explains the different boron distributions that are obtained.

It is expected that the distribution in depth of the defects is roughly independent of the crystal orientation of the implanted substrate surface. This is in agreement with the results given in fig. 2.

Summarizing, we conclude that the observed redistribution effects are mainly due to the an-isotropic properties of the crystal lattice caused by the elastic stress fields and strain, which are introduced by the implanted ions and defect structures. This elastic behaviour is systematically distributed in depth and depends strongly on the implanted ion species.

ACKNOWLEDGEMENTS

The authors wish to thank Mr. M. A. H. de Grefte for his kind co-operation in performing SIMS measurements and Ir. M. G. Collet and Drs. J. Hornstra for discussions.

REFERENCES

[1] P. Baruch, C, Constantin, J. C. Pfister and R. Saintesprit, Disc. Faraday Society 31, 76 (1961).

[2] H. Strack, J. Appl. Phys. 34, 2405 (1963).

[3] R. L. Minear, D. G. Nelson and J. F. Gibbons, J. Appl. Phys. 43, 3468 (1972).

[4] W. K. Hofker, H. W. Werner, D. P. Oosthoek and H. A. M. de Grefte, Rad. Effects 17, 83 (1973).

[5] W. K. Hofker, H. W. Werner, D. P. Oosthoek and H. A. M. de Grefte, Appl. Phys. 2, 265 (1973).

[6] A. Monfret, R.T.C. La Radiotechnique-Compelec, Caen, private communication.

[7] W.S. Johnson and J. F. Gibbons, Projected Range Statistics in Semiconductors (Stanford University Book Store, 1969).

[8] J. E. Lawrence, Brit. J. Appl. Phys. 18, 405 (1967).

IMPLANTATION OF BORON AND LITHIUM IN SEMICONDUCTORS AND METALS

J.P. Biersack, D. Fink

Hahn-Meitner-Institut

Glienicker Str. 100, Berlin 39, W.-Germany

ABSTRACT

Range and damage profiles of implanted particles in semiconductors and metals are measured, and compared with theoretical predictions up to the 3rd moment of the distribution. Mobility of implanted atoms is studied during irradiation as well as after irradiation, for the observation of radiation enhanced diffusion and of normal diffusion.

Damage, range and diffusion profiles of implanted particles are measured by a method employing the neutron induced reactions $^6Li(n,\alpha)T$ and $^{10}B(n,\alpha)^7Li$. From angular emission patterns also lattice positions of implanted atoms can be derived by means of channeling and blocking patterns.

INTRODUCTION

The present paper contains a comparison of theoretical predictions and measured impurity profiles of lithium in niobium and boron in silicon. The profiles under investigation include range distributions, damage distributions (observed through impurity atoms which are immobily bound to defects in Nb), and profiles reflecting radiation enhanced and regular diffusion.

The system Nb(Li) is relevant for fusion reactors, where liquid lithium will be in contact with the metallic vacuum vessel and other structural materials under the influence of a considerable radiation field. Earlier studies of the system Nb(Li) have already been published |1|.

The system Si(B) is of wellknown interest for semiconductor fabrication and has been investigated by several authors during the past years |e.g. 2-4, and references therein|. Most of them measured the electrically active fraction after anneal by means of the C-V method, or applied the sputter technique in connection with SIMS measurements |3,4|.

For the present study of range, damage, and diffusion profiles of Li and B, a method is applied which (i) detects all implanted particles independent of their lattice positions, including precipitated material at the surface, and (ii) leaves the original profile unchanged for further repeated measurements with the same sample. This method is based upon the thermal neutron induced (n,α) reactions of Li-6 or B-10, where the energy loss of emitted α particles is a direct measure of the depth of origination, i.e. the energy spectrum is an image of the depth profile. This method has formerly been applied independently by Ziegler et al. and by ourselves |1,5|; it is believed to be a suitable tool for following up profile changes, and for gaining some insight into the basic behaviour of implanted atoms - particularly if combined with angular resolution of α particle emission for lattice location studies by means of channeling and blocking effects.

EXPERIMENTS

The Nb and Si samples are single crystalline disks, cut perpendicular to the <111> axis, and prepared for good surface finish. The ions $^6Li^+$, $^{10}B^+$, and $^1H_2^+$ (the latter for predamaging or diffusion enhancement) are implanted with current densities of typically 1 to 10μA/cm^2 at 50 to 220 keV using the 350 keV Danfysik accelerator. Implantation is performed into a random direction of the crystals about 6° from the <111> axis, at temperatures between 20 and 700°C, in a vacuum of typically $2 \cdot 10^{-6}$ Torr.

The emitted α particles resulting from ^6Li or $^{10}B(n,\alpha)$ reactions in thermal neutron fluxes of 10^4 to 10^7/cm^2, are detected and energy analyzed by Si surface barrier detectors (100μm depletion depth), and multichannel analysers. The overall resolution of the detection system is better than 25 keV, corresponding to 0.08μm in Si, or 0.04μm in Nb (viewed at 30° towards the surface normal).

For converting energy spectra into depth profiles, the recent (and within ≤4% accurate) stopping power data of Lin, Olson, and Powers |6| are used, i.e. S = 544 keV/μm for the 2.0 MeV α particles of ^6Li in Nb, and S = 270 keV/μm for 1.47 MeV α particles of ^{10}B in Si. The increase of S over the path of ≤0.5μm is smaller than the error of S, and is neglected. Deconvolution with the detector resolution function is not carried out either. Both corrections would tend to make the profiles slightly narrower, and particularly steeper at the depth.

IMPLANTATION OF BORON AND LITHIUM

THEORY

For the calculation of the total path length of implanted ions, both electronic and nuclear stopping powers are taken into account. The electronic stopping power S_e is obtained from the recently published interpolation formula |7| $1/S_e = 1/S_L + 1/S_B^*$ which agrees with the Lindhard-Scharff theory $S_L \sim \sqrt{E}$ |8| at low energies, and then - following the experimentally determined maximum - merges into the Bethe formula $S_B^* \sim \varepsilon^{-1} \cdot \ln(\varepsilon+a+b/\varepsilon)$, $\varepsilon = 2m_0v^2/<I>$; the original Bethe formula $S_B \sim \ln(\varepsilon)/\varepsilon$ is slightly modified by adding $a+b/\varepsilon$ under the logarithm in order to remove the negative pole at $\varepsilon=0$, and to provide adjustable parameters which are effective near the maximum $\varepsilon \approx 1$ (for light ions good fits are usually obtained with $a=1$ and $b=3...5$). The contribution of nuclear stopping is calculated from the analytical expression |9|

$$S_n = \frac{4\pi a\ NM_1 Z_1 Z_2 e^2}{M_1 + M_2} \frac{\ln(x)}{2x(1-x^{-3/2})}, \quad x = \frac{aM_2 E}{Z_1 Z_2 e^2 (M_1+M_2)},$$

where x is the dimensionless energy variable, and $\underline{a}$ denotes the Thomas-Fermi screening radius. This formula represents the computational S_n data, as obtained on the basis of the Thomas-Fermi-Molière potential (very close agreement with the accurate computation |10|, good agreement with the momentum approximation |11|). From S_e+S_n the total path length of the projectiles is obtained.

Flight paths of light and relatively low energetic particles deviate from straight lines due to the nuclear collisions, thus giving rise to range and damage distributions of considerable width and skewness. Computations of the first three moments of damage and range distributions of light ions have been performed by Weissmann et al. |12| on the basis of the LSS collision cross section (approximate TF potential) |11,13|. A similar calculation for first and second moments on range and damage profiles has earlier been performed by Schiøtt |14| with the same results.

RESULTS ON Li-6 IMPLANTED IN NIOBIUM

<u>Range distributions</u> of 200 and 220 keV Li in Nb are measured after room temperature implantation. Theoretical predictions in the form of Edgeworth expansions, using the first three moments of the distribution |12|, yield perfect fits after adjusting the electronic stopping power |8| of Li in Nb to a higher value by applying a factor of 1.30 (a rather common correction). The third moment was found to greatly improve the fit, not only with respect to skewness, but also with respect to the width.

Damage distributions are obtained by implanting Li^+ into niobium at elevated temperatures around 700°C. The observation of damage profiles becomes feasable by the fact that point defects trap a certain number of Li atoms (probably 3), and form very stable configurations. The remaining Li which does not get trapped, is lost to the surface or into the bulk due to high interstitial mobility at this temperature. The trapped Li configuration is found thermally stable in post-irradiation anneal experiments up to 1040°C. Such stable defect configurations are not observed at lower implantation temperatures, where only vacancies are mobile but not the Li atoms; from this it is concluded that more than one Li atom must join the vacancy in order to form such stable defects. (Similar defect configurations have been observed for He in W |16,17|, and are supported by calculations of Wilson and Bisson |18|.

Lithium atoms implanted at room temperature to maximum concentrations of ≤4% atomic fraction show regular diffusion at anneal experiments between 550°C and 650°C in accordance with theoretical concentration profiles ($D=1.7 \cdot 10^{-12} cm^2/s$ at 600°C, $\Delta H=1.3 eV$). All Li atoms are found to be mobile, and no trapped or precipitated lithium is observed in these experiments. In one experiment the implanted Li concentration was raised to about 50% which resulted in precipitations over most of the range; in this sample the Li could be redissolved upon anneal to ≥900°C corresponding to an energy of solution slightly above 2 eV.

RESULTS ON B-10 IMPLANTED INTO SILICON

Theoretically expected range profiles are obtained after pre-damaging the Si crystal with $2 \cdot 10^{16}$ protons per cm^2 of 100 keV energy. With this proton irradiation an energy density of the order of $10^{23} eV/cm^3$ is deposited in elastic collisions in the silicon lattice over a depth of about 1.5μm, cf. Fig. 2a; this causes enough defects to trap the implanted boron atoms, but does not render the silicon amorphous. A comparison to the results of Wittmaack et al. |5|, who also predamaged the crystals, yields good agreement with the presently obtained profiles, e.g. Fig. 1a.

Without predamaging the crystal, one obtains slightly deeper boron profiles with a considerable tail extending about twice as far into the depth as theoretically predicted, Fig. 1b. The steep cut-off which is theoretically provided by the negative third moment is no longer recognizable; instead, the experimental profile resembles a Gaussian with some additional tail extending into the depth, Fig. 1b. The 100 keV boron profile, shown here, agrees well with the result obtained by Ziegler et al. |19| under similar conditions. Seidel |2| also implanted B into undamaged crystals and measured the electrically active boron profiles by means of the C-V method. His results similarly are Gaussian profiles with approximatily exponential tails, $e^{-x/\lambda}$ with $\lambda \approx 0.08 \mu m$. Such exponential

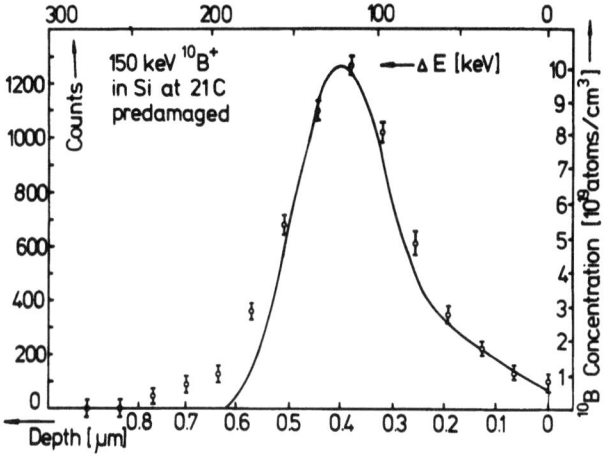

Fig. 1a: Implantation of 150 keV $^{10}B^+$ in a non-channeling direction of a silicon crystal after <u>predamaging</u> with 10μA min/cm² of 200 keV H_2^+. Agreement between theoretical range profile (full line) and experiment is satisfactory, if the 3rd moment of the theoretical distribution is taken into account.

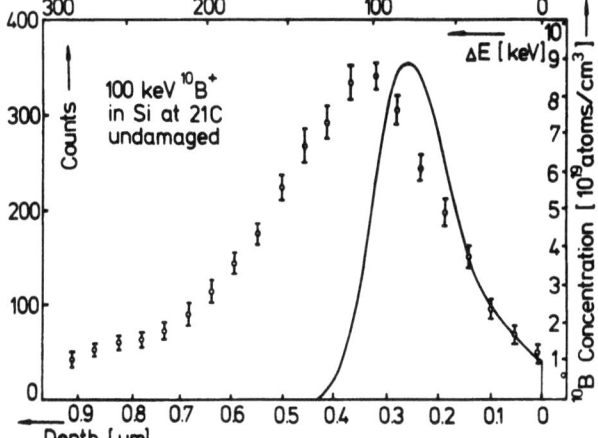

Fig. 1b: Implantation of 100 keV $^{10}B^+$ into an <u>undamaged</u> Si crystal (non-channeling direction). The experimental profile is deeper than the theoretical prediction (full line), and exhibits a tail. Both features deviate from common range distributions, but might be attributed to diffusion.

tails beyond the range distribution (i.e. extending into undamaged regions of the lattice), could possibly be explained by interstitial motion of B atoms until trapping occurs into an intrinsic lattice defect; assuming $N=2\cdot 10^{16}$ traps per cm³ with an effective trapping radius of R=6Å, one arrives ideed at $\lambda=(4\pi RN)^{-1/2} = 0.08\mu m$ (independent of D and, hence, of temperature).

For a direct study of <u>radiation enhanced diffusion</u>, a Si sample is implanted with 50 keV B^+ to a concentration of about 10^{20}at./cm³ within 0.2μm depth, and then post-irradiated with 10μA min/cm² of 200 keV H_2^+. The resulting B distribution is shifted into the depth, and exhibits a rather flat portion extending to about 1.5μm which equals the proton range. The experimental results are depicted in Fig. 2a together with the original boron range profile, and with the theoretical damage profile of 100 keV protons (dashed line accurate

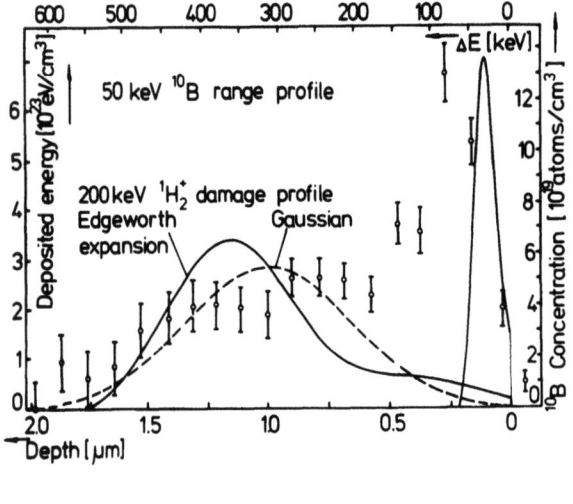

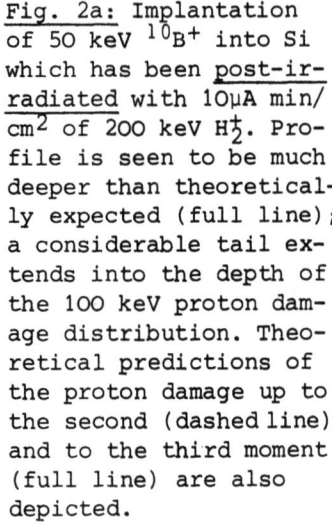

Fig. 2a: Implantation of 50 keV $^{10}B^+$ into Si which has been post-irradiated with 10μA min/cm² of 200 keV H_2^+. Profile is seen to be much deeper than theoretically expected (full line); a considerable tail extends into the depth of the 100 keV proton damage distribution. Theoretical predictions of the proton damage up to the second (dashed line) and to the third moment (full line) are also depicted.

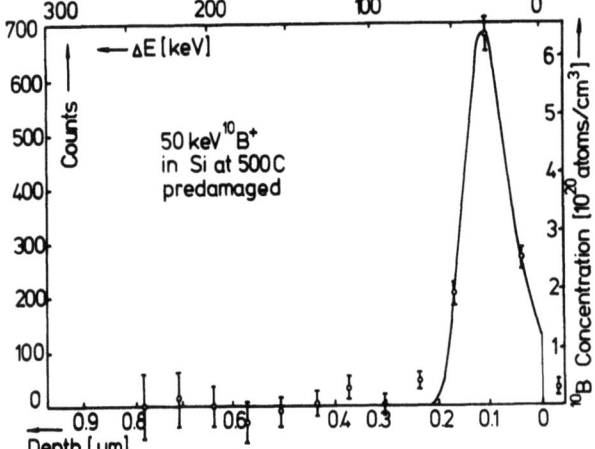

Fig. 2b: For comparison with Fig. 2a the original 50keV boron range profile is depicted together with the theoretical prediction. Crystal has been predamaged with 10μA min/cm² of 200 keV H_2^+. The absence of any tail should be noticed.

to second moment, solid line to third moment). The original boron range profile has been checked before, and was found to agree with theory as shown in Fig. 2b. Besides, this rather narrow profile is used to demonstrate that no tail or broadening occurs during the process of implantation even at the elevated temperature of 500°C, if the crystal is sufficiently predamaged.

In undamaged crystals some broadening of the boron profile is generally seen, even at room temperature, cf. Fig. 1b. This may also be attributed to radiation enhanced diffusion caused by the boron implantation itself: collision cascades may lead to a release of B previously trapped in Si vacancies, thus enabling some interstitial motion until new trapping occurs; this effect would be particularly effective at the far end of the distribution, i.e. beyond the region

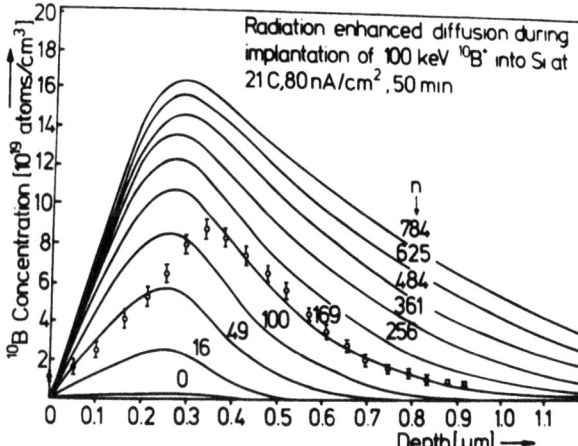

Fig. 3: Boron range profile effected by radiation enhanced diffusion. Theoretical predictions for simultaneous implantation and diffusion; curves are plotted for $Dt = n \cdot 3.37 \cdot 10^{-12} cm^2$ at indicated values of n.

of maximum damage. Regular diffusion, neglecting the influence of defect concentration, cannot account for the observed profiles, as shown in Fig. 3. The observed broadening corresponds to an effective diffusion coefficient of roughly $D \sim 10^{-13} cm^2/sec$, but the shape of the observed profile can not be reproduced theoretically under the assumption of a constant D value for radiation enhanced diffusion.

RÉSUMÉ

Range and damage profiles can very accurately be predicted by theory, when taking into account the third moment of the distribution, provided that the electronic stopping power has been determined once.

Radiation enhanced diffusion at room temperature is observed for boron in undamaged Si crystals. The boron mobility is ascribed to the implantation itself, or - more pronounced - to post-irradiation of 100 keV protons.

Radiation retarded diffusion is observed in the case of lithium implantation at 700°C in niobium. In this case very stable defect configurations are formed, probably involving more than one Li atom per defect, which remain immobile and stable in post-irradiation anneals up to 1040°C.

ACKNOWLEDGEMENT

We acknowledge the kind support of Prof. W. Heinz and his coworkers of the Physikalisch-Technische Bundesanstalt Braunschweig, W. Germany, where the experiments were carried out at the FMRB reactor facility. We greatly appreciate the help of Dipl.-Phys. E. Santner in implanting our samples.

REFERENCES

1. J.P. Biersack and D. Fink, proceedings of the international conferences on "Ion Beams to Metals", Albuquerque 1973, "Surface Effects in Controlled Thermonuclear Fusion Devices and Reactors", Argonne 1974, "Symp. on Fusion Technology", Noordwijkerhout, 1974.

2. I. Ruge, J. Graul (Eds.), "Ion Implantation in Semiconductors", Springer, Berlin 1971, T.E. Seidel, p. 47; M. Tamura et al., p. 96; G. Dearneley et al., p. 439; H. Müller et al., p. 85; S.M. Davidson, p. 79.

3. B.L. Crowder (Ed.), "Ion Implantation in Semiconductors and Other Materials", Plenum, N.Y. 1973; K. Wittmaack et al., p. 119.

4. W.K. Hofker, H.W. Werner, D.P. Oosthoek, H.A.M. de Grefte, Rad. Effects $\underline{17}$ (1973) 83.

5. J.F. Ziegler, G.W. Cole, J.E.E. Baglin, J. Appl. Phys. $\underline{43}$ (1972) 3809.

6. W.K. Lin, H.G. Olson, D. Powers, J. Appl. Phys. $\underline{44}$ (1973) 3631, and Phys. Rev. B $\underline{8}$ (1973) 1881.

7. J.P. Biersack, D. Fink, proceedings of the international conference on "Atomic Collisions in Solids", Gatlinburg, Tenn., 1973.

8. J. Lindhard, M. Scharff, Phys. Rev. $\underline{124}$ (1961) 128.

9. J.P. Biersack, Z. Physik $\underline{211}$ (1968) 495.

10. J.P. Biersack, report HMI-B37 (1964).

11. J. Lindhard, M. Scharff, E. Schiøtt, Danske Vid. Selbsk., Mat.-Fys. Medd. $\underline{33}$ (1963), No. 14.

12. R. Weissmann, P. Sigmund, Rad. Effects $\underline{19}$ (1973) 7.

13. K.B. Winterbon, P. Sigmund, J.B. Sanders, Danske Vid. Selsk., Mat.-Fys. Medd. $\underline{37}$ (1970), No. 14.

14. H.E. Schiøtt, Danske Vid. Selbsk., Mat.-Fys. Medd. $\underline{35}$ (1966), No. 9.

15. K.B. Winterbon, Rad. Effects $\underline{13}$ (1972) 215.

16. E.V. Kornelsen, Rad. Effects. $\underline{13}$ (1972) 227.

17. S.T. Picraux, F.L. Vook, in "Applications of Ion Beams to Metals", Plenum, 1974.

18. C.L. Bisson, W.D. Wilson, in "Applications of Ion Beams to Metals", Plenum, 1974.

19. J.F. Ziegler, B.L. Crowder, G.W. Cole, J.E.E. Baglin, B.J. Masters, Appl. Phys. Lett. 16, $\underline{21}$ (1972).

INFLUENCE OF ANNEALING AND RADIATION DAMAGE ON ELECTRICAL CARRIERS

PROFILES IN PHOSPHORUS IMPLANTED SILICON ALONG THE |110| AXIS

F. Cembali, R. Galloni, R. Lotti, F. Zignani [+]

Laboratorio di Chimica e Tecnologia dei Materiali e dei

Componenti per l'Elettronica. C.N.R., Bologna, Italy

ABSTRACT

Phosphorus ions at energies of 100, 200 and 300 KeV have been implanted into silicon crystals along the |110| axis. The shape of carriers profiles, as a function of annealing temperature between 100 and 900°C, has been compared with the damage distribution due to the implantation itself and correlation between phosphorus electrical activation and damage has been shown. It seems possible to single out three different regions in the carriers depth distributions with different mechanisms of phosphorus electrical activation.

INTRODUCTION

We have confined our experiments to implantations of phosphorus ions along the |110| axis of silicon samples, at different energies and doses, with the aim of abtaining some information on the correlation between radiation damage nature and distribution, and phosphorus electrical activation mechanisms. Lattice defects introduced by the implantation have strong influence on the diffusion mechanisms that produce the electrical activation of phosphorus; such mechanisms, concerning the migration of the dopant ions in the lattice or to recovery of compensating defects, act at different or partly overlapped temperature ranges at different depths inside the crystal, so that the carriers profiles assume different shapes at the different annealing temperatures.

([+]) Istituto Chimico, Facoltà di Ingegneria, Università di Bologna.

EXPERIMENTAL TECHNIQUES AND RESULTS

Phosphorus ions at doses of 5×10^{14}, 1×10^{15}, 5×10^{15} at./cm^2 have been implanted along the |110| axis of silicon wafers at energies of 100, 200, 300 KeV. The angular divergence of the beam scanned on a (2x2)cm^2 region was ± 0.1°, and the same accuracy was used in the beam-channel alignment. The characteristics of the ion-accelerator used, a 400 KeV Cockroft - Walton by Accelerator Inc. (Austin, U.S.A.) have been described in a previous paper (1). Silicon wafers |110| orientation, p-type (B doped), floating zone, dislocation free, 500-1000 ohm x cm produced by Wacker Chemitronic Co. (Germany), have been used in our experiments. Electrical carriers depth distribution have been determined by differential resistivity and Hall coefficient measurements and subsequent layers have been removed by the anodic oxidation process with the sample fixed in a special electrolitic cell (2,3). Due to the strong depth dependence of the electrical carriers concentration, produced by the low temperature post-implantation annealing of the samples, the special technique described in our previous paper (2) has been used to obtain ohmic contacts for the electrical measurements. Radiation damage distribution produced by phosphorus ions implanted along the |110| axis of silicon at 100, 200, 300 KeV have been determined by Rutherford Back-scattering using 300 KeV protons.

Electrical carrier profiles of 100 KeV $10^{15}P^+$/cm^2 after annealing at 110°C, 450°C, 900°C, are shown in fig. 1(a). At 900°C the number of carriers measured from the integral of the distribution, is 95% of the implanted dose; at 450°C the surface peak height is unchanged but its width is about half of the foregoing value and a minimum in the distribution appears at about 0.4µm. Two peaks partly overlapped are shown at depths between 0.4 and 1.7µm where only 40% of the implanted dose shows electrical activity. In the sample annealed at 110°C, the carriers concentration is very low in the 1µm thick surface region and grows to a maximum of 4×10^{16} carriers/cm^3 in the region of well channeled ions at 1.4µm depth; the fraction of electrically active phosphorus is only 0.15% of the implanted dose. Similar distributions are shown in figs. 1(b) and 1(c) for $10^{15}P^+$/cm^2 implanted at 200 and 300 KeV; also a de-channeled peak at very low concentration (<10^{16} carriers/cm^3) is here evident.

Comparison between carriers profiles after annealing at 450°C and radiation damage distribution from back-scattering analysis is shown in figs. 2(a), 2(b), 2(c), for 100, 200, 300 KeV implantations respectively. It can be seen that a good depth correspondence exists between carriers concentrations and damage peaks. Deeper inside the crystal no damage, either by back-scattering or by Transmission Electron Microscopy analysis, was observed(1).

Samples implanted with 200 KeV, $10^{15}P^+$/cm^2 have been isochronally annealed in the range 25-450°C after 1µm and 1.7µm respectively

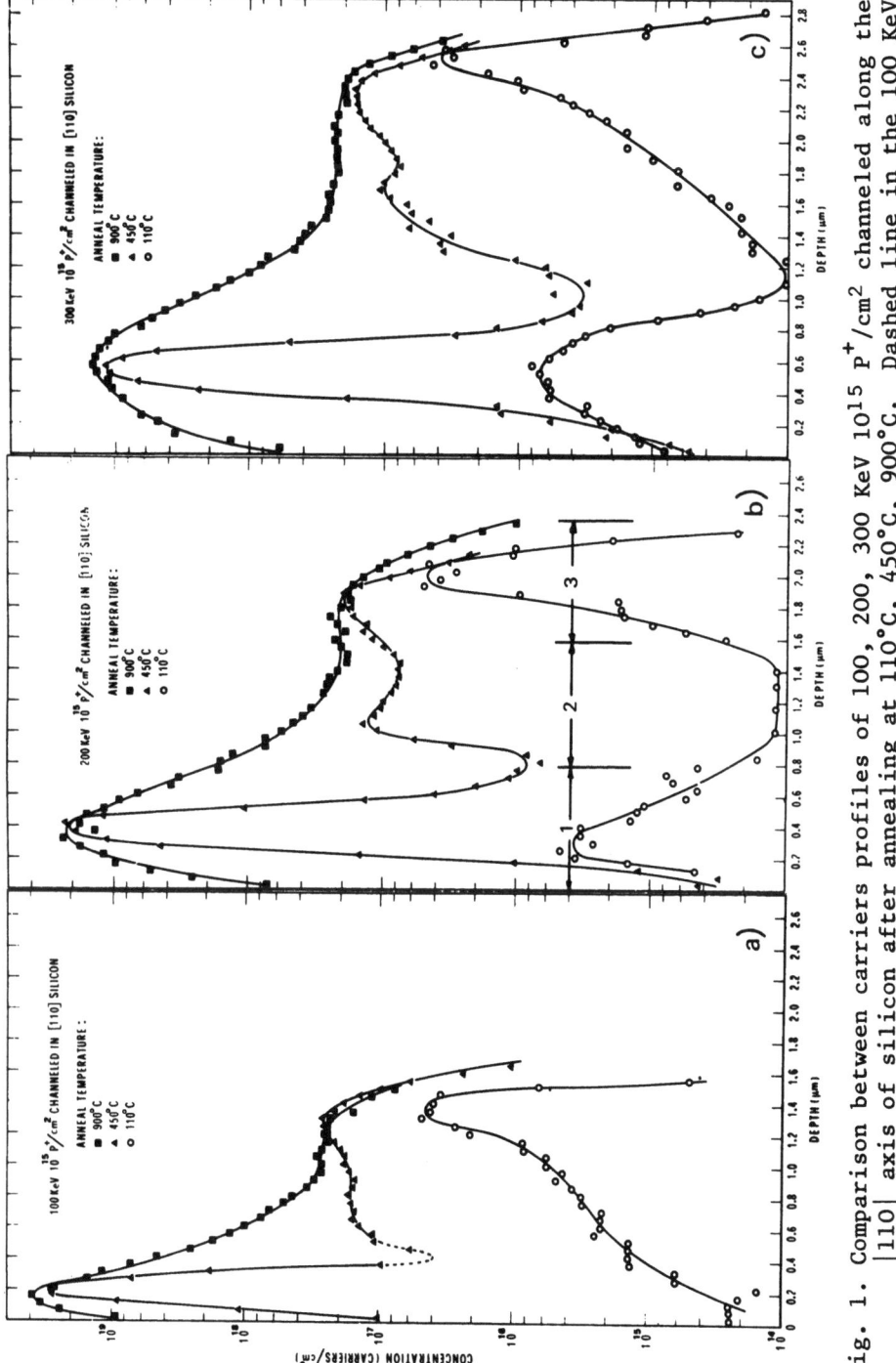

Fig. 1. Comparison between carriers profiles of 100, 200, 300 KeV 10^{15} P^+/cm^2 channeled along the $|110|$ axis of silicon after annealing at 110°C, 450°C, 900°C. Dashed line in the 100 KeV profile is a tentative interpolation of higly scattered data.

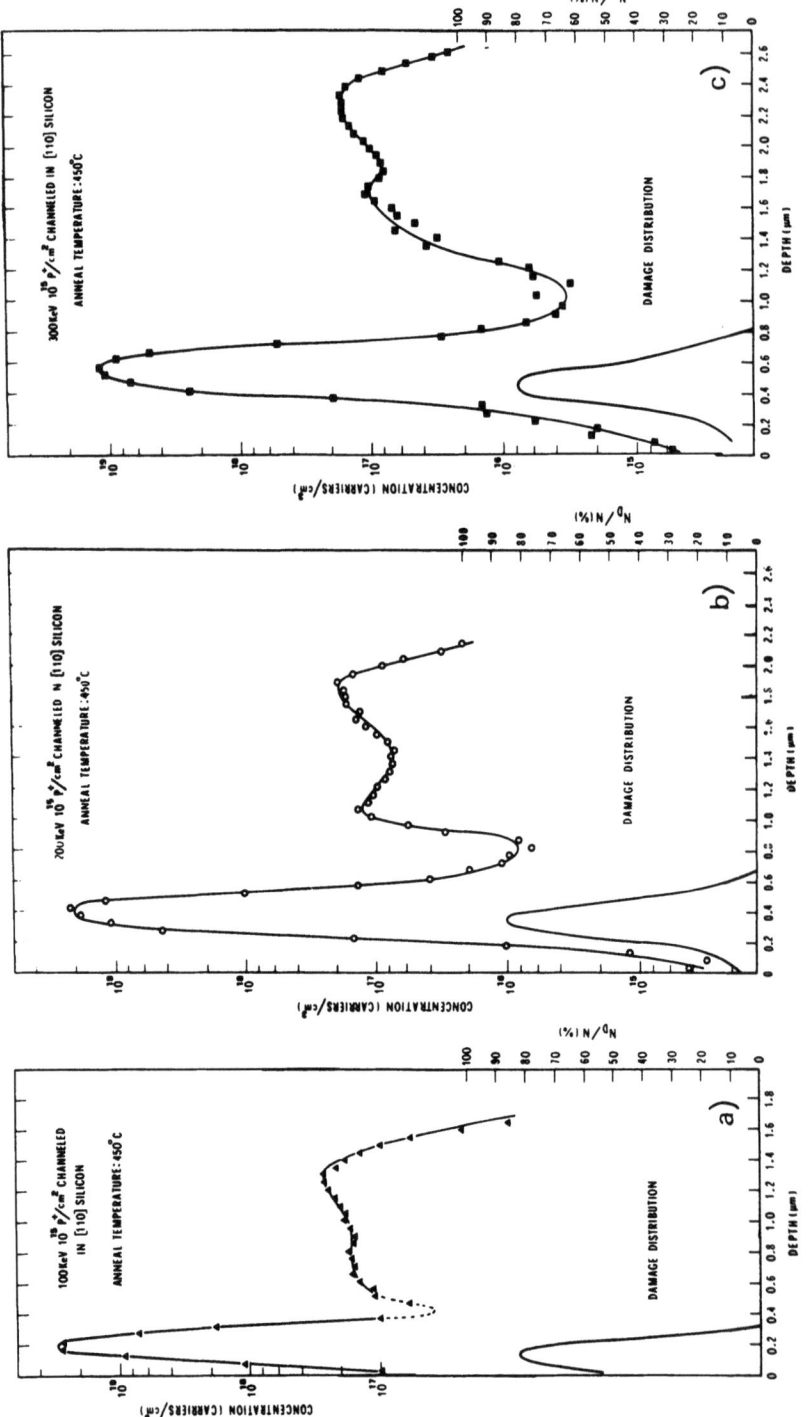

Fig. 2. Comparison between carriers profiles after 450°C annealing and relative radiation damage distributions by Rutherford back-scattering analysis. N_D/N is the fraction of displaced atoms.

had been stripped by anodic oxidation from two distinct regions of the same implanted surface. Some preliminary results are shown in fig. 3. Curve (a) and (b) are relative to the implanted sheet between 1.6µm and R_{max}, and 1µm and R_{max} respectively; curve (a) represents therefore the annealing behaviour of the well-channeled ions. The first annealing stage is between 100°C and 200°C in both curves; in curve (b) it is also possible to see a second stage starting at 400°C that can be attributed to the activation of phosphorus in the region between 1µm and 1.6µm.

DISCUSSION

On the basis of the experimental results shown, it is possible to draw a few hypoteses on the understanding of the phosphorus electrical activation mechanisms at different penetration depths. With reference to fig. 1(b), the implanted sheet can be divided schematically into three layers, with different electrical activation mechanisms; similar considerations can be applied at profiles of 100 and 300 KeV implantations.

1st. Layer. From the surface to $\sim$ 0.8µm depth, detectable damage is in the form of amorphous regions of average diameter of $\sim$ 50Å (1), but we reasonably think that isolated defects are also present. Phosphorus atoms trapped in amorphous regions, become electrically active during their recrystallisation and as the annealing temperature for this process is a function of the dimensions of the amorphous regions, as suggested by Nelson (4), we should expect a large temperature range for the electrical activation of these atoms. A small surface peak starts, infact to appear after annealing at 110°C

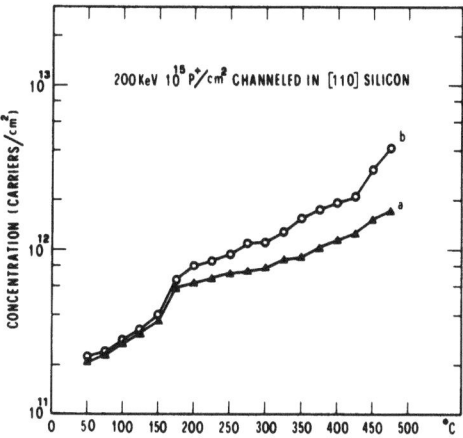

Fig. 3. Isocronal annealing of $O^{15}P^+/cm^2$ implanted at 200 KeV: (a) sheet between 1.6µm and R_{max}; (b) sheet between 1µm and R_{max}.

and after annealing at 500°C it still does not cover all the heavily damaged volume. The much higher temperatures (6) required for the complete electrical activation of the layer is probably connected with the recovery of compensating centers, due for example to dislocation, dislocation loops, staking faults etc., formed during the annealing, or already present in the crystal.

2nd. Layer. From comparison between the isocronal annealing data and the electrical profiles, it can be seen that electrical activation in this layer is between 450°C and $\sim$ 700°C. This temperature range is practically overlapped with the one we have found in the 1st. layer for the electrical activation of the regions with lower density of amorphous zones and dominated by the effect of compensating centers. Studies of the kinetics of recovery and measurements of activation energy will be necessary to confirm these interpretive hypoteses.

3rd. Layer. This is the region of well channeled ions. From the profiles after 110°C annealing and from the first step in the isocronal annealing curve, it is possible to see that electrical activation in this region occurs at temperatures below 200°C. From the tails of the distributions at 110°C of fig. 1(a), 1(b) 1(c), it can be seen that R_{max} show an $E^{\frac{1}{2}}$ dependence which suggests, in very good agreement with the data from Goode, Wilkins, Dearnaley (7), that well channeled P^+ ions loose their energy only in electronic interations within a channel; this means that they should end up their trajectories in electrically inactive interstitial positions. Moreover in the highly ionized crystal, during the implantation, the charge exchange mechanism of interstitial migration proposed by Bourgoin and Corbett (8), could be responsible for the interstitial migration of the implanted ions until they are trapped in a vacant lattice site, but as from Watkins and Corbett (9) observations, no single vacancies are present in silicon at room temperature, this proposed mechanism could be responsible only for a small fraction of the channeled ions that are captured by vacancies created during the implantation itself. In fact at room temperature, we observe that a small fraction of channeled ions is electrically active. The larger fraction of well channeled ions, becomes electrically active between 150°C and 200°C. This temperature seems to agree with the value of 170°C observed by other authors (10) in samples implanted with 20 KeV phosphorus ions along the |110| axis, and there ascribed to the migration of the E center. Moreover, temperatures of 100-180°C and activation energy of 0.94 eV have been observed by other authors (11) for the activation of phosphorus ions in silicon in agreement with the activation energy of 0.93 eV found by Watkins (9) for the reorientation of the E center. The formation of the E center, could be due to the well channeled ions that having lost their energy within a channel, during the implantation process, migrate at room temperature and are trapped by divacancies. The electrical activity of phosphorus is a consequence of the migration and trapping in sinks, like oxygen or carbon atoms, of the E center. This mechanism, feasible only in the presence of divacancies is supported by the stability of divacancies in silicon up to temperatures of 250-300°C. By the variations in our profiles from the lower to the higher annealing temperatures (see fig. 1), it is actually possible to single out a small diffusion of the tails of the order of 10^3 Å that could be ascribed to the migration of the E center. The different slope of the tails in our case and in Dearnaley's experiment (12) could be due to the bulk silicon characteristics: as already shown in fact (12), the channeled peak maximum concentration and the slope of the tails are different in samples from different stocks. This observation can also be drawn from comparison between the profiles here presented and the profiles shown in our previous work (1) in which we used silicon of nearly the same characteristics but from a different factory.

To conclude we should like to point out that these can only be tentative hypoteses, as the available experimental data are not sufficient yet to draw conclusions on the activation mechanisms. The absence of sufficiently accurate theoretical models, able to give calculated activation energies to be compared with the experimental values is also a great handicap.

ACKNOWLEDGEMENTS

Thanks are due to A. Desalvo and to R. Rosa for their help in the computer calculation of the radiation damage, to L. Pedulli for his help in the elaboration of the experimental data. To R. Angelucci for his help in the annealing treatments.

REFERENCES

(1) F. Cembali, R. Galloni, F. Mousty, R. Rosa, F. Zignani, Rad. Effects, April (1974).
(2) F. Cembali, R. Galloni, F. Zignani, J. Phys. E, Sci. Instrum. (in press).
(3) B.L. Crowder, J.M. Fairfield, J. Electrochem. Soc., $\underline{117}$, 363, (1970).
(4) R.S. Nelson, Proc. European Conf. on Ion Implantat., Reading, England, p.212 (1970).
(5) J.F. Gibbons, Proc. of the IEEE, $\underline{9}$, 1062, (1972).
(6) B.L. Crowder, J. Electrochem. Soc., $\underline{118}$, 943, (1971).
(7) P.D. Goode, M.A. Wilkins, G. Dearnaley, Rad. Effects, $\underline{6}$, 237, (1970).
(8) J.C. Bourgoin, J.W. Corbett, Phys. Lett. $\underline{38A}$, 135, (1972).
(9) G.D. Watkins, J.W. Corbett, Phys. Rev. A, $\underline{134}$, 1359, (1964).
(10) Ph. Glotin, J. de Phys. $\underline{29}$, 926, (1968).
(11) M. Hirata, M. Hirata, M. Saito, J. Phys. Jap. $\underline{27}$, 405, (1969).
(12) G. Dearnaley, M.A. Wilkins, P.D. Goode, J.H. Freeman, G.A. Gard, Atomic Collision Phenomena in Solids, (North Holland) p.633, (1970)

II-VI COMPOUND SEMICONDUCTORS AND OTHER MATERIALS

ION IMPLANTATION OF As in CdTe: ELECTRICAL CHARACTERISTICS AND RADIATION DAMAGE[†]

J. C. Bean[*], J. F. Gibbons[*], T. J. Magee[**], and J. Peng[**]

[*]Stanford University, Stanford, California USA 94305

[**]Stanford Research Institute, Menlo Park, Cal. USA 94025

ABSTRACT

The effects of arsenic and krypton implantations have been investigated in semi-insulating cadmium telluride crystals. Nearly 100 percent electrical activity was observed when As implants were performed at 300°C in samples subjected to a 24-hour 500°C cadmium vapor anneal prior to implantation. A jet thinning technique was utilized to prepare samples for transmission electron microscopy analysis and a number of micrographs obtained. In_2Te_3 and $CdCl_2$ precipitates were identified in indium and chlorine compensated materials. Implantation damage was also observed in the form of small vacancy loops.

INTRODUCTION

Cadmium telluride is noted for its relatively large direct bandgap and high average atomic number. These properties have suggested its use as a room temperature gamma ray detector material. In addition, there has recently been considerable interest in employing CdTe in solid state solar power cells. With these applications in mind, we have investigated the application of ion implantation in CdTe. This study concentrates on the formation of shallow doped layers in semi-insulating CdTe. The effects of implantation are evaluated using conventional mobility and sheet resistivity measurements. A modified jet thinning technique has been applied to obtain samples for microstructural analysis in a transmission electron microscope [1]. Using correlated data from these experiments, we have investigated the effect of preanneal, substrate temperature and post anneal temperature upon implanted ion activity.

MATERIALS CHARACTERIZATION

The initial problem was that of characterizing commercially grown CdTe gamma ray detector material. It has long been known that the presence of electrically active native defects in CdTe [2] effectively precludes its refinement to a high resistivity state. As such, typical detector material is rich in acceptor-like defects which have been compensated by adding indium or chlorine donors. To evaluate the microstructure of this material, a technique for TEM sample preparation has been developed and applied [1]. A bright field micrograph of an indium compensated sample is shown in Fig. 1. A transmission diffraction pattern is superimposed in the upper right hand corner of the figure. In addition to the expected (111) diffraction pattern, the inset shows a number of polycrystalline diffraction rings. The crystalline spacings computed from these rings are found to coincide with those of Te and In_2Te_3 crystallites. The micrograph itself shows a large number of precipitate particles, most of which display a clearly defined Moiré pattern. From the fringe spacings, one may calculate "d" spacings of the precipitate particles [3]. The computed spacings are found to be equivalent to those of In_2Te_3, thereby identifying the particles. A similar analysis resulted in the identification of $CdCl_2$ precipitates in chlorine compensated materials.

IMPLANTATION

Based on the investigation of Donneley et al. [4], arsenic was chosen for implantation. The availability of only a 60 keV machine required that we forego the use of protective sample coatings to insure adequate ion ranges in the high mass CdTe substrates. Implantations were performed initially into as-delivered commercial CdTe material. Despite varying post anneals up to 500°C, initial results were consistently poor. Implantations at room temperature and 100°C showed no measurable electrical activity. Implantations at 200°, 300°, and 400°C resulted in slight activity, increasing with implantation temperature to a maximum of about 10 percent fractional ion activity at 400°C. The 400°C implantation did, however, result in marked surface deterioration and subsequent implantations were performed at 300°C.

The microscopic effect of a 5×10^{13} As/cm^2 room temperature implant in as-received In-doped material is shown in Fig. 2. In addition to In_2Te_3 precipitates, small 50 Å diameter dislocation loops are observed. To confirm that the loops were due to radiation damage, samples were implanted to doses in the range, 5×10^{13} to 5×10^{15} $ions/cm^2$ and subsequently analyzed. The measured loop concentration exhibited an approximately linear dependence on dose, with possible saturation appearing at the higher dose levels. Additional implants were performed at room temperature to a dose of 10^{17} As/cm^2 and it was found that all samples retained single

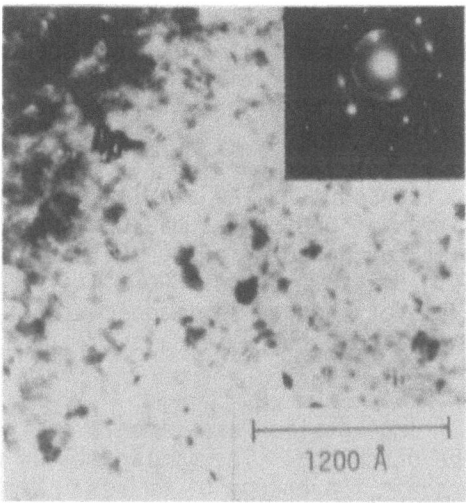

Fig. 1. Transmission electron micrograph of as-received indium doped sample. A selected area transmission electron diffraction pattern obtained in this region is shown in the right corner of the micrograph.

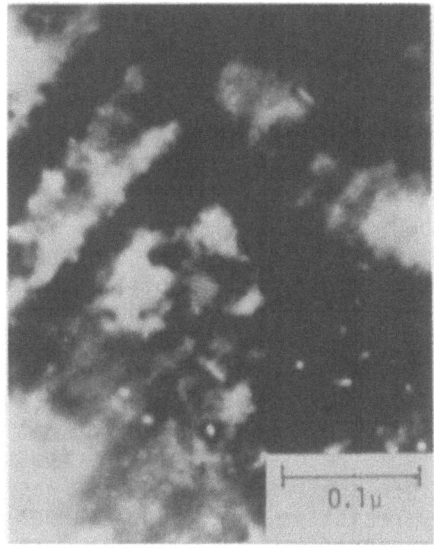

Fig. 2. Electron micrograph of unannealed indium-doped sample exposed to room temperature As-implant dose of 5×10^{13} ions/cm^2.

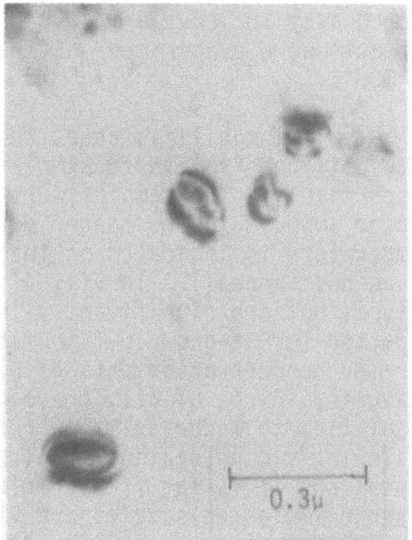

Fig. 3. Electron micrograph of sample subjected to 24 hour, 500°C preanneal in Cd vapor.

crystal spot patterns for both reflection and transmission diffraction.

It was subsequently found that a cadmium preanneal could significantly increase ion activities. Samples were sealed in evacuated quartz ampoules with pieces of 6N purity cadmium. Cd and CdTe samples were positioned at opposite ends of the ampoule and the CdTe end placed at the maximum temperature point of the furnace to inhibit Cd vapor transport to the sample surfaces. A time-temperature combination of one day at 500°C was selected to maintain the semi-insulating character of the CdTe. The effect of such an anneal is shown in Fig. 3. Large 1200 Å dislocation loops have appeared. An analysis of changes in image contrast as the sample was tilted in the microscope [5,6] reveals that the dislocation loops are indeed vacancy loops.

A recent set of implantation experiments on the preannealed material will now be described. The implantations were performed at 300°C with an arsenic ion energy of 60 keV. Samples were implanted at dose rates from 100 to 150 na/cm^2 to a total dose of about 10^{15} As/cm^2. To isolate the doping effects of outdiffusion, damage, and true chemical doping, multiple control schemes were used. One set of samples was placed in the implantation apparatus and simply run through the heating cycle. Another set was heated and implanted with krypton, and the final set with arsenic. The entire sample face was implanted and then portions were chemically etched away to leave a mesa van der Pauw [7] pattern. Nickel contacts were evaporated on the four pads and alloyed for 8 minutes at 300°C.

Typical mobility, sheet resistivity and carrier concentration data obtained from these samples are shown in Figs. 4, 5, and 6, respectively. The samples were post annealed for one hour in a closed ampoule in the manner described above. Post anneal temperature is indicated on the horizontal axis (25°C corresponds to no post anneal). In Fig. 4 it is noted that all mobilities are p-type, but are significantly less than the value of 70-90 cm^2/V-sec expected for as-grown p-type material. The sheet resistivity and carrier concentration data display several significant and unusual features. First of all, the control samples have changed from semi-insulating to moderately low resistivity (high carrier concentration) following vacuum heating. There is some recovery with higher post anneal but it is far from complete. In contrast, the krypton and arsenic implanted samples are highly resistive up to a post anneal temperature of about 400°C. At this point, the resistivities are reduced sharply by several orders of magnitude. In terms of carrier concentration, the arsenic and krypton samples peak at concentrations of approximately 100 and 30% of the total implanted dose respectively. The curves might be explained in the following manner. The heated vacuum implantation cycle may lead to the out

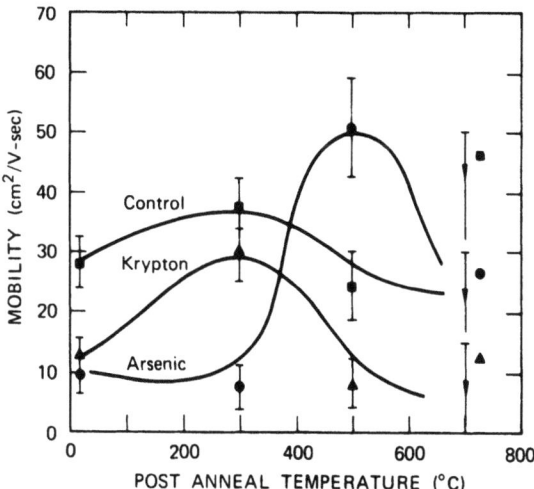

Fig. 4. Mobility versus post anneal temperature

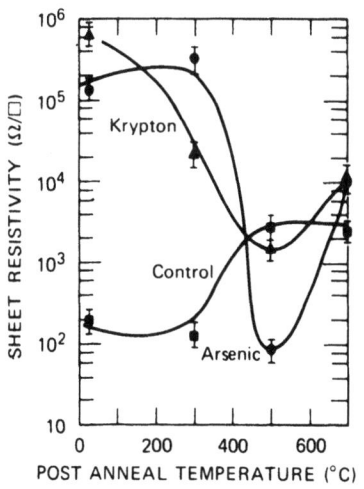

Fig. 5. Sheet resistivity versus post anneal temperature

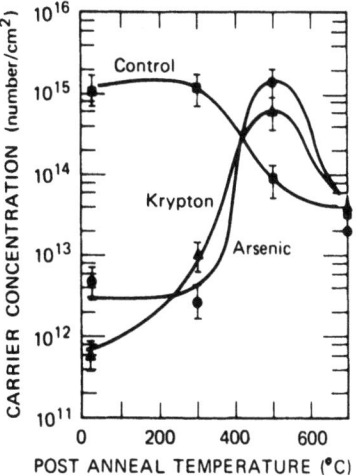

Fig. 6. Carrier concentration versus post anneal temperature

diffusion of Cd to form acceptor-like Cd vacancies or vacancy complexes. Presumably, this occurs in arsenic, krypton, and controlled samples alike, but is masked by radiation damage in the arsenic- and krypton-implanted samples at low post-anneal temperatures. At higher post-anneal temperatures, two things may occur. First, Cd may diffuse into the crystal from the vapor to annihilate vacancy acceptors and begin to form insterstitial cadmium donors. Simultaneously, radiation damage may anneal, increasing the effect of chemical doping. This would explain the initially high hole concentration in the control sample and its gradual compensation with higher post-anneal temperature. The data would then also indicate a sharp damage annealing stage at about 400°C leading to the sudden increase in arsenic doping activity.

The concurrent increase in doping seen in the Kr-implanted sample is as yet unexplained. In this and other experiments, Kr-implanted samples have displayed a carrier concentration of about 30 percent of the implanted dose. To investigate the nature of this damage-induced doping, low temperature measurements and further electron microscopy analyses are in progress.

†This work was supported in large part by the Office of Naval Research Contract No. N00014-73-C-0378, and in part by the Joint Services Electronic Program.

References

1. T. J. Magee, J. Peng, and J. C. Bean, "Preparation of Thin Cadmium Telluride Samples for Electron Microscope Studies," Proc. Electron Microscopy Soc. Amer., Claitor Publ. Co., Baton Rouge, La. (1974).

2. D. de Nobel, Phillips Res. Rpt. $\underline{14}$, 442 (1959).

3. S. Amelinckx, Direct Observation of Dislocations, p. 414, Academic Press, New York (1964).

4. J. P. Donnelly, A. G. Foyt, E. D. Hinkley, W. T. Lindley, and J. O. Dimmock, Appl. Phys. Lett. $\underline{12}$, 303 (1968).

5. G. W. Groves and A. Kelly, Phil. Mag. $\underline{7}$, 892 (1962).

6. B. Edmonson and G. K. Williamson, Phil. Mag. $\underline{9}$, 277 (1964).

7. L. Van der Pauw, Phillips Res. Rpt. $\underline{13}$, 1 (1958).

DIRECT MEASUREMENT OF IMPURITY AND DAMAGE DISTRIBUTION IN ION

IMPLANTED ZnTe BY CATHODO AND PHOTOLUMINESCENCE[+]

D. Demars[*], M. Quillec, M. Ravetto, J. Marine, G. Guernet

L.E.T.I. - Centre D'Etudes Nucléaires de Grenoble

B.P. 85 - 38041, Grenoble - France

ABSTRACT

In this paper, we present the luminescence properties of Boron and Aluminium implanted ZnTe. Once implanted and eventually annealed, ZnTe is deepened by Argon ionic sputtering to different depths (down to 2.2 µ). Then cathodoluminescence at 77°K and photoluminescence at 4.2°K are performed on each step. From the spectral distribution and the quantum efficiency measurements, it appears clearly that the effect of implantation is significant far beyond the penetration depth of impurities. On the other hand, annealing at 550°C during 30 minutes is not sufficient to eliminate the created damage. On the contrary, far from the surface, the annealing reduces the quantum efficiency. A model, based on the interaction between damage and luminescent centers associated with implantation, is proposed to explain the behaviour of implanted impurities in II-VI semiconductors. Important conclusions are also inferred in order to improve the external quantum efficiency of ZnTe implanted electroluminescent diodes.

INTRODUCTION

Zinc Telluride is the only wide bandgap II-VI semiconductor to be obtained p-type as-grown. Its luminescence properties, 1% external quantum efficiency at 300°K in the green (5500 Å), make it very interesting for electroluminescent devices purpose. In the mean time, tentatives to make useful devices by ion implantation [1,2] show that a compensated region, of several microns width, is created between the substrate and the implanted layer, whereas the impurity distribution, measured by the "ion microanalyser", is in good agreement with theoretical calculations [3]. However, if we assume that the compensated region is due to the diffusion of implanted impurities, the sensibility of the "ion microanalyser", limited to few 10^{17} cm^{-3} [4], is not sufficient to detect the phenomenon.

Luminescence measurements are widely used to characterize impurities in II-VI compounds [5]. They allow to detect impurity concentrations as low as 10^{15} cm^{-3}, in the case where the impurity is associated with a radiative transition. Furthermore, by mixing ionic sputtering and luminescence measurements, new informations about the impurity distribution can be obtained. This paper relates the results obtained, after Boron or Aluminium implantation in ZnTe, by cathodo and photoluminescence versus depth. An original method that permitted the fabrication of green electroluminescent diodes, is deduced and briefly described.

EXPERIMENTAL PROCEDURE

ZnTe samples obtained by a Bridgman technique in our laboratory have 25 mm diameter. They are mechanically polished and chemically etched in a solution of 1% Bromine in Methanol before being implanted on all the surface except on a small masked region used as a reference. Two series have been implanted. The first one with Aluminium at 200 KeV (samples temperature 77°K), and the second one with Boron at 140 KeV (samples temperatures 300°K) to obtain almost the same penetration depth. In both cases, implanted dose was 5.10^{14} at/cm^2. The annealing, when indicated, is performed at 550°C during 30 minutes under a flux of zinc vapour.

A number of pits are performed by sputtering with Argon ions at 500 eV. The current density of the Argon beam is 500 µA/cm^2 giving a sputtering rate of 12 Å/s. For each study, the different depths are carried out on the same 25 mm diameter slice. Depth measurements are taken with a "Talystep", they indicate that all the craters are flat-bottomed to within 2%.

Cathodoluminescence is performed on each step with the sample held at 77°K by sticking it to a cold copper finger. The energy of the incident electrons is 20 KeV and the current, falling on 1 mm^2 area, limited to 50 µA. Furthermore, to prevent heating the sample, the current is pulsed. Pulse duration is 100 µs every millisecond.

For photoluminescence measurements at 4.2°K, the excitation is produced by the 4880 Å line of an Argon Laser. The power used is continuous 150 mW and heating problems resolved by immersing the sample in the Helium coolant.

RESULTS AND DISCUSSIONS

Our first care was to verify that the ion sputtering did not disturb the crystal surface leading to parasitic transitions. The only effect observed, for an unimplanted sample, is a reduction of the quantum efficiency when the surface is strongly sputtered. The maximum difference between a virgin surface of ZnTe and a 2.2. µ sputtered surface is lower than 50%.

A) <u>Aluminium implantations</u>

On fig. 1, we represent the cathodoluminescence spectra before and after Aluminium implantation. The implanted sample is annealed.

IMPURITY AND DAMAGE DISTRIBUTION

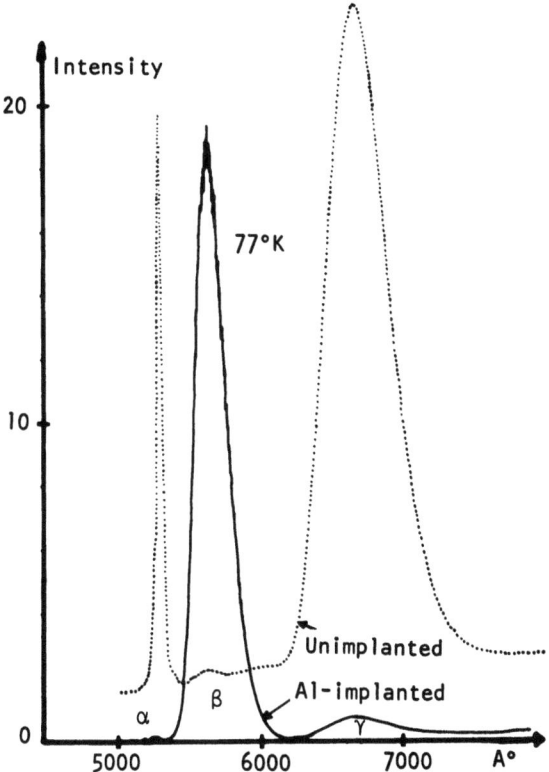

Fig. 1 - Cathodoluminescence spectra at 77°K of ZnTe before and after Aluminium implantation. Electrons energy is 20 KeV.

Three lines are detected. The α transition at 2.34 eV (5280 Å) is a narrow line, its importance, compared to the total emitted energy, depends on the analysed depth. This line is also detected before implantation as well as the γ transition at 1.85 eV (6650 Å). Both of them are identified to be due to bound excitons for the α line [6] and to oxygen for the γ line [5]. The β transition is the new one that appears after Aluminium implantation. Its identification is not as clear as for α and γ lines. It could be attributed to the defects created by the ion beam, mainly zinc vacancies, or to a luminescent center associated with the Aluminium impurity.

To clarify this point we took the photoluminesce spectra of the same sample at 4.2°K as indicated on fig. 2 where we represent also the spectrum obtained after low dose, 2.10^{13} at/cm^2, implantation. The 2.232 eV (551 Å) line and its phonon replicas, designed β-line in the cathodoluminescence spectrum, are well resolved at low dose rate. But as we increase the dose, this line broadens as shown on fig. 2 and phonon replicas disappear. MEESE [7], BRYANT [8], and RODOT [9], attribute this line to the transition between the conduction band and the second ionisation level of zinc vacancy. Considering this interpretation, the evolution of β line versus depth would give the distribution of zinc vacancy, i.e. the damage distribution. However, samples implanted with Argon or with Zinc, just to produce the damage, do not show any transition identical to those discussed above, whereas lightly Aluminium doped as-grown AnTe, about 10 p.p.m level, has the same spectral response as Aluminium implanted samples. We observed also that in these crystals, the β line is more or less important depending on the concentration of Aluminium in the melt. An explanation could be that the transition

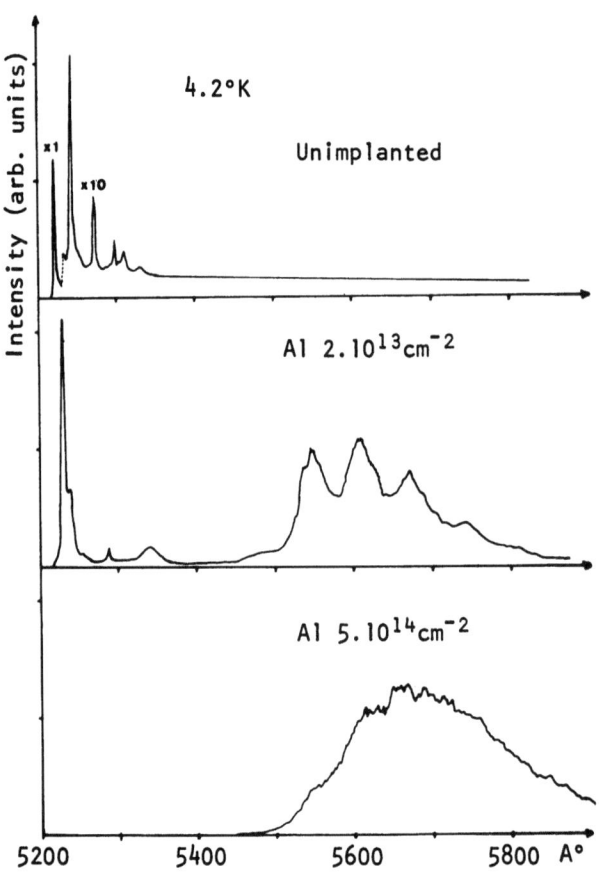

Fig. 2 - Photoluminescence spectra at 4.2°K of Aluminium implanted ZnTe. Excitation is produced by the 4880 Å line of an Argon laser.

involving Aluminium, isolated or associated with native defects, on one hand, and Zinc vacancy on the other hand occurs at the same place, or that it is a transition which is not sufficiently identified. Our experimental results as well as those available in the literature are not sufficient to remove the doubt.

In spite of this discussion, the evolution of the β-line versus depth would give the range of action of ion implantation in this material. Fig. 3 shows how the total quantum efficiency varies with sputtered depth. The measurements are carried out by cathodoluminescence at 77°K. As in all analysed spectra, taken to 2.2 μ depth, the β-line dominates, the curve represented on Fig. 3 gives the evolution of the effect associated with ion implantation. On this figure we also indicate the damage profile calculated by J.L. COMBASSON [10] according to a model proposed by K.B. WINTERBON [11]. Three zones can be defined:
- zone A with decreasing efficiency
- zone B where it grows up to a maximum
- zone C where it decreases towards a low value.

The first remark is that the effect of ion implantation extends far beyond the penetration depth of the damage or of the impurities detected by the ion microanalyser (β-line dominates, even after etching 2.2 μ).

To explain the behaviour of the efficiency reported on fig.3 let us consider a profile of luminescent centers, associated with the ion implantation, extending up to 1.5 μ as indicated on fig. 4 curve A. On the other hand, we suppose that we have a uniform generation function 1 μ width produced by the 20 KeV excitation electrons. A simple convolution of curve A with this step function would give curve B with a maximum on the surface. In the mean time, the damaged region up to 4000 Å have two effects:

1) the non radiative centers, due to defects, dominate the recombinations and reduce the absolute value of the quantum efficiency as indicated on curve C.

2) the perturbed zone introduces a strong reabsorption which explains the decrease in the external efficiency represented by curve D.

Finally, the decrease in efficiency up to 1800 Å, zone A of fig. 3, is attributed to surface recombination directly related to the local concentration of defects. This non radiative process would absorb a part of created pairs until the damaged zone has disappeared.

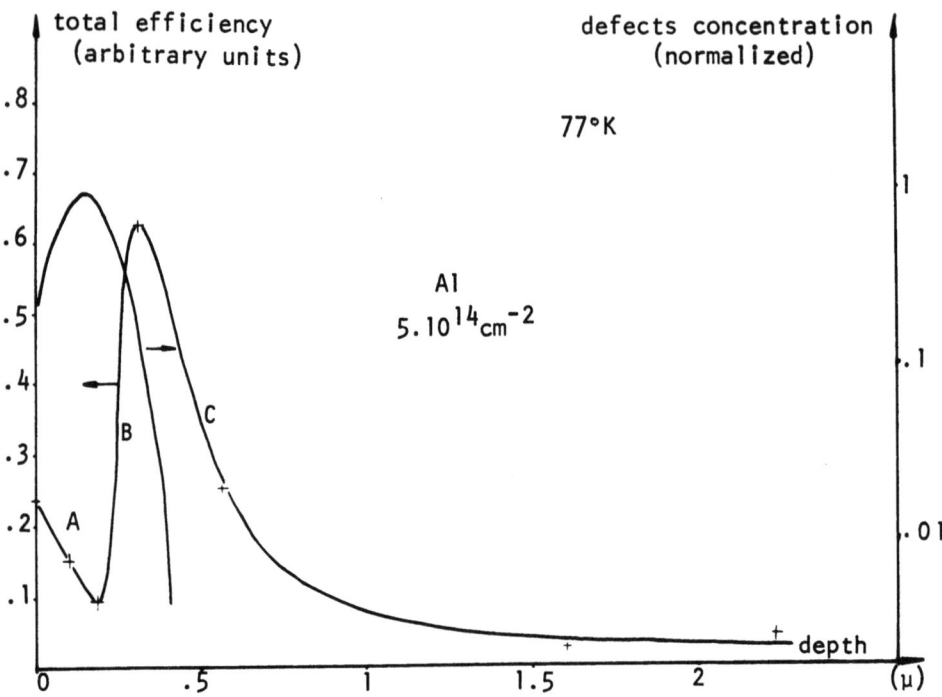

Fig.3 - Total efficiency variation with sputtered depth measured by cathodoluminescence at 77°K (Scale on the left hand side), and calculated damage profile for 200 KeV Aluminium implantation (Scale on the right hand side).

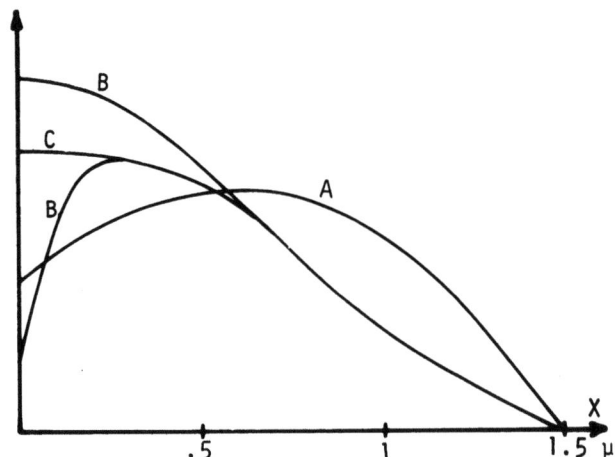

Fig. 4 - Schematic representation of our proposed model. Vertical scale is arbitrary.

IMPURITY AND DAMAGE DISTRIBUTION

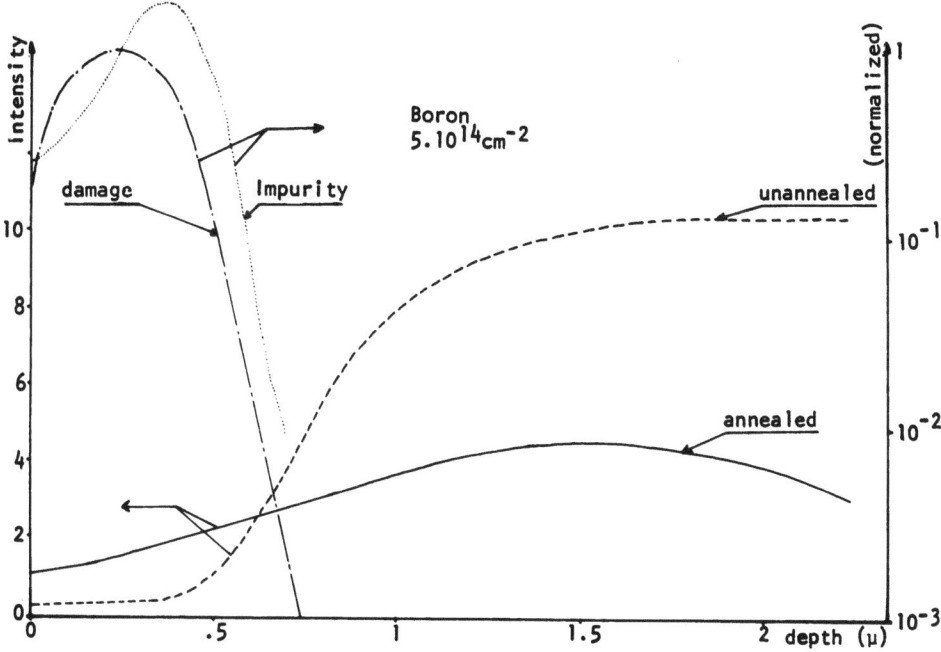

Fig.5 - Luminescence intensity variation with sputtered depth measured by photoluminescence at 4.2°K. The scale on the left is identical for annealed and unannealed samples. Calculated damage and measured Boron profiles are normalized (scale on the right).

To summarize, our simplified model brings out that even after annealing, implanted ZnTe has a weak luminescent efficiency in the damaged region associated with a strong reabsorption and a non radiative recombination process on the surface. This behaviour, different from the one observed with mono-atomic semiconductors like Silicon, could be the consequence of the fact that we have two kinds of defects, those involving Zinc atoms and those involving Tellurium atoms. When heated, the recombination kinetic of defects may generate stable complex centers which are responsible of non radiative transitions.

B) <u>Boron implantations</u>

We carried out the same measurements on Boron implanted ZnTe. Photoluminescence intensity variation versus sputtered depth is shown on fig.5 before and after annealing. The curves represent the total energy variation but the evolution of each line is identical, they are not drawn in order to make the figure comprehensive. Photoluminescence compared to cathodoluminescence is a surface phenomenon. Pairs, generated in a depth lower than 1000 Å, probably diffuse to less than 2000 Å. So the curves of fig.5, represent approximatly the real variation of luminescent centers efficiency. Impurity and damage profiles are also represented on this figure to facilitate interpretation. The annealing, whilst improving efficiency in the perturbed zone, has also the effect of reducing it in the depth. On the other hand, the annealing at 550°C is not sufficient to bring the efficiency to its pre-implantation value whereas channeling and electron microscopy studies show that such temperatures restore the cristallinity [12]. Fig.5 indicates also that the effect of implantation is significant far beyond the range predicted by L.S.S. theory. This migration is not particular to ZnTe, it seems to be the general case for II-IV compounds [13].

Another proof of this migration is brought by capacitance measurements simultaneously carried out on diodes obtained by evaporation of Indium contacts on the implanted side and electroless Gold contacts deposited on the back side of the samples. C (V) curves indicate that an insulating region almost entirely depleted at zero-bias voltage is present. Its width deduced at V = 0 is 1.5 µ before annealing and grows up to 3.8 µ after 550°C annealing during 30 minutes. These values are obtained for Boron implantation at 140 KeV and a dose of 5.10^{14} at/cm^2. The migration is enhanced by annealing but also existed before. It takes place during implantation, even at 77°K, as we saw it by capacitance measurements carried out, in situ, after Boron implantation at 77°K. As dominating defect, Zinc vacancy, would

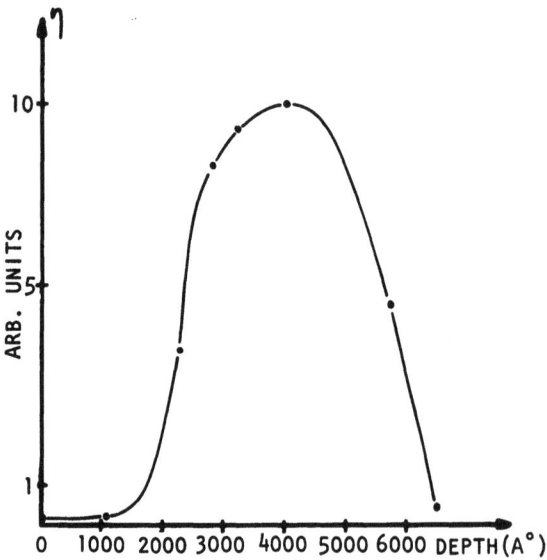

Fig. 6 - Electroluminescence external quantum efficiency variation with sputtered depth. Diodes are Boron implanted at 140 KeV.

accentuate the p-type character, the existence of a compensated region implies the presence of a donor. The nature of compensating center is still unknown to this day, it might be an association between implanted donor and Zinc vacancy.

ELECTROLUMINESCENT DIODES

The above experimental study brings out two main conclusions; the effect of ion implantation extends to several microns and the annealing reduces the quantum efficiency in the extended region. Now, the electroluminescence comes from this region where the recombination of minority injected carriers occurs, therefore annealing is not suitable for electroluminescence purpose and has to be replaced by the following procedure.

Once implanted, ZnTe surface is sputtered to a depth depending on the nature of implanted ion as well on its energy. The aim of this etch is to remove defects, because they are closer to the surface tham impurities, while keeping the major part of implanted impurities. Then, contacts are deposited on both sides and diodes set up for measurements.

Fig.6 indicate how the external quantum efficiency of these diodes varies with the sputtered depth. The curve is drawn for Boron implantation at 140 KeV and a dose of $5.10^{14} at/cm^2$. From 0 to 1800 A° surface recombination dominates, so we have a very low quantum efficiency. Then, as damaged layer is eliminated, reabsorption of emmitted ligth is reduced ; the external efficiency grows up to a maximum and falls down again because the recombinating active region becomes too small compared to the diffusion length of minority carriers injected from the bulk. After sputtering 4000 Å, the external quantum efficiency compared to the one obtained with annealed diodes is over five times higher. However, to this day, electroluminescence efficiency of implanted and sputtered ZnTe diodes is still lower than the one obtained with Gallium Phosphide green LEDS.

CONCLUSION

Combining controlled surface sputtering with cathodo and photoluminescence measurements have permitted to settle the variation of radiative recombination mechanisms versus depth. Our study points out that luminescence effects due to implantation extend far beyond the penetration depth of implanted impurities. That is to say that a diffusion occurs even at low temperature. Furthermore, annealing does not produce the same effect as in mono-atomic semiconductors case. We measured a reduction of luminescence quantum efficiency after an annealing which was sufficient to eliminate implantation damage. According to these two facts, we conceived a new technology for ZnTe LEDS fabrication where annealing is replaced by sputtering the damaged layer. So doing, electroluminescence quantum efficiency of ion implanted ZnTe diodes is considerably enhanced.

ACKNOWLEDGMENTS

We wish to express our gratitude to J.C. PFISTER and J.L. PAUTRAT for many useful discussions.

We also thank J.L. COMBASSON for impurity and damage profiles calculations as well as B. BLANCHARD and E. LIGEON for ion microanalyser measurements and sputtering experiments.

REFERENCES

[1] J. MARINE, H. RODOT - Appl. Phys. Letters, 17, 352 (1970)
European Conference on ion implantation, Reading (1970)

[2] F. CHERNOW - Private communication

[3] J.L. COMBASSON - Internal notes

[4] B. BLANCHARD, N. HILLERET, J.B. QUOIRIN - J. of Radioanalytical chemistry, 12, 85, (1972)

[5] J.L. MERZ, L.C. FELDMAN - Appl. Phys. Letters, 15, 129, (1969) and Phys. Rev. 176, 961, (1968)

[6] R.E. HALSTED, M. AVEN, H.D. COGHILL - J. of Electrochem. Soc. 112, 177, (1965)

[7] J.M. MEESE, Y.S. PARK - Conference on defects in semiconductors Reading (1972)

[8] F.J. BRYANT, A.T.J. BAKER - Physica Status Solidi (a), 11, 623 (1972)

[9] H. RODOT, P. LECLERC, N. HAMMOND - Conference on electronic materials - SAN FRANCISCO (1971)

[10] Damage profile is assimilated to the deposited energy profile

[11] K.B. WINTERBON - Rad. Effects, 13, 215, (1972)

[12] A. BONTEMPS, E. LIGEON, J. MARINE, J.C. PFISTER - Congrès du Centenaire de la S.F.P. - VITTEL (1973)

[13] P.F. ENGEL, F. CHERNOW - This conference

+ This work is supported by D.G.R.S.T. contract

* Actual address : University of OTTAWA - CANADA

PROPERTIES OF Aℓ AND P ION-IMPLANTED LAYERS IN ZnSe

Y. S. Park, B. K. Shin,* D. C. Look,† and D. L. Downing

Aerospace Research Laboratories

Wright-Patterson Air Force Base, Ohio 45433

ABSTRACT

Low-resistivity n- and p-type layers have been produced in ZnSe by room-temperature Aℓ and P implantation, respectively, and subsequent annealing. The layers have been characterized by electrical and photoluminescence measurements as functions of ion energy and dose, and annealing time and temperature.

INTRODUCTION

Ion implantation has shown much promise as a doping technique in ZnSe. Recently, we reported the successful formation of p-n junctions in Li-implanted ZnSe.[1] ZnSe p-n junction electroluminescent diodes[2] and switching and memory devices[3] were also fabricated by P-implantation. In these reports, evidence regarding type conversion and p-n junction formation was presented from thermal probe measurement, I-V characteristics, photovoltaic effect, and low-voltage injection electroluminescence. Hall-effect and sheet-resistivity measurements have also been made on n-type layers produced by Aℓ implantation.[4] In this report we present the results of a more detailed investigation of the implantation effects.

The substrate used was a melt-grown cubic single crystal of either high- ($10^8 \Omega$-cm) or low- (0.1Ω-cm) resistivity n-type ZnSe. The use of a high-resistivity substrate for Aℓ-implantation makes it possible to confine the resulting electrical characteristics to the lower-resistivity implanted layer region. In the case of P-implantation, the formation of p-n junctions provides the necessary isolation from the low-resistivity substrate. Low-resistivity substrates were prepared from the high-resistivity Aℓ-doped samples by firing in molten Zn for 24 hours. The crystals were mechanically polished, and then chemically etched (to remove mechanical damage) at 90°C for 1 min, in a mixture of 2 parts H_2SO_4 and 3 parts saturated aqueous solutions of $K_2Cr_2O_7$, followed by a 20 sec rinse in a boiling 25% solution of NaOH. The electrical measurements on the P-implanted samples were carried out using a van der Pauw configuration.[5]

RESULTS AND DISCUSSION

Aℓ IMPLANTATION

ZnSe substrates, prepared as described above, were implanted at room temperature with 90 keV Aℓ ions and then annealed for 30 minutes in evacuated quartz tubes at temperatures ranging from 700 to 1050°C. In Fig. 1 is plotted the room temperature sheet resistivity ($\rho_\Box$) for samples implanted with $10^{12} - 10^{16}$ Aℓ ions/cm^2 and then annealed at 700, 800, or 900°C. For annealing temperatures much lower than 700°C, or much higher than 900°C, $\rho_\Box$ was very high, approaching the substrate value. It is obvious from Fig. 1, especially the 900°C data, that something besides the Aℓ is contributing to the conductivity. The crystals were analyzed by mass spectroscopy to determine what residual impurities existed. The most abundant impurity was Aℓ at 1ppm ($\sim 10^{16}$/cm^3). An earlier group of crystals, from a different growth, did not exhibit this "excess" conductivity, and further studies will be required to determine the exact cause.

For the 800°C annealing data, $\rho_\Box$ is quite dependent upon the Aℓ dose for doses greater than 10^{13}/cm^2, and thus the temperature dependence of $\rho_\Box$, shown in Fig. 2, should be characteristic of the donor induced by the Aℓ implantation and subsequent annealing. Near room temperature $\rho_\Box$ becomes rapidly varying, approaching an activation energy of about 0.1 eV, while at low temperatures it appears that impurity conductivity dominates. The activation energy for the impurity conductivity decreases from about 10 meV to about 2 meV as the Aℓ dose increases from 10^{13}/cm^2 to 10^{15}/cm^2, implying that the carriers are becoming degenerate.[6]

To determine the resistivity ρ of the layer, the depth x must be known ($\rho = \rho_\Box x$). We have investigated this problem by means of a stripping technique in which $\rho_\Box$ is monitored as successive layers are removed by using the etch described previously. This etch is known to remove 50Å/sec from a polished ZnSe surface at room temperature.[7] Typical results are shown in Fig. 3 in which the inverse resistance of a sample annealed at 900°C is plotted as a function of the amount of material removed. The actual donor profile, ignoring mobility variation with depth, should be the derivative of such a

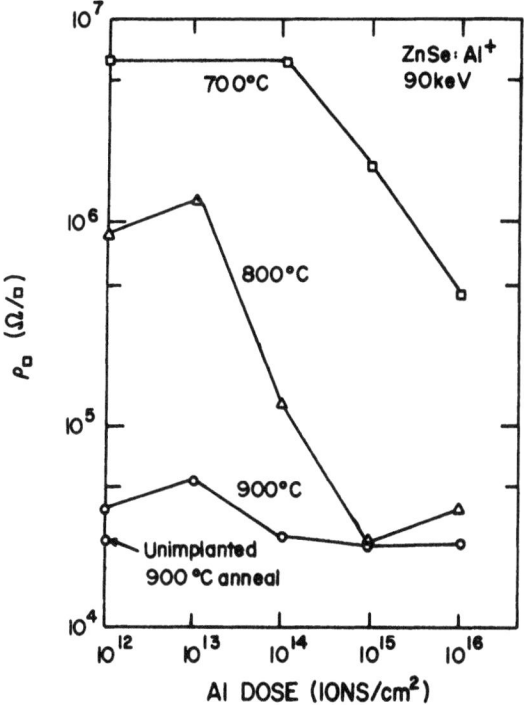

Fig. 1. The room-temperature sheet resistivity $\rho_{\square}$ vs Al dose for samples annealed at 700, 800, or 900°C.

curve but it is clear that appreciable conductivity exists at depths greater than 1μm. Since the projected range of the 90 keV Al ions is only about 0.1 μm, it appears that diffusion has taken place at 900°C. There is wide scatter in the data for different samples, evidently due to non-uniform etching of the surfaces, but in general the samples annealed at 700°C show x≲0.05μm, at 800°C, x≈0.1-0.5μm, and at 900°C, x≳1μm. Unfortunately, the present data are not good enough to calculate a diffusion constant.

The Hall mobilities of these samples are very low, typically about 1 cm^2/V sec. Using this value, the sheet carrier concentrations, given by n≈6x10^{18}/$\rho_{\square}$μ, are about the same as the implanted-Al doses for the 10^{13}, 10^{14}, and 10^{15}/cm^2 samples annealed at 800°C, although not for the 10^{16}/cm^2 sample. Since the layer depth is about 1μm or less, we would expect the ionized impurity density to appreciably limit the mobility. However, the measured Hall mobility may be lower than the true, microscopic Hall mobility if the layer is inhomogeneous.[8] This problem must be investigated further.

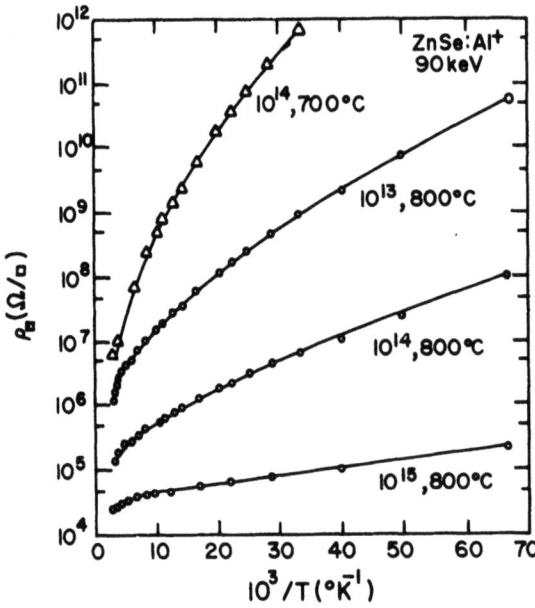

Fig. 2. The sheet resistivity $\rho_\square$ vs inverse temperature for three samples (of different Aℓ doses) annealed at 800°C, and for one sample annealed at 700°C.

To summarize the Aℓ-implantation data, we can create a conducting layer with $\rho \simeq 1\Omega$-cm by implanting with 10^{15}Aℓ/cm^2 and annealing at 800°C. By annealing at 900°C, however, many crystals, regardless of the implanted Aℓ content, exhibit a resistivity of this order of magnitude. This "excess" conductivity also displays nearly degenerate electrical characteristics, implying that a high density of donors is involved. Since mass analysis shows no impurity of greater than 1ppm density, we conclude that a native defect, perhaps Se$_V$, is involved. Another result of the present study, of technological importance, is that Au or Pt contacts sputtered onto the implanted layer are ohmic, whereas these metals usually form rectifying Schottky barriers in n-type ZnSe.

Photoluminescence spectra of as-implanted samples at 4.2°K are dominated, in general, by the characteristic phonon-assisted broad edge emission bands (Fig. 4a) which result from a recombination of a hole bound to an acceptor and an electron bound to a donor,[9] but do not show a well-known self-activated luminescence band centered around ~6300Å arising from transitions between localized states of

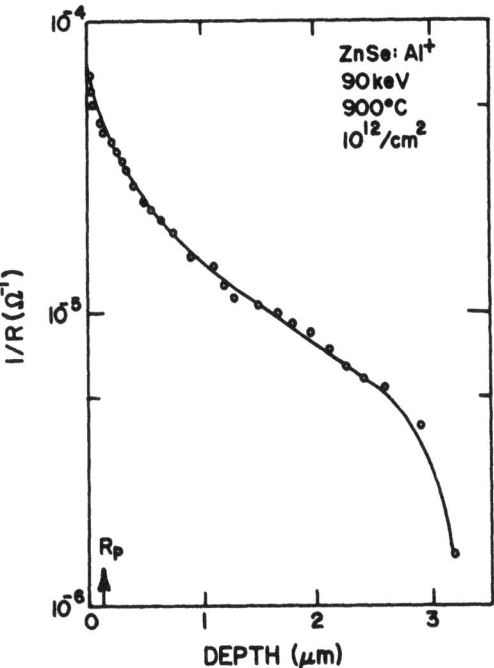

Fig. 3. The conductance (1/R) vs amount of surface removed for a ZnSe sample implanted with $10^{12}A\ell/cm^2$ at 90 keV and then annealed at 900°C. These data are typical for all samples annealed at 900°C, regardless of $A\ell$ content.

a Zn vacancy-aluminum complex center.[10] The intensity of the edge emission bands in the implanted region is inversely proportional to the ion doses. This reduction in emission intensity can be attributed to the non-radiative centers formed by implantation-induced damage which compete with the radiative process for a hole-electron pair recombination. The self-activated luminescence band does not appear in those samples annealed at up to 900°C for 30 minutes, but is dominant in those samples annealed for a prolonged period of 24 hours at above 1050°C. Apparently, complexing between the $A\ell$ dopants and Zn vacancies is taking place above this temperature. Fig. 4b shows the self-activated luminescence band seen in the $10^{15}/cm^2$ samples annealed at 1050°C for 24 hours. One of the features of this luminescence band is that as the temperature of the sample is increased from 4.2°K, thus decreasing the bandgap energy, the luminescence peaks shift toward higher energy, and the band increases in width according to a square root hyperbolic cotangent. For the

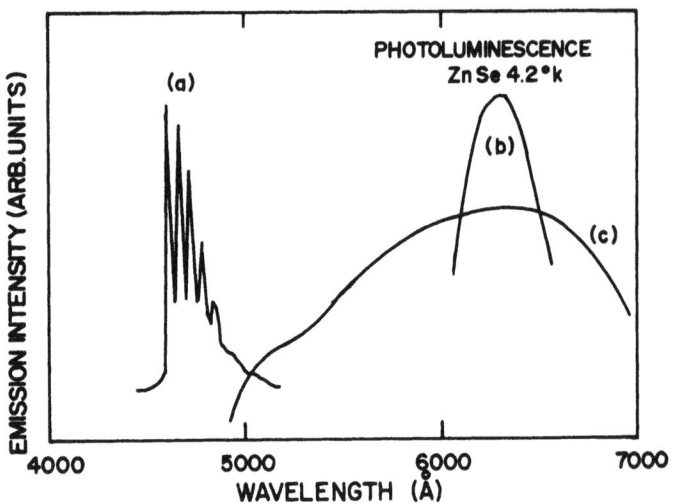

Fig. 4. Photoluminescence Spectra of Al-implanted ZnSe at 4.2°K.
(a) 90 KeV, $10^{14}/cm^2$ Al, 800°C, 30 min anneal.
(b) 90 KeV, $10^{15}/cm^2$ Al, 1050°C, 24 hr anneal.
(c) 90 KeV, $10^{16}/cm^2$ Al, as-implanted.

heavily implanted $10^{16}/cm^2$ samples, the broad band extending from 5000 to 7000Å (Fig. 4c) was seen before annealing and the characteristic edge emission bands were not observed, evidently due to this disorder.

P IMPLANTATION

P ions were implanted at room temperature in both high- (10^8 Ω-cm) and low- (0.1 Ω-cm) resistivity n-type ZnSe substrates doped with Al. For low-energy (70 keV) implants, an effective hole mobility up to 15 cm^2/V sec was observed by van der Pauw measurements. Three isothermal-annealing stages were observed. Within the first 5 min of annealing, the sheet resistivity decreased considerably; for the next 1 hour of annealing, the effective hole mobility went through a maximum, and prolonged (~24 hour) annealing suppressed the mobility. Isochronal annealing of 10^{16} P/cm^2 implants at ~450°C showed sheet-resistivity minima of ~10^6 Ω/◻ and 5 x 10^4 Ω/◻ for implants on high- and low- resistivity substrates, respectively. For ~600°C anneal, the sheet resistivity of 10^{14} P/cm^2 implants had minima of 10^7 Ω/◻ and ~5 x 10^5 Ω/◻ for implants on high- and low-resistivity substrates, respectively. When these implants are annealed above 600°C, electrical activities are reduced which suggests that enhancement of well-known compensation effects in this material takes place at higher temperatures. For moderate-energy (400 keV) implants, optimum electrical characteristics are achieved for 10^{14} P/cm^2 implants when annealed at 500°C for 1 hour; these samples exhibit a sheet resistivity ~$10^5 \Omega$/□, an effective hole mobility ~10 cm^2/V sec, and an effective sheet-hole concentration of ~10^{13}/cm^2. At high energy (~1.8 MeV), the ion dose and post-annealing conditions are virtually independent of electrical properties of the implanted samples, the sheet resistivity remaining high. This effect is believed to be due to formation of the self-induced anion vacancy complex frequently formed by high-energy positive ion implantation in binary semiconductors, which results in total compensation. Optical measurements give an implanted depth less than 1μm. This agrees with the P profiles obtained from Auger spectral analysis with sequential removal of layers by Ar sputtering. Therefore as a result of P implantation, a low-resistivity (~10 Ω-cm) p-type layer was achieved having ~10^{17} cm^{-3} hole concentration with ~10% doping efficiency. The presence of a low-resistivity p-layer suggests the formation of shallow acceptors by P implantation.

Diodes were fabricated to study the junction properties of the implanted samples. The I-V characteristics of the annealed diodes obey an ideal p-n junction relation $I = I_o \{\exp(eV/nkT) - 1\}$ similar to those previously reported.[2] The coefficient n varies considerably from diode to diode between 1.3 and 1.7 depending on post-annealing conditions. In a wide bandgap material such as ours, it is known that for the forward current in a p-n junction diode device, n = 1 when the diffusion current dominates and n = 2 when the recombination current dominates. When these currents are comparable, n had a value between 1 and 2. The breakdown voltage V_B in reverse bias occurred at ~30 V at room temperature. In general, V_B may be expressed in terms of the bandgap energy E_g and the background concentration N_B of the low-doped side of an abrupt junction, given by[11] $V_B \simeq 60(E_g/1.1)^{3/2} (N_B/10^{16})^{-3/4}$. For the implanted junction, where $V_B = 30$ V and $E_g = 2.67$ eV at room temperature, N_B is then ~10^{17} cm^{-3}. Therefore, the doping concentration in the low doped side of the junction must be ~10^{17} cm^{-3}. This is consistent with the hole concentration found in Hall measurements.

Previously, very large values of ionization energy of the known acceptors were reported,[12] ranging from 0.66 to 0.75 eV from the valence band edge. However, in a recent report of p-type conduction in undoped ZnSe,[13] a shallow acceptor level (~0.1 eV) was measured in addition to a deep level (0.65 to 0.75 eV). A shallow acceptor level of 0.05 eV was also measured in P-implanted CdS.[14] These authors concluded that a complex defect center was responsible for formation of the shallow acceptor level. The shallow acceptor level which formed in our P-implanted ZnSe samples may also be due to P-complexes. Further experimental studies are required to identify the existence and nature of the shallow acceptor level in P-implanted ZnSe.

ACKNOWLEDGEMENTS

The authors wish to thank J. A. Hutchby of NASA Langley Research Center who performed the Aℓ implantation of the samples. Thanks are also due J. E. Ehret and D. E. Johnson of our laboratory for implanting P and sample preparations, and to D. Walters for mass analyzing an unimplanted sample.

REFERENCES

1. Y. S. Park and C. H. Chung, Appl. Phys. Lett. $\underline{18}$, 99 (1971).

2. Y. S. Park and B. K. Shin, J. Appl. Phys. $\underline{45}$, 1444 (1974).

3. B. K. Shin and Y. S. Park, Proc. IEEE $\underline{62}$, 538 (1974).

4. B. K. Shin, Y. S. Park and D. C. Look, Appl. Phys. Lett. $\underline{24}$, 435 (1974).

5. P. M. Hemenger, Rev. Sci. Instrum. $\underline{44}$, 698 (1973).

6. See, e.g., N. F. Mott and W. D. Twose, Advan. Phys. $\underline{10}$, 107 (1961).

7. J. Santiago (private communication).

8. See, e.g., K. W. Böer, Jour, Non-Cryst. Sol. $\underline{2}$, 444 (1970).

9. See, for example, P. J. Dean and J. L. Merz, Phys. Rev. $\underline{178}$, 1310 (1969).

10. W. C. Holton, M. de Witt and T. L. Estle, in International Symposium on Luminescence, The Physics and Chemistry of Scintillators, edited by N. Riehl and H. Kallman (Karl Thienig, Munich, 1966), p. 454.

11. S. M. Sze and G. Gibbons, Appl. Phys. Lett. $\underline{8}$, 111 (1966).

12. Y. S. Park, P. M. Hemenger, and C. H. Chung, Appl. Phys. Lett. $\underline{18}$, 45 (1971). (Other references therein).

13. R. W. Yu and Y. S. Park, Appl. Phys. Lett. $\underline{22}$, 345 (1973).

14. W. W. Anderson and R. M. Swanson, J. Appl. Phys. $\underline{42}$, 5125 (1971).

*Permanent Address: Systems Research Laboratories, Dayton, Ohio
†Permanent Address: University of Dayton, Dayton, Ohio

N-ION IMPLANTATION INTO ZnSe[*]

C. H. Chung, H. W. Yoon and H. S. Kang

Dept. of Physics, Yonsei University

and

C. H. Tai

Dept. of Physics, Ewha Women University

Seoul, Korea

ABSTRACT

Low resistivity n-type ZnSe single crystals implanted with 60 keV or 60 and 25 keV N-ions were converted to p-type conductivity after proper annealing. The type conversion was confirmed by a thermal probe method, I-V characteristics and a photovoltaic effect. Photoluminescent and electroluminescent measurements showed the edge emission (2.68 eV), intense green (2.38 eV) and faint red (1.99 eV) bands at $77^{\circ}K$. The green emission quenched out completely at room temperature and only weak red band (1.97 eV) remained. The radiative recombination of the anomalous emission intensity at $77^{\circ}K$ can be attributed to the coorperation of the radiative recombination of the implanted nitrogen center (0.09 eV) and the intrinsic edge emission.

INTRODUCTION

Recently, the interest in ion implantation into wide bandgap II-VI compounds is increasing because this is an effective means of achieving type conversion of their conductivities despite the self-compensation effect. Many workers converted conductivity types and investigated consequent properties as reported in CdTe: As (1), CdS:Bi (2), CdS:P (3,4), CdS:N (5), ZnTe:F (6) and ZnTe: Cl (7).

[*]Work supported in part by the Graduate School, Yonsei University and the Foundation of Industry-Academy Cooperation

Park and Chung obtained p-type conductivity with high conductivity n-type ZnSe by Li-ion implantation (8). Type conversion was checked by thermal probe method, I-V characteristics and photovoltaic method. Also they discussed the type conversion in conjunction with sharp emission lines in low temperature photoluminescence. Park and Shin reported ZnSe p-n junction electroluminescent diodes, swtching and memory devices, and backward diodes fabricated from n-type ZnSe:Al by P-ion implantation (9). Chung and Kim worked on photoluminescent properties of ZnSe:P-ion (10). Santiago et al. investigated on broad band luminescence of Br, N and Ar-ion implanted ZnSe (11). They studied the behavior of green and red bands due to annealing temperature, and discussed the nature of the both bands.

In this report we will present an anomalous emhancement of the edge emission intensity in forward biased ZnSe diodes formed by N-ion implantation. The experimental method and results will be described in detail.

EXPERIMENTAL

The starting samples for implantation were cleaved from undoped n-type ZnSe single crystals grown by Eagle-Pitcher Co. As-grown crystals showed very high resistivity in the range of 10^8-$10^9 \Omega$.cm at room temperature. Molten-Zn heat treatment (12) for 24 h at 900°C decreased resistivity to the order of 1 Ω.cm. Heat treated samples were mechanically polished, optically finished and chemically etched in concentrated boiling NaOH for 1 min.

All implantations were performed at room temperature with energy up to 60 keV at a dose of 10^{15}-10^{16} ions/cm^2. Sometimes double implantation technique was adopted to insure uniform concentration with energies of 60 keV and 25 keV.

Isochronical annealing temperature was ranged from 150°C to 450°C for 20 minutes each. The type conversion was confirmed by a thermoelectric probe method, I-V characteristics and a photovoltaic effect. The Ohmic contacts to the diode were made by a gold sputtering to the implanted surface and an indium fusion to the opposite side respectively.

To observe the emission spectra, a phase-sensitive method was applied, using an 1/2 m Jarrel-Ash Monochromator and a PAR lock-in amplifier (Model 122) with a chopping frequency of 26 Hz. Samples were directly immersed in liquid nitrogen both for photoluminesence and electroluminesence measurement. The same monochromator was used for photovoltaic measurement with a tungsten lamp.

RESULTS AND DISCUSSION

A dc thermal probe method was employed for type checking in the dark between two Au-sputtered dots made on the implanted side. The signal was compared with those of known p-type ZnTe and n-type CdS to exclude ambiguities. The signal was same polarity with that of ZnTe.

The photovoltaic effect of ZnSe diode was measured at room temperature in order to check the type conversion due to nitrogen ion implantation. This sample was implanted to a dose of 10^{15} ions/cm^2 with the energy of 60 keV and the annealing was carried out at 350°C for 20 minutes. As shown in Fig. 1, the spectral response curve peaked at about 4600 Å (~ 2.7 eV) corresponding to the bandgap energy of ZnSe at room temperature. The small peak appeared at 5000 Å. The origin of this peak can not be explained at present. The polarity of the photovoltage was positive with respect to the unimplanted layer.

The I-V characteristics was measured at room temperature by a curve tracer as shown in Fig. 2. The forward knee voltage observed in the one of the best diode was 1.2 volts and the reverse breakdown voltage was occured at about 35 volts.

By means of the dc thermal probe method, the photovoltaic

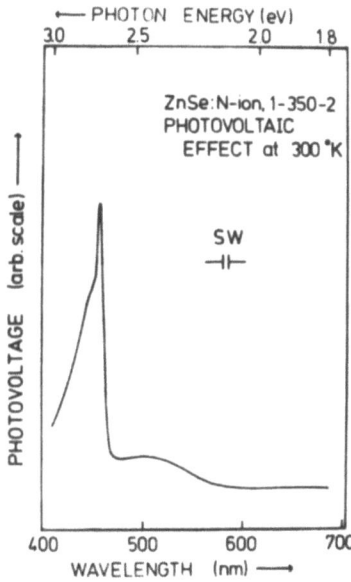

Fig. 1 Spectral response curve of the photovoltage in a N-ion implanted ZnSe diode at room temperature.

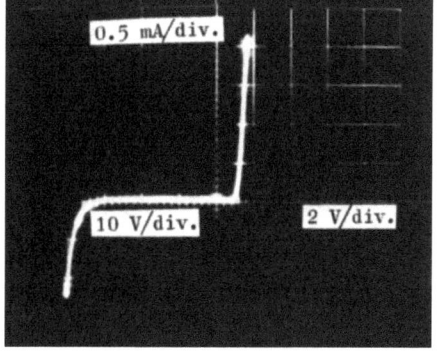

Fig. 2 I-V characteristics of the same diode with Fig. 1 measured in the dark at room temperature.

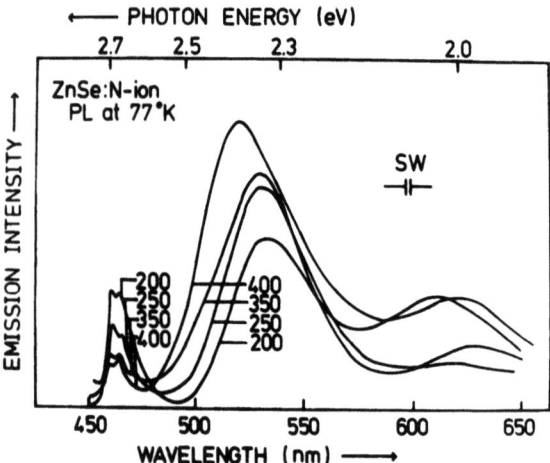

Fig. 3 Photoluminescence spectra of a N-ion implanted ZnSe at 77°K after various annealing stages. Implantation condition; 60 keV at dose of 10^{15} ions/cm^2.

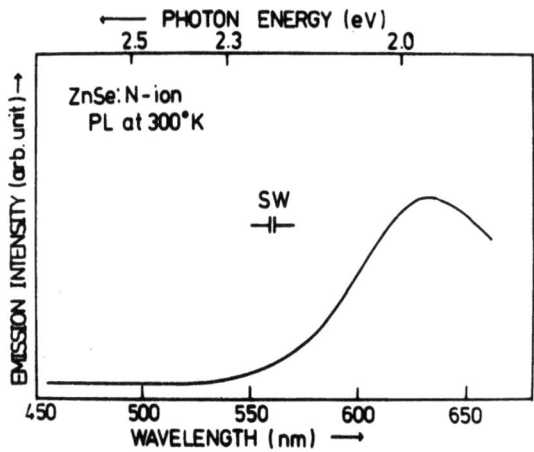

Fig. 4 Photoluminescence spectra at room temperature. Same condition with Fig. 3.

effect and the rectifying property, we have confirmed that the implanted layer was converted to p-type conductivity after proper annealing.

Photoluminescence was measured at both room and liquid nitrogen temperature by uv excitation. At 77°K, the edge emission (~2.7 eV), the strong green band and the weak red band appeared and, in the other hand, they were completely quenched out at room temperature except the red band (1.97 eV) as shown in Fig. 3 and Fig. 4. The green band showed the tendency of the peak shift to

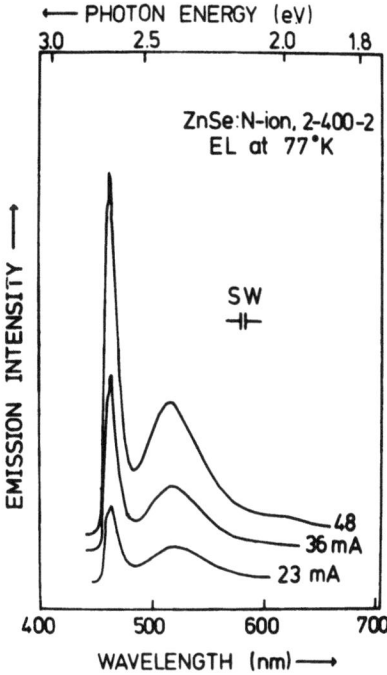
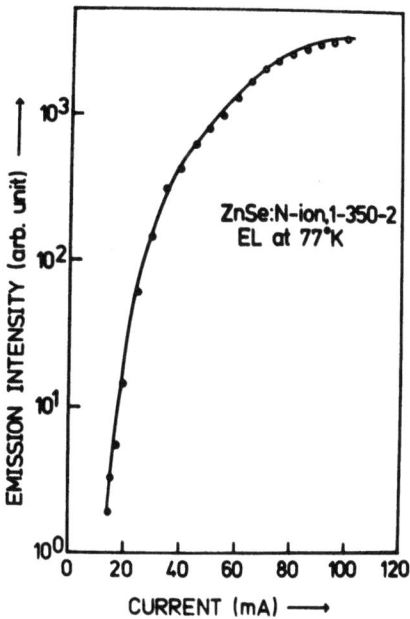

Fig. 5 Electroluminescence spectra of a N-ion implanted ZnSe diode. Annealed at 350°C for 20 min.

Fig. 6 The emission intensity dependence on the forward current in the same diode with Fig. 5.

shorter wavelength side as raising the annealing temperature as shown in Fig. 3. The optimum annealing temperature was estimated to be 350° - 400°C for 20 minutes as deduced from the luminescent intensity.

The green band observed in this study was very similar to those of ZnSe Li-ion (8) and ZnSe:P-ion (10). The green band appeared not only Li-ion implanted sample, but Li, Na or K-doped ZnSe (13). Also similar green bands were observed in Br, N and Ar-ion implanted ZnSe (14). Morever the red band observed at room temperature is similar to that of ZnSe:Al, P-ion (9). These experimental evidences suggest that the green band can be ascribed to Zn-vacancy or Zn-vacancy complex produced during ion implantation and annealing stages (14).

Electroluminescence was investigated at liquid nitrogen temperature. The edge emission, green and red bands correspond to those of photoluminiscence observed at 77°K from uv excitation, but anomalous enhancement of the edge emission intensity has been detected as shown in Fig. 5. Usually the peak intensity of the edge emission of ZnSe is comparable with that of the deep centers only at the near of liquid helium temperature (8). At liquid nitrogen temperature, normally the peak intensity of the edge emission of ZnSe is weaker than that of deep center. The anomalous enhancement of the edge emission intensity can be attributed to the coorperation of the radiative recombination between the free electrons in the conduction band and the shallow aceptors (0.09 eV from the top of the valence band) (14) with the intrinsic edge emission (15). This is very significant behavior as compared to the other ZnSe diode produced by ion implantation, because the enhancement suggests that an efficient injection takes place in the diode other than the coorperation described above.

Further studies of electrical properties are on the way to clarify the nature of the implanted nitrogen centers in ZnSe.

ACKNOWLEDGEMENTS

We would like to express our sincere thanks to Profs. C. C. Lee and Y. K. Ko and Mr. C. N. Hwang who have implanted the samples for this experiment. We are also indebted to Drs. Y. S. Park and S. Ibuki for the provision of the ZnSe sample. Also we appreciate Dr. U. Kim for his encouragement and valuable discussions.

REFERENCES

1. J. P. Donnelly, A. G. Foyt, E. D. Hinkley, W. T. Lindley and J. O. Dimmock, Appl. Phys. Lett., 12, 3030 (1968).

2. F. Chernow, G. Eldridge, G. Ruse and L. Wahlin, Appl. Phys. Lett., 12, 339 (1968).

3. W. W. Anderson and J. J. Mitchell, Appl. Phys. Lett., 12, 334 (1968).

4. Y. Shiraki, T. Shimada, and K. F. Komatsubara, J. Appl. Phys., 43, 710 (1972).

5. S. L. Hou and J. A. Marley, Jr., Appl. Phys. Lett., 16, 467 (1970).

6. S. L. Hou, K. Beck and J. A. Marley. Jr., Appl. Phys. Lett., 14, 151 (1969).

7. J. Marine and H. Rodot, Appl. Phys. Lett., 17, 352 (1970).

8. Y. S. Park and C. H. Chung, Appl. Phys. Lett., 18, 99 (1971).

9. Y. S. Park and B. K. Shin, *Proceedings of the International Conference on Solid State Devices*, Tokyo, 1973, p. 508.

10. C. H. Chung and K. B. Kim, to be published in J. Korea, Phys. Soc., vol. 7, No. 2 (1974).

11. J. J. Santiago, J. E. Ehret, W. R. Woody and Y. S. Park, *Proceedings of the International Conference on Ion Implantation*, New York, 1972.

12. M. Aven and H. H. Woodbury, Appl. Phys. Lett., $\underline{1}$, 53 (1962).

13. Y. S. Park, P. M. Hemenger and C. H. Chung, Bull. Am. Phys. Soc. II, $\underline{16}$, 374 (1971).

14. Phil Won Yu (private communication).

15. D. C. Reynolds, L. S. Pedrotti and O. W. Larson. J. Appl. Phys., $\underline{32}$, 3230 (1961).

ENERGY LEVEL ANALYSIS OF N^+ IONS IMPLANTED CdS

Yoshio Machi, and Sadao Adachi

Tokyo Electrical Engineering College

Chiyoda-ku, Tokyo, JAPAN

ABSTRACT

Energy level analysis has been studied for N^+ ions implanted and not implanted CdS diodes. Those were performed extensively by the methods of thermally stimulated current(TSC), photo-thermoelectric power(PTP), photo-capacitance(P-CAP), thermoluminescence (TL), cathodo-luminescence(CL), and other measurements. Shallow energy levels such as 0.05 eV, 0.1 eV, and 0.13 eV were found from TSC and identified as acceptor type levels. Also, the deep acceptor levels(0.77 eV and 1.2 eV from the V.B.) was found from TSC, P-CAP, and CL experiments. Isochronal anneal for 10 minutes studies were performed between 100 and 700 °C by TSC method. The best annealing temperature for the activation of the shallow levels was about 430 °C. Mn spin density of ESR signal was increased at this temperature.

INTRODUCTION

Using ion implantation technic, a great deal of work has gone into investigating the wide bandgap Ⅱ-Ⅵ compound semiconductors for making p-n junction electroluminescent diode. Especially, there are lot of reports for CdS:N^+, CdS:P^+, and CdS:Bi^+ [1]. In these ion implantation, we wish substitutional group-Ⅴ elements on sulfer site and behavior as an acceptor impurity. However, most of reports discussed only type conversion and deep acceptor levels and did not say so much the origins of the type conversion.

In this paper, we wrote mainly shallow energy level analysis for N^+ ions implanted CdS single crystals, it's type identification and it's annealing effects.

EXPERIMENTALS

The crystals used in these experiments were undoped and low resistive(ρ=0.68 Ω-cm) CdS single crystal(D-31) and high resistive ($\rho \fallingdotseq 2 \times 10^8$ Ω-cm) one(M-4) doped with Mn about less than several ppm. Ion implantation was performed at 200 KeV to a dose of 10^{15} ions/cm^2 at room temperature. For contact to the n-layer and implanted layer, indium and gold metal were evapolated respectively.

EXPERIMENTAL RESULTS

Fig. 1(a) shows typical TSC data for N$^+$ ions implanted and annealed diode and not implanted diode. From the N$^+$ implanted diode, a big bump which contained several peaks was found at low temperature. Therefore, thermally cleaning technic was used to analize these peaks. Fig. 1(b) shows TSC data after level cleanings. For calculation of the activation energies, following equation [2] was used.

$$\exp(E_t/kT) = N_c \cdot v \cdot S_T \cdot T_m^2 / \beta \cdot E_t$$

where assume negligible retraping. E_t is thermal activation energy, β heating rate, T_m TSC peak temp., v thermal velocity of the carrier, S_T capture cross section of the trap, N_c effective density of states.

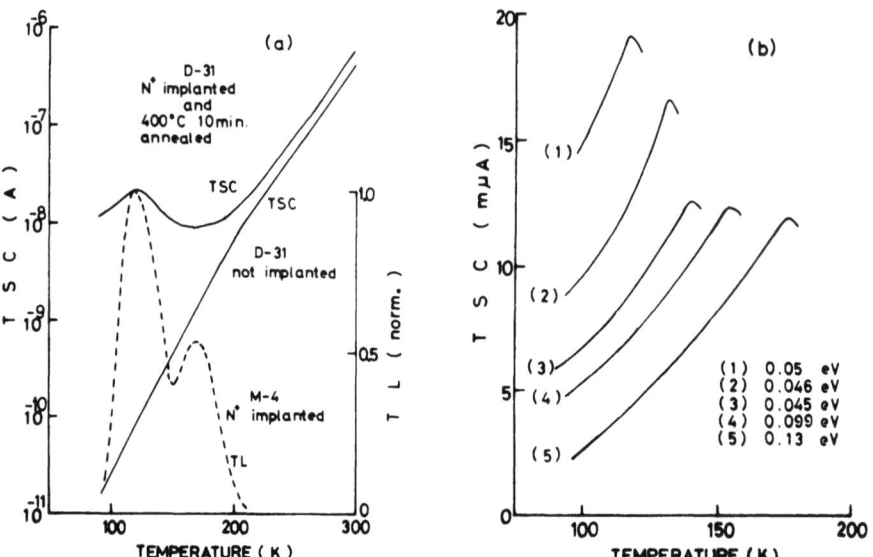

Fig. 1. Comparison of the TSC and TL curves; a) for N$^+$ implanted and annealed CdS and not implanted CdS; b) TSC curves after level cleaning.

This method was known as a variation of heating rate method. The activation energies of 0.05 eV, 0.1 eV and 0.13 eV were obtained from the separate peaks. In those shallow levels, the density of 0.05 eV was dominant than the others. Also, we had more levels such as 0.26 eV and 0.66 eV at higher temperature than those.

TL was measured simultanously when TSC measurement was carried out. The TL glow curve is shown in Fig. 1(a). Before N^+ ions implantation, extremely faint TL glow exists, but after implantation and post annealing, relatively strong TL glow appears. The energy of the two peaks calculated from a initial rise method [2] coincide with TSC shallow energy levels such as 0.05 eV and 0.1 eV. TL emission spectrum shows maximum intensity at about 0.75 μm (1.7 eV from V.B.).

The type identification for the shallow level was performed by the modified thermally stimulated capacitance measurement. Namely capacitance was measured under the two conditions of filled or empty at 77 K. As the depletion layer mainly spreads into n layer of low carrier density than the p layer under reverse bias condition, the capacitance decrease after thermally cleaning of the shallow level, means this level is an acceptor or a hole type level. In our case, the capacitance decreases about 11 % at -1 volt bias voltage. Therefore, it was determined this was an acceptor or a hole type level.

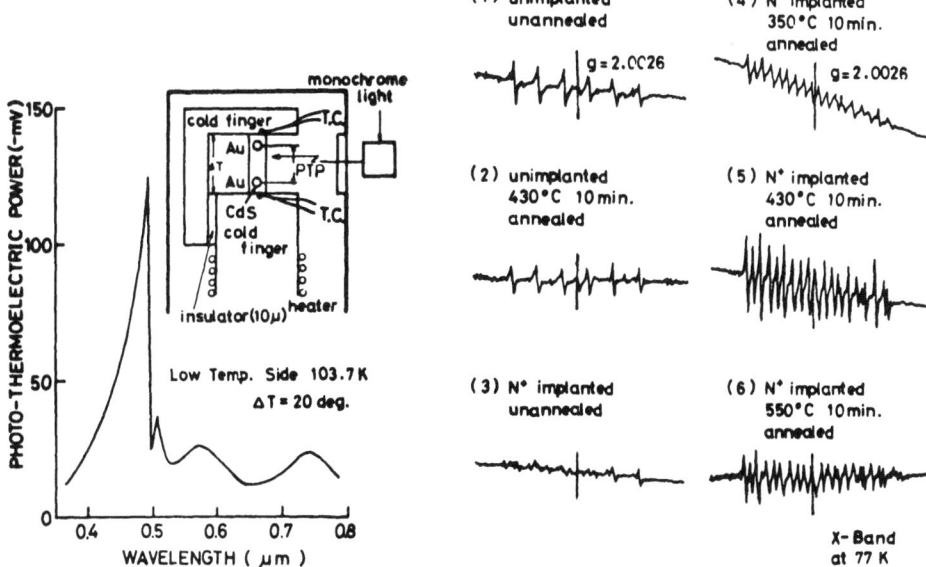

Fig.2. Photo-thermoelectric power measurement of N^+ ions implanted CdS(D-31).

Fig.3. ESR spectra for annealing effect.

Thermoelectric power measurements for the ion implanted layer were performed and N^+ ions implanted layer showed p type in the range of 100~300 K. Also, PTP measurements [3] were applied to the same sample. Fig. 2 shows one of the results. From this figure, several peaks exist and those are 0.06 eV, 0.27 eV, 0.87 eV, and 1.24 eV from the absorption edge. And this experiments were applied to identify those levels types. Those levels were confirmed as acceptor type levels.

For annealing effect study, TSC measurements and ESR measurements were applied. ESR measurements at 77 K were used for isochronal annealing experiment. As M-4 CdS was slightly doped with Mn(less than several ppm), six hyperfine lines of Mn [4] were detected before implantation as shown in Fig. 3. For as-implanted sample, these six lines went into noise level, but after annealing those signals were enhanced in intensity and each of lines showed clear five super hyperfine structure. The optimum annealing conditions for Mn spin density was at 430°C for 10 minutes. Also, isochronal anneal studies using TSC measurement were shown in Fig.4(a). From this figure, isochronal annealing affects strongly for the glow curves. Obviously, it seems suitable annealing temperature gives optimum activation of the shallow energy level such as 0.05 eV. Fig. 4(b) shows total sweep out charge from the depletion layer for the shallow levels as a function of annealing temperature. Therefore, the best annealing condition was at 430°C for 10 minutes.

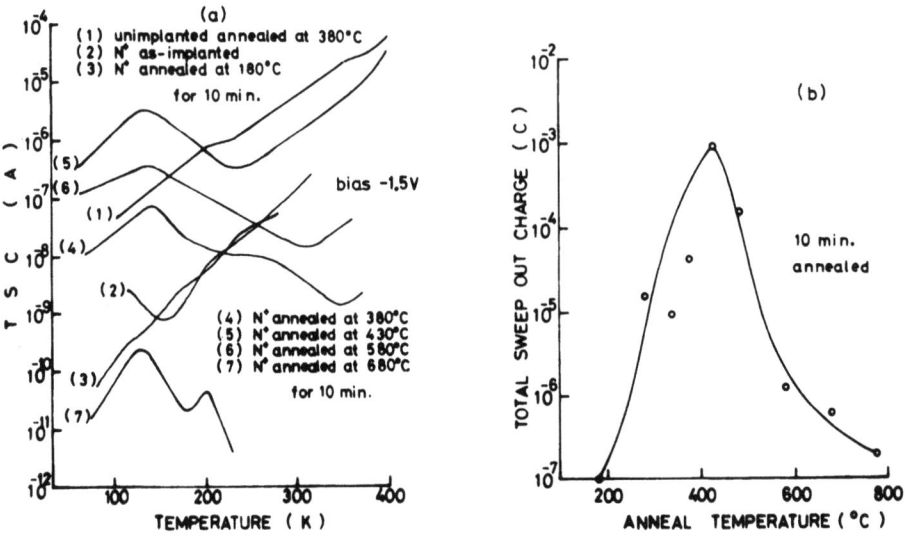

Fig. 4. Isochronal anneal characteristics; a) TSC curves; b) total sweep out charge from the shallow level of the N^+ ions implanted diodes as a function of annealing temperature (M-4).

For isothermal annealing studies at 430°C, TSC method was applied and about 10 minutes anneal gave the optimum activation of the shallow levels.

A profile of the shallow energy levels was examined by TSC experiments for N^+ ions implanted and post annealed (at 380°C for 10 min.) sample (M-4) after each etching removal process by the solution of 20 % HCl. Before ion implantation, the shallow acceptor level probably due to Mn impurity [5] exists slightly. But after ion implantation and post anneal, the density of them incleases only implanted surface side. Especially it becomes the same density to the bulk at about 0.8 μm from the surface.

For the deep level analysis, P-CAP measurements were performed at 77 K. Two deep levels such as 0.95 eV(1.3μm) and 1.3 eV(0.95μm) were found and those were identified as a donor type level and an acceptor type level respectively. Also, CL and PL experiments were performed. Only saying to the deep level, a new relatively strong emission of about 0.7μm comes out after N^+ ions implanted and post annealed crystal.

SUMMARY

We examined energy level analysis of CdS:N^+ single crystals. These levels were summarized in Table 1, also the energy band diagram was shown in it. The shallow acceptor type energy 0.05

Table 1. Energy level of CdS after N^+ ions implantation and post anneal (a), and possible band diagram (b).

(a)

Experimental methods	Energy levels (eV)						
T S C	0.05	0.09	0.13	0.26	0.66		
P T P	0.06			0.27		0.87	1.24
T L	0.05	0.1		1.7 (from V.B.)			
P - C A P						0.95	1.3
Photovoltaic effect							1.24
C L , P L				1.77 (from C.B.)			

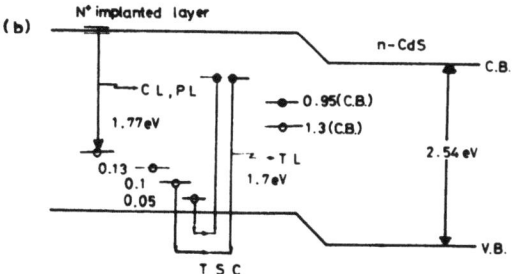

eV was found in D-31 sample or the density of it was enhanced in M-4 sample after N^+ ions implantation and post anneal. For M-4 sample, it contains Mn impurity, but for D-31, we do'nt have any information except undoped crystal. It is known in ZnS:Mn system, Mn impurity acts as shallow accepter level, and substitutional Mn on Zn site makes local expansion due to the difference of ionic radius [6]. In CdS:Mn system, local contraction occurs at Cd site. After N^+ ions implantation and post anneal, also a new contraction occur at S site due to substitutional N. We believe those contraction will create shallow acceptor type complex levels, but still we have a little possibility of an acceptor level due to simply substitutional N on S site.

The acceptor type densities of the shallow level and deep level(1.24eV), which were calculated at 1.1μm from surface using m-TSCAP and P-CAP data for D-31 were about $8 \times 10^{15}/cm^3$ and $1.5 \times 10^{15}/cm^3$ respectively. An another deep acceptor level of 0.77 eV (V.B.) was found from CL and PL by the ion implantation.

ACKNOWLEDGMENT

This work was supported by the advanced research fund of Tokyo Electrical Engineering College. One of author, Y. Machi wish to thank Dr. W.W.Anderson of the Ohio State Univ. for stimulating this work originally. Also the authors would like to acknowledge Dr.K.F. Komatsubara, Dr.Y.Shiraki, and Dr.T.Shimada of Central Research Laboratory of Hitachi Ltd for the ion implantation and valuable discussions, and Dr.T.Anbe of Matsushita Research Institute Tokyo Inc. for suppling CdS single crystal, and Dr.H.Morishima and Mr. M.Shimada of JAPAN Electron Optics Laboratory Co Ltd for ESR measurements and Mr.S.Kitano of Kitano Seisakujo Ltd for suppling a cryostat for the cathodoluminescence experiments.

REFERENCES

[1] see, for example,Y.Shiraki,T.Shimada,and K.F.Komatsubara: J. Appl. Phys. 43,710(1972);W.W.Anderson,and R.M.Swanson: J.Appl. Phys. 42 ,5125(1971); F.Chernow,G.Eldridge,G.Ruse,and L.Wahlin: Appl. Phys. Lett. 12,339(1968).
[2] K.H.Nicholas,and J.Woods: Brit.J.Appl.Phys. 15,783(1964).
[3] R.Lawrance,and R.H.Bube: J.Appl.Phys. 30,1807(1968).
[4] M.F.Deigen,V.M.Maevskii,V.Ya.Zevin, and N.I.Vitrikhovskii: Sov.Phys. Solid State,6,2193(1965).
[5] D.Curie,and J.S.Prener: in Physics and Chemistry of II-VI Compounds,ed. M.Aven and J.S.Prener(North-Holland Publishing Co.,Amsterdam 1967) p479.
[6] R.S.Title: in Physics and Chemistry of II-VI Compounds, ed. M.Aven and J.S. Prener, p290.

DEEP PENETRATION OF IMPLANTED Po IN CdS

P.F. Engel
F. Chernow

Department of Electrical Engineering
University of Colorado
Boulder, Colorado 80302

ABSTRACT

The distribution of 25 KeV Po^{210} implanted into CdS platelets was determined by energy analyzing the α-particles emitted by Po^{210}. Two samples were implanted along an axial channeling direction, and one along a planar channeling direction.

A new diffusion model was developed. The predicted distribution function as well as the distribution obtained from multistream diffusion theory were fitted to the experimental data. The present theory yielded a substantially better fit in the samples implanted along the axial channeling direction, in which there was measurable activity at a depth of 1μ. Both theories yielded comparable fits in the sample implanted along the planar channeling direction, in which there was measurable activity over a depth of 5000Å.

I. INTRODUCTION

All theories which predict the range distribution of implanted ions on the basis of their energy loss assume that an ion will cease to move once it has lost all its kinetic energy. In some cases this assumption is not correct and the ions move over distances which are large compared to their ballistic range[1-3]. One such case is that of Bi implanted in CdS. CdS crystals implanted along a channeling direction with radioactive Bi^{210} at an energy of 25 KeV retain more than 2% of the Bi activity after etching about 2500Å[4]. The maximum range of a perfectly channeled Bi ion in CdS can be estimated to be about 2400Å if its energy loss to atomic processes is neglected and if the LSS electronic stopping power is used[5]. Inclusion of energy loss to atomic processes, by using a ring model for the atoms[6] decreases the estimated range by a factor of 2, if a Thomas-Fermi potential is used. Thus the above result cannot be readily explained by a ballistic theory.

II. THEORY

The following diffusion model is a combination of single-stream (SSD)[1] and multistream diffusion (MSD)[3], and will be referred to as plural-stream diffusion (PSD).

Assume that the implantation process generates interstitial bombardants, and defects (e.g. vacancies) which are able to diffuse beyond the generation region. The interactions among defects and between defects and bombardants are the same as in MSD theory. Assume in addition that native, immobile traps are present which can trap the bombardant atoms. Thus a bombardant interstitial can recombine with either a generated defect, or with a native defect. In either case the bombardant is assumed to be immobile after trapping, and detrapping is neglected.

The models described by Sparks[3] can be readily adapted to the present theory, and this adaptation will be carried out with his "V-V-B" model. Let $n_V(x)$, $n_B(x)$ denote the concentrations of generated vacancies and bombardant interstitials, respectively, at a depth x below the surface of the sample. If $n_T(x,t)$ is the concentration of bombardants trapped at a depth x after the implantation has proceeded for a time t, then:

$$\frac{\partial n_T}{\partial t} = \beta_{VB} n_V n_B + \gamma n_B \tag{1}$$

The first term in (1) represents the rate at which the diffusing bombardants are trapped in the vacancies while the second term is the rate at which the bombardants recombine with the native defects.

Since the implanted area dimensions are large compared to the depth of penetration, it is reasonable to assume that the diffusion problem is one-dimensional. Thus the equations for n_B and n_V are:

$$\frac{\partial n_B}{\partial t} = D_B \frac{\partial^2 n_B}{\partial x^2} - \beta_{VB} n_V n_B - \gamma n_B \tag{2}$$

$$\frac{\partial n_V}{\partial t} = D_V \frac{\partial^2 n_V}{\partial x^2} - \beta_{VV} n_V^2 - \beta_{VB} n_V n_B \tag{3}$$

To solve equations (2) and (3), the following assumptions are made:
A) $n_V = n_B = 0$ at $x=\infty$
B) n_V is large compared to n_B
C) n_B and n_V reach steady state values in a time short compared to the implantation time τ
D) γ is a constant independent of t and x.

Assumptions A-C are examined in Sparks'[1] paper. Assumption D implies that the concentration of native defects is independent of position and time, since γ is proportional to this concentration. From equations (2) and (3) it follows that:

$$n_B = cz^{\frac{1}{2}} K_\nu(z) \qquad (4)$$

where
$$z = \alpha(x+d) \qquad (5a)$$
$$\alpha = (\gamma/D_B)^{\frac{1}{2}} \qquad (5b)$$
$$d = [6D_V/\beta_{VV} n_V(\ell)]^{\frac{1}{2}} - \ell \qquad (5c)$$
$$\nu = (6\beta_{VB} D_V/\beta_{VV} D_B + 1/4)^{\frac{1}{2}} \qquad (6)$$

where ℓ is the depth, over which there is generation and K_ν is the modified Bessel function of the second kind[7]. c is a constant proportional to the dose-rate of implantation. n_T can be obtained from (1) by using assumption C, equation (4) and Sparks' result for n_V:

$$n_T = D_B \tau \alpha^2 c(\nu^2 - 1/4 + z^2) K_\nu(z)/z^{3/2} \qquad (7)$$

The limiting cases of (7) are:

$$n_T \propto (x+d)^{-\nu-3/2}, \text{ for } z \ll (\nu^2-1/4)^{\frac{1}{2}} \qquad (8)$$

$$n_T \propto e^{-\alpha x}, \text{ for } z \gg (\nu^2-1/4)^{\frac{1}{2}} \text{ and } z \gg (\nu^2-1/4)/2 \qquad (9)$$

$z \ll (\nu^2-1/4)^{\frac{1}{2}}$ implies that the dominant mechanism trapping the diffusing bombardants is recombination with generated defects and corresponds to the MSD case. $z \gg (\nu^2-1/4)^{\frac{1}{2}}$ and $z \gg (\nu^2-1/4)/2$ implies that the recombination with native defects dominates the trapping of bombardants, corresponding to the SSD case. In order to compare the PSD and the MSD theories, the MSD distribution function will be taken to be

$$n'_T \propto (x+d')^{-\nu'-3/2} \qquad (10)$$

III. EXPERIMENTAL PROCEDURES

CdS platelets grown from the vapor phase were implanted with 25 KeV Po^{210}, Bi^{210} and Bi^{209}. The Bi contaminants were unavoidable. The typical total dose was 5×10^{13} ions/cm^2. Po^{210} emits monoenergetic α-particles with an energy of 5.3 MeV, Bi^{210} emits β-particles with a maximum energy of 1.16 MeV, and Bi^{209} is stable[8]. Samples 1 and 2 were implanted with the ion beam perpendicular to the $(2\bar{1}0)$ plane while sample 3 was implanted with the ion beam perpendicular to the $(3\bar{1}0)$ plane. The orientation of samples 1 and 2 corresponds to an axial channeling direction and the orientation of sample 3 corresponds to a planar channeling direction.

Successive energy spectrum measurements (of the α-particles emitted by the implanted Po^{210}) and etches were performed. The energy spectra were measured using a silicon surface barrier detector connected to standard pulse-handling electronics and a 1024-channel pulse-height analyzer. The solid angle subtended by the detector was 0.2 Sr. and the resolution full width at half maximum was typically 21 KeV. Residual activity data were obtained by adding the counts in each channel of the spectrum.

IV. DATA REDUCTION

In order to compare the measured α-particle energy spectra to the theoretical depth distribution of the implanted Po^{210} atoms, a theoretical energy spectrum corresponding to this distribution was calculated. The relationship between the depth distribution of α-particle emitting atoms and the energy distribution of the emitted α-particles has been deduced elsewhere[10,11]. The stopping power for the α-particles was taken to be 170 KeV/μ[10]. To convert this distribution to the analog of a measured spectrum, the broadening effect of the α-particle detector and associated electronics was taken into account. This was accomplished by assuming that the resolution line shape was Gaussian and by convolving this Gaussian with the theoretical distribution in energy of the emitted α-particles. Using the data from the experimental energy spectra, and the residual activity, least squares fits of the PSD and the MSD theories to these data were performed. The channel number corresponding to the energy edge of each spectrum was allowed to vary during the least squares fits. The degree to which each theory fit the data was expressed quantitatively in terms of the rms error defined as

$$\varepsilon = [\frac{1}{N} \sum_{i,j} (Y_{ij}-y_{ij})^2/Y_{ij} + \frac{1}{N} \sum_{K} (R_K-r_K)^2/R_K]^{\frac{1}{2}} \quad (11)$$

where Y_{ij} and y_{ij} are the experimental and theoretical counts, respectively, in channel i of spectrum number j; R_K and r_K are the experimental and theoretical residual activity counts, respectively, after the Kth etch, and N is the total number of data points.

V. RESULTS AND DISCUSSION

Fig. 1 shows two spectra obtained from sample 1. Curve A was obtained after etching the sample 490Å and curve B was obtained after etching 925Å. The spectra provide data over approximately 1μ depth from the surface of the sample. The PSD and MSD theories were fitted to both of the experimental spectra, resulting in the theoretical spectra shown in Fig. 1. The parameters of the fit were as follows: for the PSD theory d=1340±150Å, ν=1.97±0.18, α=(5.93±0.20)×10^{-4}Å^{-1} and ε=6.5. For the MSD theory d'=3620±230Å, ν'=4.62±0.22 and ε=16.

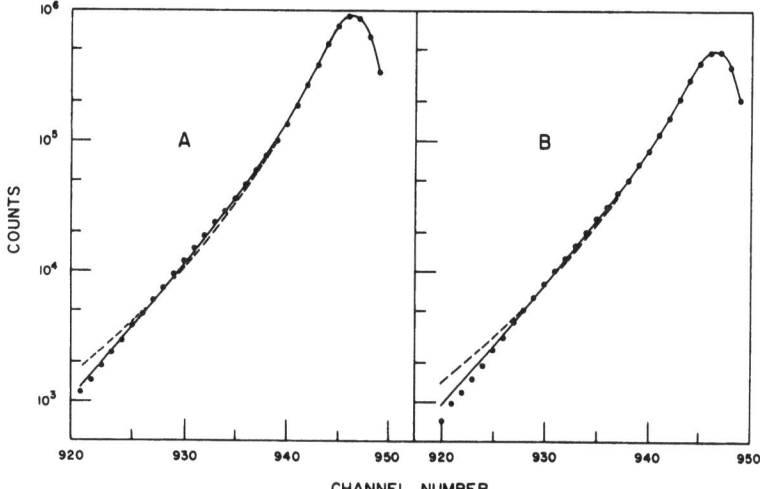

Figure 1. α-particle energy spectra from sample 1. Curve A: spectrum after etching 490Å. Curve B: spectrum after etching 925Å. o = experimental points. Solid lines = theoretical PSD spectra. Dashed lines = theoretical MSD spectra.

Fig. 2 shows the distribution of Po activity implied by the fit just discussed. The amplitudes of the two distributions were taken from this fit.

The two theories discussed so far were fitted to two spectra from sample 2, which were obtained after first etching 600Å and then 420Å. The residual activity predicted by the resulting fits was calculated, and is shown in Fig. 3. The parameters for this fit were: for the PSD theory, $d=1760\pm140$Å, $\nu=2.46\pm0.18$, $\alpha=(5.5\pm0.2)\times10^{-4}$Å^{-1} and $\varepsilon=2.1$; for the MSD theory $d'=3800\pm300$Å, $\nu'=4.95\pm0.30$ and $\varepsilon=3.0$.

Three spectra obtained after etching to a depth of 450Å, 580Å and 1060Å and all the residual activity points obtained after etching 450Å or more were used as data for sample 3. The results obtained from this sample have some uncertainty owing to the rather high background levels experienced during measurement. The residual activity data were background corrected but such a correction was not feasible for the spectra. The parameter α of the PSD theory was held fixed during the least squares fit because the experimental data did not reach deeply enough into the sample. The value chosen for α was 5.7×10^{-4}Å^{-1}, which is the average of the values obtained from samples 1 and 2. The values obtained from the fit for the other parameters were $d=475\pm41$Å, $\nu=3.34\pm0.13$ and $\varepsilon=2.11$ for the PSD theory and $d'=526\pm43$Å, $\nu'=3.54\pm0.13$ and $\varepsilon=2.04$ for the MSD theory.

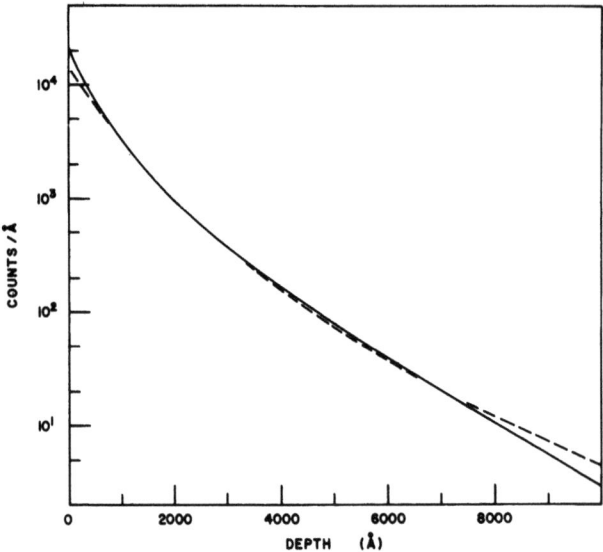

Figure 2. Theoretical distribution of Po^{210} activity in sample 1. Solid line = PSD theory. Dashed line = MSD theory.

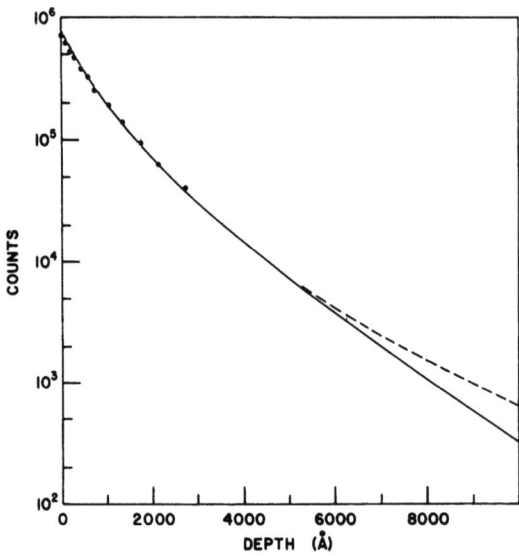

Figure 3. Residual activity from sample 2. o = experimental points. Solid line = PSD theory. Dashed line = MSD theory.

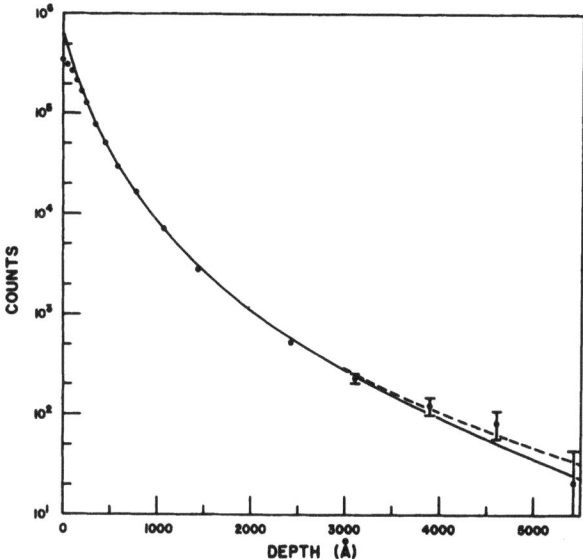

Figure 4. Residual activity from sample 3. o = experimental points. Vertical bars represent one standard deviation due to statistical errors. Solid lines = PSD theory; Dashed lines = MSD theory.

Fig. 4 shows the residual activity data for sample 3 and the corresponding theoretical curves obtained from the fit. The data deviate significantly from the theoretical curves at depths of less than 200Å. Since the LSS projected range is approximately 100Å[5], this deviation must be due to generation processes occurring over distances comparable to this LSS range. The good agreement between theory and experiment obtained near the surface for sample 2 (Fig. 3) can be ascribed to the fewer close encounters between bombardant and substrate atoms in an axial channeling direction than in a planar channeling direction. The same conclusion follows from the fact that d for sample 3 is smaller than d for samples 1 and 2, since Eq. (5c) implies that n_V is larger for sample 3 than for samples 1 and 2, if ℓ is assumed to be approximately the same for all samples. These implications cannot at present be pursued quantitatively, in view of the uncertainties in ℓ and in the results for sample 3.

It is to be noted that there exist effects that are not within the scope of the PSD theory. For example, saturation of the native traps may occur at high doses if the concentration of native traps holding bombardant atoms becomes comparable to the total concentration of native traps present. This effect would violate assumption D in section II. For sample 2 the native trap density was estimated to be roughly $10^{17} cm^{-3}$, by using the value of α from the fit[1,12].

The concentration of trapped bombardants at 2600Å was estimated to be about $3 \times 10^{17} cm^{-3}$ for sample 2. At this depth the two trapping mechanisms of section II were equally important. Thus assumption D was invalid in this case. The detailed quantitative consequences of modifying assumption D to allow for this effect have not yet been investigated.

The PSD theory has been quantitatively compared only to the MSD theory. The other theories known to the authors have features which prevent them from yielding as good a fit as the above two. Thus the SSD distribution, which is exponential, has zero curvature when plotted on semilog paper, in contrast to the experimental data. Extensive numerical evaluation of the distribution predicted by the theory due to Bourgoin et al.[13] always led to the wrong curvature in the logarithm of this distribution when compared to the experimenta data.

REFERENCES

1. J.A. Davies, P. Jespersgard, Can. J. Phys. 44, 1631 (1966).
2. E.V. Kornelsen, F. Brown, J.A. Davies, D. Domeij, G.R. Piercy, Phys. Rev. 136, A849 (1964).
3. M. Sparks, Phys. Rev., 184, 416 (1969).
4. G. Eldridge, F. Chernow, G. Ruse, J. Appl. Phys. 44, 3858 (1973).
5. W.S. Johnson, J.F. Gibbons, Projected Range Statistics in Semiconductors (Stanford University Bookstore, Stanford, Calif., 1969).
6. H.H. Anderson, P. Sigmund, Mat. Fys. Medd. Dan. Vid. Selsk, 34, No. 15 (1966).
7. Handbood of Mathematical Functions, edited by M. Abramovitz and I.A. Stegun, National Bureau of Standards, Applied Mathematics Series 55 (U.S. GPO, Washington, D.C., 1964).
8. C.M. Lederer, J.M. Hollander and I. Perlman, Table of Isotopes (Wiley, New York, 1967).
9. G. Eldridge, P.K. Govind, D.A. Nieman and F. Chernow, Proceedings of the European Conference on Ion Implantation (Peter Peregrimes, Stevenage, Hertfordshire, England, 1970), p. 143.
10. P.F. Engel, J.A. Borders and F. Chernow, J. Appl. Phys. 45, 38 (1974).
11. B. Domeij, I. Bergstrom, J.A. Davies, J. Uhler, Arkiv Fysik 24, No. 29, 399 (1963).
12. G. Eldridge, Ph.D. Thesis (unpublished).
13. J. Bourgoin, D. Peak and J.W. Corbett, J. Appl. Phys. 44, 3022 (1973).

ZnS:Mn DC ELECTROLUMINESCENT CELLS

BY ION IMPLANTATION TECHNIQUES

T. Takagi, I. Yamada, and A. Sasaki

Department of Electronics, Kyoto University

Kyoto, Japan

ABSTRACT

New approaches to fabricate ZnS:Mn DC El cells by low-energy ion implantation process which combined with deposition process have been proposed. Mn ions as a activator are implanted during or after deposition of ZnS. By simultaneous or alternative operation of deposition and implantation process, dose distribution of Mn element can be controlled externally by adjusting ion beam current and accelerating voltage. Wide depth distribution of luminescent centers can be controlled by this method with less lattice disorder or deffect. Usual ZnS:Mn thin film EL cells fabricated by thermal processing show little or no emission under DC excitation without any coactivator. But, the cells by ion implantation have shown strong emission under both DC and AC excitations and low impedance characteristics under DC excitation. The ionized cluster beam deposition method which is new deposition technique proposed by authors using newly designed "Vapourized Metal Cluster Ion Source" may be one of the special case of simultaneous operation of implantation and deposition above mentioned. In the technique, ZnS:Mn vapour aggregate (cluster) is used instead of atomic or molecular ions in ordinal ion implantation techniques. Under DC excitation, these cells showed good luminescence in spite of no coactivators.

INTRODUCTION

For a long time, most of the effort on a thin-film DC EL have been devoted to ZnS:Mn, Cu, Cl and ZnS:Cu, Cl [1]. But the maintenance characteristics under DC operation must be improved considerably. DC EL in ZnS:Mn film containing no coactivator shows little or no electroluminescence. New approaches to fabricate ZnS:Mn DC

EL cells by low-energy ion implantation techniques have been studied. One approch is the simultaneous or alternative operation method. Amount and distribution profile of activator could be controlled exactly, and implantation effects of activator for the substrate material gave much influence upon the substitution of activator elements to substrate crystal. As the results, a low-impedance type DC EL cell could be obtained and long maintenance characteristic could be expected.

THE SIMULTANEOUS OR ALTERNATIVE METHOD OF DEPOSITION AND IMPLANTATION

Fig.1 shows the schematic illustration of this method. Activator elements (such as Mn, Cu, Ag, etc.) are implanted from ion source during or after deposition of substrate materials (such as ZnS). One example of EL cells by the alternative operation of deposition and by implantaion has been obtained by electron beam deposition and by ion implantation of Mn (10 kV) [2]. After ZnS is deposited on the NESA glass to the thickness of 1.0 μm, Mn is implanted from 0.5×10^{15} to 1×10^{15} ions·cm^{-2} for a single layer type. For double layer type, 0.5 μm deposition of ZnS and Mn implantation are repeated twice. After annealing, Al electrode is deposited. Fig.2 shows the X-ray diffraction pattern of the ZnS film deposited by electron beam deposition, which shows polycrystalline state. Fig.3 shows the spectrum of electroluminescence from the cell. The peak of 5850 Å from Mn center obtained. Fig.4 shows the brightness-voltage characteristics where we can see that the luminescent intensity of the double layer type is stronger than that of the single layer type with the same total does of activator. The brightness-voltage relation is expressed by the familiar $V^{-1/2}$ law for the excitation of Mn center. Figs.5 and 6 show typical characteristics by DC excitations. The cell shows the low impedance characteristics as compared with usual ZnS:Mn EL cell, and does not show the rectification property which is seen in the DC EL fabricated by the thermal diffusion processing. Fig.7 shows the maintenance characteristics. Long life cells could be expected by the ion-implantation method.

Mn deposition on the ZnS by electron beam and its annealing has been made to study the difference between ion implantation and thermal diffusion of Mn into ZnS. The quantity of same order of implanted Mn was deposited on the film, and cell was annealed. But the cell did not show the electroluminescence.

IONIZED-CLUSTER BEAM DEPOSITION METHOD

The ionized-cluster beam deposition method which is new deposition technique proposed by authors using newly designed "Vapourized Metal Cluster Ion Source" [3, 4, 5] may be one of the special case

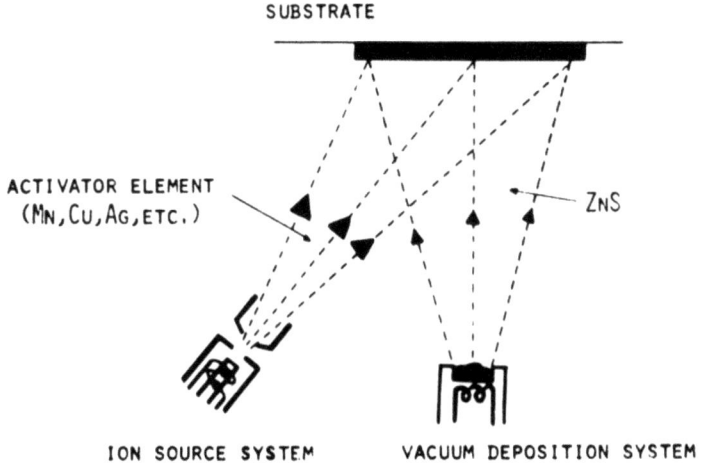

Fig.1 Schematic illustration of the system of the simultaneons or alternative method of deposition and implantation.

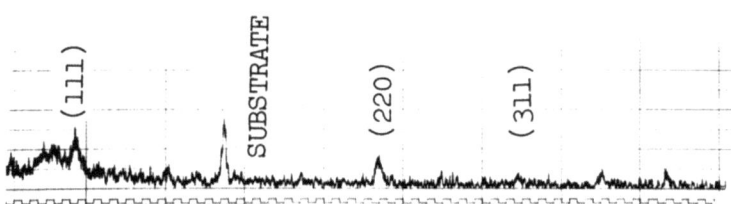

Fig.2 X-ray diffraction patterns of ZnS film by electron beam deposition.

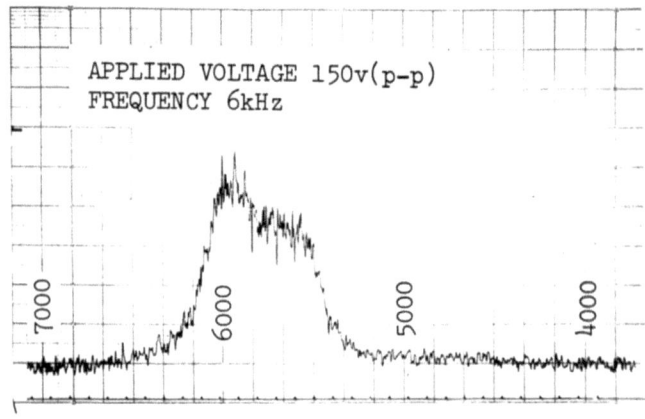

Fig.3 Luminescent spectrum of the cell

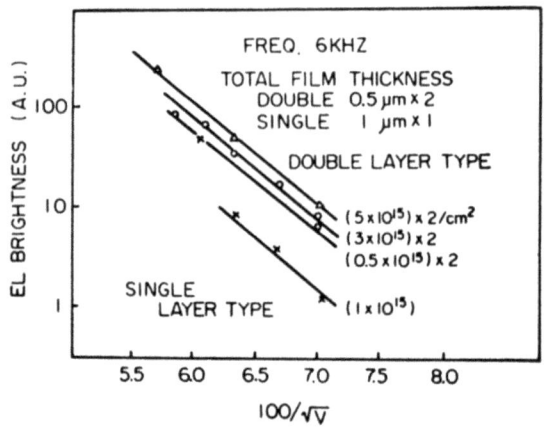

Fig.4 Brightness-Voltage characteristics

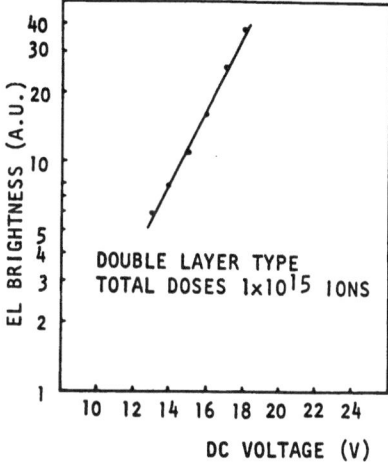

Fig.5 B-V characteristic

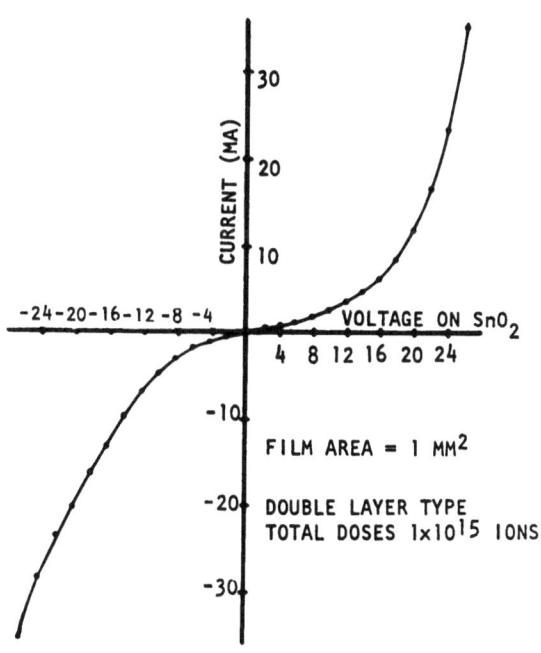

Fig.6 V-I characteristic

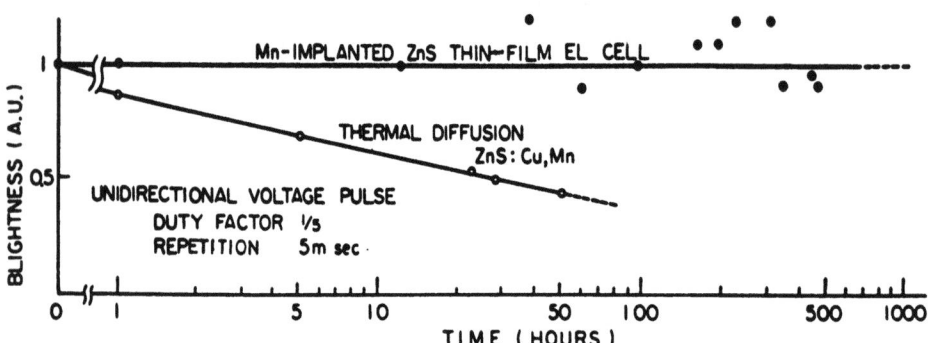

Fig. 7 Maintenance characteristic

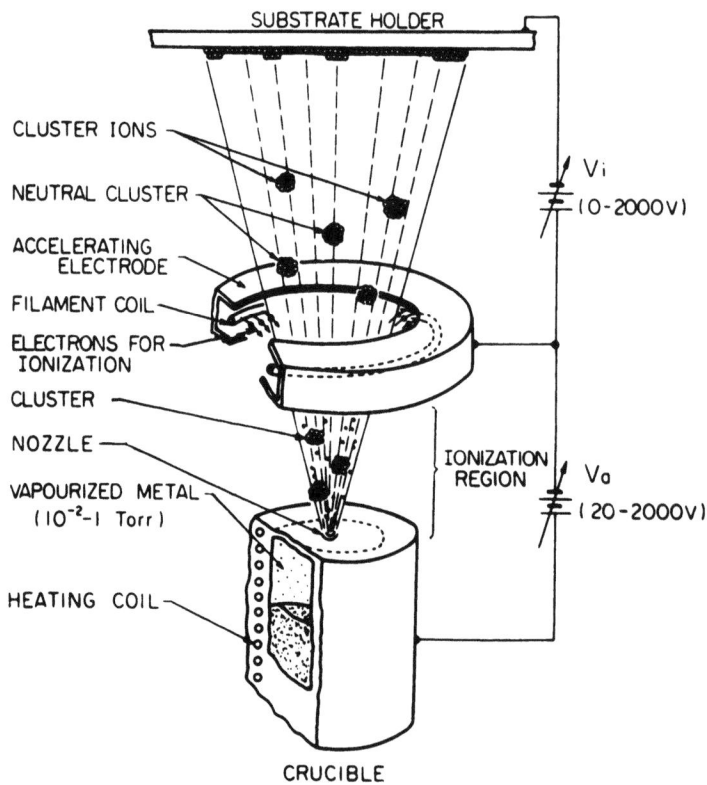

Fig. 8 Ionized cluster beam deposition system.

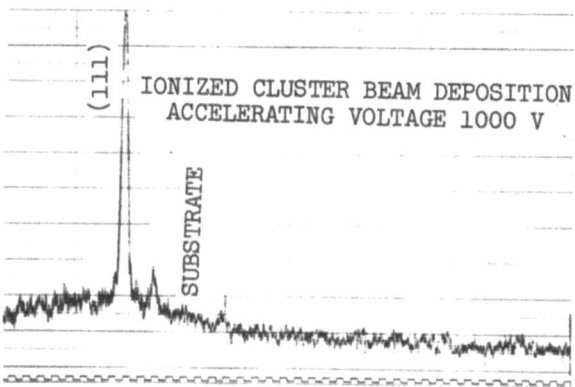

Fig. 9 X-ray diffraction pattern of the film by the ionized cluster beam deposition method.

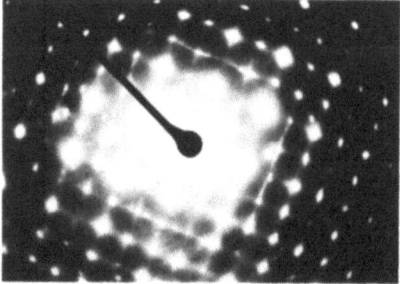

Fig. 10 Electron diffraction pattern of ZnS:Mn deposited on cloven rocksalt by the ionized cluster beam deposition method.

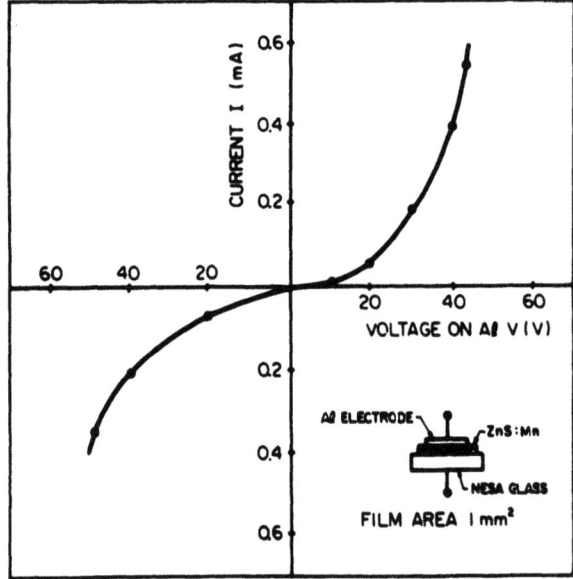

Fig. 11 V-I characteristic from the cell fabricated by ionized cluster beam deposition method.

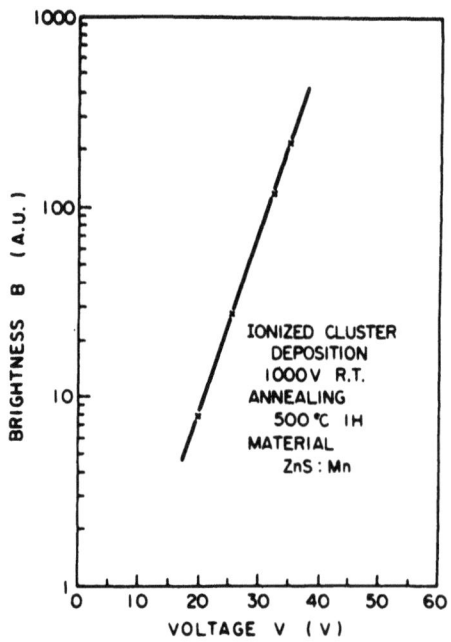

Fig.12 B-V characteristic

of simultaneous operation of implantation and deposition above mentioned. In the technique, ZnS:Mn vapour aggregate (cluster) is used instead of atomic or molecular ions in ordinary ion implantation techniques. A cluster beam consisting of 10^2-10^3 atoms is formed by adiabatic expansion and ionized by electron bombardment. The atoms which consist a cluster are bound by the Van der Waals' force. Therefore they are not so tight coupled each other but act as one aggregate. Only one atom in an ionized cluster is ionized. The ionized clusters are accelerated by the accelerating boltage (1-2 kV) and impact on to the substrate with neutral clusters. When the ionized and neutral clusters bombard to the substrate, they are scattered and separated to each atom. Scattered atoms migrate on the surface by thermal and kinetic energy and condense with good crystal state.

In this experiment, the substrate temperature is kept at 250°C during the deposition. After deposition, film is annealed at the temperature of 500°C 1h in the vacuum. Fig.9 shows the X-ray diffraction pattern of ZnS:Mn on the NESA glass by the cluster method. By the ionized cluster beam deposition, the films are improved in crystal state as compared with the method by usual electron beam deposition. The ionized cluster beam deposition method is seemed to be suitable to obtain fine crystal state deposited film. As one example diffraction pattern of ZnS:Mn film on cloven rocksalt is shown in Fig.10. The single crystal state could be obtained as shown in this figure.

Under DC excitation, the cell shows good luminescence. Fundamental characteristics of the cells are similar with the cell fabricated by simultaneous method of deposition and implantation. The current voltage characteristic is shown in Fig.11. The typical dependence of brightness on DC voltage is shown in Fig.12. The efficiency of the cells fabricated by the cluster method is still low as compared with that of previous method. Futher investigations in improving the characteristics are under way.

REFERENCES

1. A.Vecht, N.J.Werring; J.Phys.D.Appl. Phys, B, p.105, 1970.
2. T.Takagi, I.Yamada, A.Sasaki, and T.Ishibashi; IEEE Trans. on Electron Devices, ED-20, 11, p.1110, 1973.
3. T.Takagi, I.Yamada, M.Kunori, and S.Kobiyama; Proc. of the 2nd International Conf. on Ion Source, Vienna, Sept., p.790, 1972.
4. T.Takagi, I.Yamada, K.Yanagawa, M.Kunori, and S.Kobiyama; Presented at the sixth International Vacuum Congress, March 1974, Kyoto.
5. T.Takagi, I.Yamada, and K.Yanagawa; Ionized Cluster Beam Deposition, Presented at this Conference.

ION-BOMBARDMENT-INDUCED SURFACE EXPANSION OF SOLIDS

H. TANOUE and T. TSURUSHIMA

Electrotechnical Laboratory

5-4-1 Mukodai, Tanashi, Tokyo, JAPAN

ABSTRACT

Surface expanding phenomena in GGG($Gd_3Ga_5O_{12}$) bombarded with various ions have been observed by an interferometer. An expansion model is proposed to explain the phenomena taking into account present and other investigators' results. The effect of surface expansion on the profile of implanted ions in the substrate is discussed.

INTRODUCTION

It is well known that the increase of lattice constant or the volume expansion is observed in solids bombarded with high energy particles. The factors to explain the volume expansion may be attributed to radiation products, in short, empty spaces including vacancies, divacancies etc., as well as implanted ions such as in interstitial sites. In the usual ion implantation processes that the number of ions introduced into the target is relatively small (less than a few percent of the substrate atoms in density), and the target during implantation is near at room temperature, the volume expansion can be attributed mainly to the existence of the empty spaces. In this paper, considerations are given on the correlation between the volume expansion and the effective energy to make an quantitative explanation of the surface elevation due to the radiation induced volume expansion.

EXPERIMENTAL

GGG substrates covered with a Mo-mask are bombarded at room temperature with H, He, N, Ne, Ar, Zn, Kr and Xe ions of $5 \cdot 10^{14}$ ∿

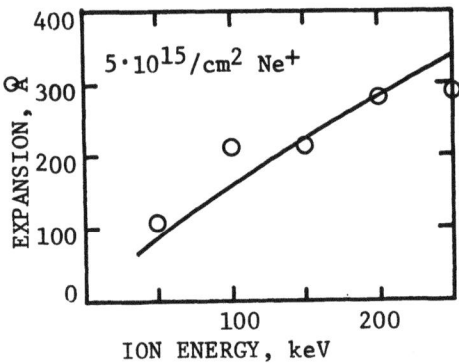

Fig.1 Energy Dependence of Surface Step Height

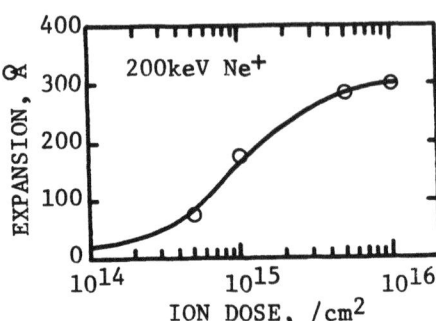

Fig.2 Ion Dose Dependence of Surface Step Height

$5 \cdot 10^{16}$/cm^2 in the energy range 50 $\sim$ 250 keV. Beam line is tilted 7 degrees from the normal of the target surface whose orientation is <111>. The surface elevation in the region bombarded with ions through windows whose size is approximately 350X500μm in the mask is observed by an interference contrast microscope, and measured quantitatively by Newton Ring method.

RESULTS AND DISCUSSIONS

Ion energy dependence of the step heights by $5 \cdot 10^{15}$/cm^2 Ne ion bombardment is shown in Fig.1 with open circles. A solid line in this figure is calculated from an equation based on a model which will be dicussed in the following section. The figure shows that the expansion step height increases with incident ion energy, which is reasonable from the view point that the expansion is caused by the radiation products. Though the surface elevation on GGG bombarded with $5 \cdot 10^{15}$/cm^2 protons is too small to be measured, it is to be noted that the step by low energy bombardment is more clearly distinguished than that by high energy bombardment. Deep damage may produce smaller surface elevation than shallow one. If there is a threshold value of energy deposition rate to create volume expansion, the value may invalidate the energy deposition near the surface region in high energy proton bombardment. Thus the introduction of the threshold energy deposition may also explain the results of proton bombardment.

Dose dependence of the surface elevation in GGG by 200keV Ne ion bombardment is shown with circles in Fig.2. The step increases with ion dose and tends to saturate.

Most bombardments are carried out under the dose rate about 1 μA/cm^2. No dependence of the step heights on the dose rate was observed.

Circles in Fig.3 show the step heights created on GGG surface

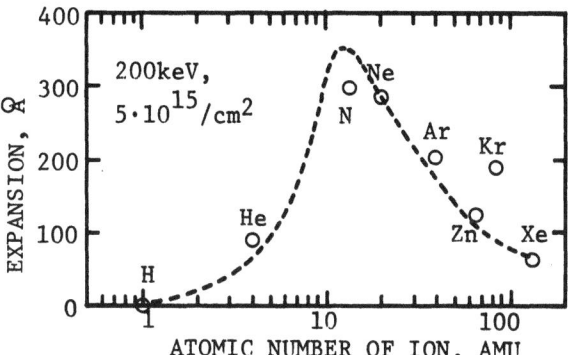

Fig.3 Ion Mass Dependence of
Surface Step Height

by various ion bombardment of 200keV $5 \cdot 10^{15}$/cm². Under this condition the maximum step is created by nitrogen ion bombardment. Though the decrease of step height in heavy ion range may be explained by surface atom sputtering effect, we consider the effect can be neglected. Sputtering must be taken into account if the ion energy is less than a few keV and the beam line is tilted as large as 60 degrees to the normal of target surface. The results of Fig.3 can be explained by the fact that the damaged region comes to be shallower as the ion mass increases, and the volume expansion saturates when the energy deposition exceeds a limiting value.

SURFACE ELEVATION MODEL

A surface elevation model taking into account the experimental results is proposed, and an equation to calculate the surface step heights is deduced. Basic idea is that the energy deposited from incident ions causes volume expansion, which in turn creates the surface elevation.

Assume a small region at depth x. The energy deposited into this region from incident ions creates radiation defects in this region, and because of those defects including vacancies and other empty spaces that region will expand, and compressive stress is generated in the region. It may be proper to assume that the strength of the stress is proportional to the strength of the expanding force. Then the force is transmitted towards the surface attenuating exponentially with distance. The surface elevation is made up with all these contributions of expanding force from each depth where volume expansion is induced. The surface elevation S will be calculated from the following equation:

$$S = \int_{x_2}^{x_1} A \cdot f(x) \cdot \exp(-C \cdot x) dx$$

where $f(x)$ represents energy deposition rate into atomic processes.[†]

Two values of depth x_1 and x_2 are to define the region where the values of $f(x)$ exceed a specified value f_L. This value is estimated from the dose under which the surface elevation is not observed, and is assumed to be considerably smaller than the energy deposition density corresponding to amorphizing dose. A constant f_0 is introduced as a limiting value of $f(x)$ to represent the saturation phenomena in dose dependence. The value of f_0 is expected to be larger than the value corresponding to the amorphizing dose. C is the attenuation coefficient of lattice expansion propagation. Another constant A is taken as a factor to convert the energy to volume expansion. For example, A=0.022 $Å^3$/eV in GGG means that the volume expands 2.2% by 1 eV/$Å^3$ energy absorption. J. C. North et al.[1] reported that the lattice parameter expands 1 % by 1 eV/$Å^3$ energy absorption in magnetic garnet $YGdTmFe_{4.2}Ga_{0.8}O_{12}$.

MODEL APPLICATION TO EXPERIMENTAL DATA

If we assume the following values for the constants A, C, f_L and f_0 to explain the experiments of surface elevation on GGG with the equation proposed in the preceeding section, solid curves in Fig. 1 and Fig.2 and broken curves in Fig.3 are obtained. A=0.0219 $Å^3$/eV, 1/C=2.5μm, f_L=0.0035 eV/$Å^3$ and f_0=4 eV/$Å^3$. Further we can estimate the ion dose dependence of the surface step height on GGG by 200keV $5 \cdot 10^{15}$/cm^2 proton bombardment. The result is consistent with the experimental result. Thus the present model gives a good estimation of the surface elevation on GGG substrates.

Values of the constants for silicon are assumed as follows after the experimental data by R. E. Whan et al.[2] and W. Beezhold [3]. A=0.0025 $Å^3$/eV, 1/C=2.5 μm, f_L=0.003 eV/$Å^3$ and f_0= 27 eV/$Å^3$. Broken curves in Fig.4 are obtained with these constants. The present calculation gives the value 272 Å for the surface elevation on silicon bombarded with 200keV 10^{16}/cm^2 Si ions, which is comparable with the

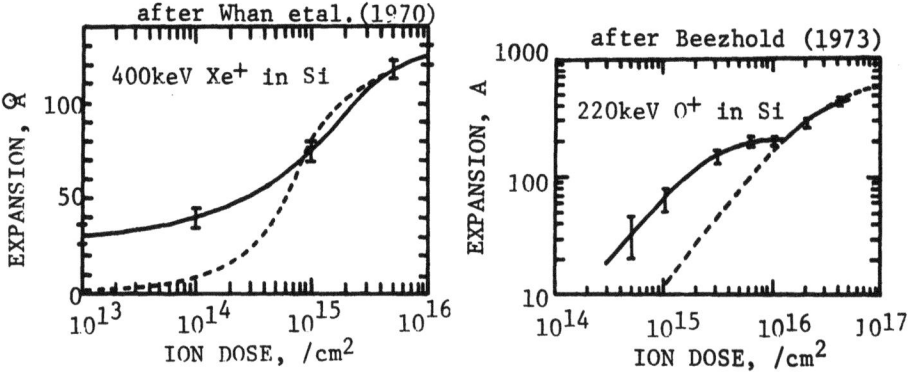

Fig.4 Ion Dose Dependence of Surface Step Height on Si
(Broken curves: calculated, Solid curves: experimental data)

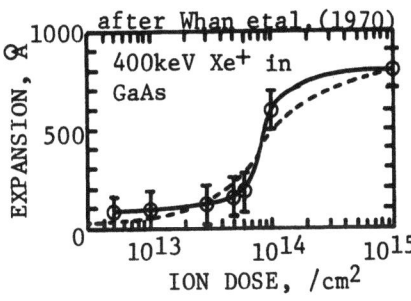

Fig.5 Surface Expansion on GaAs

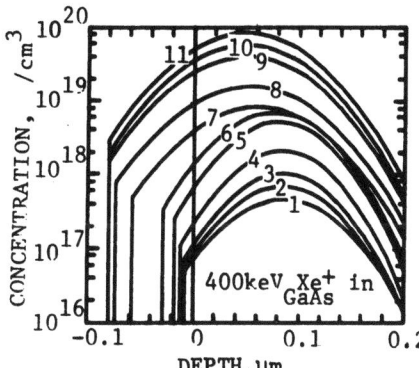

Fig.6 Modified Profiles of Implanted Ions in Expanded GaAs
(on the data of [2]) 1:$5 \cdot 10^{12}$, 2:$7 \cdot 10^{12}$, 3:10^{13}, 4:$2 \cdot 10^{13}$, 5:$5 \cdot 10^{13}$, 6:$7 \cdot 10^{13}$, 7:10^{14}, 8:$2 \cdot 10^{14}$, 9:$5 \cdot 10^{14}$, 10:$7 \cdot 10^{14}$, 11:10^{15} /cm^2.

value 220± 20 Å obtained by K. N. Tu etal.[4]. The present model can also estimate the mass dependence of the surface steps on silicon created by 200keV $5 \cdot 10^{15}$/cm^2 ion bombardment. The maximum step is created by argon ions.

Assuming the following constants for GaAs, we can make a comparison between the data of Whan etal.[2] and the calculated values as in Fig.5. A=0.18 Å^3/eV, 1/C=2.5μm, f_L=0.025eV/Å^3 and f_0=3.5eV/Å^3.

PROFILE OF IMPLANTED IONS IN EXPANDED REGION

Recently the measurements and the theoretical calculations of the distribution of implanted ions have been carried out considerably precisely. In the sample bombarded with ions less than about 10^{13}/cm^2, the surface elevation will be negligible. If the surface elevation is observed, however, some modification of range parameters usually obtained by simple calculation will be needed. To make a simplified estimation, we assume that the expansion is caused by the increase of empty spaces, the stopping power in those spaces can be neglected, and the expansion is uniform over the damaged region. Then the range parameters in expanded regions will be greater than those in unexpanded regions. As an example, the modified profiles based on the experimental data of R. E. Whan etal.[2] are illustrated in Fig.6.††

ANNEALING

An appropriate annealing treatment is expected to take out the surface elevation. Isochronal annealing of GGG bombarded with 250 keV $2 \cdot 10^{15}$/cm^2 Ar ions are carried out. The result tells that the surface expansion is almost disappeared by 900°C, 15 minutes annealing.

STRESSED REGION IN ION BOMBARDED MAGNETIC GARNET

It is known that in magnetic garnets bubbles tend to adhere to the edge of the ion implanted area. Observation by polarization microscope reveales the existence of the region unique in polarization characteristics surrounding ion bombarded area, and that bubbles are caught in this unique region[5]. The most probable explanation is that this unique region is the stressed region due to the expansion by ion bombardment. The width of this unique region is observed to be 10∼15 μm, if we use the value of the attenuation factor C assumed in this paper to obtain the position where the stress is attenuated to 1 % of the peak value, we get the width of stressed region to be 11.5 μm, which gives a fairly good agreement with the observed width.

REFERENCES

[1] J. C. North and R. Wolfe; *Ion Implantation in Semiconductors and other materials*, Plenum Press, (1973) p.505

[2] R. E. Whan and G. W. Arnold; Appl. Phys. Lett. <u>17</u> 378 (1970)

[3] Wendland Beezhold; *Ion Implantation in Semiconductors*, Springer-Verlag, (1971) p.267

[4] K. N. Tu, P. Chaudhari, K. Lal. B. L. Crowdwer and S. I. Tan; J. Appl. Phys. <u>43</u> 4262 (1972)

[5] T. Tsurushima etal.; *Proc. 5-th Conf. on Solid State Devices, Tokyo* (1973) p.464

† For the procedure to get f(x), see a paper published by the authors; J. Phys. Soc. Japan <u>31</u> 1968 (1971).

†† Though the projected ranges measured from the expanded surface seem to increase largely with ion dose in Fig.6, those expressed in μg/cm² have only small differences among themselves as shown in the following table. The projected range of 400keV Xe ions in unexpanded GaAs substrate is estimated to be 47.7μg/cm².

R_p \ Dose/cm²	5E12	7E12	1E13	2E13	5E13	7E13	1E14	2E14	5E14	7E14	1E15
in Å	936	940	943	952	981	1035	1187	1266	1294	1296	1296
in μg/cm²	46.9	47.0	47.1	47.2	47.4	47.4	47.5	48.1	48.8	49.1	49.3

NITROGEN IMPLANTATION IN SiC: LATTICE DISORDER AND

FOREIGN-ATOM LOCATION STUDIES

A. B. Campbell, J. Shewchun and D. A. Thompson
McMaster University, Hamilton, Ontario, Canada

and

J. A. Davies and J. B. Mitchell
CRNL, Chalk River, Ontario, Canada

ABSTRACT

The implantation behaviour of 25-80 keV nitrogen in α-SiC has been investigated for doses of 2×10^{15} and 2×10^{16} N atoms/cm^2 as a function of implantation temperature (20-450°C) and subsequent anneal treatments up to 1485°C. Backscattering yield measurements (using a 2 MeV helium beam) and $^{15}N(p,\alpha)$ nuclear reaction yield measurements (using a 1 MeV proton beam) have been used to study the amount of residual damage in the crystal and the lattice location of the implanted nitrogen atoms, respectively. The results clearly indicate the advantage of high temperature implantation in producing a high "substitutional" fraction (∼ 70%) and a minimum damage level. In the 450°C implants, a maximum substitutional nitrogen concentration of ∼ $1 \times 10^{21}/cm^3$ was obtained. Evidence for an anomalous "inverse annealing" stage at ∼ 800°C is observed for the 350 and 450°C implants, with the residual damage level increasing markedly and the "substitutional" nitrogen fraction decreasing to around the 50% level.

INTRODUCTION

Silicon carbide has good possibilities as a material from which to fabricate high power and high temperature semiconductor devices[1]. Standard diffusion doping techniques, however, require temperatures in excess of 2200°C and are difficult to perform due to surface dissociation. Ion implantation provides an alternative doping technique, provided the associated radiation damage can be annealed at reasonable temperatures and the implanted atoms become electrically active.

Previous work[2,3] on the nitrogen/silicon carbide system has shown that an n-type layer can be formed in p-type material by nitrogen implantation at room temperature and a subsequent 750°C anneal. Annealing at temperatures in the range 1100-1700°C results in 40% to 50% electrical activity. Although the donor concentration remains relatively constant over this anneal temperature range, the mobility of the carriers increases with increasing anneal temperature, indicating a decrease in the density of defects.

Earlier work in these laboratories[4] and elsewhere[5] has concentrated on room temperature implantation, and has shown that the damaged layer anneals from the surface inward, thus producing a buried damage region which is then quite difficult to anneal completely.

Damge studies in other binary semiconductors[6,7] such as GaAs and GaP have shown that implantation into heated substrates, where annealing of each collision cascade can occur during implantation, usually produces less residual damage than a comparable post-implantation anneal of a room temperature implant. In the present investigation of SiC we have therefore investigated the dependence of lattice disorder on implantation temperature (20-450°C) as well as on subsequent anneal treatment.

EXPERIMENTAL CONDITIONS

The silicon carbide crystals were α-type, aluminum doped with a nominal acceptor concentration of 10^{18} cm^{-3}. They were kindly provided by R.B. Campbell of the Westinghouse Astronuclear Laboratory. Neutron activation analysis[8] indicated that the total Al concentration was about 10^{19} cm^{-3}, i.e. about an order of magnitude higher than the net acceptor concentration.

Before implantation the crystals were degreased and etched in HF acid to remove any surface oxide layer and then annealed in vacuum (10^{-6} torr) at 1485°C for 5 minutes. This method of surface preparation was found to give the minimum amount of surface disorder, as shown by backscattering analysis.

Except for a preliminary series of 80-keV room temperature implants, all implantations were double implants at 45 keV and 25 keV in order to provide a relatively uniform nitrogen atom density in the implanted region. The ^{15}N isotope was selected so that we could use the ^{15}N(p, α)^{12}C nuclear reaction$^{(9,10)}$ in conjunction with channeling techniques to study the lattice location of the implanted nitrogen atoms. This particular reaction is sensitive enough to measure the amount of nitrogen implanted; also, the background from competing reactions is acceptably low (i.e. equivalent to $\sim 3 \times 10^{14}$ atoms of ^{15}N/cm^2).

One series of samples were implanted to a dose of $\sim 2 \times 10^{15}$ atoms/cm^2 at substrate temperatures of room temperature ($\sim 20°$C) and 350°C. A second series were implanted to a dose of $\sim 2 \times 10^{16}$ atoms/cm^2 at substrate temperatures of 250, 350 and 450°C. The 450°C sample had both faces implanted. From theoretical predictions of range profiles in silicon carbide, we estimate the peak concentrations to be $\sim 2 \times 10^{20}$ and $\sim 2 \times 10^{21}$ nitrogen atoms/cm^3 in these two series of implants. These peak concentrations are sufficiently high to overcome the compensating effects of the ($\sim 10^{18}$ cm^3) acceptor impurities. This is a necessary condition if the nitrogen implanted samples are subsequently to be used for device applications. Anneals of 5-minute duration were performed in a vacuum of 10^{-6} torr to a maximum temperature of 1485°C.

Lattice disorder measurements were conducted using a beam of 2 MeV He$^+$ ions. The backscattered particles were detected and energy analysed by a silicon surface-barrier detector with a 15-keV resolution (FWHM). Backscattered energy spectra were collected with the sample c-axis both aligned and non-aligned with the incoming beam. A comparison of these two spectra provides a quantitative measure of the amount of radiation damage$^{(11)}$.

A 1 MeV proton beam was selected as the optimum energy for the nitrogen-atom location measurements. We define the "substitutional" fraction (S) measured in this experiment as the percentage of implanted nitrogen atoms located on sites in the SiC lattice that are shadowed along the c-axis. Wherever the crystal contains a large amount of residual disorder, a discussion of substitutional percentage is of doubtful significance. For this reason, we do not attempt to evaluate S for those cases where the observed damage peak exceeds $\sim 20\%$ of the non-aligned backscattered yield.

RESULTS AND DISCUSSION

Figure 1A shows a portion of the He$^+$ backscattering spectrum for a sample implanted at 20°C with a dose of 2×10^{15} atoms of 80-keV ^{14}N/cm^2. The effect of subsequent annealing to temperatures as high as 1400°C is also shown. Note that the damage anneals predominantly from the surface inward, thus producing a buried damage region that becomes progressively narrower with increasing anneal temperature. Marsh and Dunlap[2] found evidence of a buried semi-insulating layer, probably due to the presence of defects, in their study of the electrical properties of nitrogen implanted SiC diodes. They also found that this layer decreased in thickness with increasing anneal temperature.

Figure 1B shows similar backscattering spectra for the corresponding double implant (45 and 25 keV). Note that, for these lower energy implants, the damage extends completely to the surface and that this damage subsequently anneals from the damage-crystalline interface out towards the surface. No buried damage region was observed in this case. Also the α yield from the ^{15}N(p, α) reaction showed that approximately 50% of the implanted nitrogen atoms had diffused out of the target during the 1485°C

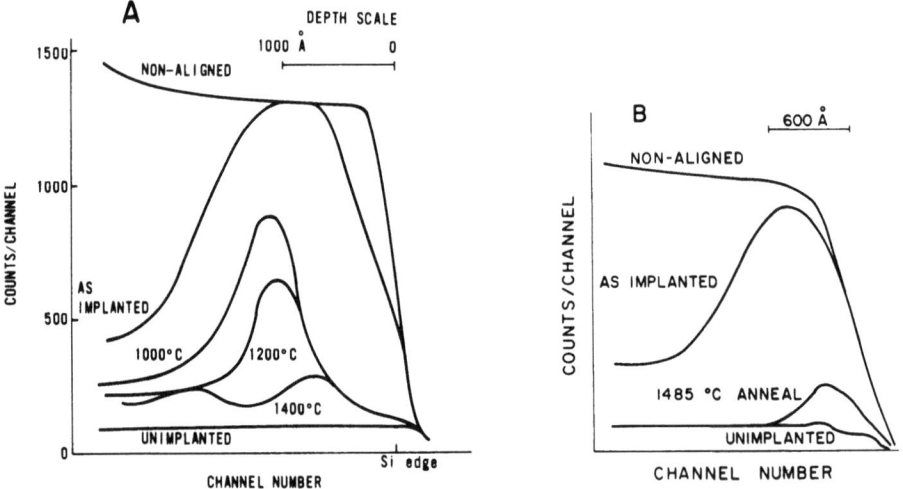

Fig. 1 Backscattering spectra of 2.0 MeV He$^+$ for aligned (c-axis) and non-aligned directions of incidence, before and after implantation and subsequent anneals. Implantation conditions: 20°C, 2×10^{15} N atoms/cm^2 at A) 80 keV and B) 45 and 25 keV (double implant).

anneal. Presumably, some of the nitrogen becomes trapped at the damage-crystalline interface and thus gets swept out toward the surface as this low energy implant anneals.

Figures 2A and 2B show the observed damage spectra for the high dose (2×10^{16} N atoms/cm^2) implants at 350 and 450°C respectively. In each case the damage was also measured after various anneal temperatures and was found to increase markedly following an 800°C anneal and then drop following higher temperature anneals. This "reverse annealing" phenomenon is not understood at present. Within the accuracy of the measurement, the hot implants do not lose any nitrogen even after the 1485°C anneal. Note too that the damage in the "as implanted" cases is much lower than for the room temperature implants.

Table I lists the observed Si damage peak heights (figs. 1, 2) normalized to the non-aligned yield at the same depth. The data show that the normalized damage levels for the room temperature implants and for the high dose heated implants up to 350°C approach unity, indicating nearly complete destruction of crystalline order. Note that the damage in the "as implanted" case decreases markedly with increasing implantation temperature, as expected. An implant temperature of 350°C in the low dose case and 450°C in the high

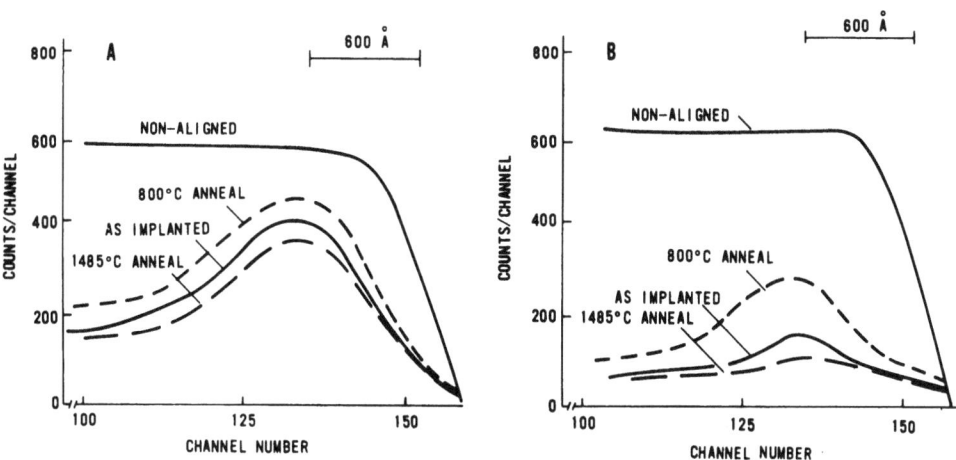

Fig. 2 The backscattering spectra of 2.0 MeV He$^+$ following implantation of 2×10^{16} N atoms/cm^2: A) at 350°C, and B) at 450°C.

TABLE I

Normalized Damage Peak Heights

Dose	Anneal Temp. (°C)	Implantation Temperature			
		20°C	250°C	350°C	450°C †Side 1 Side 2
$2 \times 10^{15}/cm^2$	As Implanted	.88	-	-	- -
	1485	.225	-	.18	- -
$2 \times 10^{16}/cm^2$	As Implanted		.82	.76	.26 .45
	800		.82	.82	.47 .45
	1485		.70	.64	.19 .35

† Side 1 showed a considerably thicker oxide film growth than side 2, when the sample was subjected to 1200°C oxidation in O_2 gas. Presumably, side 1 is the carbon face, and side 2 is the Si face.

dose case is sufficient to maintain the aligned yield near that of the unimplanted sample, indicating almost complete annealing of radiation damage during implantation.

Only in the case of the 450°C high dose implant is the damage level low enough for an investigation of the substitutional nitrogen content to be justified. The results of these nitrogen-location measurements, together with those for the low dose implants, are shown in Table II. The highest substitutional percentage (~ 70%) is

TABLE II

Dose	Implantation Temperature	Anneal Temperature	χ^*_{min} (side 1/side 2)†	S(%) (side 1/side 2)
2×10^{15}	20°C	1485°C	0.09	39
2×10^{15}	350°C	As implanted	-	-
		1485	.06	47
2×10^{16}	450°C	As implanted	-	-
		600	.11/.17	70/45
		800	.13/.17	52/45
		1000	.09/.12	53/54
		1200	.09/.13	49/52
		1485	.09/.13	51/56

† see footnote to Table I.

found for the high dose sample following a 600°C anneal. However, the damage increases following an anneal to 800°C, as shown earlier in figure 1B and is accompanied by a drop in the substitutional component from 70% to 52%. Subsequent anneals to temperatures of 1000, 1200 and 1485°C do not raise the percentage of substitutional nitrogen, although the damage level decreases considerably. It would appear that a high temperature implant with little or no subsequent anneals results in the largest substitutional component with a tolerable damage level. Unfortunately, however, the reverse annealing observed with these heated implants produces an unstable condition which could not be tolerated in high temperature device applications. Consequently, implantation conditions yielding a stable, high substitutional component over a wide range of anneal temperatures must be found. Perhaps even higher implant temperatures than 450°C would result in a low damage level, while still yielding a high substitutional percentage.

The 450°C implant data in Tables I and II show that a significant difference in damage level and in substitutional nitrogen level exists between the carbon face (side 1) and the silicon face (side 2). The cause of this difference is not understood and further work is required to clarify whether or not it is a general result.

CONCLUSIONS

1) Implants of nitrogen into heated silicon carbide result in much less damage and a higher substitutional nitrogen fraction than if the implants were performed at room temperature.

2) A reverse annealing effect around 800°C is observed for the hot implants in that the substitutional nitrogen component decreases and at the same time, the normalized damage peak height increases markedly. Further annealing to 1485°C reduces the damage level again but with little change in the substitutional nitrogen component.

ACKNOWLEDGEMENTS

We thank O.M. Westcott for performing the ^{15}N implantations and F. Brown and T.A. Eastwood for a critical review of the manuscript.

REFERENCES

1) Robert W. Keyes, "Silicon Carbide - 1973", edited by R. C. Marshall, J. W. Faust, Jr., and C. E. Ryan (University of South Carolina Press, 1974) p. 534.

2) O. J. Marsh and H. L. Dunlap, Rad. Effects 6, 301 (1970).

3) H. L. Dunlap and O. J. Marsh, Appl. Phys. Letters 15, 311 (1969).

4) R. R. Hart, H. L. Dunlap and O. J. Marsh, Rad. Effects 9, 261 (1971).

5) A. B. Campbell, J. B. Mitchell, J. Shewchun, D. A. Thompson and J. A. Davies in "Silicon Carbide - 1973" (cf. ref. 1) p. 486.

6) J. L. Whitton and G. Carter, "Proceedings of International Conference on Atomic Collision Phenomena in Solids", (North Holland Publishing Co., 1970), p. 615.

7) A. B. Campbell, W. A. Grant and G. A. Stephens, Rad. Effects 17, 19 (1973).

8) We are particularly grateful to W. D. Mackintosh and H. H. Plattner of CRNL for carrying out these neutron activation analyses.

9) F. B. Hagedorn and J. B. Marion, Phys. Rev. 108, 1015 (1957).

10) A. V. Cohen and A. P. French, Phil. Mag. 44, 1259 (1953).

11) J. W. Mayer, L. Eriksson and J. A. Davies, "Ion Implantation in Semiconductors", Academic Press, New York (1970).

METALS

ENHANCEMENT OF THE SUPERCONDUCTING TRANSITION TEMPERATURE BY ION IMPLANTATION IN ALUMINIUM THIN FILMS

O. Meyer

Institut für Angewandte Kernphysik

Kernforschungszentrum Karlsruhe

ABSTRACT

Selected ions have been implanted at room temperature into thin aluminium films. The superconducting transition temperature T_c has been measured and was found to depend on implanted ion species and ion concentration. The largest increase in T_c was found for atoms which have large size factors and electronegativity values. The T_c behaviour, as observed for aluminium based ion implanted alloys differs considerably from results observed for some conventionally prepared alloys. Heavy ion radiation damage was found to affect T_c only slightly and an increase of T_c from 1.20 to 1.21 K has been observed.

INTRODUCTION

It is well known that a close relationship exists between metallurgy and superconductivity. The superconducting transition temperature T_c for example is a very specific property of a given metal, element, alloy or compound. T_c is affected by the general metallurgical condition of the sample, and can therefore provide information about what has happened to the sample. T_c - measurements have been used to determine the purity of samples, the solid solubility levels, the composition of samples, the phases which are present and the appearance of new phases[1]. Ion implantation is a simple tool to introduce impurities over a wide range of concentrations into the near surface region of solids. As the maximum obtainable concentration is not controlled by equilibrium processes there is a possibility to produce new stable structures, as well as metastable phases which may have unexpected superconducting properties.

In this preliminary study the simple (s-p) metal aluminium has been used as a target material mainly because numerous studies have been performed about the influence of impurities[2,3], pressure,[4] strain[5] and damage[6] on T_c. The influence of the impurity concentration in Al based solid solutions on T_c has been studied by Chanin, Lynton and Serin[2]. They found that at low concentrations (< 0.1 at %) soluble impurities produced a decrease of a few percent in T_c of Al, independent of the nature (electronegativity, valence, atomic size) of the solvent. This decrease in T_c is explained by the removal of the anisotropy effect[7]. A further increase in concentration C, yields a positive curvature in dT_c/dC, the magnitude of this curvature seems to depend on the valence of the solute.

Seraphim, Chiou and Quinn[3] extended the experiments to higher (> 1 at %) solvent concentrations and described the T_c dependence on the concentration by a single equation: $T_c = C(k_1 + k_2 \ln C)$ where the parameters k_1 and k_2 seem to depend on the valence of the solvent and also on the mean free path of the electrons. They believe that clustering will effectively remove the solute from the solution.

Hydrostatic pressure p applied to Al was found to decrease T_c.[4] At low pressures dT_c/dp shows a linear variation for $p \to 0$, indicating that a negative pressure effect (decrease in density) may result in an increase in T_c. This assumption is supported by the observation of T_c - enhancement obtained by applying tensile strain on Al films evaporated on mylar[5]. In addition some work on ion implantation in Al[8,9] has been performed and some information on radiation damage and precipitate growing during bombardment is available.

Therefore it is hoped that T_c measurements will provide new information about the metallurgical nature of ion implanted Al layers.

EXPERIMENTAL AND ANALYSIS

Al-films with thicknesses between 2000 and 4000 Å were prepared by electron beam evaporation in an ultra-high vacuum system. These films were implanted with selected ions from each group of the periodic system. Ion energies used were chosen in order to obtain a mean projected range of about 1000 Å, for very heavy ions the maximum available energy of 400 KeV has been used. The fluxes were maintained between 5 and 10 µA/cm² and fluences were used from 10^{12} to 10^{17} ions/cm². Layer thickness, sputtering during implantation and the depth distribution of implanted ions were recorded by use of nuclear backscattering technique. The usefulness of this technique is demonstrated in Fig. 1 where the backscattering spectra, obtained with 2 MeV α-particles from a thick Al layer are shown prior to and after implantation of $1.3 \cdot 10^{17}$ S⁺/cm² at 100 KeV. By analysis of such spectra, which is described elsewhere[10] one can obtain the sputter yield and the maximum concentration of the implanted species; together with additional thickness measurements of the Al-layer and of the step between implanted and unimplanted layer with a stylus equipment, lattice parameter changes in the implanted layer can be determined.

The superconducting transitions were measured resistively, with T_c being defined as the temperature at which the resistance decreased to half of its normal value. During implantation the layers were partially covered in order to restrict the implanted area to 5 x 10 mm². For the T_c measurements the voltage contacts were placed at equal distances from the boundary between the implanted and unimplanted areas. With this contact arrangement the resistance drop reveals a step as it is shown in Fig. 2 for a Al layer, implanted with $2 \cdot 10^{16}$ Cs$^+$/cm², 320 KeV at room temperature. The contact arrangement is shown in the inset of Fig. 2. For comparison the resistance drop of the unimplanted Al layer at 1.2 K is included and can be used as a check to show that the shielded part of the layer is not influenced during the implantation procedure.

RESULTS AND DISCUSSION

1) Influence of Radiation Damage on T_c

In a first step we studied the influence of heavy ion radiation damage on T_c by implanting high fluences of Ar$^+$ ions (10^{17} Ar$^+$/cm²) at room

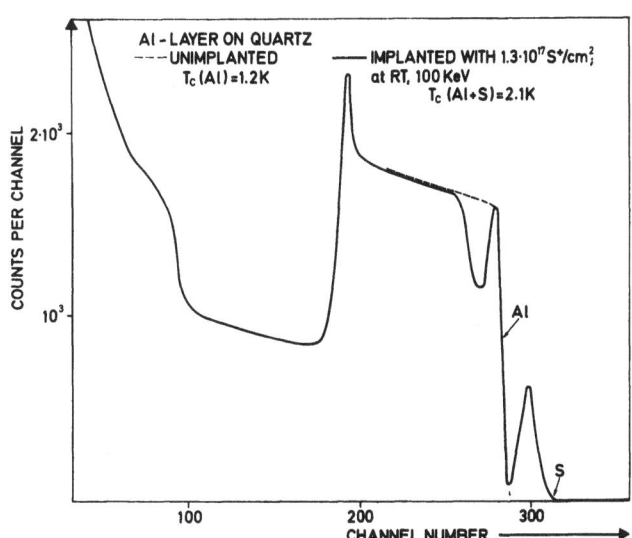

Fig. 1 Backscattering spectra from a thick aluminium layer before and after implantation of $1.3 \cdot 10^{17}$ S$^+$/cm², 200 KeV.

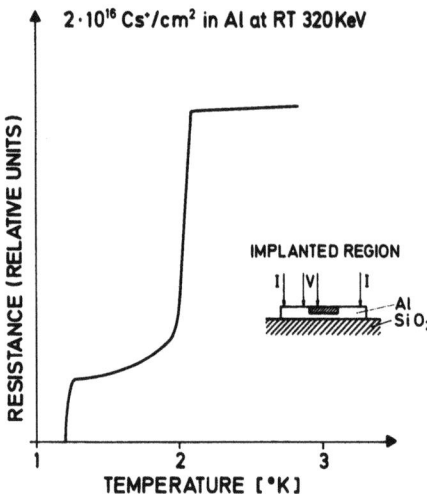

Fig. 2 Resistance drop as a function of temperature for an Al layer implanted with $2 \cdot 10^{16}$ Cs^+/cm^2, 320 KeV; contact arrangement is shown in the inset of the figure.

temperature. A slight increase in T_c from 1.20 → 1.21 K has been observed. This result shows that radiation defects produced at room temperature by inert gas ions only weakly affect T_c and this result is in agreement with observations, that defects produced by cold working at liquid He-temperature do not influence T_c[6].

2) Doping Effects

In order to compare T_c - values measured for Al based alloys produced by conventional alloying and produced by ion implantation we implanted the elements Ge, Zn and Mg in the fluence range between 10^{12} to 10^{17} ions/cm^2 and with mean projected ranges of about 1000 Å. The concentration in the peak maximum of the Gaussian distribution thus varied between 0.001 and 20 at %. The solid solubility levels for these elements in Al are rather high (Ge 2 at %, Zn 25 at %, Mg 15 at %).

In Fig. 3 we have presented the results on T_c obtained for conventionally produced alloys together with our results on ion-implanted alloys (Ge, Zn, Mg in Al) produced at room temperature. For Ge and Zn implants an increase in T_c, ΔT_c is observed in the concentration region between 0.001 and 0.01 at %). The slope dT_c/dC decreases and is found to be nearly constant between 0.1 and 20 at %. The conventionally produced alloys[2] show a decrease in T_c for concentrations between 0.01 and 1 at %. The T value given for the

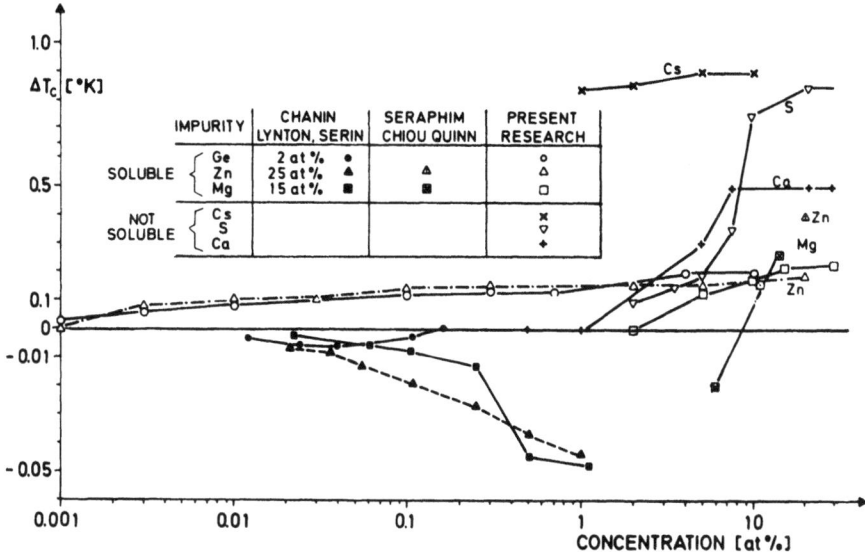

Fig. 3 Changes of the superconducting transition temperature ΔT_c in Al layers after implantation of different ions as a function of concentration compared with results obtained with conventional alloying techniques from other authors.

system Al - Zn (20 at %)([3]) is found to be larger than the corresponding value for an implanted system. For Mg on the other hand an increase in T_c is observed for concentrations above 2 at % and a saturation level in T_c, ΔT_c (max) is found for concentrations above 20 at %. In this case the behaviour of the ion implanted alloy is not too different from results obtained for conventional alloys ([3]) although the different slope and the increase in ΔT_c at smaller concentrations indicate that an additional effect is influencing the increase in T_c in implanted alloys.

It is obvious that our results obtained for as implanted samples can not be described by the equation given by Seraphim, Chiou and Quinn([3]) and we may conclude, that valence effects and the influence of impurities as scattering centers for electrons on the mean free path are not important. The results on the influence of pressure and strain as mentioned in the introduction leads us to the hypotheses that localized strain fields associated with the impurity ions may be responsible for the observed increase in T_c. The possibility that local strain fields of the impurity atoms could influence the energy gap of superconducting Al-based alloys has already been treated by Claiborne([11]). He suggested that the sign of the difference between the atomic radii of the impurity and the host atoms will be responsible for a possible increase or decrease of T_c on alloying. This statement is not verified by our observations however. For the above mentioned elements this effect will only occur if they do not occupy substitutional sites after implantation. The saturation in the T_c increase may be due to precipitation effects, as precipitate growing during implantation at room temperature has been observed for Cu^+, Pb^+ and Sb^+ ions in Al[8].

In order to test the assumption that strain fields connected to the impurity atom or to an impurity-defect complex are important we implanted ions known to have large size effects in Al. The size mismatch between solvent and solute atoms results in a deviation from Vegard's law and is proportional to the elastic strain energy ([12]). The size factor Ω_{sf} is defined by $(\Omega^x - \Omega_{Al})/\Omega_{Al}$, where Ω_{Al} is the atomic volume of Al and Ω^x is the value of the volume for a 100 at % impurity concentration extrapolated from the slope of the lattice parameter versus concentration obtained at low impurity concentrations. Table 1 shows the results for such impurities in Al, where the size factors are known([13]):

Table 1: Size factor and ΔT_c(max) in ion implanted Al based alloys.

Dopant	Ca	Mg	Ge	Ga	Ag	Zn	Cu	Pb
Ω_{sf}	+ 177	+ 41	+ 13	+ 5	+ 0.12	- 5.7	- 38	- 53
ΔT_c(max)	+ 0.5	+ 0.23	+ 0.2	+ 0.22	+ 0.1	+ 0.18	0	+ 0.2

In the limited number of Al based alloys for which size factors are available([13]), Ca undoubtely has the biggest value and this value correlates with the largest value of ΔT_c(max) in this special selection of elements. As an example the dependence of T_c for Ca on the concentration is also shown in Fig. 3. For the other elements the correlation is not obvious and Cu, for example, which has a negative size factor does not show any increase in T_c at all. This observation may indicate that Cu completely precipitates during implantation at room temperature([8]). The strain fields connected to the precipitates are not believed to have any influence on T_c otherwise such influence should have been observed in conventional alloys with high impurity concentration levels.

In a further study we selected ions of different electronegativity, χ in order to see if the chemical nature of the element is important for T_c enhancement.

In Table 2 the results from this systematic study are summarized:

Table 2: Electronegativity and ΔT_c(max) in ion implanted Al based alloys.

Dopant	Cs	K	Ba	La	In	Pb	Sb	B	Au	C	S	N
χ	0.7	0.8	0.9	1.2	1.7	1.8	1.9	2	2.4	2.55	2.58	3
ΔT_c(max)	+0.9	0	+0.25	+0.1	+0.2	+0.2	+0.25	+1.1	0	+0.6	+0.9	+0.65

The T_c-enhancements seem to be pronounced for elements with large electronegativity values but there are two important exceptions: Au has a large electronegativity value but it does not show any T_c - enhancement, this is similar to the result obtained for Cu. Cs has a very low electronegativity value but a large increase in T_c is observed. Here we believe that similar to the results obtained for Ca a large size mismatch is responsible for this observation. From this study on electronegativity one may conclude that the chemical behaviour of the implanted ion itself does not play a major role for the T_c - enhancements observed in ion implanted Al base alloys produced at room temperature. Large electronegativity values on the other hand may reduce the migration probability and prevent from precipitation. The T_c dependence on the concentration of implanted S^+ and Cs^+ ions are also included in Fig. 3.

SUMMARY AND CONCLUSION

Only 4 species (Ar, Cu, Au and K) out of 24 ions implanted in Al films at room temperature were found not to increase T_c above 1.2 K. For all other dopants under consideration an increase of T_c and saturation levels, T_c(max) have been observed. T_c(max) was found to depend on the implanted ion species and on their concentration. The largest increase in T_c was found for atoms with large size factors and with large electronegativity values. The T_c dependence for the implanted Al-Ge and Al-Zn alloys is completely different compared with results for conventional alloys, and can not be explained by a removal of the anisotropy effect, by a mean free path effect or valence effects. Vice versa the results from the T_c measurements indicate, that the as implanted ions are not on clean substitutional sites. This statement will be investigated by channeling measurements. Hauser[14] studied mixtures of Al and 3.6 at % Ge sputtered at room temperature and observed a maximum T_c of 2.5 K. He explained his results by an observed 5 - 10 % increase in volume. This explanation is in accordance with a possible negative pressure effect and the strain effect discussed above and therefore we assume that the density effect is also responsible for our results. Thus the conditions we want to impose on an impurity in order to obtain a **maximum increase** in T_c of Al are:

a) low precipitation probability, i. e. low diffusibility even in a stream of mobile vacancies during implantation.

b) large size factor in order to produce large strain energies.

Al based ion implanted alloys produced at room temperature have provided unexpected results concerning the superconducting transition temperature. Vice versa the T_c measurements clearly indicate the complex behaviour of implanted species in the matrix.

ACKNOWLEDGEMENTS

The author greatfully adknowledges the assistance of M. Kraatz and R. Smithey in the experimental part of this work. He also wished to thank J. Geerk and G. Linker for valuable discussions.

REFERENCES

[1] B. T. Matthias, Metallurgy from Superconductivity

[2] G. Chanin, E. A. Lynton and B. Serin, Phys. Rev. 114, 719 (1958)

[3] D. P. Seraphim, C. Chiou and D. J. Quinn, Acta Metallurgica 9, 861 (1961)

[4] M. Levy and J. L. Olsen, Solid State Communications 2, 137 (1964)

[5] H. A. Notary, Appl. Phys. Letters 4, 79 (1964)

[6] G. von Minnigerode, Z. Physik 154, 442 (1959)

[7] D. Markowitz and L. P. Kadanoff, Phys. Rev. 131, 563 (1963)

[8] P. A. Thackery and R. S. Nelson, Phil. Mag. 19, 169 (1969)

[9] G. J. Thomas and S. T. Picraux, Proc. of the Internat. Conf. on Appl. of Ion Beams to Metals Oct.2 - 4, 1973, Albuquerque, New Mexico

[10] M. A. Nicolet, J. W. Mayer and I. V. Mitchell, Science 177, 481 (1972)

[11] L. T. Claiborne, J. Phys. Chem. Solids 24, 1363 (1963)

[12] J. Friedel, Phil. Mag. 46, 514 (1955)

[13] H. W. King, J. of Mat. Science 1, 79 (1966)

[14] J. J. Hauser, Phys. Rev. B3, 1611 (1971)

THE INFLUENCE OF HEAVY ION BOMBARDMENT ON THE SUPERCONDUCTING

TRANSITION TEMPERATURE OF THIN FILMS

G. Linker and O. Meyer

Institut für Angewandte Kernphysik

Kernforschungszentrum Karlsruhe

ABSTRACT

Thin layers of the transition metals vanadium, niobium and tantalum have been bombarded with Ne^+ ions. Energies and fluences were chosen such that the ions penetrated the layers and that equal amounts of energy lost in nuclear collisions were deposited in the different layers. A strong decrease of the superconducting transition temperature T_c after bombardment has been observed and similar relative changes $\Delta T_c/T_c$ occured in V, Nb and Ta. These changes are assigned to the disorder produced on the paths of the penetrating ions. An enhanced decrease in T_c was induced in the A-15 compound Nb_3Ge by He^+ ion irradiation.

INTRODUCTION

There has been considerable interest in the effects of irradiations on superconductors since 1960 and many studies were concerned with the influence of radiation damage on the critical parameters I_c, H_c. Irradiations were performed using mainly neutrons, protons and deuterons and striking effects were reported for the critical current density I_c, however little influence of radiation damage on the superconducting transition temperature T_c has been observed. A summary of this work till 1968 is given by Cullen[1].

Recently also heavy ions have been used for damage production in superconductors[2,3] and substantial decreases in T_c as a result of heavy ion bombardment has been found for Nb by Crozat et al.[4] and for V, Nb, Ta and Nb_3Sn by Meyer et al.[5]. Kübler[6] found a dependence of T_c reductions in V and Nb layers on mass, energy and fluence of the bombarding ions and thus assigned the reductions to defects produced in the layers. A threshold effect in T_c reduction of some A-15 compounds with neutron irradiation has been observed by Schweitzer and Parkin[7]. In this present study which is a continuation of ([5],[6]) an attempt is made to study more quantitatively damage production and distribution from heavy ions and its influence on T_c of some transition metal superconductors. The results of such measurements are important when ion implantation is used as a doping technique for superconductors, when superconducting devices are used in a radiation environment, and may be of interest when the mechanism of superconductivity is studied in disordered layers as it is known that considerable changes in T_c are observed in disordered layers of transition metals obtained by evaporation onto cryogenic substrates[8].

EXPERIMENTAL

Layers of the transition metals V, Nb and Ta have been prepared by electron gun evaporation onto heated quartz substrates to thicknesses of about 1000 Å and a pressure in the 10^{-8} torr range was maintained during evaporation. Layer thickness was monitored with a quartz oscillator during evaporation and then controlled by stylus instrument measurements and by backscattering of 2 MeV He^+ ions, the latter technique also giving information on layer homogeneity, sputtering effects during implantation and possible impurity content[9]. Transition temperatures were measured resistively with a standard four point probe method to an accuracy of about 0.05 K. The layers have been bombarded with Ne^+ ions at room temperature with a scanned ion beam with energies and fluences in the range 170 KeV to 360 KeV and $2 \cdot 10^{14}$ ions/cm^2 - $2 \cdot 10^{17}$ ions/cm^2 respectively. Fluxes were kept at about 1 - 3 $\mu A/cm^2$, the pressure in the implantation chamber prior to implantation was $< 10^{-7}$ torr and increased slightly during bombardment to values $< 10^{-6}$ torr due to the reduced pumping efficiency of the vacuum system for noble gases.

DAMAGE PRODUCTION

The damage in the layers has been produced by Ne^+ ion bombardment. The energy of the ions was chosen such that they penetrated the layers and came to rest in the substrates thus effecting layer properties only by the energy lost on their paths through the layers. The condition for the minimum bombardment energy was: $Rp > d + \Delta Rp$ where d is the layer thickness and Rp and ΔRp are the mean projected range and the average fluctuation in projected range respectively.

On their way through a layer the particles loose energy by two processes, namely by electronic excitations and by elastic nuclear collisions. Both processes determine the range of the particles but only nuclear scattering leads to displacements of target atoms. In order to obtain a quantitative measure for the distribution and the total energy lost in nuclear collisions by the Ne^+ ions in different layers the primary energy deposition profile for the projectiles has been calculated from universal $(d\varepsilon/d\rho)_n$ data (ρ and ε are dimensionless range and energy parameters introduced by Lindhard et al. (LSS) ([10])) tabulated by Schiott([11]) by converting these data from an energy to a depth scale for a particular ion-target system under consideration of electronic energy losses. In the actual computations an analytical expression for $(d\varepsilon/d\rho)_n$ from Sigmund([12]) which is a good fit to Schiott's data and electronic stopping values from the LSS theory were used.

The conversion was performed by calculating the nuclear stopping $(d\varepsilon/d\rho)_{n,i}$ at energy positions ε_i. The ε_i values were obtained successively by calculating the total energy losses $\Delta\varepsilon_i$ in small constant depth intervals $\Delta\rho (\approx \Delta x)$ starting at the energy of the incident projectile $\varepsilon_o (\approx E_o)$ thus giving $\varepsilon_i = \varepsilon_{i-1} - \Delta\varepsilon_i, i = 1, 2...$; the total stopping power (electronic + nuclear) has been assumed to be constant in $\Delta\rho$. The conversion procedure stops when $\varepsilon_i < 0$ and the total energy lost in nuclear collisions E_n in a layer is obtained from $\sum_i \Delta\rho (\frac{d\varepsilon}{d\rho})_{n,i}$, the summation taken on the particle path over the layer thickness in steps of $\Delta\rho$. The penetration depth of the particles and the conversion of depth intervals $\Delta\rho$ to projected depth intervals $\Delta\rho_p$ has been estimated from LSS projected range statistics computed by Johnson and Gibbons([13]).

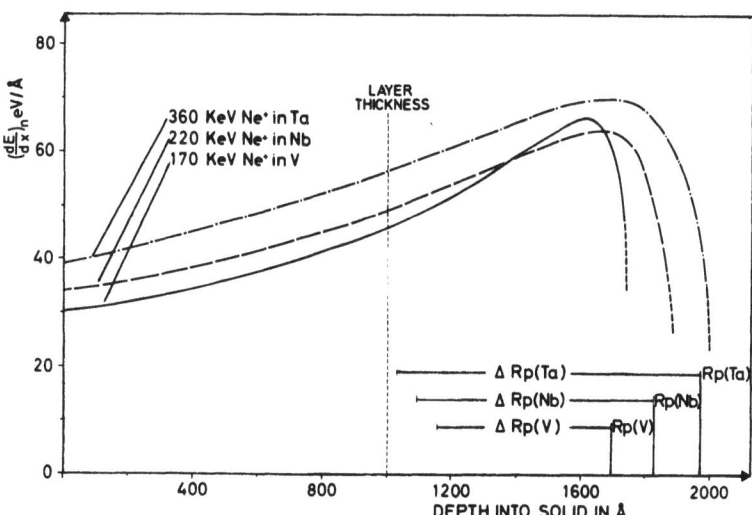

Fig. 1 Calculated primary energy deposition profiles for an average Ne ion in V, Nb and Ta; R_p and ΔR_p values from LSS theory are included in the figure.

Table 1. Characteristic values for the bombardment of V, Nb and Ta with Ne$^+$ ions. The symbols are: E - energy of the incident particles; R, R$_p$, ΔR$_p$ -range, projected range and fluctuation in projected range respectively; Δx - depth intervals used in the profile calculations and E$_n$ - total energy lost in nuclear collisions by a Ne$^+$ ion in a layer of 1000 Å thickness.

	E(KeV)	R (Å)	Rp(Å)	ΔRp(Å)	Rp/R	ΔX(Å)	E$_n$(1000Å) (KeV)
V	170	2484	1691	535	0.6808	38	37.04
Nb	220	3359	1834	734	0.5460	48	41.43
Ta	360	4577	1973	948	0.4311	61	47.55

Some characteristical values obtained for the bombardment of 1000 Å thick V, Nb and Ta-layers with Ne$^+$ ions are given in Table 1 and calculated deposition profiles are shown in figure 1. These profiles are rather homogeneous in layers of about 1000 Å thickness as compared to profiles when particles come to rest in the layer, it should be emphasized however that these are profiles for an average particle as fluctuations in range were neglected in the calculations. Nevertheless they are considered to be a reasonable approximation on the path of the particles as influence of fluctuations on primary deposition profiles is evident mainly in the range where the particles come to rest([14]).

Similar densities of energies deposited in nuclear collisions in different materials were obtained by choosing appropriate fluences for the Ne$^+$ ions (Fluence x Energy/Å = const.).

RESULTS AND DISCUSSION

The transition temperatures of the as-evaporated V, Nb and Ta layers defined as the midpoints of the transition curves from the superconducting to the normal state were close to the values of bulk material and usually were in the range 5.2 - 5.3 K for V, 9.0 - 9.3 K for Nb and 4.0 - 4.2 K for Ta while the transition full widths for all samples were < 0.2 K. Layer bombardment with Ne$^+$ ions resulted in a decrease of the transition temperatures the maximum absolute value being obtained for Nb with a T$_c$ reduction of about 40 %. This result is in accord with our previous measurements([5]) and with the results of Crozat et al. ([4]) but is in marked contrast to results reported for high integrated fast neutron fluences (> 10^{19}n/cm^2) where T$_c$ reductions observed were always below 1 % ([15]). Relative T$_c$ reductions ΔT$_c$/T$_c$ from our present measurements are shown in

INFLUENCE OF HEAVY ION BOMBARDMENT

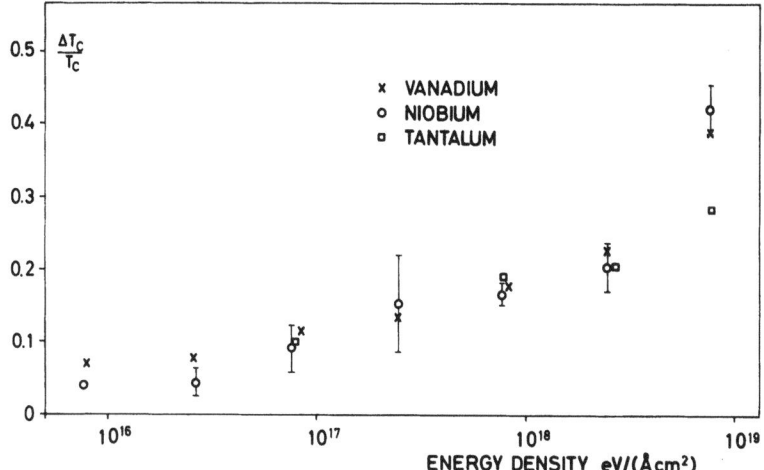

Fig. 2 Relative decrease of the superconducting transition temperature $\Delta T_c/T_c$ as a function of energy density deposited in nuclear collisions (ded).

figure 2 as a function of the density of energy deposited in the layers by nuclear collisions (ded) with error bars being included for the Nb measurements. A slight increase in the relative reductions (0.05 < $\Delta T_c/T_c$ < 0.2) is observed in the ded-range from 10^{16} eV/(Å cm^2) to 10^{18} eV/(Å cm^2) and similar values are obtained for V, Nb and Ta. Beyond 10^{18} eV/(Å cm^2) corresponding to roughly 50 - 100 dpa (atom displacements per atom), this value depending on the assumption of an average displacement energy E_d, a steeper slope in the dependence of $\Delta T_c/T_c$ on ded is observed showing a lower value for Ta than for V and Nb. This pronounced change in slope indicates that a different process influencing the superconducting transition temperature must occur in these heavily bombarded layers as compared to those with lower energy densities deposited.

The shape of the transition curves and backscattering analysis of the bombarded layers gave some indication to the nature of damage occuring at high fluences. The width of the transition curves for the bombarded samples with low ded-values (< 10^{18} eV/(Å cm^2)) was a little broader than for the as evaporated layers but was always below 0.5 K, for high fluences however transition widths greater 1 K were observed. The former result indicates homogeneous distortions of the bombarded layers in agreement with the calculated damage profiles, the latter suggesting inhomogeneities in the layers. These conclusions were confirmed by the backscattering results. No changes in layer thickness and in the shape of the spectra from layers before and after bombardment were observed for low ded values

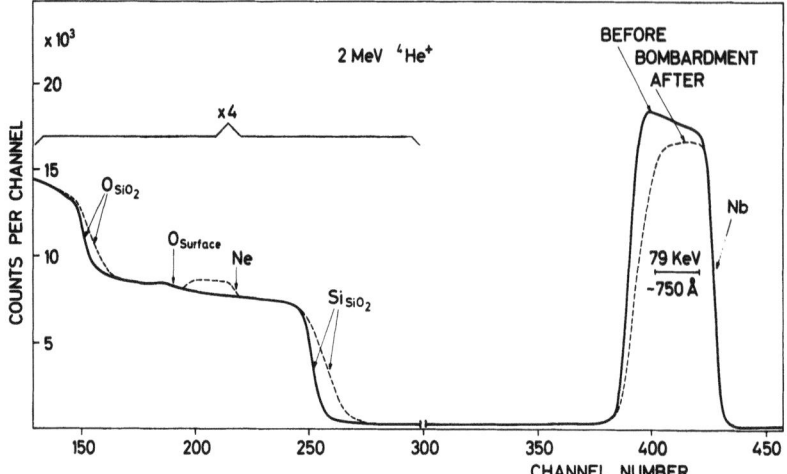

Fig. 3 Backscattering spectrum from a heavily bombarded Nb layer ($5.9 \cdot 10^{17}$ Ne$^+$/cm^2, 240 KeV) compared with a spectrum from a non bombarded layer on quartz substrates.

demonstrating that sputtering effects were negligible in this region. For high ded values however changes were observed and an example with pronounced effects is shown in figure 3. Here spectra from an undamaged and from a heavily bombarded ($5.9 \cdot 10^{17}$ Ne$^+$/cm^2) Nb layer on quartz substrates are compared. It is demonstrated that the bombardment due to sputtering effects leads to an overall reduction in layer thickness and distortions in the shape of the spectrum. From these distortions in shape and also from the reduction in sharpness of the silicon and oxygen edges from the quartz substrates it can be concluded that inhomogeneities arose in the bombarded layer may be due to discontinues material removal([16]).

It should be noted that the presence of small amounts of oxygen dissolved in all V, Nb and Ta results in a drastic reduction of T_c ([17]). In some of our backscattering spectra as also indicated in figure 3 we observed small oxygen peaks due to a surface oxide layer. These surface oxides however did not influence the T_c's as they were observed also in high T_c as evaporated layers. Nevertheless it can not be completely excluded that during Ne$^+$ ion bombardment knock on oxygen atoms could be implanted into the layers and thus may have effected our quantitative results (large error bars). Difficulties in the quantitative reproducibility of such measurements due to film preparation and protection were also pointed out by Crozat et al.([4]).

Backscattering analysis however within its limits of accuracy did not show an increase of impurity content during ion bombardment, thus we conclude that in the low ded region T_c reduction occurs due to heavy ion radiation damage resulting in defect clusters deteriorating the bcc-structure of the investigated metals and thus reducing the density of states at the fermi level N(0) and/or the average phonon frequency as radiation hardning is known to occur in bcc metals. Of course these are only two possible factors for an microscopic explanation of the T_c - reductions and no solid arguments can be given from our present experiments. The enhanced reduction in layers where high densities of energy were deposited we ascribe to macroscopic distortions of the layers like particle size effects rather than to defect aggregates. Results from thin film X-ray analysis support this argument as from the heavy distorted layers only the two strongest reflections were recorded and these showed distinct broadening as compared with the equivalent reflections from not bombarded layers.

Recently Schweitzer and Parkin[7] have reported a high reduction in T_c of some A-15 compound superconductors as a threshold effect with high fluence fast neutron irradiations. In accordance with this result in respect to its magnitude we have observed in a preliminary experiment a reduction of T_c in a Nb_3Ge layer from 20.7 to 7.1 K after irradiation with 10^{17} He^+/cm^3 and 300 KeV.

ACKNOWLEDGEMENTS

The authors would like to thank M. Kraatz for performing the implantations and R. Smithey for sample preparation and T_c - measurements.

REFERENCES

[1] Cullen, G. W. 1968 Proc. Summer Study on Superconducting Devices and Accelerators, Brookhaven National Laboratory, p. 437

[2] Tsypkin, S. I., Chudnova, R. S., Soviet Physics-Solid State, 13 (1972), 2588

[3] Ischenko, G., Mayer, H., Voit, H., Besslein, B., Haindl, E., Z. Physik 256 (1972), 176

[4] Crozat, P., Adde, R., Chaumont, J., Bernas, H., Zenatti, D., Int. Conf. Appl. Ion Beams Metals, Albuquerque N. M., 1973, Oct. 2 - 4

[5] Meyer, O., Mann, H., Phrilingos, E., ibid.

[6] Kübler, G., Diplomarbeit, Karlsruhe 1973

[7] Schweitzer, D. G., Parkin, D. M., Appl. Phys. Lett., 24 (1974), 333

[8] Crow, J. E., Strongin, M., Thompson, R. S., Kammerer, O. F., Phys. Lett., 30 A (1969), 161

[9] Linker, G., Meyer, O., Gettings, M., Thin Solid Films, 19 (1973), (1973), 177

[10] Lindhard, J., Scharff, M., Schiott, H. E., Mat.-Fys. Medd. 33 (1963), Nr. 14

[11] Schiott, H. E., Mat.-Fys. Medd. 35 (1966), Nr. 9

[12] Winterbon, K. B., AECL - 3194, Nov. 1968

[13] Johnson, W. S., Gibbons, J. F., LSS Projected Range Statistics in Semiconductors, Stanford, Calif., 1970

[14] Kulcinski, G. L., Laidler, J. J., Doran, D. G., Rad. Eff. 7 (1971), 195

[15] Kernohan, R. H., Sekula, S. T., J. Appl. Phys., 38 (1967), 4904

[16] Meyer, O., Mann, H., Linker, G., Appl. Phys. Lett. 20 (1972), 259

[17] DeSorbo, W., Phys. Rev., 132 (1963), 107

CHEMICAL ASPECTS OF ION IMPLANTATION

G. K. Wolf, W. Fröschen, U. Sahm

Lehrstuhl für Radiochemie, Universität

Heidelberg, 75 Karlsruhe, Postfach 3640

Until now in contrast to solid state physicists only very few chemists used the technique of ion implantation (1-4). The purpose of my paper is to show ion implantation being a very useful method for studies in very different fields of chemistry.

The first figure gives a survey on the activities of our group in Karlsruhe and Heidelberg in this field.

The figure covers very different disciplines of chemistry, from diffusion (physical chemistry) and radiation damage (radiation chemistry) studies to experiments on chemical reactions of energetic ions with solids (solid state chemistry) and gaseous compounds (ion molecule reactions). The first two subjects will be treated later in detail. Concerning the last two I want to mention only that we investigate mainly the mechanism and probability of reactions between accelerated ions and solid inorganic compounds on one hand and gaseous compounds on the other hand in dependency of dose, ion energy etc. (4-8).

The chemical syntheses of compounds (inorganic chemistry) I will describe later in detail. Of more technical interest are the catalytic properties of materials after ion implantation. Here Baumgärtner and Burkhardt (9) could show the hydrogenation of ethylene with metal catalysts proceeding with a higher reaction rate using ion bombarded catalysts compared with untreated ones.

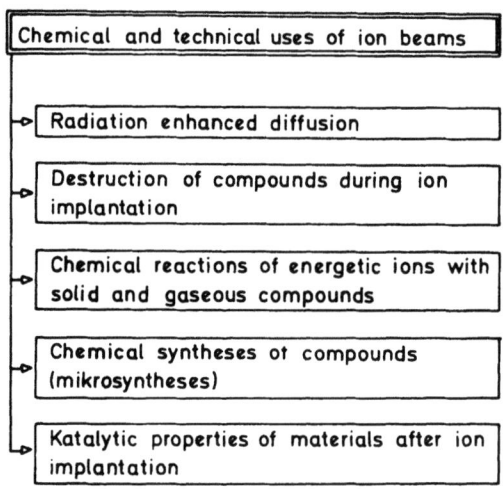

Fig. 1: Main subjects studied by the Heidelberg-Karlsruhe Group

All our experiments were performed with a Danfysik 100 keV ion accelerator which we modified considerably at the gas inlet and target side. The target chamber enables us to irradiate up to 24 targets, metals or pressed powders at temperatures ranging from liquid nitrogen to ≉200°C. For the gas phase reactions we constructed a differentially pumped reaction chamber, equiped with a quadrupol mass spectrometer for the analysis of the reaction products (7,8).

Some of the fields, shown in the first figure I will treat now in detail.

1. Radiation enhanced diffusion

We made an approach to this problem different from most other studies on these subjects (10). We measured the diffusion of Cadmium in Silver foils (10^4 Å thick) determining the speed of evaporation of the Cadmium at ≉900°C. The radioactive Cadmium was created in the Silver via the Ag (α,xn) Cd reaction bombarding the foil with 30 MeV deuterons. Thus we had an uniform distribution of Cd in the Ag. Afterwards the foils were bombarded with $10^{15}-10^{16}$ ions/cm^2 of different chemical elements and the evaporation time of the radioactive Cadmium out of this foils compared with the case of no bombardement. It turned out that
- irradiations with Ag$^+$ ions did not change the evaporation time,
- irradiating with Ge$^+$, O$^+$, Kr$^+$ ions the evaporation speed increased by a factor of 5 - 10.

One explanation for this increase could be a change in the surface properties of the Ag due to the bombardement, facilitating the removal of the Cadmium. This possibility we could rule out because
- the big difference between the effect of Ag$^+$ and other ions cannot be understood in that way;
- the activation energy deducted from the temperature dependency of the evaporation speed was in the range of 1 eV, being typical for diffusion but not for the removal from the surface.

Thus we have a second explanation or rather working hypothesis: The ions penetrating during the implantation at liquid nitrogen temperature < 1000 Å into the Ag diffuse during the heating period together with the Cadmium through the whole Ag foil stabilizing the defects by preventing them from annealing. Thus the

Cadmium may diffuse much faster via a defect mechanism compared with the untreated case. After Ag^+ implantation Ag anneals fast together with the defects, thus causing no increased evaporation speed.

2. Destruction of compounds during ion implantation

There exist various methods for the determination of the destruction of compounds or the amount of lattice disorder in ion bombarded elements and compounds as backscattering, X-ray analysis, optical reflection measurements etc. For a number of chemical compounds one may also use a chemical method: "The reactivity of implanted ions with the compound". I want to outline this possibility for two examples:
If one bombards an organometallic compound as $Cr(CO)_6$ or $Mo(CO)_6$ or $W(CO)_6$ with a low dose of Cr^+ labelled with radioactive ^{51}Cr a certain amount of the $^{51}Cr^+$ will react chemically forming $^{51}Cr(CO)_6$. This amount depends on the ion/target combination and is dose independent at low doses. In this region the disorder spikes, caused by individual ions are well separated from each other. At higher doses they start to overlap and this behavior implies a decrease of the chemical reactivity of the $^{51}Cr^+$. Thus, the decrease of the $^{51}Cr(CO)_6$ formation is reflecting the growing decomposition or lattice disorder. Fig. 2 shows this behavior for the reaction $^{51}Cr^+$ (60 keV) + $Mo(CO)_6 \rightarrow {}^{51}Cr(CO)_6$ (11). At low doses ≈ 50 % of the $^{51}Cr^+$ is converted to $^{51}Cr(CO)_6$. At a dose of ≈10^{12} ions/cm² the destruction of the compound starts and at ≈10^{14} ions/cm² the $Mo(CO)_6$ is thoroughly damaged.

The second example (6) is the decomposition of the complex Cobalt compound $[Co(ethylendiamin)_2Cl_2]NO_3$ during Co^+ bombardement. Here we have again a decrease for the probability of the reaction
$^{57}Co + [Co(en)_2Cl_2]NO_3 \longrightarrow [^{57}Co(en)_2Cl_2]NO_3$.
Between 10^{12} and 10^{13} ions/cm² the compound formation drops from ≈70 % to ≈30 %. But in contrary to the carbonyl case the decrease does not approach nearly zero applying higher doses. This fact was confirmed by Moessbauer measurements with a probe bombarded with 5 x 10^{15} ions/cm² showing a distinct amount of ^{57}Co in the form of the complex compound (12). Our interpretation of this tail in the destruction curve is a certain amount of $^{57}Co^+$ penetrating into less damaged lattice regions due to (radiation enhanced) diffusion . There it has the possibility to react even very high doses having been applied to the sample.

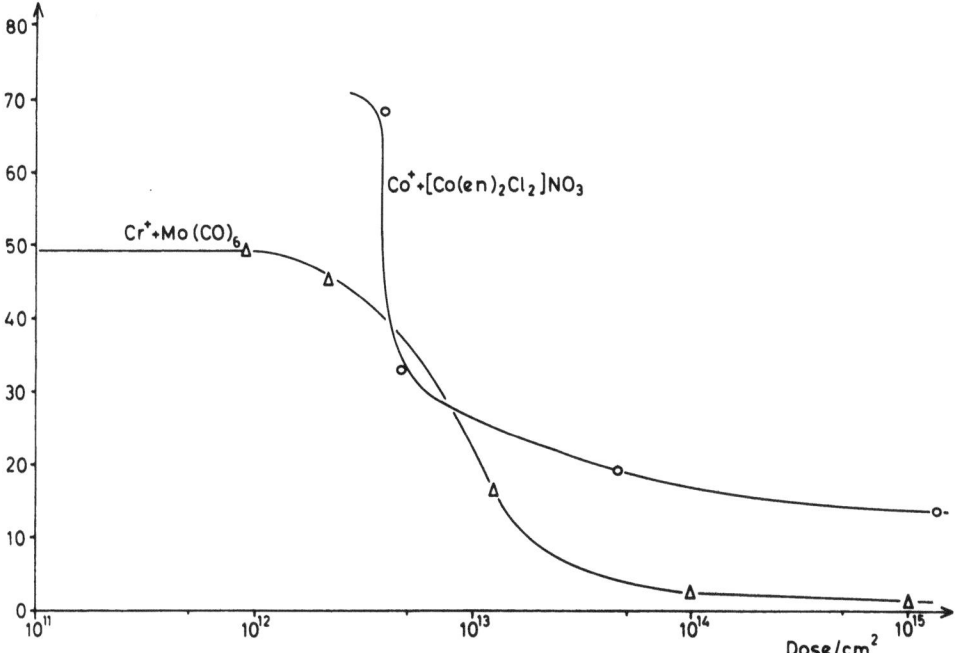

<u>Fig. 2</u>: Destruction of the chemical compounds $Mo(CO)_6$ and $[Co(en)_2Cl_2]NO_3$ during implantation with increasing doses of Cr^+ (labelled with ^{51}Cr) and Co^+ (labelled with ^{51}Co) respectively.

This method may be also used in a different way the destruction of the compound being done by Gamma or electron radiation or bombardement with rare gases followed by a low dose implantation of a suitable reactive ion ($^{51}Cr^+$ or $^{57}Co^+$ in our examples). It is interesting to compare the results presented here with results obtained with other methods for different types of lattices. Our doses for total destruction of the compounds are only a little lower than the critical doses for the amorphization of semiconductors (Si and Ge) determined with various methods (10,13) but ≉2 orders of magnitude lower than the critical doses for metals as Vanadium (14) or compounds as CdS (15) measured with the backscattering technique and optical reflection respectively.

3. Chemical syntheses of compounds

The possibility of using ion implantation for chemical syntheses is due to the fact that different from conventional chemical reactions where both reaction partners have about the same concentration and temperature we deal with reactions of single atoms having very high energies with partners (targets) being present in high excess at arbitrary temperatures. Thus unconventional reactions may proceed leading to products being yet unknown or difficult to synthetize. Unfortunately, the amount of the products is very low, typically 10^{13} - 10^{17} molecules and thus the analysis difficult to perform.

We tried to detect unusual oxidation states of Krypton and rare earth, studies I don't want to treat in this paper and to synthetize heteronuclear carbonyl compounds.

Implanting for example radioactive Mn^+ into $Re_2(CO)_{10}$ a fraction of the Mn^+ is transformed into a volatile and extractable form. We came to the conclusion $MnRe(CO)_{10}$ having been formed because of the following reasons:

- a Mn carbonyl with one metal atom does not exist,
- a formation of $Mn_2(CO)_{10}$ is not possible because of the statistical probability two Mn atoms coming close enough together being to low.
- other volatile and extractable stable Mn compounds are not known under the experimental conditions.

Table 1: Formation of heteronuclear metal carbonyls by ion implantation

Reactions	Product	% Yield	otherwise synthetized
$^{51}Cr^+ + Re_2(CO)_{10}$	$Cr(CO)_6$ or Re-Cr-Carbonyl	17	-
$^{56}Mn^+ + Re_2(CO)_{10}$	Mn-Re-Carbonyl ($MnRe(CO)_{10}$)	7	$(OC)_5MnRe(CO)_5$
$^{57}Co^+ + Re_2(CO)_{10}$	Co-Re-Carbonyl ($ReCo(CO)_9$)	6,5	$(OC)_5MnCo(CO)_4$
$^{57}Co^+ + Ru_3(CO)_{12}$	Co-Ru-Carbonyl	8	-
$^{57}Co^+ + Rh_4(CO)_{12}$ $^{57}Co^+ + Ir_4(CO)_{12}$	Co-Rh-Carbonyl Co-Ir-Carbonyl ($CoRh_3(CO)_{12}$)	4	$Co_2Rh_2(CO)_{12}$
$^{59}Fe^+ + Re_2(CO)_{10}$	?	1	$[Re_2(CO)_{10}]Fe(CO)_4$
$^{59}Fe^+ + Ru_3(CO)_{10}$	Fe-Ru-Carbonyl ($FeRu_2(CO)_{12}$)	20	$FeRu_2(CO)_{12}$
$^{59}Fe^+ + Rh_4(CO)_{12}$ $^{59}Fe^+ + Ir_4(CO)_{12}$	Fe-Rh-Carbonyl Fe-Ir-Carbonyl	10 6,5	-

Afterwards we found out that this particular compound has been already synthetized by Russian workers (16) in macroscopic amounts. Later we started to examine a number of other ion/target combinations and the table 1 shows the results. One can see a great number of heteronuclear carbonyls having been formed with different yields. Some of them are already known others are entirely new compounds. Certainly the identification of the compounds is based only on indirect proofs but the arguments are rather strong as we think. Unfortunately a determination of the structure of the compounds is not possible because of the small amounts available with the exception of the cobalt compounds. They may be studied via Moessbauer spectroscopy - an experiemnt being very difficult to perform. Recent results indicate that it is even possible to insert copper into heteronuclear carbonyls. If this proofs to be true it would be the first carbonyl compound containing copper ever synthetized.

Summarizing one can say ion implantation being a useful tool for basic and applied research in chemistry and physical chemistry, deserving broader distribution in these disciplines.

REFERENCES

1) T. Andersen, G. Sørensen, Trans.Farad.Soc. 62, 3427 (1966)
2) T. Andersen, T. Langvad, G. Sørensen, Nature 218, 1158 (1968)
3) G.K.Wolf, T. Fritsch, Radiochim.Acta 11, 194 (1969)
4) A. Maddock, Proceedings of the 8th Intern.Conf. on Low Energy Ion Accelerators and Mass Separators, Billingehus, Sweden, 12.-15.6.1973
5) G.K. Wolf, KFK-Nachrichten 3, 14 (1973)
6) E. Mohs, Diplomarbeit Heidelberg 1972
7) M. Becker, private communication 1974
8) G.K. Wolf, Proc. of the 8th Int.Conf.on Low Energy Ion Accelerators ..., p. 307
9) F. Baumgärtner, H.G. Burkhardt, private communication (will be published)
10) G. Dearnaley, J.H. Freeman, R.S. Nelson, J. Stephen, "Ion Implantation", North Holland, Amsterdam, London (1973)
11) U. Heinstein, Diplomarbeit Heidelberg 1973
12) J. Fleisch, P. Gütlich, E. Mohs, G.K. Wolf 7th Intern. Hot Atom Chemistry Symposium Jülich (1973) AED-Conf.-73402-080
13) V.M. Gusev, M.I. Guseva, C.V. Sarinin Radiation Effects 15, 251 (1972)
14) G. Linker, M. Gettings, O. Meyer Ion Implantation in Semiconductors and other Materials, B.L. Crowder (Ed.), Plenum Press N.Y. 1973, p. 465
15) J.A. Hutchby, Radiation Effects 16, 189 (1973)
16) E.W. Abel, F.G.A. Stone, Quart.Review 23, 343 (1969)

ION IMPACT CHEMISTRY IN THIN METAL FILMS; ARGON, OXYGEN AND NITROGEN BOMBARDMENT OF TANTALUM

K.H. Goh, K.G. Stephens and I.H. Wilson

University of Surrey

Department of Electronic and Electrical Engineering
Guildford, Surrey, England

ABSTRACT

Nitrogen bombardment of tantalum thin films is shown to produce large variations in resistivity (ρ) and temperature coefficient of resistivity (TCR) with ion dose, and electron diffraction shows that these changes are probably due to the formation in the films of Ta_2N and TaN. Films were prepared by evaporation in two quite different vacuum systems and by sputter deposition in an argon environment and the amount of oxygen in the as deposited tantalum films seems to play an important part in the magnitude of the variations of ρ and TCR.

Argon bombardment produces large changes in ρ with little change of TCR. Electron micrographs show bubble formation.

Oxygen bombardment produces a similar change in ρ to that with argon, but the TCR goes increasingly negative with dose.

1. INTRODUCTION

Tantalum thin films are the basis of most thin film high performance hybrid passive networks in use today. All films in commercial use are sputter-deposited and the films giving the best stability are either tantalum nitride or tantalum aluminium alloy with resistivities in the region 200 to 300 $\mu\Omega$ cm and a thermal coefficient of resistivity (TCR) of -100 to -150 ppm/°C. Higher resistivities necessary to minimise area are obtained with films containing oxygen but usually at the expense of a high TCR.

We are investigating the properties of ion implanted tantalum films in the hope that we can control the electrical characteristics in a way that may be of use in producing passive circuits. It is important to characterise the starting material and in this paper results for specimens produced by evaporation in a UHV evaporator are presented and compared with results reported earlier[1,2] for specimens evaporated in a commercial evaporator, and specimens that have been sputter deposited. The UHV evaporated films have been analysed by Rutherford backscattering[3] and transmission electron microscopy both before and after bombardment. Effects of bombardment with nitrogen have been studied in detail and contrasted with argon bombardments to establish the role of damage and sputtering in isolation from chemical effects.

2. THE 'AS DEPOSITED' FILMS

The most recent batch of films have been produced at the University of Surrey by electron beam evaporation. The vacuum system consists of a bakeable work chamber with a titanium sublimation pump, a Vacuum Generators Ltd. combination nitrogen trap and a diffusion pump charged with poly-phenyl-ether oil. The base pressure of this system is 2×10^{-10} torr. During the evaporation of the films for which results are quoted the pressure rose to a maximum of 5×10^{-7} torr, but stayed below 1×10^{-7} torr for most of the run. Most of the substrates used were one inch squares of Corning 7059 glass and all the normal cleaning procedures were followed in a clean-room environment. Some films were deposited at the same time onto polished vitreous carbon substrates for use in analysis by Rutherford backscattering, and onto Corning 7059 substrates coated with aluminium. The tantalum films deposited on the aluminium were floated free by dissolving the aluminium layer and fragments placed on a grid for analysis in the transmission electron microscope.

As reported earlier[1,2] one batch of films was prepared in a conventional system, at pressures of up to 10^{-5} torr by Edwards High Vacuum Research Laboratories. Another batch of films was prepared by sputter deposition at Ultra Electronics Ltd.[2]

A summary of the electrical properties, structure and estimated oxygen content of the different films is given in Table 1.

The Transmission electron micrograph (Fig. 4) of the 'Surrey' films reveals four diffuse rings indicating a semi-amorphous structure with some indication from the d-spacing that the tantalum exists as β-tantalum. The analysis by backscattering[3] indicates an oxygen content of about 30 atomic % and this is fully consistent with published data on the effect of oxygen on the electrical properties[4]. Electron diffraction of the Edwards films reveals one very diffuse

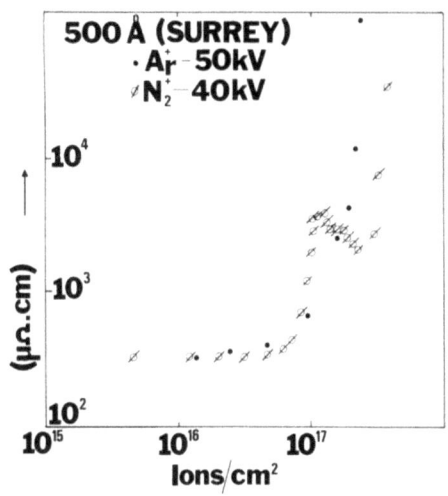

Fig. 1: Resistivity v. dose
(N_2^+ & Ar^+ ions)

Fig. 2: TCR v. dose
(N_2^+ & Ar^+ ions)

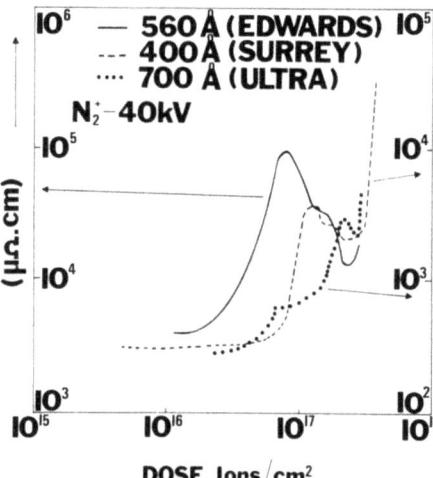

Fig. 3: Resistivity v. dose for different films (N_2^+)

TABLE 1

Deposition Method	Resistivity (μΩcm)	TCR (ppm/°C)	Structure	Oxygen conc. (distributed)
Evaporated at < 10^{-5} torr (Edwards)	3000 to 9500	-500	Amorphous (TEM)	> 55% (electrical properties)
Evaporated at < 5×10^{-7} (Surrey)	360	-170	Semi-amorphous β-Ta (TEM)	30% (RBS)
Sputter Deposition (Ultra Electronics)	235	+145.0	Mixture b.c.c. and β-Ta (X-ray)	~12% (RBS) 40 Å surface oxide (ESCA)

ring indicating an amorphous structure. The electrical properties indicate an oxygen content in excess of 55%[4]. Analysis of the Ultra films by X-ray diffraction indicated that the films are a mixture of b.c.c. and β-tantalum[2]. A depth profile obtained by sputter etching and analysis in a photoelectron spectrometer (ESCA) revealed that there existed a fully oxidised layer to a depth of 40 Å below the surface. Backscattering analysis of a different batch of Ultra films deposited by the same method[3] indicated that the oxygen content was in the region of 12% which again is consistent with the published relationship between resistivity and oxygen content.

Selected films were vacuum annealed at 250°C for several hours before bombardment, but showed little change in resistance as a result of this treatment. However large changes in resistance occurred when films were heated in vacuum to above 300°C when the resistance showed a linear dependence on $(time)^{\frac{1}{2}}$ suggesting parabolic oxidation growth kinetics.

3. THE ION BOMBARDMENT

The resistors were in the form of a strip typically 8 mm × 25 mm with evaporated titanium/gold contact pads on each end. The bombarded area was 12 mm across the film strip × 8 mm along the axis so that the change in resistance of one square was measured.

The ion bombardment was carried out in a 600 keV heavy ion accelerator with magnetic mass analysis. Uniformity of dose in the bombarded area was ensured by electrostatically scanning a de-focused beam over a much larger area. The pressure in the ion pumped target chamber was always < 1×10^{-6} torr and usually 5×10^{-7} torr during bombardment.

A Marconi bridge type TF 2700 (frequency 1 kHz) was used to measure the resistance in some cases but usually a Wheatstone bridge

ION IMPACT CHEMISTRY IN THIN METAL FILMS

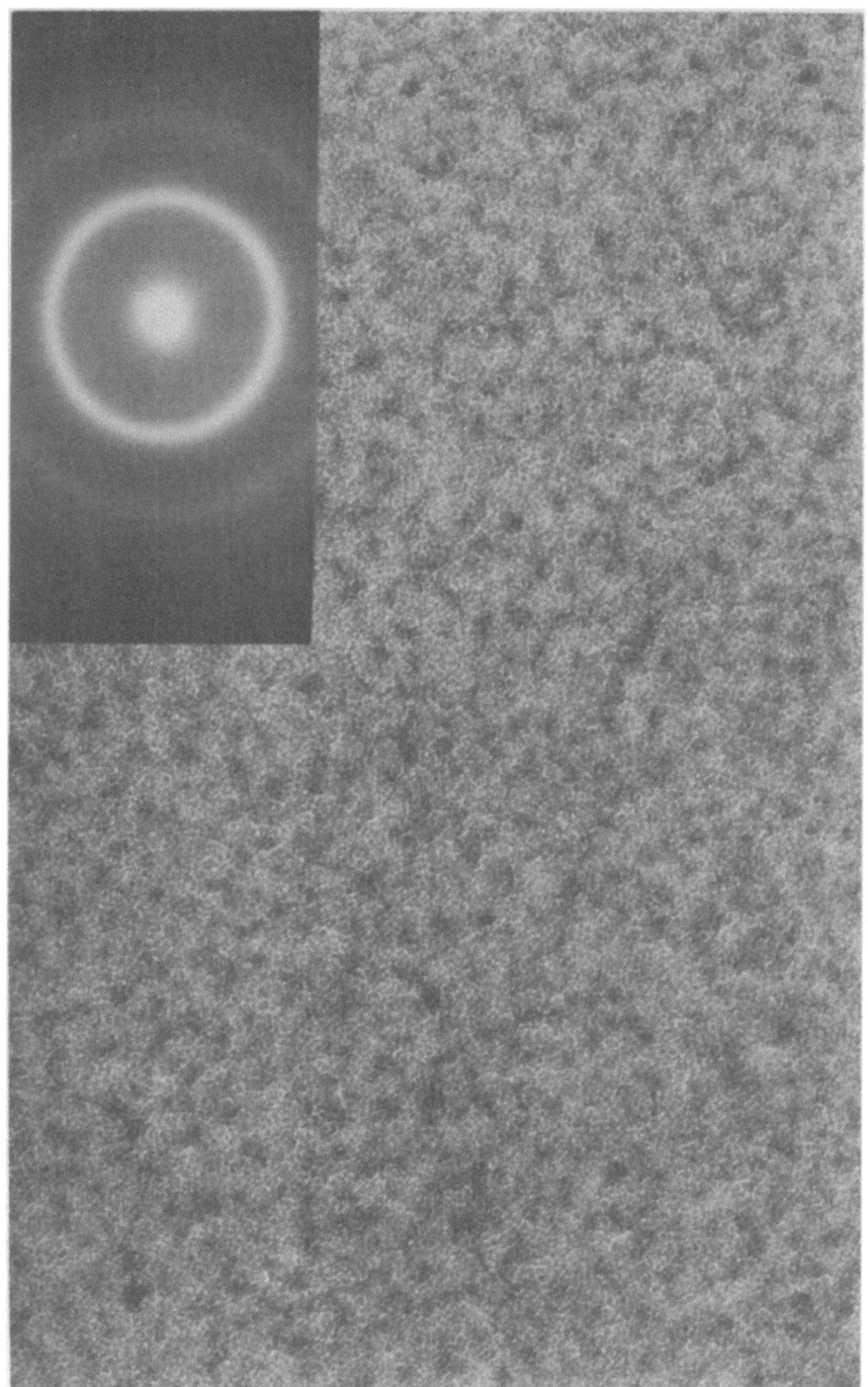

Fig. 4: Electron diffraction & micrographs (a) as deposited, Long axis: 1.7μ

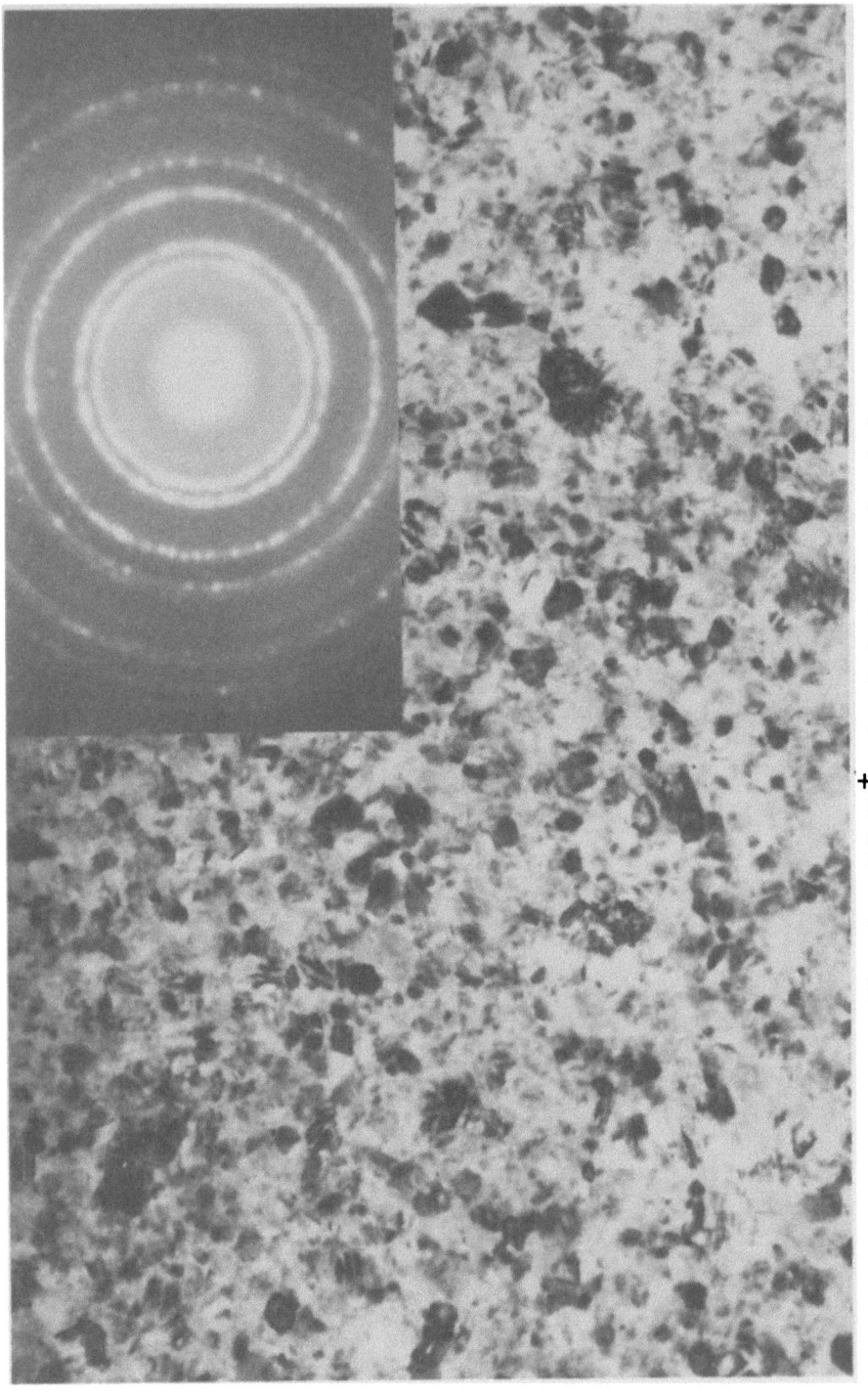

Fig. 4: (b) N_2^+ bombarded, Long axis: 1.4μ

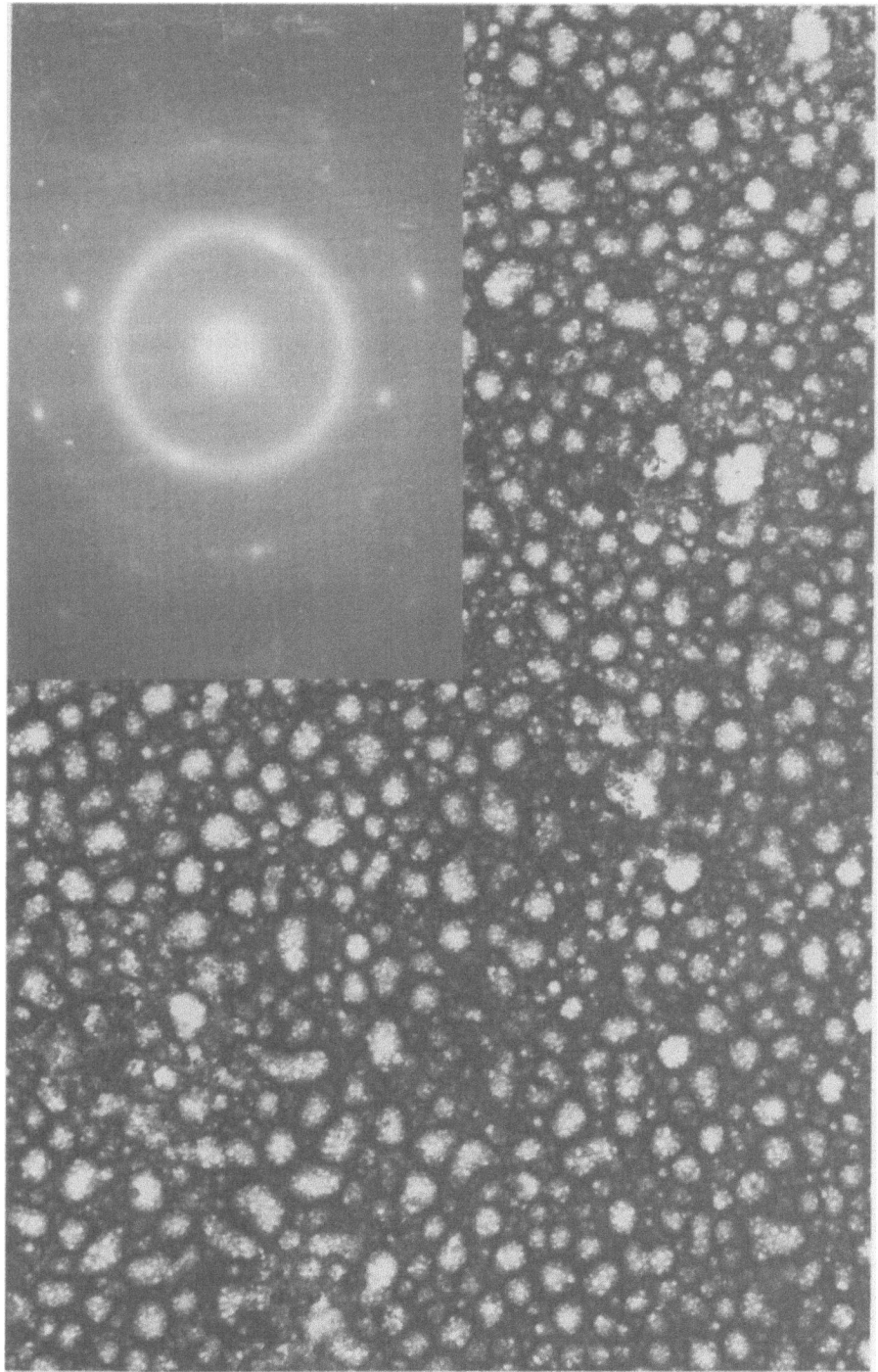

Fig. 4: (c) Ar$^+$ bombarded, Long axis: 2.7µ

accurate to 0.03% or a constant current source and digital voltmeter were used.

In the latest runs using Surrey evaporated films the specimen temperature could be varied in situ between 10°C (water cooling) and 400°C. The specimens were annealed at 200°C after each bombardment (temperature during bombardment was always < 200°C) and then cooled for determination of TCR.

4. RESULTS

Figures 1 and 2 show the dose dependence of resistivity and TCR for nitrogen and argon bombarded 'Surrey' films. Figure 3 shows a comparison of the dose dependence of resistivity for nitrogen bombardment of the three types of film.

Figures 4a, b and c show the diffraction patterns and transmission electron micrographs of Surrey films:- a) 'as-deposited' b) after bombardment with nitrogen to a dose of 9.4×10^{16} ions/cm^2 c) after bombardment with argon to a dose of 9.4×10^{16} ions/cm^2.

5. DISCUSSION AND CONCLUSIONS

With nitrogen bombardment all the films show peaks in the resistivity versus dose plots. However the magnitude of the resistivity change at the peaks is greatest for the Edwards films, smallest for those from Ultra, with the Surrey films intermediate between the two. The difference between the three types is even more dramatic when the dose dependence of TCR is compared. All again show at intermediate doses a large negative TCR which gets less negative at higher doses, but this effect is not very marked in the case of the Ultra films whereas the TCR changes dramatically to positive at a critical dose for the Edwards films. Again the behaviour of the Surrey films is somewhere between the other two types. This ranking list for ρ and TCR coincides with that of the oxygen content originally in the films (Table 1), suggesting that a high oxygen content is necessary to achieve very low TCR with a high resistivity as suggested by Hardy et al.[5] Clearly this is a topic for further investigation by implantation of both nitrogen and oxygen.

It can be speculated that the role of oxygen in the case of nitrogen bombarded films is to enhance the diffusion of nitrogen and encourage the precipitation and growth of the nitride phases. It has been shown[7] that very pure tantalum nitrides deposited by chemical vapour deposition have a positive TCR. Thus the Edwards films could be composed of connected islands of single crystal nitrides in a matrix of tantalum oxide, the formation of these precipitates being helped by radiation enhanced diffusion.

For Surrey films the micrographs reveal that the tantalum, which is very fine grained before bombardment, is recrystallised during nitrogen bombardment with a growth in grain size and the appearance of hexagonal platelets which may be the proposed precipitates of Ta_2N and TaN. Preliminary analysis of the diffraction pattern confirms the presence of Ta_2N and TaN. Bombardment with argon appears to produce a highly damaged film as the diffraction pattern becomes more diffuse, but spots are also seen so some ordering takes place. The micrograph reveals the presence of bubbles. It can be seen from Figures 1 and 2 that a very high resistivity with low TCR can be achieved by bombarding with argon. The TCR gradually becomes less negative with argon dose in the way expected from the reduction in film thickness and the resulting dominance of surface scattering in determining the resistivity. The sputter-thinned films although having properties that are attractive may show the same instability due to oxidation that led, in the case of sputter deposited films, to the development of tantalum nitride, and tantalum/aluminium films[6].

Oxygen bombardment of the Surrey films results in similar behaviour reported previously[1,2], i.e. no resistivity peaks and an increasing negative TCR with dose. Electron diffraction studies of these bombarded films are now being made.

REFERENCES

1) M. Deery, K.H. Goh, K.G. Stephens and I.H. Wilson. Thin Solid Films, 17, 59, (1973).
2) I.H. Wilson, K.H. Goh and K.G. Stephens. Int. Conf. on Applications of Ion Beams to Metals, Albuquerque, 1973, Paper V8.
3) P.L.F. Hemment (Private Communication).
4) L.G. Feinstein and R.D. Huttemann. Thin Solid Films, 20, 103, (1974).
5) W.R. Hardy, J. Shewchun, D. Kuenzig and C. Tam. Thin Solid Films, 8, 81, (1971).
6) R.G. Duckworth. Thin Solid Films, 10, 337, (1972).
7) K. Hieber. Institute of Physics Conference on Thin Films (University of Sussex,(April 1974), Paper C21.

ACKNOWLEDGEMENTS

The authors gratefully thank Mr. I. Sheikh, who deposited the Surrey films, Mr. R. Duckworth of Ultra Electronics Ltd., Dr. P. Hemment, members of the Accelerator Laboratory and Mr. M.J. Hepburn for help with the TEM work.

IRON SURFACE TREATMENT BY BORON IMPLANTATION

T. Takagi, I. Yamada, and H. Kimura*

Department of Electronics, Kyoto University
Kyoto, Japan
*Yamasaki Denki Kogyo Co.
Sakato-machi, Iruma-gun, Saitama, Japan

ABSTRACT

As an application of ion implantation technique outside semiconductor fields, the technique may be useful for metal surface treatments. Because, the depth distribution of treated surface, dose of element, and selection of doping element can be easily controlled externally by changing the ion beam energy, current, and kinds of ions, respectively. Moreover, this technique has the potentiality in the fabrication of new metal compounds which cannot be expected under the conditions of the conventional thermal processing. As one of the application of surface treatment, surface hardening by boron has been studied. Some metallugical properties, wear properties, and magnetic properties of implanted substrates have been obtained.

INTRODUCTION

Metal surface treatment by ion implantation seems to offer more benefical results than the treatment by thermal diffusion processing. Many studies about corrosion, low-friction, and wear resistant meatl surfaces have been done by ion-implantation techniques [1, 2]. As application of the ion-implantation to the surface treatemnt, surface hardening has been studied by boron implantation with a low accelerating voltage (10-25 kV). Even by this low accelerating voltage, boronide could be formed on the surface of iron substrate and significant increase in hardening could be observed without any digradation of the magnetic properties.

EXPERIMENTAL

Boron ions have been implanted with 10-25 KeV into the iron substrates at room temperature by using the newly designed ion source [3]. Ion doses to be implanted have ranged from 5×10^{13}-10^{15} ions.cm^{-2}. Fig.1 shows the transmission electron diffraction patterns of B implanted Fe. From the analysis of the pattern, formation of Fe_2B and FeB compounds could be confirmed after 500°C 1h annealing. The amount and penetration depth of implanted atoms were measured by B^{11}(p, α) Be^8 reaction and ion backscattering methods. The depth of the maximum concentration of B in the mirrored iron substrate is about 300 Å from the surface. After the annealing of the substrate in the vaccum chamber, surface state of implanted sample have been observed by optical microscope. Fig.2 shows the surface patterns of the substrates for different implantation conditions and annealing temperatures. The pattern from the sample fabricated by thermal processing is shown in Fig.3(a) for comparison. Thermally prepared one is fabricated by gas method. In this method the substrate is placed in the BCl_3+H_2 atmosphere at the temperature of 700-1000°C. FeB is formed dominantly near the surface and Fe_2B is dominant under the surface. They are distributed each other complicately in the interface as shown in Fig.3(b). In Fig.3(a), the surface pattern of the interface where Fe, FeB, and Fe_2B are mixed complicately is shown by etching of a thin surface layer. In the case of ion implanted substrates, depth distribution of boronide region is so shallow that the surface pattern will be simillar to that of Fig.3(a). In fact, it is seen in Fig.2 that the surface pattern approaches to that of Fig. 3(a) according to the increase of boron dose.

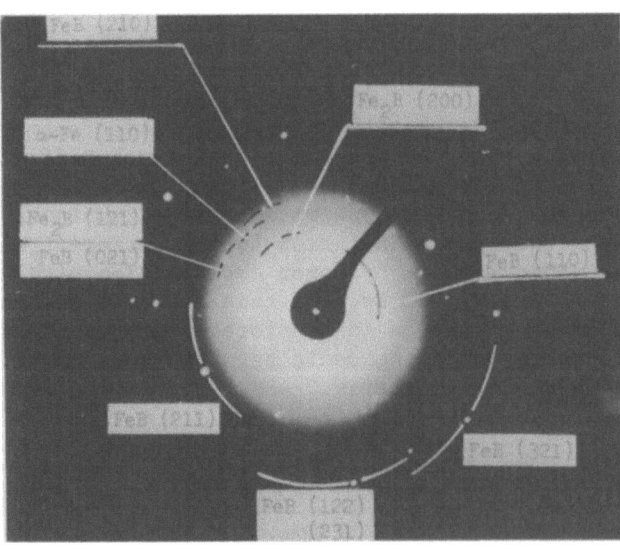

Fig.1 Electron diffraction pattern of boron implanted iron.

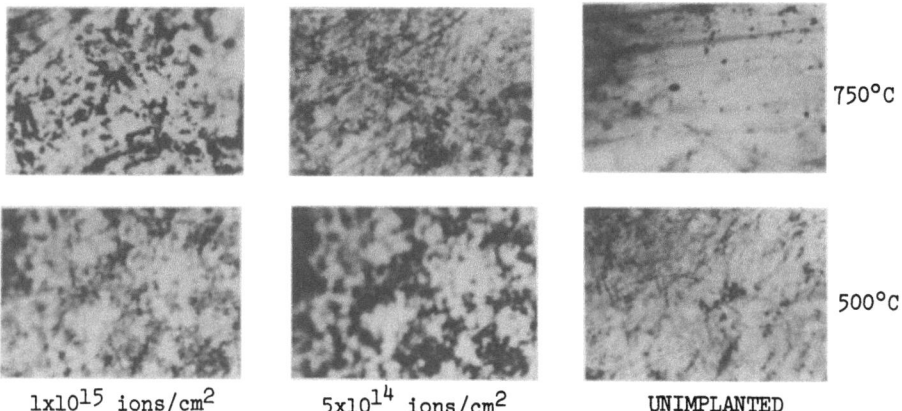

Fig.2 Surface states of boron implanted iron substrates.

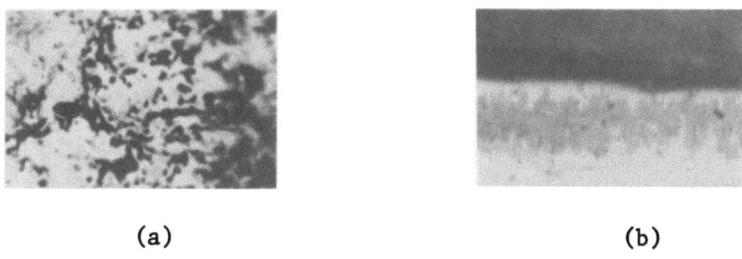

(a) (b)

Fig.3 Boronized patterns observed on (a) the surface and (b) the vertical section.

MECHANICAL PROPETIES

A microhardness tester has been used to measure the Vickers hardness. By measuring the depth of indentation by tester, Vickers hardness number can be obtained. Since the depth of indentation by the tester is about 7 μm and the depth of implanted surface layer is about 1000 Å, only the effect of ion implanted surface cannot measure separately. Fig.4 shows the change in the hardness of the implanted substrates including the effects of substrate material for different annealing temperatures. The figure shows that the hardness increases with the doses of implanted boron atoms. The decrease of the hardness with annealing temperatures is due to the decrease of the hardness of substrate materials.

Abrasion characteristics have been measured by the tape friction test. The substrate of boron-implanted permalloy was used as the test pieces. The surface of the test piece has the curvature to fit the sliding tape. Fig.5 shows the abrasion characteristics by the tape test. In the 10^{15} ions·cm^{-2} implanted substrate, signifi-

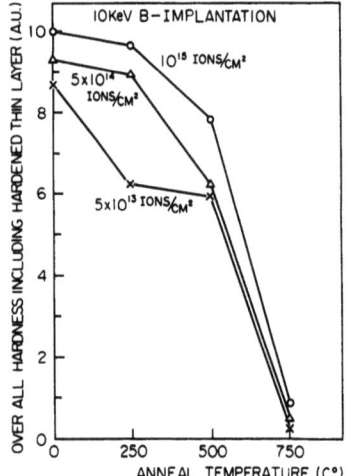

Fig.4 Change in the hardness of implanted substrates.

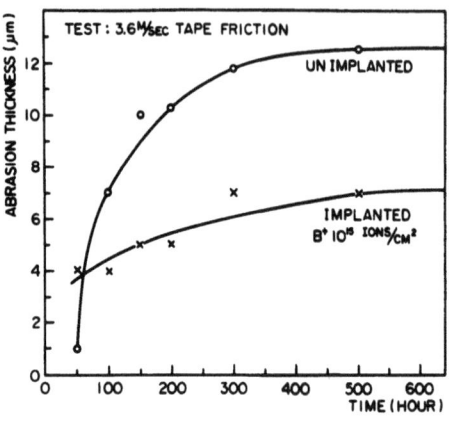

Fig.5 Abrasion characteristics by tape test.

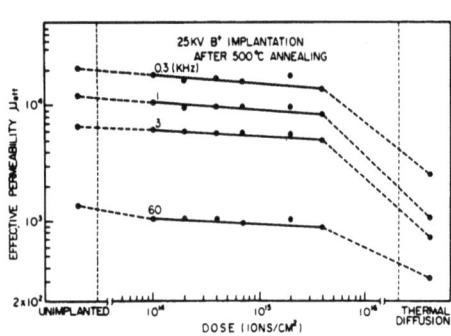

Fig.6 Effective permeability as a function of B^+ ion dose.

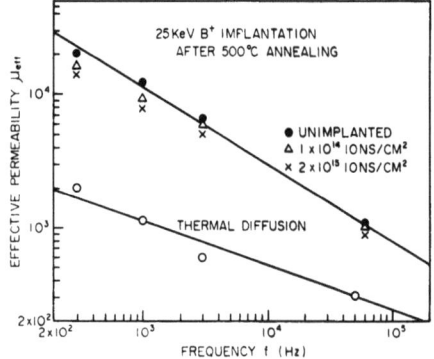

Fig.7 Frequency characteristics of effective permeability.

cant decrease in abrasion could be observed as compared with the unimplanted substrate.

MAGNETIC PROPERTIES

Frequency characteristics of the effective permeability have been measured. The dose dependencies of the permeability for the boron-implanted permalloy after annealing are shown in Fig.6. After the implantation and annealing, a little digradations in the effective permeability with dose of implanted boron could be observed, but in the case of thermal processing, significant degradation in permeability was observed.

IRON SURFACE TREATMENT BY BORON IMPLANTATION

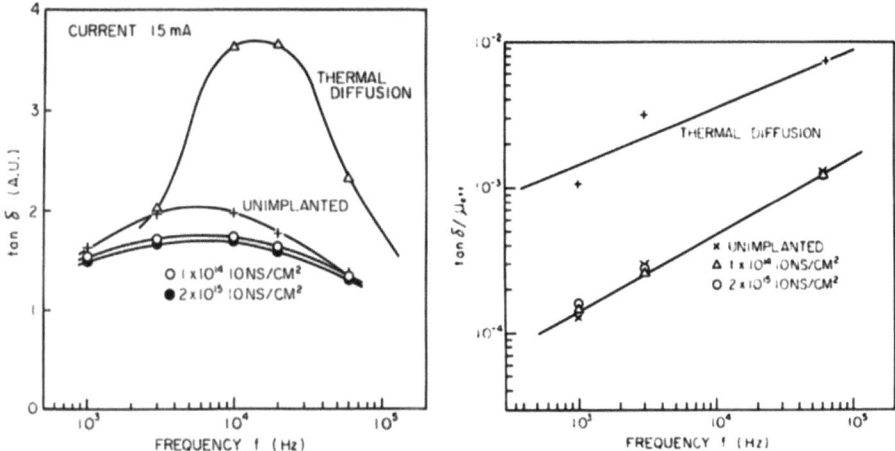

Fig.8 Frequency characteristics of tan δ.

Fig.9 Equivalent loss factor.

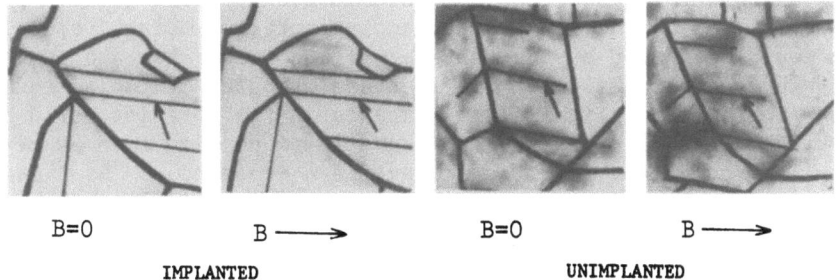

Fig.10 Displacement of magnetic domain wall.

Fig.7 shows the frequency characteristics of the permeability for the ion implanted and the thermally prepared substrates. The figure shows that the change in the permeability for the thermally prepared one is much greater than that of the ion implanted one.

Fig.8 shows the frequency characteristics of the loss factor, tan δ. The loss factor decreases with the dose of boron. But, in the sample by thermal processing, the loss factor increase. Equivalent loss factor (tan δ/μ_{eff}) which shows the quality of magnetic material will be defined by the ratio of the loss factor to the effective permeability. Fig.9 shows the equivalent loss factor as a function of frequency after annealing. The figure shows that the equivalent loss factor does not change by implantation in contrast to remarkable change in thermal processing. This result shows that tan δ decreases although μ_{eff} decreases by the boron implantation.

Therefore, it can be seen that by the boron implantation, deterioration of μ_{eff} was compensated by the decrease of the tan δ of the materials. Study about the difference between loss factor by ion implantation and thermal processing are under way.

Fig.10 shows the displacement of magnetic domain wall by magnetic field. The bold line in the figure shows the grain boundaries and thin line shows the magnetic domain walls. By comparison with the observation of displacement of magnetic domain wall under magnetic field in both unimplanted and implanted substrates, it is seen that hardening of the magnetic domain wall occurs by ion implantation. Because the thin line in implanted substrate does not shift whereas thin line in unimplanted substrate shift clearly. After implantation hardening of domain wall could be observed. One of the reasons for the degradation of the μ_{eff} may be due to the hardening of domain walls.

CONCLUSIONS

Surface treatment by boron implantation with low accelerating energy has been proposed, and some fundamental properties have been studied. Even by the low energy ion implantaion, effective boronide layer could be formed on the surface of iron substrates and increasing of surface hardness with the increase of doses of implanted atoms could be observed. Significant decrease in the abrasion could be seen by the tape friction test. Magnetic properties such as permeability and equivalent loss factor have been measured. No change in the equivalent loss factor could be observed after implantation. This ion implantaion method is the useful way to harden a whole or a part of the device surface without any change in the magnetic properties after assembling it by machine process.

ACKNOWLEDGEMENTS

We should like to express our thanks to Mr. H.Mori and Mr. H. Morimoto for the measurments of magnetic properties.

REFERENCES

1. Picraux; Application of Ion Beams to Metals, Plenum, 1973.
2. Dearnaley; Ion Implantation, North-Holland, 1973.
3. T.Takagi, I.Yamada, and J.Ishikawa; IEEE Trans. NS-19, 2, 1972.

IONIZED-CLUSTER BEAM DEPOSITION

T. Takagi, I. Yamada, and K. Yanagawa*

Department of Electronics, Kyoto University

Kyoto, Japan

ABSTRACT

A new deposition technique has been developed by using "Vapourized-Metal Cluster Ion Source" by authors. This new technique may be named "Ionized-Cluster Beam Deposition". The fundamental effects of the formation of deposited films such as surface cleaning, deep etching, local heating, and ion implantation effects by injecting cluster ions have been studied. This technique is the one of the special case of simultaneous operation method of implantation and deposition which is proposed by author. By the technique, not only metals but also insulating materials can be depsited on metal substrates or insulating substrates with fairly strong adhesion and in good crystalline state.

INTRODUCTION

A new type of the ion source which has a simple structure with a high deposition rate has been developed [1, 2]. This may be named "Vapourized-Metal Cluster Ion Source". In the source, metal vapour aggregates formed by ejecting from a crucible are ionized by electron bombardment. In cluster ion, one of the atoms which consist of a cluster is ionized, and thus cluster ion is resulted in intense macroparticle with a small charge-mass ratio. A use of the cluster ion source has many advantages over the ion plating, ion beam deposition and other applications, because of their higher energy, small space-charge effects, and a small charge-mass ratio. The increase in the surface mobility of the incident ions, local temperature rise

* Present adress : Yokogawa Hewlett Packard Co., Hachiōji, Tokyo.

on the surface, and sputtering of surface layers can lead to fabricate a fine quality film with strong adhesion. The small charge-mass ratio reduces the trouble caused by charging up on the surface due to a positive charge of a depositing material, when a thin film is deposited on an insulating substrate. This new technique may be named "Ionized-Cluster Beam Deposition". In the paper, some characteristics of ion source and properties of deposited film will be reported.

CHARACTERISTICS OF ION SOURCE

The illustration of our ion source is shown in Fig.1. Metal vapour is ejected into a high vacuum region through a small aperture of a special crucible in which metal is vapourized by heating of coil surrounding the crucible or by electron bombardment. Then metal vapour aggregate, that is, cluster is formed by adiabatic expansion due to the ejection into a high vacuum region through a nozzle of a heated crucible. The clusters are ionized by electrons. They are emitted from a filament coil in an accelerating electrode assembly which is located coaxially in front of the crucible and kept at negative potential V_a with respect to the crucible. The cluster ion beam is effectively produced by electron impact, because the mass in the cluster state is larger than that in the atomic vapour state, and the cross-section for the ionization is roughly proportional to the mass. Fig.2 shows the difference between atomic ions and cluster ions. The cluster ions have a very small value of the charge-mass ratio and they produce an intense equivalent current. The cluster ions are accelerated to the direction opposite to the flow of injected electrons by the potential difference V_a between the accelerating electrode assembly and the crucible. A negative

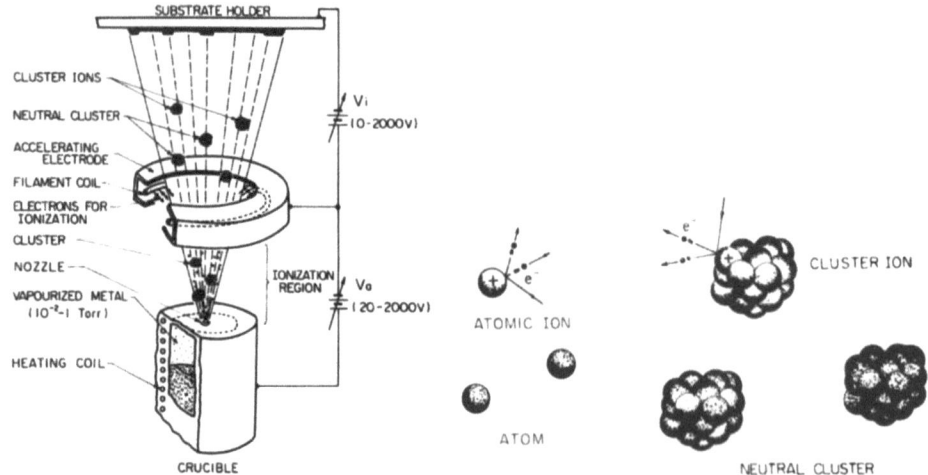

Fig.1 Schematic diagram of the vapourized-metal cluster ion source.

Fig.2 Schematic illustration of the cluster.

potential Vi can be applied to the substrate to attract the positive ions. The energy of ions incident to the substrate can be adjusted by the potential difference between the substrate and the crucible. One of the reasons why fine quality depsition film with strong adhesion can be obtained is that the cleaning effect by sputtering and the deposition function on a substrate are done simultaneously during the deposition. The kinetic energy of an accelerated cluster ions is converted partially to the energy which sputters substrate materials.

Fig.3 shows typical chracteristics of the deposition. In this experiment, censor of a thickness meter is located at the edge of the substrate to eliminate the effects by the deposition of the neutral cluster. A deposition rate is increased with an accelerating voltage. But, over a certain range, deposition rate is decreased with an ion accelerating voltage due to serious sputtering effects. In practice, cluster ion beam with neutral cluster which are not ionized in the ionization region is utilized to obtain a high deposition rate and a desirable quality of films. Therefore, one of actual deposition characteristics is shown in Fig.4. At a low deposition rate, sputtering effects appear clearly at a higher accelerating voltage. But at a high deposition rate, sputtering effects are often masked, because of the compensation of sputtering by deposition. However, cleaning effects are not reduced.

The time-of-flight method for measuring the cluster size has been used. In the time-of-flight experiment, a shot pulse voltage is applied to the shutter electrode between accelerating electrode assembly and substrate electrode. The flight time is measured on a synchroscope. The cluster size of $5 \times 10^2 - 10^3$ is observed.

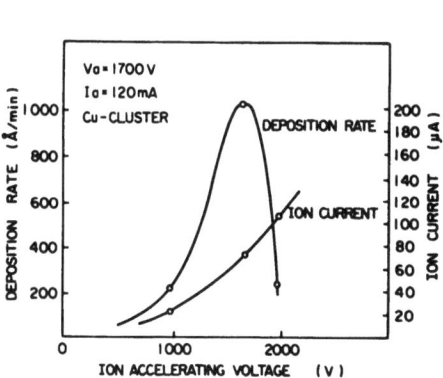

Fig.3 Dependencies of the ion current and the deposition rate on accelerating voltage.

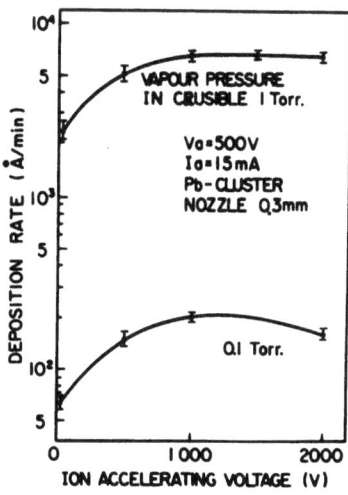

Fig.4 Typical deposition rate

Cluster size is seemed to be constant against the fluctuations of the crucible temperature, that is, the metal vapour pressure in the curcible, in spite of remarkable change in a deposition rate.

PROPERTIES OF DEPOSITED FILMS

The atoms which consist a cluster are bound by the Van der Waals' force. Therefore, they are not so tightly coupled each other, but act as one aggregate. When cluster ions are bombarded with neutral clusters, they are scattered and separated to each atom. Scattered atoms migrate on the surface by the kinetic and thermal energies converted from the incident energy, and condense in a good crystal state. Beside these effects, other fundamental effects which give influences on the formation of deposited films are considered as surface cleaning, deep etching, local heating, and ion implantation effects by injecting cluster ions. These effects give the significant contribution to film properties such as crystalline state, interfacial properties between substrate and deposited film, adhesion strength, etc.

Transmitting electron diffraction patterns and electron micrographs from Pb layers deposited onto the (100) plane of a single crystal of NaCl are shown in Fig.5 for different accelerating voltages and substrate temperatures. As shown in Fig.5(a), the film which was obtained by ordinary vacuum deposition technique is an polycrystalline film. Figs.5(b) and 5(c) show the patterns obtained at the accelerating voltage of 500 volts and 1000 volts at the room temperature, respectively. The patterns show that crystallization and size of crystallines increase with an accelerating voltage. Fig.5(d) shows the patterns obtained by ordinary vacuum deposition technique at 200°C of the substrate temperature. The deposited films are im-

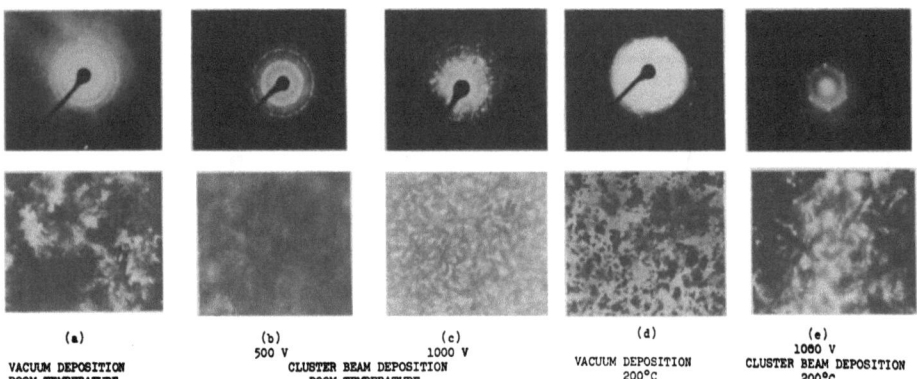

Fig.5 Transmitting electron diffraction patterns and electron micrographs of the Pb films for various deposition conditions. Original magnification 1100X.

proved in the crystalline state by increasing of accelerating voltage and substrate temperature in the region of this experiment.

The bombardment of cluster ion with kinetic energy to a substrate surface may introduce some defects and implantation effects in a surface layer. It is seemed that the defects enhanced the diffusion of deposited material at a high temperature on a substrate due to the thermal energy converted from the kinetic energy. The interfacial layer may be formed by these processes and deep etching process of the surface. The profile of the interfacial layer is analyzed by ion backscattering technique as shown in Fig.6. The figure shows the spectra for 1.5 MeV He^+ incidence onto a Cu films

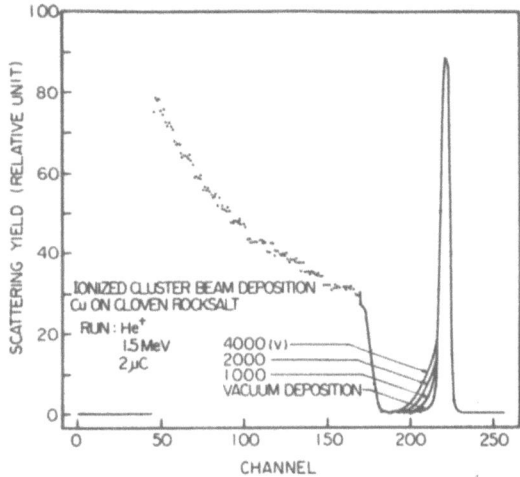

Fig.6 Backscattering spectrum for 1.5 MeV He^+ incidence on to the Cu film deposited by cluster ion beam.

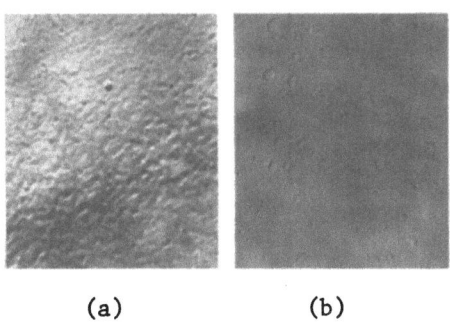

Fig.7 Surface conditions of the glass substrate (a) before deposition and (b) after removing the Cu film deposited by cluster ion beam.

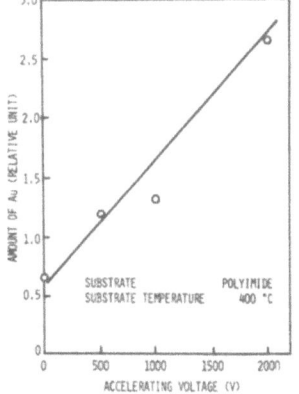

Fig.8 Ion-implantation effect by cluster ion beam deposition.

deposited by cluster ion source for different accelerating voltages. The front edge at 230 channel corresponds to Cu atoms at the surface, and the back edge near 200 channel which has long slope in this figure shows the deposited layer. Changing of the long slop shows the formation of interface. The interface is increased with the increase of incident energy of cluster ions. Sputtering effect on the substrate was observed by the electron microscope. Fig.7(a) shows the surface of substrate before deposition, and Fig.7(b) shows the surface after removing of the film deposited by cluster ions. Significant etching effect by cluster ions could be oberved. One of the reasons why fine quality deposition film with strong adhesion can be obtained is that the cleaning function by sputtering and the deposition function on a substrate are done simultaneously during deposition process.

Ion-implantation effect was measured by the method of radioactivation analysis. After deposition of Au, film was removed by chemical etching completely. Amount of implanted Au in the Polyimide substrate was measurd by γ-ray counter. Fig.8 shows the dependencies of implanted Au on the accelerating voltage. The implanted atoms increase with the increase of the accelerating voltage of cluster ions.

CONCLUSIONS

Deposition mechanisms by cluster ion beam and fundemental effects such as sputtering, etching, and ion implantation have been studied. In our study, the ionized cluster beam deposition has many features as follows.
  i) Good crystal films can be deposited by using the cluster ion beam and neutral clusters with strong adhesion.
  ii) The structure of the ion source is simple and suitable to industrial applications and epitaxy by physical processing instead of conventional chemical processing.
  iii) Since the working temperature of substrate does not increase, it is available for low melting temperature substrate such as plastic plate.
  iv) The deposition speed can be adjusted in a wide range from a few hundreds Å/min to a few μm/min.

REFERENCES

1) T.Takagi, I.Yamada, M.Kunori and S.Kobiyama, Proc. of the Second International Conference on Ion Sources (SGAE, Vienna, Sept., 1973) p.790.
2) T.Takagi, I.Yamada, M.Kunori and S.Kobiyama, Presented at the Sixth International Vacuum Congress, Kyoto, March, 1974.

EFFECTS OF ION BOMBARDMENT ON METAL-SILICON INTERFACE

H. Nishi, T. Sakurai, and T. Furuya

Fujitsu Laboratories Limited, Kawasaki, Japan

ABSTRACT

Backscattering analysis with 1.5-MeV He ions is used to investigate the effects of Ar and As ion bombardments on thin Mo and Cr films evaporated on silicon. Ion bombardment induces the migration of metal atoms across the metal-silicon interface at room temperature. Migration is caused mainly by an atomic recoil process (knock-on effects), and not by radiation enhanced diffusion. It is also affected by the interface interaction between metal film and silicon substrate. The interface interaction is enhanced by lattice defects produced in the substrate near the interface during ion bombardment, especially in the lower dose range where lattice defects are pronouncedly created.

Migration of metal atoms depends strongly on the thickness of metal film, and occurs most noticeably when the value of the film thickness is fairly smaller than the projected range of incident ions in the metal film. Experimental results show that in the most efficient case, about four Mo-atoms are recoil-implanted into silicon by every 150-KeV Ar ion, and about ten Cr-atoms by every As ion.

INTRODUCTION

Recoil implantation was first proposed theoretically by Nelson as a useful technique to introduce impurities into the quite thin layer of the solid surface [1]. Parkins reported that Ar ion bombardments on Al films evaporated on glass substrate induce the recoil implantation of Al atoms, and which can enhance the adhesion of Al films to the substrate [2]. It is quite interesting to apply this technique to a metal evaporated silicon on the view points of electric contact between metal and semiconductor, and of studying experimentally the secondary collision phenomena at the metal-silicon interface. However, it is reported that radiation enhanced interdiffusion and intermixing occur in Pd-Si case [3,4]. So, both types of atomic migration, i.e. recoil implantation and enhanced diffusion, are expected to occur across the interface during ion bombardments.

The authors have reported the experimental results on the effects of Ar ion bombardment on Mo-Si interface [5]. The purpose of the present work is to investigate the effects of ion bombardments on metal-silicon interface in more detail, and to obtain the rate of recoil implantation experimentally using backscattering analysis with 1.5-MeV He ions.

EXPERIMENTAL

Single-crystal silicon wafers used were p-type 1-2 Ω-cm, oriented in <111> direction. The method for sample preparation is found in the previous letter [5]. The thicknesses of Mo films were ranged from 200 to 1100 Å, and Cr films 100-800 Å. Their thicknesses were monitored by an oscillating quartz crystal method during evaporation. They were also checked by the backscattering technique [6]. The experimental error in estimating film thickness was less than 50 Å for Mo, and 100 Å for Cr films.

150-KeV Ar and As ions were implanted into the Mo-Si and Cr-Si structures, respectively. Implantations were carried out at room temperature, in the dose range 1×10^{13} to $2 \times 10^{16}/cm^2$. The Mo and Cr films were subsequently removed chemically. The Mo etchant consisted of H_3PO_4, HNO_3, and CH_3COOH, and Cr films were etched off by ($K_3F(CN)_6$ + NaOH + H_2O). Then backscattering analysis with 1.5-MeV He ions was performed to investigate the migration phenomena of metal atoms into the substrate.

RESULTS AND DISCUSSIONS

The energy spectra of He ions backscattered from the specimens covered with Mo films indicate that the formation of Mo-silicides was not induced by the Ar ion implantations. This result is quite different from the Pd-Si case [3,4]. After Ar implantation and subsequent Mo-etching, a significant amount of Mo atoms ($\lesssim 3 \times 10^{16}/cm^2$) were detected in the backscattering spectra. The surface of Mo-etched specimen was checked by Auger electron spectroscopy. Strong electron emission from Si atoms was observed, and this means that Mo films were removed completely. When the surface

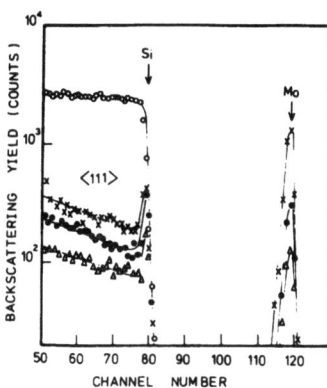

Fig. 1 Effects of thermal treatments on Mo evaporated silicon observed in <111> aligned spectra. Evaporated Mo films were etched off, after the heat treatment at 700°C for 10 min (●), and 30 min (×). A spectrum for an ordinarily prepared specimen (△) is also shown.

layer of 100-Å thick silicon was removed by anodic oxidation and HF stripping method, the Mo atoms were still detected, but the number of them was reduced to about 10% of that before the stripping. In a word, the Mo atoms detected in the backscattering spectra have migrated really into the substrate, and besides the range of them is smaller than 100 Å.

Even in the surfaces of unimplanted specimens a small number of Mo atoms, which corresponds to about atomic mono-layer of molybdenum, were detected in the backscattering spectra, after the evaporated Mo films were etched off. These unremovable Mo atoms were not found at the surface of 2000 Å thick SiO_2 which was thermally grown on silicon, after the Mo-etching by the same etchant. These facts suggest the strong interaction of Mo and Si atoms at the interface, which originates from the active characteristics of the surface of single-crystal silicon.

The interface interaction is enhanced by thermal treatments after evaporation or pre-bombardments to the silicon crystal before evaporating Mo film on it. For example, pre-bombardment with 150-KeV Ar ions to a dose of $1 \times 10^{15}/cm^2$ created about two times of the number of unremovable Mo atoms compared to the ordinarily prepared specimens. The variation of <111> aligned spectra due to heat treatments at 700°C is shown in Fig. 1. The unremovable Mo atoms increase with heating time. At the same time, larger surface peak of silicon was observed in the spectra. The unremovable Mo atoms and the displaced Si atoms at the crystal surface caused by heat treatments form a interface layer with chemically different properties from either Mo film or silicon substrate. This layer can not be etched off by the usual Mo-etchant described above.

In the case of Cr evaporation, unremovable Cr atoms were also detected by backscattering analysis. However, the number of them was about two or three times smaller than that of the Mo case. The interface interaction between Mo and Si should be stronger than that between Cr and Si.

In the present study, therefore, the number of metal stoms migrating into silicon is defined as the number having actually moved as a result of ion bombardment, so the number of metal atoms detected before implantation is subtracted from that after the implantation.

Dose and Dose Rate Dependence

Figure 2 shows the dose rate dependence of the number of Mo atoms migrating into silicon after Ar ion implantation to a total dose of $1 \times 10^{15}/cm^2$. The Mo films used were 300 Å thick. As seen from Fig. 2, no distinct dependence on dose rate was observed. Migration of Mo atoms occurred also into SiO_2 substrate as well as into silicon substrate, as described below. These facts suggest that the migration of Mo atoms was not caused by enhanced diffusion as a result of bombardment-generated defects [7], at the interface (in the surface region of silicon crystal), or heating effects during bombardments. In the case of As ion bombardment onto Cr-Si structure, the effects of radiation enhanced diffusion were also not recognized throughout the experiments.

Dose dependences of the number of migrating Mo and Cr atoms due to Ar and As implantation are shown in Fig. 3 (a) and (b), respectively. Mo films of 300-Å thick, and 550-Å Cr films were used. In both cases, the migrating number increases with ion dose, but shows non-linear relation. If the migration of Mo and Cr atoms occur only as a result of atomic recoil process (knock-on effects), their numbers are expected to increase in proportion to ion dose in the full dose range. The linear relation is observed in the higher dose range. Considering the strong interaction between metal and silicon at their interface, and that the interaction is enhanced by prebombardment to silicon substrate as described above, it is probably expected that in addition to recoil process, effects of bombardment-enhanced interaction give a larger increasing rate of the migrating number of Mo and Cr atoms in the lower dose range.

In practice, this expectation is supported by the experimental results on dose dependence of the lattice defects generating in silicon near the interface, as is shown in Fig. 4. The dose dependences of both migrating number and lattice defects were obtained from the same specimens. That is to say, in the lower dose range, ion bombardment causes continually lattice defects, such as dangling bonds, in silicon near the interface. These lattice defects enhance the interface interaction of metal atoms and the substrate. As the increase of the number of displaced silicon atoms saturates at around the dose of $5 \times 10^{14}/cm^2$ in both cases, so effects of enhanced interaction are no more strengthened at around this dose. Therefore, the dose dependences of the migrating numbers of Mo and Cr atoms show higher increasing rate in the lower dose range.

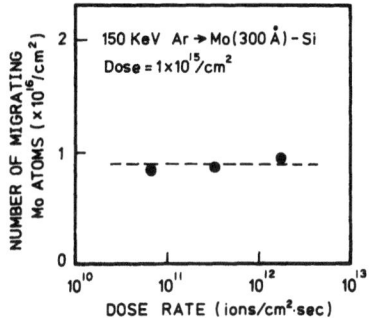

Fig. 2 Dose rate dependence of the number of Mo atoms migrating into silicon due to 150-KeV Ar ion implantation to a total dose of $1 \times 10^{15}/\text{cm}^2$.

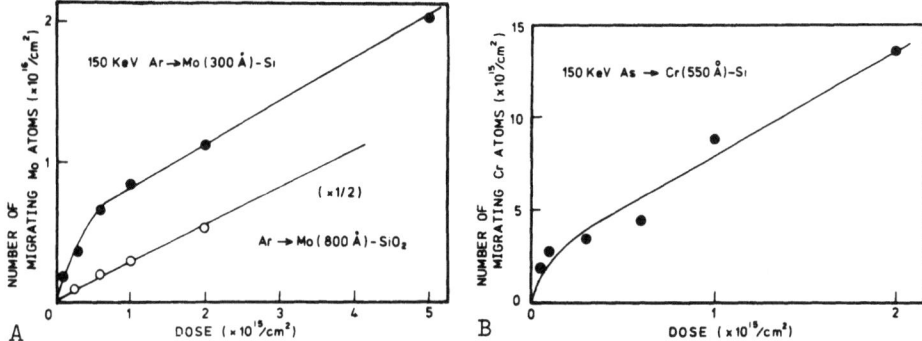

Fig. 3 Dose dependences of the numbers of migrating Mo atoms (a), and Cr atoms (b), induced by 150-KeV Ar and As ion implantations, respectively. Mo evaporated SiO_2 was used for comparison, and the result is shown in Fig. 3 (a).

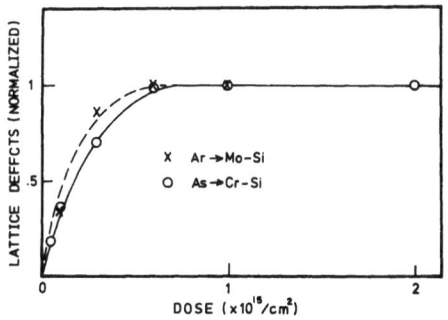

Fig. 4 Dose dependence of the lattice defects generating near the interfaces of Mo-Si (×) and Cr-Si (○) as a result of Ar and As ion implantations. The thicknesses of Mo and Cr films were 300 and 550 Å, respectively.

Dose dependence shows a linear relation, in fact, when Mo evaporated substrate is not crystal silicon but SiO_2, and their interface interaction is much weak, as is shown in Fig. 3 (a).

The rate of recoil-implantation is obtained from the slope of the number of migrating Mo or Cr atoms versus ion dose in the higher dose range. For example, from the data in Fig. 3 (a) and (b), recoil-implantation rate is evaluated to be 3.0 for Mo and 5.8 for Cr, respectively.

Thickness Dependence

The migration phenomena of Mo and Cr atoms into silicon depend strongly on the thickness of the metal film. In Fig. 5, the dependence of the number of migrating Mo atoms on the thickness of Mo film, is shown. In this case, the dose of implanted Ar ions was $5 \times 10^{15}/cm^2$. Solid line in Fig. 5 indicates the depth distribution of Ar atoms implanted in Mo, which was obtained at the same time by backscattering technique as follows.

The average concentration of Ar atoms implanted in Mo between thicknesses z_i and z_{i+1} is given by,

$$n_i = \frac{N_i - N_{i+1}}{z_{i+1} - z_i}$$

where N_i is the number of Ar atoms implanted into silicon substrate through Mo film with thickness z_i.

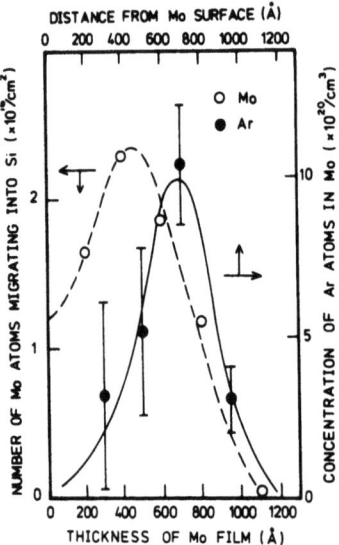

Fig. 5 Dependence of the number of migrating Mo atoms into silicon due to 5×10^{15} Ar ions/cm^2 implantation at 150-KeV on the thickness of Mo film. The depth distribution of Ar atoms in Mo obtained by backscattering analysis is also shown.

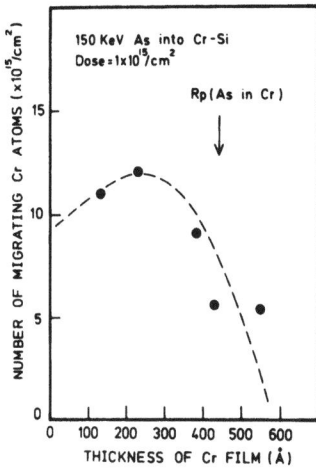

Fig. 6 Thickness dependence of the number of Cr atoms migrating into silicon due to 150-KeV As implantation to a dose of $1 \times 10^{15}/cm^2$. Projected range of As in 800-Å thick Cr-film is calculated by the absolute energy shift of As peak from As surface energy edge using stopping power data [6].

The migration of Mo atoms occurred most remarkably at a film thickness of 450 ± 50 Å, where about five Mo atoms for every incident Ar ion were introduced into the substrate. It is quite interesting that this value of thickness was by about 35% smaller than the projected range of 150-KeV Ar ions in Mo, which was 700 ± 100 Å, as is found from Fig. 5. Thickness dependence of the number of migrating Cr atoms due to 1×10^{15} As ions/cm^2 implantation is shown in Fig. 6. The projected range of As ions in Cr was 450 ± 100 Å, which was obtained by usual method to calculate the peak depth of As in Cr from backscattering spectrum using stopping power data [6]. In the case of As implantation into Cr-Si structure, the error of Cr-film thickness was large, and thickness dependence was not so distinguished as the Mo case. But it is clear that the migration of Cr atoms occurs pronouncedly also at film thicknesses shallower than the As range in Cr.

For the present both cases, i.e. Ar into Mo-Si and As into Cr-Si, the relation between the depth distribution of implanted ions in the metal and the thickness dependence of migrating metal atoms is similar to that between the range and energy deposition of ions in material [8]. Considering that the range of most recoiling atoms is at most a few tens of atomic distance [1,9] (this fact is checked experimentally for Mo case and described above), the curves indicated by the dashed line in Fig. 5 and in Fig. 6 are understood to correspond fairly well to the depth distribution of the deposited energy for Ar ions in Mo, and As ions in Cr, respectively.

From these experimental results, the maximum value for Mo atoms to be recoil-implanted into silicon by 150-KeV Ar ion implantation is expected to be about four. In As-Cr system, this value is about ten.

CONCLUSIONS

When energetic ions are implanted through metal evaporated films of thickness comparable to ion range, migration of metal atoms into the substrate occurs. The migration of metal atoms is caused mainly by atomic recoil process (knock-on effects), and not by enhanced diffusion due to bombardment-generated defects during ion bombardments, in the case of Mo and Cr. Migration is also affected by the atomic interaction between metal and silicon at the interface which is enhanced by lattice defects generating in silicon near the interface due to ion bombardment, especially in the lower dose range.

Migration phenomena of Mo and Cr atoms show strong dependence on the thickness of the metal film. In both case, the value of thickness where the migration occurs remarkably is fairly smaller than the projected range of incident ions in the metal film.

The efficiency of recoil implantation depends on atomic masses both ion and metal. The maximum value for Mo atoms to be recoil implanted into silicon is about four by 150-KeV Ar implantation, and for Cr atoms about ten by As implantation.

ACKNOWLEDGEMENTS

The authors wish to thank Professor A. Hiraki of Osaka University for helpful discussions, Dr. T. Oshida for his encouragement, and Mr. T. Akamatsu for his available discussions and preparing samples. They are also indebted to Dr. M. Maeda and Mr. S. Tokuhira for valuable suggestions on this work.

REFERENCES

1. R. S. Nelson, Radiation Effects $\underline{2}$, 47 (1969).
2. J. G. Parkins, J. Non-Crystalline Solid $\underline{3}$, 349 (1972).
3. D. H. Lee, R. R. Hart, D. A. Kiewit, and O. J. Marsh, Phys. stat. sol. (a) $\underline{15}$, 645 (1973).
4. W. F. Van der Weg, D. Siguard, and J. W. Mayer, International Conference on the Application of Ion Beam to Materials, New Mexico, 1973 (unpublished).
5. H. Nishi, T. Sakurai, T. Akamatsu, T. Furuya, Appl. Phys. Lett. $\underline{25}$, No. 6 (1974), in press.
6. J. F. Ziegler and W. K. Chu, IBM Research Report RC 4288 (1973).
7. R. L. Minear, D. G. Nelson, and J. F. Gibbons, J. Appl. Phys. $\underline{43}$, 3468 (1972).
8. D. K. Brice, Appl. Phys. Lett. $\underline{16}$, 103 (1970).
9. T. Tsurushima and H. Tanoue, J. Phys. Soc. Japan $\underline{31}$, 1695 (1971).

LATTICE LOCATION OF DEUTERIUM

IMPLANTED INTO W AND Cr*

S. T. Picraux and F. L. Vook

Sandia Laboratories

Albuquerque, New Mexico   87115

ABSTRACT

Ion implantation and ion channeling are used to study hydrogen in metals. Deuterium implanted at room temperature into Cr and W is found to be localized at distinct positions, predominately near the octahedral and tetrahedral interstitial sites respectively. Irradiation of the implanted W by the 750 keV ^{3}He analysis beam results in the movement of the deuterium to a new location.

INTRODUCTION

Many of the physical properties of metals are modified by the presence and lattice location of hydrogen. Recently ion channeling has been applied to obtain the lattice location of hydrogen in single crystal metals [1,2]. The controlled introduction of hydrogen into the lattice by ion implantation allows the study of metal systems with low hydrogen solubility [2]. Such systems are difficult to investigate by traditional techniques such as neutron diffraction or NMR which require high concentrations throughout the crystal. In contrast, the ion channeling technique requires moderately high concentrations (~ 0.1 at.% hydrogen) only near the surface, and this can easily be achieved using ion implantation.

This paper discusses lattice location studies of ^{2}H (referred to as D hereafter) in the VIB transition metals W and Cr. Preliminary results for D in W were reported previously [2]. No other information is available for the lattice location of D in W or Cr

---

*This work supported by the U. S. Atomic Energy Commission.

due to the low hydrogen solubility. Experiments were made at room temperature since hydrogen diffusivities in W and Cr are relatively low. For higher diffusivities it would be necessary to introduce the hydrogen at low temperatures for *in situ* analysis.

EXPERIMENTAL TECHNIQUE

The nuclear reaction $D(^3He,p)^4He$ was used to detect the D with a 750 keV ^{3}He analysis beam having a full width angular divergence $\leq 0.06°$. Implants and analyses were done *in situ* on single crystals with the $\langle 100 \rangle$ axis normal to the surface. The crystals were of good quality for ion channeling as indicated by low $\langle 100 \rangle$ minimum yields of 1.7 and 1.1% for Cr and W, respectively, prior to implantation. Implants using D_3^+ molecular beams were performed 7 to 12° from the $\langle 100 \rangle$ direction. The emitted protons were detected at $\sim 135°$ using a 300 mm^2 surface barrier detector. The detector was covered by a 12 μm alluminized mylar foil to prevent the intense backscattered ^{3}He signal from entering the detector except through several pin holes. The ^{3}He signal was used to monitor the Cr or W channeling dips at depths corresponding to the D ion range.

RESULTS AND DISCUSSION

A. As-Implanted Location in W and Cr

Previous results [2] identified implanted D as being predominantly located near the tetrahedral interstitial site in W. In striking contrast to the W results, as-implanted D in Cr is observed to occupy a distinctly different lattice location, the octahedral interstitial site. Angular channeling distributions for the $\langle 100 \rangle$ axis and the $\{100\}$ plane are shown for W and Cr in Figs. 1 and 2. The backscattering and nuclear reaction yields are normalized by the random (non-channeled) yields. The insets show a part of the projected bcc unit cell with the circles indicating the atom rows, the lines indicate the atom planes, and the boxes show the relative number of projected locations for the tetrahedral and octahedral interstitial sites.

For the $\langle 100 \rangle$ axis the large central flux peaks of angular width $\approx 0.2°$ for W and $\approx 0.3°$ for Cr indicate significant fractions of the impurity are relatively near the center of the channel. In addition for the case of Cr the D signal drops below the random level with a minimum yield ≈ 0.73 and angular width of the order of that for the Cr lattice dip, suggesting that nearly 1/3 of the D atoms are located along the $\langle 100 \rangle$ rows. For W the distribution is consistent with the majority of the implanted D occupying the tetrahedral interstitial site where all the projected positions for the interstitial are off lattice rows and a 1/3 component in the channel center

LATTICE LOCATION OF DEUTERIUM

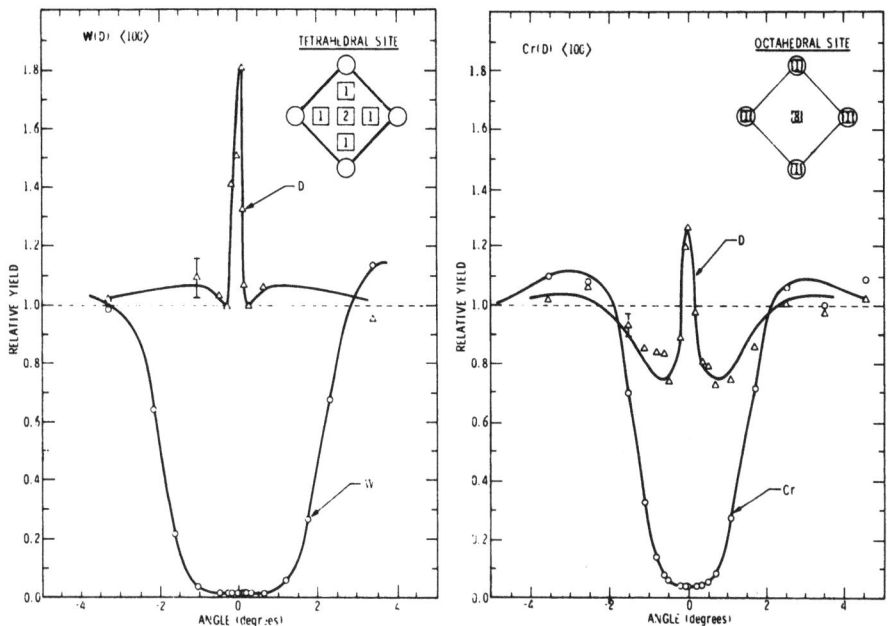

Fig. 1 ⟨100⟩ scans: 3×10^{15} D/cm^2; 30 keV D in W, 15 keV D in Cr.

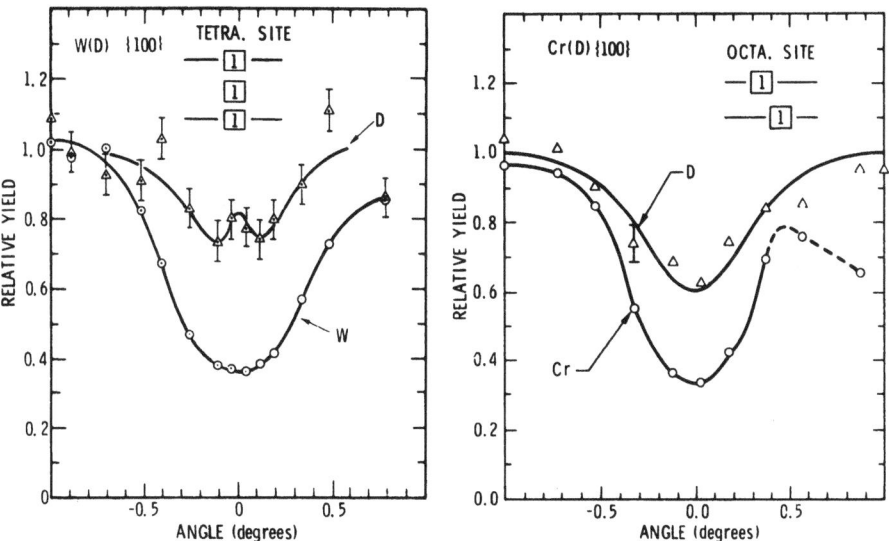

Fig. 2 {100} scans for W and Cr. Implant conditions as in Fig. 1.

gives rise to a central flux peak. For the octahedral interstitial site in Cr the 1/3 of the interstitial atoms located along lattice rows would give a 1/3 dip and the other 2/3 in the channel center would give rise to a central flux peak.

The {100} planar data in Fig. 2 are also qualitatively consistent with this identification of the lattice sites. For the tetrahedral site in W a 2/3 dip with a small central flux peak superimposed is expected, and for the octahedral site in Cr a simple dip is expected. The magnitudes of the D planar dips relative to the lattice are less however, than would be expected if all D atoms were located exactly on the assigned interstitial sites and if the vibrational amplitude of the D atoms was no larger than that of the lattice atoms. For example in Cr the observed D dip is 60% of the Cr lattice dip and the angular width has been narrowed $\sim$ 10%. Possible reasons for the weaker dips include the greater vibrational amplitude of D relative to the host atoms as well as possible small distortions of the time averaged position from the center of the interstitial sites.

A general feature of ion channeling is that greater resolution in determining the localization of impurities is obtained from angular scans if the impurity is located along a crystal axis or plane than if it is in a central flux peak location. This occurs because the average potential which gives rise to steering of the channeled ions changes much more rapidly near the rows or planes than in the center of the channel. The threshold sensitivity for localization near an axis or row is of the order of the vibrational amplitude of the lattice atoms; whereas the threshold for interstitials located in the center of the channel would be appreciably greater. For the $\langle 100 \rangle$ in W we estimate from the continuum model [3] that to resolve increased vibrational amplitudes by an increase over the natural width of experimental flux peaks ($\approx 0.2°$) would require amplitudes ≥ 0.3 Å. This is much greater than either the W atom vibrational amplitude which is 0.09 Å or the estimated D vibrational amplitude in W which (assuming the same force constant) is 0.14 Å.

B. Irradiation Induced Change in D Location

Continued irradiation by the 750 keV ^{3}He analysis beam induces the implanted D in W to move to a new lattice location which is still interstitial in nature. Figure 3 shows angular scans for a 15 keV D implantation in W with the amount of D retained in the W lattice being $\approx 1 \times 10^{15}$/cm^2. The points for the initial scan are taken sequentially from the center out to either side with a fluence of 2 μC per point. The D scan labeled "after bombardment" was taken after a ^{3}He fluence of 74 μC along the $\langle 100 \rangle$ axis. The initial scan is consistent with that shown previously in Fig. 1a, however, the greater detail gives some indication of side band flux peaks located at approximately 0.6° from the center of the dip. Continuum model estimates correlate this peak position with a D component located 1.0 Å from the center of the $\langle 100 \rangle$ channel toward the $\langle 100 \rangle$ rows. This value should be compared to the predicted location for the

LATTICE LOCATION OF DEUTERIUM

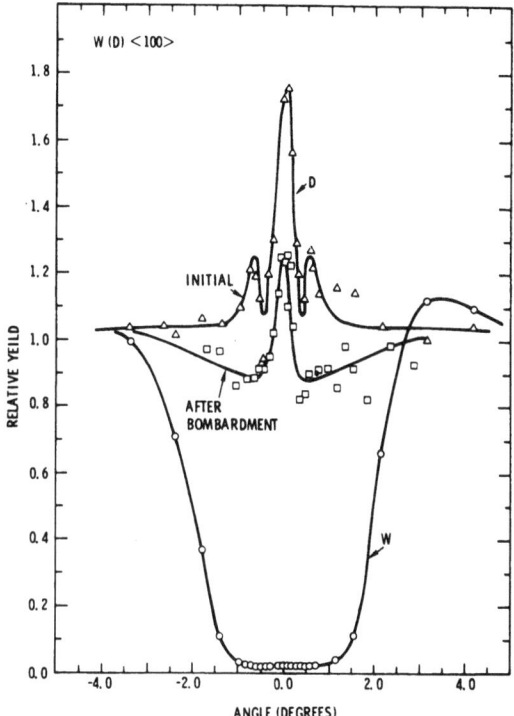

Fig. 3 Angular scan through ⟨100⟩ for $1 \times 10^{15}/cm^2$ 15 keV D implanted W before and after 74 μC irradiation with 750 keV ^{3}He$^+$.

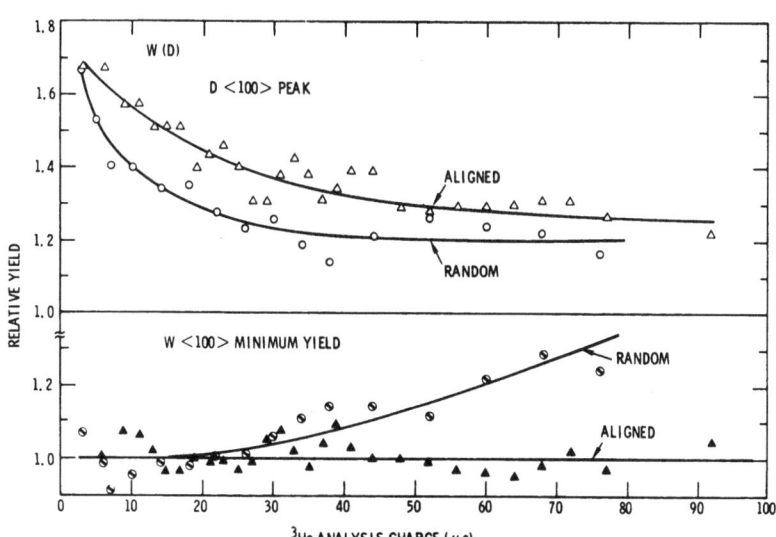

Fig. 4 Relative yield of central flux peak and W minimum vs. 750 keV ^{3}He irradiation fluence for same implant conditions as Fig. 3.

1/3 component for tetrahedral site occupation which is 0.79 Å from the center (see inset in Fig. 1a). The angular scan after bombardment gives some indication of a small dip (∼ 0.9) along the ⟨100⟩ direction as well as the central flux peak of somewhat lower magnitude. Additional angular distributions are needed to determine the new D location; however, the ⟨100⟩ scan suggests that the D is distorted off of the tetrahedral interstitial site appreciably, perhaps towards the octahedral interstitial site.

The rate at which the change in lattice site is induced by the ^{3}He beam for a channeled and a random (non-channeled) irradiation is shown in Fig. 4, where the relative heights of the D flux peak and the W minimum yield are plotted vs. ^{3}He fluence. For random bombardment, the random irradiation is alternated with aligned measurements of the flux peak and the total fluence plotted. The flux peaks decrease fairly rapidly and reach a relatively stable level ≈ 1.2 to 1.3 for both aligned and random irradiation, even though for random irradiation the W minimum yield increases appreciably. The data of Fig. 4 suggest that the change in location is induced fairly rapidly by a mean ^{3}He fluence of 20 to 30 μC (≈ 1-2 x 10^{16}/cm^2), and that the new position remains relatively stable. The normalized value of 1 for the W minimum yield scale corresponds to χ_{min} ≈ 1.4%. Since the backscattering yield is a relative measure of close impact parameter events, the rate of near surface defect production is about 70 times greater for random than for aligned irradiation. However, the rate of decrease in the D flux peak is relatively insensitive to irradiation direction. This implies that if the site change is caused by direct interaction with irradiation induced defects then these defects must be migrating from the end of range of the 750 keV helium ions (i.e.>1 μm). The decrease in the flux peak observed during irradiation does not result in any change in the total D yield for a random orientation indicating the D does not leave through the surface or migrate to the damage at the ^{3}He end-of-range.

In summary ion implantation combined with ion channeling is a valuable technique to study implanted hydrogen location and stability under irradiation. This technique is particularly applicable to transition metals such as Cr and W which exhibit low hydrogen solubility.

Expert technical assistance was given by R. G. Swier.

REFERENCES

1. H. Fischer, R. Sizmann and F. Ball, Z. Physik **224**, 135 (1969).

2. S. T. Picraux and F. L. Vook, Applications of Ion Beams to Metals, ed. by S. T. Picraux, E. P. EerNisse and F. L. Vook (Plenum Press, New York, 1974) p. 407.

3. See, for example, D. Van Vliet, Radiation Effects **10**, 137 (1971).

THE FORMATION OF SUBSTITUTIONAL ALLOYS IN fcc METALS BY HIGH

DOSE IMPLANTATIONS*

J. M. Poate, W. J. DeBonte and W. M. Augustyniak
Bell Laboratories, Murray Hill, New Jersey 07974

J. A. Borders
Sandia Laboratories, Albuquerque, New Mexico 87115

ABSTRACT

The lattice location of Au implanted to high doses at room temperature into single crystal Ni, Cu, Pd and Ag has been determined using ^{4}He ion channeling and backscattering. In all cases the Au atoms are found to be 100% substitutional for room temperature implants without annealing. Au implants in Cu at 15°K are found to be 100% substitutional. High dose implants of W in Cu, a reportedly insoluble impurity, are also found to be highly substitutional.

INTRODUCTION

High dose ion implantation in metals offers the possibility of forming alloys under non-equilibrium conditions, such as with materials which are normally immiscible. A parameter of importance in understanding the ion implantation process is the lattice site location of the implanted species. Lattice location [1] can be determined using the channeling technique which also provides information on the amount of damage or strain produced by the implant. Most lattice site location measurements on impurities in metals have been performed on low dose implants and none produced an implant that was 100% substitutional [1]. One possible inference from this behavior was that the implantation process introduced damage that precluded 100% substitutionality. However we report here the formation of 100% substitutional Au alloys produced by room temperature Au

*The work done at Sandia Laboratories was supported by the U.S. Atomic Energy Commission.

implantations of 10^{16} cm^{-2} in single crystal Ni, Cu, Pd and Ag. Furthermore we have observed a high substitutionality of a reportedly [2] insoluble impurity (W) in Cu. These results will be discussed within the context of the Hume-Rothery criteria [3] for solid solutions.

EXPERIMENTAL TECHNIQUES

The single crystals used in this work were cut with a face perpendicular to a <110> axis and Syton polished to a mirror finish. Some Cu crystals were electropolished in a 2:1 phosphoric acid-water mixture. The crystals were analyzed before implantation with 1.8-2.0 MeV ^{4}He channeling and backscattering. The measured minimum yields (χ_{min}) for the <110> axis were typically 5% for the Syton polished samples and 2-3% for the electropolished Cu samples. Implants of Au or W were usually performed at room temperature with energies between 70-200 keV and doses between 5×10^{14} - 10^{17} cm^{-2}. Those dose levels correspond to peak atomic concentrations up to 15 at .%. Angular scans across the <110> axis were then carried out using 1.8-1.9 MeV ^{4}He ions. Most measurements were performed at room temperature but in situ implantation and channeling measurements for Au in Cu were also made using a goniometer capable of cooling the sample to 12-15°K.

RESULTS AND DISCUSSION

Figure 1 shows random and aligned 1.9 MeV ^{4}He backscattering spectra for a 2×10^{16} cm^{-2} <110> implant of 200 keV Au$^+$ ions into Cu. The extended depth distribution of the Au is due to channeling. It is seen immediately from the large reduction of the Au scattering yield, when the beam is channeled, that the Au is highly substitutional. The appearance of a surface peak in the Au <110> spectra also indicates the high degree of Au substitutionality in the Cu lattice. If the usual definition of the substitutional fraction, $S = (1-\chi_{min}(\text{Impurity}))/(1-\chi_{min}(\text{Host}))$, is assumed then the Au implant is 100% substitutional. However channeling angular scans [4] need to be performed for unambiguous site determination.

In Fig. 2 are shown <110> angular scans for implants of 10^{16} cm^{-2} 150 keV Au ions into Ni, Cu, Pd and Ag. The host and Au scattering yields have been normalized and correspond to scattering from atoms at an average depth of 200 Å from the surface. The Ni, Cu and Ag scans were taken with an incident ^{4}He energy of 1.9 MeV and the Pd with an energy of 1.8 MeV. Within the accuracy of the measurements the normalized scans are identical and it can be concluded that the implanted Au resides 100% on substitutional lattice sites in Ni, Cu, Pd and Ag.

SUBSTITUTIONAL ALLOYS IN fcc METALS

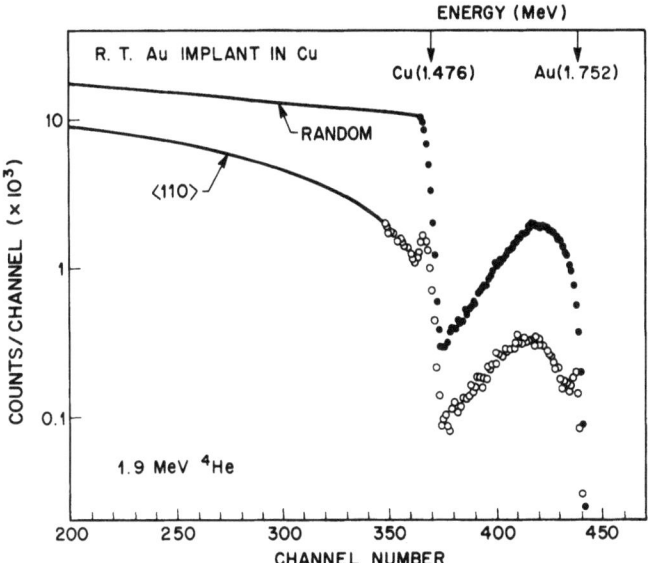

Figure 1. Random and <110> aligned spectra for a single crystal Cu sample implanted with 2×10^{16} Au cm^{-2}.

Ni and Pd show (Fig. 2) higher χ_{min} after implantation than Cu or Ag. This suggests that the damage produced by the incident Au anneals less readily for the case of Cu and Ag. It is not unreasonable to expect the annealing of the host lattice damage to be related to the self-diffusion coefficients. This is consistent with the observation that the measured activation energies for self-diffusion coefficients in Ni and Pd (60-70 kcal/mole) are higher than those for Cu and Ag (40-50 kcal/mole).

The present lattice site location results can be compared with those of Alexander and Poate [5] who found that Au is 100% substitutional in a grown alloy crystal of Cu-Au, containing 2 at .% Au. The importance of the present work, however, is that the Au which was implanted at room temperature, is 100% substitutional before any high temperature processing. If equilibrium conditions applied one could predict a good likelihood of high substitutionality for the case of Au in Cu. Au and Cu are completely soluble and the three criteria of Hume-Rothery [3] for solid solutions are all satisfied; similar considerations apply to Au in Ag and Au in Pd. Au and Ni, however are outside the Hume-Rothery size criteria and the electro negativities have the largest difference of the four systems. In addition the Au-Ni binary phase diagram [2] shows a miscibility gap indicating

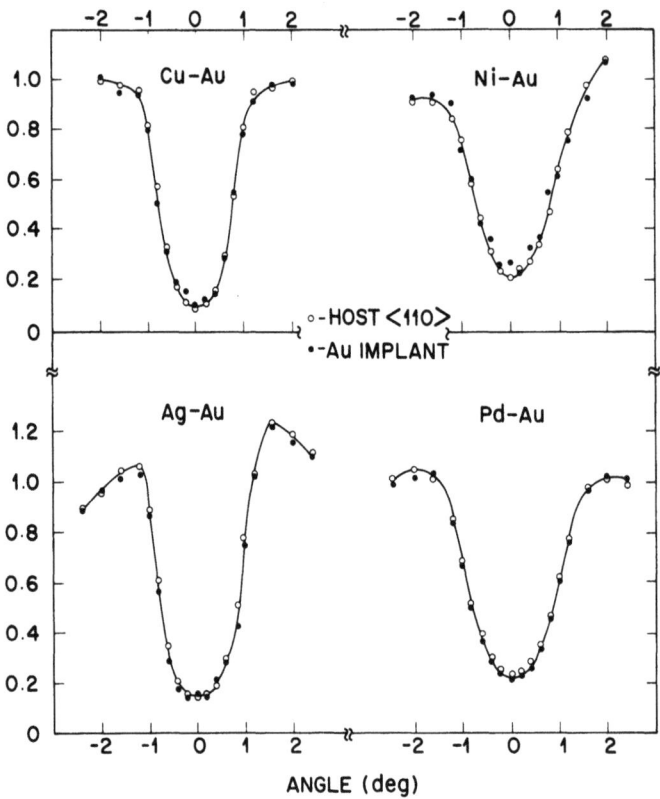

Figure 2. Normalized <110> angular distributions for implants of 10^{16} Au cm^{-2} in Ni, Cu, Pd and Ag. The solid circles indicate the scattering yield for the implanted Au and the open circles show the yield from the host crystal.

that the two elements are not completely soluble. Nevertheless in spite of the failure of these criteria, Au implanted at room temperature also shows 100% substitutionality in Ni.

We have presented results of Au implanted systems which, if normal equilibrium considerations applied, might be expected, a priori, to demonstrate substitutionality. However implants of W in Cu also show a high degree of substitutionality. This is surprising if there is any relationship between substitutionality of implanted species and equilibrium solubility as W is reportedly [2] insoluble in Cu. Figure 3a shows angular scans for a room temperature 80 keV W implant into <110> Cu at a dose of 10^{16} cm^{-2}. The width of the W and Cu angular distributions are the same and a comparison of the host and impurity χ_{min} indicate that 90% of the W atoms are substitutional in Cu. Figure 3b shows the results of an angular scan on the same sample after a vacuum anneal for

SUBSTITUTIONAL ALLOYS IN fcc METALS

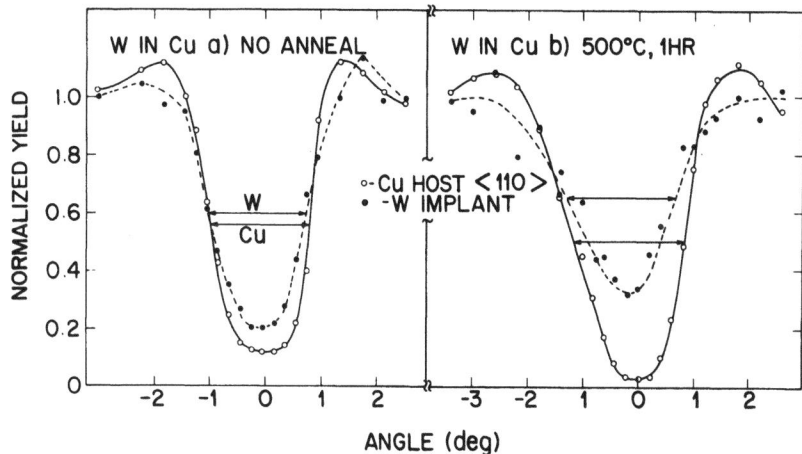

Figure 3. Normalized <110> angular distributions of 10^{16} W cm^{-2} in Cu a) before annealing and b) after annealing at 500°C for 1 hr. in vacuo. The substitutional fractions are a) 0.9 and b) 0.7.

1 hr. at 500°C. The decrease in yield for the Cu lattice to its pre-implant level shows that the damage caused by the implant has been annealed but the W shows an increased yield over that taken after the room temperature implant. The W is still highly substitutional however and there is no sign of the complete aggregation or precipitation observed for implanted Hf in Ni [6] on annealing.

Au implants in Cu and backscattering analysis were also performed at 15°K. The Au was found to be 100% substitutional. These results are noteworthy in that at 15°K both vacancies and interstitials are expected to be immobile [7]. This suggests a model where the Au undergoes a replacement collision at the end of its range and has no interaction with the lattice damage. These low temperature results however are not conclusive as the localized temperature near an ion track may be quite high for a short time (thermal spike model [8]).

We have presented results of high dose implants in fcc metals. It is possible to produce Au substitutional alloys in Ni, Cu, Pd and Ag but W also occupies substitutional sites in Cu even though equilibrium solubility considerations could be thought to rule against this eventuality. de Waard and Feldman [1] note high substitutional fractions for low dose implants in Fe although in no case do they report 100% substitutionality. They suggest that vacancies stabilize these configurations allowing fairly large atoms to occupy substitutional sites. Additional measurements are in progress to extend the type of ion implanted so that we can investigate further the relationships between lattice site location, damage and chemical nature of the implanted species. These high dose implants are also suitable for more specific metallurgical characterization in that x-ray diffraction and x-ray photoelectron spectroscopic measurements should be feasible.

ACKNOWLEDGEMENTS

We are grateful to W. L. Brown and F. L. Vook for encouraging this work, to W. F. Flood for the crystal polishing and the accelerator staffs at Bell and Sandia for aid with the implants.

REFERENCES

1. H. de Waard and L. C. Feldman, Applications of Ion Beams to Metals (edt. by S. T. Picraux, E. P. EerNisse and F. L. Vook, Plenum, New York 1974) p. 317.

2. M. Hansen, Constitution of Binary Alloys (McGraw-Hill, New York, 2nd Edition, 1958).

3. W. Hume-Rothery, R. E. Smallman and C. W. Haworth, The Structure of Metals and Alloys (Institute of Metals, London, 5th Edition, 1969).

4. R. B. Alexander, P. T. Callaghan and J. M. Poate, Phys. Rev. 9, 3022 (1974).

5. R. B. Alexander and J. M. Poate, Radiation Effects, 12, 211 (1972).

6. E. N. Kaufmann, J. M. Poate and W. M. Augustyniak, Phys. Rev. B7, 951 (1973).

7. W. Schilling et al., Vacancies and Interstitials in Metals (edt. by A. Seeger et al. North Holland, Amsterdam, 1970).

8. M. W. Thompson, Defects and Radiation Damage in Metals (Cambridge University Press, 1969).

THE EFFECT OF ION IMPLANTATION ON THE CORROSION BEHAVIOUR OF Fe

V Ashworth, D Baxter,[*] W A Grant,[*] R P M Proctor and
T C Wellington. Corrosion and Protection Centre, UMIST,
Manchester. (* Dept of Elec Eng, University of Salford
Salford M5 4WT, Lancashire, U.K.)

ABSTRACT

Previous work[2] indicated that the corrosion and passivation
behaviour of metals such as Al, Cu and Fe can be influenced by high
doses of various low energy ions. This paper presents further
investigations of the Fe system mainly following implantation with
Cr and Ar but also with Fe and Ta. Cathodic reduction tests show
that ion implanted Fe has an oxide layer of approximately 60Å, a
thickness twice that of a normal air formed oxide. The film
formed consists of both Fe_3O_4 and γ-Fe_2O_3, as in air formed films
and the extra oxide thickness is contributed by the Fe_3O_4 layer
with the outer γ-Fe_2O_3 layer remaining constant. Potentiostatic
polarisation tests on Cr implanted Fe showed a large increase in
sample passivity. In order to assess the quality of the steels
produced, ion implanted Cr/Fe surface alloys were compared to
conventional alloys of similar bulk composition. Unconventional
alloys formed by implantation of Ta were also investigated by
potentiostatic polarisation measurements.

INTRODUCTION

Ion implantation is a useful technique for altering and
investigating the surface or para-surface properties of materials.
It has been extensively applied in the semiconductor field for
producing electronic devices and much of the early basic research
effort has also involved semiconductors[1]. In recent years ion
implantation has become increasingly employed in investigating a

wider range of materials and properties.[2,3] We present here the
results of further work on the corrosion behaviour of ion implanted
iron. The results of our previous work may be briefly summarised
as follows. The injection of various dopants (e.g. B, Ar, Mo)
into Al, Cu and Fe influenced the corrosion and passivation of
these metals in various aqueous solutions. Aluminium for example
was implanted with a high dose ($> 10^{16}$ ions.cm^{-2}) of 20keV Ar$^+$
ions and examined potentiostatically. The anodic section of the
polarisation curve indicated that passivity (due to the formation
of Al_2O_3 surface film) was retained to higher voltages than for
unimplanted samples. In chloride solutions the implanted specimens
showed a slightly longer passive region before breakdown. The
breakdown of passivity in aluminium is mainly due to the presence
of weak spots and pores in the oxide film so that the reason for a
reduction in breakdown was attributed to the oxide film being
thicker or more coherent or both. Argon implanted Fe samples
showed no differences from pure iron when polarisation was carried
out first in the cathodic direction but samples polarised initially
in the anodic direction showed an increase in polarisability of the
anodic reaction. Cathodic polarisation of Fe electrodes would be
expected to electrochemically reduce any oxide film (unlike Al)
and it was concluded that the implantation process altered the
composition or thickness of the natural oxide present at
Fe surfaces.

Since some of the effects observed appeared to be independent
of the ion species injected further work was needed to ascertain
the nature and thickness of any oxide film present at metal surfaces
following implantation. The present work has been restricted to
the Fe system whose oxides are electrochemically reducible and
falls naturally into two parts namely (i) cathodic reduction
measurements to measure the thickness and composition of both
unimplanted and ion-implanted specimens and (ii) potentiostatic
polarisation tests on Cr implanted Fe and a comparison with
conventional Cr/Fe alloys of similar bulk composition. In
addition, the unconventional ion implanted alloy Ta/Fe was
also investigated.

In aqueous solutions the behaviour of Fe is best described by
reference to pH value. Within the acid range (pH $<$ 5) the reaction
produces hydrogen evolution and no protective oxides are formed.
In the neutral range (pH 5-10) Fe initially has relatively poor
resistance[4] to corrosion, yet if undisturbed a protective oxide
film forms, considerably slowing the rate of reaction. Alkaline
solutions have only a slight corrosive action on iron because of
the development of the protective oxides and this resistance
increases with alkalinity. Iron forms three stable oxides;[5]
Wustite FeO, Magnetite Fe_3O_4 and Haematite Fe_2O_3. Wustite only
exists above 570°C. The thickness of air formed oxide films on

Fe have been measured by various techniques including gravimetric, optical and electrochemical. It is generally found to be 10-30Å depending on surface treatment, oxygen pressure and time of exposure. The outer section of the oxide if γ-Fe_2O_3 with a thicker layer of Fe_3O_4 below.

EXPERIMENTAL METHODS AND RESULTS

Specimens were implanted in the Salford University isotope separator at room temperature and at a pressure of < 10^{-6} torr. High doses (> 10^{16} ions.cm^2) of various ions at 20keV including Fe^+, Cr^+, Ar^+ and Ta^+ were implanted at dose rates sufficient to keep the rise of sample temperature to < $10°C$. The surface layers of the ion implanted layers were investigated by two basic methods (i) by cathodic reduction in borate/HCl solutions to obtain a measurement of the type and quantity of oxide film(s) present and (ii) by potentiostatic polarisation tests in acetic acid/sodium acetate buffer solutions to examine both the behaviour of the as-implanted specimens and that of the alloys exposed following reduction of any surface oxide. Bulk Cr/Fe alloys were also examined potentiostatically for comparison with those prepared by implantation.

Analysis of the results of cathodic reduction measurements on both unimplanted Fe and on Fe samples implanted with 5×10^{16} ions.cm^{-2} of 20keV Ar^+ indicated that the oxide haematite γ-Fe_2O_3 is reduced first followed by magnetite Fe_3O_4. Using values for the current efficiencies of these two reduction reactions, together with oxide densities, the thickness of the two oxide films may be calculated. For unimplanted samples the average thicknesses were 13Å of Fe_3O_4 with an outer layer of 11Å of γ-Fe_2O_3. Samples implanted with Ar were found to have substantially thicker oxide layers namely 33Å Fe_3O_4 with 11Å γ-Fe_2O_3. The oxide thickness and composition for unimplanted Fe compares satisfactorily with previous measurements by Sewell, Stockbridge and Cohen.[6] On implanted Fe the oxide has approximately doubled in total thickness, mainly by an increase in the Fe_3O_4 layer. The outer layer of γ-Fe_2O_3 is of identical thickness on both unimplanted and implanted specimens. Both types of specimen were also aged for two months in a desiccator before being examined by cathodic reduction tests; the results are given in Table I. In each case the oxide film has increased by 10Å, the extra thickness being mainly contributed by the Fe_3O_4.

A typical potentiostatic polarisation test for pure Fe is shown in Figure 1. The specimen was first polarised anodically from the corrosion potential to +1000mv and during this first anodic sweep the oxide film is thickened. The sample was next polarised cathodically to -1400mv which results, amongst other

TABLE I. Thickness of Fe_3O_4 and $\gamma\text{-}Fe_2O_3$ present on unimplanted Fe and on Fe samples implanted to a dose of 5×10^{16} ions cm^{-2} 20keV Ar^+.

	Fe_3O_4 Å	$\gamma\text{-}Fe_2O_3$ Å	Total Å
UNIMPLANTED	13	11	24
UNIMPLANTED AND AGED	20	14	34
IMPLANTED	33	11	44
IMPLANTED AND AGED	43	11	54

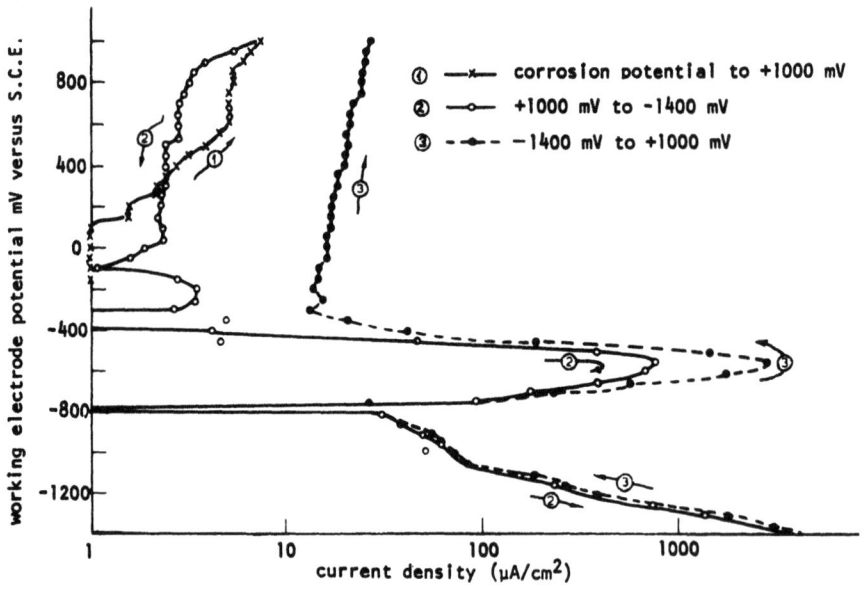

FIGURE 1. Potentiostatic polarisation curves for pure Fe in sodium acetate/acetic acid solution, pH 7.3.

reactions, in the reduction of any oxides present on the metal surface. Finally, the sample was polarised anodically again from -1400mv to +1000mv. It can be seen from Figure 1 that between -700mv and -200mv the anodic current density rises rapidly to a peak value of I_c, the critical current density for passivation. This peak value is greater in the second anodic polarisation sweep than that in the previous cathodic sweep since the metal is now oxide free and the anodic dissolution reaction can reach its full magnitude. The formation of the metal oxide occurs also during this range of potential so that the current density falls rapidly from I_c and the surface is said to be passivated. The value of the critical current density I_c can be used as a measure of the ease of passivation.

Conventional Cr/Fe alloys were tested as above and analysis of the data indicates that during the second anodic sweep the critical current density I_c is considerably reduced compared to the pure Fe and the sample passivates more easily, as expected for the Cr/Fe alloy. The results of polarisation measurements on alloys of various compositions are summarised in Figure 3 where the critical current for passivation I_c is plotted against the atomic percentage of chromium in the alloy. Fe samples implanted with various doses of 20keV Cr^+ ions were investigated in the same manner and Figure 2 shows a representative result. Following implantation with a dose of 5×10^{16} ions cm^{-2} the samples were reduced to the oxide free condition by the cathodic polarisation sweep and then polarised anodically as previously. The ion implanted Cr/Fe alloy behaved similarly to conventional alloys; I_c in Figure 2 is again reduced compared to that for pure Fe. The implanted alloy composition was varied by varying the total implanted Cr^+ dose and I_c was noted in each case. Reference to Figure 3 then enables the equivalent conventional alloy composition to be estimated. Doses of 5×10^{16} Cr^+ ions cm^{-2} and 2×10^{17} Cr^+ ions cm^{-2} produced implanted surface alloys corresponding to conventional bulk alloys of 4.2% and 6.2% respectively.

Finally, Fe samples implanted with Fe^+ and Ta^+ were also investigated potentiostatically. The Fe^+ implanted specimens behaved identically to unimplanted samples once the first cathodic sweep had removed the surface oxide film and Ta^+ implanted samples showed an increase in the degree of passivation.

CONCLUSIONS

The results of cathodic reduction and polarisation tests enable some clear conclusions to be drawn concerning the corrosion behaviour of Fe implanted with various ions. The air formed oxide film always present on Fe is increased in thickness as a result of Ar^+ implantation. It is believed that this oxide thickening is a

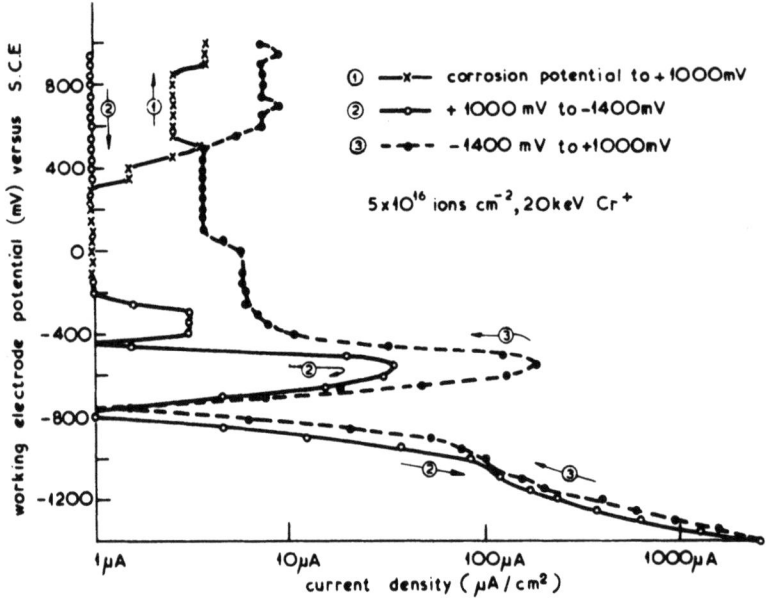

FIGURE 2. POTENTIOSTATIC POLARISATION CURVES FOR Cr^+ IMPLANTED Fe IN SODIUM ACETATE/ACETIC ACID SOLUTION pH7.3.

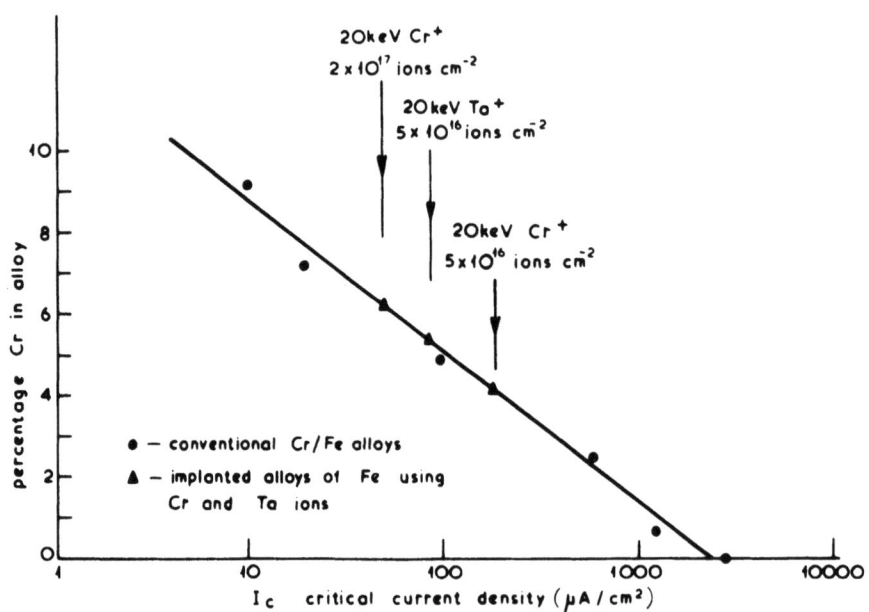

FIGURE 3. CRITICAL CURRENT DENSITY I_c FOR PASSIVATION AS A FUNCTION OF PERCENTAGE OF Cr IN Cr/Fe ALLOY

general effect of the ion implantation process and occurs with ion species other than Ar^+. An increase in surface oxide thickness would explain the increased corrosion resistance shown in polarisation measurements on Al as reported previously[2]. Removal of this oxide layer, in the case of Fe, by cathodic polarisation exposes the underlying implanted metal surface for analysis.

Ar^+ and Fe^+ implanted iron samples have no increased corrosion resistance other than that due to the increased oxide film mentioned above. In potentiostatic polarisation tests, reduction of this thicker oxide film exposes an Fe surface that behaves identically to unimplanted material.

Implantation of Fe with 20keV Cr^+ ions increases the passivation of the metal. This effect is clearly distinguishable from the passivation due to the thicker air formed oxide present following implantation. Reduction of this initial oxide reveals an implanted Cr/Fe surface alloy that behaves, in the aqueous environments used, similarly to conventional Cr/Fe bulk alloys. The implanted alloy composition can be altered by varying the Cr^+ ion dose. The behaviour of a particular implanted alloy can be compared to conventional alloys by measuring I_c, the critical current density for passivation. The degree of passivation (as measured by I_c) produced by Ta^+ implantation is greater than that produced by an equivalent dose of Cr^+.

REFERENCES

1. Ion Implantation in Semiconductors. J W Mayer, L Eriksson and J A Davies. Academic Press (1970).
2. Ion Implantation in Semiconductors and Other Materials. Edited by B L Crowder. Plenum Press (1973).
3. Application of Ion Beams to Metal. Edited by S T Picraux, E P Eernisse and F L Vook. Plenum Press (1974).
4. F L Laque and H R Cobson. Corrosion Resistance of Metals and Alloys. Reinhold (1963).
5. O Kubaschewski and B E Hopkins. Oxidation of Metals and Alloys. Butterworth (1962).
6. C D Stockbridge, P B Sewell and M Cohen. J Electrochem Soc 108 928 (1961).

A RUTHERFORD BACKSCATTERING AND CHANNELLING STUDY OF Dy IMPLANTED INTO SINGLE CRYSTAL Ni

G A STEPHENS, E ROBINSON AND J S WILLIAMS[+]

DEPT OF PHYSICS. ([+]DEPT OF ELEC ENG.) UNIVERSITY OF SALFORD, SALFORD M5 4WT, LANCASHIRE, U.K.

ABSTRACT

Lattice disorder, atom location and implant migration during isochronal annealing of the 20keV Dy^+ implanted Ni system have been examined by Rutherford backscattering and channelling techniques. Two annealing stages for the disorder are identified. The first between 300-600°C is associated with a reduction in the channelling disorder peak, the second corresponds with a sharp decrease in dechannelling rate between 650-800°C. High depth resolution backscattering indicates that the lattice disorder anneals from the bulk towards the Ni surface and that marked outdiffusion of Dy to the Ni surface occurs above 600°C. The lattice location of Dy in Ni appears to be most sensitive to the implant conditions.

INTRODUCTION

The motivation for choosing the Dy implanted Ni system arises from an increasing interest in the magnetic properties exhibited by alloys of rare earth and ferromagnetic materials.[1] Over the past decade and a half ion implantation has been used successfully to introduce low concentrations of an active impurity into the surface layers of semiconductor materials for the fabrication of solid state devices. More recently, ion implantation has been applied with similar success to metals, as a means of improving the physical and chemical properties of metal surfaces,[2] and would therefore seem to be ideally suited for the introduction of controllable concentrations of rare earths into the near surface of ferromagnetic materials.

As a precursor to magnetic property measurements we decided to examine the anneal characteristics of the Dy implanted Ni system by means of Rutherford backscattering and channelling. In this study we have measured lattice disorder, implant migration and atom location at various stages during an isochronal anneal cycle up to 800°C. A glancing incidence Rutherford backscattering geometry[3] has been employed to provide increased depth resolution for monitoring the disorder distribution and implant profile as a function of temperature. The results indicate two apparent disorder anneal stages and marked outdiffusion of Dy at high temperatures. In addition, the atom location of Dy in the Ni lattice seems to be extremely sensitive to the implant conditions.

EXPERIMENTAL

Thin slices of single crystal Ni, cut several degrees off the <100> axis, were mechanically and chemically polished to remove surface damage. Care was taken to ensure a large area flat surface suitable for low angle Rutherford scattering. Prior to implantation the samples were set up for 2MeV He$^+$ backscattering analysis both to locate the available low angle channels for subsequent post-implant analysis and to obtain a measure of residual damage from sample preparation. Both near-normal <100> and <110> axes and glancing incidence <100> and <110> axes (inclined at $\theta \approx 84°$ and $\theta \approx 80°$ to the Ni surface normal respectively) were employed during analysis. The samples were scribed before removal from the goniometer to facilitate post-implant relocation of the various aligned directions. The Ni samples were then implanted off axis with 20keV Dy$^+$ ions at a dose rate of $\sim 2 \times 10^{13}$ ions.cm^{-2}.sec^{-1} to a total dose of about 5×10^{15} ions.cm^{-2}. The samples were analysed by 2MeV He$^+$ backscattering following each isochronal anneal in an inert Ar atmosphere at temperatures in the range 25-800°C.

RESULTS

Lattice Disorder

From a comparison of aligned spectra before and after implantation, the remnant lattice disorder is manifested by an increased near-surface disorder peak and a high dechannelling rate behind the peak. Two disorder anneal stages were evident from monitoring of the disorder peak area and dechannelling rate during the anneal cycle. The first anneal stage associated with a decrease in the disorder peak area is shown in Figure 1. Here the disorder peak area, as observed along the <100> ($\theta \approx 12°$) and <110> ($\theta \approx 40°$) axes, is plotted as a function of anneal temperature. The results of both the <100> and <110> measurements are in good agreement; the disorder appearing to anneal out to the pre-implant disorder level in the temperature range 300-600°C.

RUTHERFORD BACKSCATTERING AND CHANNELING STUDY

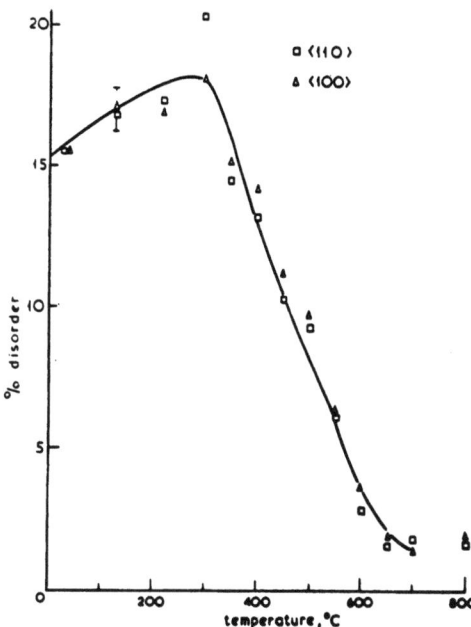

FIGURE 1. Peak disorder annealing behaviour.

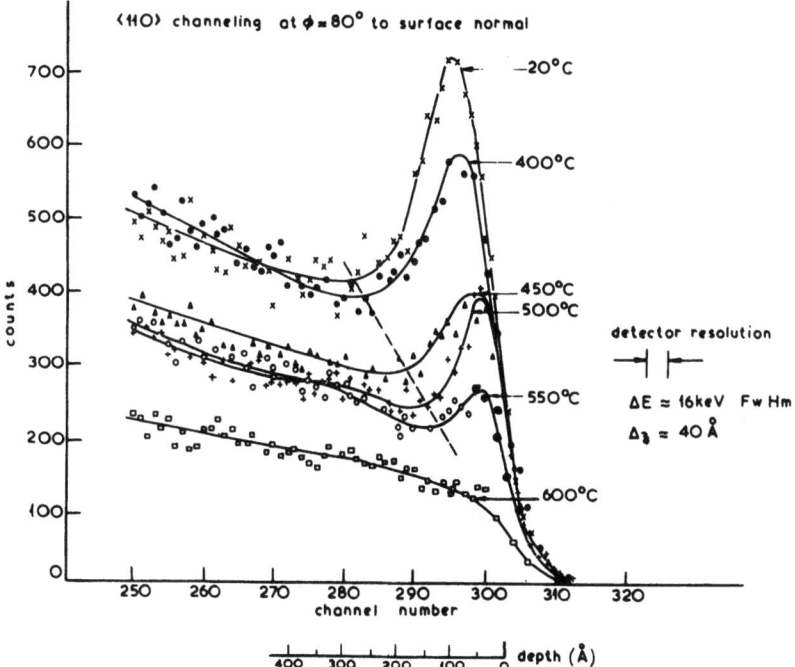

FIGURE 2. Glancing incidence channelled spectra from Ni showing the annealing of the disorder profile.

Glancing incidence channelling along the <100> and <110> directions provided additional information on the disorder distribution as a consequence of the improved depth resolution. Figure 2 is a typical family of glancing incidence aligned spectra taken during the disorder peak annealing stage. Here, the <110> ($\theta \approx 80°$) geometry provides a depth resolution enhancement of 4.2 compared with normal incidence backscattering. Upon fitting a depth scale[4] to the spectra, the depth resolution of the detection system was found to be approximately 40Å. The post-implant (20°C) disorder profile peaks at $\approx$ 110Å with the "tail" extending to about 270Å. It is clear that with increasing anneal temperature the disorder anneals from the bulk towards the Ni surface.

A second anneal stage between 650-800°C was evident in the aligned spectra as a sharp reduction in the rate of dechannelling behind the disorder peak. During the first anneal stage associated with the reduction in disorder peak area (300-600°C) only a slight reduction in the dechannelling rate was apparent (see Figure 2). Similarly, during the second anneal stage no further reduction in the disorder peak was observed.

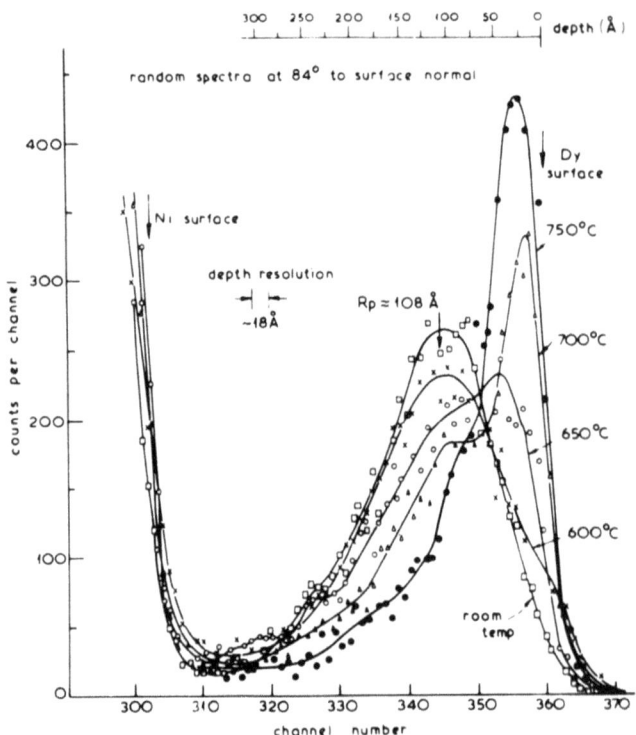

FIGURE 3. Glancing incidence random spectra (Dy region only) showing migration of Dy above 600°C.

Dy Implant Distribution

Using a glancing incidence geometry with $\theta \approx 84°$ to the surface normal and with the probe beam incident in a random crystal direction the family of curves shown in Figure 3 was obtained from the resultant backscattered spectra. The resolution enhancement here is 6.4 compared with normal incidence backscattering, leading to a depth resolution[4] of about 20Å. The post-implant (20°C) Dy profile is near Gaussian in shape with a measured most probable range of ≈ 108Å. The shape of the distribution does not change substantially up to 600°C (i.e. during the first annealing stage). Above 600°C rapid outdiffusion of Dy occurs until by 750°C it is virtually all residing at the crystal surface. No substantial loss of Dy from the system was observed.

Atom Location

The random and aligned integrated Dy yields were used to obtain a measure of the Dy atom position within the Ni lattice. In some cases up to a 60% attenuation in Dy yield was measured along both <110> and <100> **aligned directions** after annealing at 600°C. Angular scans about <100> and <110> directions gave dips for the Dy and Ni yield with similar FwHm (1.46° for 1.5MeV He$^+$) which can be interpreted to indicate a significant substitutional component of Dy atoms in such cases. There is further evidence to suggest a marked reduction in substitutional component above 750°C. However, some implanted samples exhibited almost no indication of preferred Dy lattice position during the entire anneal cycle. Indeed it would seem that the atom location of Dy is extremely sensitive to the implant conditions. Preliminary glancing incidence random and aligned spectra obtained from samples exhibiting high substitutional component indicated a significant difference in the shape of the Dy profile between random and aligned measurements. Such behaviour may arise from variation in the substitutional Dy component throughout the implant profile.

DISCUSSION

Recent studies of hyperfine structure measurements combined with ion channelling and atom location for the Yb implanted Fe system[5,6] have shown a strong correlation between the number of Yb atoms on substitutional sites and the number of high magnetic sites. Moreover, these authors mention an apparent lack of agreement between absolute Yb substitutional content for measurements under different implant conditions and attribute this to different disorder distributions and states of Yb in the Fe lattice (as precipitates, alloys etc).[5] Such behaviour underlines the importance of obtaining a detailed knowledge of lattice disorder, atom location, implant migration and precipitation and alloying effects of rare earth species within the metal lattice in order to better understand the magnetic properties of the composite system. We suspect that the sample-to-sample variations in Dy substitutional content in our case may be a result of some subtle difference between the various implant conditions. At present we are examining the effect on Dy substitutional content of such parameters as dose rate, total dose and energy of the implanted Dy ions.

In contrast to the atom location results, the post-implant lattice disorder measurements and subsequent anneal characteristics are in excellent agreement from sample-to-sample; a summary of the results is given below:-

(i) 25-300°C. No significant reduction was evident in either the disorder peak area or the dechannelling rate behind the peak. Similarly no change in shape was observed in either the near Gaussian disorder profiles (peak ≈ 110Å) or the Dy implant profiles (R_p ≈ 108Å).

(ii) 300-600°C. Complete annealing of the disorder peak to the pre-implant level occurred, the disorder annealing from the bulk towards the surface. A corresponding slight reduction in the dechannelling rate was observed but no significant change in the Dy implant profile.

(iii) 650-800°C. A sharp reduction in dechannelling rate to near the pre-implant level was observed. Simultaneously, marked outdiffusion of the implant took place, Dy building up at the Ni surface.

The two apparent anneal stages (namely the annealing of the disorder peak between 300-600°C and the sharp reduction in dechannelling rate between 650-800°C) have also been observed by Gettings and co-workers[7] for the Bi implanted Ni system. The presence of two specific anneal stages has been identified in other implanted metal systems (e.g. Sb implanted Al[8]) although the anneal regions have not usually been associated with both a decrease in the disorder peak and a reduction in dechannelling rate.

From TEM studies of radiation induced defects in Ni by other workers[9-11] we suggest that the 300-600°C reduction in the disorder peak is associated with the annealing of implantation induced defect clusters and loops. For Bi implanted Ni[7] the disorder peak was considerably closer to the surface than the Bi ion-range whereas our disorder and implant profiles have similar depth distributions. Our glancing incidence channelling spectra indicate that the disorder anneals from the bulk towards the surface: presumably, any beam annealing of the disorder which takes place during implantation (at high dose rates) would result in similar behaviour. It would therefore seem reasonable to expect partial beam annealing of disorder during implantation to result in a remnant disorder peak nearer the surface than the ion range.

The simultaneous occurrence of the second annealing stage between 650-800°C and the migration of Dy to the surface may well be more than coincidental. For instance, the second anneal stage in Sb implanted Al[8] was explained as a break up of Sb clusters (identified by TEM) and the observed long range migration of Sb in the Al lattice. We suggest that the Dy in the Ni lattice prior to 650°C (possibly as Dy clusters, some Dy-Ni alloy, etc) may directly (or indirectly as Ni lattice strain) contribute to a high dechannelling rate which is reduced as the Dy migrates to the surface.

Finally, for positive identification of the nature of the disorder and the various annealing mechanisms operative in the Dy implanted Ni system, a correlated TEM and channelling study would seem to be the obvious next step.

ACKNOWLEDGEMENTS

We are most grateful to W A Grant for communicating this paper and for discussions. M J Nobes and P Cardwell are acknowledged for technical assistance.

REFERENCES

1. See for example, K H J Buschow, Proc Conf on "Rare Earths and Actinides", Durham 1971. Publ. Inst of Phys, London (1971) p.140.
2. See for example, R S Nelson, Vacuum $\underline{23}$, 79 (1972).
3. J S Williams, Rad Effects, in press.
4. Stopping powers for He$^+$ in Ni according to L C Northcliffe and R F Schilling, Nuclear Data Tables A7, 233 (1970) have been used to fit depth scales to the spectra.
5. R B Alexander, E J Ansaldo, B I Deutch, J Gellert and L C Feldman in "Applications of Ion Beams to Metals", Eds S T Picraux et al. Publ. Plenum, New York (1974) p.365.
6. F Abel, M Bruneaux, C Cohen, H Bernas, J Chaumont and L Thome, Sol State Comm $\underline{13}$, 113 (1973).
7. M Gettings, K G Langguth and G Linker, in Ref.5 p.241.
8. G J Thomas and S T Picraux, in Ref.5 p.257.
9. A Seeger and M Mehrer, in "Vacancies and Interstitials in Metals", Eds A Seeger et al. Publ. North Holland, Amsterdam (1970) p.1.
10. D I R Norris, Phil Mag $\underline{19}$, 653 (1969).
11. K Urban, Phys Stat Sol (a) $\underline{4}$, 761 (1971).

RADIATION DAMAGE I

DEFECT PRODUCTION IN SEMICONDUCTORS[++]

J. BOURGOIN

Groupe de Physique des Solides de l'E.N.S.[+]
Tour 23, 2 Place Jussieu, 75221 Paris Cedex 05

ABSTRACT

The different stages which govern the defect production are briefly described. Emphasis is placed on particular problems encountered in ion implantation such as the nature of the damage in highly disordered regions and the influence of the ionization on the damage production. Electronic mechanisms which can play a role in defect production and annihilation are discussed.

I. INTRODUCTION

The process of defect production in a material by energetic particles can be divided in several stages: energy is transmitted to the atoms of the material by the particles; these atoms are displaced from their substitutional position giving rise to (stable or metastable) vacancy-interstitial pairs; vacancies and interstitials interact with each other and with impurities to form complex defects. We shall briefly discuss these three stages in the second section. Then, in the third section, we shall describe the influence of various parameters which characterize the implanted material as well as the irradiation itself, on the production and the nature of the defects. And because many aspects of the defect production depend on the electronic state of the defects, we shall develop the effects of the ionization on their behaviour in a last section.

+ Laboratoire associé au CNRS.
++ This work was supported by the Délégation Générale à la Recherche Scientifique et Technique under contract 73.7.1650.

II. DEFECT PRODUCTION MECHANISM

2.1 The transmission of an energy T from an energetic particle to an atom occurs through Coulombic interaction in case of charged particles; in case of neutral particles the interaction is described by elastic collisions (1). For a charged particle the differential cross section $d\sigma$ is given by the Mott formula (2), or by the Mott-Rutherford formula (5) (approximated by the McKinley-Fesbach formula (6)) when the energetic particle is relativistic. Hard spheres scattering corresponds to an isotropic distribution of the transmitted energy T and therefore to a constant differential cross-section. For charged particles the differential cross-section decreases rapidly with T up to the maximum transmitted energy T_m (Fig. 1); because of the electron-proton mass ratio, T_m is several orders of magnitude smaller in the case of electrons than in the case of protons, even taking into account the relativistic correction for the electrons (Fig. 2).

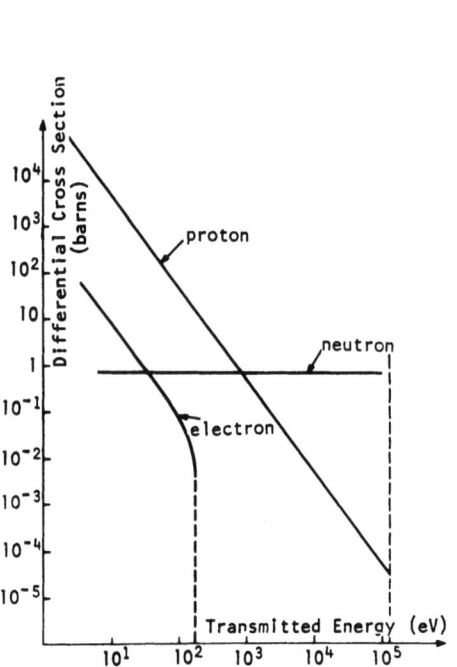

Fig. 1. Differential cross-section vs transmitted energy for 1 MeV protons, electrons and neutrons in silicon

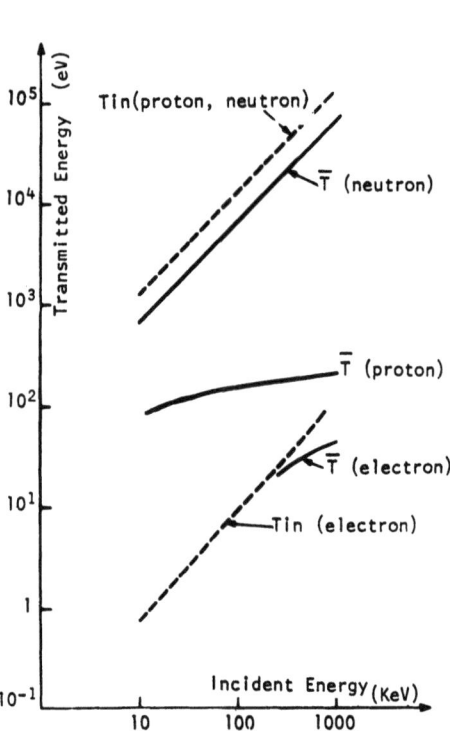

Fig. 2. Maximum transmitted energy (T_m) and average transmitted energy ($\bar{T}$) vs incident energy for protons, electrons and neutrons in silicon

2.2 The number of primary defects produced by an incident particle of energy E is given by the cross-section for displacement:

$$\sigma(E) = \int_{T_d}^{T_m} g(T) \cdot d\sigma$$

where $g(T)$ is the displacement probability function. The threshold energy T_d, i.e. the minimum transmitted energy which results in the displacement of an atom, has been evaluated by Kohn [7] and Bauerlein [8] assuming that the atoms in the lattice are in isotropic square wells of depth T_d. Although these evaluations are adiabatic [9], they give threshold energies in reasonable agreement with the experimental results [4] for germanium and silicon. In this approximation of an isotropic square well, the displacement probability is: $g(T) = 0$ for $T < T_d$ and $g(T) = 1$ for $T_d \leq T < 2 T_d$. Such an approximation is incorrect since the probability to displace an atom depends not only on T but also on the direction in which it is transmitted, compared to the crystallographic orientation.

The nature of the defects produced depends only on the transmitted energy: particles of different nature and of different energy which transmit the same energy to a crystal produce the same defects. We shall therefore consider the nature of the defects produced as a function of the average transmitted energy $\bar{T}$. The average transmitted energy $\bar{T}$ is larger for neutrons than for charged particles (Fig. 2) for which $d\sigma$ decreases rapidly with T; $\bar{T}$ is smaller for electrons than for ions for which $d\sigma$ and T_m are several orders of magnitude larger.

2.3 When $T_d \leq \bar{T} < 2 T_d$, only one lattice atom is displaced by the incident particle and a vacancy-interstitial pair is created. Are vacancy-interstitial pairs observed? No, in most cases, even when the irradiation is performed at helium temperature. In silicon the defect creation rate observed at 4.2 °K is very low, compared to the rate expected from the calculation; the isolated vacancy is observed [10], but not the interstitial. In aluminum doped silicon Watkins [11] has shown that the appearance of aluminum interstitials after 4 °K electron irradiation can only be the consequence of the mobility of silicon interstitials at 4 °K (this point will be discussed in the fourth section). In semiconductor compounds the nature of the recovery processes remains to be fully established: in III-V compounds the lower temperature stages are generally attributed to close pair annihilation; in II-VI compounds chalcogen and metal vacancies have been reported but no direct measurements on interstitials have been made; it is also argued [12] that interstitial migration occurs at the low temperature stage (around 60 °K). Only in n-type germanium strong indications of the presence of vacancy-interstitial pairs can be found [13]; it is thought that the annealing stage at 65 °K corresponds to the annihilation of the pairs through the diffusion of the interstitials.

2.4 When $\bar{T}$ is on the order of several T_d, the primary knock-on atom (PKA) can in turn displace other atoms; as a result, when neighbouring atoms of the PKA are displaced, complex defects are formed. Some of these complex defects have been identified in silicon (they have been classified by Vook and Stein (14) as "ITI" defects - defects whose production is temperature independent - by opposition to ITD defects - defects whose production is temperature dependent). These defects are the divacancy (15), the trivacancy and the quadrivacancy (16) in several charge states. The concentration of impurities being to low, complex defects involving impurities cannot be created directly by the irradiation in a non-negligible concentration.

2.5 In case of neutron and ion irradiation, $\bar{T}$ is very large compared to T_d and the secondary displaced atoms can in turn displace other atoms, resulting in a cascade of displacements. The problem of the defects produced in a cascade can be considered in two ways: whether it is assumed that the distance between adjacent knock atoms is large enough so that the defects produced are only point defects or it is assumed that the energy lost in nuclear collisions in the volume of the cascade is so large that it results in a "spike" which leaves this region in a highly disordered state. In the first case a simple theory, due to Kinchin and Pease (3), gives the number of displaced atoms: $g(T) = T/2\, T_d$ where T is the energy transmitted by the incident particle to the PKA. This formula assumes that each atom transmits half of its energy. If it is assumed that an energy (T_d) is lost for each collision, then (17): $g(T) = (T+T_d)/2\, T_d$. Such formulas suppose elastic collisions. A more general expression has been proposed later by Sigmund (18). In the second case the cascade is described as a thermal spike (19), a displacement spike (20) or a depleted zone (21). In both cases the concentration and the nature of the defects formed in the cascade is difficult to calculate because the process of the showing down of energetic atoms in a lattice, as well as the process of the return to an equilibrium state of a highly excited lattice, are not well known. Only the number of initial collisions can be evaluated. This number can be very high; for instance, for 1 MeV neutron: $T \simeq 4 \times 10^3\, T_d$ in silicon and the number of collisions is $\simeq 10^4$. The cascade will therefore correspond to a highly disordered region surrounded by point defects (such as divacancies (22)).

In semiconductors such disordered zones have been first described by Gossick (23), Crawford and Cleland (21) and Bertolotti et al. (24). It does not seem that the studies which have been performed so far allow to know the nature of these zones. On one hand, using X-ray observations, Gonser and Okkerse (25) concluded that the disordered zones formed in gallium arsenide and indium antimonide by 12 MeV deuteron irradiation are no more crystalline: electron microscopy (26-28) seems to indicate the presence of "amorphous" zones in ion implanted silicon but, even in the case of low temperature implantation, only a small fraction of the disordered zones is amorphous. On the other hand, microscopic measurements, such as electron paramagnetic resonance (EPR) [29] and infrared absorption [22], as well as macroscopic observations, such as X-ray diffraction [30] and electron microscopy [31], tend to indicate that disordered zones formed by fast neutrons in silicon and germanium are still crystalline.

2.6 When the dose of irradiation increases, the concentration of the disordered regions increases until they overlap; a continuous disordered layer is then formed. This overlapping corresponds to a saturation of the amount of disorder (Fig. 3). The saturation is observed whatever is the technique used to monitor the damage (Rutherford scattering, EPR, electron microscopy, Raman scattering, optical absorption, resistivity). A phenomenological description of the formation of these "amorphous" layers has been given by Morehead and Crowder [32] which illustrates their dependence upon ion mass, dose and temperature. What is the dose ϕ_c for which this saturation occurs? In order to make a comparison between different irradiations, we consider the total energy deposited into atomic processes E_t [33]. In silicon, for room temperature implantations, the energy E_t corresponding to the saturation of the disorder is on the order of 10^{24} eV cm^{-3}, i.e. 20 eV per atom (but the exact value seems to be dependent upon the technique of measurements [34]). Since T_d is on the order of 10 to 25 eV this means that, in order to produce a continuous disordered layer, it is necessary to displace each atom.

From the variation of the disorder with the dose of implantation it is possible to get a measure of the lateral spread ΔR_{pp}, i.e. of

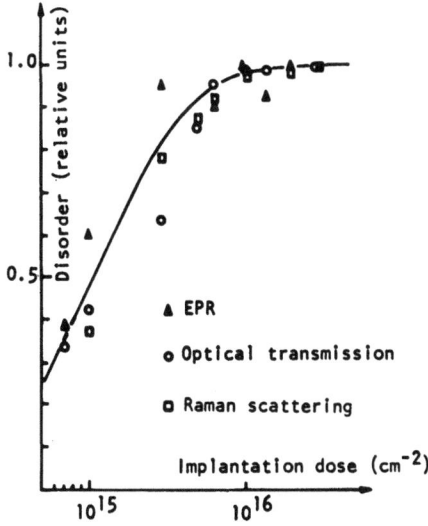

Fig. 3. Disorder vs dose for 70 keV nitrogen ions in diamond

the diameter of the disordered region created by an ion (35). (An approximate value of ΔR_{pp} is $\emptyset_c^{-1/2}$ (32).) One gets values of ΔR_{pp} which are considerably smaller than the values expected from the calculation, even when they are extrapolated towards zero temperature (35); for instance, for 40 keV antimony ions implanted in silicon one measures ΔR_{pp} = 18 Å; the values measured can be even lower, on the order of the interatomic distance (for 200 keV boron ions in silicon or for 70 keV nitrogen ions in diamond). This strongly suggests that only a small part of a disordered region is "amorphous" and that this part depends upon the nature of the implanted ion. Previous measurements of the lateral spread, by junction decoration (36) and backscattering (37), could not distinguish between the lightly and heavily disordered parts of the disordered region.

There are some measurements which indicate that the disordered layer is amorphous when the saturation of the disorder is reached: electron diffraction shows that there is no long range order in the layer (38). Crowder (39) has reviewed some of the physical properties (optical absorption edge, EPR, Raman scattering) of amorphous layers made by implantation in silicon and he concluded that these layers were identical to amorphous layers prepared by evaporation.

III. THE PARAMETERS WHICH INFLUENCE THE DEFECT PRODUCTION

3.1 The parameters which define an irradiation are characteristic of the irradiation itself as well as of the irradiated material. The parameters which characterize the irradiation are: the nature of the energetic particle, its energy, the dose and the intensity of the irradiation; those which characterize the irradiated material are: the nature of the material, the temperature, the type and the concentration of the impurities and of the free carriers, the lifetime of the minority carriers. All these parameters are not independent: thus, during irradiation, the carrier concentration is dependent upon the temperature, the intensity of the irradiation, the doping and the lifetime; the carrier lifetime, function of the concentration and of the nature of the impurities, varies with the irradiation dose (because the defects produced are effective recombination centers). In this section we shall briefly discuss the effects of three of these parameters: the temperature, the intensity of irradiation and the electronic state of the material, since the other parameters which are dependent upon the transmitted energy have been considered in the second section.

3.2 Since defects are stable only below a characteristic temperature, the temperature plays an important role on the nature of the defects created and on their production rate. As illustrated on Fig. 4, the annealing of primary defects results in the appearance of more complex defects, association of the primary defects between themselves or with impurities. For room temperature irradiations, when $\bar{T}$ is on the order of T_d, only mobile interstitials and vacancies are created; since their instantaneous concentration is small compared to the impurity concentration, they associate with impurities, giving rise to A-centers (association with oxygen), E-centers (association with group III and group V impurities), etc. But when $\bar{T}$ is large compared to T_d, the nature of the defects which are created changes because the local concentration of the primary defects is very large compared to the impurity concentration; the defects formed are intrinsic defects (vacancy clusters). In this case the degree of mobility of the primary defects, i.e. the temperature, is an important factor: the higher the temperature, the larger the primary defect mobility and the better the chance that primary defects will form complexes with impurities. This is also illustrated by the effect of the temperature on amorphous layer formation: the critical dose $\emptyset_c$ required to form an amorphous layer depends directly on the mobility of the vacancies since the characteristic dimension of the amorphous region produced by an ion is decreased by the diffusion length of the vacancies; as a result the dose $\emptyset_c$ necessary to produce overlapping of the amorphous regions increases with the temperature [32].

3.3 The influence of the intensity of the irradiation on the defect production has not been systematically studied; it has been observed mainly in the case of ion implantation [40-46]. The intensity of irradiation can have two effects: it varies, during the irradiation, the instantaneous concentrations of the primary defects and of the electron-hole pairs. An increase of the instantaneous concentration of vacancies (or of interstitials) can induce the enhancement of impurity migration; it can also decrease the flux of vacancies which come out of amorphous regions and therefore decrease the critical dose $\emptyset_c$ for amorphous layer formation. In case of electron irradiation the intensity used is not sufficient for such ef-

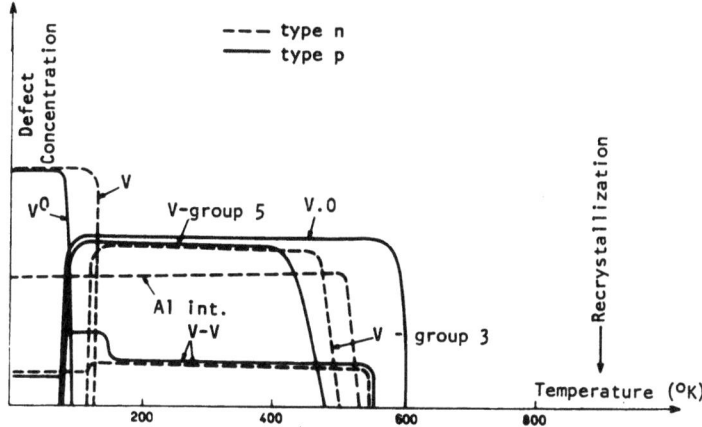

Fig. 4. Recovery stages associated with identified defects in silicon

fects to occur, but the intensity of irradiation can have an effect through the influence of the electron-hole concentration which determines the electronic state of the defects.

3.4 The influence of the charge state of a defect on its behaviour is qualitatively known (47). The fact that the stability and the mobility of a defect is charge state dependent can be easily understood: the interaction energy of a defect with the host atoms depends on the bonding and therefore on the electronic state of the defect.

When the charge state of a metastable defect influences its stability, i.e. when the barrier for annihilation and (or) for formation of a stable defect is charge state dependent, then the concentration of the stable defects is Fermi level dependent. In such case the production rate of the stable defects varies exponentially with the inverse of the temperature. Such behaviour has been observed in germanium and silicon (48,49).

The influence of the charge state of a defect on its mobility can be described phenomenologically in the following way (50): let D_B and D_S be the diffusion coefficients of a defect in two electronic states, B and S; the apparent diffusion coefficient of the defect will be the sum of these diffusion coefficients balanced by the respective occupation of $B(\alpha)$ and $S(1-\alpha)$ states:

$$D_{eff} = \alpha D_B + (1-\alpha) D_S.$$

D_{eff} varies from D_B to D_S when the Fermi level goes through the level associated with the transition $B \rightleftarrows S$. Such a behaviour has been observed in many cases, the variation of the Fermi level being obtained by temperature variation, irradiation, light illumination and carrier injection in a junction (51).

The presence of electron-hole pairs produced by the irradiation can have another effect: it can induce athermal migration of defects and of impurities (i.e. the migration is not thermally activated). The ionization enhanced diffusion (IED) which leads to such athermal migration has been recently reviewed by Bourgoin and Corbett (51). In the next section we shall briefly recall the principle of the IED mechanisms and discuss the indications which can be found for their existence in the case of ion implantation.

IV. ATHERMAL IONIZATION ENHANCED DIFFUSION MECHANISMS

4.1 There are two IED mechanisms which can lead to athermal migration of defects or of impurities. The first mechanism (the energy release mechanism) occurs when an energy is supplied to a defect site. Trapped carriers can release their energy under the form of a cascade of phonons, adding to thermal phonons; a strain energy, such as the Jahn-Teller energy, can also be released. The second mechanism (the Bourgoin mechanism) occurs when defect configuration is so strongly charge state dependent that the equilibrium configuration of the defect in one charge state corresponds to the saddle point configuration of the defect in another charge state; then, as illustrated on Fig. 5a, successive trapping of carriers of opposite sign results in the athermal migration of the defect. In a more general case the Jahn-Teller distortion produced by the change of charge state does not correspond to the saddle point, the symmetry of the defect configuration is lowered and more than one equilibrium configuration are possible in a lattice cell; as illustrated on Fig. 5b the mechanism still applies, but it is then thermally activated.

The phenomenology of these mechanisms has been developed (50); the diffusion coefficient is: $D_{eff} = a^2/4\tau$ where a is the lattice parameter and $1/\tau$ is the frequency of change in charge state (in the case of the Bourgoin mechanism) or the number of jumps experienced by the defect (in the case of the energy release mechanism). The existence of the Bourgoin mechanism has been justified to some extent theoretically for interstitials using classical (52) and LCAO-MO treatments (53). The mechanism has been extended later to vacancy and vacancy-type defects (4); in the case of the vacancy it is argued that the addition of an electron on a negative vacancy, localized on this vacancy, will result in such electron-electron interaction that the doubly negative vacancy will assume the split-vacancy configuration (which is the saddle point for vacancy migration).

4.2 The applicability of these mechanisms in the case of electron, γ and X-ray irradiations has already been discussed (51). In particular it has been shown that the IED mechanisms provide possible explanations for the interstitial mobility at 4 °K in silicon and germanium due to electron irradiation or light illumination, for the behaviour of the vacancy and of E-centers in silicon. Are there additional indications for the existence of these IED mechanisms in the case of ion implantation? Picraux and Vook (40,41) suggested that the ionization which accompanies an implantation induces the annealing of defects within clusters in a similar way as for electron created defects. They argued that ionization stimulation of the annealing of the disorder is responsible for the differences observed between lattice reordering during implantation at a given temperature and post-implantation lattice reordering at that temperature. Norris et al. (54) studied the behaviour of aluminum interstitials produced by ion implantation in silicon under ionization; because they did not observe any change in the concentration of these interstitials, they concluded that no substantial migration of silicon interstitials was taking place. But, even if isolated interstitials exist after implantation, their migration can have been prevented by the large concentration of the defects present after implantation. Bourgoin et al. (55) proposed that the tail, deviation from the Gaussian shape of an ion profile, could be due to the migration,

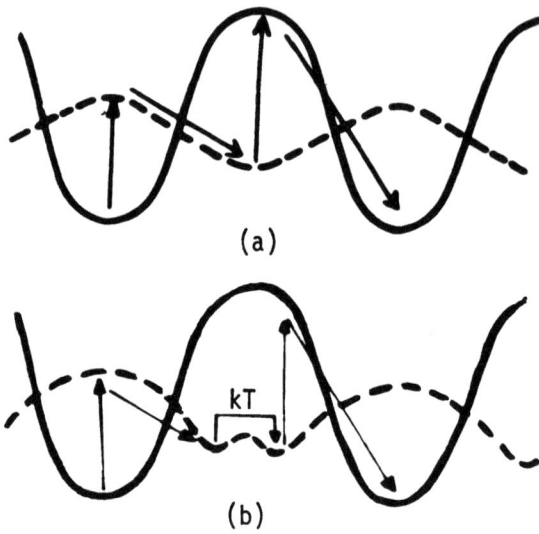

Fig. 5. Potential energy vs distance in the lattice for a defect whose equilibrium configuration is charge state dependent (b). In case a, when the equilibrium position for a given charge state corresponds to the saddle point in another charge state, the IED mechanism is athermal.

through an athermal IED mechanism, of some of the implanted ions in the unperturbed region below the implanted layer. But, according to Blood et al. (56), it seems now more likely that this tail is due to a rechanneling process although such process cannot explain the fact that the slope of the tail varies with the dose and the intensity of the implantation (57)-which an IED mechanism could explain. Other anomalous diffusions of defects could be related to IED mechanisms; for instance, post-implantation in boron implanted layer can produce, in some cases, an enhanced annealing stronger than expected by a defect enhanced diffusion mechanism (58). Other observations, related to the effect of the intensity of the implantation on damage production could also be interpreted in terms of IED mechanisms (41). Changes in the nature of defects with increasing intensity of irradiation (44,59) can be accounted for by the variation of the instantaneous defect concentration, but also by the variation of the instantaneous electron-hole pair concentration. More work needs to be done before any conclusion can be reached.

4.3 IED mechanism could also possibly play a role in the formation of amorphous layers, during which damage creation and annealing compete. We argue that, in addition to the recrystallization of amorphous material which can occur in the region of a thermal spike, crystallization can also be caused by an ionization process. We propose that photocrystallization, which has been observed in amorphous compounds such as As_2S_3 (60), could also take place in amorphous regions created by implantation because of the ionization present during the irradiation. IED mechanisms provide a way through which photocrystallization could take place: namely, vacancies which exist at the interface between crystalline and amorphous regions could diffuse by an ionization mechanism towards the amorphous region when the crystal phase is more stable than the amorphous phase. The fact that photocrystallization is not temperature independent implies that an intermediate metastable configuration such as illustrated on Fig. 5b is involved (such a metastable configuration is observed at low temperature in As_2S_3 (61)). Some indications that crystallization processes are occurring because of implantation can indeed be found, for instance in amorphous germanium (62). Parsons and Balluffi argue that the crystallization caused by ion implantation is due to local heating in the displacement spikes. But such explanation does not agree with the fact that epitaxial growth of silicon can be assisted by ion implantation (63). We favor an explanation base on an IED mechanism which can also account for photocrystallization and crystallisation under electron irradiation (64,65).

REFERENCES

1. The calculation of T can be found in references 2-4.

2. F. Seitz and J.S. Koehler, Solid State Phys. 2, 305 (1956).

3. C.H. Kinchin and R.S. Pease, Report Prog. Phys. 18, 1 (1955).

4. J.W. Corbett and J.C. Bourgoin, Point Defects in Solids, ed. J.W. Crawford and L.M. Slifkin, Plenum Press, New York (1974),vol. 2, chap. 1.

5. N.F. Mott, Proc. R. Soc. A 124, 426 (1929) and A 135, 429 (1932).

6. W.A. McKinley,Jr. and H. Fesbach, Phys. Rev. 74, 1759 (1932).

7. W. Kohn, Phys. Rev. 94 A, 1409 (1954).

8. R. Bäuerlein, Z. Naturforsch. a 14, 1069 (1959) and Z. Phys. 176, 498 (1963).

9. J.W. Corbett, J.C. Bourgoin and C. Weigel, Radiation Damage and Defects in Semiconductors, The Institute of Physics, London (1973), p. 1.

10. G.D. Watkins, J. Phys. Soc. Japan 18, Supp. II, 2 (1963).

11. G.D. Watkins, Effets des Rayonnements sur les Dispositifs à Semiconducteurs, ed. F. Cambou, Journées d'Electronique, Toulouse (1968), p. A 1.

12. G.D. Watkins, Radiation Effects in Semiconductors, ed. J.W. Corbett and G.D. Watkins, Gordon and Breach, New York (1971), p. 301.

13. J. Bourgoin and F. Mollot, Phys. Stat. Sol. 43, 343 (1971).

14. F.L. Vook and H.J. Stein, Radiation Effects in Semiconductors, ed. F.L. Vook, Plenum Press, New York (1968), p. 99.

15. G.D. Watkins and J.W. Corbett, Phys. Rev. 138, A 543 (1965).

16. K.L. Brower, in Ref. 12, p. 189.

17. W.S. Snyder and J. Newfeld, Phys. Rev. 99, 1636 and 1326 (1955).

18. P. Sigmund, Appl. Phys. Letters 14, 114 (1969).

19. F. Seitz, Disc. Faraday Soc. 5, 271 (1949).

20. J.A. Brinkman, J. Appl. Phys. 25, 961 (1954).

21. J.H. Crawford, Jr. and J.W. Cleland, J. Appl. Phys. 30, 1204 (1959).

22. H.J. Stein, F.L. Vook, D.K. Brice, J.A. Borders and S.T. Picraux, Rad. Eff. 6, 19 (1970).

23. B.R. Gossick, J. Appl. Phys. 30, 1214 (1959).

24. M. Bertolotti, T. Papa, D. Sette, V. Grasso and G. Vitali, Nuovo Cimento 29, 1200 (1963).

25. U. Gonser and B. Okkerse, Phys. Rev. 105, 737 (1957) and 109, 663 (1958).

26. J.R. Parsons, Phil. Mag. 12, 1159 (1965).

27. J.A. Davies, J. Denhartog, L. Erikson and J.W. Mayer, Can. J. Phys. 45, 4053 (1967).

28. J.D. Mazey, R.S. Nelson and R.S. Barnes, Phil. Mag. 17, 1145 (1968).

29. D.F. Daly and H.E. Noffke, in Ref. 12, p. 179.

30. M.C. Wittels, J. Appl. Phys. <u>28</u>, 921 (1957) and <u>40</u>, 2909 (1969).

31. M.L. Swanson, J.R. Parsons and C.W. Hoelke, in Ref. 12, p. 359.

32. F. Morehead and B.L. Crowder, Rad. Effects <u>6</u>, 27 (1970).

33. F.L. Vook, in Ref. 9, p. 60.

34. J.C. Bourgoin, J.F. Morhange and R. Beserman, Rad. Eff., to be published.

35. J.L. Combasson and J.C. Bourgoin, to be published.

36. Y. Akasaka and K. Horie, Appl. Phys. Letters <u>21</u>, 128 (1972).

37. F. Furukawa and H. Matsumura, *Ion Implantation in Semiconductors and Other Materials*, ed. B.L. Crowder, Plenum Press, New York (1973), p. 193.

38. L.N. Large and R.W. Bicknell, J. Math. Sci. <u>2</u>, 589 (1967).

39. B.L. Crowder, *Ion Implantation in Semiconductors*, ed. S. Namba, Japan Soc. for the Promotion of Science (1972), p. 63.

40. S.T. Picraux and F.L. Vook, *Ion Implantation in Semiconductors*, ed. I. Ruge and J. Graul, Springer-Verlag, Berlin (1972), p. 1.

41. S.T. Picraux and F.L. Vook, Rad. Effects <u>11</u>, 179 (1971).

42. F.L. Vook and H.J. Stein, Rad. Effects <u>2</u>, 23 (1969).

43. N.G. Blamires, in Ref. 40, p. 119.

44. F.H. Eisen and B. Welch, European Conf. on Ion Implantation, The Institute of Physics, Peter Peregrinus, Stevenage (1970), p. 227.

45. F.H. Eisen and B. Welch, Rad. Effects <u>7</u>, 143 (1971).

46. J.S. Harris, in Ref. 40, p. 157.

47. J.W. Corbett and J.C. Bourgoin, IEEE Trans. Nucl. Sci. <u>NS-18</u>, 11 (1971).

48. R.E. Whan, in Ref. 14, p. 195.

49. F.L. Vook and H.J. Stein, in Ref. 14, p. 99.

50. J.C. Bourgoin, J.W. Corbett and H.L. Frisch, J. Chem. Phys. <u>59</u>, 4042 (1973).

51. J.C. Bourgoin and J.W. Corbett, Proc. Int. Conf. on Lattice Defects in Semiconductors, Freiburg (1974).

52. J.C. Bourgoin and J.W. Corbett, Phys. Letters $\underline{38\ A}$, 135 (1972).

53. C. Weigel, D. Peak, J.W. Corbett, G.D. Watkins and R.P. Messmer, Phys. Rev. $\underline{B\ 8}$, 2906 (1973).

54. C.B. Norris, K.L. Brower and F.L. Vook, Rad. Effects $\underline{18}$, 1 (1973).

55. J. Bourgoin, D. Peak and J.W. Corbett, J. Appl. Phys. $\underline{44}$, 3022 (1973).

56. P. Blood, G. Dearnaley and M.A. Wilkins, in Ref. 51.

57. W.A. Grant and J.N. Baruah, Rad. Effects $\underline{14}$, 261 (1972).

58. L.O. Bauer, Appl. Phys. Letters $\underline{20}$, 107 (1972).

59. N.G. Blamires, in Ref. 40, p. 119.

60. E. Finkman, A.P. DeFonzo and J. Tauc, Proc. 12th Int. Conf. on the Physics of Semiconductors, Stuttgart (1974).

61. J. Cernogora, F. Mollot and C. Benoit à la Guillaume, in Ref. 60.

62. J.R. Parsons and R.W. Balluffi, J. Phys. Chem. Solids $\underline{25}$, 263 (1964).

63. T. Itoh and T. Nakamura, Rad. Effects $\underline{9}$, 1 (1971).

64. N. Baltateanu and I. Spinulescu, Rad. Effects $\underline{20}$, 149 (1973).

65. I. Spinulescu and M. Baltateanu, Rad. Effects $\underline{14}$, 257 (1972).

TRANSPORT OF ION DEPOSITED ENERGY

BY RECOILING TARGET ATOMS*

David K. Brice

Sandia Laboratories

Albuquerque, New Mexico 87115

ABSTRACT

A previous method for directly calculating the spatial distribution of energy deposition into damage or ionization for ions implanted into solid targets is extended to account for the energy transport by recoiling target atoms. The new calculations extend the applicability of the method to lower incident ion energies. Good agreement is obtained between experiment and theory using the improved procedure.

INTRODUCTION

The kinetic energy, E, of an ion incident on a solid target becomes partitioned into the kinetic energy of recoiling target atoms and into ionization and excitation of the electrons of the target as the ion slows to a stop. The partitioning of energy into atomic and electronic processes also occurs for the recoil products so that finally the total incident kinetic energy may be regarded as partitioned into atomic processes (damage) and electronic processes. In a previous publication [1] a method was presented for calculating directly the initial depth distribution of the energy deposited into these two categories. For energy deposited into atomic processes the method basically involves evaluating the integral

---

*This work was supported by the U.S. Atomic Energy Commission.

$$Q(E,x) = \int_0^E P(E,E',x)\, \Sigma(E')\, (d\bar{R}/dE')\, dE', \tag{1}$$

where

$$\Sigma(E') = N \int q(T)\, d\sigma(E',T). \tag{2}$$

In these equations $P(E,E',x)$ is the depth distribution of the ions when the incident energy has been reduced from E to E', $d\bar{R}$ is the average distance traveled by an ion of energy E' while losing energy dE', N is the atomic density of target atoms, $d\sigma(E',T)$ is the differential atomic cross section for the transfer of energy T (In the interval T to T + dT) to a recoiling target atom by an ion of energy E', and $q(T)$ is that portion of T which will ultimately be deposited into atomic processes. (The quantity $q(T)$ is conventionally referred to as the damage energy of the recoiling target atom.) The initial depth distribution of all energy deposited into atomic processes is then given by $Q(E,x)$. Similar equations apply to energy deposition into electronic processes.

The quantity $Q(E,x)$ may also be identified as the _final_ depth distribution of energy deposition into atomic processes, providing there is no significant energy transport by the recoiling target atoms. This approximation is valid [1-3] for incident ion energies above a lower (ion-target dependent) energy limit. Because of this limit energy deposition distributions for many ion-target combinations at energies of interest in the ion implantation field cannot be obtained from Eq. (1). It is therefore of interest to extend the technique to lower energies by taking into account energy transport by recoiling target atoms. The purpose of this paper is to describe a procedure for accomplishing this.

THEORY

Consider an ion with energy E', traveling parallel to the x-axis, and located at a depth x in the target. The ion transfers damage energy into the lattice at a rate $\Sigma(E')$. The energy does not ramain at x, but becomes distributed in depth. Let $D(E',x'-x)$ be the final _normalized_ distribution in depth x' of the energy initially deposited at x. For a distribution, $P(E,E',x)$, of ions having energy E' the relative depth distribution of damage energy is then given by $S(E,E',x')$, where

$$S(E,E',x') = \int_{-\infty}^{\infty} P(E,E',x)\, D(E',x'-x)\, dx. \tag{3}$$

The final depth distribution of <u>all</u> energy deposited into atomic processes is then obtained from

$$Q(E,x) = \int_0^E S(E,E',x) \, \Sigma(E') \, (d\bar{R}/dE') \, dE' \ . \tag{4}$$

An approximation included in Eqs. (3) and (4) is the assumption that all ions travel parallel to the x-axis. This approximation is quite good for heavy incident ions for which the lower energy limit of the earlier technique [1] was a particularly severe restriction. For light incident ions the recoil effects tend to be small at incident energies of interest, except in the immediate vicinity of the target surface where the approximation again is quite good. As before, however, the limits of applicability of the new procedure will be evaluated by comparing the distributions with experiment and by comparing moments, M_{nQ}, of the calculated energy deposition distributions, with exact moments obtained from solutions to the transport equations [1-3].

To evaluate Eq. (4) the function $S(E,E',x)$ is required. Since $\Sigma(E')$ is a slowly varying function of E', and S is a relatively narrow function of x, compared with Q, the details of the precise shape of S will be lost in the integration. Thus, an exact representation of S is not necessary for an accurate determination of Q. For this reason, the first few spatial moments of S will be used to construct an approximate representation and the constructed distribution will be used in evaluating Eq. (4). The same procedure was used in the earlier [1] evaluation of Eq. (1).

For a distribution $F(x)$ the spatial moments, M_{nF} are defined as

$$M_{nF} = \int_{-\infty}^{\infty} x^n F(x) dx \ . \tag{5}$$

For the moments of the distributions of interest above replace F by P, D, S, or Q. Combining Eqs. (3) and (5) one obtains

$$M_{nS}(E,E') = \sum_{k=0}^{n} C_{nk} M_{kP}(E,E') M_{n-kD}(E') \ , \tag{6}$$

where the C_{nk} are binomial coefficients. The moments M_{nS} thus depend upon the moments of the ion depth distribution, M_{nP}, and the moments of the recoil atom damage distribution, M_{nD}. Values for the M_{nP} can be obtained by a method which is described elsewhere [4], and the M_{nD} are evaluated by an average over the interaction cross section $d\sigma(E',T)$ through

$$M_{nD}(E') = \int M'_{nQ}(T,\eta) \, q(T) \, d\sigma(E',T) / \int q(T) \, d\sigma(E',T) \qquad (7)$$

In Eq. (8) $M'_{nQ}(T,\eta)$ is the contribution to $M_{nD}(E')$ by a target atom recoil of energy T, whose initial velocity has direction cosine η with respect to the x-axis. For purposes of evaluating Eq. (7) in the present paper the damage energy, $q(T)$, is assumed to be uniformly distributed along a straight line from the initial location of the recoil atom in the target to the position at which it ultimately comes to rest. The moments M'_{nQ} are then simple functions of the moments of the recoil atom range distribution which can be evaluated by previously described techniques [4].

DISCUSSION OF RESULTS

Figure 1 shows a comparison of the depth distributions of energy into atomic processes obtained from Eqs. (1) and (4) for 100 keV boron ions incident on a silicon target. The distribution resulting from Eq. (1) is shown as the dashed curve while the solution from Eq. (4) is shown as a solid curve. Moments through second order (n = 2) were used in evaluating Eqs. (6) and (7) and $S(E,E',x)$ was assumed gaussian. For this ion-target combination it was earlier

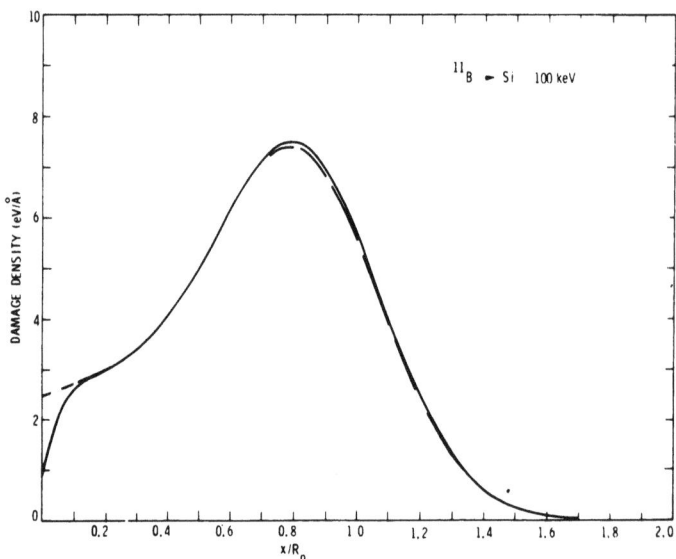

Fig. 1 Damage density versus depth for 100 keV boron ions incident on a silicon target. Recoil energy transport is included in the solid curve, neglected in the dashed curve. Depth scale normalized by the average projected range of the incident ions, R_p = 2890 Å.

estimated [1-3] that Eq. (1) should give valid distributions for ion energies down to ∼ 5-10 keV, and it is seen from this comparison that the inclusion of energy transport by target atom recoils has little effect on $Q(E,x)$ except in the immediate vicinity of the surface of the target, where a significant reduction in energy deposition is noted.

Figure 2 shows a similar comparison for 100 keV antimony ions incident on silicon. The inclusion of recoil energy transport clearly causes a significant change in the distribution of energy deposition for this case over its entire depth, since for antimony ions incident on silicon Eq. (1) results in accurate distributions only for ion energies above ∼ 5000 keV [1-3]. Also shown for comparison in Fig. 2 are the experimental results of Bøgh, et. al. [5] (filled circles). Their experimental depth scale was obtained [5] by analysis of the channeling-backscattering experiment using a channeling stopping power which was 0.5 times the random stopping power. It is now known [6] that a larger value for the channeling stopping power should be used in the analysis of such experiments. A corrected analysis moves the experimental points nearer the surface by 10-15%, as indicated by the bars to the left of the circles, and brings the experimental points almost into coincidence with

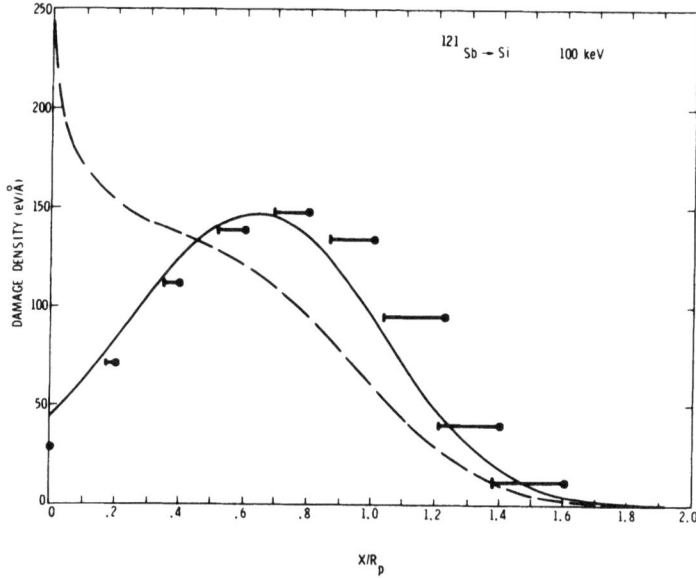

Fig. 2 Damage density versus depth for 100 keV antimony ions incident on a silicon target. Recoil energy transport is included in the solid curve, neglected in the dashed curve. Experimental points are from Bøgh, et. al. [5]. Depth scale normalized by the average projected range of the incident ions, R_p = 430 Å (theoretical), = 500 Å (experimental).

the distribution obtained as a solution to Eq. (4). It thus appears that energy transport by target atom recoils is adequately accounted for by Eq. (4) using a gaussian representation for $S(E,E',x)$, and the approximate moments M'_{nQ} described above.

In order to estimate the energy range of applicability of the present procedure for including the energy redistribution by recoils, the approximate moments M_{nQ} of energy deposition distributions calculated from Eq. (4) have been compared with exact moments determined from the transport equations [3,7]. The results of this comparison, which will be presented in more detail in a later publication, indicate that for <u>all</u> ion-target combinations the distributions Q are accurately given ($\sim 5\%$) by solutions to Eq. (4) for incident ion energies down to $\epsilon = 0.01$, where ϵ is the incident ion energy expressed in Lindhard's dimensionless units [8]. When recoil energy transport is <u>not</u> included the lower energy limit [1-3] is $\sim \epsilon = 1$ for light incident ions and $\sim \epsilon = 10$ for heavy incident ions. Thus the present method of accounting for the transport of damage energy by recoil target atoms extends to much lower values the range of energies for which energy deposition distributions can be calculated by the direct method.

REFERENCES

1. D. K. Brice, Rad. Effects <u>6</u>, 77 (1970).

2. P. Sigmund, M. T. Matthies, and D. L. Phillips, Rad. Effects <u>11</u>, 39 (1971).

3. K. B. Winterbon, Proceedings of the Fifth International Conf. on Particle Solid Interactions, Gatlinburg, Tenn., Sept. 23-28, 1973 (to be published).

4. D. K. Brice, Rad. Effects <u>11</u>, 227 (1971).

5. E. Bogh, P. Hogild, and I. Stensgaard, Proceedings of the Symposium on Radiation Damage in Reactor Materials, <u>77</u>, Vienna, Austria (1969).

6. J. Bøttiger and F. H. Eisen, Thin Solid Films <u>19</u>, 239 (1973).

7. K. B. Winterbon, Rad. Effects <u>13</u>, 215 (1972).

8. J. Lindhard, M. Scharff, and H. E. Schiøtt, Kgl. Danske Videnskab. Selskab, Mat.-Fys. Medd. <u>33</u>, No. 14 (1963).

ION IMPLANTATION THROUGH SURFACE LAYERS:

A TRUNCATED GAUSSIAN MODEL

A. V. S. Satya and H. R. Palanki

IBM System Products Division

Hopewell Junction, New York 12533 U.S.A.

ABSTRACT

An analysis of the effects of surface layers, used to obtain modified ion-implanted profiles, is presented as an extension of the LSS-approach. The results obtained from the APL analysis, which combines a truncated Gaussian model with a simple iteration scheme, are in substantial agreement with Furukawa and Ishiwara's quasi-Monte Carlo results as well as with the published experimental data. These findings are contrary to the previously published opinion that the LSS approach is not applicable to multi-layer implantation problems.

INTRODUCTION

The implanted profiles in single-material substrates can be predicted through Lindhard-Scharff-Schiott (LSS)[1,2] theory, or from Monte Carlo simulation. The interest in predicting the implanted profiles in multi-layered structures is evident from recent publications and the success of the ion-implantation technique as a viable tool for unique micro-electronic device applications. Apart from the traditional surface layers, such as the passivation (thin) and the masking (thick) surface layers, used in the semiconductor processing with ion implantation, "partial mask" type of surface layers can be gainfully employed to obtain simultaneous multiple profiles and in other unique applications.

The first attempt to predict the implanted profiles penetrating the surface layer is by Furukawa and Ishiwara[3] using quasi-Monte Carlo simulation technique. It is so far considered impossible to obtain such distributions by applying the LSS theory. In this paper a truncated Gaussian model is presented for implantation into multi-layered materials as an extension of the LSS theory.

TRUNCATED GAUSSIAN MODEL

For an implantation with energy E_0 and dose D, through a partial-mask layer of thickness t (up to $t = R_p \pm 3\times\sigma$), the profile trapped in the surface layer is given by:

$$N(x)_{SL} = \frac{D}{\sigma_{SL}\sqrt{2\Pi}} \exp\left[-\{x-R_{pSL}\}^2 \div 2\sigma_{SL}^2\right] \qquad \ldots 1$$

$$:-\infty \leq x \leq t$$

If we assume that this trapped profile is not influenced by the presence of the second layer, then the balance dose of

$$D - A = D - \int_{-\infty}^{t} N(x)_{SL}\, dx \qquad \ldots 2$$

should penetrate into the substrate.

If these particles also suffer random collisions in the substrate material, then the penetrating profile is a truncated Gaussian in form, resulting from the same energy distribution at the interface, which is also obtainable with an equivalent energy E^* when the surface layer is absent. The penetrating profile can now be:

$$N(x)^* = \frac{D}{\sigma^*\sqrt{2\Pi}} \exp\left[-(x-R_p^*)^2 \div 2\sigma^{*2}\right] \qquad \ldots 3$$

$$: t \leq x \leq \infty$$

with (R_p^*, σ^*) corresponding to E^*, as determined by iterating on energy from E_0, $E_0 - \Delta E$, $E_0 - 2.\Delta E$, ... to E^*, till the equation:

$$\frac{D - A}{D} = \int_t^\infty \frac{1}{\sigma_{sub}\sqrt{2\Pi}} \exp\left[-(x-R_{p\,sub})^2 \div 2\sigma_{sub}^2\right] dx \qquad \ldots 4$$

is satisfied. The implanted profiles in multiple surface layers can also be predicted by successively extending the same procedure.

The truncated Gaussian model predicts that there exists a discontinuity in concentration at the surface-layer-substrate interface, which depends on t, E_0, D, and the stopping power of the materials. If the R_p^* is within the LSS-bounds of R_{pSL} and R_{pSUB}, then the peak concentration of the penetrating

ION IMPLANTATION THROUGH SURFACE LAYERS

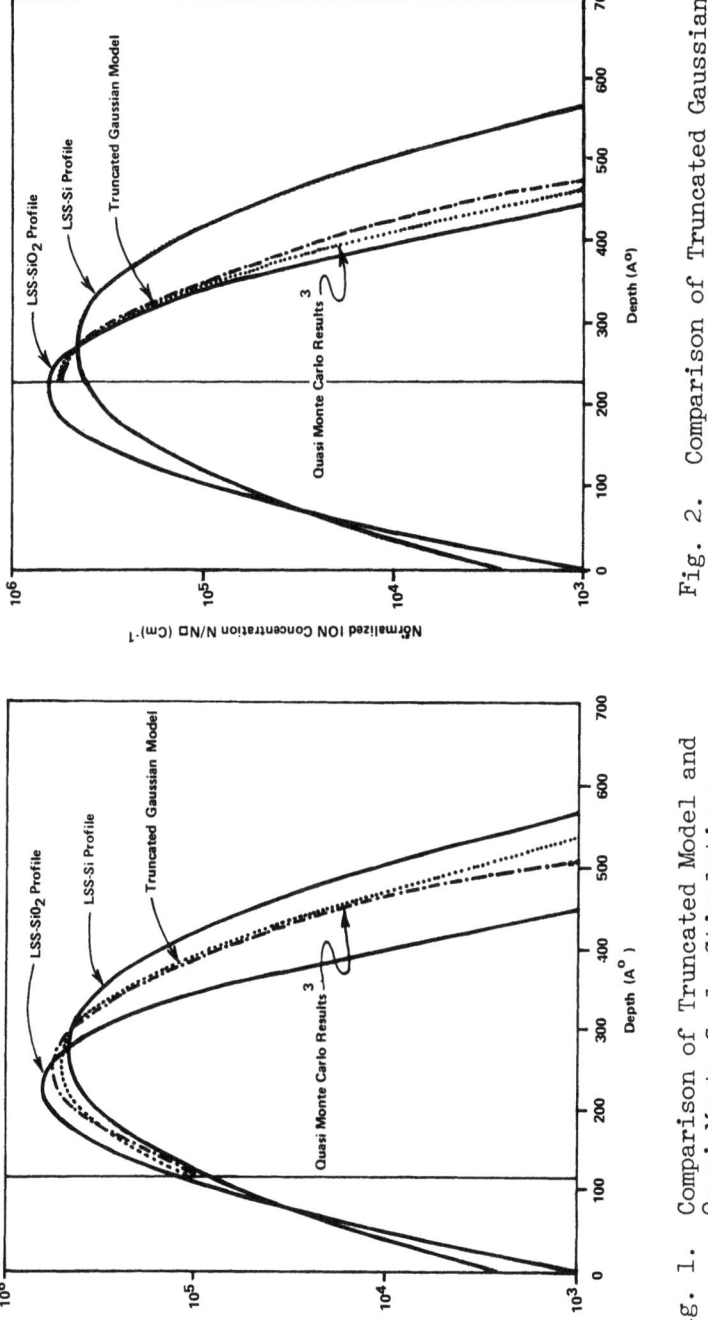

Fig. 1. Comparison of Truncated Model and Quasi Monte Carlo Stimulations.

Fig. 2. Comparison of Truncated Gaussian Model and Quasi Monte Carlo Stimulations

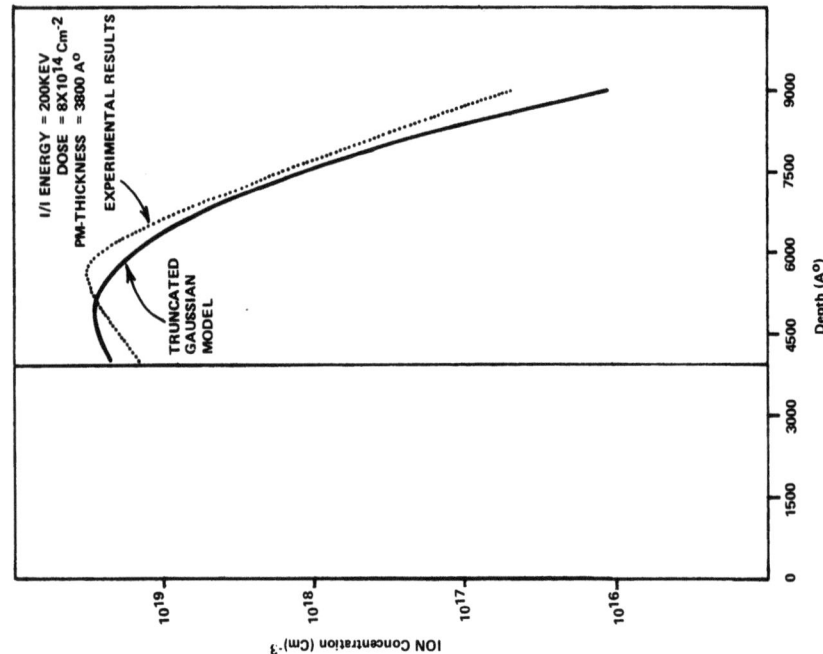

Fig. 4. Ion Implantation of B^+ into Si Through SiO_2 Partial Mask

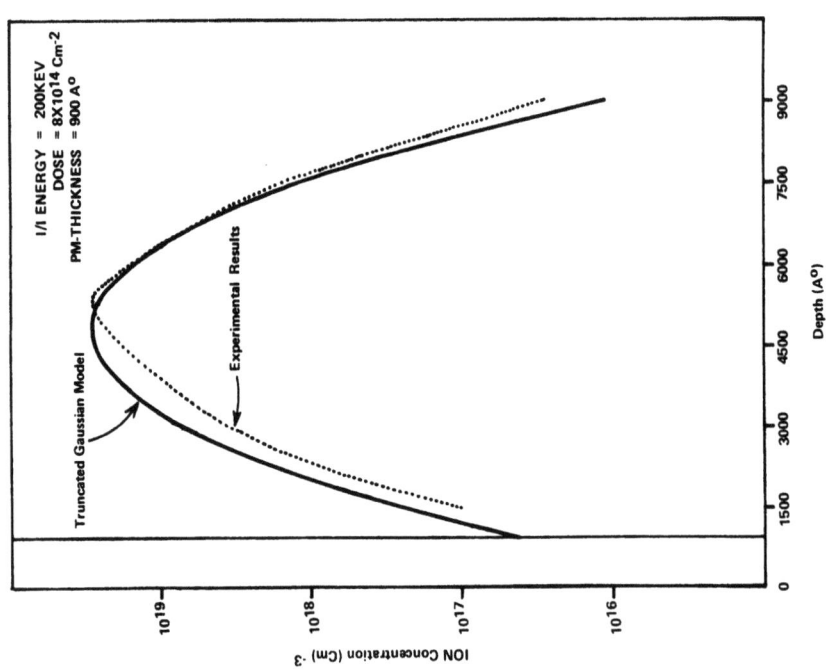

Fig. 3. Ion Implantation of B^+ into Si Through SiO_2 Partial Mask

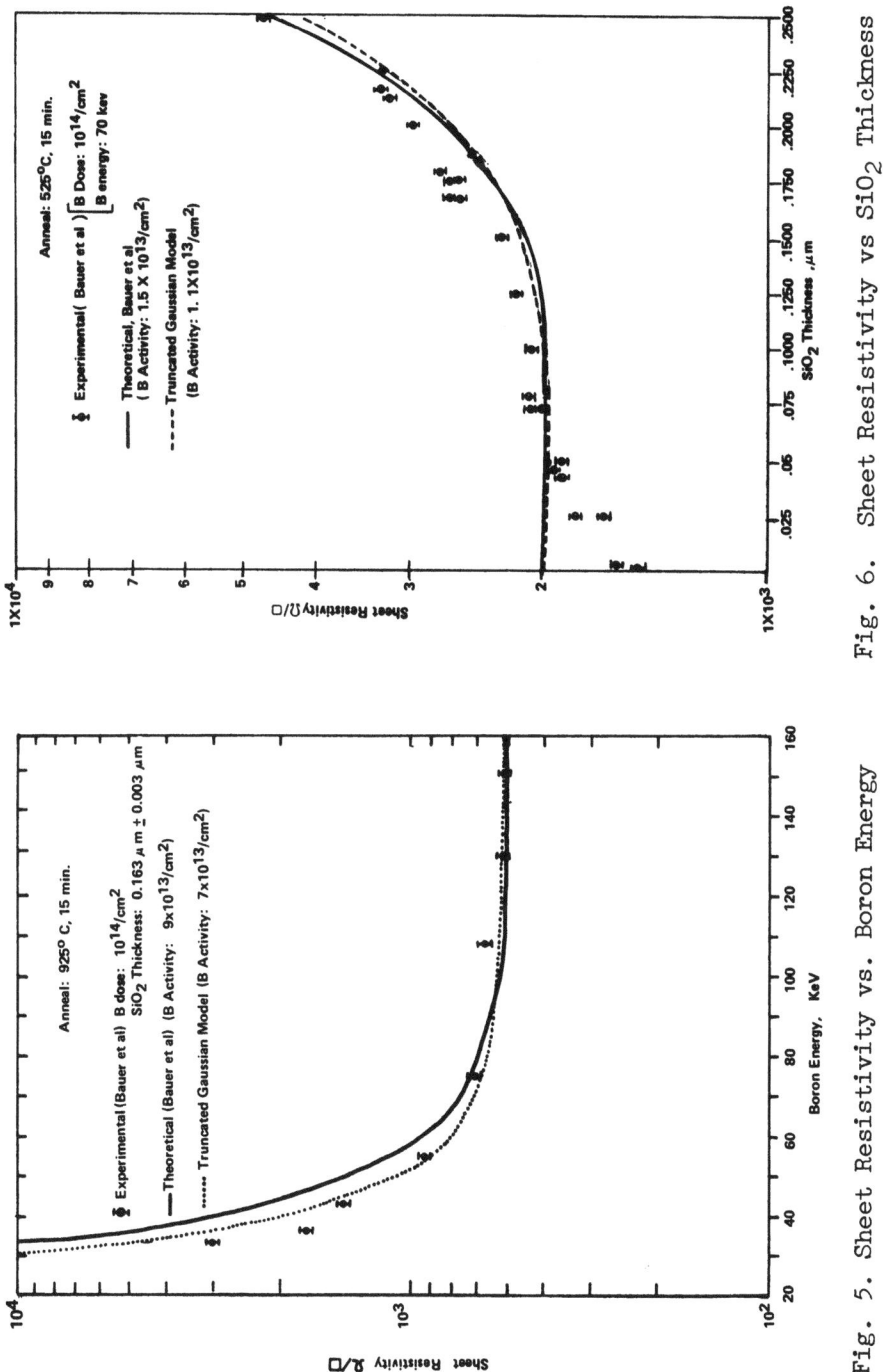

Fig. 5. Sheet Resistivity vs. Boron Energy

Fig. 6. Sheet Resistivity vs SiO_2 Thickness

profile occurs away from the interface. In addition, if the partial-mask material has a higher stopping power for the ions than the substrate has, then the peak concentration of the penetrating profile will be higher than that for an implantation without the partial mask for the same implantation conditions; and the peak concentration will be less than the LSS-peak-concentration when the partial-mask stopping-power is less than that of the substrate. These features of the penetrating profile are also found by Furukawa and Ishiwara in their quasi-Monte Carlo simulation.

EXPERIMENTAL RESULTS

The truncated Gaussian model serves as a simple extension of the LSS theory, from the LSS (or experimental) Projected Range Statistics in terms of the Gaussian parameters R_p and σ as a function of energy for the surface- and the substrate- materials, without considering the detailed energy distributions as in the quasi-Monte Carlo simulation. Figures 1 and 2 compare the truncated Gaussian model predictions with the simulation results by Furukawa and Ishiwara. The predicted results from the two techniques are in fair agreement for these two examples.

To examine the most common application in semiconductor processing with ion implantation, Boron implantation at 200 KeV into silicon substrates with 900 and 3800Å thick oxide surface layers was considered for experimental verification. Figures 3 and 4 compare the predicted penetrating profiles with those obtained by Ion Probe Analysis. The truncated Gaussian model is further extended to include the sheet resistivity calculation as by Bauer et al.[4], for verification against their published results. Figures 5 and 6 show that the agreement is fair between our prediction and the experimental data from Bauer et al.

CONCLUSIONS

There is a 20 to 30% disagreement between the experimentally determined profiles and projected range statistics from different analysis techniques[5]. As and when established experimental data on the projected range statistics are available, the Gaussian approximation used in the LSS-theory and the truncated Gaussian model can be improved to predict implanted profiles in multi-layered structures more accurately by considering higher moments in the range distributions.

The truncated Gaussian model presented here is a first approximation to treat the multi-layer problem as an extension of the LSS-theory and further experimental evaluations need be performed to verify and improve this model.

ACKNOWLEDGMENTS

The authors gratefully acknowledge the experimental support received from Messrs. J. Winnard and F. Anderson's groups, in terms of ion implantation and ion-probe analyses. It is a pleasure to thank Drs. H. N. Ghosh, W. S. Johnson, H. F. Quinn, E. Valsamakis, and B. L. Crowder for the many valuable suggestions and discussions.

REFERENCES

1. J. Lindhard, M. Scharff and H. E. Schiott, Kgl. Danske Videnskab., Mat. Fys. Medd. Dan. Vid. Selsk., $\underline{33}$, #14, 1963.

2. W. S. Johnson and J. F. Gibbons, "Projected Range Statistics in Semiconductors," Stanford University Book Store, 1969.

3. S. Furukawa and H. Ishiwara, J. Appl. Phys., $\underline{43}$, #3, 1972.

4. L. O. Bauer, M. R. Macpherson, A. T. Robinson, and H. G. Dill, Solid State Electronics, $\underline{16}$, #3A, 1973.

5. International Symposium on Ion Implementation, Yorktown Heights, Dec. 11 through 14, 1972; unresolved problem of Rump Session.

THE EFFECTS OF NON-GAUSSIAN RANGE STATISTICS ON ENERGY DEPOSITION PROFILES

S. W. Mylroie and J. F. Gibbons

Stanford Electronics Laboratories, Stanford University

Stanford, California 94305   USA

ABSTRACT

The technique for calculating low order moments of the projected range distribution previously described by these authors [1] has been extended to allow calculation of intermediate range statistics. The results obtained show that at intermediate energies the skewing can be much larger than observed in the final range distribution. The effect of this skewing on energy deposited in atomic processes is computed by the two step method of Brice [2], from which it is found that computations including the skewing provide a substantially better fit for experimental data.

INTRODUCTION

A problem of considerable interest for applications of ion implantation is that of computing the depth distribution of energy deposited into atomic processes as an implanted ion comes to rest in a solid. Two independent techniques have been used to treat this problem, one due to Lindhard et al. [3]; and one due to Brice [2]. In the technique pioneered by Lindhard and subsequently refined by Sigmund and Sanders [4] and Winterbon [5], an integral equation for the probability of energy deposition is formulated and then the statistical features of the solution are obtained by finding the moments of the probability function. The procedure is fundamentally similar to that used to obtain the statistical features of the projected range distribution function; however, an even larger number of moments is required to construct the energy deposition profile with reasonable accuracy, and because of computational difficulties, this technique is effectively restricted to rather low values of the initial energy (e.g., $E_0 \lesssim 100$ keV for B in Si if Thomas-Fermi stopping powers are used).

To circumvent the computational problems that arise in the method described above, Brice [2] and Gibbons [6] have proposed what are called two-step methods for calculating the depth distribution of deposited energy. In these methods, one first calculates the range distributions of the ion as a function of energy, after which an integration over energy is performed across the distributions at a fixed depth to determine the energy loss rate at that depth. The integrand for this calculation is the probability that a particle is in an interval ΔX about X with energy in an interval ΔE about E, multiplied by a weighting factor which accounts for the (energy-dependent) rate at which that particle loses energy to atomic processes.

In its simplest form, the two-step method neglects the transport of energy by knock-ons and is therefore only applicable for light ion implantation into heavy substrates. In the work reported here this restriction is still operative, though recent calculations by Brice [7] effectively remove this restriction; and with this new correction the method described here should give satisfactory results for most ion-substrate combinations of interest for ion implantation. In effect, our work removes the restriction that the intermediate range distribution functions be Gaussian, while that of Brice [7] improves the computation of the weighting factor to include energy transport by knock-ons.

COMPUTATIONAL STRATEGY

As mentioned above, the two-step method for computing energy deposition begins with a calculation of the parameters that are required to construct the range distribution function at any intermediate energy $\mathcal{E}(0 \leq \mathcal{E} \leq E_0)$. The basic equation for calculating these parameters is obtained from a generalization of of the basic LSS integral equation

$$1 = \int_0^{\gamma E} \{p(r,\theta,E) - p(r,\theta,E-T)\} \, d\sigma(T) \ . \tag{1}$$

This equation describes the final probability distribution of ions injected at initial energy E into a target. γ is the mass ratio, defined by $\gamma = 4M_I M_T/(M_I+M_T)^2$ where M_I = ion mass and M_T = target mass.

Brice [6] has shown that this same equation with minor modification will also give the spatial probability distribution of implanted ions at intermediate energies $\mathscr{E}$, where $0 < \mathscr{E} < E_0$. The appropriate equation is

$$1 = \int_{E'}^{\gamma E} p(r,\theta,\mathscr{E},E) \, d\sigma(T) + \int_0^{E'} \{p(r,\theta,\mathscr{E},E)-p(r,\theta,\mathscr{E},E-T)\} \, d\sigma(T) \quad (2)$$

where $E' =$ minimum of γE and $(E-\mathscr{E})$. The first term is needed to account for the ions which lose more energy than $(E-\mathscr{E}_i)$ before reaching the depth associated with a point which has coordinates (r,θ).

The solution technique for final range statistics previously described by the present authors [1] can also be used to solve the intermediate range Eq. (2). One first expands the probability function in a Legendre polynomial cosine series in θ and takes moments with respect to r. This results in a coupled system of integral equations in the variables $\mathscr{E}$ and E for the coefficients of the Legendre polynomial series for the moments m_ℓ^n. Those equations have the general form

$$g_\ell^n(E) = \int_0^E \{m_\ell^n(\mathscr{E},E) - m_\ell^n(\mathscr{E},E-T)\} \, d\sigma(T) \quad (3)$$

For any fixed value of $\mathscr{E}$, each of the integral equations can be expanded in a system of linear first order differential equation of the form

$$g_{\ell,j}^n = \beta_{\ell,0}^n m_{\ell,j}^n + \beta_{\ell,j}^n \frac{dm_{\ell,j}^n}{dE} \quad (4)$$

where the β coefficients are defined by

$$\beta_{\ell,0}^n = N \int_0^{E'} [1 - P_\ell(\cos\theta[E-T])] \, d\sigma_n(T) + \int_{E'}^{\gamma E} d\sigma_n(T) \quad (5a)$$

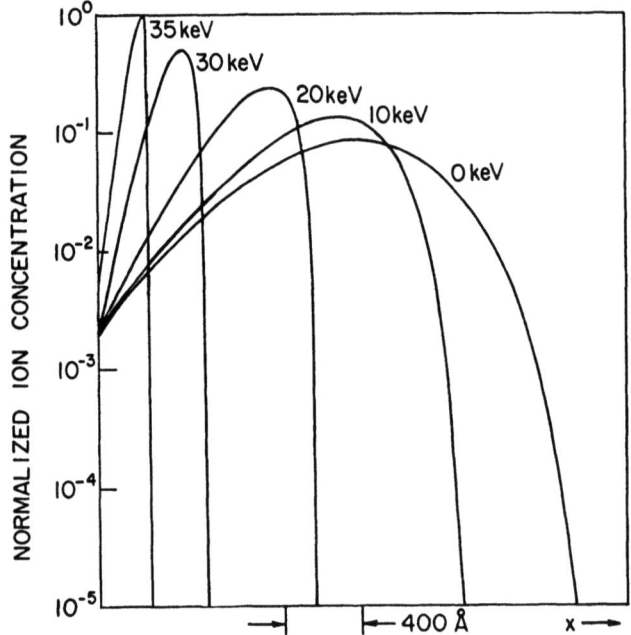

FIG. 1 Intermediate range profiles for 40 keV B in Si and 10 keV steps in intermediate energy.

$$\beta_{\ell,j}^{n} = N \int_{0}^{E'} T^{j} P_{\ell}(\cos\theta [E,T]) \, d\sigma_{n}(T) \tag{5b}$$

The variable $m_{\ell,0}^{n}$ is the first order solution for the ℓ coefficient in the Legendre series for the nth moment, and $m_{\ell,j}^{n}$ is the jth order correction to the first order solution. The solution method (to be explained in greater detail in a forthcoming publication) consists in first evaluating the integrals for $\beta_{\ell,j}^{n}$ and $g_{\ell,j}^{n}$ and then using these values to solve for $m_{\ell,i}^{n}$ in Eq. (4).

PROFILE CONSTRUCTION

The general procedure described above permits one to calculate moments to any desired order. Both the present authors [1] and Winterbon [9] have shown that the third central moment is very substantial and cannot be ignored in the construction of the final range distribution. Using the method outlined above we also find that the skewness factor (ratio of third central moment to σ_p^3) is even larger at intermediate energies than it is in the final range distribution.

A general idea of the importance of the skewing can be obtained from Fig. 1, which shows projected range distributions for a 40 keV B implant in Si at 10 keV intervals. A Pearson type III distribution [11] was used to construct the profiles. Of particular interest is the substantial increase in skewing with intermediate energy. As a result of this extreme skewing, any estimate of energy deposition based on Gaussian intermediate range profiles will markedly underestimate the depth to which energy is transported.

Since we have been unable to find a totally satisfactory distribution profile that can accommodate large values of the third central moment, we have used several distribution functions: the joined half-Gaussian [10], the Pearson type III, the Gram-Charlier, and the Edgeworth [11]. The first two of these are limited to skewing ratios (i.e., third central moments/σ_p^3) of 1 and 1.8 respectively. The Gram-Charlier and Edgeworth expansions permit larger values of the skewing ratio, but may suffer from the problem of (unphysical) negative values and oscillations.

They also require for their construction either a value for the fourth moment ratio (not calculated), or an assumption about the relationship between the third and fourth central moments.

For qualitative purposes it is frequently useful to assume the distribution functions have a fourth central moment ratio of 3 (this

TABLE I

Moments of Energy Deposition Profiles for B in Si

Energy	Gaussian Profiles			
	Mean Å	Standard Deviation Å	Skewness Coeff.	Kurtosis
40	900	540	.25	1.95
70	1520	840	.03	1.86
100	2100	1090	-.10	1.86
200	3770	1770	-.41	1.96
400	6440	2720	-.69	2.24

Energy	Skewed Profiles			
	Mean Å	Standard Deviation Å	Skewness Coeff.	Kurtosis
40	900	530	.09	1.66
70	1520	840	-.17	1.57
100	2100	1090	-.32	1.57
200	3750	1750	-.60	1.73
400	6400	2680	-.86	2.09

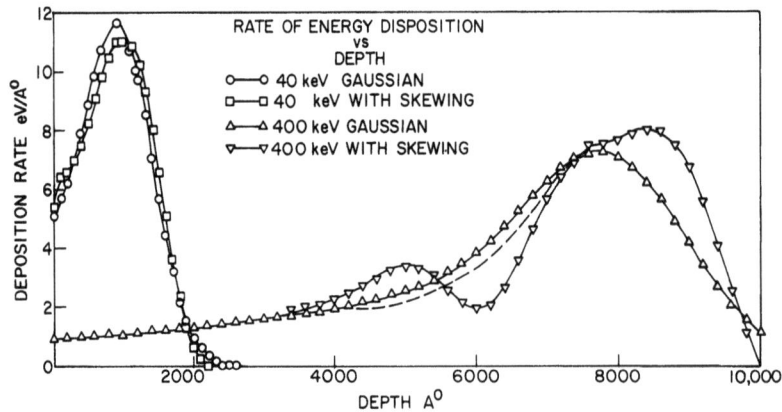

FIG. 2 Rate of energy deposition by 40 and 400 keV implants of B in Si.

being the value for a true Gaussian distribution); or else use the fourth moment values calculated for power law interatomic potentials with neglect of electronic stopping [9]. A better fit to experimental data can be obtained by choosing the fourth moment so that the distribution function has the value predicted by the Gaussian approximation at $x = R_p$.

ENERGY DEPOSITED INTO ATOMIC PROCESSES

As mentioned earlier, the final step in the calculation of energy deposition involves evaluation of the integral

$$\int_0^E \left\{ p(E,\mathcal{E},x) \right\} \left\{ \frac{1}{\bar{r}(\mathcal{E})} \int_0^{\mathcal{E}} \bar{\nu}(\eta) d\sigma(\eta) \right\} d\mathcal{E} \qquad (6)$$

for fixed values of initial energy E and depth x. $\bar{r}(\mathcal{E})$ is the average distance along its path traveled by a particle in losing an amount of energy $d\mathcal{E}$ and

$$\int_0^{\mathcal{E}} \bar{\nu}(\eta) d\sigma(\eta)$$

accounts for the subsequent partition of energy transferred to a knock-on into its nuclear and electronic components [12].

To evaluate the integral given in Eq. (6), moments of the range distribution were calculated for B in Si, using 5 keV increments of intermediate energy, for a set of initial energies between 40 and 400 keV. The damage profiles were then obtained by performing an integration using energy increments of 1/16 keV and interpolation of the calculated moments.

Table I summarizes the principal features of these calculations. Of major note is a shift of the peak towards the bulk and a change in the width of the distribution as compared to results based on Gaussian profiles. A better idea of this effect is given in Fig. 2 which compares our results with those of Brice [13] for 40 and 400 keV implants. The oscillation in mid-depth range is an artifact that is due to the oscillations in range profile models. The dotted curve is probably a better estimate to the true curve.

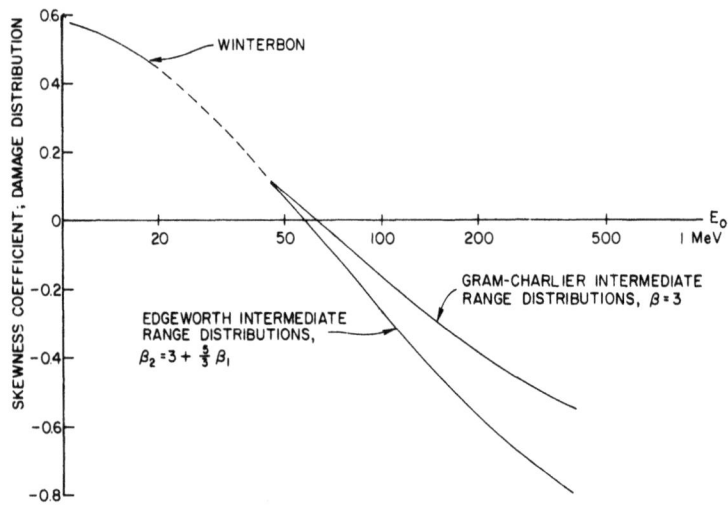

FIG. 3 Mean and FWHM vs energy for calculated profiles and experimental measurements of Eisen.

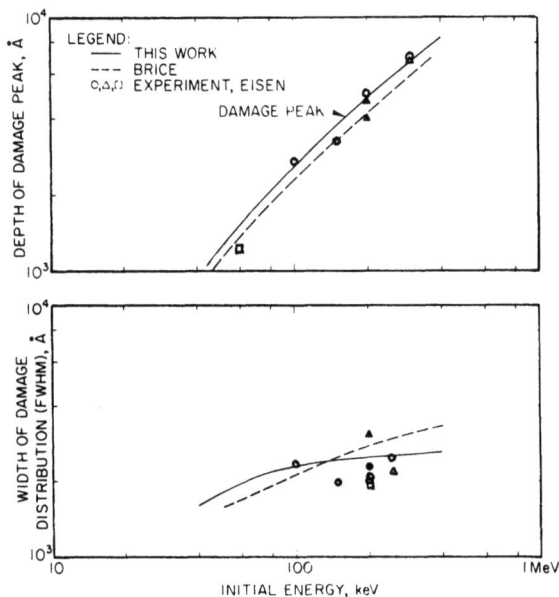

FIG. 4 Skewing coefficients of various calculated energy deposition profiles for B in Si as a function of energy.

In Fig. 3, the results are compared with experimental data on the mean depth and FWHM of the damage distribution as measured by Eisen [14]. The skewing coefficient of the calculated deposited energy profile also extrapolates well to the low energy values of Winterbon [15] as can be seen in Fig. 4. The divergence of the results for two probabilities models at higher energies again point up the need for a better probability model and calculation of the fourth and maybe even higher moments in case of very large skewing ratios in range distributions.

Acknowledgements

The authors gratefully acknowledge their indebtedness to the National Science Foundation for support of this work.

References

1. S. W. Mylroie and J. F. Gibbons, "Computation of Third Central Moments for Projected Range Distributions of Common Ion-Implanted Dopants in Silicon," Third International Conf. on Ion Implantation, Yorktown Heights, New York (December 1972).

2. D. K. Brice, Rad. Effects 6, 77-78 (1970).

3. J. Lindhard, V. Nielsen, M Scharff, and P. V. Thomson, Kgl. Danske. Videnskab, Mat. Fys. Medd. 33, No. 10 (1963).

4. P. Sigmund and J. B. Sanders, Proc. Int. Conf. on Application of Ion Beams in Semiconductor Technology, Grenoble 1967.

5. K. B. Winterbon, Rad. Effects. 13 (1972).

6. J. F. Gibbons, Proc. of U.S.-Japan Seminar on Ion Implantation in Semiconductors, Kyoto 1971.

7. D. K. Brice, Proc. 4th International Conference on Ion Implantation in Semiconductors, Osaka 1974.

8. D. K. Brice, Sandia Labs Research Report SC-RR-710599.

9. N. L. Johnson and S. Kotz, Continuous Univariate Distributions, Vol. II, Houghton-Mifflin, New York 1970.

10. J. F. Gibbons and S. W. Mylroie, Appl. Phys. Letters 22, No. 3, (February 1973).

11. F. Eisen, B. Welch, J. E. Westmoreland, and J. W. Mayer, Proc. Int. Conf. Atomic Collision Phenomena in Solids, U. of Sussex (1969).

PROJECTED RANGE DISTRIBUTION OF IMPLANTED IONS IN A DOUBLE-LAYER SUBSTRATE

H. Ishiwara, S. Furukawa, J. Yamada, and M. Kawamura

Dept. of Physical Electronics, Faculty of Engineering

Tokyo Institute of Technology, Tokyo 152   JAPAN

ABSTRACT

A theoretical method to derive the projected range distribution in a double-layer substrate such as a SiO_2-Si substrate is presented. Validity of the analysis is checked by the C-V measurement. One of the conclusions by this analysis is that a discontinuity in the ion concentration generally exists at the interface of the layers.

Next, the optimum implant conditions to control the gate threshold voltage of MOSFET's are theoretically predicted using the above analysis. Some of the optimum conditions show good agreement with what have been found empirically.

INTRODUCTION

Recently, ion implantation technique has been widely applied to fabrication of MOS devices, especially to adjustment of the gate threshold voltage [1]-[3], fabrication of high speed [4] and high voltage devices [5], doping of complementary wells [6], and so on. In most of these applications, doping ions are implanted through an oxide layer over the surface of the substrate. So, it can be said that theoretical studies on the projected range distribution of implanted ions in a double-layer substrate are important for exact design of the MOS devices. However, no theoretical study on this problem has been reported, except the study on the total range [7].

In this paper, a theoretical method to derive the projected range distribution in a double-layer substrate is proposed, and validity of the analysis is checked by the C-V measurement. Next, the optimum implant conditions to control the gate threshold voltage of MOSFET's are theoretically predicted using the above analysis.

PROJECTED RANGE DISTRIBUTION IN A DOUBLE-LAYER SUBSTRATE

a. Theoretical Consideration on a Special Double-Layer Substrate

We consider the projected range distribution in a special double-layer substrate such as a diamond on graphite substrate, each layer of which is composed of identical elements but has different densities. According to the assumption by Lindhard et al. [8] that each collision between ions and atoms is independent, the projected range distributions in a diamond and a graphite substrate are perfectly equal in a space normalized by the average projected range Rp, in case when the identical ions are implanted with the same energy and dose. In the real space, the distribution in graphite is broader than that in diamond, since the atomic density n is lower in graphite and Rp is inversely proportional to n as shown in Fig. 1 (a) and (b).

Here, two films composed of diamond and graphite and with thicknesses equal to Rp in each substrate are considered. In this case, ions implanted into the films are considered to experience the same collision processes except that spatial distances between collisions are shorter in diamond than those in graphite. Then, it can be said that the energy distributions of ions passing through the films are equal in both cases.

Using the above consideration and the assumption that the ions having an energy distribution at the end of the first layer are implanted into the second, the projected range distribution in a diamond (thickness=Rp) on graphite substrate can be expressed as shown in Fig. 1 (c), in which the left half of the distribution in Fig. 1 (a) and the right half in (b) are connected at the interface. The same discussion holds for the case when the thickness of the first layer isn't equal to Rp.

b. Application to a More General Double-Layer Substrate

We discuss the projected range distribution in a more general double-layer substrate composed of such two materials as the distributions in the normalized space are not perfectly but nearly equal.

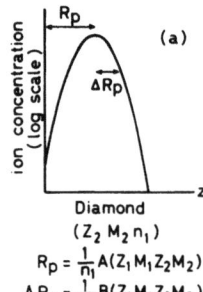
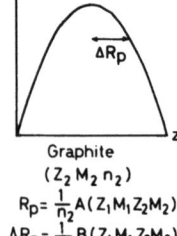
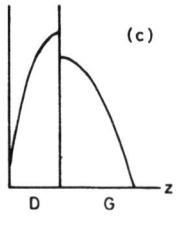

Fig.1 Ion distribution in a diamond on graphite substrate

This condition is realized in materials whose average atomic numbers and masses are nearly equal, and the ratio between Rp and the projected standard deviation ΔRp is considered to give a measure of this condition. As an industrially important example, any double-layer substrate composed of Si, SiO_2, Si_3N_4, and Al_2O_3 satisfies the above condition.

A graphical method to construct a projected range distribution in a SiO_2-Si substrate is shown in Fig.2, in which the gaussian distribution is not assumed. In Fig.2(a) and (b), the distributions of ions implanted with the same condition are shown. In Fig.2(c), the distribution in the SiO_2 layer is the same with that in (a), and the distribution in the Si layer is equal to a deeper portion of that in (b) truncated so that the remainder number of ions N_2' is equal to N_2 in (a). In this method, the total number of ions in the double-layer substrate is equal to the dose N_0.

In case when the gaussian distribution is assumed, the distribution in (c) can be expressed by the four values of Rp_1, ΔRp_1, Rp_2', and ΔRp_2, where Rp_1 and ΔRp_1 are the average projected range and the standard deviation in Fig.2(a), ΔRp_2 is that in (b), and Rp_2' is given by the next equation.

$$Rp_2' = d + (Rp_1 - d) \cdot \Delta Rp_2 / \Delta Rp_1 \quad (1)$$

where d is the thickness of the first layer.

This method is considered an extension of the exact one shown in Fig.1, since both methods coincide in case when the ratios ΔRp/Rp are equal in the first and the second layer.

The range distribution in a SiO_2-Si substrate calculated by this analysis (solid line in SiO_2 + broken line in Si) well agrees with that by the method in Ref. (7) (solid line) as shown in Fig.3. One of the conclusions by this analysis is that a discontinuity in ion concentration generally exists at the interface of layers.

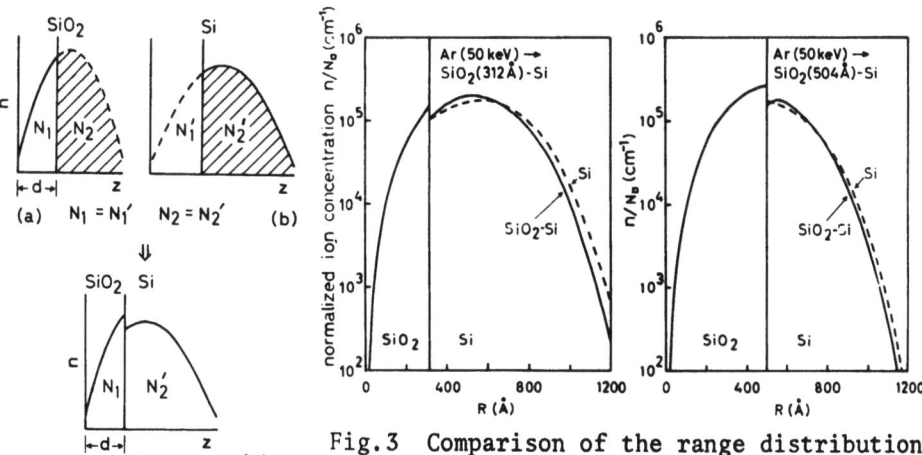

Fig.3 Comparison of the range distribution

Fig. 2 Construction of distribution in a SiO_2-Si substrate

c. Experimental Check of the Analysis

Validity of the analysis is checked by the C-V measurement. Specimens used were n on n^+ epitaxial wafers with 10 Ωcm layer resistivity. Oxide layers with thickness between 800 and 2000 Å were thermally formed in dry oxygen atomosphere. Thickness of the oxide layer was measured by the backscattering technique. Implantation was carried out near room temperature with 150 keV P ions for both SiO_2-Si and bare Si substrates. The dosage was between 1×10^{12} and 10^{13} cm^{-2}. After implantation, samples were annealed at 800°C for 10 min. and removed the oxide layer. Schottky barrier contacts for the C-V measurement were formed by vaccum evaporation of gold.

A typical result is shown in Fig.4, in which circles represent concentration of P ions implanted through 850 Å SiO_2 layer. The solid line in the figure shows the semi-theoretical result constructed using Fig.2(c), where the experimental distribution in a bare Si substrate is used instead of the theoretical one. Agreement between both results is excellent.

It is concluded that this analysis is substantially valid, from our experiment and the work by French researchers [9] in which the discontinuity in ion concentration at the interface has been observed using the ion micro-analyzer.

OPTIMUM IMPLANT CONDITIONS TO CONTROL GATE THRESHOLD VOLTAGE OF MOSFET

In case when the ion implantation technique is used to control the gate threshold voltage V_T of MOSFET's, the next conditions are considered desirable.
(1) To localized impurity ions near the semiconductor surface.
(2) To minimize the total dose.
(3) To minimize the fluctuation in the threshold voltage shift ΔV_T against that in the thickness of the oxide layer.

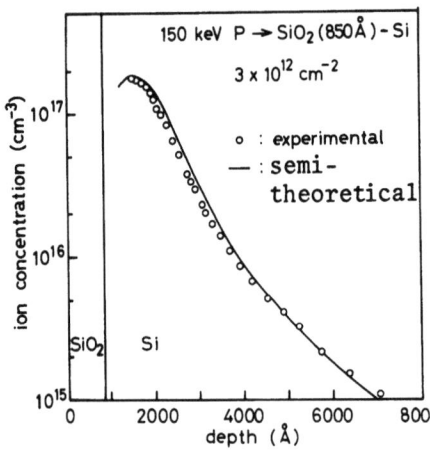

Fig.4
150 keV P ion concentration in SiO_2-Si substrate.

The gate threshold voltage shift ΔV_T of MOSFET is known to be given by the next relation [2], provided implant ions are localized near the surface of the semiconductor and the number of ions $N'_\square$ in the semiconductor is less than 1×10^{12} cm^{-2}.

$$\Delta V_T = qN'_\square / C_{ox} \tag{2}$$

where, q is the electronic charge and C_{ox} is the capacitance of the oxide layer per unit area. Since $N'_\square$ is related to the total dose $N_\square$ as Eq.(3) in gaussian approximation, the relations among $N_\square$, ΔV_T, and d are uniquely determined for a given implant energy.

$$N'_\square = \frac{N_\square}{\sqrt{2\pi}} \int_{(d-Rp_1)/\Delta Rp_1}^{\infty} \exp(-t^2/2) \, dt \tag{3}$$

A chart to determine the optimum implant conditions for control of the V_T by Boron ions is shown in Fig.5, in which the density and the dielectric constant of SiO$_2$ are assumed equal to 2.25 and 3.4, respectively, and electrically activated fraction equal to 100 %. The solid lines labelled energy values are the calculated results by Eq.(2) and (3) and the thin solid line labelled Rp=d shows a condition that Rp of B ions in SiO$_2$ is equal to the thickness of the SiO$_2$ layer. The broken line shows the optimum implant energy to satisfy the conditions (1) and (2) simultaneously, which is determined so as to minimize the product between $N_\square$ and the distance measured from the interface of the center of gravity of the distribution in Si. On the dotted and dashed line all lines labelled energy values become flat, and the condition (3) is best satisfied on it.

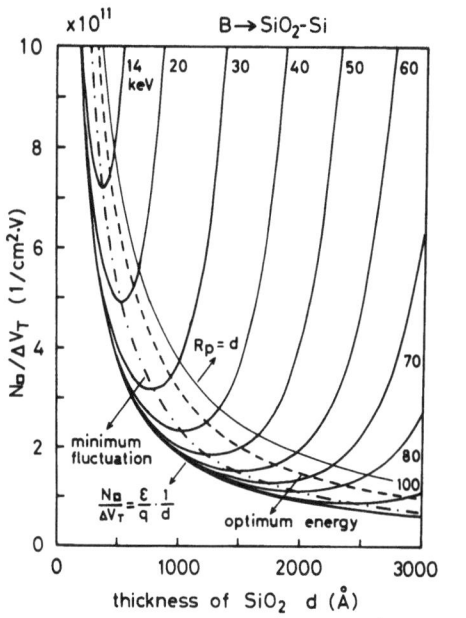

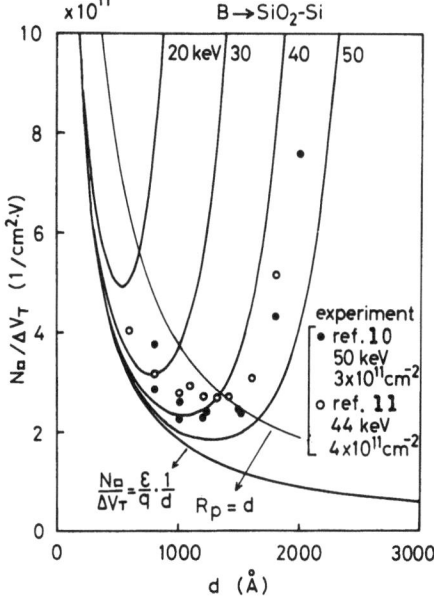

Fig.5 Chart to determine optimum implant conditions for V_T control.

Fig.6 Comparison with the experimental results.

From this figure, it is concluded that (1) the optimum implant energy to control the gate threshold voltage by B ions is several keV higher than the energy satisfying the condition that Rp=d, (2) there exists such a thickness of the SiO_2 layer as to minimize the dosage for a certain threshold voltage shift, in case when ions are implanted with a constant energy, and (3) the minimum fluctuation in ΔV_T is obtained at the energy satisfying the minimum dose condition stated in (2).

Fig.6 shows comparison of the chart with the experimental values reported [10],[11]. Both results show qualitatively good agreement, though the experimental values of $N\square/\Delta V_T$ is larger by 20~50 %. This discrepancy will be reduced by assuming other values of the density of SiO_2 or of the electronic stopping power in SiO_2. From this figure, it can be said that validity and usefulness of the chart are experimentally shown.

CONCLUSIONS

A theoretical method to derive the projected range distribution in a double-layer substrate was presented, and its validity was experimentally shown. The optimum implant conditions to control the gate threshold voltage of MOSFET were theoretically predicted using the above analysis. The calculated results showed quantitative agreement as well as qualitative with the experimental values reported.

REFERENCES

1. M.R.MacPherson, Appl. Phys. Letters, 18, 502 (1971)
2. T.Warabisaco, I.Yoshida and T.Tokuyama, Proc. 4th Conf. Solid State Devices, Tokyo, Suppl.Japan Soc.Appl.Phys. 42, 181 (1973)
3. M.Kamoshida, Appl. Phys. Letters, 22, 404 (1973)
4. R.W.Bower, H.G.Dill, K.G.Aubuchon and S.T.Thompson, IEEE Trans. ED-15, 757 (1968)
5. I. Yoshida, T.Matsuhara, M.Kubo and T.Tokuyama, will be presented in 6th Conf. Solid State Devices, Tokyo, 1974 No.B-6(5)
6. P.J.Coppen, K.G.Aubuchon, L.O.Bauer and N.E.Moyer, Solid-State Electron. 15, 165 (1972)
7. S.Furukawa and H.Ishiwara, J. Appl. Phys. 42, 1268 (1972)
8. J.Lindhard, M.Scharff and H.E.Schiott, K. Danske Vidensk. Selsk. mat.-fys. Medd., 33, No.14 (1963)
9. J.L.Combasson, J.Bernard, G.Guernet and M.Bruel, Proc. 3rd Intern. Conf. Ion Implantation in Semiconductors and Other Materials, edited by B.L.Crowder (Plenum Press, New York, 1973) p.285
10. K.Nakamura and M.Kamoshida, Tech. Report of No.350 Appl. Electronics and Solid State Physics Meeting in Japan Soc. Appl. Phys. No.3 (1973)
11. R.B.Palmer, C.C.Mai and M.Hswe, J. Electrochem. Soc.,Solid-State Sci. and Tech., 120, 999 (1973)

GENERATION OF KNOCK-ONS IN SOLIDS BOMBARDED WITH ENERGETIC IONS AND ENERGY PARTITION RELATIONS

T. Tsurushima and H. Tanoue

Electrotechnical Laboratory

5-4-1 Mukodai, Tanashi, Tokyo, JAPAN

ABSTRACT

Theoretical considerations are given on the energy partition relation of collision sequences in solids bombarded with energetic ions, and a step by step method is proposed to calculate the statistically averaged total number of knock-on atoms. For assigning the initial energy of each knock-on atom, a basic equation governing radiation effects is modified to a simple integral form, and iterative numerical computations are carried out. The results suggest that a modification factor ξ for the Kinchin-Pease formula (*(total knock-ons)*$=\xi \cdot E_n/2E_d$) comes up to about 0.6, which is theoretically acceptable as a consequence employing a realistic (Thomas-Fermi) interaction potential. Approximations for reducing the computational burden and the accuracy of those results are discussed. Effects of the displacement threshold energy and the electronic stopping constant on the number of knock-ons and the energy partition relations are illustrated to be compared with experimental data.

INTRODUCTION

As the Kinchin-Pease formula [1] to predict the average number of knock-on atoms in a solid bombarded with energetic ions appeared to overestimate the damage rate by a factor of two to ten [2], the validity of the formula has been widely examined both in theoretical and experimental ways, and many refinements to the formula have been proposed to push the predicted number down to smaller values [2-4]. A confusion arised, however, because the discrepancy between the theoretical prediction and the experimental (backscattering) results in some typical semiconductors implanted with ions came to be negative, that is, the formula "underestimated" the damage rate at half or less [5-10]. These facts may suggest that many problems are still there both in theoretical and experimental treatments. This paper is to show a simple step by step method for estimating the number of knock-ons with energy partition relations in collision sequences. The method is developed in a sense of statistical average, and a basic principle of energy conservation is confirmed at every step of the method.

OUTLINE OF THE METHOD

As we usually consider the stopping of an ion incident on a target material results from the sum of the two components, the electronic stopping and the nuclear stopping, we can estimate quite easily the energy dissipated in electronic collisions in primary processes (processes in which primary knock-ons are produced) from

$$E_{e1} = \int_0^{E_0} dE \{ S_e(E)/(S_e(E)+S_n(E)) \}, \qquad (1)$$

where E_0 is the initial energy of the incident ion, $S_e(E)$ and $S_n(E)$ are the electronic and nuclear stopping powers, respectively. So we get the energy which is devoted to the nuclear collisions in primary processes as $E_{n1} = E_0 - E_{e1}$.

If we define the displacement stopping power [11]

$$S_{nd}(E,E_d) = \int_{E_d}^{\gamma E} T_n d\sigma_n(T_n), \qquad (2)$$

the energy lost thermally in primary processes can be calculated by

$$E_{l1} = \int_0^{E_0} dE \{ (S_n(E) - S_{nd}(E,E_d))/(S_e(E)+S_n(E)) \}, \qquad (3)$$

where E_d is the displacement energy of the substrate atoms, and γ is given by $\gamma = 4M_1M_2/(M_1+M_2)^2$, where M_1 and M_2 are the atomic masses of the incident ions and the substrate atoms, respectively.

Equations (1) and (3) deduce that the energy which is used to produce primary knock-ons is given by

$$\begin{aligned} E_{nd1} &= E_{n1} - E_{l1} = E_0 - (E_{e1} + E_{l1}) \\ &= \int_0^{E_0} dE \{ S_{nd}(E,E_d)/(S_e(E)+S_n(E)) \}. \end{aligned} \qquad (4)$$

Furthermore, the average number of primary knock-ons is obtained from [11]

$$<\nu_1> = \int_{E_d/\gamma}^{E_0} dE \{ \sigma_{nd}(E,E_d)/(S_e(E)+S_n(E)) \} \qquad (5)$$

where $\sigma_{nd}(E,E_d)$ is the cross section for the production of displaced atoms, and is given by

$$\sigma_{nd}(E,E_d) = \int_{E_d}^{\gamma E} d\sigma_n(T_n) . \tag{6}$$

Equation (5) is a modified integral form of the basic equation governing radiation effects corresponding to that given by Lindhard et al. in general differential form [12]. It is to be noted that equation (5) only describes a series of collisions, and to solve the problem completely, iterative applications of the equation to every series of collisions, primary, secondary, teritary etc., will be needed.

Since we assumed a sharp threshold displacement energy E_d, the total energy absorbed in primary atomic displacement collisions (not transferred to the primary knock-ons, but absorbed in permanent lattice deformation) is simply given by $<\nu_1>E_d$, and so the sum of the initial kinetic energy of each primary knock-on is obtained from

$$\sum_{j=1}^{<\nu_1>} \binom{initial\ energy}{of\ j\text{-}th\ knock\text{-}on} = E_{nd1} - <\nu_1>E_d$$
$$= E_0 - (E_{e1} + E_{l1} + <\nu_1>E_d) . \tag{7}$$

If we can assign the initial energy of each primary knock-on in some way, we are able to iterate the procedure mentioned above for the secondary knock-ons, and in the same way, for the tertiary, quaternary, etc., knock-ons. The problem, therefore, arrives at how to get the initial energy of each knock-on atom. These are summarized as follows (suffix 1 indicating the primary process is generalized to (i)):

$E_e^{(i)}$: Energy dissipated in electronic processes.

$$E_e^{(i)} = \int_0^{E_0^{(i-1)}} dE\{S_e(E)/(S_e(E)+S_n(E))\}$$

$E_l^{(i)}$: Energy lost thermally.

$$E_l^{(i)} = \int_0^{E_0^{(i-1)}} dE\{(S_n(E)-S_{nd}(E,E_d))/(S_e(E)+S_n(E))\}$$

$<\nu^{(i)}>E_d$: Energy absorbed in permanent lattice deformation.

$$\sum_{j=1}^{<\nu^{(i)}>} (Initial\ energy\ of\ j\text{-}th\ knock\text{-}on)$$

$E_0^{(i-1)} \to [E_n^{(i)} \to [E_{nd}^{(i)} \to$

$i = 1, 2, \ldots$

(Iteration)

In such a manner summarized above, the initial energy E_0 is separated into three terms after all, the total energy absorbed in permanent lattice deformations, the total energy dissipated in transient lattice vibrations, and the total energy inelastically lost into e.g. electron exitation or ionization.

From equation (5) we can calculate the energy E_j^* which satisfies the following equation:

$$j = \int_{E_j^*}^{E_0} dE \{\sigma_{nd}(E,E_d)/(S_e(E)+S_n(E))\} \tag{8}$$

for $j = 1, 2, \cdots, \bar{\nu}_1$, where $\bar{\nu}_1$ is the maximum integer not more than $\langle \nu_1 \rangle$ obtained from equation (5). The energy difference $E_0 - E_j^*$ gives the energy that the incident ion loses, in a sense of statistical average, until it produces the j-th knock-on. In other words, the incident ion produces the j-th knock-on somewhere on the way it reduces its energy from E_{j-1}^* to E_j^*.

If we assume that the j-th primary knock-on is produced by the incident ion when its energy is reduced just to E_j^*, we can assign the initial energy of the j-th primary knock-on as $\langle T_d(E_j^*, E_d) \rangle - E_d$, and obtain the lowest average number of total knock-ons [11], where $\langle T_d(E, E_d) \rangle$ is the average transferred energy in an atomic displacement collision given by

$$\langle T_d(E, E_d) \rangle = S_{nd}(E,E_d)/\sigma_{nd}(E,E_d) . \tag{9}$$

If we assume, on the other hand, that the j-th primary knock-on is produced when the incident ion energy falls down just to E_{j-1}^*, we can assign the initial energy of the j-th knock-on as $\langle T_d(E_{j-1}^*, E_d) \rangle - E_d$, and obtain the highest average number of total knock-ons. Thus, for assigning the energy to the j-th knock-on, it may be appropriate to give it an energy $\langle T_d(E', E_d) \rangle - E_d$, where $E' = E_{j-\eta}^*$, for $0 \leq \eta \leq 1$. For a first order approximation we can choose $\eta = 0.5$ independent of the energy as well as of the value of j. This will be discussed later.

As is evident from equations (8) and (9), the following equation is obtained to describe the relation between the total energy loss of the incident ion from E_{j-1}^* to E_j^* and the average energy transferred to the j-th primary knock-on:

$$1 = \int_{E_j^*}^{E_{j-1}^*} dE\{S_{nd}(E,E_d)/(S_e(E)+S_n(E))/\langle T_d(E,E_d) \rangle\} . \tag{10}$$

The average transferred energy $\langle T_d(E, E_d) \rangle$ is of course a function of energy in the energy range between E_{j-1}^* and E_j^*, but if we assume that it can be replaced by a substitute value $\langle \bar{T}_d \rangle$, equation (10) gives the following relation:

$$\langle \bar{T}_d \rangle = \int_{E_j^*}^{E_{j-1}^*} dE\{S_{nd}(E,E_d)/(S_e(E)+S_n(E))\} . \tag{11}$$

In a similar manner, the average energies dissipated in lattice vibrations without permanent atomic displacement and in electronic collisions during the process in which the colliding particle decreases its energy from E_{j-1}^* to E_j^* are given by

$$\langle\bar{T}_l\rangle = \int_{E_j^*}^{E_{j-1}^*} dE\{(S_n(E)-S_{nd}(E,E_d))/(S_e(E)+S_n(E))\} \tag{12}$$

and

$$\langle\bar{T}_e\rangle = \int_{E_j^*}^{E_{j-1}^*} dE\{S_e(E)/(S_e(E)+S_n(E))\} , \tag{13}$$

respectively. It is quite natural that the transferred energies given by equations (11), (12) and (13) satisfy the following relation of energy conservation:

$$\langle\bar{T}_d\rangle + \langle\bar{T}_l\rangle + \langle\bar{T}_e\rangle = E_{j-1}^* - E_j^* . \tag{14}$$

$\langle\bar{T}_d\rangle$, $\langle\bar{T}_l\rangle$ and $\langle\bar{T}_e\rangle$ are the functions of energy only through the value of j. The initial kinetic energy of the j-th knock-on will quite reasonably be assigned to $\langle\bar{T}_d\rangle - E_d$.

RESULTS AND DISCUSSIONS

Two approximations for estimating the initial energy of a knock-on atom corresponding to the cases in which the values of η defined before are equal to 0 and 1, respectively, have been examined to compare the obtained lowest and highest values of the average number of knock-ons. The results are summarized in Table 1. In most cases, the differences between the two were about 15% of the highest.

The number corresponding to the initial energy obtained from equation (11) (hereafter we call this category of calculations ECM, energy coservation method) gives the value very close to the mean of the highest and the lowest as shown also in Table 1. The ratio between the numbers obtained by the ECM and by the Kinchin-Pease formula comes to about 0.6 for the present substrate. This seems theoretically agreeable because we use the Thomas-Fermi potential as the base for calculating collision cross-sections, stopping powers and other parameters in equations (1)-(14). According to an analysis made by Robinson [4], or by Sigmund [13], a modification factor ξ for the Kinchin-Pease formula (*(total knock-ons)*$=\xi \cdot E_n/2E_d$) will be smaller than $(12/\pi^2)\ln 2 = 0.84$ for realistic interaction potentials.

The numbers thus obtained, however, are not immediately enough to explain most of the experimental data obtained by backscattering as shown in Table 1, while that obtained by ESR [14] looks to fit the present result. In any case, many questions are still there on the correspondence between the theoretical predictions and the experimental results [15].

Brice [16] has pointed out that the authors incorrectly assigned the initial energy of the knock-ons in their previous paper [11]. Fig. 1 shows a typical example of the initial energy of the knock-ons obtained by the ECM and the two approximation methods, $\eta=0$ and $\eta=1$,

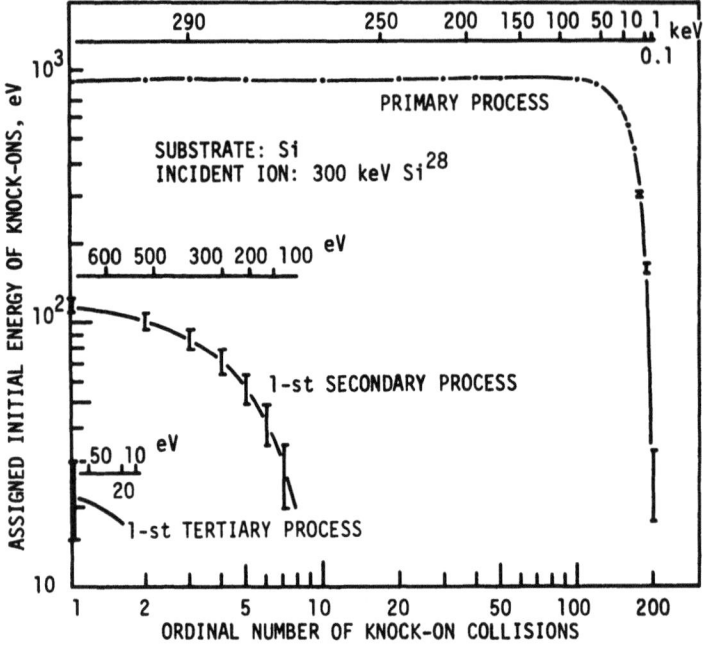

Fig. 1 Initial energy of knock-on atoms. Scales in the figure show the energy that a colliding particle possesses when it produces the j-th knock-on.

as a function of ordinal number of collisions. This figure includes only the energy of each knock-on produced by the incident ion (primary process), by the 1-st primary knock-on (1-st secondary process) and by the 1-st secondary knock-on due to the 1-st primary knock-on (1-st tertiary process), but it may present a simple picture of the knock-on processes. Discrepancy among the initial energies assigned by the ECM and the two approximation methods can be ignored as long as a colliding particle has the energy more than about 10 keV. This is more evident in Fig. 2, which plots the differences of the assigned initial energies obtained by the approximation methods from that obtained by the ECM. The differences increase and saturate to about ±8 eV, as the colliding energy decreases to less than about 1 keV. They cumulatively affect, however, the total number of knock-ons up to about ±7.5 %. Another feature concerned with Fig. 2 is that the approximation employing $\eta=0.5$ gives quite close solutions (plots of open circles) to those given by the ECM. This approximation is useful for reducing the computation time.

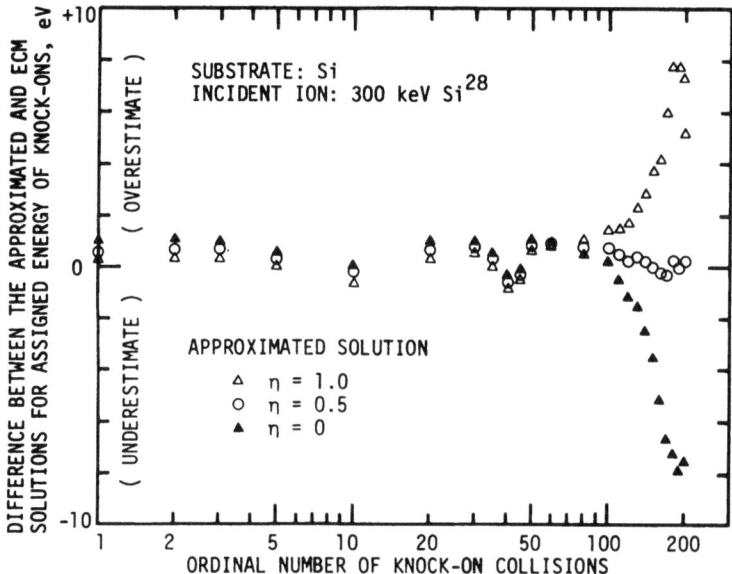

Fig. 2 Degree of accuracy of approximations.

Table 1 Number of Knock-on Atoms in Si

Incident Ion	Si^{28}				Ne^{20}	P^{31}	Sb^{121}	Hg^{202}
Energy (keV)	30	100	300	1000	80	280	40	75
Number of Knock-ons								
Lowest ($\eta=0$)	407	1132	2390	4155	806	2466	612	1149
Highest ($\eta=1$)	474	1308	2751	4771	927	2841	713	1333
ECM	437	1208	2544	4416	859	2623	656	1228
K-P Formula	726	2009	4233	7357	1430	4361	1087	2037
Factor ξ	.60	.60	.60	.60	.60	.60	.60	.60
Experiments					3000*	2800†	3000*	5000*
							3300*	

* Backscattering [5-10], † ESR [14]

The value of the displacement energy E_d strongly affects the total number of knock-ons. We have so far used $E_d = 15.8$ eV for silicon [17], while, assuming $E_d = 7.9$ eV (a half) and 31.6 eV (twice), we get the total number of knock-ons, e.g., for the case of 300 keV Si-ion into Si-substrate as 5084 and 1277, respectively, which are strictly proportional to E_d^{-1} as expected from the Kinchin-Pease formula. Another interesting parameter is the electronic stopping constant, that is, the k-value in $S_e(E) = k \cdot E^{1/2}$. Table 2 shows the changes in the number of knock-ons and the total energy dissipated in electronic collisions, when the k-value changes from that in

Table 2

Initial Energy (keV)	k_{LSS} Total N	k_{LSS} Total E_e (keV)	$(1.5) \times k_{LSS}$ Total N	$(1.5) \times k_{LSS}$ Total E_e (keV)
30	437	7.06	390	9.61
100	1208	36.53	1020	46.71
300	2544	166.2	2010	194.9
1000	4416	767.5	3284	828.1

the LSS theory to the 50% increased. The total number of knock-ons decreases to 90% (at 30 keV) $\sim$ 75% (at 1 MeV), while the total energy dissipated in electronic collisions which is closely related to the ionization in the substrate increases by 8% (at 1 MeV) $\sim$ 36% (at 30 keV). Sattler's experimental data [18] on the energy dependence of the ionization produced in silicon by energetic silicon atoms will be well explained, if we simply assume the electronic stopping constant in this case to be (1.3$\sim$1.5) times larger than that obtained from the LSS theory. One of the questions deduced from this fact is whether the LSS stopping constant in the Si-on-Si case should be replaced by some larger value just as in the B-on-Si case [19].

REFERENCES

1. G. H. Kinchin et al., Rep. Progr. Phys. 18, 1 (1955).
2. P. Sigmund gave a short summary for recent experiments in his paper: Appl. Phys. Lett. 14, 114 (1969).
3. O. S. Oen et al., Appl. Phys. Lett. 2, 83 (1963). See also the references of this paper.
4. M. T. Robinson, Phil. Mag. 12, 741 (1965).
5. D. J. Mazey et al., Phil. Mag. 17, 1145 (1968).
6. J. A. Davies et al., Canad. J. Phys. 45, 4053 (1967).
7. J. W. Mayer et al., Canad. J. Phys. 46, 663 (1968).
8. L. Eriksson, in *Radiation Effects in Semiconductors* ed. F. L. Vook, Plenum Press, New York (1968).
9. S. T. Picraux et al., Appl. Phys. Lett. 14, 7 (1969).
10. G. Della Mea et al., Radiation Effects 3, 259 (1970), and G. Della Mea et al., Appl. Phys. Lett. 16, 382 (1970).
11. T. Tsurushima et al., J. Phys. Soc. Japan 31, 1695 (1971).
12. J. Lindhard et al., Mat. Fys. Medd. Dan. Vid. Selsk. 33 (1963).
13. P. Sigmund, Radiation Effects 1, 15 (1969).
14. B. L. Crowder et al., Radiation Effects 6, 87 (1970).
15. For example, P. Sigmund discusses on this problem in ref. [2].
16. D. K. Brice, in *Ion Implantation in Semiconductors and Other Materials* ed. B. L. Crowder, Plenum Press, New York (1973).
17. H. Flicker et al., Phys. Rev. 128, 2557 (1962). See also J. E. Westmoreland et al., Radiation Effects 5, 245 (1970).
18. A. R. Sattler, Phys. Rev. 138, A1815 (1965).
19. F. H. Eisen et al., *Int. Conf. Atomic Collision Phenomena in Solids*, University of Sussex, (1969).

DEFECTS IN ION IMPLANTED SiO$_2$ LAYERS ON Si[*]

E. P. EerNisse and C. B. Norris

Sandia Laboratories

Albuquerque, New Mexico 87115

EXTENDED ABSTRACT

The introduction rates and annealing behavior of structural defects created by ion implantation into or through thermally grown SiO$_2$ layers on Si have been investigated by measuring the induced volume compaction of the SiO$_2$. In contrast to conventional capacitance-voltage techniques, the present measurements produce information about the structural defects independent of the defect electrical state. Thus, defect annealing is monitored, not charge redistribution. The induced compaction was measured by monitoring the curvature of 1.5 cm × 0.25 cm × 0.025 cm oxidized Si samples introduced by implantation of Ar$^+$, H$^+$, and low energy electrons into one major surface. Curvature was determined in situ with a cantilever beam technique and after isochronal anneals with a Dec Tac instrument.

The structural defects are shown to be produced by two distinct mechanisms, one dependent upon ion energy deposited in the SiO$_2$ as atomic collisions, the other dependent upon the ion energy deposited in the SiO$_2$ as ionization events. Implants of 500 keV Ar^{++} (projected range 5000 Å) into the 6500 Å SiO$_2$ layers are used to produce the defects related to atom collisions. Implants of 250 keV H$^+$ (projected range $\gg$ 6500 Å SiO$_2$ thickness) through the SiO$_2$ to leave only ionization events in the SiO$_2$ and 18 keV electrons (projected range $\gg$ 6500 Å SiO$_2$ thickness) which cause only ionization are used to isolate the defects related to ionization.

---

[*]This work was supported by the U. S. Atomic Energy Commission.

It is found that the atomic collision compaction effect has saturated by 2×10^{14} cm^{-2} 500 keV Ar^{++} (2×10^{20} keV/cm^3 deposited into atomic collisions). Twenty-minute isochronal anneals in 100°C steps to 900°C in dry N$_2$ produced the following anneal behavior. The atomic collision related compaction produced by a 2×10^{13} cm^{-2} 500 keV Ar^{++} implantation (well below the damage saturation level) begins annealing at 400°C in a gradual manner leading to 10% of the damage remaining after the 900°C anneal. The before-implant density of the SiO$_2$ is attained only after annealing at temperatures comparable to the growth temperature of the SiO$_2$. The atomic collision related compaction for fluences high enough to cause saturation anneals in a featureless manner between 300 and 700°C, being essentially complete by 700°C.

The ionization induced structural damage is found to saturate at 2×10^{23} keV/cm^3 deposited into ionizing events. Isochronal annealing of SiO$_2$ layers damaged to levels below and above the saturation dose produced the same results, one well-defined stage centered at 650°C. The before-implant density of the SiO$_2$ is attained after an 800°C anneal.

From an ion implanted device point of view, any oxide which has had ions implanted through it should be annealed to at least 800°C to remove the ionization related structural damage. Any oxide which has atomic collision related structural damage in it must be annealed to above 900°C. Measurements are underway to identify the electrical effects of the damage from the two separate mechanisms.

SECONDARY DEFECTS IN BORON IMPLANTED SILICON

G.P. Pelous, D.P. Lecrosnier, P. Henoc

Centre National d'Etudes des Télécommunications

22301 LANNION, FRANCE

ABSTRACT

On 1 MeV Boron implanted silicon, we have drawn the depth distribution of secondary defects by using a combination of transmission electron microscopy and anodic stripping. From 1.6 μm to 2 μm, perfect dislocation loops are the predominant defects. These loops lie into the {111} plane parallel to the surface. They are of vacancy type with a Burger vector of $\frac{a}{2}$ <110>. The density presents a maximum value located at the peak of boron distribution.

This results suggest a long range interaction between primary defects and impurities, as an adsorption of boron atoms along the dislocations. To check this assumption, lattice site location of boron measurements have been performed, using $^{11}B(p, \alpha)2\alpha$ nuclear reaction and angular scans through the major axial and planar channeling dips. This results seem to indicate that almost of implanted boron lie on well-defined sites into {111} planes.

INTRODUCTION

The aim of this paper is to present transmission electron microscopy experiments and channeling analysis performed on 1 MeV boron implanted silicon. The use of such a high energy, leading to well buried layers, gives very suitable samples for investigating the depth distribution of defects. Moreover, for channeling analysis, it could be though that the channeled proton flux is in quasi-equilibrium, so deep in the crystal.

EXPERIMENTAL PROCEDURE

After implantation, the samples were annealed at 1000°C in a nitrogen ambiant during 30 minutes. This temperature is high enough to ensure a 100 % yield of electrical activity [1]. In such conditions, it was found that diffusion was not completely negligeable ($Dt \simeq \Delta R_p$) but didn't change greatly the shape of the distribution of boron as indicated on fig. 1. Using previous results [2] on non anneal 1 MeV boron profile a value of 7.10^{-14} cm^2/s for the coefficient diffusion was found in applying the method given by SEIDEL [3]. This value agrees well with those obtained by HOFKER [4] for lower energies and is larger than those measured in conventional diffusion ($\sim 1,5.10^{-14}$ cm^2/s). In addition, it must be noted that for the chosen dose, no precipitation is expected because the higher concentration of the profile does not reach the solubility limit.

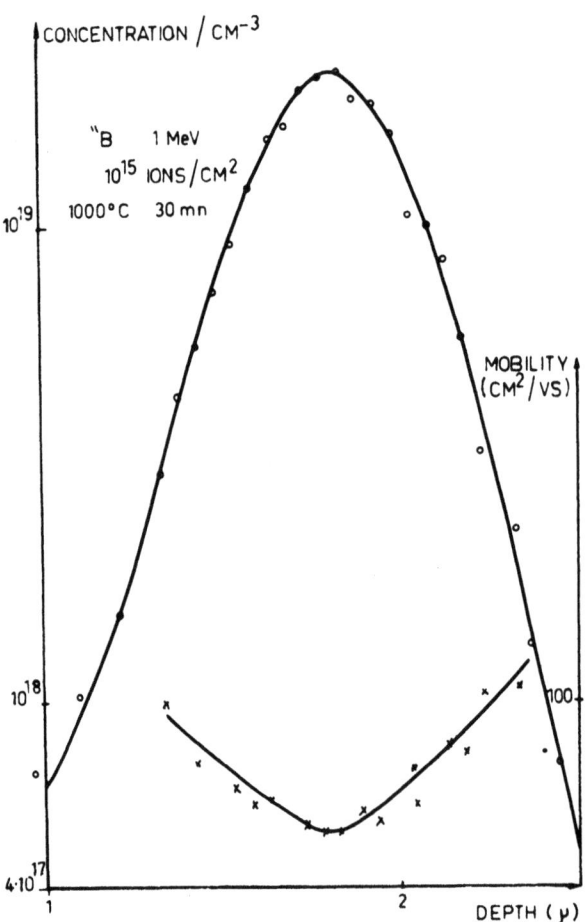

Fig. 1. Carrier concentration profile of 1 MeV Boron implanted silicon. The distribution has been obtained by Hall effect measurements combined with anodic stripping.

Channeling analysis

To determine lattice location of boron, the yield of $^{11}B(p, \alpha)^8Be$ reaction near 670 keV was compared to the yield of elastically scattered protons by silicon-substrate at the same energy.

The scattered protons were detected by a small surface-barrier detector at 135° and the detector pulses were treated by standard electronics, in such a way as only the protons scattered by silicon atoms located at the depth of boron concentration peak would be registered in a counting scale.

The α particles from $^{11}B(p, \alpha)^8Be$ were detected by a large annular surface barrier detector, covered with 8 -thick mylar in to stop elastically-scattered protons but not the reaction product [5]. So, it is possible to detect α particles with very large solid angle and high beam current without any pulse pile-up.

In order to obtain reliable and reproductible data, great care was taken in the orientation of the crystal. At first, a scan in the chosen tilt direction was made to check that dip was not influenced by high order planes or axis and to measure the part of scan vital to the determination of dip shape. After these ajustements, beam impact was changed and angular scans were performed in 0.1° or 0.05° steps beginning by the bottom of the dip, so minimizing the displacement effect of proton bombardement, as previously described by SIGURD and BJORWVIST [6].

Microscopy observations

Distribution of defects is determined by TEM observations after removing a calibrated surface layer by anodic oxydation and HF stripping. After the front surface being removed and carefully protected by wax, the samples were back-thinned to about 1000 Å by chemical etching technique for microscopy observation.

The observations were carried out at 100 keV, using a goniometer to orientate the specimen. In this case, when only the incident beam and one of the diffracted beam have a noticeable intensity, the observed contrasts are more easy to explain [7,8].

Generally, we found no oscillations in diffracted intensities along defects. Therefore these are nearly parallel to the entry face. Because a dislocation cannot appear or vanish in the bulk of the crystal, the observed defects are dislocation loops with their larger dimensions parallel to the surface ({111} plane). In other hand, when dipoles are resolved, no interference fringes appear inside them, excluding the assumption of loops surrounding a stacking fault. To strenghten this assumption, we have tried to determine Burgers vectors of these loops by looking for lighting conditions as defect contrast vanishes, using the non-contrast rule

$$\vec{g} \cdot \vec{b} \wedge \vec{u} = 0$$

where $\vec{g}$ is the reciprocal lattice vector
$\vec{b}$ the Burgers vectors of the dislocation
$\vec{u}$ the unit vector along positive direction
of the dislocation line. We find $\vec{b} = \frac{1}{2} <110>$, value which characterize perfect dislocations. In order to determine if these loops are vacancy type or intersticial type, we have chosen to study the variation of anormal absorption in thick enough foils. The "black-white" variation of diffracted intensity in the direction of observation vector g on dark field pictures, allows us to find the defects to the vacancy type [9].

EXPERIMENTAL RESULTS

Channelling

At first, it should be noted that at 1.5 μm deep in the silicon crystal, quasi statistical equilibrium of the channeled proton flux is obtained.

Scans across major axis (111, 110, 112) and major planes (110, 111) were performed. Some of these are presented on figures 2 a to c. Across {111} plane, we find that the widths of Si-dip and B dip are the same. Across <111> and <110> axis, we observe a slight narrowing in the bottom of dip is observed [10] and across {110} plane, well defined shoulders appear in boron yield.

We observe on fig. 2 that the minimum yields are quite high but it should be kept in mind that a large fraction of incoming proton beam is dechanneled at this depth in the crystal. This part of beam reacts with boron atoms in a "random" way. So, we can estimate that about 80 % of boron atoms occupie the same position but it seems very hard to do a more detailed quantitative analysis.

Microscopy

Fig. 3 (a) to 3(f) show TEM micrographs taken at specimen depth of respectively 1,4 ; 1,6 ; 1,7 ; 1,8 ; 1,9 ; 2 microns. From surface to 1,5 μm, no defects are seen at any magnification and only extinction contours are visible. This region looks like a perfect crystal indicating that in this part defects are created in a negligeable amount or they have been annealed during the heat treatment. It must be noted that the energy deposited in this region by the incoming ions is essentially dissipated in electronic processes.

At 1,6 μm and 1,7 μm elongated dislocations loops are observed with increasing density with depth. In addition to loops, long rod-shaped defects are also seen. These linear defects have been previously observed by several authors [11, 12] in boron implanted silicon but only for anneal temperature in the range 600°C - 800°C.

SECONDARY DEFECTS IN BORON IMPLANTED SILICON

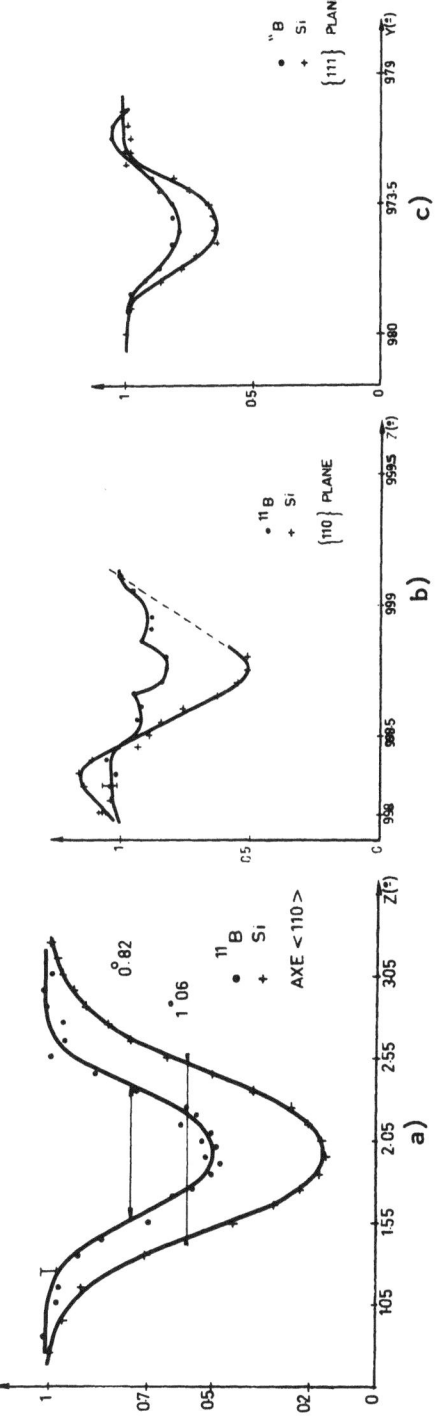

Fig. 2a, b, c. Comparison between angular dependence of the yield of $^{11}B(p,\alpha)^2$ reaction and $^{28}Si(p,p)^{28}Si$ reaction at the same depth
a) <110> axis, b) {110} plane, c) {111} plane

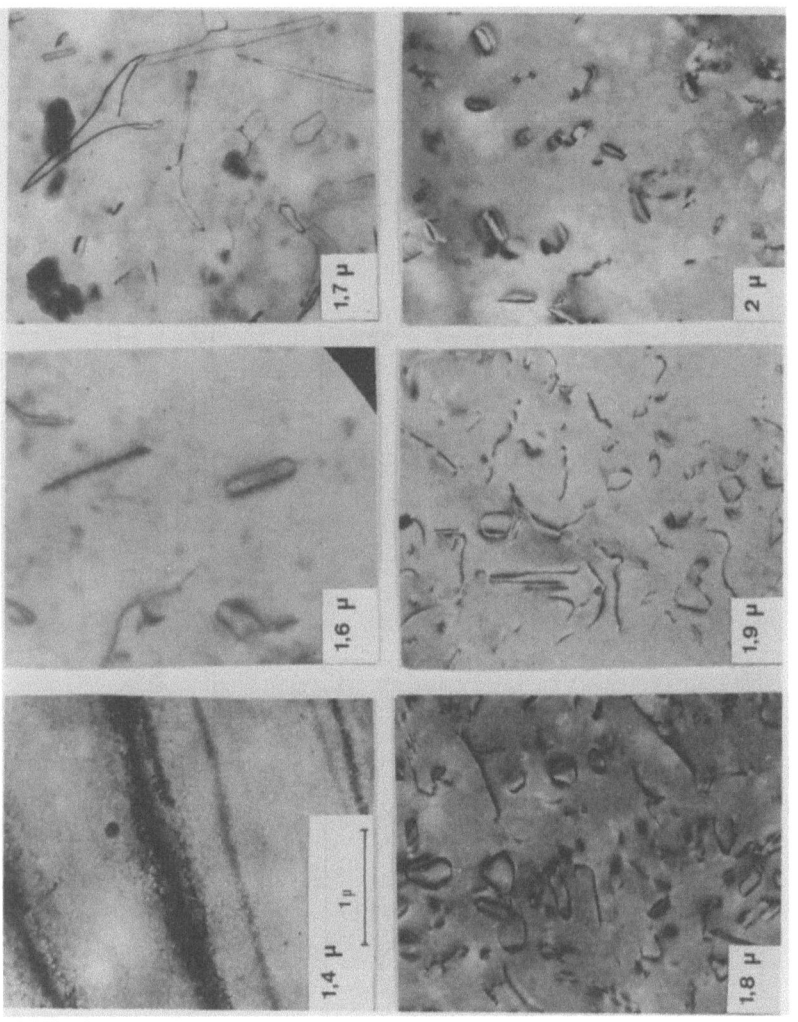

Fig. 3. Micrographs of the defects seen at various depths for sample implanted with 10^{15} 1 MeV Boron ions/cm^2 and annealed at 1000°C during 30 minutes.

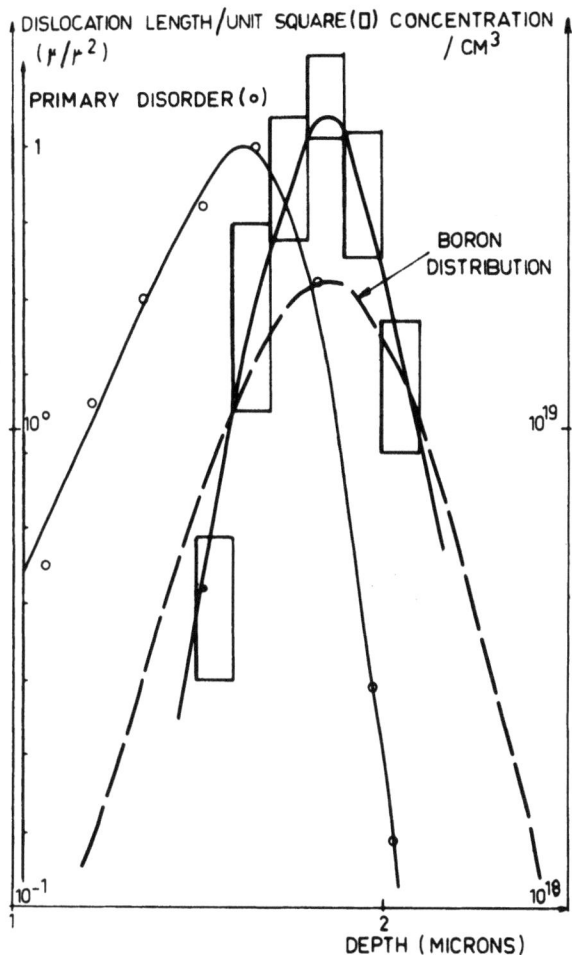

Fig. 4. Distribution of scattering center density (primary defects defects) after implantation; and disolcation loops length measured after anneal at 1000°C during 30 minutes. (The scattering center density has been measured on a more heavily implanted sample (4.10^{15} Boron.cm^{-2} at 77°K).

The micrograph 3 (d) shows clearly the gradual change from a small number of large loops seen at 1,7 μm to a large number of small loops. In conjunction with these small loops, few rod-like defects and complicated networks are also seen. At 1,9 and 2 μm, the density of dislocation is decreasing and their general feature is similar to that observed at 1,8 and 1,6 μm respectively. At 2,1 μm, very few loops are detectable.

DISCUSSION

Before any heat treatment, back-scattering experiments (fig. 4) shows the defect concentration is maximum at 1,6 m from the surface with a FWHM of about 3500 Å in quite good agreement with calculation of SIGMUND-SANDERS [13]

After 1000°C anneal during 30 minutes, we plot (fig. 4) dislocation length measured from fig. 3 versus depth and we find clearly a shift of defects towards the impurity distribution from 1,6 μm to about 1,8 μm. This change has been observed by AKASAKA [19] with slightly different experimental condition. The "secondary defects" observed by AKASAKA [14] in backscattering experiments may be explained as boron precipitates. In our case, boron precipitates are not expected but the coïncidence between dislocations loops and impurities distribution suggest a strong interaction boron-defects. As regard to the large size of loops, this interaction may explain the increase of diffusion coefficient at 1000°C. We also point out that for higher diffusion temperature the defect distribution and the boron distribution diffuse in the same way with normal diffusion coefficient. Following the assumption of BICKNELL [11] and calculating the number of atoms that could be accomodated in a single layer within the area of loops we find a quite good agreement with boron concentration. BICKNELL [11] has suggested that the enclosed atoms within the loops are silicon which have left their substitutionnal site for boron impurities. This assumption would mean that boron atoms would occupy a perfect substitutionnal site in the lattice. But this is not supported by channeling analysis. From those experiments, using same interpretation as PICRAUX [10] or SIGURD [15] with some symetry considerations, we could conclude that almost all the boron atoms lie in {111} planes but slightly displaced from lattice sites.

Taking into account all these results, we are lead to propose tentatively the following mecanism :

- In a first phase, point defects condense into vacancy clusters where boron atoms precipitate by adsorption mecanism. This produce loops surrounding a stacking fault with Burger vector equal to $\frac{1}{3}$ <111>. This could also explain the fast diffusing boron atoms.

— In a second phase, as described by THOMSON [16] the resulting stacking fault is removed by a shearing in the {111} plane of silicon neighbouring plane by a vector $\frac{1}{6}$ <112>. This process is energetically favourable and results in a loop of full dislocation with b = $\frac{1}{2}$ <110> as experimentally found. During this shearing, boron atom would be paired with either a silicon atom or a silicon vacancy, leading in both cases to a lattice dilatation, and would occupy a position shifted in <112> direction from a regular site. All the channeling results could be explain with such a position but quantitative calculations had to be done to acertain this assumption.

CONCLUSION

The experimental results obtained on primary and secondary defects in 1 MeV boron implanted silicon make obvious the interaction between defects and impurities during the anneal treatment. However, the proposed mecanism would need further experimental support, specially in order to determine the position of boron, i.e the nature the secondary defects observed here.

ACKNOWLEDGEMENTS

The authors would to express their thanks to J. FLOCH and G. JOURDE for their careful help in performing transmission electron micrographs and R. DESPRETZ and A. GUIVARC'H for channeling analysis.

REFERENCES

1. T.E. Seidel, A.U. Mac Rae, Proc. 1 st Int. Conf. on Ion Implantation, pp. 149-154, 1971

2. D.P. Lecrosnier, G.P. Pelous, European Conf. on Ion Implantation pp. 102-106, 1970

3. T.E. Seidel, A.U. Mac Rae, Transaction of the metallurgical Society of AIME, Vol. 245, 1969

4. W.K. Hofker, H.W. Werner, D.P. Oosthock, H.A.M. de Grefte, Appl. Phys. 2, 265-278, 1973

5. E. Ligeon, A. Bontemps, J. Radio Anal. Chem. 12, pp. 335-351, 1972

6. G. Fladda, K. Björkqvist, D. Sigurd, Applied Physics Letters, 1970, Vol. 16, n° 8, 313

7. Hirsch et al., Electron microscopy of thin crystals, Butterworths London, 1965

8. G. Saada, Microscopie électronique des lames minces cristallines, Masson, Paris, 1966

9. B. Jouffrey, Méthodes et Techniques nouvelles d'observation en métallurgie physique, S.F.M.E., Paris 1972

10. S.T. Picraux, W.L. Brown, W.M. Gibson, Phys. Rev. B, 6, 1383, 1972

11. R.W. Bicknell, Proc. Royal Soc., A, Vol. 311, p. 75, 1969
12. L.T. Chadderton, F.H. Eisen, Proc. 1st Int. Conf. on Ion Implantation, pp. 445-453, 1971
13. P. Sigmund, J.B. Sanders, Int. Conf. on Application of Ion Beams to Semiconductor Technology, 1967
14. Y. Akasaka et al., J. Appl. Phys., Vol. 44, N° 1, January 1973
15. D. Sigurd, K. Björkqvist, Radiation Effects, 1973, Vol. 17, pp. 209-220
16. M.W. Thompson, Defects and radiation damage in metals, Cambridge University press - 1969

TERNARY DEFECTS RESULTING FROM THE IMPLANTATION OF B, F, BF, AND BF_2 IONS INTO SILICON, THEIR FORMATION AND EFFECT UPON DEVICE PROPERTIES

S. Prussin, consultant

TRW Semiconductors and TRW Systems

Lawndale, Ca.   90260 U.S.A.

ABSTRACT

Ions of B, F, BF and BF_2, implanted with fluences of 10^{14} and $10^{15} cm^{-2}$, were subjected to several thermal treatments. Wet oxidation at $1100°C$ resulted in a heavy generation of ternary defects. Annealing in dry N_2 at $1100°C$ prior to wet oxidation resulted in the effective solution of defect nuclei while annealing at $1000°C$ was only partially effective. N^+p diodes, prepared by chemical phosphorus indiffusion were evaluated electrically. Diodes prepared from implanted surfaces which underwent the $1100°C$ dry N_2 anneal exhibited low leakage under reverse bias. For the $1000°C$ anneal, only the BF_2 implantation exhibited low leakage, the B and BF implantations resulting in leakages several orders of magnitude greater.

INTRODUCTION

Annealing processes which result in complete return of implanted ions to electrically active substitutional positions usually leave a concentration of microdefects of a size requiring transmission electron microscopy for their individual distinction[1]. These have been referred to as secondary defects[2]. Initial studies have shown that the thermal oxidation of silicon surfaces containing B and P implanted at room temperature causes the small extrinsic dislocation loops to expand by several orders of magnitude into dislocation loops large enough to be easily detected by chemical etching and optical microscopy[3]. These are here referred to as ternary defects.

The oxidation technique permits us to evaluate the effectiveness of various annealing treatments in reducing defect concentrations. This is particularly important since ternary macrodefects are developed during the oxidation step in the preparation of bipolar devices. In a previously described work[4] an ion implantation of 1.2×10^{15} B cm^{-2} was substituted for a conventional BN base deposition in the preparation of bipolar I.C.s. This was followed by a standard PH_3

emitter diffusion. The reverse leakage of the n^+p^+ test diodes was found to depend upon the nature of the base drive. Using the conventional base drive of 15 min wet N_2 followed by 30 min dry N_2, both at $1140°C$, a reverse leakage three orders of magnitude greater than normally found with BN deposition was obtained. This was attributed to the generation of ternary defects arising from the use of an oxidizing atmosphere in the first stage of the base drive. These ternary defects interacted with the subsequent emitter phosphorus diffusion resulting in diffusion-induced dislocations. The greatly increased reverse leakage of the p^+n^+ diodes was attributed to these diffusion-induced defects. Using a base drive of 15 min dry N_2, 15 min wet N_2, and 15 min dry N_2, all at $1140°C$, ternary defects were avoided. The reverse leakge as a result was reduced to values at or below those found with BN base deposition.

The need of a high temperature anneal to avoid the generation of ternary defects limits the use of B implantation for shallow junctions. In this study the use of lower annealing temperatures as well as the use of molecular implant species which might permit such lower temperatures was investigated.

Müller et al. [5] compared the use of BF_2 molecules with elemental B as a means of implanting fluences of 10^{14} and 10^{15} B cm^{-2} into silicon. By using implant energies proportional to the molecular and atomic weights, they were able to achieve identical projected ranges for the implanted B ions. Isochronal annealing and He backscattering studies indicated that high implant damage caused the formation of amorphous layers at the silicon surface for the $10^{15} cm^{-2}$ BF_2 implantation.

EXPERIMENTAL PROCEDURES

Implantations of B, F, BF and BF_2 were carried out at fluences of 10^{14} and $10^{15} cm^{-2}$ into 2" diameter (100) n-type dislocation free pedestal-grown silicon wafers of 30 to 50 Ωcm resistivity. The implant energies, 30 keV for B, 52 keV for F, 82 keV for BF and 134 for BF_2 were chosen proportion to the molecular weights to achieve equal range and range straggling for the B ions as well as for the F ions in the various implantations. This permitted us to compare B implantations with varying amounts of superimposed damage as well as to compare the defects generated by an implanted ion such as F which diffuses out at the surface with those generated by B ions which remain imbedded in the lattice. Sets of eight specimens, consisting of four implant species at the two fluences were subjected to three thermal treatments. The first consisted of a 1h wet oxidation at $1100°C$, the second of a 1h dry N_2 anneal at $1100°C$ followed by a 1h wet oxidation at $1100°C$, while the third consisted of a 1h anneal at $1000°C$ followed by a 1h wet oxidation at $1100°C$. At the conclusion, the $6300 A°$ thick oxide was stripped in HF and the wafers were Sirtl etched (50g Cr_2O_3, 100 ml H_2O, 75 ml 48-50% HF) and examined by optical and scanning electron microscopy.

Table I Structures resulting from various thermal treatments

Spec. #	Fluence cm^{-2}	Energy keV	Species	Thermal Treatment	Appearances
1A	10^{14}	30	B		SF, clear background
6A	10^{14}	52	F		DL, clear background
9A	10^{14}	82	BF		SF, discrete etch pits
15A	10^{14}	134	BF$_2$	1h wet O$_2$	SF, discrete etch pits
21A	10^{15}	30	B	@ 1100°C	SF, etch pit clusters
26A	10^{15}	52	F		Heavy dislocation loop con.
29A	10^{15}	82	BF		No SF, etch pit clusters
35A	10^{15}	134	BF$_2$		Few SF, etch pit clusters
1B	10^{14}	30	B		Clear
6B	10^{14}	52	F		Clear
9B	10^{14}	82	BF	1h dry N$_2$	Clear
15B	10^{14}	134	BF$_2$	@ 1100°C +	Clear
21B	10^{15}	30	B	1h wet O$_2$	Few SF, clear background
26B	10^{15}	52	F	@ 1100°C	Clear
29B	10^{15}	82	BF		Clear background
35B	10^{15}	134	BF$_2$		Few SF, clear background
2A	10^{14}	30	B		Few SF, clear background
7A	10^{14}	52	F		Clear
10A	10^{14}	82	BF	1h dry N$_2$	Few SF, clear background
16A	10^{14}	134	BF$_2$	@ 1000°C +	Fewer SF, clear background
22B	10^{15}	30	B	1h wet O$_2$	SF, etch pit clusters
27A	10^{15}	52	F	@ 1100°C	Reduced dislocation loop con.
30A	10^{15}	82	BF		Few SF, etch pit clusters
36A	10^{15}	134	BF$_2$		No SF, etch pit clusters

Two sets of 6 wafers were used in device fabrication. For one set, after implantation of B, BF, and BF_2 with fluences of 10^{14} and 10^{15} cm^{-2}, a base drive of 1h at 1100°C in dry N_2, 1h at 1100°C in wet O_2, and 30 min in dry N_2 at 1140°C was used. For the other set, the base drive consisted of 1h at 1000°C in dry N_2, 1h at 1100°C in wet O_2, and 40 min in dry N_2 at 1140°C. The difference in time for the last step of these two drives was chosen to obtain similar junction depths. Emitter windows were opened in the oxide and emitters were generated by a 15 min PH_3 diffusion at 1020°C. Metallization of contacts permitted us to measure leakage currents for 5 differently sized diodes on each wafer.

EXPERIMENTAL RESULTS

Table I lists the various structures obtained in this study. For the 1h wet oxidation we expect the generation of the greatest number of ternary defects. The defect structure for the 10^{14} B cm^{-2} fluence, consisting of uniformly distributed stacking faults (SF) on a clear background has been previously described and illustrated [3]. Figure 1 illustrates the discrete etch pit background found for 10^{14} cm^{-2} BF and BF_2 implantation, while Fig. 2 illustrates the etch pit cluster background found for 10^{15} cm^{-2} B, BF, and BF_2. The defect structure for 10^{15} cm^{-2} F is distinctly different, Fig. 3.

The 1h anneal in dry N_2 at 1100°C appears to play an effective role in dissolving defect nuclei. Sequential wet oxidation fails

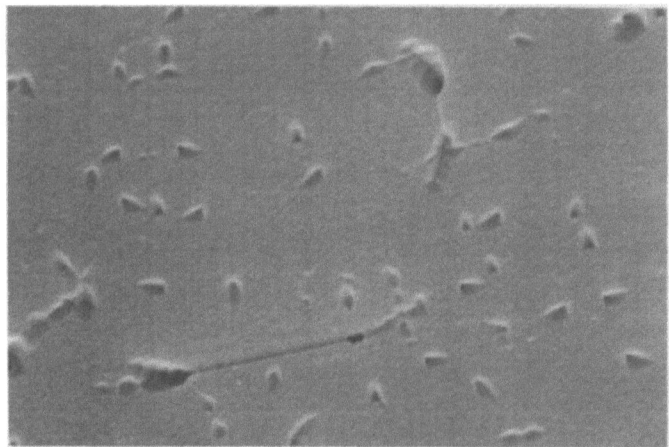

Fig. 1 Scanning electron micrograph, specimen 9A, 10^{14} BF cm^{-2}, 82 keV, wet oxidized at 1100°C for 1h, Sirtl etched 30 sec, magnification 5400x.

TERNARY DEFECTS 453

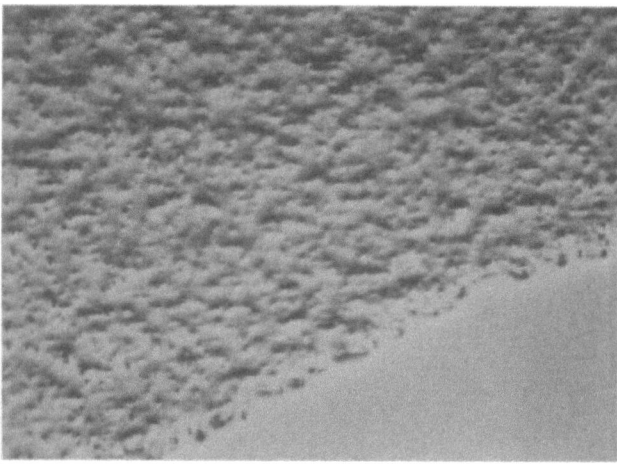

Fig. 2 Scanning electron micrograph, specimen 35A, 10^{15} BF_2 cm^{-2}, 134 keV, wet oxidized at $1100°C$ for 1h, Sirtl etched 15 sec, magnification 5400x.

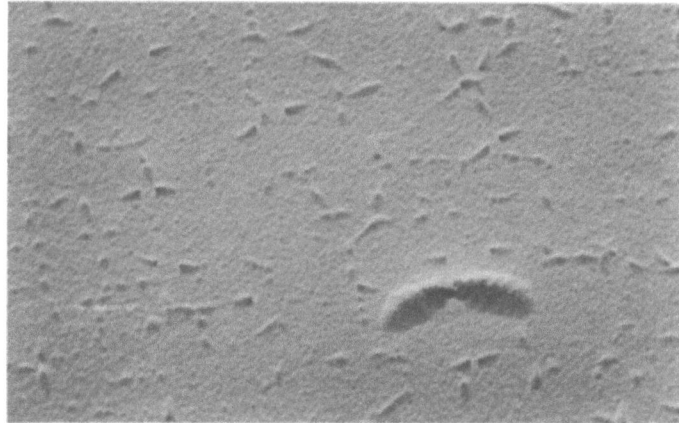

Fig. 3 Photomicrograph of ion-implant boundary, specimen 26A, 10^{15} F cm^{-2}, 52 keV, wet oxidized at $1100°C$ for 1h, Sirtl etched 60 sec, Nomarski interference contrast optics, magnification 430x.

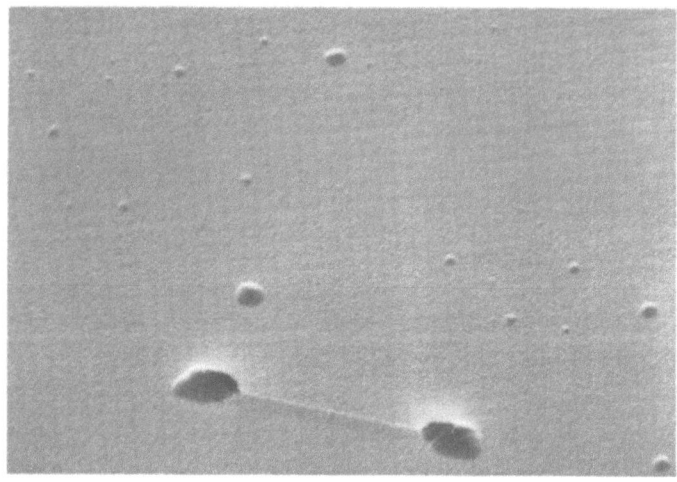

Fig. 4 Scanning electron micrograph, specimen 35B, 10^{15} BF_2 cm^{-2} 134 keV, annealed in dry N_2 at $1100°C$ for 1h, wet oxidized at $1100°C$ for 1h, Sirtl etched 15 sec, magnification 2700x.

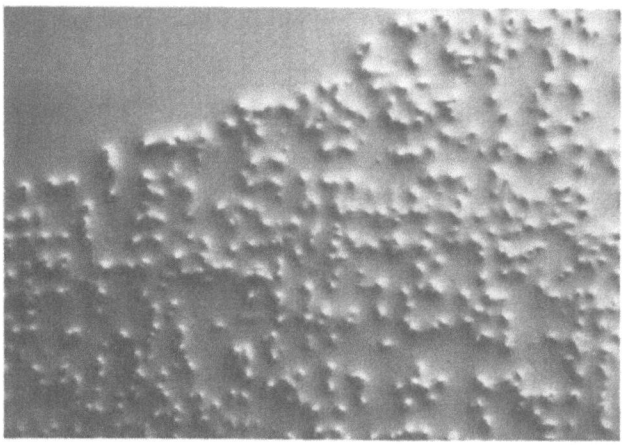

Fig. 5 Photomicrograph of ion implant boundary, specimen 27A, 10^{15} F cm^{-2}, 52 keV, annealed in dry N_2 at $1000°C$ for 1h, wet oxidized at $1100°C$ for 1h, Sirtl etched 15 sec, Nomarski interference contrast optics, magnification 430x.

Table II Diode leakage currents, reverse bias of 3V applied to n^+ region of a n^+p^+ diode, p^+ region by implantation, n^+ region by PH_3 diffusion, 15 min at $1020°C$. Annealing treatment, 1h dry N_2, followed by 1h wet oxidation at $1100°C$. Diode areas, D_1 = .1 mil^2, D_2 = .5 mil^2, D_3 = 2 mil^2, D_4 = 65 mil^2, D_5 = 150 mil^2.

Fluence cm^{-2}	Species	Anneal. Temp.	Reverse Current In Picoamps				
			D_1	D_2	D_3	D_4	D_5
10^{14}	B	1100	.05	.27	.14	11.4	22.8
10^{14}	BF	1100	.01	.09	.11	15.3	23.
10^{14}	BF_2	1100	.036	1.24	.042	2.98	61.2
10^{15}	B	1100	.080	.142	.402	32.0	69.6
10^{15}	BF	1100	.205	.322	2.22	516	386
10^{15}	BF_2	1100	.102	.135	.72	82	146.2
10^{14}	B	1000	.12	1.08	4.92	232	592
10^{14}	BF	1000	1.22	3.4	225	1450	5×10^6
10^{14}	BF_2	1000	<1.0	<1.0	<1.0	–	36.4
10^{15}	B	1000	416	3.96×10^4	4.86×10^4	4.6×10^6	4.04×10^6
10^{15}	BF	1000	24.6	36.8	9.5×10^2	3.83×10^4	1.39×10^6
10^{15}	BF_2	1000	.01	.05	.145	3.0	17.6

to cause the development of any ternary defects for all four species implanted with a fluence of 10^{14} cm^{-2}. For 10^{15} cm^{-2} B, BF, and BF$_2$, a few stacking faults are found and the background is essentially clear of the etch pit clusters encountered in specimens subjected to the 1h oxidation alone. As seen in Fig. 4, only shallow circular pits are found. It is not certain whether they are related to lattice defects.

The use of a 1h anneal in dry N_2 at $1000°C$ results in complete solution of defect nuclei only for the 10^{14} F cm^{-2} implantation. For the 10^{14} cm^{-2} B, BF, and BF$_2$ implantations the stacking fault concentration was significantly reduced and the presence of discrete etch pits in the background was eliminated. For the 10^{15} F cm^{-2} implantation we had a marked reduction in dislocation loop concentration, Fig. 5. For the 10^{15} B, BF, and BF$_2$ cm^{-2} implantations, there was noticable reduction in concentration of stacking faults and background defects. On the basis of this evidence it did not appear likely that an annealing treatment of 1h at $1000°C$ would be successful in eliminating reverse leakage.

Table II summarizes the electrical measurements made on diodes of five different areas: Each value listed represents the average obtained from measurements on five identical diodes. We find that the leakage current appears to be significantly greater for the lower annealing temperature with two notable exceptions. These are the diodes prepared with 10^{14} and 10^{15} cm^{-2} BF$_2$ implantations. Previous work on elemental B implantation [4] correlated ternary defects with emitter diffusion-induced dislocations and these in turn with leakage currents. Further studies are contemplated to determine why the BF$_2$ implanted structures which do appear to develop ternary defects, do not exhibit high leakage currents when processed into n p diodes.

ACKNOWLEDGEMENTS

The author thanks Dr. Arthur Hochberg, Ralph Miller and Anthony Fern for their support and useful discussions, and Leonard Braun and Jay Levine for preparing the ion-implanted specimens.

REFERENCES

1. J.F. Gibbons, Proc. IEEE **9**, 1062 (1972)
2. Masao Tamura, Appl. Phys. Letters, **23**, 651 (1973)
3. S. Prussin, J. Appl. Phys. **45**, 1635 (1974)
4. S. Prussin, Spring Meeting, E.C.S. San Francisco May 1974, abstract no. 85 (submitted for publication, J.E.C.S.)
5. H. Müller, H. Ryssel, and I. Ruge, Proceedings of the Second International Conference on Ion Implantation in Semiconductors, edited by I. Ruge and J. Graul (Springer Verlag, New York, 1971)

TECHNIQUES FOR STUDYING ION IMPLANTATION IN DIAMOND

J. F. Morhange, R. Beserman

Laboratoire de Physique des Solides, Université Paris VI,

Tour 13, 4 place Jussieu, Paris 5e, France

J. C. Bourgoin

Groupe de Physique des Solides de l'E.N.S. Université

Paris VII, Tour 23, 2 place Jussieu, Paris 5e, France

P. R. Brosious, Y. H. Lee, L. J. Cheng and J. W. Corbett*

Department of Physics, State University of New York at

Albany, Albany, New York 12222, U.S.A.

ABSTRACT

Measurements using electron paramagnetic resonance, optical absorption, Raman scattering and luminescence have been performed in ion implanted type IIa diamonds. Results are described which show how these techniques can be used to monitor the total amount of damage as well as the behaviour of point defects introduced by implantation.

INTRODUCTION

The aim of this work is to investigate the applicability of different techniques: electron paramagnetic resonance (E.P.R.),

---

*This work was supported in part by the Office of Naval Research under Contract No. N00014-70-C-0296.

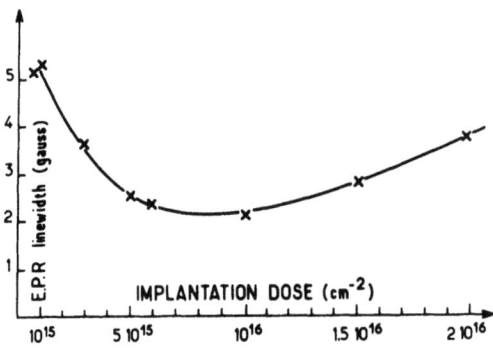

Fig. 1. Variation of the linewidth of the EPR isotropic signal with the dose of nitrogen ions implanted at 70 keV.

optical absorption, Raman scattering (R.S.) and luminescence to the measurements of the total amount of disorder, as well as to the detection of point defects, introduced by ion implantation in diamond. In this paper we shall describe the use of these techniques and illustrate them by providing results concerning the introduction rate of defects and their annealing in natural IIa diamonds implanted at room temperature and high temperature with ions (mostly nitrogen) of moderate (70 keV) and high energy (0.5 to 3 MeV).

ELECTRON PARAMAGNETIC RESONANCE

The EPR spectrum observed in diamond implanted at room temperature with ions of moderate or high energy is an isotropic line of Lorentzian shape at g = 2.0023. The linewidth of this line ranging from 2 to 5 gauss varies, as shown in Fig. 1, with the dose of implantation. Because this spectrum is very similar to the spectrum observed on diamond surfaces [1], Brosious et al. [2] suggested that it is associated with dangling bonds on internal surfaces.

In case of implantation (10^{16} N^+/cm^2) at high temperature (600°C) with the incident energy 1.7 MeV the spectrum appears to be a superposition of signals coming from two different centres: one is the isotropic line described above; the other line appears to have a small anisotropy with a minimum linewidth with a maximum intensity along the <100> direction and a maximum linewidth with minimum intensity along the <111> direction. Because the orientation-dependent linewidth is identical at X and Q bands, we think this new anisotropic center is most likely a spin-1 center arising from a small D tensor having a <111> axial symmetry.

Figure 2 shows that the introduction of spins for the iso-

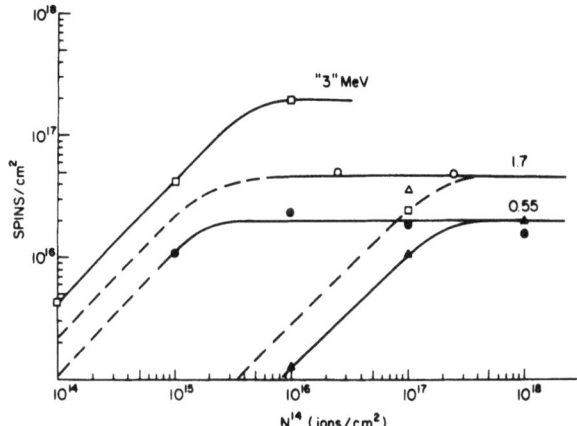

Fig. 2. Defect spin concentration versus ion implantation fluence at several nitrogen ion energies and diamond temperatures (□ 25°C, 3 MeV; o 325°C, 1.7 MeV; Δ 25°C, 1.7 MeV; ■ 600°C, 1.7 MeV; ● 20°C, 0.55 MeV; ▲ 650°C, 0.55 MeV).

tropic line increases linearly with the dose of implantation until it saturates. This saturation occurs for the same spin concentration, whatever is the ion energy. The critical dose for which the saturation occurs increases with the temperature of implantation (indicating that annealing occurs during high temperature implantation).

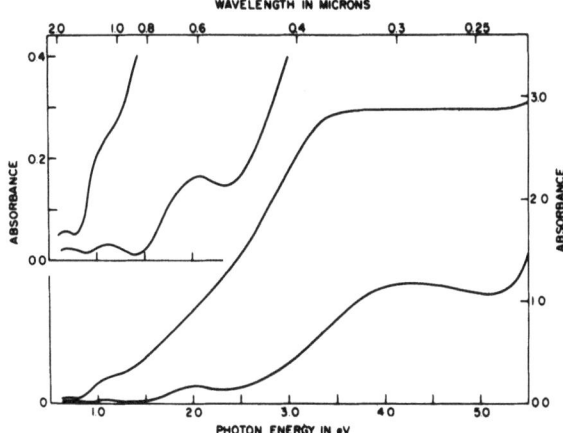

Fig. 3. Optical absorbance versus photon energy (top curve 10^{16} N+/cm^2, 18°C, 0.55 MeV; bottom curve 10^{15} C+/cm^2, 17°C, 0.55 MeV).

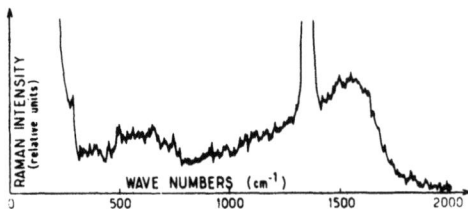

Fig. 4. Raman spectrum of a diamond implanted with 10^{16} N^+/cm^2 at 70 keV.

OPTICAL ABSORPTION

The implanted samples undergo visible color changes as a function of implantation dose. An increase in the absorption coefficient at 4880 Å is observed with increasing doses, until it reaches saturation. Preliminary optical absorption measurements in the visible and UV range show (Fig. 3) that implantation introduces a broad absorption band (for energies higher than 3 eV) and two small bands at respectively 2 eV and 1.12 eV. The broad UV band, also observed in neutron or electron irradiation diamond [3], is thought to arise from nitrogen aggregates; the 2 eV band is the GR1 center, interpreted as being associated with the vacancy.

RAMAN SCATTERING

As shown in Fig. 4, the Raman scattering cross-section of implanted diamond is composed of two spectra: the intense line at 1331 cm^{-1} is the Raman active zone-center mode of diamond; the continuous spectrum, which presents a maximum at 1550 cm^{-1} is due to the disorder layer produced by the implantation [4]. When the dose of implantation increases, the intensity of the crystalline peak decreases and the intensity of the broadband increases. The variation of the crystalline intensity I_c at 1331 cm^{-1} can be used to monitor the introduction of the disorder; the quantity $1-(I_c/I_0)$, where I_0 is the crystalline intensity prior to implantation, increases with the dose of implantation and saturates near 10^{16} ions/cm^2 in the case of 70 keV N^+ implantation.

LUMINESCENCE

Prior to implantation natural IIa diamonds often present an emission band at 19868 cm^{-1} with a phonon replica which is associated with centers involving nitrogen [5]. No change in the spectrum is observed after ion implantation. But after annealing (800°C for 1 hour) new lines appear which indicate the appearance of point defects (see Fig. 5).

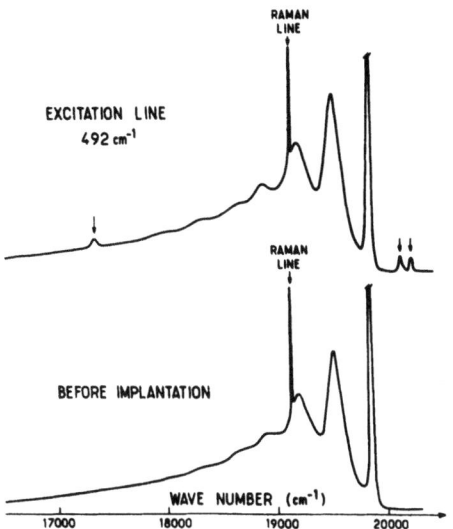

Fig. 5. Luminescence spectrum of a sample implanted with 10^{16} cm^{-2} 70 keV nitrogen ions and annealed at 800°C.

REFERENCES

1. G. K. Walters and T. L. Estle, J. Appl. Phys. 32, 1854 (1961).

2. P. R. Brosious, J. W. Corbett and J. C. Bourgoin, Phys. Stat. Sol. (a) 21, 677 (1974).

3. C. D. Clark and E. W. J. Mitchell, Rad. Effects 9, 219 (1971).

4. J. F. Morhange, R. Beserman and J. C. Bourgoin, (to be published).

5. P. J. Dean, Phys. Rev. 139, A588 (1965).

THE CHARACTERIZATION OF IMPLANTED LAYERS USING THE CONDUCTIVITY MODULATION EFFECT

M.J. Howes, D.V. Morgan, P. Ashburn

Electrical Engineering Department, Leeds University

Leeds LS2 9JT, England

ABSTRACT

In this paper we have considered in detail how the introduction of radiation damage centres into the depletion region of silicon p-n junctions effects the small signal terminal admittance of the diodes. The results are interpreted in terms of conductivity modulation arising from a trapping level 0.31eV above the conduction band which gives rise to a hole capture type recombination process. Damage profiles have been extracted from the forward bias capacitance voltage curves.

INTRODUCTION

Capacitance/voltage measurements on reverse biassed p-n junctions or Schottky barriers have been widely used to characterize ion implanted layers (1). In this paper we report some preliminary results on the forward biassed, small signal admittance of silicon p^+n junctions in which damage centres have been introduced into the depletion region.

The diodes used for this work were silicon p^+n diodes formed by the implantation of boron into a slice of n-type silicon followed by a $900^\circ C$ anneal to remove the radiation damage. Recombination centres were selectively introduced into the depletion region of the diodes by implanting with carbon ions. The profile of the damage could be accurately controlled by varying the energy and dose of the implantation.

The defects resulting from the carbon implantations have been characterized using the techniques of thermally stimulated current and thermally stimulated capacitance and reported in previous publications (2,3).

THEORY

It has been shown by a number of workers (4,5) that an approximate small signal equivalent circuit for forward biassed asymmetrical p-n junction diodes consists of a parallel GC network in series with a parallel GL network (Fig. 1). The GC network arises as a result of the depletion layer capacitance C_D the diffusion capacitance C_d and the diffusion conductance G_d whilst the GL network models conductivity modulation effects in the bulk region of the device.

The inductive properties of p-n junctions may be understood by considering a device under steady state forward biassed conditions and subject to a small increase δV in terminal voltage at $t = 0$. At $t = 0$ a fraction of δV will appear across the bulk region of the device as a result of the finite conductivity of the bulk region. As time elapses, however, this fraction will decrease due to the increase in conductivity resulting from hole injection (i.e. the current lags the voltage). As far as sinusoidal terminal voltage perturbations are concerned this phenomenon may be modelled in an equivalent circuit by an admittance in which the susceptive part is inductive. It is of interest to note at this point that the magnitude of inductive susceptance is proportional to the rate of change of conductivity of the bulk region and therefore to the rate at which the hole concentration changes at the edge of the depletion region. Thus the effect of conductivity modulation on the terminal admittance of the device is expected to be influenced by recombination centres at the edge of the depletion region. We consider the role of hole capture type centres in this respect later.

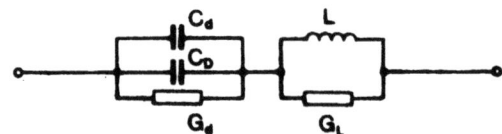

Fig. 1 The equivalent circuit model for a forward biassed, asymmetrical p-n junction.

RESULTS AND DISCUSSION

Experiments were carried out to determine the effect of multiple trapping centres introduced by carbon implantation on the small signal admittance of the ion implanted diodes described earlier. Measurements of terminal admittance were made using the Wayne Kerr B201 bridge at 1MHz (signal level 20mV). The results of these measurements are shown in Figs. 2(a) and 2(b). Computer calculations using the methods described in a previous publication [7] are also presented in the case of diode 1. The terminal C-V characteristic of diode 1 is qualitatively as expected with the depletion layer capacitance dominant at low forward bias, diffusion capacitance causing a rapid increase in terminal capacitance as the injection level increases until at sufficiently high injection levels the inductive nature of conductivity modulation in the bulk region results in the terminal capacitive susceptance reaching a maximum. Further increase in forward bias results in a rapid decrease in effective terminal capacitance until the terminal susceptance becomes inductive. The results of the computer analysis are in quantitative agreement with these results.

However, for the low dose carbon implanted diodes (diodes 2-4 shown in Fig. 2(a) the experimental results show significant deviations from this behaviour. The C-V characteristics of these devices exhibit pronounced minima at approximately the same value of forward bias. The form of minima indicates a deepening, a broadening and an extension to lower forward bias as the carbon dose increases.

The proposed explanation of the terminal admittance characteristics of these diodes is as follows. In previous papers [2,3] we have shown that carbon implanted diodes are characterised by multiple trapping centres and in particular we have identified a trapping level 0.31eV from the valence band. In these diodes this level will result in hole capture type recombination processes.

At a forward bias corresponding to the initial peak in the C-V curves hole capture mechanisms slow down the rate of increase in conductivity of the bulk region and the capacitive susceptance of the device decreases rapidly. The capacitive susceptance reaches a minimum at a forward bias of $\sim$ 0.53V for all the diodes corresponding to a depletion layer width of approximately 0.24μm. This coincides with the point at which the density of trapping sites is expected to be a maximum [8]. Further increase in forward bias results in the density of traps at the edge of the depletion region decreasing and the effect on the inductive properties of the device becoming rapidly less pronounced. The result is that the terminal susceptance increases until at sufficiently high injection levels conductivity modulation effects once more dominate the device small signal behaviour.

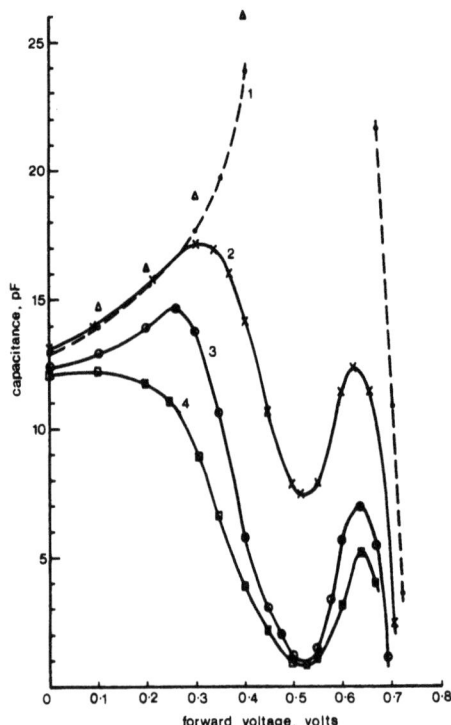

(a)

Fig. 2

The terminal capacitance C vs forward bias voltage V.

(a) unimplanted diode (diode 1), and low dose diodes:- diode 2 (1×10^{11}cm^{-2}), diode 3 (5×10^{11}cm^{-2}) and diode 4 (1×10^{12}cm^{-2}).

(b) high dose diodes:- diode 5 (5×10^{12}cm^{-2}), diode 6 (1×10^{13}cm^{-2}), and diode 7 (5×10^{13}cm^{-2})

(Δ theoretical results for the unimplanted diode)

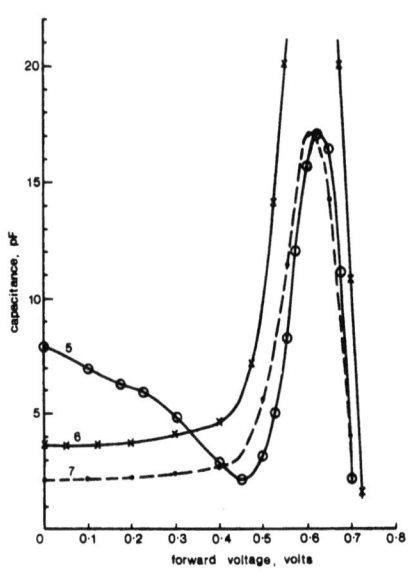

(b)

The diode which does not show the effects described above was the one subjected to the heaviest dose of carbon ions (diode 7 shown in Fig. 2(b)). In this case it is suggested that charge compensation is significant (as suggested by Ashburn and Morgan[2]) and the diode is effectively transformed into a PIN structure. This is supported by the C-V plot for this diode which shows that the terminal capacitance is essentially constant up to moderate forward bias voltages. This is precisely the expected characteristic of a PIN diode structure. Also a simple calculation of the capacitance of the forward bias structure yields a value of 2.2 pF in good agreement with experiment. Charge compensation is also apparent from the C-V curves appropriate to diodes (2-4). This manifests itself in a small decrease in the zero bias depletion layer capacitance. However, the compensation is a small effect ($\sim \frac{2\Delta C}{C} = 8\%$) for carbon implantation doses up to 10^{12} ions cm^{-2}.

Diodes 5 and 6 (see Fig. 2(b)) have properties which are intermediate between those of the high dose and low dose diodes. The characteristics exhibit pronounced minima, but at voltages which differ from those obtained in the characteristics of the low dose diodes. Diode 5 has a minimum at 0.45 volts, and diode 6 a minimum at 0.07 volts. These devices also show the presence of the charge compensation effect which was observed in the high dose diode.

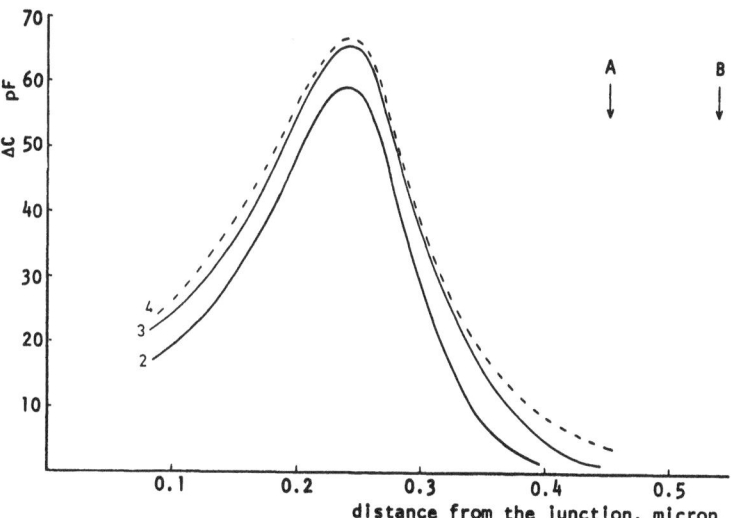

Fig. 3   The damage profile derived from C/V measurements (diodes 2, 3 and 4). A is the position of the peak in the carbon profile, and B is the position of the edge of the depletion width at zero bias.

Consider now the diodes which have received relatively light implantation doses (diodes 2, 3 and 4). If we assume that the reduction in capacitance ΔC is directly proportional to the density of defects at the edge of the depletion region and is not dependent on the injection level* then the conductivity modulation data can be converted into a damage profile. In earlier publications [2,3] it has been shown that, in diodes 2, 3 and 4, all the defects are in the neutral charge state at room temperature. It is therefore possible to convert the voltage scale in Fig. 2 into a depth scale (see Fig. 3) using the following relationship

$$x = \left[\frac{2\varepsilon\varepsilon_o}{e\,N_D} (\phi_o - v) \right]^{\frac{1}{2}}$$

where $N_D = 3 \times 10^{15}$ ions cm^{-3} and $\phi_o = 0.67$ volts

Implicit in the use of this equation is the assumption that the damage level is not great enough to destroy the abrupt nature of the junction and thus we have only attempted to obtain profiles for diodes 2, 3 and 4. The point A shows the position of the projected range of the carbon implantation and B indicates the position of the depletion edge at zero bias.

The general shape of the profiles shown in Fig. 3 is in good qualitative agreement with thos predicted by Brice[8]. Furthermore our projected range ($\sim$0.63μm) compares well with his theoretical figure of $\sim$0.60 bearing in mind the uncertainty in our values resulting from the uncertainty in the thickness of the surface oxide film (SiO$_2$).

REFERENCES

(1) J.W. Mayer, L. Eriksson, J.A. Davies: "Ion Implantation", Academic press (1970).

(2) P. Ashburn, D.V. Morgan: Solid state electronics, vol.17, pg.689 (1974).

(3) M.J. Howes, D.V. Morgan, P. Ashburn: Submitted to Solid State Electronics.

(4) I. Ladany: I.R.E. Trans. on Electron Devices, pg.303, October 1960.

* A detailed theoretical study is at present underway to investigate the validity of this approximation.

(5) H. Melchior, M.J.O. Strutt: I.E.E.E. Trans. on Electron Devices, pg.47, February 1965.

(6) J.L. Moll: "Physics of Semiconductors" pg.128, McGraw-Hill, 1964.

(7) P. Ashburn, D.V. Morgan, M.J. Howes: Accepted for publication in Solid State Electronics.

(8) D.K. Brice: Radiation Effects, vol.6, pg.77 (1970).

RADIATION DAMAGE II

BACKSCATTERING AND ESR STUDIES IN HEAVILY DAMAGED LAYER

Kohzoh Masuda

Faculty of Engineering Science, Osaka University

Toyonaka, Osaka, Japan

ABSTRACT

Two important technologies (Backscattering and ESR) in measurements of physical properties in implanted material are noticed as complementary technologies.

A typical feature of ESR of the amorphous centers is described and inhomogeneous character which was found recently by microwave frequency dependence at low temperature in spin system is mentioned.

INTRODUCTION

At the very high implantation dose such as that over 6×10^{14} phosphorus-ions/cm^2 with acceleration voltage of 200keV in silicon, one can obtain perfectly disordered lattice (normally this state is called as amorphous state). Annealing behavior of the amorphous layer is quite different from that of single crystal. For example, electrical properties of amorphous layer is recovered at only about 550°C annealing [1].

An energy level diagram for the amorphous state for tetrahedral configuration is propsoed by Mott and Davies [2] as shown in Fig. 1 (d). They suggested that some kind of voids are necessary to contain some unpaired electrons in the sample which are contributed in conductivity.

The change in amorphous peak position and height with annealing was shown in Fig. 2 by Mayer et al. [3]. This means that the recovery of crystallinity is started from interface between amorphous layer and crystal just behind it.

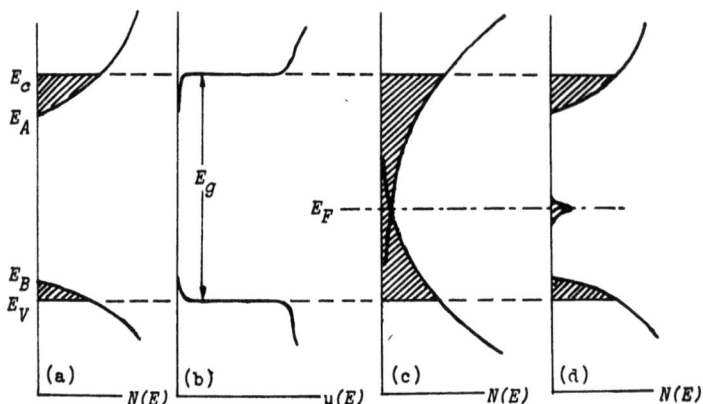

Fig. 1. Density of states and mobility as functions of energy in amorphous semiconductors. The energy difference (Ec - Ev) defines a mobility gap Eg{sec(a) and (b)}. In certain glasses the density of states is shown by Cohen, Fritzsche, Ovshinsky model as (c).

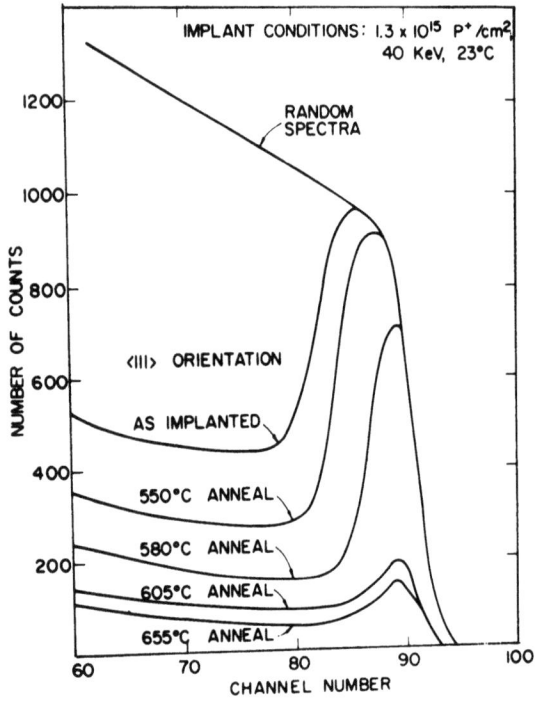

Fig. 2 Effect of annealing on the aligned backscattering spectra from a silicon sample implanted at room temperature. The analyzing beam was 1.0-MeV helium ions.

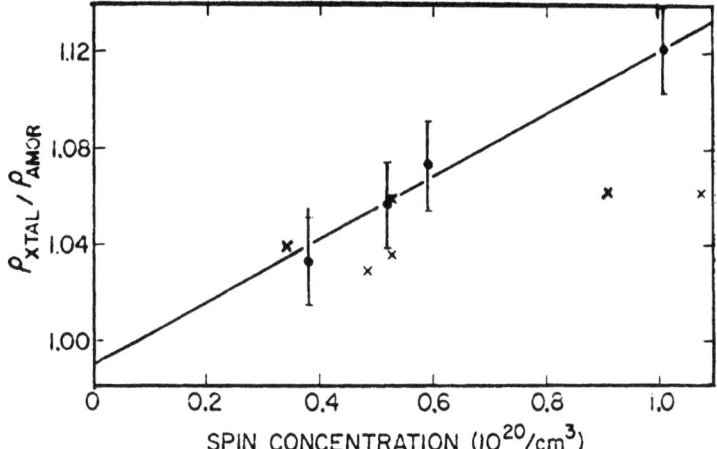

Fig. 3 Ratio of the density of crystalline to the density of amorphous layer against to the spin concentration in amorphous silicon layer.

A change of density of weight has been found by Brodsky et al. [1] by means of backscattering measurements as shown by Fig. 3. The backscattering yield is proportional to the total cross sections of target atoms. Therefore this method is suitable to measure the absolute number of target atoms and hence the target density. This is one of the most superior features of backscattering method.

The backscattering method has small sensitivity for vacancies. If the defect has only vacancies and no displaced atoms and impurities few backscattered particles are obtained. On the other hand, ESR measurement is very suitable and sensitive method to detect local structure of the vacancies or voids when they have unpaired electrons. Oscillator strenghts of unpaired electrons are always the same. So that the measurement of absolute number of unpaired spin is always accurate.

Backscattering and ESR methods are comlementary techniques to understand the structure of implanted solid.

THREE REGIONS OF IMPLANTATION DOSE

With increasing of implantation dose ESR spectrum is changed depending on low, intermediate and high dose region. For example, ESR spectrum in the sample implanted by 200keV phophorus ions can be divided into three region of implantation dose. [4].

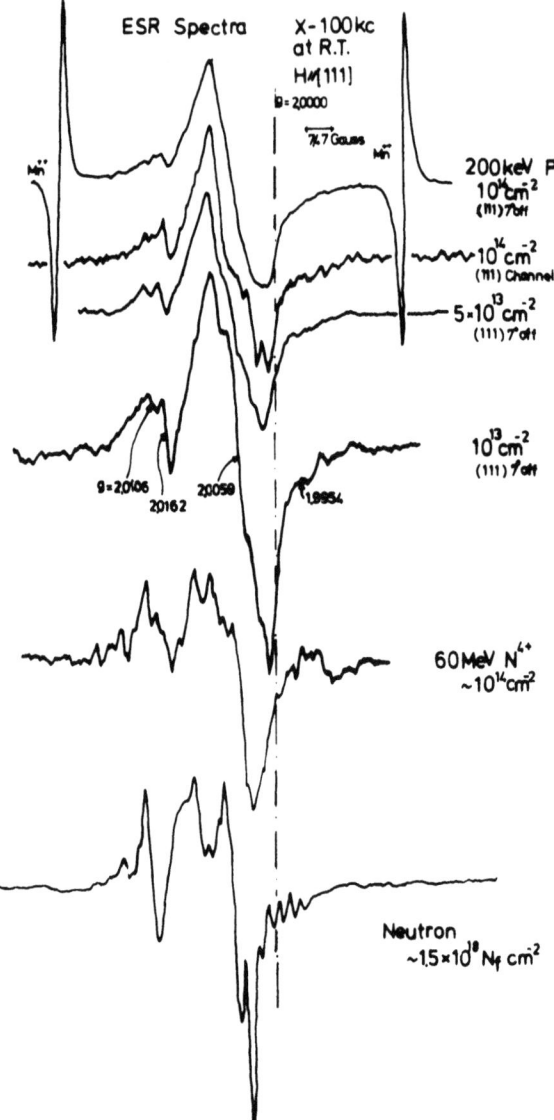

Fig. 4 Concentration dependence of ESR signal of silicon implanted by 200keV phosphorus. Measurement was done at room temperature. Bottom figure is signal from the sample bombarded by reactor neutron.

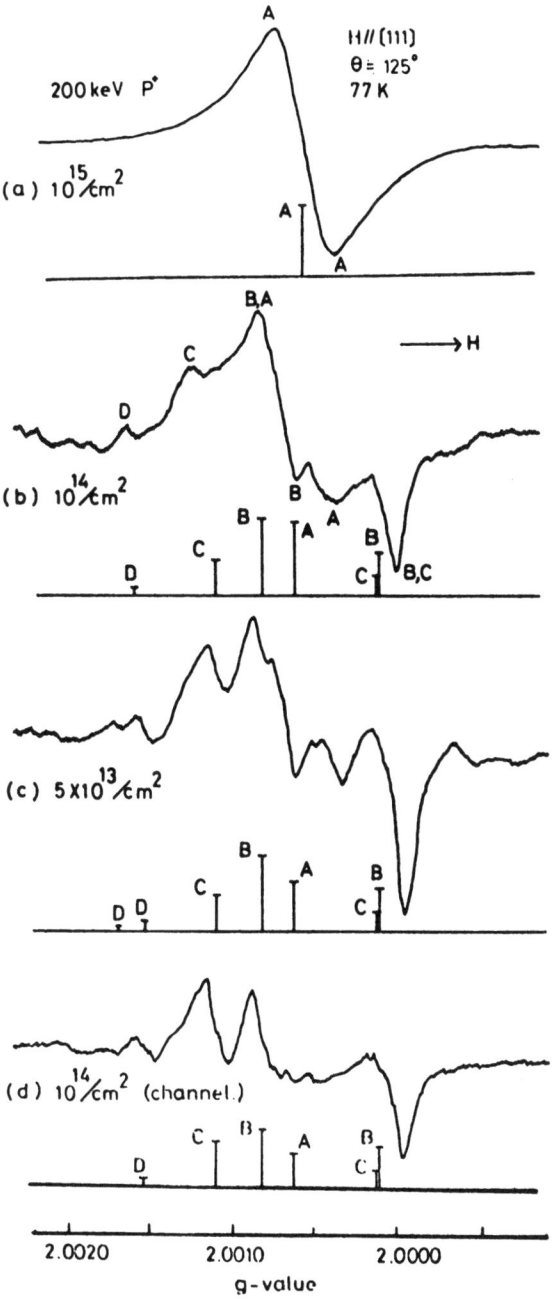

Fig. 5 Concentrate dependence of ESR signal of silicon implanted by 200keV phosphorus. Measurement was done at 77°K. Bottom figure is signal from the sample bombarded at channeling direction.

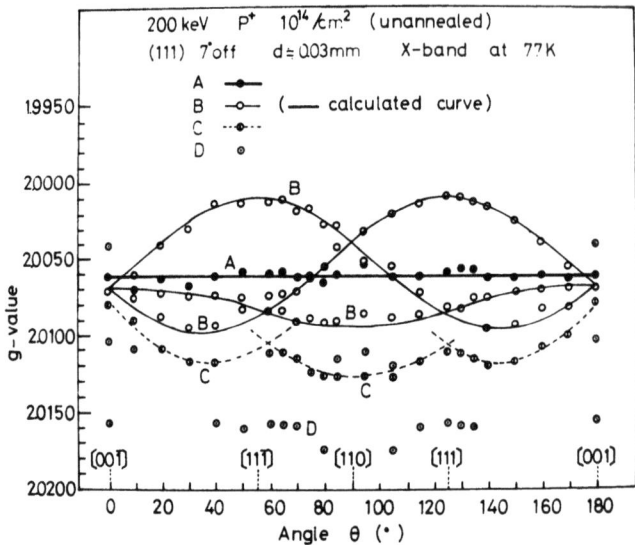

Fig. 6 Angular dependence of g values in region 2)

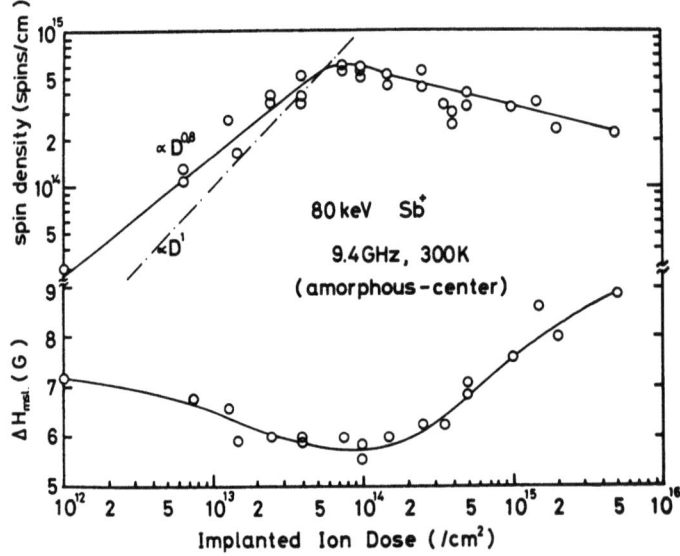

Fig. 7 Dose dependence of spin density and line width in gauss (distance between maximum slopes)

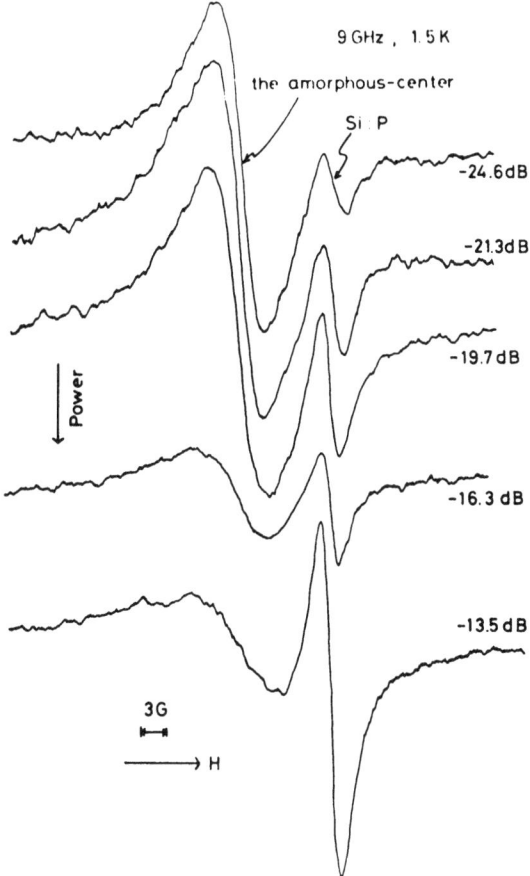

Fig. 8 Power dependence of ESR signal.

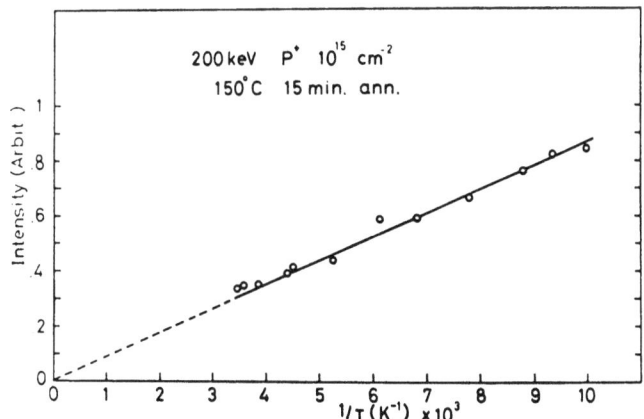

Fig. 9 Temperature dependence of ESR signal between 325K to 77K.

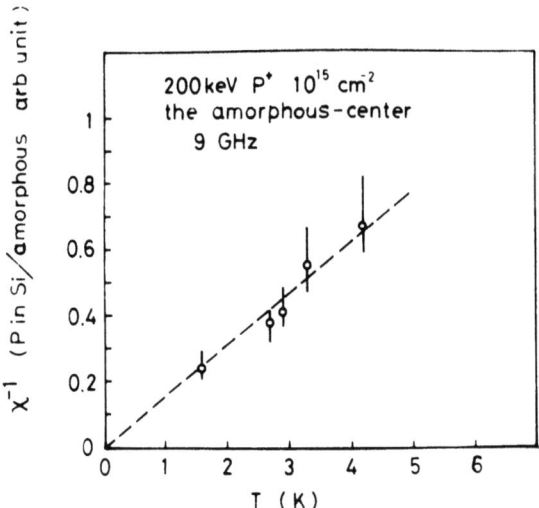

Fig. 10 Temperature dependence of ESR signal between 1.6K and 4.2 K.

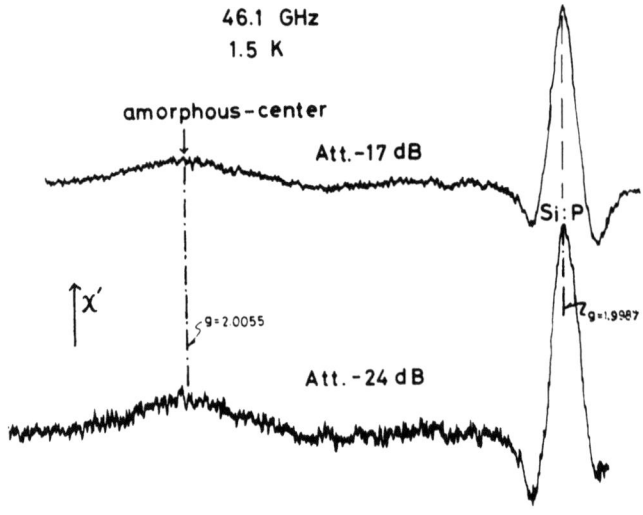

Fig. 11 Line width of the amorphous signal at 46.1GHz 1.5K.

Region 1) Dose $\leq 1 \times 10^{13}$ ions/cm^2
In this region isolated vacancies are observed. They are
negative divacancy[3] (Si-G7)
neutral 4-vacancy[5] (Si-P3) produced by O implantation
(Si-B2)
and (Si-P3) produced by N and P implantation
Those centers have anisotropic g values.

Region 2) Dose between 1×10^{13} ions/cm^2 and 6×10^{14}/cm^2
In this region, both isolated centers and amorphous state centers are observed.
Isolated centers are Si-B2
Si-O1
and Si-P3 produced by P implantation
Beside those centers amorphous center is detected.

Region 3) Dose over 6×10^{14} ions/cm^2.
Isotropic singel Lorentzian of line width of 5~6 gauss is obtained in this region. Spin density of this spectrum is saturated at 2×10^{20} spin/cm^2.

RESULTS AND DISCUSSIONS

Changes of spectrum depend upon the dose are shown in Fig. 4 and Fig. 5 at different temperature. One can observed easily that the fine structure disappeared at high dose.

Isotropic signal and anisotropic signal are easily separated by changing crystal axis about the static magnetic field as shown in Fig. 6. Signal from the amorphous centers is clearly separated from others.

Spin density in silicon implanted with 80keV Sb is shown in Fig. 7. Dose of 10^{14} ions/cm^2 corresponds to both saturation point of spin density and appearance of amorphous signal of backscattering spectrum

The main interest of this paper is to find some fine structure of amorphous center. Unpaired electron located at voids in amorphous center interacts with its sorroundings. They may contribute to the electronic hopping conduction. If the electron system includes two different spin systems one can obtain two different saturation effects by changing the power of measurement. It is not so easy to distinguish two groups of spin because of homogeneous saturation as shown in Fig. 8. Temperature dependence of intensity shows that this spin system follows Curie's law as shown in Fig. 9 and Fig. 10. This means the level at which spin exists is a singly occupied level.

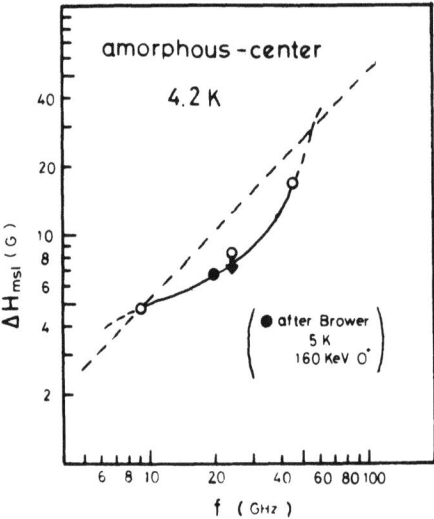

Fig. 12 Frequency dependence of line width (distance between maximum slopes) at 4.2K.

When the electron system has many groups of spins (inhomogeneous system) the line width will increase depending upon the increase of observing frequency. But if exchange of electron is faster than observing frequencies (homogeneous system) then line width will not be changed by changing frequency. Line width measurement at 46.1 GHz is shown in Fig. 10 and frequency dependence is also shown in Fig. 11. So that this spin system has two characters which are inhomogeneous and homogeneous.

SUMMARY

In amorphous phase voids model is quite reasonable in view points of backscattering and ESR experiments. In amorphous phase the isotropic ESR signal is observed which can be used for criterion of formation of amorphous layer. Those are already established.

Spin system in voids in amorphous layer is consisted of at least more than two groups. This is a problem and also a key to solve a problem in the future.

ACKNOWLEDGMENTS

Thanks are due to Professor S. Namba for his valuable discussion.

Thanks are also due to Dr. K. Gamo and Mr. K. Murakami for performing experiments and to Mr. K. Mino and Mr. K. Kawasaki for assistance during experiments.

REFERENCES

[1] F.F. Morehead and B.L. Crowder: J.A.P. 43 (1972) 1112.
[2] N.F. Mott and E.A. Davis: Electronic Processes in Non-Crystalline Materials, Clarendon Press, Oxford, 1971. p.199.
[3] J.W. Mayer, L. Ericksson, S.T. Picraux and J.A. Davies; Can, J. Phys. 46 (1968) 663.
[4] K. Murakami, K. Masuda, K. Gamo and S. Namba: J.J.A.P. 12 (1973) 1307.
[5] G.D. Watkins and J.W. Corbett: Phys. Rev. 123 (1962) 1605.

ON THE DETERMINATION OF DEFECTS DISTRIBUTION IN IMPLANTED

LAYERS BY MEANS OF BACKSCATTERING TECHNIQUE

NORIAKI MATSUNAMI and NORIAKI ITOH

Department of Nuclear Engineering, Nagoya, Japan

Furo-cho, Chikusa-ku, Nagoya, Japan

The various methods of deriving the random fraction caused by defects and of obtaining the depth profile of defects in Si are examined, such as the single scattering, multiple scattering, plural scattering and diffusion models. It is found that the diffusion model gives the appropriate result over the most wide range of defect concentration. Methods are suggested for deriving both the random fraction caused by defects and the depth profile of defects experimentally.

I. INTRODUCTION

Dechanneling measurements have been widely used for extraction of the depth profile of defects in ion-implanted layers.[1] In deriving the depth profile of defects, the relation between the increment of the random fraction Δx caused by defects and depth z from the surface is necessary. The single scattering,[2] multiple scattering,[2] plural scattering[3] and diffusion models[4] have been used to obtain the relation between Δx and z. The depth profiles derived from backscattering experiments depend on the method of analysis and do not always agree with the theoretical depth profile.[5] A typical experimental depth profile[6] of defects derived with the single scattering model in Si implanted with 175 keV B is compared with the theoretical depth profiles of defects in Fig. 1. From the figure, it is seen that the experimental depth profile has a small value near the surface and shows a rapid increase at large depth: being not in agreement with the theoretical depth profiles. The theoretical depth profiles agree well with the depth profiles directly determined from EPR and optical experiments. Moreover, Eisen[7] pointed out that the

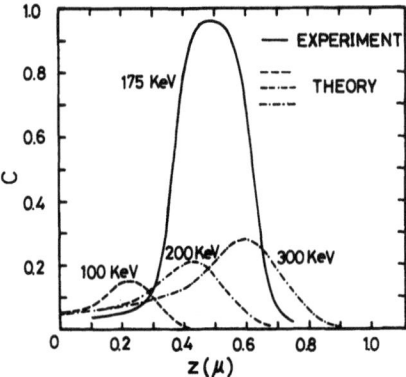

Fig. 1. Comparison between a typical experimental depth profile of defects derived with the single scattering model and the theoretical depth profiles for B implanted Si. The solid curve is the experimental result obtained by Ziegler, with 2.5 MeV He$^+$. The broken, double dash-dot and dot-dash curves are the theoretical depth profiles calculated by Brice. The energies in the figure are the implantation energies.

half-width of the depth profile of defects derived with the single scattering model increases by a factor two as the implantation dose increases; which is rather improbable.

In order to solve these difficulties, the carefull evaluation of the relation between $\Delta\chi$ and z is necessary, since the fractional concentration c of defects is very sensitive to the value of $\Delta\chi$ used to evaluate c. It has been shown that the diffusion model, which takes the modification of particle distribution by defects into consideration, provides proper $\Delta\chi$ over the wide range of defect concentration for KCl,[8] therefore it is of interest to calculate $\Delta\chi$ in Si with the diffusion model and compare the result with $\Delta\chi$ obtained with the other models.

In the present paper $\Delta\chi$ for 1.0 MeV He$^+$ incident along a ⟨110⟩ channel of Si was obtained with various models. It was found that the $\Delta\chi$ vs z curve obtained with the diffusion model gives the most appropriate results, and that the depth profile similar to the theoretical depth profile may be obtained from the backscattering experiments.

II. COMPARISON OF VARIOUS MODELS

Methods of calculation of $\Delta\chi$ with the single scattering, multiple scattering, plural scattering and diffusion models have

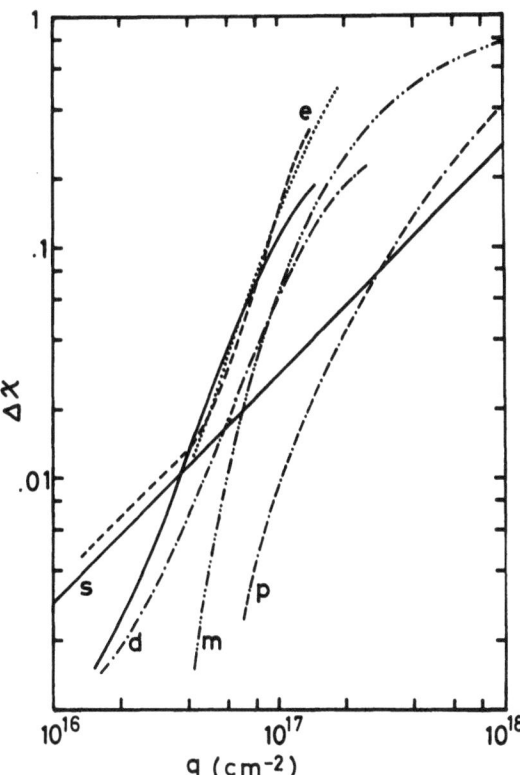

Fig. 2. The increment of the random fraction $\Delta\chi$ caused by defects, calculated with the single scattering (s), multiple scattering (m), plural scattering (p) and diffusion (d) models, as a function of surface concentration q of defects, for 1 MeV He$^+$ incident along a <110> channel of Si. The curves denoted by d were obtained for the defect concentration c=1.5 % (———) and c=5 % (—·—·). The curves denoted by e are derived from the experimental backscattering spectrum of Si, implanted with B at room temperature (broken line) and at liquid nitrogen temperature (dotted line), measured by Eisen[7] with 1 MeV He$^+$, and from the depth profile calculated by Brice.

been given in references.[3] For the plural scattering the distribution suggested by Meyer[9] was used. For the diffusion model, the diffusion coefficient[8] was calculated taking the upper limit of integration as $10\varepsilon_\perp^c$, where $\varepsilon_\perp^c$ is the critical transverse energy, which equals to the particle energy times the square of the experimental critical angle.

The $\Delta\chi$ obtained with the various methods for 1 MeV He$^+$ incident on a $\langle 110 \rangle$ channel of Si are shown in Fig. 2. Since the $\Delta\chi$ obtained with single, multiple scattering models depends on z through a function of total defect concentration or surface concentration q of defects, given by $q = N\int_0^z c(z')dz'$, where N is the number of the atoms per cm^3, $\Delta\chi$ is plotted as a function of q in Fig. 2. The curves denoted by d in Fig. 2 were obtained with the diffusion model for the different defect concentrations, it is to be noted that the $\Delta\chi$ obtained with the diffusion model is almost independent on c and only a function of q. It is seen from the figure that the $\Delta\chi$ calculated with the diffusion model increases slowly for small q ($q \lesssim 10^{16}$ cm^{-2}), similarly to the $\Delta\chi$ obtained with the single scattering model. For large q ($q \gtrsim 10^{17}$ cm^{-2}), however, the $\Delta\chi$ calculated with the diffusion model increases more rapidly than the $\Delta\chi$ calculated with the single scattering model and finally converges to the $\Delta\chi$ obtained with the multiple scattering model. This behavior that the $\Delta\chi$ calculated with the diffusion model is larger than those calculated with other models at the intermediate value of q has been ascribed to the modification of the particle distribution caused by defects.[8] The plural scattering model yields the value of $\Delta\chi$, which is smaller than those calculated with other models.

III. DEPTH PROFILES OF DEFECTS

The fact that the $\Delta\chi$ obtained with the diffusion model begins to increase at a relatively low surface concentration of defects q ($q = 2 \sim 4 \times 10^{17}$ cm^{-2}) may explain the difference between the experimental and theoretical depth profile of defects shown in Fig. 1. The fractional concentration of defects c is given by

$$c(z) = (\chi_2(z) - \chi_R(z))/(1 - \chi_R(z)),$$

and

$$\chi_R(z) = \Delta\chi(q) + \chi_P(z), \quad (1)$$

where $\chi_2(z)$ is normalized aligned yield obtained from the backscattering experiments and $\chi_P(z)$ is the random fraction for pure crystal. In the single scattering model, in which $\Delta\chi$ is nearly a linear function of q, the rapid increase in $\chi_2(z)$ gives the rapid increase in c(z). On the contrary, according to the diffusion model, it is

DETERMINATION OF DEFECTS DISTRIBUTION

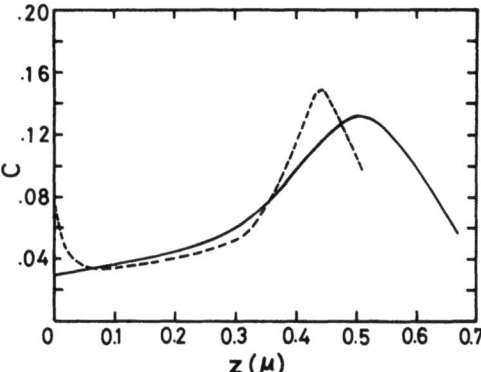

Fig. 3. Brice's theoretical depth profile and the depth profile derived from experimental data for two different implantation doses (see text).

suggested that the increase in $\chi_2(z)$ represents the increase in $\Delta\chi$ rather than $c(z)$. In this case the increase in $c(z)$ with the increasing depth is not so rapid, and the depth profile of defects similar to that calculated by Brice[5] may be obtained.

The $\Delta\chi$ vs q curve denoted by the broken line in Fig. 2 was derived from the experimental backscattering spectrum from 200 keV B implanted layer in Si measured by Eisen with 1 MeV He^+ and from the depth profile of defects calculated by Brice, which was adjusted so that the depth profile has the maximum at the same depth as the backscattering spectrum. The maximum fractional concentration was determined as 0.13, so that the $\Delta\chi$ at low q agrees with the result of the single scattering model. This value is smaller by a factor 4 than that determined by Eisen. The $\Delta\chi$ begins to increase similarly to the $\Delta\chi$ calculated with the diffusion model. The $\Delta\chi$ vs q curve denoted by the dotted line was obtained from result of implantation at $-150°C$, with the maximum concentration of 0.36 and with the same depth profile. It is to be noted that these two curves are in good agreement. This result suggests that if one uses $\Delta\chi$ vs q curve similar to that obtained with the diffusion model one gets the depth profile which agrees well with the theoretical profile.

Although the diffusion model indicates that $\Delta\chi$ calculated with the single and multiple scattering models is not appropriate, it is not very easy to calculate the $\Delta\chi$ vs q curve with the diffusion model precisely. In fact the comparison of curves d and e shows that the diffusion model is erroneous for large $\Delta\chi$, where the particle distribution is highly distorted. We propose the following two methods of deriving the depth profile of defects. The most convenient method is to derive the depth profile near the surface with the single scattering model and assume the depth profile calculated by Brice.

Another method is to obtain the depth profile under two assumptions: (i) $\Delta\chi$ is only a function of q and (ii) the shape of the depth profile does not depend on the implantation dose. According to the assumption (ii) $c(z)$ can be written as

$$c(z) = f c_0(z). \qquad (2)$$

Here f is a factor depending on the implantation dose and $c_0(z)$ is the normalized depth profile. We rewrite the eq. (1) as the following:

$$\Delta\chi_2(z) = \Delta\chi(q) + c(z)\{1 - \Delta\chi(q) - \chi_p(z)\}, \qquad (3a)$$

and

$$\Delta\chi'_2(z) = \Delta\chi(Fq) + Fc(z)\{1 - \Delta\chi(Fq) - \chi_p(z)\}, \qquad (3b)$$

where $\Delta\chi'_2(z)$ is the normalized aligned yield obtained from the backscattering spectrum with larger implantation dose, that is, $F > 1$. Eq. (3a) and (3b) lead to the equation:

$$\Delta\chi(Fq) = \frac{\Delta\chi(q)\{F(1-\chi_p(z)) - \Delta\chi'_2(z)\} + (1-\chi_p(z))\{\Delta\chi'_2(z) - F\Delta\chi_2(z)\}}{1 - \chi_p(z) - F\Delta\chi_2(z) + (F-1)\Delta\chi(q)}, \qquad (4)$$

Eq. (4) gives $\Delta\chi$ at Fq, if $\Delta\chi$ at q is known, and thus $\Delta\chi$ at any values of q can be obtained by interpolation. The factor F is determined such that c near the surface equals to that obtained with the single scattering model. Figure 3 shows the depth profile of defects obtained with the above two methods. It is seen that the depth profiles obtained with two different methods agrees well, indicating that the above two methods are reasonable. The second method gives the $\Delta\chi$ vs q curve almost the same as curves e in Fig. 2. It follows that the curves e are the most appropriate $\Delta\chi$ vs q curves.

In conclusion, the extraction of the depth profile of defects with the single scattering model or multiple scattering model is not appropriate. A similar shape of the depth profile as calculated by Brice should be obtained if proper $\Delta\chi$ vs q curve is used. The one of the most convenient method would be to measure the backscattering spectrum at various implantation doses and to use the eq. (4) to obtain $\Delta\chi$ and the depth profile.

The authors express their gratitude to K. Morita for valuable discussions. The calculation was made with the Facom 230-60 at Nagoya University.

REFERENCES

1. J. W. Mayer, L. Eriksson and J. A. Davies, Ion Implantation in Semiconductors, (Academic Press, New York, 1970).
2. E. Bøgh, Canadian J. Phys. 46, 653 (1968).
3. J. E. Westmoreland, J. W. Mayer, F. H. Eisen and B. Welch, Radiation Effects 6, 161 (1970).
4. N. Matsunami and N. Itoh, Physics Letters 43A, 435 (1973).
5. D. K. Brice, Radiation Effects 6, 77 (1970).
6. J. F. Ziegler, J. Appl. Phys. 43, 2973 (1972).
7. F. H. Eisen, B. Welch, J. E. Westmoreland and J. W. Mayer, Atomic Collision Phenomena in Solids (North-Holland, Amsterdam, 1970) p.111
8. N. Matsunami, to be published. see also N. Matsunami and N. Itoh, Proceeding of the International Conference on Atomic Collisions in Solids (Gatlinburg, 1973).
9. L. Meyer, Phys. Stat. Sol. (b) 44, 253 (1971).

THE USE OF MOLECULAR IONS FOR IMPLANTATION STUDIES IN Si AND Ge

J. B. Mitchell, J. A. Davies, L. M. Howe,

R. S. Walker and K. B. Winterbon

CRNL, Chalk River, Canada

and

G. Foti (University of Catania, Italy)

J. A. Moore (Brock University, Canada)

ABSTRACT

Results of an investigation of the lattice disorder resulting from equal atom dose implants of molecular and atomic ions in Si and Ge are presented. In each case, the molecular ion implants had the same energy per atom and were performed at the same atomic flux and fluence as the atomic ion implants. With heavy ions (As, Sb, Te and Bi), the molecular beam produces roughly 50% more damage than the atomic beam, indicating that damage production depends not only on the amount but also on the localized concentration of deposited energy. Preliminary experiments with lighter ions (H, D) in both Si and Ge do not show this molecular effect. The significance of these results is discussed in relation to the average energy density within the collision cascade.

INTRODUCTION

The amount and nature of the lattice disorder created around an ion track depends primarily on the amount of energy deposited into nuclear processes by both the incident ion and the secondary or knocked-on particles. This deposited energy produces small damage clusters within the cascade volume surrounding the ion track. The resulting total damage depends strongly on the extent to which these damage clusters overlap (and thus inhibit annealing) during the bombardment. It has been realized for some time that implantation at low temperatures (e.g. 40°K) reduces markedly the rate of cluster annealing and therefore produces more damage than would be present for the same implantation dose at higher temperature (e.g. 300°K) where annealing can take place during implantation.

Even at high temperatures, it should be possible to inhibit annealing during implantation, and thus increase the observed damage, provided we can find a technique for increasing the density of energy deposition within each cascade volume. One simple way of accomplishing this is to use molecular ion beams. Consider a diatomic ion with the same energy per atom as the corresponding monatomic ion: the two cascade volumes are identical, but the nuclear energy deposited into this cascade volume in the molecular ion case is double that of the monatomic ion. Since the energy deposition within a heavy-ion cascade is almost instantaneous ($< 10^{-12}$ secs), this doubling in energy density occurs on a much shorter time scale (by $\sim 10^8$) than can be obtained by merely increasing the dose rate of the monatomic ion.

The first direct comparison of the damage created by monatomic and diatomic ions has been reported recently by Moore, Carter and Tinsley[1], who injected 40-keV As_2^+ and 20-keV As^+ into GaAs crystals at 20°C and found that the damage created per arsenic <u>atom</u> was about 50% greater in the diatomic case.

In the present study we have investigated the molecular ion effect (i.e. the difference in total damage created by equal atom-dose implants of monatomic and diatomic ions) in silicon and germanium at room temperature, using helium-ion backscattering. Several heavy ions (As, Sb, Te and Bi) in the energy region 15-60 keV have been used. For comparison, we have also investigated the damage created by much lighter, higher energy ions (300-keV D^+ vs 600-keV D_2^+) where the damage cascades are so large and dilute that (in the D_2^+ case) the subcascades from each D atom should not appreciably overlap, and the molecular ion effect should be negligibly small.

EXPERIMENTAL

The heavy ion implantations were performed at room temperature using the Chalk River isotope separator. The doses were selected to produce an aligned yield of 10-30% of the random value in the subsequent backscattering analysis. This damage level is sufficiently high to permit accurate analysis, but not high enough to cause saturation effects. Typical doses ranged from 0.3-1 x 10^{13} atoms/cm^2. All implantations were performed ~8° away from the surface normal, in order to minimize channeling along the <111> direction. In each monatomic-diatomic comparison, the implantation time and the total number of implanted <u>atoms</u> were kept constant. Furthermore, the ions had the same energy per atom for both the monatomic and diatomic implants - i.e. the diatomic ion had twice the energy of the monatomic ion. In this way, the total amount of deposited energy, the cascade dimensions, the mean range and the straggling of the implanted atoms were all held constant. Two or more duplicate runs were performed for each implantation condition and the results were averaged. All duplicate analyses agreed within ± 5%.

ANALYSIS

The total number of displaced atoms per unit area was obtained by summing the damage peak area in the aligned backscattered spectrum, after normalizing it by the corresponding random or non-aligned spectrum. These techniques have been fully described in the literature by Eisen[2]. A linear dechanneling correction takes account of those particles that are scattered out of the aligned direction by the damage and subsequently backscatter from the regular lattice atoms. Although this is an oversimplification, the error introduced is negligible for the shallow implants considered in the present study. Appropriate (random) stopping power values have been used for the incident and scattered parts of the helium ion trajectory[3].

A typical set of damage spectra is shown in figure 1. The molecular Sb_2^+ implant has obviously displaced at least 50% more Si atoms than the corresponding Sb^+ case, even though the same dose of (25-keV) Sb <u>atoms</u> is involved.

From the peak areas in figure 1 and from similar data for the other heavy ion implants, we may calculate directly the total number (ΣN_D) of displaced lattice atoms, as shown in Table I. Note that the normalized value (N_D^*) of displaced atoms per incident ion in the diatomic case is much greater than twice the corresponding monatomic value. The magnitude of this molecular effect N_D^*(molecular)/$2N_D^*$(atomic) is tabulated in the final column of Table I. Both in Si and Ge we observed a large enhancement factor (1.4-1.8) for the molecular beam.

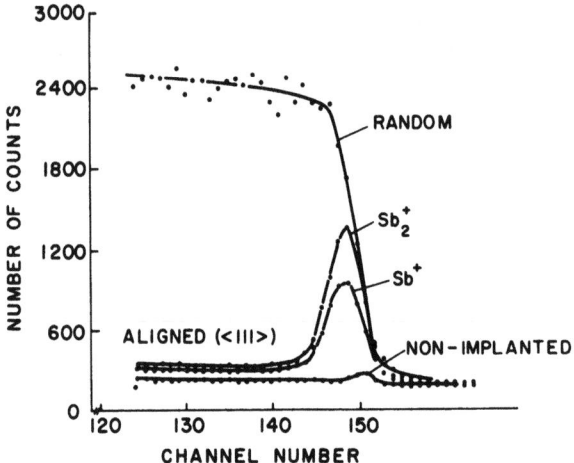

Fig. 1 Backscattering energy spectra for aligned (<111>) and random incidence from two silicon crystals implanted at room temperature with the same atomic dose (1.65×10^{13} Sb atoms/cm^2) of 25-keV Sb$^+$ and 50-keV Sb$_2^+$ ions, respectively. The analysing beam was 1.0 MeV helium.

TABLE I

Observed Number of Displaced Atoms, N_d

Ion	Energy (keV)	Ion Dose (ions/cm^2)	ΣN_d (total number)	$N_d^* = \dfrac{\Sigma N_d}{\text{ion dose}}$	$\dfrac{N_d^* \text{(mol)}}{2N_d^* \text{(atomic)}}$
As$^+$ in Si	25	2.20×10^{13}	28.7×10^{15}	1306	1.7
As$_2^+$	50	0.91	41.3	4504	
Sb$^+$	25	1.76	43.1	2450	1.6
Sb$_2^+$	50	0.81	61.7	7610	
Te$^+$	20	1.44	29.2	2024	1.4
Te$_2^+$	40	0.79	44.9	5660	
Bi$^+$	15	0.94	23.7	2532	1.6
Bi$_2^+$	30	0.50	40.9	8196	
As$^+$ in Ge	25	1.79	52.3	2921	1.6
As$_2^+$	50	0.62	57.9	9340	
Sb$^+$	25	.22	8.35	3800	1.8
Sb$_2^+$	50	.175	24.5	14,000	

A set of damage spectra for the D^+/D_2^+ comparison in silicon is shown in figure 2. This experiment was performed at 40°K in order to minimize annealing effects and thus permit us to observe measurable damage at fairly low deuteron doses. Because of the much larger depths involved, it is much more difficult to extract an accurate value of N_D^* from these D-bombardment data. However, we see clearly that the aligned backscattering yield curves for the diatomic and monatomic D implants are identical. A similar experiment at room temperature (not shown) required a considerably larger deuteron dose to produce observable damage but again the molecular effect was negligible. This is not surprising because, as noted in the Introduction, the damage cascade for 300-keV D atoms is so large and so dilute that one would not expect significant overlap to occur between the subcascades along the two D-atom trajectories.

DISCUSSION

For each of the implants, we have evaluated in Table II the deposited energy density ($\bar{E} = 0.35\, E/N_v$) in the cascade and the corresponding damage density ($F_D = 0.35\, N_D^*/N_v$). To do this, we first estimate the number of lattice atoms N_v contained within the cascade volume: i.e. within a spheroid whose axes are determined

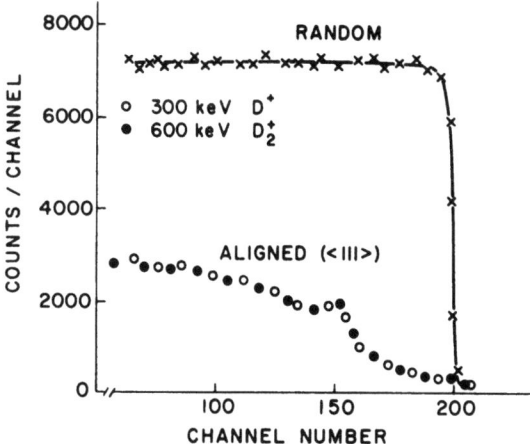

Fig. 2 Backscattering energy spectra for aligned (<111>) and random incidence from two silicon crystals implanted at 40°K with the same atomic dose (1×10^{15} D atom/cm²) of 300-keV D^+ and 600-keV D_2^+ ions, respectively. The analysing beam was 1.0 MeV protons.

TABLE II

Effective Displacement Energies, E_d

Ion	Energy (keV)	$\nu(E)$ (eV)	$\bar{E}$ (eV/Atom)	F_D	$E_d = \dfrac{0.8\,\nu(E)}{2N_d^*}$ eV
As^+ in Si	25	17,350	.2	.01	5.3
As_2^+	50	34,700	.4	.04	3.0
Sb^+	25	17,900	.4	.04	2.9
Sb_2^+	50	35,800	.8	.13	1.9
Te^+	20	14,550	.6	.06	2.9
Te_2^+	40	29,100	1.2	.17	2.1
Bi^+	15	11,350	1.7	.29	1.6
Bi_2^+	30	22,700	3.5	.94	1.0
As^+ in Ge	25	18,000	.4	.05	2.5
As_2^+	50	36,000	.8	.15	1.5
Sb^+	25	18,500	.8	.13	1.9
Sb_2^+	50	37,000	1.7	.47	1.0

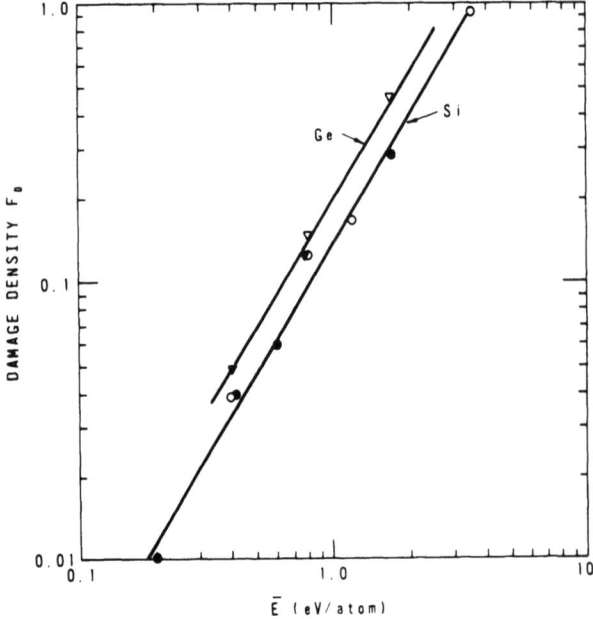

Fig. 3 Relationship between F_D and the deposited energy density $\bar{E}$ for various collision cascades in Si and Ge (data from Table II). The open set of symbols (○,▽) refer to molecular ions, and the filled set (●,▼) to the corresponding atomic ion implantations.

by the longitudinal $(\langle\Delta x^2\rangle)^{\frac{1}{2}}$ and transverse $(\langle y^2\rangle)^{\frac{1}{2}}$ components of the damage straggling, as calculated using the procedures discussed in reference (4). Note that our definition of N_v includes only that portion of the cascade volume extending out to one standard deviation from the region of maximum damage: i.e. the central region containing (if the distribution were Gaussian) roughly 35% of the total cascade damage. Hence, we introduce a factor of 0.35 in evaluating F_D and $\bar{E}$.

Figure 3 shows the existence of a strong correlation between the damage density F_D and the average energy $\bar{E}$ per atom in the cascade: viz. $\log F_D \simeq 1.6 \log \bar{E}$ for both Si and Ge crystals. For the highest $\bar{E}$ values in fig. 3, the value of F_D approaches unity (i.e. the 100% disorder level), indicating that the central region of each individual cascade is completely disordered. We plan shortly to test the validity of this conclusion by transmission electron microscopy. Note too that the F_D vs $\bar{E}$ curves in fig. 3 show no indication of levelling off as the damage level approaches 100% of the calculated cascade volume. This suggests that, in such high-density cascades, a thermal spike concept may provide a more appropriate interpretation of the observed damage behaviour, as was recently suggested by Sigmund[5].

Table II also contains the effective displacement energies E_d, calculated by substituting our observed N_D^* values (Table II) into the standard Kinchin-Pease formula[6]

$$N_D^* = 0.8 \, \nu(E)/2 \, E_d \tag{1}$$

where $\nu(E)$ is the portion of the incident ion energy that ends up in nuclear collision events. The values of E_d thus obtained are not constant, but decrease markedly as the damage density F_D increases. Furthermore, in all cases, our E_d values are much smaller than the typical threshold value (~ 14 eV for Si and Ge) obtained from e^- bombardment studies. This discrepancy is even greater than the data in Table II indicate, because (as pointed out by Sigmund[7]) the Kinchin-Pease formula gives only an upper limit to E_d. Such extremely low experimental values for E_d again suggest that a thermal spike concept may provide a more useful basis for interpreting heavy ion cascade phenomena, since in the "spike" model one expects the heat of melting (or vaporization) to be a more relevant quantity than the threshold displacement energy. In both Si and Ge, the heats of melting and vaporization are smaller than 1 eV per atom and hence are not inconsistent with our observed range of E_d values (Table II).

CONCLUSIONS

1) Considerably more damage occurs for diatomic heavy ion implants than for the corresponding monatomic implants. This molecular effect is of similar magnitude in both silicon and germanium. The effect is not observed for high energy light ions, presumably because the cascade is so dilute that negligible interaction of subcascades occurs.

2) For the highest density cascades studied - viz. 30-keV Bi_2^+ in Si and 50-keV Sb_2^+ in Ge - the observed amount of damage corresponds to almost 100% of the calculated cascade volume, indicating that each cascade is almost entirely amorphous. Furthermore, there is no evidence of the damage levelling off as it approaches this 100% limit.

3) The effective displacement energy E_d depends strongly on the damage density within the cascade, and is always much smaller than the threshold values (~ 14 eV) obtained from e^- bombardment.

4) Points 2 and 3 are interpreted as evidence for a thermal-spike type of model.

ACKNOWLEDGEMENTS

We thank O. M. Westcott for performing the implantations for this study and F. Brown and T. A. Eastwood for a critical review of the manuscript.

REFERENCES

(1) J. A. Moore, G. Carter and A. W. Tinsley, Radiation Effects, in press.
(2) F. H. Eisen in "Channeling", Ed. by D. V. Morgan, John Wiley and Sons, London, Chapter 14, (1973).
(3) J. F. Ziegler and W. K. Chu, Thin Solid Films 19, 281 (1973).
(4) K. B. Winterbon, Rad. Effects 13, 215 (1972).
(5) P. Sigmund, Appl. Phys. Lett. 25, 169 (1974)
(6) G. H. Kinchin and R. S. Pease, Rept. Prog. Phys. 18, 1 (1955).
(7) P. Sigmund, Appl. Phys. Lett. 14, 114 (1969).

DAMAGE PRODUCTION AND ANNEALING IN IMPLANTED SILICON

AS STUDIED BY OPTICAL REFLECTIVITY PROFILING

E. T. Yen, B. J. Masters, and R. Kastl

IBM System Products Division

Hopewell Junction, New York 12533

ABSTRACT

An optical reflectivity technique was devised to investigate ion implanted damage profiles and their annealing behavior. A set of damage profiles for 400 keV implanted ^{75}As^{+} was obtained. The results are in good general agreement with theoretical approximations. The anomalous excess damage layer induced by the thru-oxide arsenic implantation was briefly studied.

INTRODUCTION

The understanding of implanted damage distribution and its annealing behavior in semiconductor materials is of fundamental interest. Experimental techniques such as electron diffraction and electron microscopy [1], radiotracer [2] and backscattering [3] analyses had been employed to study the damage profiles. However, few of them could avoid sophisticated data acquisition and/or analytical equipment. Simpler techniques such as optical reflectivity measurements [4,5,6,7] and optical transmission study of absorption [8,9] were mainly used for determination of fluence dependence of damage production. Other techniques such as electron spin resonance [10] and Raman scattering [11] measurements were also used primarily for fluence dependence studies. The ellipsometric technique [12] is still under development in its application for implanted damage study.

In this work, we have attempted to modify the existing optical reflectivity techniques and develop a simple and effective method that can determine the ion implanted damage distribution, its fluence dependence, and its annealing characteristics. We are presently concentrating our study on implanted damage caused by a relatively heavy ion, i.e., arsenic.

It is well known [13,14] that the optical reflectivity of crystalline Si peaks at wavelengths of 2750Å and 3420Å arising from direct interband transition. The suppression of these characteristic peaks provides a measure of the degree of macroscopic disorder in the surface layer that the light penetrates. This property has been utilized by McGill et al. [6] who related the fractional change in reflectivity ($\Delta R/R_o$) at 2750Å wavelength to the fluences (ϕ) and compared it with backscattering results by Davies [15]. The two experiments found good qualitative agreement. However, the absolute reflectivity (R) is hard to reproduce accurately due to factors such as sample polishing, doping levels, and instrumental instabilities, etc. Thus, Sell and MacRae [7] employed a different approach. They converted their reflectivity curves for GaAs in the 3500Å ~ 4500Å wavelength region into the energy derivative of the reflectivity curve ((dR/dE)/R). They felt this to be a more direct and reliable measure of damage than absolute reflectivity itself, since the scale factors of the absolute reflectivity mentioned above can no longer contribute to these converted curves. However, they did not derive any quantitative fluence versus damage relationship, presumably because of the difficulty of the interpretation of these converted curves.

The scheme used in the present study has the merits of the method of Sell and MacRae but is simpler and easier to interpret. A minimum in the reflectivity curve occurring at 3300Å (Fig. 1) was found to be essentially independent of the neighboring peaks. The reflectivity at this point changes only 1-3% for a structural change from the crystalline to amorphous state. Therefore, this reflectivity (R_c) may be used as a reference point for all measurements. Further, the net reflectivity ($R'=R-R_c$) was observed to be essentially free of the reproducibility problems mentioned above, and was used throughout.

We have further considered the amplitude change in R' at 2750Å wavelength from unimplanted crystal state (R'_o) to implanted amorphous state (R'_a) as the measure of silicon crystalline order. Therefore, the degree of disorder (DD) of the surface area being studied can be obtained from:

$$DD = 1 - (R' - R'_a) / (R'_o - R'_a) \tag{1}$$

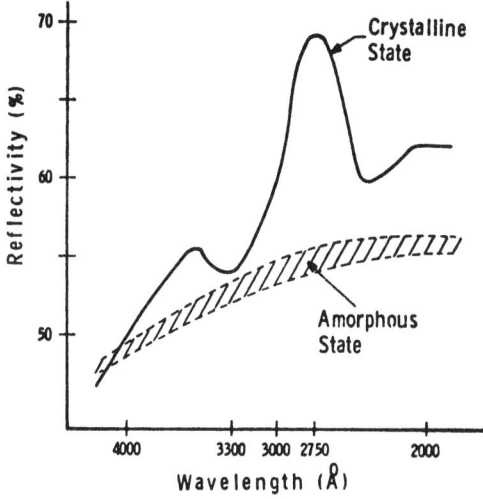

Fig. 1. Optical reflecting spectrum of silicon in ultraviolet light region.

The combined indeterminacy of R'_o and R'_a is within 10%. The total systematic error of the absolute value of DD is thus less than 10%.

By combining this reflectivity method with the successive surface layer removal by anodization and stripping, implanted damage profiles in silicon are thus determined.

EXPERIMENTAL

The optical reflectivity measurements were obtained using a Beckman DB spectrophotometer with a specular reflectivity attachment. The incident light beam was 10 degrees off the normal to the sample surface. The total area exposed to the incident light was about 3mm by 5mm.

Chemi-mechanical polished $\langle 100 \rangle$ silicon wafers were used, except the thru-oxide implants where $\langle 111 \rangle$ oriented specimens were used.

A Van de Graaff accelerator was used to implant 400 keV $^{75}As^+$ ions at a flux of about 0.5 $\mu A/cm^2$. All wafers were implanted with the ion beam 7 degrees off the major axis in order to minimize channeling effects. All implants were done at room temperature. The annealing of samples was conducted in a tube furnace under a flowing Ar atmosphere.

Anodizations were performed in a N-Methylacetamide solution using well known techniques for the calibrated removal of silicon layers. The anodized

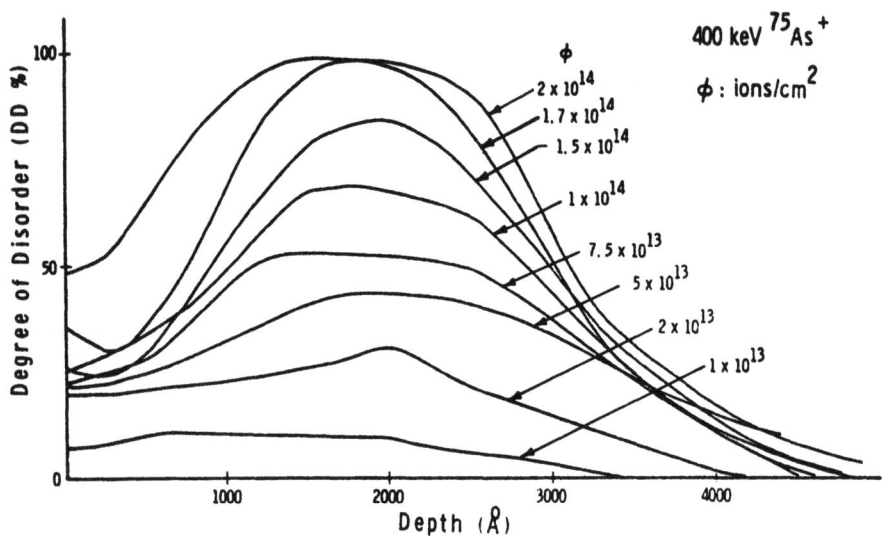

Fig. 2. Damage depth distribution curves - Part I.

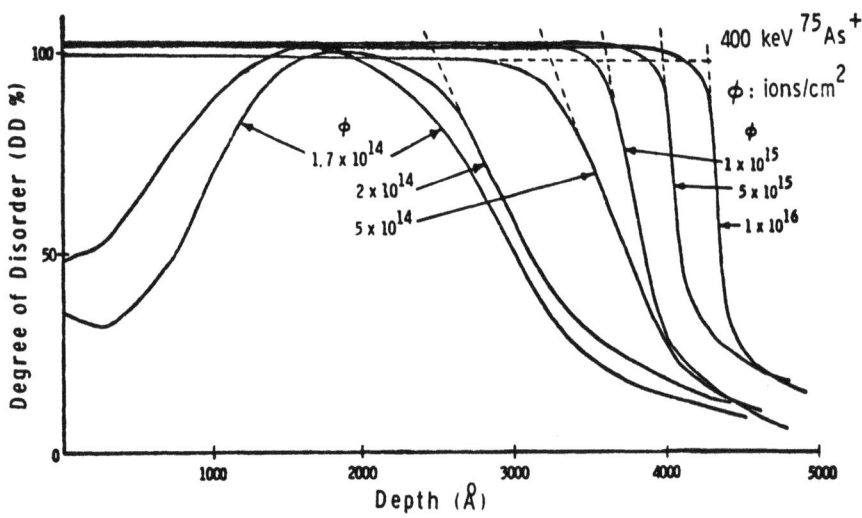

Fig. 3. Damage depth distribution curves - Part II.

film was removed in a dilute HF solution prior to each reflectivity measurement.

RESULTS AND DISCUSSION

The damage profiles of 400 keV $^{75}\text{As}^+$ ions implanted into silicon at doses ranging from 1×10^{13} to 1×10^{16} ions/cm^2 are shown in Figs. 2 and 3. As seen in Fig. 2, for fluences of 1.7×10^{14} ions/cm^2 and less, the damage distributions are approximately of Gaussian shape, particularly in the dose range of $1 \times 10^{14} - 1.7 \times 10^{14}$. The damage projected range and its straggling are estimated as $R_D \simeq 1900$Å and $\Delta R_D \simeq 1000$Å. Sigmund and Sanders [16] have theoretically approximated the relationship between projectile range distribution and its damage range distribution. According to their tabulation for arsenic implant at moderate energy, R_D is about $0.83 R_p$ and ΔR_D is about $1.30 \Delta R_p$. From LSS theory with Eisen correction [17], the projected range data are given as $R_p = 2200$Å and $\Delta R_p = 700$Å. Theoretical estimates of damage distribution can thus be obtained as $R_D \simeq 1830$Å and $\Delta R_D \simeq 910$Å, which are in good agreement with our experimental estimates.

Both the peaks of the degree of disorder (DD) at about R_D and the total disorder produced (area ratio under the damage profile) over the dose range $2 \times 10^{13} - 1.5 \times 10^{14}$ ions/cm^2 are plotted against ion fluence (ϕ) in Fig. 4. The observed curve is approximately a straight line which follows the amorphous cluster model [18]. This model predicts that

$$1 - DD = \exp(-A_i \phi) \qquad (2)$$

where A_i is the average area of an amorphous cluster projected on the surface of the sample. From the slope of our data, the projected amorphous cluster area for arsenic implant is about $(11\text{Å})^2$.

At the critical dose of 1.7×10^{14} ions/cm^2, a buried amorphous layer begins to form. With continued implantation, this buried layer broadens until at about 5×10^{14} ions/cm^2 a continuous amorphous layer extends to the surface. Upon further increasing the ion fluences, the flat-topped Gaussian-like damage profile advances toward greater depth. Each advancing edge of the amorphous layer may be located by the position of the extension of its slope to 100% DD line (see Fig. 3) and the values thus estimated are tabulated in Table 1, column 1.

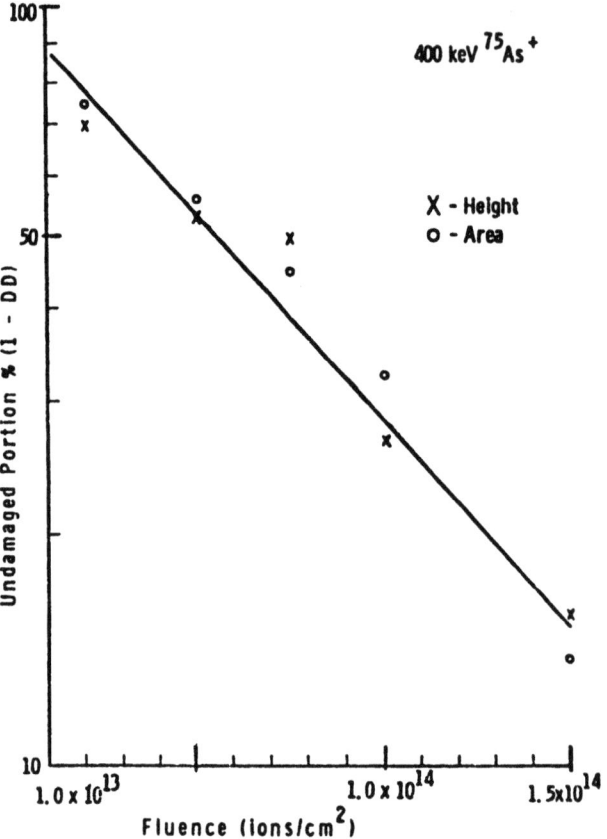

Fig. 4. Dose dependence of damage distribution.

Table 1. Estimates of the amorphous layer edge locations

Fluences (ions/cm^2)	Experimental Estimate (Å)	Theoretical Estimate (Å)
2×10^{14}	2475	2300
5×10^{14}	3200	3150
1×10^{15}	3600	3530
5×10^{15}	4000	4150
1×10^{16}	4300	4400

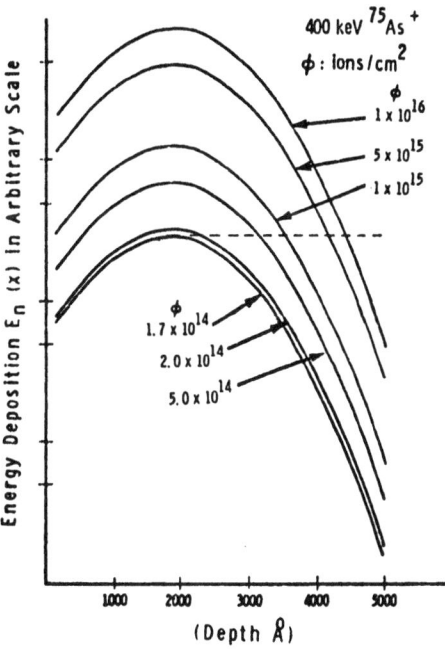

Fig. 5. Computer simulated energy deposition curves by using Gaussian approximation.

We assume the total energy deposited in silicon due to nuclear collision is proportional to the ion fluence if the implantation energy is held constant. Then we may convert the Gaussian range distribution function into an energy distribution function $E_n(x)$ on an arbitrary scale as

$$E_n(x) = \frac{\phi}{\sqrt{2\pi}\,\Delta R_D} \exp\left[-(x - R_D)^2 / \Delta R_D^2\right] \qquad (3)$$

The results of this distribution for various doses are plotted in Fig. 5. The peak of the energy distribution of fluence 1.7×10^{14} ions/cm^2 is assumed as the saturation energy dose for amorphous phase formation. Thus, in this figure, an equal energy line drawn horizontally from this peak should intersect the curves for other fluences at the amorphous layer depth. These values are listed in Table 1, column 2. As shown in this table, our experimental estimates are in general agreement with this first order theoretical approximation.

This optical reflectivity profiling technique has also been used to study the annealing behavior of some continuous amorphous layers which were produced by 400 keV arsenic implantation upon bare substrates. As shown

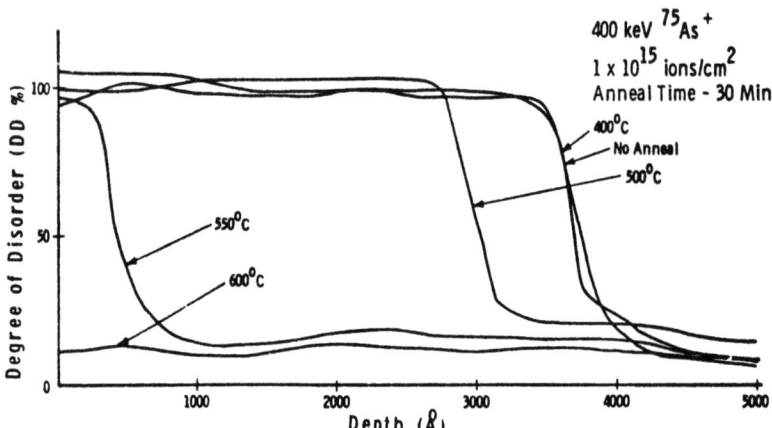

Fig. 6. Annealing behavior of arsenic implanted damages.

in Fig. 6, epitaxial regrowth of these layers starts at the amorphous/crystalline interface and proceeds toward the surface upon a 500°C 30 min anneal. Most of the implanted damage is gone for a 600°C 30 min anneal.

For some samples, half of the substrate was masked by thermal oxide several hundred angstroms thick and the whole substrate was implanted with 400 keV $^{75}As^+$ of dose 1×10^{16} ions/cm^2. The continuous amorphous layers generated showed a drastic difference between the two halves of the substrate after a 600°C 30 min anneal. The surface remained amorphous in the thru-oxide implanted region but not in the bare implanted region. A brief sectioning step revealed a residual amorphous layer of 200 ~ 300Å at the surface of the masked region. This amorphous layer disappeared only after a 850°C 30 min anneal. However, considerable damage (DD ≈ 20%) still remained in this region even after a 1000°C 30 min anneal. This confirms an earlier finding by electron microscopy [19]. Further investigation is being directed toward determining whether this anomaly is due to the forward sputtering of oxygen atoms from the oxide into silicon. This sputtering process may produce a highly oxygen-doped amorphous silicon layer retarding epitaxial regrowth during subsequent anneal.

CONCLUSION

This study shows the optical reflectivity measurement to be a simple and useful technique for profiling implanted damage in silicon. A series of damage profiles of 400 keV $^{75}As^+$ implantation is established. Good agreement has been found in this study between experimental results and theoretical approximations. The existence of excess damage layer in the case of thru-oxide implantation has been confirmed.

REFERENCES

1. S. M. Davidson and G. R. Booker, Proc. IEEE, pp. 51-61, Sept. 1972.
2. B. J. Masters, J. M. Fairfield and B. L. Crowder, Proc. 1st. Int. Conf. on Ion Implantation, New York: Gordon and Breach, pp. 81-87, 1971.
3. J. W. Mayer, L. Eriksson, S. T. Picraux and J. A. Davis, Can. J. Phys. 46 663 (1968).
4. S. Kurtin, G. A. Shifrin and T. C. McGill, Appl. Phys. Letters 14 223 (1969).
5. R. R. Hart and O. J. Marsh, Appl. Phys. Letters 14 225 (1969).
6. T. C. McGill, S. L. Kurtin and G. A. Shifrin, J. Appl. Physics 41 246 (1970).
7. D. D. Sell and A. U. MacRae, J. Appl. Physics 41 4929 (1970).
8. H. J. Stein, F. L. Vook, D. K. Brice, J. A. Borders and S. T. Picraux, Proc. 1st. Int. Conf. on Ion Implantation, New York: Gordon and Breach, pp. 17-24, 1971.
9. E. C. Baranova, V. M. Gusev, Yu V. Martynenko, C. V. Starinin and I. B. Hailbullin, Ion Implantation in Semiconductor and Other Materials, B. L. Crowder, Ed., Plenum, pp. 59-71, 1973.
10. J. G. DeWitt and C. A. J. Ammerlan, Proc. 2nd Int. Conf. on Ion Implantation, I. Ruge and J. Graul, Eds., New York: Springer, pp. 39-46, 1971.
11. J. E. Smith, Jr., M. H. Brodsky, B. L. Crowder and M. L. Nathan, Phys. Rev. Letters 26 642 (1971).
12. M. M. Ibrahim and N. M. Bashara, Surface Science 30 632 (1972).
13. W. C. Dash and R. Newman, Phys. Review 15 1151 (1955).
14. H. P. Philipp and E. A. Taft, Phys. Review 120 37 (1960).
15. J. A. Davies, J. DenHartog, L. Eriksson and J. W. Mayer, Can. J. Phys. 45 4053 (1967).
16. P. Sigmund and J. B. Sanders, Proc. Int. Conf. on Applications of Ion Beams to Semiconductor Technology, P. Glotin, Ed., France, pp. 215-233, 1967.
17. S. Furukawa, H. Matsumura, H. Ishiwara, Proc. U.S.-Japan Sem. on Ion Implantation in Semiconductors, S. Namba, Ed. Kyoto, Japan, (1972).
18. W. S. Johnson, Technical Report No. K701-3, pp. 64-68, Stanford University, 1969.
19. T. R. Cass and V. G. K. Reddi, Appl. Phys. Letters 23 268 (1973).

Displacement Damage in Ne Implanted

Magnesium Oxide

    B. D. Evans

    Naval Research Laboratory

    Washington, D. C. 20375 U. S. A.

ABSTRACT

Displacement damage profiles measured for 1-to-4.8 MeV Ne^+ implantation of MgO agree with LSS energy deposition profiles for doses near 10^{14} cm^{-2}. For higher doses, damage profiles appear to penetrate further than anticipated on the basis of LSS. Below 3×10^{13} cm^{-2}, for room temperature implants, approximately 100 stable F^+ centers are formed per incident 3 MeV Ne^+. This defect formation efficiency quickly decreases for high dose, falling to 0.1 per Ne^+ by 10^{17} cm^{-2}. Following F^+-center anneal at $600°C$, a large deformation related band at 5.8 eV remains, the intensity of which suggests massive deformation and plastic flow within and near the damage layer. Such plastic flow, releaving stress built-up by high-dose implantation, may be related to the distorted damage profiles.

INTRODUCTION

Atomic displacement damage has been studied in high purity crystalline MgO in order to lucidate the type effects that may accompany ion implantation of insulating oxide materials in general. Such studies will further the understanding of implant doping of MOS gate insulator structures, implant fabrication of planor waveguides and possibly of heavy-ion bombardment simulation of fast-neutron irradiated insulators. MgO was chosen because much fundamental knowledge is already known about the optical and ESR spectroscopy of the primary defect centers - an

electron trapped at an anion vacancy (the F^+ center) and a hole
localized near a cation vacancy (the V^- center).[1] Hardening
effects introduced by particle irradiation has been discussed by
W. A. Sibley et al.[2] and implantation produced stress effects have
been presented by W. A. Primak et al.[3]. We are here primarily
interested in the influence of ion energy, dose and dose rate upon
the spatial distribution of the volume concentration of defects introduced by Ne^+ implantation and in a comparison of LSS theory[4]
for the energy deposition profile with experimentally determined
profiles. The number of stable, primary defects observed per
incident ion as a function of dose and target temperatures has also
been determined.

RESULTS AND DISCUSSION

High purity crystalline magnesia was implanted with 1, 2, 3, 4
and 4.8 MeV Ne^+ to doses ranging from 10^{13} to 10^{17} ions/cm^2 at
dose rates of 0.7 to 3.5×10^{12} ions/cm^2 sec. To avoid the possibility of channeling effects samples were cut and polished into
platelets whose normal was away from any major crystallographic
direction. Implantation results in an intense optical absorption
band near 5.0 eV (Fig. 1) and a strong ESR signal near g=2.0023.[5]
A single electron trapped at an anion vacancy is associated with
this paramagnetic center. These results, taken with Smakula's
equation, yields the oscillator strength for this defect system,
$f = 0.7 \pm 2$.[5]

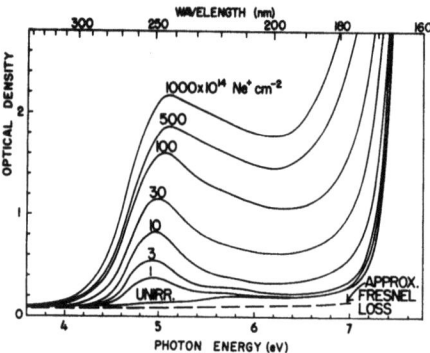

Fig. 1. Room-temperature optical absorption spectra of 0.15
mm thick samples of MgO implanted to doses as indicated
by 3-MeV Ne^+.

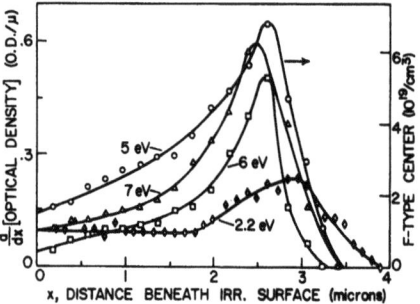

Fig. 2. Spatial profiles of
absorption at the indicated
photon energies induced by 3-
MeV Ne^+ implants to 3×10^{15}
cm^{-2}; vertical scales for the
2.2 eV profiles should be
divided by 200.

Table I. Dose-Dependent F^+-center production by a 3-MeV Ne^+ implant.

Dose (x 10^{14} cm^{-2})	F^+ centers/Ne^+
.1	100
.3	80
1	31
3	19
10	8.2
30	3.0
100	0.77
500	0.12
1000	0.090

By a chemical layer-removal technique the spatial profile of the defects associated with the 5.0 eV band may be determined.[5] This is shown in Fig. 2 together with profiles of absorbance at 6.0 eV and 7.0 eV, and the profile of a weak band near 2.2 eV that, infact, is responsible for the pale blue appearance of the implanted material.[6] A comparison of Fig. 2 with the shape of the electronic and nuclear stopping power distributions clearly suggests that most of the induced abosrption is related to displacement events,[5] in agreement with previous reports by Chen et al.[7] for the electron-irradiation induced 5-eV band. Very little effect of dose rate upon profile shape was observed for the range measured.

Thru a combination of optical and ESR spectroscopy, the number of stable, F^+ centers resulting after implants to various doses may be determined (see Table I). Saturation at about 10^2 centers/Ne^+ appears below a dose of 3×10^{13}. Damage regions surrounding particle tracks may begin to overlap for doses above 10^{14} cm^{-2}, and certainly overlap above 10^{15} cm^{-2}. Therefore, it is not surprising that the apparent damage production efficiency should decrease markedly for doses above 10^{13} cm^{-2} (see Table I). Similar studies have been performed on samples held near 80°K during implant. The resulting 5-eV band is enhanced about 2.5 times after warming to room temperature. This enhancement suggests that thermally assisted vacancy-in-

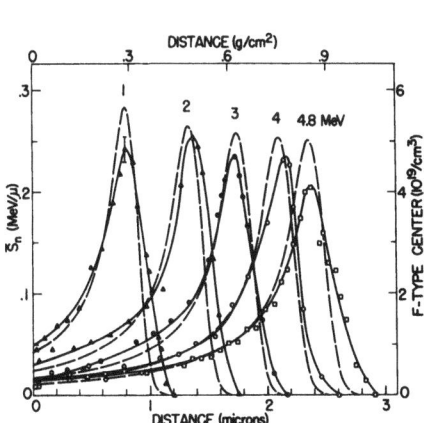

Fig. 3. A comparison of F^+-center concentration profiles (solid lines, data points) with $S_n(x)$ based on LSS (dashed lines).

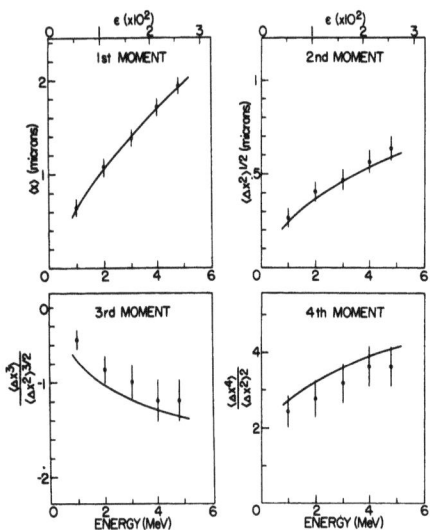

Fig. 4. Comparison of the moments of profiles in Fig. 3; solid line, LSS theory, and data points for experimental profiles.

terstitial recombination destroys many Frenkel pairs during implant. This observation is supported by further determinations of the dose dependence of the intensity of the 5-eV band.[5] Above a dose of 10^{13} cm^{-2} the slope of the OD vs dose curve decreases; above 10^{16} the 5-eV band nearly saturates at a concentration of approximately 10^{20} cm^{-3}.

Figure 3 compares the experimental spatial profiles of the volume concentration of F$^+$ center with the profiles for the nuclear stopping power, $\overline{S}_n(x)$, obtained from a Manning-Mueller code for an LSS treatment[8] for 1, 2, 3, 4, and 4.8 MeV Ne$^+$ incident onto MgO. Figure 4 compares the moments of the experimental and theoretical profiles shown in Fig. 3, where the moments are defined in the usual manner: $\langle x^n \rangle = \int x^n N(x) dx$, and $\langle \Delta x^n \rangle = \int (x - \langle x \rangle)^n n(x) dx / \langle x^0 \rangle$. The experimental centroid data points agree with theory well within the estimated ±15% experimental error bars. However, the experimental profile width, $\langle \Delta x^2 \rangle^{1/2}$, appears consistently larger than expected from theory by about 7%. The data points for the third moment suggest we consistently observed proportionally more damage than expected from theory on the deep side of the damage peak than on the surface side. Again, the fourth moment suggests proportionally more damage in the wings of the distribution than expected from theory. These general trends are further born out by damage profiles resulting from higher dose implants for a fixed ion energy, as shown in Fig. 5. At higher dose, both the damage at the surface increases and the damage peak appears further from the surface. We note that this dose dependent profile peak shift appears to begin to saturate

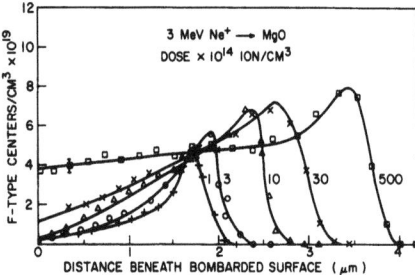

Fig. 5. Damage profiles for 3-MeV Ne$^+$ implants at various doses.

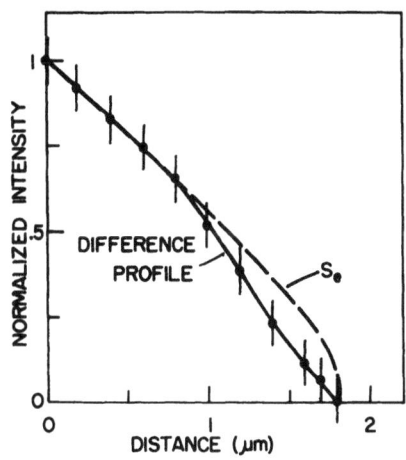

Fig. 6. Normalized comparison of the electronic stopping power, S_e, with the slope of the additional damage observed in going from 3×10^{15} to 5×10^{16} cm^{-2}.

above 10^{16} cm^{-2}. It is also above this dose that the 5.0-eV band shows saturation. Large stresses may be introduced by implantation by both ion stuffing and by void formation produced by separate anion and cation aggregation. The fact that the dose dependence of the 2.2-eV band[6] for which there is evidence to suggest is associated with some type of aggregate defect, possibly a divancy, saturates at 5×10^{15} cm^{-2}, indicates that aggregation may have proceeded to such an extent that voids are present at doses above 10^{16} cm^{-2}. Also, it has been suggested that the presence of ionizing radiation may alter the charge state of defects or impurities and thus influence their migration and diffusion.[9]

Figure 6 shows a tentative comparison between the electronic stopping power, S_e, obtained from the same code and the increase in defect concentration observed in going from the 3×10^{15} profile to the 5×10^{16} cm^{-2} profile (labeled difference profile). The resemblance is striking, and indeed suggests the influence of ionizing radiation (about 10^{13} rads) on the alteration of the damage profile.

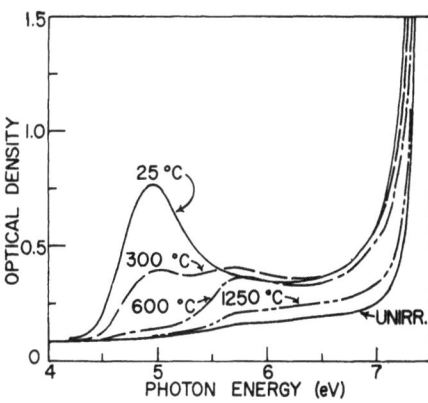

Fig. 7. Effect of annealing on the uv absorption spectra resulting from a 3-MeV implant to 10^{15} cm^{-2}.

In order to investigate stress introduced in and near the damage layer by implantation samples were annealed in an Ar atmosphere to 600°C reducing the F^+ band and revealing a band at 5.8 eV as shown in Fig. 7. Sibley et al.[10] and Turner et al.[11] find this band in stressed MgO crystals. They have further shown the absorption coefficient, α, at 5.8 eV to be proportional to the deformation. From the spectrum in Fig. 7 and the 5.8-eV profile in Fig. 2, α, may be determined. The result suggests well over 10%

deformation in the damaged layer. Pristine material normally begins plastic flow near 1% compression[2] corresponding to a flow stress of 2.0 kg/mm^2. McGowan and Sibley[2] have reported hardening of MgO with fast neutron irradiation; the incremental increase in flow stress was found proportional to the square root of 5.0-eV band intensity. The flow stress could be increased to 30 kg/mm^2 with 3×10^{18} F-type centers cm^{-3}, although most samples underwent brittle fracture above these values. At Ne$^+$ doses near 10^{15} cm^{-2}, F-type center concentrations[5,6] in the damage layer approach 5×10^{19} cm^{-3}; assuming no fraction, flow stresses approaching 100 kg/mm^2 may be realized. We have not observed blistering or other near-surface fraction on samples exposed up to 10^{17} cm^{-2}; nor was blistering observed after high dose samples were partially etched to reveal the buried damaged layer. However, Primak[3] has reported spontaneous surface exfoliation of crystalline MgO samples implanted with 140 keV He$^+$ to 5×10^{17} cm^{-2}. For our case, in the absence of observed exfoliation, plastic flow in and adjacent to the buried damage layer may release stress built-up by ion stuffing and void formation resulting from high dose implants. These effects may be related to the deviations of the damage profiles from LSS energy deposition distributions at high dose.

In Figure 8 we show the isochromal anneal curves for optical absorption at several photon energies. The deformation-related band at 5.8 eV anneals at much higher temperatures than the F$^+$-center, suggesting that substantial deformation remains in the lattice after the primary lattice defect has been removed. Also shown is the anneal curve for the 5-eV band in additively colored MgO [samples heated in a Mg vapor] taken from the data of Chen, Williams and Sibley.[12] By this process, it is believed only the anion vacancy is introduced, without the corresponding anion interstitial which surely accompanies particle irradiation. Figure 8 shows that nearly half the melting temperature, T_m, is required to allow diffusion of the single anion vacancy through the lattice, where the Frenkel defect recombination takes place

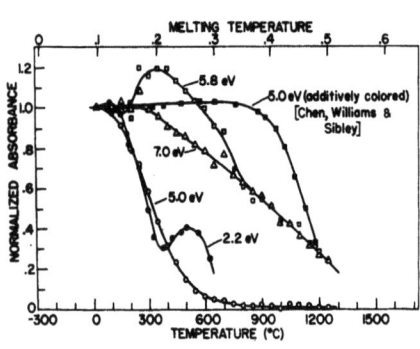

Fig. 8. Isochromal anneal curves for the several bands resulting from 3-MeV Ne$^+$ implant to 10^{15} cm^{-2}.

by 0.2 T_m. Similar high anneal temperatures (i.e., 0.5 T_m), were observed by Primak³ in order to remove dialation produced by He^+ implants in MgO crystals.

ACKNOWLEDGMENTS

The author wishes to thank P.R. Malmberg for implanting the samples. We are indebted to W.R. Hunter for use of the VUV spectrometer. We thank I. Manning for discussions and access to early results from his energy deposition code for LSS, thank J.E. Westmoreland for discussions regarding profiles and J. Comas for technical assistance.

REFERENCES

1. A.E. Hughes and B. Henderson, in Point Defects in Solids, ed. by J.W. Crawford, Jr. and L.M. Slifkin (Plenum, New York, 1972).
2. W.C. McGowan and W.A. Sibley, Phil. Mag. 19, 967 (1969).
3. W. Primak and J. Luthra, Phys. Rev. 150, 551 (1966).
4. J. Lindhard, M. Scharff, and H.E. Schiott, K. Dan. Vidensk. Selsk. Mat.-Fys. Medd. 33, No. 14 (1963).
5. B.D. Evans, Phys. Rev. B9, 5222 (1974).
6. B.D. Evans, J. Comas, and P.R. Malmberg, Phys. Rev. B6, 2453 (1972).
7. Y. Chen, D.L. Trueblood, O.E. Schow and H.T. Tohver, J. Phys. C3, 85 (1974).
8. I. Manning and G.P. Mueller, Comput. Phys. Commun. 7, 85 (1974).
9. J.C. Bourgoin and J.W. Corbett, Phys. Letters A38, (1973).
10. W.A. Sibley, J.L. Kolopus, and W.C. Mallard, Phys. Status Solidi 31, 223 (1969).
11. T.J. Turner, C. Murphy, and T. Schultheiss, Phys. Status Solidi B58, 843 (1973).
12. Y. Chen, R.T. Williams, and W.A. Sibley, Phys. Rev. 182, 960 (1969).

AN EPR STUDY ON HIGH ENERGY ION IMPLANTED SILICON*

Y. H. Lee, P. R. Brosious, L. J. Cheng and J. W. Corbett

Department of Physics

State University of New York at Albany

Albany, New York 12222, U.S.A.

ABSTRACT

EPR is used to study the lattice disorder produced by the C^+ and N^+ implantation in silicon in the high-energy range 0.55-3.50 Mev. Two new EPR spectra, Si-A12 (anisotropic) and -A13 (isotropic) are resolved for E > 1.65 Mev. Isochronal annealing shows that the amorphous line does not completely anneal at 700°C. The depth of the amorphous layer is roughly proportional to $E^{1/2}$ in the high-energy range.

INTRODUCTION

High-energy ion implantation, particularly along the channeling directions, recently became important both from the point of view of understanding the nature of defect production in the energy range where the electronic energy loss is dominant and from the possible application to a high-power device fabrication. The early EPR work [1-3] on defects in heavy-ion implanted silicon has been mostly limited to the low or intermediate energies ($\leq$ 500 kev). High-energy ion implantation, however, has been studied by using other experimental techniques such as transmission electron microscopy, X-ray diffraction topography and electric property measurements [4]. Dose-dependence, annealing and depth profile of the amorphous layer have been studied by monitoring an intensity variation of the amorphous line.

---

*This work was supported in part by the Office of Naval Research under Contract No. N00014-70-C-0296.

EXPERIMENTAL PROCEDURES

The magnetic field-analyzed beam from the Dynamitron at SUNY/Albany was utilized to implant N^+ and C^+ into silicon with the incident energy varying from 0.55 to 3.5 Mev. The silicon sample was oriented by the X-ray diffraction to within an accuracy of ± 1° and then cut to a size of 2 mm x 0.5 mm x 2.5 cm. Although various silicon samples (P- or B- doped and Czochralski or Float-zone grown) were used to study impurity-dependence, particular emphasis was given to the intrinsic float-zone silicon ($\rho = 10^4$ Ω-cm). Implantation was carried out at room temperature as well as at high temperature near 600°C; the vacuum in the target chamber was less than 3.0×10^{-6} Torr. The annealing of samples was conducted in a closed quartz-tube with an ambient gas (N_2 or Ar), and temperature was monitored by a thermocouple located inside the tube.

RESULTS AND DISCUSSION

Figure 1 shows the actual EPR signal which was observed in the C^+ - implanted silicon after subsequent anneals up to 400°C. The sample was originally implanted at room temperature to the total fluence 2×10^{17} ions/cm^2, with incident energy 1.65 Mev. In addition to the "a" line at g = 2.0055 corresponding to the amorphous state [2], two new spectra, labelled as Si-A12 and -A13, are resolved. The A12 spectrum has an anisotropic g-tensor: $g_{[100]}$ = 2.0079 (±0.0003), $g_{[011]}$ = 2.0030 and $g_{[01\bar{1}]}$ = 2.0131. The A13 spectrum is an isotropic line at g = 2.0023 (±0.0003) with the peak-to-peak width, ΔH_{pp} = 1.6G. These new spectra were observed over the incident energy range of 1.65 - 3.5 Mev. In the 550 kev implantation, on the other hand, only the "a" line was observed,

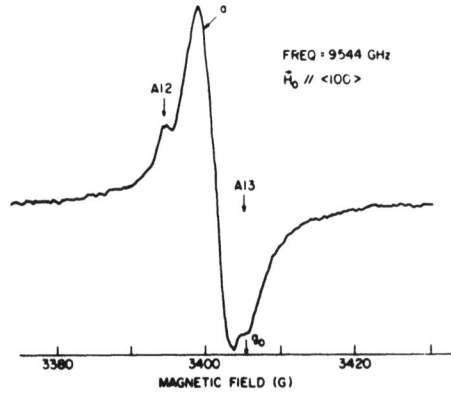

Fig. 1. Actual EPR spectra taken at 300°K from the intrinsic silicon after implanted 2×10^{17} C^+/cm^2 and then annealed at 400°C. "a" indicates the amorphous line and both A12 and A13 are the new spectra.

regardless of the annealing temperature and of the ion dose between $10^{14} - 10^{18}$ ions/cm^2. Both A12 and A13 spectra disappear at 500°C. Further details on the spectra will be published elsewhere.

We have performed detailed studies on the "a" line of the amorphous state produced by the high-energy ions. This line has been studied in detail for low energy implantation [2,3]. In Fig. 2, the intensity of the "a" line is plotted with respect to the total ion fluence for the two different incident energies, 550 kev and 1.65 Mev. For E = 550 kev, the amorphous state is saturated at 1.0×10^{17} ions/cm^2. The total spin density at this fluence was estimated to be 4.0×10^{19} spins/cm^3 for the N^+ - implantation, and half that for the C^+ - implantation. The production rate is almost linear without regard to the implanting ions, inconsistent with the channeling experiment done by Feldman and Rodgers [5]. They observed in the C^+ - implantation at E = 500 kev a non linear production rate in the number of displaced atoms. A sub-linear production of the amorphous state was actually observed in the case of the 1.65 Mev implantation, presumably due to the fact that point defects would diffuse away from the heavily damaged region preventing amorphization. Although the total spin density starts to saturate near 5×10^{17} ions/cm^2, a complete saturation should not occur at that ion dose, because the A12 spectrum exists at 5×10^{17} ions/cm^2.

Isochronal annealing was carried out up to 1000°C with a 30 min. step. Intrinsic silicon was chosen for the study so as to lock the Fermi level at the middle of the forbidden-band throughout the experiment. Figure 3 shows the annealing results obtained in the 550 kev and 1.65 Mev implantations, as the amorphous state is saturated. The annealing was conducted under an argon environment.

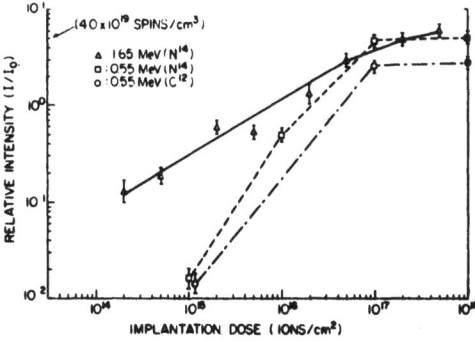

Fig. 2. Dose dependence of the amorphous state: Δ represents the N^{14}-ions at E = 1.65 Mev, ◻ the N^+-ions at E = 550 kev and ○ the C^{12} ions at E = 550 kev. Both implantation and EPR measurement were performed at room temperature.

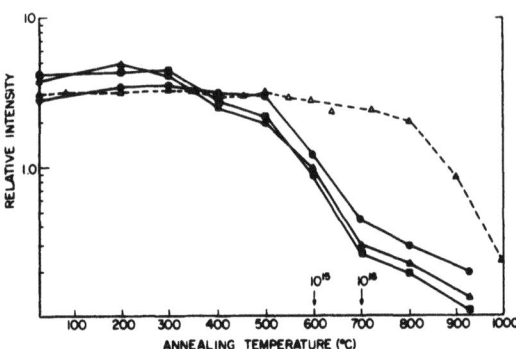

Fig. 3. Isochronal annealing of the amorphous state: ▲ = 2×10^{17} N^+/cm^2 (E = 1.65 Mev), ■ = 1×10^{18} N^+/cm^2 (E = 550 Kev), ● = 1×10^{18} C^+/cm^2 (E = 550 kev). These three samples were annealed in an argon ambience; Δ (2×10^{17} C^+/cm^2, E = 1.65 Mev) was annealed under the nitrogen atmosphere, and the delay of annealing temperature requires further studies to unravel its origin. The amorphous line at the total ion dose 10^{15} and 10^{16} ions/cm^2 anneals near 600 and 700°C, respectively.

The amorphous state starts to recrystalize near 500°C but its residue (about 10 percent of the original concentration) remains even at 930°C. The existence of the residue was manifested by annealing a sample at 750°C for 2 1/2 hr. in vacuum (1.5×10^{-6} Torr.); we observed the "a" line with an intensity comparable to the one in the sample annealed in the Ar - ambience. We also indicate in Fig. 3 that the amorphous state at the dose 10^{15} and 10^{16} ions/cm^2 disappears at 600°C and 700°C, respectively. We note that Schwuttke, et al. [4] observed the two annealing stages at 600° and 1000°C in the 2 Mev N^+ - implanted silicon with the dose of 10^{16} ions/cm^2.

We performed implantation at 490° and 650°C with the 550 kev ions. In the 650°C implantation no EPR signal was observed at the dose of 10^{17} ions/cm^2, but the amorphous line could be resolved at the dose of 10^{18} N^+/cm^2, with the intensity being three orders of magnitude smaller than that of the room temperature implantation after annealing at 720°C. In the case of the 490°C implantation with the total dose of 4×10^{17} N^+/cm^2, the intensity of the amorphous line is two orders of magnitude larger than that at 650°C, although it is much weaker compared with the one in the room temperature implantation. We presume that the residue of the amorphous state may be due to such defects as dislocations and stacking faults which can trap a small point defect diffusing away from the amorphous region.

The depth distribution of the amorphous layer was measured by

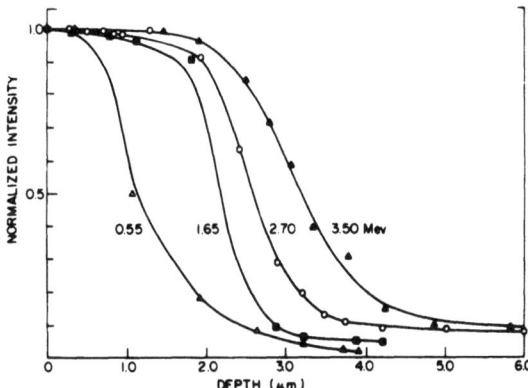

Fig. 4. Depth profile of the amorphous layer for E = 550 kev (Δ), 1.65 Mev (■), 2.70 Mev (○) and 3.50 Mev (▲). Ions were implanted along the ∿ <110> axis at room temperature.

monitoring the intensity variation of the "a" line, while grinding the implanted surface with diamond powder. The results for the four different energies, 0.55, 1.65, 2.70 and 3.50 Mev of the N^+ ions are shown in Fig. 4. The integrated intensity of the "a" line versus depth is normalized to the initial total intensity. A general feature of the damage profile is that near the surface the silicon lattice is virtually free of paramagnetic defects and underneath this undamaged layer is the heavily disordered layer. This was also observed in the previous work [4] on the 2 Mev N^+ - implantation along the <111>-axis.

The projected range of the incident ions which create the amorphous state is estimated at the depth where the total concentration of the "a" line is reduced to 20 percent of the initial value, assuming that incident ions stop mostly at the end of the amorphous layer. The results are given in Table I along with theoretical estimations; LSS and Firsov represent calculations based on the

TABLE I

Projected Range (R_p in μm) of Nitrogen Ion into Silicon

Incident Energy (Mev)	0.55	1.65	2.70	3.50
LSS	1.24	2.62	3.55	4.15
Firsov	1.07	2.22	2.99	3.49
Experiment	1.85	2.55	3.2	4.0

LSS and Firsov theories, respectively, using the Brice's computor program [6]. The LSS theory predicts the ranges remarkably well in good agreement with the experimental data, even though the calculation is for a randomly oriented target. This indicates that the layer of amorphous state is created by those ions reflected from the channeling direction, because we know from the Al3 spectrum that a considerable amount of the incoming ions actually travel deep into the crystal along the channel. The width of energy deposition obtained from Brice's calculation is approximately 1 μm, consistent with the thickness of the amorphous layer in Fig. 4.

SUMMARY

The following conclusions on the amorphous state are obtained: (1) at high-energy implantation, the amorphization process is delayed by the out-diffusion of point defects from the heavily damaged region; (2) approximately 10 percent of the initial spin concentration remains after the 930°C anneal. This is presumably caused by the presence of dislocation and stacking fault which may be created by the relaxation of clusters of vacancies and interstitials; (3) the depth of the amorphous layer in high-energy implantation is in good agreement with the projected range predicted by the LSS theory.

REFERENCES

1. K. L. Brower, F. L. Vook and J. A. Borders, Appl. Phys. Lett. 15, 208 (1969); D. F. Daly and K. A. Pickar, Appl. Phys. Lett. 15, 267 (1969).

2. B. L. Crowder, R. S. Title, M. H. Brodsky and G. D. Pettit, Appl. Phys. Lett. 16, 205 (1970).

3. K. Masuda, S. Namba, K. Gamo and K. Murakami, in Ion Implantation in Semiconductors, S. Namba, Ed., Soc. for Promotion of Sci., Japan (1972), p. 19.

4. G. H. Schwuttke, K. Brack, E. E. Gardner and H. M. DeAngelis, in Radiation Effects in Semiconductors, F. L. Vook, Ed., Plenum Press, New York (1968), p. 406.

5. L. C. Feldman and J. W. Rodgers, J. Appl. Phys. 41, 3776 (1970).

6. D. K. Brice, Sandia Report (SLA-73-0416), (1973).

ESR LINE WIDTH OF CONDUCTION ELECTRONS IN P⁺ ION IMPLANTED Si

T. Shimizu, S. Hasegawa and H. Karimoto

Department of Electronics, Faculty of Technology

Kanazawa University, Kanazawa 920, Japan

ABSTRACT

In order to investigate the nature of lattice imperfections remaining after annealing, ESR due to conduction electrons and electrical measurements are carried out in parallel with a successive layer removal for P⁺ ion-implanted Si annealed at above 600°C. The similar experiments are done for P-diffused Si with various diffusion depths and samples implanted with Ne⁺ ion into the P⁺ ion-implanted Si annealed at sufficiently high temperature for comparison. As a result, the following interesting results are found: (1) ESR line width (ΔH) for samples which have a large gradient of a carrier density does not depend on a carrier density in contrast with those for bulk samples. (2) ΔH depends sensitively on some imperfections remaining after annealing the sample damaged heavily by ion implantation with high doses and broadens largely even when those imperfections do not have an appreciable influence on a carrier density and mobility. (3) On the other hand, point defects produced by ion implantation with low doses which make carrier density decrease have no influence on ΔH. Such an imperfection broadening ΔH appears to be random strain due to dislocation loops which have been observed by electron microscope.

INTRODUCTION

The investigation on the nature of lattice imperfections in ion-implanted materials remaining after annealing is important in connection with a device application. In the present work, the ESR method is applied to the problem. The sample used is Si implanted with P^+ ion. The ESR signal with g-value of 2.0055 from lattice defects produced by ion implantation disappears after annealing at above 600°C[1], then the Lorentzian type ESR signal with g-value of 1.9989 from conduction electrons is newly observed[2]. The variations of the line width (ΔH) of the ESR signal with a successive layer removal are investigated. Electrical measurements are also carried out at the same time. From the results, the nature of the ESR signal of conduction electrons for ion-implanted samples is found to be different from that for bulk materials as reported previously by present authors [2].

In order to clarify the origin of the difference, the ESR and electrical measurements are carried out for Si wafers diffused with P under various diffusion conditions which have a high donor density only in the surface layer as in the ion-implanted sample. Similar experiments are also done for two kinds of samples implanted with Ne^+ ion at low and high doses, respectively, into the P^+ ion-implanted Si annealed at sufficiently high temperature.

SAMPLES AND EXPERIMENTALS

The samples used for the present work are shown in Table I. The substrate materials for ion implantation were polished slices of p-type $\langle 111 \rangle$ oriented Si with resistivities of 5~30 Ωcm. The direction of the incident ions is tilted by 7 degrees from the $\langle 111 \rangle$ direction to avoid channeling effects. The ion implantations are carried out at room temperature. In order to compare with the results for the ion-implanted samples, Si diffused with P under various diffusion conditions and powdered Si samples having a donor concentration (N_D) of 3.4×10^{19} cm^{-3} with various dimensions (a minimum linear dimension is around 1 μm) are also investigated.

The isochronal annealing was performed in vacuum of about 10^{-2} Torr, and the annealing time for each stage was 10 min. ESR was measured at liquid nitrogen temperature ($\approx$77 K) by an x-band spectrometer. The method of stripping the surface layer and that of determining the thickness of the layer are described in the previous paper [2]. Electrical measurements were carried out at room temperature.

Table I. Characteristics of the samples used.

Sample	Ion	Implantation energy (KeV)	Dose (cm^{-2})
P50-15	P$^+$	50	1x 10^{15}
P100-14	"	100	1x 10^{14}
P100-15	"	"	1x 10^{15}
P200-15	"	200	"
Ne50-1·12[a]	Ne$^+$	50	1x 10^{12}
Ne50-5·12[a]	"	"	5x 10^{12}
Ne50-1·15[b]	"	"	1x 10^{15}
SD1~4	Si diffused with P under various diffusion conditions		
SP	Powdered sample of bulk Si doped with P (N_D is 3.4x10^{19} cm^{-3})		

a) The starting material before Ne$^+$ ion implantation is a sample P100-15 annealed at 1060 °C for 30 min.

b) The starting material is a sample P100-15 annealed at 1000 °C for 10 min.

The carrier density n(x), the resistivity $\rho(x)$ and the Hall mobility $\mu(x)$ at a ceatain depth x from the surface are determined by the method previously reported [2].

EXPERIMENTAL RESULTS

Ion-Implanted Sample

The results of the ESR and electrical measurements obtained by stripping the surface layer successively are shown in Fig. 1. Figures 1-(i) and (ii) show the results for sample P100-15 annealed at 950 and 1050°C, respectively. The annealing at those temperatures makes most of implanted P^+ ions electrically active. Similar results are also obtained for other P^+ ion-implanted samples. Figure 1-(iii) shows the result for sample Ne50-1·15 annealed at 700°C. For this sample before annealing no ESR signal from conduction electrons can be observed. However, for samples Ne50-1·12 and Ne50-5·12, ESR signals from conduction electrons can be observed without annealing and the result for the latter sample before annealing is shown in Fig. 1-(iv). The value of ΔH for sample Ne50-1·12 and Ne50-5·12 does not vary with a layer removal in contrast with other samples, and it has a constant value of about 2.8 G which is same as that observed before Ne^+ ion implantation, i.e. that for sample P100-15 annealed at 1060°C, although a carrier density decreases largely by Ne^+ ion implantation.

For the samples after annealing heavily damaged samples in which the ESR signal can not be observed without annealing, ΔH's decrease markedly to a constant value (2~3 G) by stripping the surface layer of the depth corresponding to the damaged regions before annealing as can be seen from Figs. 1-(i)~(iii). The defect distributions which have been obtained from ESR measurements for the sample before annealing[1] are shown by dashed lines in (a) in Figs. 1-(i)~(iii). ΔH obtained from n shown in (a) in Figs. 1-(i)~(iv) by comparing with the relation of n and ΔH obtained for the bulk materials[3] are shown together with the observed ΔH in (b) in Figs. 1-(i)~(iv), respectively. As shown in (b) in Fig. 1-(iii), the obvious difference is present between the observed and the calculated ΔH, with regard to the depth at which ΔH decreases markedly and ΔH appears to be independent of n in contrast with bulk materials.

Values of mobility calculated using n by comparing with the results for bulk materials at room temperature [4] are shown by dashed lines in (c) in Figs. 1-(i)~(iv) together with the observed ones. Agreement between the observed and the calculated values appears to be fairly good.

Diffused Sample

In order to clarify the origin of peculiar features seen in the ion-implanted samples, the ESR and electrical measurements are car-

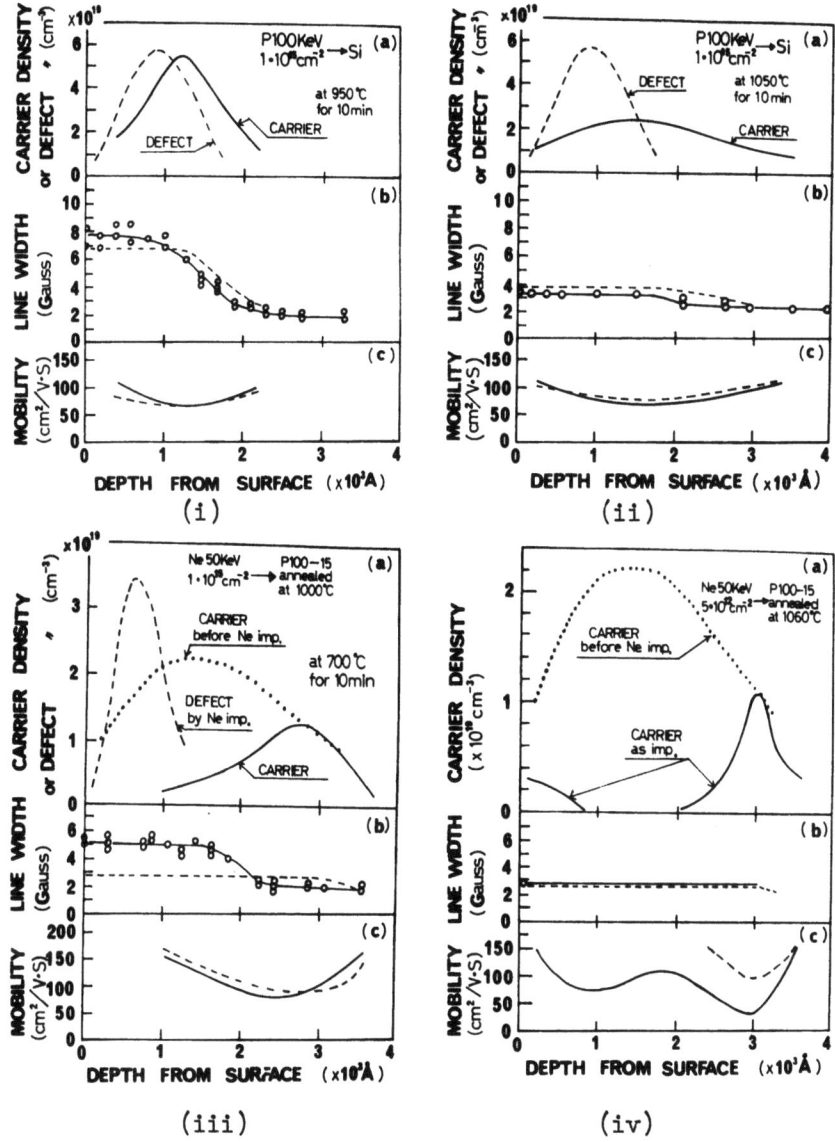

Fig. 1. Results of ESR and electrical measurements obtained by stripping the surface layer successively for the following samples: (i) P100-15 annealed at 950°C for 10 min, (ii) P100-15 annealed at 1050°C for 10 min, (iii) Ne50-1·15 annealed at 700°C for 10 min and (iv) Ne50-5·12 as implanted. Defect distributions before annealing are also shown by dashed lines in (a) for comparison. Dashed lines shown in (b) and (c) show the calculated values using the carrier density shown in (a) by comparing with the results for bulk materials.

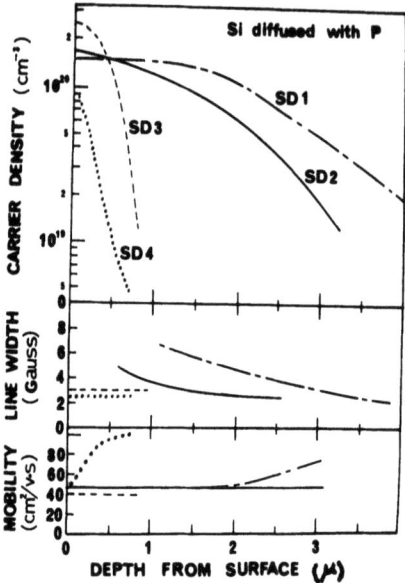

Fig. 2. Results for ESR and electrical measurements obtained by stripping the surface layer successively for samples SD1~4.

ried out for Si wafers diffused with P which have a high donor density only in the surface layer as in the ion-implanted samples. The results are shown in Fig. 2. For sample SD1, the apparent ΔH is broader than the real ΔH because of the Dyson effect, so that the value of ΔH corrected by the Dyson theory is shown in Fig. 2(b) [5]. From Fig. 2, it is found that ΔH does not depend on n for the sample with a diffusion depth smaller than about 1μm while an increase of a diffusion depth makes ΔH broader and dependent on n. Accordingly the absence of the dependence of ΔH on n in ion-implanted and diffused samples is considered to be ascribed to the presence of carriers only near the surface or to a large gradient of the carrier distribution.

Powdered Bulk Sample

In order to distinguish the above stated two possibilities, we investigated the dependence of ΔH on particle size (a minimum linear dimension is around 1μm) for powdered bulk samples with N_D of 3.4×10^{19} cm^{-3} which have no gradient of n. As a result, it is found that ΔH is about 5~6 G and independent of the particle size.

DISCUSSION AND CONCLUSION

The following conclusion can be derived from the experimental results. (1) ΔH does not depend on n when carriers are distributed within 1μm from the surface and has a constant value of 2~3 G at

77 K in the absence of imperfections, while ΔH is known to depend largely on n for bulk samples with a homogeneous carrier distribution. Although the physical mechanism for the absence of the dependence of ΔH on n is not understood, phenomenologically it is found that ΔH does not depend on n for samples with a large gradient of n. (2) ΔH depends sensitively on some imperfections remaining after annealing the sample damaged heavily by ion implantation with high doses and broadens largely even when those imperfections do not have an appreciable influence on n and mobility. So ΔH might be used as a useful tool to detect such imperfections. On the other hand, the imperfection produced by ion implantation with low doses does not broaden ΔH even when n decreases by the presence of such defects (see(a) and (b) in Fig. 1-(iv)).

Therefore the imperfection remaining after annealing the heavily damaged sample should be different from that produced by ion implantation with low doses. The former appears to be dislocation loops which have been observed by electron microscope for the sample after annealing the heavily damaged sample [6]. Although the mechanism for the broadening of ΔH in the heavily damaged sample is not clear at present, the random strain due to the above stated dislocation loops should be considered as one of the possibilities.

ACKNOWLEDGEMENT

The authors are grateful to Drs. T. Tokuyama and T. Ikeda of Hitachi Central Research Laboratory for providing the implanted samples.

REFERENCES

1. S. Hasegawa, K. Ichida and T. Shimizu, Japan. J. appl. Phys. <u>12</u>, 1181 (1973).

2. S. Hasegawa, H. Karimoto and T. Shimizu, Japan. J. appl. Phys. <u>12</u>, 1190 (1973).

3. H. Kodera, J. Phys. Soc. Japan <u>21</u>, 1040 (1966).

4. C. Yamanouchi, K. Mizuguchi and W. Sasaki, J. Phys. Soc. Japan <u>22</u>, 859 (1967).

5. F. J. Dyson, Phys. Rev. <u>98</u>, 349 (1955).

6. M. Tamura, Appl. Phys. Letters <u>23</u>, 651 (1973).

ESR STUDIES ON ANNEALING BEHAVIOR OF HEAVILY DAMAGED SILICON

K. Murakami, K. Masuda, K. Gamo and S. Namba

Faculty of Engineering Science, Osaka University

Toyonaka, Osaka, Japan

ABSTRACT

We have studied the influence of an isochronal annealing process on paramagnetic defect centers produced by implanting $1 \cdot 10^{14} P^+/cm^2$ and $1 \cdot 10^{15} P^+/cm^2$ in Si at room temperature with an energy of 200keV. For the intermediate dose implant ($1 \cdot 10^{14}/cm^2$) four types of defect centers have been observed which revealed a complicated annealing behaviour. In addition other defect centers appeared at different temperature stages during the annealing process and some centers were found to be stable up to 600°C. For the high dose implant ($1 \cdot 10^{15}/cm^2$) only the amorphous-center ($g \fallingdotseq 2.006$) has been observed and was found to anneal out above 500°C. No other defect centers appeared in this temperature range, in contrast to the results from the intermediate dose implant. In both dose ranges, ESR spectrum of implanted phosphorus donors was observed for temperature above 500°C. We also discuss the annealing behavior of both the paramagnetic susceptibility and the line width of conduction electrons from phosphorus donor atoms.

INTRODUCTION

Ion implantation introduces a great variety of defects. In general the average concentration of defects increases with increasing ion dose and, above a certain threshold dose, a continuous amorphous layer is produced. Morehead et al.(1) divided the dose range into three parts (low(I), intermediate(II), and high(III) dose ranges) on the basis of the differences in the electrical annealing behavior. In our previous work(2), we used the ESR technique in order to investigate unannealed samples implanted with intermediate and high doses of P ions. For the intermediate dose ranges(II) it is known that localized amorphous regions and heavily damaged crystalline regions coexist(2) and the complete electrical recovery needs high temperature($800°-900°C$)(1). For the high dose range (III), a continuous amorphous layer is produced(1,2) and complete electrical recovery occurs after annealing at $550°-600°C$(1). Therefore it is important to investigate both the processes of the recrystallization of a continuous amorphous layer and the annealing of heavily damaged crystallite.

In the present work, we have studied the annealing processes of paramagnetic defect centers in ranges (II) and (III) in order to find correlation between paramagnetic defect centers and implanted phosphorus donors.

EXPERIMENTAL PROCEDURE

The silicon samples were implanted in a nonchanneling direction with a intermediate($1·10^{14}/cm^2$) or high($1·10^{15}/cm^2$) dose of 200keV phosphorus ions at room temperature and subjected to a 15 minutes post implantation anneal in the temperature range between 100-900°C. Annealing has been performed under three different conditions i) in a vacuum of about 10^{-6}Torr, ii) evacuated to $\sim 10^{-6}$Torr and sealed in ESR quartz ampoule, and iii) in pure N_2 gas. The silicon samples used were B-doped and has $\sim 25\Omega$-cm(c.z.) with the (111) surface for range (III), and B-doped and i) $\sim 25\Omega$-cm(c.z.) with the (100) surface and ii) $\sim 120\Omega$-cm(c.z.) with the (111) surface for range (II).

ESR measurements at 9.2-9.4GHz were carried out at 77K and 300K under the conditions of slow passage. For $1·10^{15}P^+/cm^2$ implanted sample, it was difficult to measure ESR spectrum of conduction electrons after annealing at 500-700°C because of large Q-losses due to high conductivity of implanted layer and a very broad line width. Therefore many measurements have been performed at 77K above 500°C annealing temperature.

EXPERIMENTAL RESULTS AND DISCUSSION

[1] Intermediate dose range (II) ($1·10^{14}P^+/cm^2$)

Figure 1 shows the ESR spectra with H//[111] axis of defect centers for samples annealed at different temperature. We have observed four types of paramagnetic defect centers for unannealed samples(2): i.e. the amorphous-center, the Si-B2, -O1, and -P3 centers. The Si-P3 center was annealed out at 200°C, which is in agreement with Brower's result(3). The Si-O1 center was also annealed out at 200°C. At 200°C a center appeared which seems to be the Si-P1 center(4) as was indicated by the angular dependence of the resonance lines. At 350°C anneal temperature the amorphous- and the Si-B2 centers disappeared, but the Si-P1 center remained. A new center appeared at this temperature which has [100] axially symmetric g-tensor:i.e.g_1=2.0005± 0.0005, g_2=g_3=2.0020±0.0005, and g_1//[100] axis at 300K. Both the

ESR STUDIES ON ANNEALING BEHAVIOR

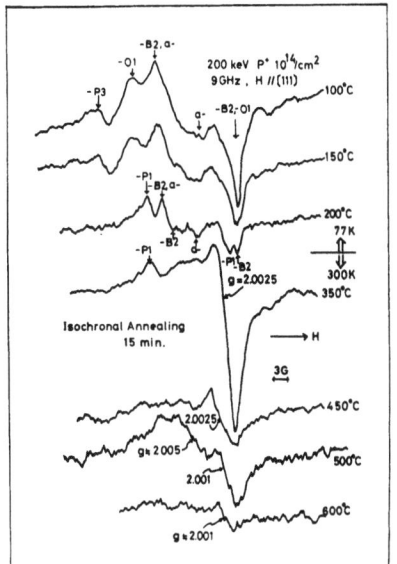

Fig.1 ESR spectra taken at 77K and 300K with H//[111] axis.

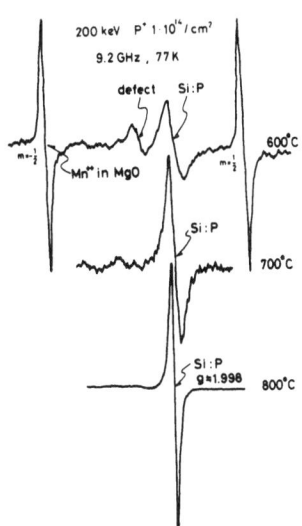

Fig.2 ESR spectra of conduction electrons from P donor atoms.

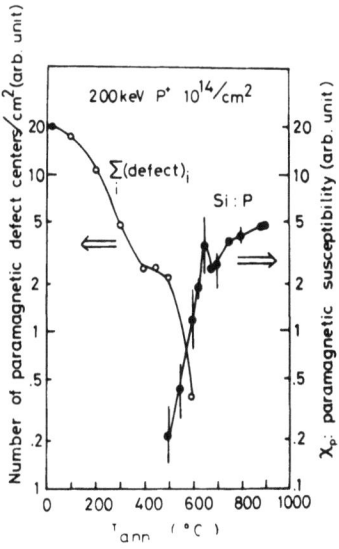

Fig.3 Isochronal annealing (15 min.) of samples for range (II).

Si-P1 and the new centers disappeared and other unidentified centers with anisotropic g values have been found between 450° and 500°C. Some defect centers remained after annealing at 600°C. One of those had the value of $g \doteq 2.001$ for H//[111] and $g \doteq 2.004$ for H//[100] at 300K. For annealing temperature above 600°C no defect centers have been observed.

Figure 2 shows the ESR spectra of conduction electrons from phosphorus donor atoms. It is possible to observe this spectrum with $g=1.998-1.999$ above 500°C anneal.

Figure 3 shows the total number, $\sum_i (defect)_i$, of all defect centers and the paramagnetic susceptibility χ_p of conduction electrons. χ_p gradually increased up to 900°C anneal, which corresponds to the recovery of surface carrier density N_s obtained by Crowder et.al.(5) except for a plateau of χ_p at 650°C. The line width ΔH_{ms1}. of conduction electrons was found to depend on the three kinds of annealing methods only between 500° and 700°C. In particular, the sealed method had an abnormal broadening of ΔH_{ms1}.

at around 650°C. At 700°C and above the line width has decreased to a constant value of 1.8G for all annealing conditions. In the same temperature range χ_p was found to increase with increasing annealing temperature. This indicates that above 700°C increase of phosphorus electrical activity, or χ_p, must be caused by annealing of compensating centers which are nonparamagnetic and not scattering centers.

[2] High dose range (III) ($1 \cdot 10^{15} P^+/cm^2$)

Figure 4 shows the ESR spectra of the amorphous-center(2) with $g \doteqdot 2.006$ and $\Delta H_{ms1} \doteqdot 6G$ which has an isotropic Lorentzian lineshape for samples annealed at different temperature. During the anneal process, only the number of the dangling bonds or voids decreased up to 500°C and no other defect centers appeared. This indicates that up to 500°C an amorphous state with voids gradually changes to an ideal amorphous state. The number of the amorphous-center is proportional to the internal surfaces of voids, because the Fermi level may be fixed to the nearly middle level of mobility gap(6) up to 500°C. At 500°C anneal, a broad and distorted spectrum was observed at 300K because implanted ions begin to act as donor atoms. No other defect centers were detected above 500°C annealing temperature.

Figure 5 shows the total number of the amorphous-center together with the paramagnetic susceptibility χ_p (arb. unit) of conduction electrons. In contrast to the results observed in range (II), χ_p recovered completely between 500° and 600°C and was almost constant up to 900°C, which corresponds to the recovery of surface carrier

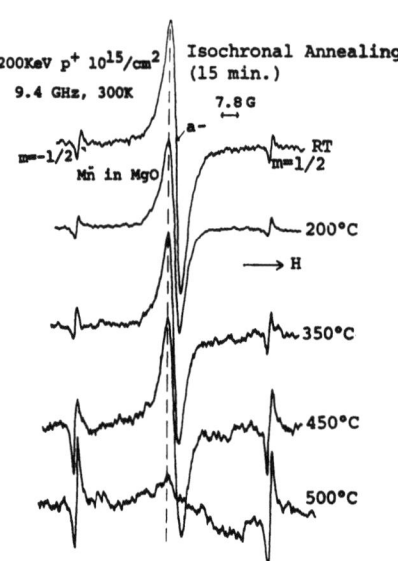

Fig. 4 ESR spectra taken at 300K.

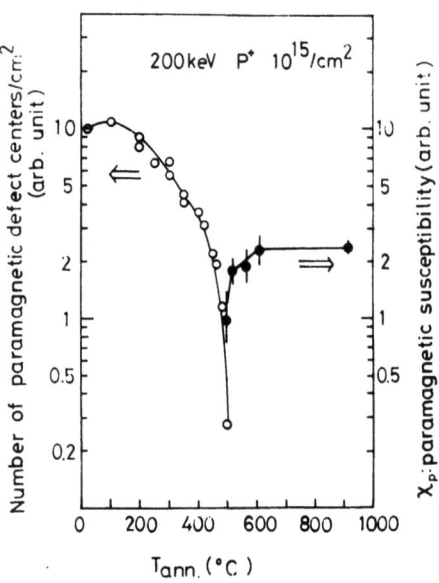

Fig. 5 Isochronal annealing(15min.) of samples for range (III).

density obtained by Crowder et.al.(1,5). The line width ΔH_{msl}. was very broad (12±2G) up to 700°C and decreased to ~6G by annealing at 900°C. From this it is concluded that main imperfections(7) which broaden the line width of conduction electrons but are not compensating centers seem to remain up to ~900°C.

The annealing behavior of the 200keV P^+ implanted silicon differs from that of 80keV Sb^+ ($1 \cdot 10^{14}$ and $1 \cdot 10^{15}$/cm^2) implanted silicon(8), where the amorphous-center remains at about 600°C anneal. There is no explanation for it as yet.

SUMMARY

The results obtained from ESR measurements on P implanted Si sample are summarized in Table 1.

(1) In the intermediate dose range(II)($1 \cdot 10^{14} P^+$/cm^2), four types of paramagnetic defect centers have been observed and they showed very complicated annealing behavior. With increasing annealing temperature some centers were found to anneal out whereas others were found to appear. Some defect centers were stable for temperature at 600°C.

Table 1 The annealing behavior for two dose ranges.

	the intermediate dose range (1×10^{14} P/cm^2)	the high dose range (1×10^{15} P/cm^2)
unannealed	four types of defect centers: the amorphous-center*), the Si-B2**), -O1***), and -P3****) centers.	only the amorphous-center ($g \approx 2.006$, $\Delta H_{msl}. \approx 4.8-6G$, and Lorentzian lineshape.)
R.T. 100°C 200°C 300°C 400°C 500°C 600°C 700°C 800°C 900°C	Si-P3 and -O1 centers / amorphous- and S-B2 centers; Si-P1****) and a new defect center with [100] axially symmetrical g tensor / other new defect centers ($g \approx 2.001$ for H//[111], $g \approx 2.004$ for H//[100]); conduction electrons. ($g=1.998-1.999$); main scatterer (or imperfections) annealed out. ← complete recovery	only the amorphous center (no other defect centers); an amorphous state with voids ↓↓; an ideal amorphous state; conduction electrons. ($g=1.998-1.999$); ← complete recovery; main scatterer (or imperfections) annealed out.

*) The amorphous-center originates from the isolated amorphous regions.
**) The g-tensor of the Si-B2 center is approximately [111] axial symmetry.
***) The Si-O1 center shows a marked motional effect at around 200K.
****) The structure of the Si-P3 and -P1 centers are thought to be tetra-vacancies in (110) plane and penta-vacancies not in (110) plane, respectively(4).

(2) In the high dose range(III)($1 \cdot 10^{15} P^+/cm^2$), only the amorphous-center($g \eqsim 2.006$) was observed up to 500°C. No other defect centers appeared in contrast to range(II).

(3) In both dose ranges, ESR spectrum of conduction electrons from implanted phosphorus donors has been observed above 500°C anneal. The annealing behavior of the paramagnetic susceptibility χ_p was almost consistent with the behavior of the recovery of surface carrier density, except at about 650°C in range(II) where χ_p had a plateau. At range(II), at 700°C and above the line width has decreased to a constant value of 1.8G. In the same temperature range, χ_p was found to increase with increasing annealing temperature. At range(III), χ_p is completely recovered between 500-600°C but the line width was found to be broader and to decrease to a constant value (~6G) at 900°C annealed temperature.

(4) The defect centers remaining at 600°C may cause the results found for range(II) that the carrier density as well as χ_p do not show complete recovery between 500°C and 600°C. One of those centers which were stable at 600°C were found to have g-values of 2.001 for H//[111] and 2.004 for H//[100].

ACKNOWLEDGEMENT

We would like to thank Drs. T. Tokuyama and T. Ikeda of Central Research Lab. of Hitachi Ltd. for preparing P^+ implanted silicon samples, and also to thank Mr. S. Yoshioka for aid in experiments.

REFERENCES

(1) F.F. Morehead,Jr., B.L. Crowder and R.S. Title: J. Appl. Phys. 43 (1972) 1112.
(2) K. Murakami, K. Masuda, K. Gamo, and S. Namba: Japan.J. Appl. Phys. 12 (1973) 1307.
(3) K.L. Brower, F.L. Vook, and J.A. Borders: Appl Phys. Letters 15 (1969) 208.
(4) Y.H. Lee and J.W. Corbett: Phys. Rev. B8 (1973) 2810.
(5) B.L. Crowder and F.F. Morehead,Jr.: Appl. Phys. Letters 14 (1969) 313.
(6) N.F. Mott and E.A. Davis: Electronic Processes in Non-Crystalline Materials (Clarendon Press, Oxford, 1971).
(7) S. Hasegawa, H. Karimoto and T. Shimizu: Japan. J. Appl. Phys. 12 (1973) 1190.
(8) K. Masuda, S. Namba, K. Gamo, and K. Murakami: Proc. of U.S.-Japan Seminar on Ion Implantation in Semiconductors, Kyoto, S. Namba ed., 1972, p.19.

ESR STUDIES OF ION IMPLANTED Si-SiO$_2$ STRUCTURE

T. Izumi and T. Matsumori

Department of Electronic Engineering, Faculty of

Engineering, Tokai University, Shibuya, Tokyo, Japan

ABSTRACT

Electron spin resonance spectra of radiation damage centers (named Pa, Pb and Pc center) in Si-SiO$_2$ structure have been observed following implantation with Ne$^+$ and O$^+_2$ ions. Pa and Pc centers with g-values of 2.0013 and 2.008, respectively, located in SiO$_2$ films and Pb with g-value of 2.006 is located in silicon. Pa center is maximum density at about 0.9Rp (projected range) in the SiO$_2$ films. The peak of spin density shifts towards the surface with elevating annealing temperature and disappeared at 400°C. The activation energy for the process is ~0.2eV in the range of 100°C — 350°C. Properties of Pb center are similar to those from the amorphous region in implanted silicon. Pc center is an asymmetric broad line, which has the maximum spin density in the deeper layer than Rp. The g-values of Pa and Pc center in the SiO$_2$ suggest that Pa is an electron resonance and Pc a hole.

INTRODUCTION

Ion implantation into MOS structures has been extensively studied by many researchers [1-3]. Interface damage generated by ion implant into Si-SiO$_2$ interface would affect the electrical properties of MOS device such as threshold voltage, carrier mobility and frequency noise. Although it has been generally mentioned that the origin of the interface-state is the structual imperfections on Si-SiO$_2$ interface [4], these physical structures are not clear till now. Although it has been generally considered that ESR measurements are very powerful to study these imperfections, no ESR data on ion-implanted Si-SiO$_2$ has been prsented.

We investigated first the annealing properties and the depth distributions of defect centers produced by ion implantation into Si-SiO$_2$ structure using an ESR technique, in order to make clear the properties of the interface states induced by ion implantation.

EXPERIMENTAL PROCEDURES

The substrates for implantation were P-type, ~20Ω-cm silicon wafers with (100) surface chemically polished. O$^+$ and Ne$^+$ of ion energies (50keV) and integrated dosage (10^{12} to 10^{14}ions-cm^{-2}) were used for implantation in SiO$_2$ films thermally grown in dry oxygen at 1050°C on Si substrates of <100> orientation. Annealing (room temperature to 600°C) was carried out in N$_2$ ambient. The ESR in these specimens were measured at 77 K using a X-band microwave spectrometer.

RESULTS AND DISCUSSIONS

Fig. 1 shows the ESR absorption derivative curve obtained by the ion (O$^+$, Ne$^+$) implanted Si-SiO$_2$ structure, measured at 77 K. Apparently, the results indicate that three kinds of paramagnetic centers named Pa, Pb and Pc center have been found in the Si-SiO$_2$ structure. Pa and Pc are located in the SiO$_2$ films and Pb is distributed in the silicon. Properties of Pb center are similar to those from the amorphous region in the implanted silicon [5]. Implantation with 10^{13}O/cm^2 at 50keV produced Pa center with a spin density of 10^{19}cm^{-3}, and increases in spin densities of Pa and Pc were observed with increasing implantation dosage.

Fig. 2 shows the depth distributions of Pa and Pc in the SiO$_2$ films. Implantation of O$^+$ and Ne$^+$ at 50keV creates the Pa distributions which have the peak positions at the depths of 460 Å and 700 Å, respectively, from SiO$_2$ surface. These peak positions correspond to the depth of ~0.9Rp (projected range). This value (0.9Rp) is very different from the peak position (2/3Rp) of the defect distribution produced in ion-implanted Si [6,7]. Pa center exhibits the same g-value (g=2.0013) and line shape as observed by S. Lee and P. J. Bray in the neutron-irradiated alumino-silicate glasses [8], except ΔHmsl (4 gauss in the case of implanted Si-SiO$_2$ and 1 gausses the irradiated glasses). On the other hand, the spectrum of Pc center is an asymmetric broad line with g-value of 2.008 and ΔH=17 gauss which is located at 900 Å from the surface in the SiO$_2$. This spectrum of Pc center is very similar as due to a hole trapped on a non-bridging oxygen, as described by S. Lee and P. J. Bray[8].

Fig. 3 shows the effect of annealing on the distribution of Pa in the SiO$_2$ films implanted at room temperature with 10^{13}O/cm^2. As it is clear from this figure, the density of Pa center decreases slightly at 100°C annealing, but decreases rapidly above 200°C annealing and the peak position of the Pa distribution shifts

ESR STUDIES OF ION IMPLANTED Si-SiO$_2$ STRUCTURE

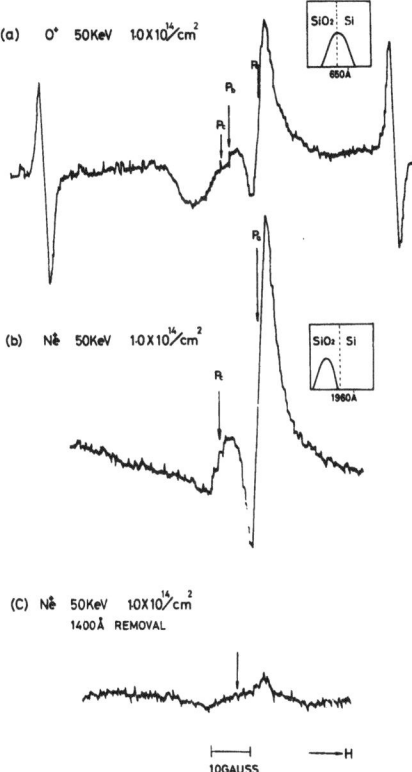

Fig. 1 Typical example of the centers on ion implanted Si-SiO$_2$ structure. Thicknesses of SiO$_2$ films were (a) 650 Å for O$^+$ implantation and (b) 1960 Å for Ne$^+$ implantation. Inset show schematically the distribution of ion species in each implantation condition. (c) When etch-off of SiO$_2$ films was performed in approximately 1400 Å, Pa disappeared and paramagnetic center with g=2.005 apeared.

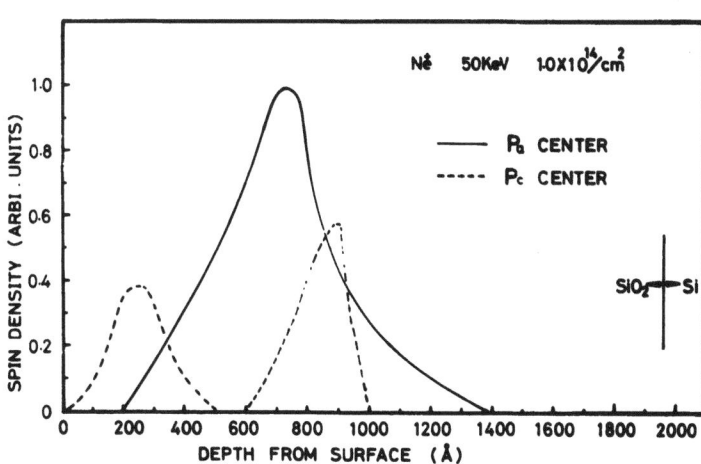

Fig. 2 Depth distribution of Pa and Pc centers in the SiO$_2$ films for 50 keV implantation of 10^{14}Ne ions/cm^2.

towards the surface with increasing annealing temperature. The result indicates that defects (Pa) in the inner part of SiO_2 films disappear at 200°C annealing, whereas the defect density on the surface side does not change remarkably. With increasing annealing temperature (above 300°C), the defects distribution at the surface also decrease. That is, the disappearance of the defects produced by implantation takes place at first in the inner part of the oxide, and proceed gradually towards the suface with inceasing annealing temperature. A similar annealing characteristics for the depth distribution of Pa produced in Ne^+ implanted specimens has been observed.

Since annealing up to 130°C has been usually done on measuring C-V characteristics, the samples were annealed at 130°C in nitrogen ambient in order to satisfy the same condition [9]. After that, the ESR measurements were done and 40% of Pa has been removed.

Fig. 4 indicates an isochronal annealing behavior of Pa center. The density of Pa center decreases exponentially above 100°C and disappears at 400°C. Activation energy of ∼0.2 eV is obtained in the annealing processes of Pa between 200°C and 350°C.

R. A. Weeks observed the two kinds of paramagnetic centers with g-values of 2.0013 ± 0.0006 and 2.0090 ± 0.0007, and half-width of 1.7 gauss and 40 gauss, respectively, in neutron irradiated quartz [10]. And it was concluded that these centers were caused by defects in the basic SiO_4 tetrahedra generated by fast neutrons and primary and secondary knockons. Two groups of lines (Pa and Pc) generated by implantation agree well with lines by R. A. Weeks, except $\Delta Hmsl$. Furthermore, T. W. Hickmtt [11] observed an ESR line (g=2.008 - 2.009) with the same g-value as described by Weeks, in rf-sputtered SiO_2 films, which is eliminated at 350°C.

As described above, it should be pointed out that Pa and Pc center produced by ion implantation have strong correlations with the defects generated in irradiated glasses judging from g-values and line shape. On the other hand, the magnitude of the g-values of the two lines suggests that Pa is electron resonance and Pc a hole. The peak position of Pa is located in the shorter depth than Rp and Pc in the longer depth than Rp. When ion-implanted into SiO_2 films, an amorphous SiO_2 lattice makes up of SiO_4 tetrahedra which include some defects and are slightly oxygen-deficient. In SiO_2 bonding scheme, each silicon in a normal structure contributes an (sp^3) electron to each of its four Si-O bonds. Ne^+ and O^+ ions implanted into $Si-SiO_2$ structure damage these Si-O bonds and sputter an oxygen atom plus an electron, and leave behind a single unpaired electron and two empty (sp^3) orbitals (Fig. 5). The spectra which were observed in the implanted $Si-SiO_2$ would be due to these defect structures.

When 1400 Å of SiO_2 films was eliminated, very weak line with the g-value of 2.005 appeared, and the distribution was located at the $Si-SiO_2$ interface (Fig. 1 (c)). Judging from g-value of 2.005, it is considered that the center is equivalent to Pb center.

ESR STUDIES OF ION IMPLANTED Si-SiO$_2$ STRUCTURE

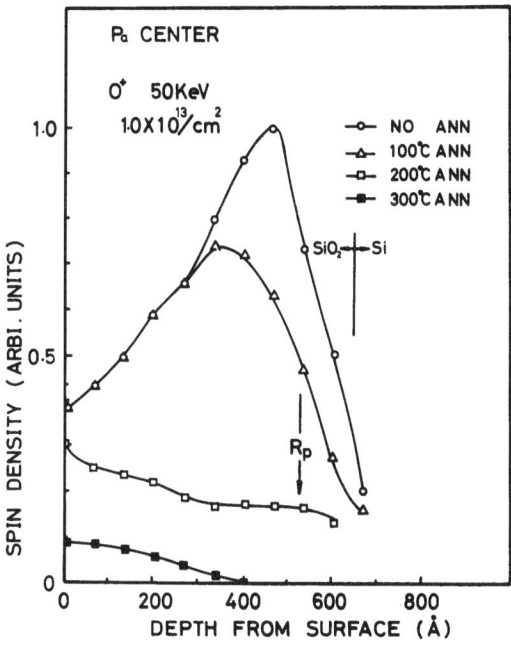

Fig. 3 Effect of annealing of the distribution of Pa center in the SiO$_2$ films implanted at room temperature with 10^{13} O ions/cm^2. Thickness of films were 650 Å.

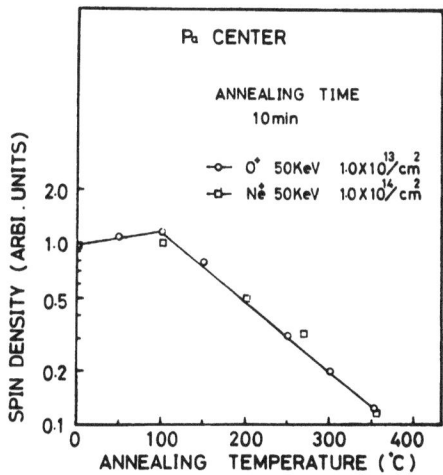

Fig. 4 Isochronal annealing behavior of Pa center.

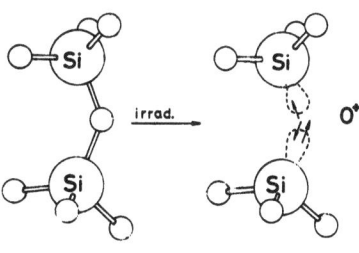

Fig. 5 Schematic representation of an amorphous SiO$_2$ lattice defects arising from ion implanted Si-SiO$_2$ structure [12].

However we could not separate this line from Pc line in as-implanted samples, for this signal is too weak to observe. It can also be considered that Pc shifts from g=2.008 to g=2.005 after eliminating Pa center (1400 Å removal). However, there are some structual defects in general, which are caused by excess Si with oxygen-deficiency in the Si-SiO$_2$ interface [13]. ESR spectra with g=2.005 could not be observed in unimplanted samples, whereas ESR signal with g=2.005 could be obtained only in implanted samples. Therefore, it can be considered that the typical paramagnetic center is generated in the SiO$_2$ very near at the Si-SiO$_2$ interface by ion implantation. Since a projected range Rp for 50 keV Ne$^+$ implantation in SiO$_2$ is 800 Å, the implanted ion should not reach to the interface (1900 Å from SiO$_2$ surface). Nevertheless, the ESR spectra can be measured in the implanted specimens (50 keV 10^{14} ion/cm^2 Ne). Therefore, it is necessary to investigate in detail why the implanted ions affect the interface distant from the Rp. Since the annealing behavior of Pa and Pc center is similar to the annealing behavior of charged interface state density induced by Ne$^+$ and O$^+$ ions implantation, it is necessary to investigate in detail how these defects may play a role on interface state generation.

SUMMARY

Electron spin resonance study has been made, in order to clear the origin of interface states induced by ion (O$^+$, Ne$^+$) implants into Si-SiO$_2$ structure. There are three kinds of centers, named Pa, Pb and Pc whose g-value are as follows; g=2.0013, ΔHmsl=4 gauss for Pa, g=2.006, ΔHmsl=6 gauss for Pb and g=2.008, ΔHmsl=17 gauss for Pc. Pa and Pc are located in the SiO$_2$, while Pb is inside of the silicon. The properties of Pb are similar to those of the amorphous Si produced by ion implanteation. The density distribution of Pa has the maximum value at about 0.9Rp. Pa disappeared at 400°C annealing. The activation energy for the annealing process is ~0.2 eV for Pa in the temperature range of 100°C - 350°C. It is suggested that Pa is due to the electron in ruptured Si-O bond. On the other hand, Pc is weak in general, so that informations on this center are not sufficient. The peak position of the Pc distribution is located at the longer depth than Rp. Paramagnetic center with g-value of 2.005 is observed in the oxide very near the interface. Futher investigations of these centers, especially paramagnetic center generated in the Si-SiO$_2$ interface, must be rquired.

ACKNOWLEDGEMENTS

The authors would like to thank Dr. T. Tokuyama, T. Yoshida and T. Warabisako of Hitachi Central Risearch Laboratory, for supplying the samples and for helpful discussions. We also thank Prof. K. Hiraoka and Y. Kimura for their advice.

REFERENCES

[1] T. Tokuyama, Proc. 5th Conf. of Solid State Devices, Tokyo 1973, p. 499.

[2] N. J. Chou and B. L. Crowder, J. Appl. Phys. <u>41</u>, 1731 (1970).

[3] T. Tokuyama, et al, 3rd Intern. Conf. Ion Implantation in Semicond. and Other Materials, (1973), p. 159.

[4] T. W. Sigmon and R. Swanson, Solid State Electro. <u>16</u>, 1217 (1973).

[5] B. L. Crowder, et al, Appl. Phys. Letters <u>16</u>, 205 (1970).

[6] B. L. Crowder and R. S. Title, Radiation Effects <u>6</u>, 63 (1970).

[7] T. Matsumori, et al, Proc. 3rd Intern. Conf. Ion Implantation in Semicond. and Other Materials, (1973), p. 33.

[8] S. Lee and P. J. Bray, Phys. and Chemi. Glasses, <u>3</u>, 37 (1962).

[9] T. Warabisako, et al, Ohyo Butsuri, <u>42</u>, 174 (1974) (in Japanese)

[10] R. A. Weeks, J. Appl. Phys. <u>27</u>, 1376 (1956).

[11] T. W. Hickmott, J. Appl. Phys. <u>45</u>, 1050 (1974).

[12] F. J. Feigl, et al, Solid State Communi. <u>14</u>, 225 (1974).

[13] Y. Nishi, Japan J. Appl. Phys. <u>10</u>, 52 (1971).

RECOVERY OF SILICON LAYERS DAMAGED BY LOW ENERGY ION BOMBARDMENT

R. Prisslinger, S. Kalbitzer and H. Kräutle
Max-Planck-Institut für Kernphysik, Heidelberg
J.J. Grob and P. Siffert
CRN Strasbourg

Abstract
Silicon has been implanted at room temperature with silicon and xenon ions to introduce lattice disorder up to the amorphous level. Annealing characteristics have been measured as a function of implanted dose. This data was analyzed with a defect model relating annealing temperature to cluster size.
Elements with masses close to silicon and xenon, respectively, have been implanted in order to search for chemical effects on the annealing behavior of damaged silicon.

Introduction
Since radiation damage, the inevitable and important main product of ion bombardment, plays a crucial role in determining the electrical properties of implanted semiconductor material, it has been of great interest to study its formation and removal.
It has been observed that the annealing characteristics do not show sharp thresholds and shift to higher temperatures with increasing disorder.
Point defects, especially interstitials, are highly mobile defects in silicon. Consequently, considerable recombination and defect clustering takes place during the transient temperature history of displacement and temperature spikes and even at room temperature or below.
These remaining aggregate centers, ranging from divacancies to many defect clusters, anneal with different activation energies. A corresponding continuum model has been proposed by Nelson (1). Different models have been discussed by Morehead and Crowder (2) and by Baranova et al. (3). The present work is an attempt to make a more quantitative check of this model by assigning annealing

temperatures to cluster sizes. In addition, by varying the ion species the role of chemically active impurity ions as potential nuclei for recovery processes has been investigated.

Experimental

Homogeneous implantations were achieved by moving the silicon wafers through the ion beam. Since the beam intensity was monitored throughout the run, the doses are estimated to be accurate to at least 5%. The annealing temperatures were controlled with a precision of a few degrees. The backscattering statistics contribute an error of about 3%.

Since the evaluation of the relative disorder and especially of the "amorphous dose" is problematic for layers whose thickness is smaller than the detector resolution, it was necessary to precisely measure the damage growth curves. The amorphous state was defined to be reached if a further substantial increase in dose, say by a factor of three, did not increase the amount of damage by more than 5%. Taking this dose level as the 100% mark the relative disorder concentrations of lower dose implants were derived.

Results and discussion

Fig.1 shows the damage growth curves published by others and our present data. In general, the data are in reasonable agreement, although the amorphous dose quoted by the individual authors lie in a bracket of a factor of 10. Methods, based on the detection of low degree cluster defects, may tend to give lower values. On the other hand, since the process is asymptotic, definitions based on backscattering data are a matter of convention. As will be seen later, our annealing data suggest that the amorphous dose for ions with mass 30 amounts to not less than $3 \times 10E15/cm \times cm$. In the investigated range of doses, the growth curves are saturating exponentials of the type $y = 1 - \exp(-x)$ where y denotes the relative disorder and x is a measure of the implanted dose.

Figs. 2a and b show the annealing characteristics for Si(Si) and Si(Xe) implants of different disorder concentrations.

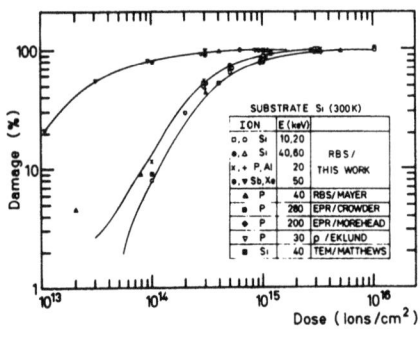

Fig.1:
Growth curves of damage, i.e. relative disorder vs. dose, for Si(M~30) and Si(M~120) implants. The values of the amorphous doses used in the present work are $3 \times 10E15$ and $8 \times 10E14$, respectively. The other authors (4-8) give the following values:
$4 \times 10E14$, $10 \times 10E14$, $6 \times 10E14$, $3 \times 10E14$, $10 \times 10E14$, M~30 ions/cm×cm.

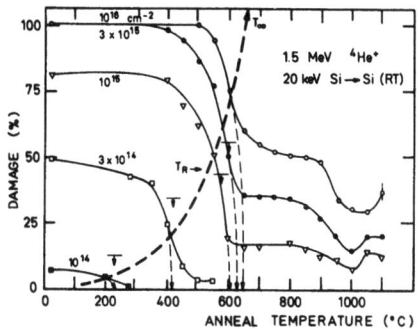 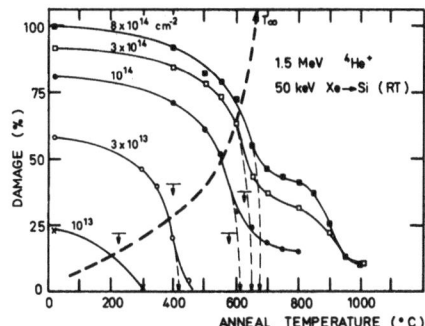

Fig. 2a and b:
Thermal recovery of single crystal silicon damaged with different doses of silicon and xenon ions, respectively. Stage I annealing is complete for damage levels up to about 50%; at higher levels additional stages comprising considerable amounts of disorder are observed. The line, crossing the stage I annealing characteristics, connects reference points (T_R) of the individual curves.

Three annealing stages are clearly distinguished:
I: T < 650 °C, II: 650 < T < 900 °C, III: T > 900 °C.
First we turn to stage I to which the cluster annealing model shall be applied. We recognize the features mentioned above: 1.) the annealing pattern extends over a temperature interval of about 200°C or more; 2.) the annealing temperatures increase with higher damage levels.
The analysis of stage I recovery, according to the model presented before, is complicated by the uncomplete annealing of the high damage implants. Whereas for the implants with damage ≲ 50% a reference point - denoting the reduction to 50%, the inflection, or the extrapolation to zero level - can be derived easily, the adequate procedure is more uncertain in the other case. By assuming an error of about ± 25 °C in the reference temperature the assignment appears to be justified. The temperature T_∞, corresponding to infinite cluster size, is about 660 °C. The further analysis proceeds as follows.
From the original formula, relating annealing temperature T_R with a spherical cluster of radius R,

(1) $T_R = T_\infty (1 - \text{const.}/R)$,

we derive:

(2) $\Delta T/T_\infty = R_0/R$,   const. $= R_0$, $\Delta T = T_\infty - T_R$.

Since R must be a monotonously increasing function of dose D, we try a relation $R \sim D^n$:

(3) $T/T_\infty = (D_0/D)^n$.

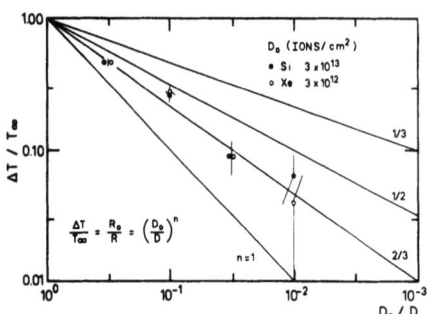

Fig. 3:
Lines representing different values of the exponent n of eq. 3 are shown together with the data which fall on the n = 2/3 line. The uncertainty in T_R is 25°C.

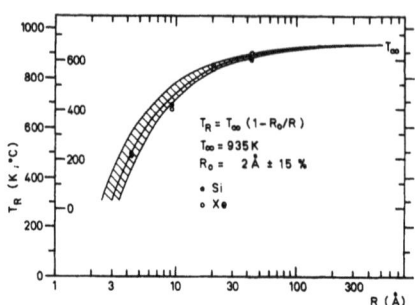

Fig. 4:
Cluster size vs annealing temperature.

This choice is suggested by the fact that in the simplest case – i.e. number of created defects = number of remaining defects, clustering uniformly in spherical geometry – n would assume the value 1/3. Fig. 3 shows the data replotted according to eq. 3.

Whether the value of n = 2/3 is caused by particular recombination kinetics or by a different cluster geometry, cannot be decided at the present time. The dose D_0 has the meaning of uncorrelated growth due to the disordering events for $D < D_0$; for $D > D_0$ the existing clusters act as nuclei for generated point defects. The numbers of D_0 = 3 x 10E13 Si/cmxcm and of 3 x 10E12 Xe/cmxcm represent the number of cylinders of radius 10 and 30 Å, respectively, needed to cover unit area. Hence, it is clear that all further displacement events take place in the immediate vicinity of these predamaged microscopic volumes. Since these volumes contain about 5000 and 40000 atoms, respectively, it follows that without recombination losses the energy input per incident ion allows for only ~10% local damage concentration; recombination effects reduce these numbers further by a substantial factor. Finally, since the displacement cascades have a much larger lateral spread than the radii, it must be concluded that the damage cylinders are essentially empty and should only be regarded as an effective target area. Indeed, to obtain the amorphous state about 10 - 100 hits into the same microarea are required, which is in agreement with the observed amorphous doses, e.g. 1 - 3 x 10E15 ions/cmxcm for Si (M~30).

If R_0 were known, absolute assignments of annealing temperature and cluster radius could be made. If we calculate the number of displaced atoms from the amount of disorder in a 1:1 relation, we find that a 20 keV Si-ion roughly displaces 120 atoms. This number is valid for the 10E14 Si/cmxcm implant; it is 250 at the threefold dose.

A Xe-ion of 50 keV is found to displace about 3000 atoms. Taking the 120 atoms per event the following combinations of cluster number/cluster radius/number of defects per cluster are derived: 1/9/120; 3/6/40; 5/5/24; 10/4/12; 20/3/6; 60/2.25/2.

Since divacancies constitute only a small fraction of the total damage (10), and since the probability of producing very few clusters of relative large size along the 300 Å (projected range)long ion track should be quite low, clusters of the size of 3-5 Å appear to be most likely numbering to about 10.

R_o can now be derived from eq.2 by relating the annealing pattern of the 10E14 Si/cmxcm implant with these estimated cluster radii; we find R_o = 2Å with an error of about 15%. The temperature-radius correlation curve based on this data is shown in fig.4. A value of R_o = 2.5 Å follows from the theoretical estimate given by Nelson (1) which is in reasonable agreement with the present data.

In figs.5 and 6 the annealing characteristics are shown for implants of nearly identical damage produced by chemically different ions of about the same mass.

Although the initial damage concentrations of the various implants are the same within the experimental limits of the measurements (±5%), the following differences are seen in the annealing characteristics. Si(Si) and Xe(Si) are described by similar patterns although the latter is somewhat smoother, as it is seen at the 400 and 600 °C points. The chemically active elements differ from these two inactive elements by a 10 -20% higher annealing over the whole measured temperature range. Individual differences among this group

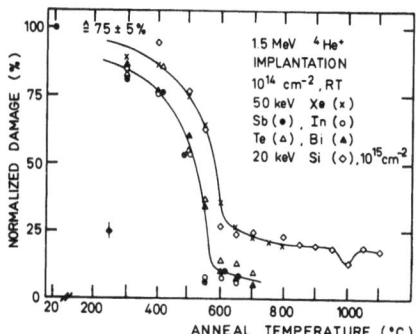

Fig. 5:
Annealing characteristics for 75% disorder implants of 115-In, 122-Sb, 131-Xe, i.e. M~120, and 204-Tl, 209-Bi, i.e. M~200, the dose being 10E14 ions/cmxcm. For comparison, a 10E15 Si/cmxcm implant is shown.

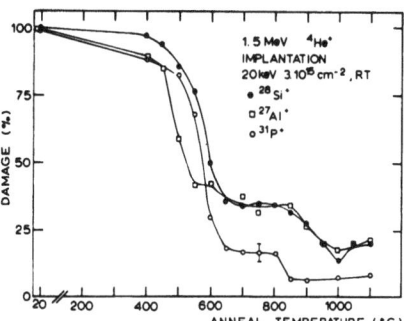

Fig. 6:
Annealing characteristics for amorphous doses of 27-Al, 28-Si, 31-P in silicon.

are indicated, e.g. In, Sb vs. Tl, Bi at 550°C. Note the high level of apparent damage present even after annealing Si(Si) and Si(Xe) at 1100 °C and 800 °C, respectively. The amorphous implants of Al, Si, P in Si show different annealing behavior over the whole temperature range.

Chemical effects based on solubility conditions seem to play a role. Precipitation of Al at relatively low temperatures may first enhance the recrystallization of the damaged silicon layer which assumes a state similar to the Si(Si)-implants at higher temperatures. The concentration of P does not exceed the solubility limit; the P atoms may serve as nuclei for recrystallization which cannot be expected for the implanted Si-ions. These observations are in general agreement with TEM data (10) where different forms of dislocation disorder in the recrystallized material above 650 °C have been observed for a number of different ion species.
Since RBS fails to give the correct amount of disorder in polycrystalline material, the interpretation of the annealing stages above 650°C is difficult. Reorientation of microcrystallites would cause an apparent reduction or increase of disorder.

Conclusions

The damage production in ion bombarded silicon at room temperature can be described by the initial formation of small defect clusters which act as nuclei for further point defects. The annealing pattern can be explained by a distribution of activation energies related to cluster size.
Chemical effects due to different projectile ions cause modifications of the annealing behavior.

References

1) R.S. Nelson, Proc.Int.Conf. on Radiation Damage and Defects in Semiconductors, The Institute of Physics, London, p. 212
2) F.F. Morehead et al., Proc.I.Int. Conf. on Ion Implantation, Thousand Oaks May 1970, p. 25
3) E.C. Baranova et al., Rad.Effects 18 (1973) 21
4) J.W. Mayer et al., Can.J.Phys. 46 (1968) 663
5) B.L. Crowder et al., Appl.Phys.Letters 16 (1970) 205
6) F. Morehead et al., J.Appl.Phys. 43 (1972) 1112
7) K.H. Eklund and A. Andersen, Proc.II.Int.Conf. on Ion Implantation, Garmisch-Partenkirchen May 1971, p. 103
8) M.D. Matthews, Rad. Effects 11 (1971) 167
9) H.J. Stein et al., see ref.2, p. 17
10) S.H. Davidson, Proc. European Conf. on Ion Implantation, Reading Sept. 1970, p.238

HIGH DOSE IMPLANTATION

ELECTRICAL BEHAVIOUR OF HEAVILY-DOPED ION IMPLANTED

LAYERS IN SILICON

J.H. Freeman, D.J. Chivers. G.A. Gard,
G.W. Hinder, B.J. Smith and J. Stephen

Atomic Energy Research Establishment
Harwell, United Kingdom

ABSTRACT

Some of the practical consequences of carrying out high fluence implantation ($\geqslant 5 \times 10^{14}$ ions/cm^2) on an industrial scale have been examined.

The annealing behaviour of heavily doped ion implanted layers of boron, phosphorus, arsenic and antimony have been studied as a function of doping rate, dose and ion-beam energy. The experiments, which were all carried out on 5 cm diameter wafers, involved the measurement of the spatial distribution of resistivity across the surface of the implanted specimens at a series of annealing temperatures. Under certain conditions, it was found that the uniformity and the reproducibility of the dopant-induced electrical activity, which are such an important characteristic of ion implantation, appear to be significantly degraded. The loss of uniformity was found to be associated with the variations in temperature which can occur across the surface of the wafers during such high-fluence implantations.

Persistent implantation temperature memory effects were observed in the electrical behaviour of antimony even after prolonged annealing at 1200°C. Boron, phosphorus and arsenic layers exhibit similar non-uniformities after implantation. For these important dopants, however, there is much less evidence of the implantation history after annealing in the temperature range 950° - 1200°C and good uniformity and reproducibility of the electrical resistivity have been obtained after doping with high-intensity ion beams under conditions corresponding to large temperature variations across the surface of the wafers.

Measurements of the heating effects of ion beams on both thermally-bonded and thermally-isolated silicon wafers have also been made and compared with theory. It is shown that the cumulative heating effect of implanting with even modest beam intensities can be significant. For example, at a typical implantation energy of 100 keV a beam current density of only 1 $\mu A/cm^2$ may result in a temperature rise of over $100°C$. The concept of IRRADIANCE is introduced as a measure of the ion-beam power density. It is seen that it is this parameter, in conjunction with the fluence and the nature of the sample mounting, which determine the temperature rise. It is also shown that in certain circumstances the beam irradiance can be reduced without a commensurate decrease in the throughput of implanted wafers.

INTRODUCTION

The low-level ion doping of silicon is widely established as a manufacturing technique and it now seems possible that the next major advances in the industrial use of implantation will involve significantly higher fluences of a somewhat wider range of ions.

It is already evident for example that increasing attention is being given to the use of arsenic as a dopant, and it can readily be seen that a variety of potentially important device applications could require ion doses in the range $\sim 10^{15}$ to 10^{16} ions/cm^2.

An initial measure of confidence in the possible technical success of such applications is provided by the now extensive literature on both the thermal annealing behaviour and the electrical quality of a wide variety of high fluence implantation experiments. Much of the reported work has, however, been carried out using relatively modest ion beams under precisely defined experimental conditions. In particular the ion-beam heating effects which are a feature of high-fluence doping are sometimes minimised by carrying out the implantations slowly, or by thermally bonding the silicon wafers to a heat sink. This measure of sample temperature control is difficult to maintain on an industrial scale of operation and we have found that heavily-doped wafers which have been implanted under production conditions may exhibit an apparent loss of uniformity and reproducibility.

It is important to stress that it is not the primary purpose of this article to consider the detailed consequences of carrying out implantation at elevated, but controlled temperatures. Extensive studies which have previously been made of this subject have established that both the damage annealing characteristics and the electrical properties may be a function of the implantation temperature (1). In general, the effects, which are reasonably consistent, and quite well understood, become more marked at high fluences.

We are concerned instead with the consequences of the <u>marked variations</u> in temperature which we have found may occur across the surface of implanted silicon wafers. These may manifest themselves initially as clearly visible coloration patterns across the implanted surfaces (see Fig.9(a)) and they are subsequently mirrored as marked spatial variations in electrical resistivity. However, as we shall see below, it is fortuitous that for the most important dopants (B^+, P^+ and As^+) these implantation thermal-memory effects, as evidenced by the electrical resistivity, can be essentially annealed out at sufficiently high temperatures, thus restoring the doping uniformity and reproducibility which are such important features of the ion-implantation process.

The implantation temperature non-uniformities arise to a large extent from the variations in thermal conduction from different areas of the implanted specimens to the mounting surfaces. The several thermal-bonding techniques which may be used to ensure good conduction from the sample to a heat sink in experimental doping studies cannot be readily adapted to implantation on a production scale. Complete thermal isolation of the wafers is equally difficult to achieve and the conventional and convenient target loading techniques tend to result in localised - but ill-defined - thermal contacting at the sample-retaining positions, whereas the heat conduction to the mounting surface from the remainder of the wafer is much more indifferent. We find, as would be anticipated, that this loss of visual and electrical uniformity tends to be somewhat irreproducible and there may be significant differences from wafer to wafer, even when these are implanted under ostensibly identical conditions.

If, as is frequently the case, the thermal contacting cannot be conveniently improved, it is apparent that both the rate of temperature rise and the final equilibrium temperature of a bombarded specimen will be determined by the ion-beam parameters.

These considerations of ion-beam heating effects lead us to introduce the concept of IRRADIANCE as an important factor in industrial-scale application. We define the irradiance as the ion-beam power density, in W/m^2:-

$$\text{Irradiance} = \frac{IE}{A}$$

where I is the beam current in amperes, E is the ion energy in

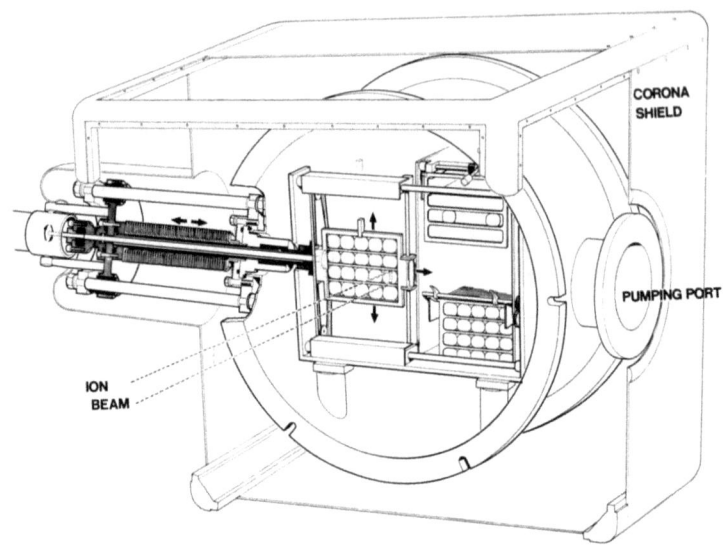

Fig.1: Harwell Implantation Chamber.

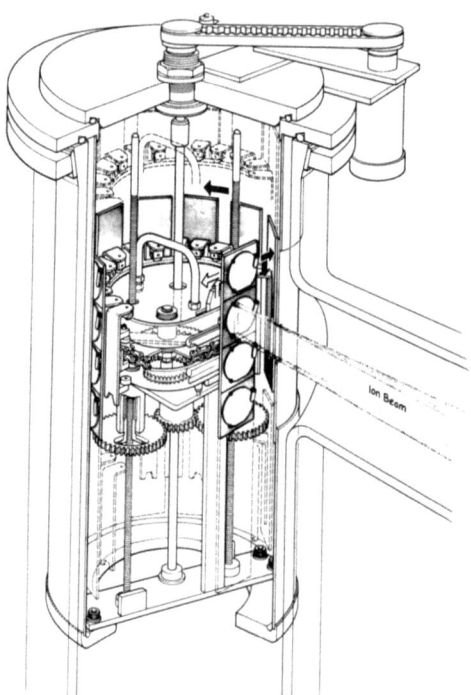

Fig.2: Lintott Process Chamber.

electron volts and A is the "effective area"* in square metres, over which the ion beam is implanted.

For a given wafer processing rate, at a defined fluence, the irradiance can only be minimised by reducing the beam energy, E, or by increasing the effective doping area, A. There may, for example, be some scope for a reduction in the implantation energy in pre-deposition applications where the final dopant profile is to be achieved by a "drive-in" diffusion. The effective doping area of a particular implantation facility is determined by the design of the target chamber. Thus, the irradiance of an electrostatic scanning facility could be reduced by the simultaneous implantation of multiple wafers although this might introduce problems in maintaining the angular beam alignment. In the case of mechanical-scanning systems, it is somewhat easier to achieve a large effective doping area and thus to reduce the temperature effects. For example, the chamber at Harwell (Fig.1), used for the experiments which are described later, has an effective doping area of about 50 cm^2 whereas the mechanical scanning principle of the processing chamber of the Lintott implanter (Fig.2) results in a substantially increased area (about 200 cm^2) for a similar ion beam. An alternative mechanical-scanning technique involving the use of high-speed rotation has recently been described by Robertson (2). This appears to be intended primarily to increase the wafer throughput but the 20" diameter disc which holds the wafers also has a quite large effective doping area. The beam heating situation is, however, presumably somewhat complicated by the need to vary the disc translation speed in order to achieve radial doping uniformity.

The experimental programme which was undertaken to study the ion beam heating effects comprised measurements of -

(a) the temperature rise of thermally-bonded and thermally-isolated silicon wafers;

(b) the spatial resistivity of uniformly implanted 5 cm diameter wafers at a series of annealing temperatures. Comparisons were made with four dopants as a function of fluence, energy, doping rate and thermal bonding;

(c) radiation damage by electron channelling.

*In implantation machines utilising two-axis electrostatic deflection, this "effective area" is essentially the scanned area of the ion beam. In implantation target chambers which utilise rapid and repetitive mechanical scanning of the specimen through a fixed ion beam, the situation is rather more complex. It is nevertheless possible to formulate an analogous area which is defined by the height (or length) of the ion beam, and the distance which might be traversed by the sample between successive scans through the beam.

These are described in more detail below.

EXPERIMENTAL

Wafer handling procedures

The 5 cm (2" diameter) Czochralski silicon wafers used for the implantation studies were supplied by Texas Instruments Ltd. (Bedford, U.K.). They were all (111) crystal orientation and were doped either p or n type to 5 to 10 ohm cm resistivity. The wafers which were polished on one face by the supplier were given an organic clean followed by a deionised water wash before implantation.

After implantation, the wafers were given an organic clean and then immediately coated with a protective layer of radio-frequency sputtered silicon dioxide to a thickness of about 1,000Å. The layer served to protect the silicon surface from impurities and attack, when exposed to high temperature in the annealing furnace. After annealing, the wafers were stripped of silicon dioxide using buffered HF and measured almost immediately.

Implantation procedure

The implantations were all carried out on a 40 kV Lintott/Harwell separator (3) equipped with a 160 kV target chamber (4). The principle of double-axis mechanical scanning used with a stationary ion beam to obtain high doping uniformity is illustrated in Fig.1.

The implantation studies involved the use of ion beams of B^+, P^+, As^+ and Sb^+ at intensities of up to 1 milliampere and over an energy range of 40 to 160 keV. The doping uniformity was $\pm$ 1% or better. These basic requirements were conservatively set in the context of several years operating experience of the implantation facility.

For most of the experiments, each sample carrier plate was loaded with four identical wafers to provide a check on the implantation reproducibility. The usual Harwell sample loading technique in which each wafer is mechanically held at three positions around the circumference was used and no attempt was made to obtain particularly good, or reproducible, thermal contact to the carrier plate. For comparison, a small number of wafers were, however, carefully thermally bonded to loading plates with "silver dag" (Acheson Colloids No. 915).

In the experimental comparisons of doping uniformity at constant fluence the irradiance (or beam-power intensity) was varied either by implanting at constant beam current, but at a series of

HEAVILY-DOPED ION IMPLANTED LAYERS IN SILICON

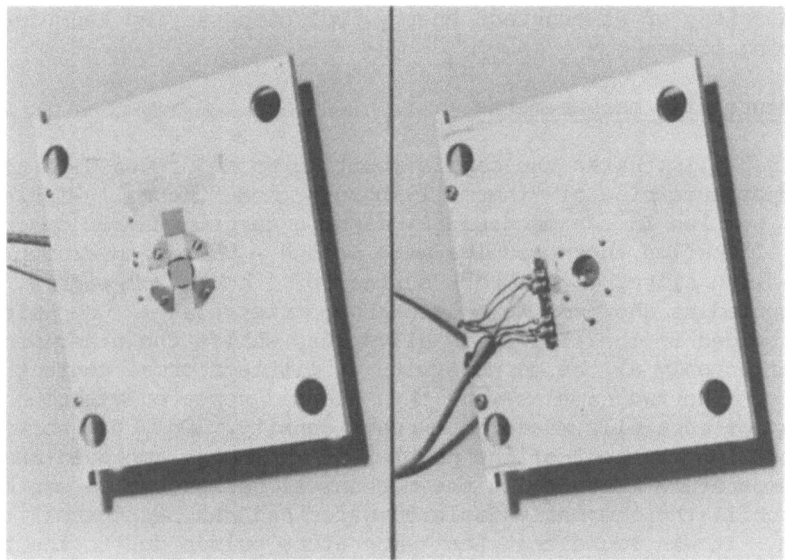

Fig.3: Sample and thermocouple mountings.

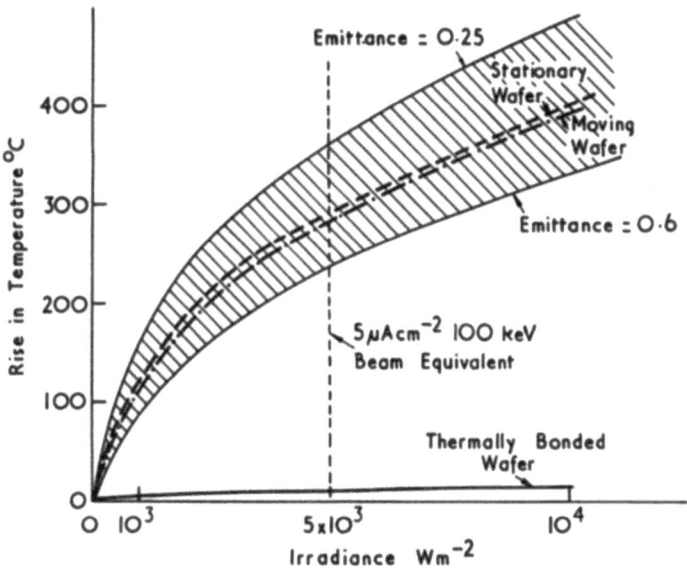

Fig.4: Temperature rise of wafers subjected to high irradiance.

beam energies; or at constant energy, but with varying beam currents and doping times.

Temperature rise measurements

Fig.3 illustrates the sample mounting procedure used to study the temperature rise of "thermally-bonded" and "thermally-isolated" silicon samples in the mechanical-scanning target chamber described above. Ultrafine thermocouples were welded - using a hydrogen flame - with silver to the back surface of 1 cm square wafers, which were mounted as shown on to a standard carrier plate. One wafer was firmly bonded to the plate with silver dag whilst the other was held in position only at the corners with as little thermal contact as possible. The two samples were then scanned repeatedly through an ion beam of carefully measured current density. Only the horizontal reciprocation movement of the double axis scan was employed and for each temperature measurement the mechanical reciprocation was continued until the thermally-isolated wafer had reached thermal equilibrium. It was found that the temperature pulses due to the movement of the sample through the ion beam were insignificant compared to the cumulative rise in temperature.

In a similar experiment intended to provide a reference comparison with a stationary sample, the thermally-isolated wafer was mounted in a fixed position in a specially designed Faraday cage, where it was irradiated with a uniformly swept ion beam.

Sheet resistivity uniformity measurements

The spatial distribution of sheet resistivity over the surface of the implanted and annealed wafers was made by the four point probe technique described elsewhere in these proceedings. For each determination the average value of sheet resistance and the standard deviation was calculated. Additionally, a histogram was drawn and an iso-resistance map was plotted by the computer using the graphical output facility.

This technique is used routinely at Harwell for implantation assessments and has been found to be accurate and reproducible to at least 1%. For example, the iso-resistance maps of implanted wafers shown in Figs.5 and 6 have measured standard deviations of only 0.7% over the entire surface. This includes both the implantation and the four point probe instrumental errors.

The particular value of the technique for these studies lay in the relative ease with which the uniformity of each wafer could be measured over a series of annealing stages up to temperatures of $1200^{\circ}C$.

HEAVILY-DOPED ION IMPLANTED LAYERS IN SILICON

Fig. 5. 5×10^{15} ions/cm.^{2} As$^+$
Beam current 100 µA
Energy 40 keV
Irradiance 9.2×10^2 W/m^2
Anneal temperature 950°C.
Mean resistivity 45.77 ohms/cm.^{2}
Standard deviation 0.76%
Good thermal bond

Fig. 6. 5×10^{15} ions/cm.^{2} Sb$^+$
Beam current 100 µA
Energy 80 keV
Irradiance 1.84×10^3 W/m^2
Anneal temperature 1200°C. only
Mean resistance 27.51 ohms/cm.^{2}
Standard deviation 0.76%

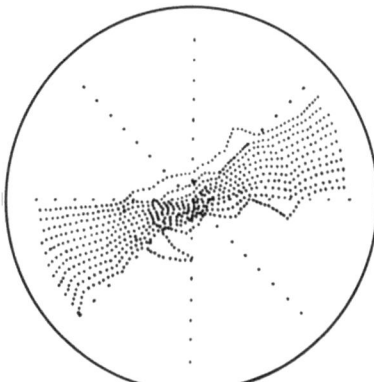

Fig. 7. 5×10^{15} ions/cm.^{2} As$^+$
Beam current 500 µA.
Energy 160 keV
Irradiance 1.84×10^4 W/m^2
Anneal temperature 950°C.
Mean resistivity 27.87 ohms/cm.^{2}
Standard deviation 6.41%

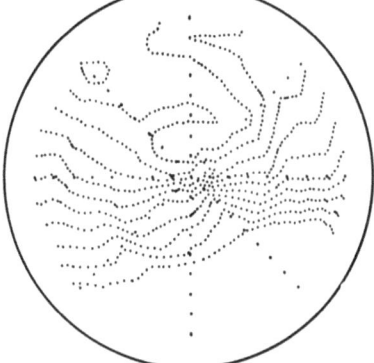

Fig. 8. 5×10^{15} ions/cm.^{2} As$^+$
Beam current 500 µA
Energy 160 keV
Irradiance 1.84×10^4 W/m^2
Anneal temperature 1100°C.
Mean resistivity 22.21 ohms/cm.^{2}
Standard deviation 4.26%

Iso-resistance maps

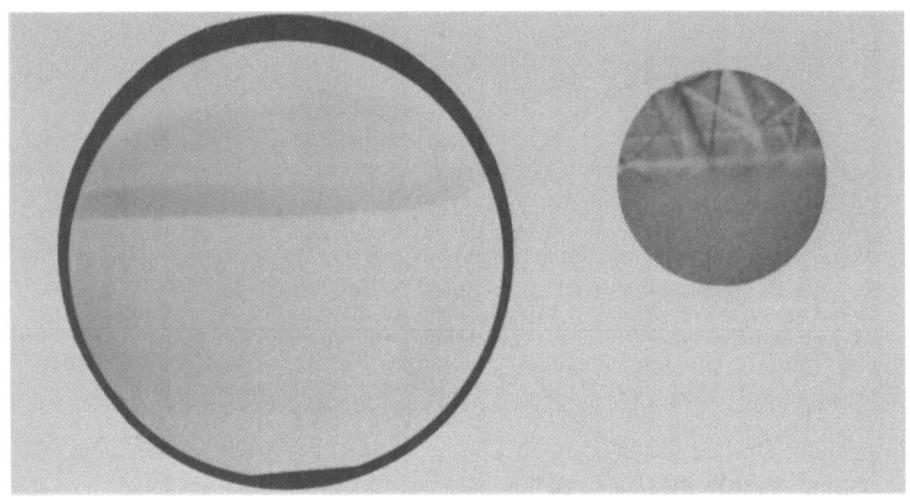

Fig.9(a): Visible pattern on wafer subjected to
5 x 10^{15} ions/cm^2 Sb$^+$ at high irradiance

(b): Photomicrograph of (a) by reflected electrons. The channelling pattern appears in the coloured band indicating crystallinity.

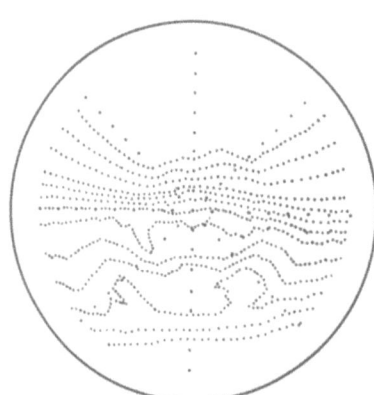 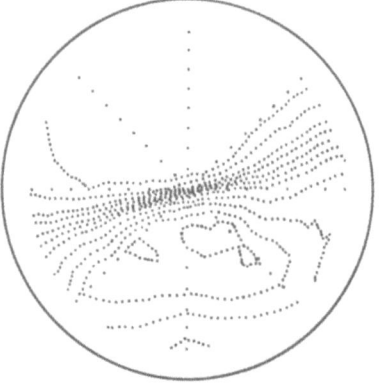

Fig. 10. 5 x 10^{15} ions/cm.^{2} Sb$^+$
Beam current 250 μA
Energy 160 keV
Irradiance 9.2 x 10^3 W/cm.^{2}
Anneal temperature 950°C.
Mean resistivity 100.88 ohms/cm.^{2}
Standard deviation 17.86%

Fig. 11. 5 x 10^{15} ions/cm.^{2} Sb$^+$
Beam current 250 μA
Energy 160 keV
Irradiance 9.2 x 10^3 W/cm.^{2}
Anneal temperature 1200°C.
Mean resistivity 18.30 ohms/cm.^{2}
Standard deviation 18.06%

Iso-resistance maps

RESULTS AND DISCUSSION

Temperature rise measurements

It was not the purpose of this particular study to undertake a detailed investigation of ion beam heating effects – an account of the factors governing both the rate of temperature rise and the equilibrium bombardment temperature has been given elsewhere (5). In particular, we have restricted ourselves to high-fluence applications where the implantation normally continues for a sufficient time to allow any thermally isolated region of the sample to reach (or at least approach) its equilibrium temperature.

The results of the beam-heating measurements are shown in Fig.4. As expected, the sample which was thermally bonded to the relatively massive loading plate showed only a very small temperature rise in the time taken for each measurement (typically ten minutes). In contrast, much more significant beam-heating effects were observed for both the stationary and the mechanically scanned samples which were "thermally isolated". These two curves fall within the shaded area which represents the calculated temperature rise (4) for silicon with an assumed emissivity in the range 0.25 to 0.6. The small discrepancy between the two experimental curves probably reflects the increased difficulty of maintaining good thermal isolation on the mechanically scanned sample.

The agreement is nevertheless sufficient to justify the validity of the "effective beam area" concept, described earlier, which was used to calculate the irradiance for the measurements made on the scanned sample.

The results confirm that ion-beam heating effects may result in significant temperature variations when high-fluence implantations are carried out at typical production rates. It also follows that over those areas of the wafer which become significantly heated, the implantation will be effected under conditions of progressively increasing temperature. Under certain high fluence doping conditions in the target chamber described, these temperature variations result in clearly visible coloured patterns across the wafer surface, as illustrated in Fig.9(a). Whilst these patterns reflect the anticipated variations in the accumulated density of radiation damage, the overall effect is evidently quite complex. Thus, whilst the reflected electron pattern in Fig.9(b) shows a clear demarcation in crystallinity corresponding to the visible pattern of Fig.9(a), it is interesting to note that it is the crystalline and not the amorphous region which is most highly coloured.

Table I

Ion	Dose ions/cm²	Beam Current mA	Energy keV	Irradiance W/m	\bar{R}(ohms/cm²) 950°C	σ(%) 950°C	\bar{R}(ohms/cm²) 1050°C	σ(%) 1050°C	\bar{R}(ohms/cm²) 1100°C	σ(%) 1100°C	\bar{R}(ohms/cm²) 1200°C	σ(%) 1200°C
B^+	2.5×10^{15}	0.25	40	2.3×10^3	47.57	1.17			42.05	0.79	37.66	0.79
					47.65	0.90			42.13	0.86	38.09	1.13
					48.03	1.22			42.61	1.01	38.16	1.28
					48.19	1.27			41.72	1.11	38.07	1.19
			80	4.6×10^3	43.83	0.86			40.34	1.18	36.25	1.07
					43.19	0.95			40.67	0.95	37.46	1.01
					43.69	0.96			40.31	0.86	36.48	1.07
					39.67	1.56			36.70	1.03	34.26	1.10
			120	6.9×10^3	40.09	1.13			36.66	1.13	33.45	1.57
					40.91	0.82			37.85	0.84	35.04	0.92
					40.73	0.91			37.96	0.80	34.27	0.94
					41.59	1.39			38.53	1.22	35.89	1.12
			180	1.04×10^4	41.42	1.83			37.89	1.39	35.06	0.99
					43.00	1.06			39.77	0.93	35.61	0.86
P^+	5×10^{15}	0.50	40	4.6×10^3	19.60	0.79					14.14	1.48
		0.50	80	9.2×10^3	19.36	1.92					15.02	2.98
		0.50	160	1.84×10^4	16.98	2.60					13.87	1.34
		0.10	40	9.2×10^2	19.12	1.76					14.66	2.39
		0.10*	40	9.2×10^2	20.28	1.02					14.81	0.82
As^+	5×10^{15}	0.50	40	4.6×10^3	50.33	1.91	36.23	1.94	29.55	2.23	26.04	2.22
					46.20	1.30	28.96	1.63	28.59	1.90	24.99	1.87
		0.50	160	1.84×10^4	27.87	6.41	22.60	4.16	22.21	4.26	19.49	4.36
		0.10	40	9.2×10^2	46.13	1.26	29.43	1.11	28.83	1.35	25.27	1.39
		0.10*	40	9.2×10^2	45.77	0.76	29.33	1.79	28.86	2.12	25.06	2.24
		0.10*	40	9.2×10^2	46.48	1.23	28.88	1.58	28.35	1.83	24.51	1.71
	1×10^{16}	0.50	160	1.84×10^4							8.96	0.93
											8.89	1.15

* = Good thermal contact \bar{R} = Mean resistivity σ = Standard deviation

Table II

Ion	Dose ions/cm²	Beam Current mA	Energy keV	Irradiance W/m²	950°C R(ohms/cm²)	σ (%)	1050°C R(ohms/cm²)	σ (%)	1100°C R(ohms/cm²)	σ (%)	1200°C R(ohms/cm²)	σ (%)
Sb⁺	1×10^{16}	0.25	140	8.05×10^3	118.2	10.64	121.3	6.62	81.51	4.30	33.32	6.58
		0.25	140	8.05×10^3	121.3	15.63	118.2	9.31	85.38	5.85	33.42	6.05
		0.50	40	4.6×10^3	359.8	2.35	376.5	9.64			137.5	7.77
			80		330.6	4.07	288.6	5.32			104.2	5.51
					329.5	2.75	296.3	6.62			107.3	4.47
				9.2×10^3	204.8	5.96	168.1	2.66			43.31	2.42
					215.7	5.51	170.3	4.77			48.02	5.36
					207.9	5.94	168.8	3.40			47.99	4.17
					234.6	4.48	177.0	2.08	111.7	1.99	63.29	2.70
					228.2	5.82	168.2	2.60	116.4	3.55	58.47	5.96
	5×10^{15}	0.25	80	4.6×10^3							30.39	1.43
					196.3	5.09	193.5	4.45	143.3	4.42	60.49	3.39
					194.6	5.52	120.5	2.88	136.7	3.06	60.55	2.87
		0.10	80	1.84×10^3							28.39	1.30
					195.4	1.71	178.0	1.51	174.7	2.58	59.80	1.13
					194.6	1.58	182.5	1.65	151.6	2.55	59.97	1.42
				• • •							27.51	0.76
					236.2	1.91	194.7	3.07	152.8	3.77	62.03	2.06
					237.3	1.45	198.6	1.64	145.5	3.38	26.87	0.83
	5×10^{14}	0.50	160	1.84×10^4	102.0	20.09	104.4	13.18			18.30	18.06
					105.2	17.02	106.3	11.53			20.60	16.77
		0.05	40	4.6×10^2	378.4	2.93	588.4	3.30			213.1	3.04
					367.2	2.47	477.3	3.96			176.0	3.10
					360.7	2.95	475.8	4.24			166.1	3.52
					369.9	2.25	487.3	3.96			171.2	3.79
	5×10^{13}	0.005	40	4.6×10	1339	6.71	1425	11.63			584.2	11.24
					1270	10.42	1352	6.92			647.8	9.26
					1363	7.15	1449	6.56			697.5	8.34
					1422	5.50	1465	4.72			787.1	6.10

• = Good thermal contact $\bar{R}$ = Mean resistivity   σ = Standard deviation

Sheet resistivity uniformity measurements

The results of a series of sheet resistivity measurements with variations of beam current, energy and irradiance are listed for boron, phosphorus and arsenic ions in Table I and for antimony in Table II. Typical iso-resistance maps illustrating both good and bad electrical uniformity for arsenic and antimony implanted wafers are shown in Figs. 5, 6, 7, 8, 10 and 11.

Although the low temperature annealing behaviour of boron is well known to be sensitive to the implantation temperature (6), no significant thermal memory effects have been observed for this dopant under a wide range of implantation conditions after annealing at $950°C$, which was the lowest temperature used in this series of experiments. This is illustrated in the boron data listed in Table I which shows both the reproducibility and the uniformity of resistivity for some typical implantation conditions.

In the case of phosphorus and arsenic high-fluence implantations, the visible patterns and the gross non-uniformities have largely disappeared at $950°C$. The residual non-uniformity is, however, somewhat larger than that observed for boron and in the case of the highest irradiance arsenic implantation is over $\pm 6\%$. In this particular case, only a marginal improvement in uniformity was subsequently observed on further annealing of the specimen to 1100 and $1200°C$, but it was found that other samples implanted at the same irradiance gave very much better uniformities (~ 1%) when annealed directly at $1200°C$.

The results in the case of antimony ions (Table II) are very different. Here, for a given dose, the electrical activity after a $950°C$ anneal is very dependent not only on irradiance but also on depth of implantation (energy). These effects still persist after a succession of anneals at increasing temperatures up to $1200°C$ and are accompanied by anomalously large variations in sheet resistivity across individual wafers. This probably results from precipitation due to the low antimony solubility and the short ion range. It can be seen from the table, however, that, as in the case of arsenic, good electrical uniformity was obtained when wafers were annealed directly at $1200°C$.

The sheet resistances of arsenic and antimony implantations given a single anneal at $1200°C$ for one hour are higher than those reported by Drum (7) who used a higher annealing temperature ($1250°C$) for up to 10 hours.

ACKNOWLEDGEMENTS

We would like to thank Mr. D. C. Marshall of Harwell for obtaining the channelling pattern.

REFERENCES

(1) Mayer, J.W. et al. - Ion Implantation in Semiconductors, Chapter 5. Academic Press, 1970.

(2) Robertson, G.I. - Rotating scan for ion implantation. Abstract No. 223, Extended Abstracts Vol. 74-1, Electrochemical Society, May, 1974.

(3) Freeman, J.H. - A variable geometry separator and low energy heavy ion accelerator. Part I, General Principles, AERE-R 6254.

(4) Freeman, J.H., Caldecourt, L.R., Done, K.C.W., and Francis, R.J. - An industrial scale ion implantation facility. AERE-R 6496.

(5) Dearnaley, G., Freeman, J.H., Nelson, R.S., and Stephen, J. - Ion Implantation, p.421. Defects in Crystalline Solids Series, North-Holland, 1973.

(6) Davies, D.E. - Applied Physics Letters, Vol.14, p.227, 1969.

(7) Drum, C.H. - Diffusion of ion implanted antimony and arsenic in silicon. Abstract No. 84, Extended Abstracts Vol. 74-1, Electrochemical Society, May, 1974.

HIGH DOSE PHOSPHORUS-GERMANIUM DOUBLE IMPLANTATION IN SILICON

N. Yoshihiro, M. Tamura, and T. Tokuyama

Central Research Laboratory, Hitachi, Ltd.

Kokubunji, Tokyo, Japan

ABSTRACT

The formation of the dislocation networks in the well annealed high dose phosphorus implanted silicon can be suppressed if phosphorus-germanium double implantation is applied. The dose of germanium less than a half of that of phosphorus is required. A wet oxidation at about 800°C before high temperature anneal enhances the effect.

INTRODUCTION

The defect structures found in ion implanted silicon depend on implantation procedures and annealing schedules [1]-[5]. In the low dose implants the defects anneal at a low temperature, returning the implanted atoms to full electrical activity. This fact has been used for MOS device manufacture.

In the higher dose implants necessary to fabricate bipolar devices, growth and interaction of the secondary defects occurs during higher temperature anneal, and leaves high density of unannealed defects.

The present work is part of a larger programme to examine anneal behavior of the defects in high dose implants which are used as a means of predeposition of dopant impurities.

In the high dose (order of $10^{15}/cm^2$ and above) implants of dopant ions, network structure of dislocations forms after sufficient anneal [5]. They are essentially the same as the "misfit dislocations" found in silicon diffused with high concentration of dopant atoms.

Generation of the misfit dislocations in the thermally diffused layer is known to be suppressed by diffusing another appropriate impurity and thus "compensating" the lattice strain caused by the dopant atoms [6]. The particular object of the present work is to examine if the "stress compensation" mechanism works effectively in the heavily damaged ion implanted layers.

Germanium (atomic radius = 1.22Å) was chosen to compensate the elastic strain caused by phosphorus (1.10Å) in silicon (1.17Å).

EXPERIMENTAL

The substrate was (111) oriented p-type silicon with the resistivity of 5 to 10 ohm·cm.

Phosphorus was implanted at 25 keV, to a dose 1.0 to 3.0×10^{16} ions/cm^2. Germanium was implanted at 50 keV, to a dose equal to or less than a half of that of phosphorus. The energies were chosen so that the projected range of both ions coincided. Both ions were implanted at room temperature.

Two systems of anneal were studied. One was a conventional anneal in nitrogen or oxygen atmosphere at 1050 to 1200°C. The other was a two-stage anneal. The samples were first wet oxidized at 800°C and then annealed at 1050 to 1200°C in dry nitrogen atmosphere.

RESULTS

Fig.1 shows the defect structures in phosphorus, phosphorus-germanium, and germanium implants annealed at 1100°C in dry nitrogen. The density of dislocations is lower and the network structure is imperfect in the phosphorus-germanium double implants as compared with the phosphorus implants.

Germanium implants do not fully anneal and complex tangle of dislocations remains.

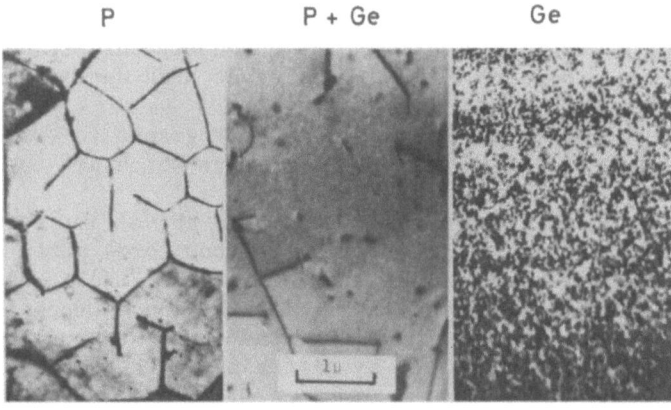

Fig.1. Electron micrographs taken after 1100°C, 40 min. anneal in N$_2$. P:2.0×10^{16}/cm^2, Ge:0.6×10^{16}/cm^2.

It was considered that if the samples were oxidized at such a low temperature as the growth of the secondary defects was negligibly slow, the damage layer formed by the implantation could be effectively removed in the oxide layer. The line of dashes in Fig.2 shows the oxide growth during wet oxidation at 800°C of phosphorus $3.0 \times 10^{16}/cm^2$ implants.

The rate of oxidation of phosphorus implants is much larger than that of non-dope silicon (thin solid line). The effect is caused by high phosphorus concentration [7], and the effect of germanium is small. Thus about 900Å thick of silicon can be oxidized in 40 min. wet oxidation at 800°C.

Fig.3 shows electron micrographs of the two-stage annealed samples. It is shown that after wet oxidation at 800°C, high density of defects, consisting of precipitates and entangling dislocations still remains in the surface layer. The double implantation seems to have little effect on the defect structure at this stage.

After the second anneal at 1100°C, imperfect networks of dislocations are seen in the phosphorus implants. Whereas in the double implants no particular defects remains at deeper than a thousand angstroms depth. The irregular contrast structure shown in the micrograph exists only within a very shallow surface layer.

Early Stage of the Second Anneal

Annealing behavior of defects in 800°C oxidized samples during short anneal at 1100°C was examined to study the difference between phosphorus- and phosphorus-germanium double- implants.

The samples were layed on the quartz plate and were inserted to and/or pulled out from the furnace within about five seconds.

The structure of defects did not change in 30 seconds' anneal.

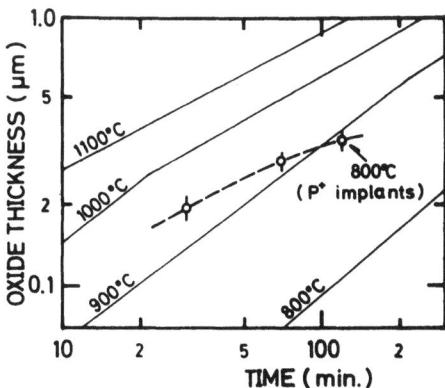

Fig.2. Oxide growth in an atmospheric steam. Thin solid lines denote those of non-dope silicon [7]. The line of dashes denotes that of phosphorus $3.0 \times 10^{16}/cm^2$ implants.

They consisted mainly of precipitates and complex tangle of dislocations. In 1 minute's anneal short straight dislocations running parallel to the surface appeared. They runned in <211> directions and were edge dislocations with Burgers vectors parallel to the surface. They grew and formed dislocation networks within another 1 minute of anneal. During the longer anneal the size of the cell of the networks grew larger as the anneal proceeded.

In the phosphorus-germanium double implants the initial structure of defects was similar to that of phosphorus implants. However, no growth of defects was observed at any stage of 1100°C anneal. The random structure seen in Fig.3 prevailed throughout anneal of longer than 1 minute. The diffraction patterns showed no sign of precipitates or other phases.

DISCUSSION

A summary of the electron microscopy studies on $3.0 \times 10^{16}/cm^2$ phosphorus implants are shown in Table 1.

In the samples wet oxidized at 800°C, precipitates and tangle of dislocations were observed.

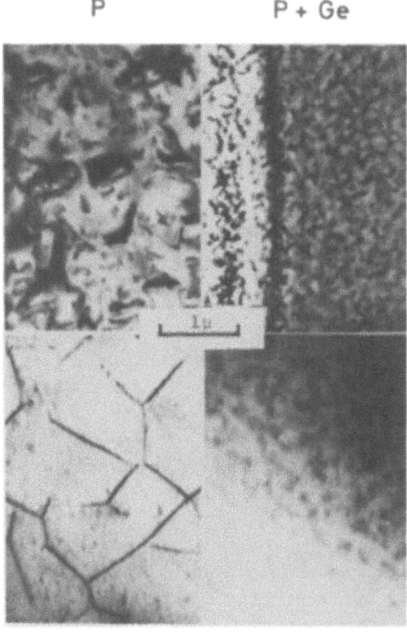

Fig.3. Electron micrographs of two-stage annealed samples, taken after 800°C, 40 min. wet oxidation (above), and after the second annealing of 1100°C, 40 min. in dry N_2.
P:$3.0 \times 10^{16}/cm^2$, Ge:$1.0 \times 10^{16}/cm^2$.

In the 1100°C annealed phosphorus implants, dislocation networks of hexagonal shape consisting of edge dislocations lying parallel to the surface formed through growth and interaction of the secondary defects. Whereas in the double implants annealed at 1100°C, density of dislocations decreased considerably. When annealed in N_2, dislocation networks still formed but were not perfect and were discontinued at here and there. In the samples annealed in O_2, the dislocation density was still lower and most of dislocations were short and half loop like in shape, with a few long ones running parallel to the surface. Some precipitates were observed when the dose of germanium was high.

In the two stage annealed phosphorus implants, imperfect networks of dislocations formed, whereas no defects were observed in the double implants at deeper than 1000Å depth.

The lattice strain can be ideally compensated if the change in the lattice constant by an impurity species is compensated at any depth by a proper amount of the second impurity species.

To achieve this condition, a rough estimation reveals that about 1.5 times as much germanium as phosphorus is required, and that the diffusion coefficients of both species must be roughly equal.

The diffusion coefficient of phosphorus, however, is about ten times as large as that of germanium at 1100°C. So the surface concentration of phosphorus drops more rapidly, and after a short anneal lattice contraction near the surface by phosphorus is compensated by high local concentration of germanium, even if the total amount of germanium is smaller than that required for ideal compensation, and thus suppress growth of secondary defects into misfit dislocations. Existence of dislocation networks in samples annealed at 1100°C in N_2 suggests that total of germanium 1/2 as much as that of phosphorus is not sufficient for this mechanism to work effectively.

Table 1. Summary of electron microscopy observations on phosphorus $3.0 \times 10^{16}/cm^2$ implants. P=precipitates, D.N.=hexagonal networks of dislocations, d.n.=imperfect networks of dislocations, S.D.=short half-loop like dislocations.

Anneal \ Ge^+/cm^2	0	7.5×10^{15}	1.0×10^{16}	1.5×10^{16}
800°C/O_2	P	P	P	P
1100°C/N_2	D.N.	d.n.	d.n.	d.n.
1100°C/O_2	D.N.	S.D.	S.D.	S.D.+P
800°C/O_2 + 1100°C/N_2	d.n.	—	—	—

Lower defect density in the samples annealed in O_2 suggest that higher surface concentrarion of germanium is effective to suppress the growth of secondary defects. Higher germanium concentration may increase the equilibrium concentration of vacancy in the region and suppress climbing motion of edge dislocations (with extra planes on the surface side) by emitting vacancies. This is another possible explanation of suppression of the network growth by germanium implantation.

CONCLUSIONS

The formation of dislocation networks in annealed high dose phosphorus implants can be suppressed by the double implantation of phosphorus and germanium. Two-stage oxidizing-non-oxidizing anneal enhances the effect.

REFERENCES

1. M. Tamura et al: Proc. 2nd Int'l Conf. on Ion Impl. in Semicond. Garmisch-Partenkirchen, (Springer-Verlag, 1971)
2. L.N. Large and R.W. Bicknell: J. Materials Sci. 2, 589 (1967)
3. R.W. Bicknell and R.M. Allen: Proc. Int'l Conf. on Ion Impl. in Semicond., Thousand Oaks, 1970
4. M. Tamura: Appl. Phys. Letters, 23, 651 (1973)
5. T. Ikeda et al: to be published in, Proc. 6th Conf. on Solid State Devices, Tokyo, 1974
6. T. Yeh and M.L. Joshi: J. Electrochem. Soc., 116, 73 (1969)
7. R.M. Burger and R.P. Donovan ed.: Fundamentals of Silicon Integrated Device Technology Vol. 1, p.39-, (Prentice-Hall Inc., N.J., 1967)

CONTROL OF SECONDARY DEFECTS BY TIN DIFFUSION IN ION IMPLANTED

SILICON CRYSTALS

G. Nakamura, Y. Yukimoto and Y. Hirose

Kitaitami Works, Mitsubishi Electric Corporation

4-1, Mizuhara, Itami, Hyogo 664, Japan

ABSTRACT

We have found that the generation of the secondary defects and the ionization ratio of the implanted ions in the ion implanted layers were affected by tin atoms present prior to the implantation. This paper reports the electrical properties and the electron microscopic observation of the secondary defects in the boron implanted layers.

INTRODUCTION

It is well known that the secondary defects observed in boron implanted silicon crystals are rod-like precipitates at the initial stage of annealing and dislocation loops and three dimensional dislocation networks at the following relatively high temperature annealing stages.[1]-[4]

Recently, Karatsyuba et al.[1] reported that rod-like precipitates and polycrystalline second phase included in dislocation loops were ascertained to be SiB_6, and Bicknell[2] measured the activation energy to convert the implanted ion from an interstitial site to an electrically active substitutional site at various annealing stages under the conditions in which no reverse annealing behavior was observed. These authors suggested that the annealing behavior of secondary defects was reduced to one aging process of boron-silicon solid solution. But they paid no regard to the effect of the strain field accompanying to the precipitates on the defect formation or the ionization of the impurity atoms.

In the present works, the nature of the strain field caused by the precipitates is investigated and how the precipitate formation is affected by the presence of additional impurities such as tin and phosphorus atoms is investigated by the observation of such defects and the electrical measurements.

EXPERIMENTAL

Silicon, (111) orientation, n-type, 2-3 ohm-cm substrates were used in this study. Boron ions were implanted at 30 keV and 80 keV with a dose ranging from $1 \times 10^{14}/cm^2$ to $3.2 \times 10^{14}/cm^2$. The implantation was performed in a random direction by tilting the substrates 8° toward the ion beam in order to reduce the channeling effect. The ion implanted specimens were annealed in nitrogen ambient for 30 minutes at the temperature range between 400 °C and 1000 °C. Boron implantations were also performed into tin diffused silicon substrates. Tin diffusion was accomplished by doped oxide source and its depth profiles were measured by 1.5 MeV He ion backscattering technique.[5] Some of the ion implanted and annealed specimens were then diffused with phosphorus atoms at the surface concentration of $7 \times 10^{20}/cm^3$ by using doped oxide source.

Depth distributions of defects and carrier concentrations have been measured by a combination of a transmission electron microscopic observation technique and an anodic stripping, method and by a sheet resistivity profile measurements, respectively.

RESULTS AND DISCUSSION

The nature of the strain field

Figure 1a shows the electron micrograph of the rod-like precipitates observed in the specimen which is implanted with boron ions at 100 keV with a dose of $1 \times 10^{15}/cm^2$ and annealed at 800 °C for 20 minutes. Figure 1 b shows a dark field image of Fig. 1a with a condition of a single strong diffraction beam (two beam case). The white contrast in the image is situated always on the negative side of g (diffraction vector) irrespective to the depth of the precipitates in thin foil specimen. The rod-like precipitates, therefore, have compressive stress $\epsilon > 0$, indicating therby that the precipitates are of interstitial type. This interpretation is based on the work of Ashby and Brown.[6]

Figure 2 shows the profiles of the carrier concentration and the defect density in the specimen implanted with boron at 100 keV with a dose of $1 \times 10^{15}/cm^2$. Open circles show the profiles obtained in the specimen annealed at 1000 °C for 20 minutes and solid circles in the specimen annealed at 800 °C for 20 minutes. From the observation of carrier profiles for 800 °C and 1000 °C, it is noted that the ionization rate of the implanted ions in a surface region are faster than that in an inner region. This result may be interpreted as the effect of the strain field accompanying to the precipitates. It is known that the concentration of vacancy surrounding the precipitates, $\epsilon > 0$, is reduced through their annihilation at the interface between the matrix and the precipitates to relax the strain field. This results in the appearance of an undersaturation of vacancy in the neighborhood of the precipitates.[7] So, the vacancy migration to the precipitates to liberate the precipitated impurity atoms to substitutional sites achieved more easily at the surface than at the inner region. Similar carrier profiles have been reported by Ryssel et al. [8]

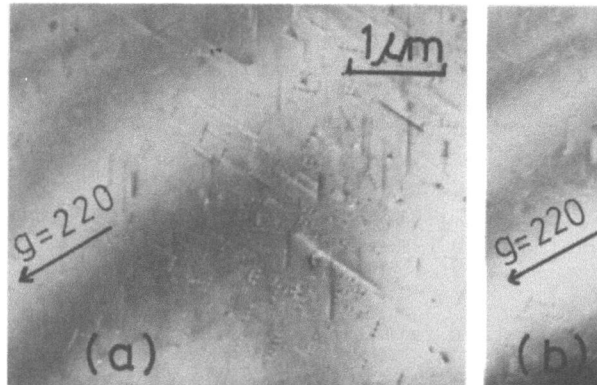
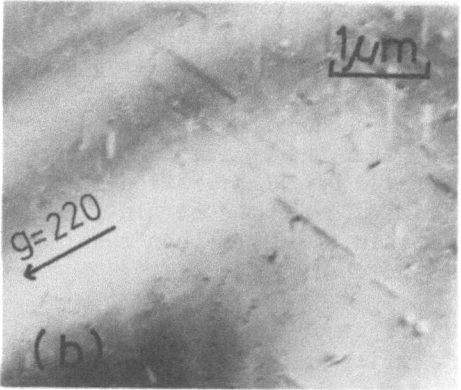

Fig. 1 Electron micrographs of B$^+$ implanted (100 keV, 1 x 10^{15}/cm^2 and annealed (800 °C, 20 min.) specimen. (a) Bright field image, (b) dark field image.

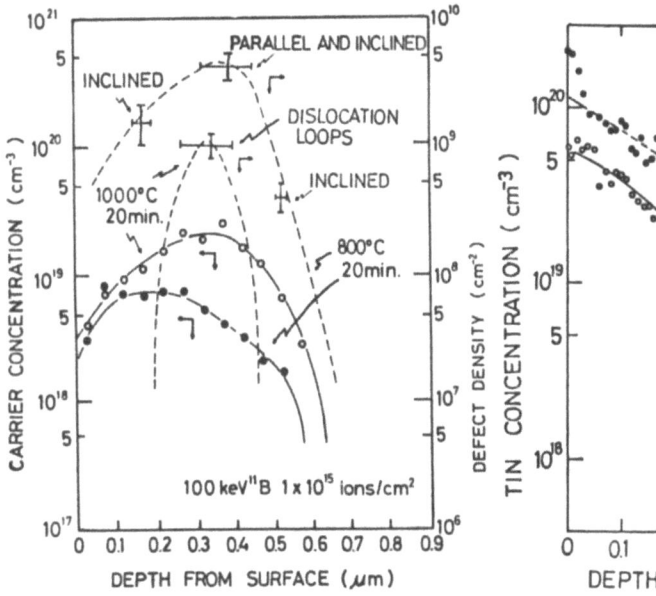
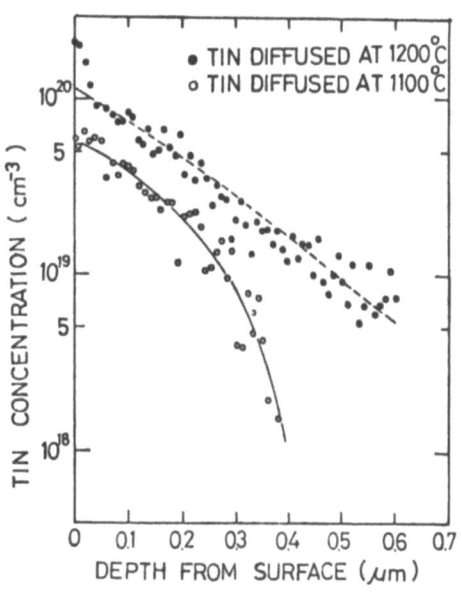

Fig. 2 Profiles of carrier concentration and defect densities in B$^+$ implanted silicon specimens.

Fig. 3 Tin profiles in silicon crystals diffused at 1200 °C, for 6 hours (●) and at 1100 °C for 20 hours (o).

In the case of annealing at 1000 °C for 20 minutes, it is observed that most implanted ions occupy the substitutional sites from the electrical measurement. Also in this case, dislocation loops were observed nearly at the depth of the implanted ion projection range. The dislocation loops would be generated from the rod-like precipitates to reduce the strain field of the precipitates.[7]

Bicknell [2] reported that the activation energy of the conversion of an interstitial boron atom to an substitutional boron atom was 1.1 ± 0.1 eV. This activation energy may be determined by thermal vacancy migration energy reported by Swalin et al.[9] We think that the ionization of the implanted boron ions will be activated by the migration of the vacancies attracted to the precipitates by the strain field. Similar vacancy attracting effect of strain field has been considered by Tsai[10] in the case of phosphorus diffusion.

The effect of tin atoms on the defect formation and electrical properties

One of the purposes of tin diffusion in silicon crystal is to relieve the strain induced by high concentration diffusions of phosphorus and boron, as described previously [11] and also it is known that tin atom acts as a vacancy sink. The action of tin atom as a vacancy sink is a useful fact which can be applied to suppress the emitter dip effect in a conventional process using double diffusions of boron and phosphorus.[5]

Figure 3 shows a typical depth profile of tin atoms diffused at 1100 °C for 20 hours and at 1200 °C for 6 hours. Channeling analysis indicated that almost all tin atoms in silicon crystal occupy the substitutional sites. Since tin atom has a larger covalent radius than that of silicon atoms, substitutional tin atoms generate positive strain field in silicon crystals. Figure 4 shows the isochronal annealing behavior of the ratio of sheet conductivity in tin diffused specimen to the sheet conductivity in the normal specimen without tin diffusion. Annealing time was 30 min. Boron ions were implanted at 30 keV with a dose of 1 and 3.2×10^{14}/cm^2. Under this condition, most boron ions were situated in the tin diffused region. Rapid increase of sheet conductivity of tin diffused specimen is observed near at 600 °C. This increase of sheet conductivity is likely proportional to tin concentration.

Figure 5 shows an example of carrier concentration and Hall mobility of the specimen descrived above. Tin atoms were diffused at 1200 °C for 6 hours and boron ions were implanted at 30 keV with a dose of 3.2×10^{14}/cm^2. Reverse annealing behavior was observed at the annealing temperature range between 500 and 600 °C for the specimen without tin atoms. On the other hand, a rapid increase of active carrier concentration in tin diffused specimen was observed at the annealing temperature range between 500 and 600 °C and became about three times larger than the normal specimens. This temperature is nearly equal to the annealing temprature

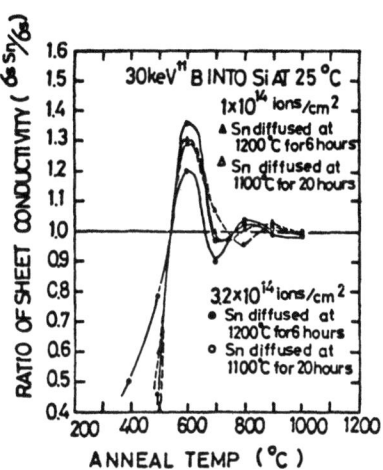

Fig. 4 Ratio of sheet conductivities between tin diffused specimens and without tin specimens vs. annealing temperature.

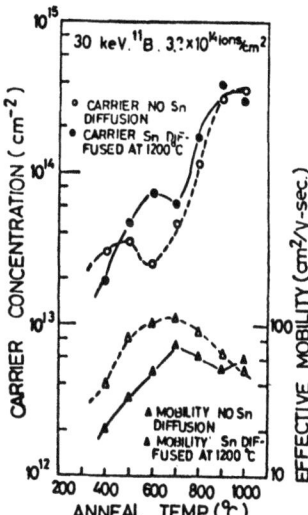

Fig. 5 Annealing behavior of carrier concentration and mobility vs. annealing temperature.

at which the small dots and rod-like precipitates appear in the electron micrograph. The tin diffused specimen shows slightly reverse annealing behavior at above 700 °C. After annealing at above 900 °C, the carrier concentration and effective Hall mobility of tin diffused specimens have the same values with those of the normal specimens.

Reverse currents I_R of P-N junctions made in previously tin diffused and without tin diffused samples have been compared at room temperature. Figure 6 shows typical characteristics of the reverse current at reverse voltage of 20V as a function of anneal-temperature. The annealing behavior of I_R for the samples without tin diffusion was similar to the previous report.[1] The I_R of tin diffused specimen has about two order low value compared to the I_R of the normal specimen, while it approaches to a same value in both specimens when annealed at above 900 °C. The increase of carrier concentration and decrease of I_R in tin diffused sample annealed at low temperature may be related to the decrease of the concentration of precipitates created in silicon crystals after boron implantation. It has been reported that the precipitates in boron implanted layer acted as deep level defect centers.[1]

Figures 7a and 7b show a comparison of electron micrographs between tin diffused samples and without tin diffused samples. Tin atoms were diffused at 1200 °C for 6 hours and boron ions were implanted at 80 keV with a dose of $1 \times 10^{15}/cm^2$. Annealing was carried out at 700 °C for 20 minutes. In this case, the ratio of

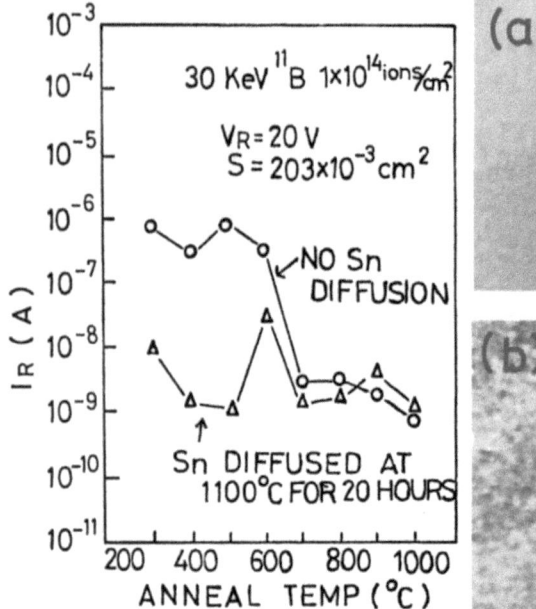

Fig. 6 Reverse current I_R vs. annealing temperature. S is an area of junction.

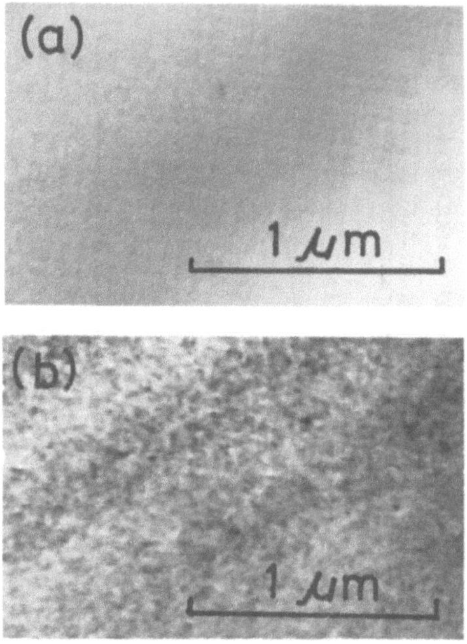

Fig. 7 Electron micrographs of B^+ implanted specimens, (a) tin diffused at 1200 °C for 6 hours, and (b) no tin diffusion.

sheet conductivities between specimens diffused with and without tin atoms is 1.42. It is observed clearly that, in the case of annealing at 700 °C, tin diffused specimen has less number of small rod-like precipitates than the specimen without tin atoms. The difference in the density and size of dislocation loops between tin diffused specimen and normal specimen is not discernible for the case of annealing at above 900 °C.

The effect of phosphorus diffusion on the defect formation

The enhancement in carrier concentration and suppression of the secondary defects in the form of precipitates have been observed in tin diffused specimens. To examine the effect of tin doping prior to the boron implantation on the defect generation during the following processes, we have diffused phosphorus into the boron implanted specimens with and without tin doping. It is known that phosphorus atoms in silicon crystal generate negative strain field, $\varepsilon < 0$, and that phosphorus diffusion creates excess vacancies which result in the emitter dip effect.[12]

Figure 8 shows the effect of phosphorus atoms on the secondary

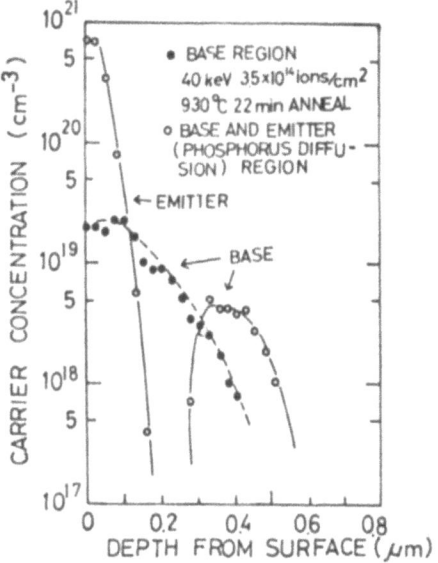 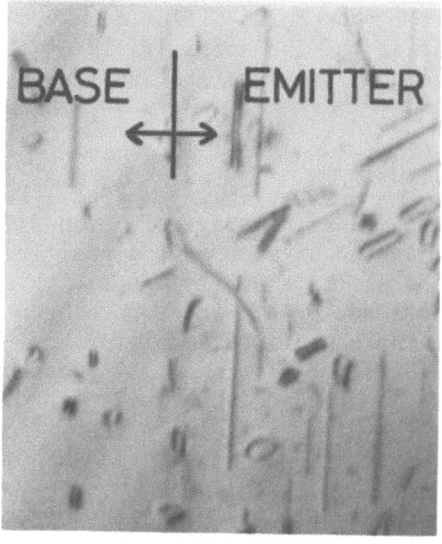

Fig. 8a Profiles of carrier concentration, before phosphorus diffusion (●) and after phosphorus diffusion (o).

Fig. 8b Electron micrograph of the left hand side sample.

defect generation in boron implanted specimen. Figure 8a shows the depth profiles of carrier concentration after phosphorus diffusion. Solid circles indicate carrier concentration in the base region and open circles indicate that in the emitter region. Figure 8b shows the electron micrograph of the base and the emitter regions. It is clearly observed that high density rod-like precipitates are reprecipitated in the emitter region, while most defects are dislocation loops in the base region. Also it has been observed that the density of the rod-like precipitates reprecipitated by phosphorus diffusion in tin diffused specimen is smaller than that in the specimen without tin diffusion.

For the interpretation of the above results, it is important to notice that the strain fields caused by tin atoms and phosphorus atoms have opposite signs each other. Tin atoms have been considered to compensate the strain caused by the impurities which have a smaller covalent radius. If tin atoms which have a compressive strain to the surrounding silicon matrix are present prior to the boron implantation, the concentration of the precipitates which have a compressive nature is reduced compared to the case in which no tin atom is present. The reduction in the concentration of the precipitates may cause the increase of the substitutional boron atoms. Contrary to the case of tin atoms, the precipitate density after phosphorus diffusion may be enhanced by the compensation effect of the strain field of the precipitates ($\varepsilon > 0$) on the strain field of the substitutional phosphorus atom ($\varepsilon < 0$).

CONCLUSION

The precipitates (dot like and rod-like precipitates) of initial aging stage of boron-silicon solid solution in boron implanted silicon have a positive strain field ($\varepsilon > 0$). The enhancement of the electrical activity and the suppression of the above mentioned precipitates have been observed in tin diffused specimen at the annealing temperature around 600 °C. Contrary to the effect of tin atoms, phosphorus diffusion into boron implanted layer enhanced reprecipitation of the above mentioned precipitates.

ACKNOWLEDGEMENTS

We want to express the sincere gratitudes to J. Shimizu, Y. Watari and I. Inoue for their helpful stimulations and discussions. We also thank Dr. Y. Akasaka, K. Horie, and K.Tsukamoto for helpful discussions and backscattering measurement of tin concentration, and thank S. Higaki and M. Kato for their experimental assistances.

REFERENCES

1. A.P. Karatsyuba, V. I. Kurinny, S. V. Rychkova, Y. P. Timashova and V. V. Yudin, Proceeding of the international Conference on Defect in Semiconductor, Reading, (1972) p. 81
2. R. W. Bicknell and R. M. Allen; "Ion Implantation" edited by F. H. E isen and L. T Chadderton (Gordon and Breach, 1971)
3. M. Tamura, T. I keda and N. Yoshihiro, Japan. J. appl. Phys., 40, 9 (1971
4. R. S. Nelson, Proceeding of the International Conference on Defect in Semiconductor, Reading (1972) p. 140
5. Y. Yukimoto, G. Nakamura, Y. Watari, K. Horie and Y. Akasaka, "Semiconductor Silicon" edited by Howard R. Huff and Ronald R. Burgess (The Electrochem. Soc. Inc., New York 1973)p. 692
6. M. F. Ashby and L. M. Brown, Phil. Mag., 8, 1649 (1963)
7. J. M. Mitchell, J. appl. Phys., 33, 406 (1962)
8. H. Ryssel, H. Muller and K. Schmid, Appl. Phys., 3, 321 (1974)
9. R. A. Swalin, J. Phys. Chem. Solids, 18, 290 (1961)
10. Joseph C. C. Tsai, Proceeding of IEEE, 57, 1499 (1969)
11. T.H. Yeh, S. H. Hu and R. H. Kastel, J. appl. Phys., 39, 4266 (1968) and T. H. Yeh and M. L. Joshi, J. Electrochem. Soc., 116, 73 (1969)
12. R. Gereth, P. G. G. Van Loon, and W. Williams: J. Electrochem. Soc., 112, 323 (1965)

CHEMICAL COMPOSITION OF HIGH DOSE IMPLANTS IN SILICON AND GERMANIUM

H.Kräutle, A.Feuerstein, H.Grahmann and S.Kalbitzer
Max-Planck-Institut für Kernphysik, Heidelberg
F. Hasselbach and M. Prager
Institut für Angewandte Physik der Universität Tübingen

Abstract
High doses of various ion species have been implanted into single crystals of silicon and germanium at energies corresponding to projected ranges of about 200Å, in order to produce binary systems of varying fractional concentration. Backscattering analysis and secondary ion mass spectrometry have been applied to measure amount and distribution of the implanted ions with a precision of a few 10 Å.

Introduction
It has been shown that the ion implantation technique may be used, within certain limits, to produce binary layers of type $A(1-x)B(x)$ by bombarding a substrate material A with ions of type B (1,2) Depending on the particular conditions the maximum fractional concentration x of the implanted ion species ranges between a few percent up to about 100%.
This process may be of interest for a number of applications, especially when conventional techniques subjected to thermodynamic limitations cannot be applied. Metallization of non-conductors and anticorrosive doping of metals are two examples of the possibilities of this low temperature ion beam chemistry (1) In some cases it may suffice to accumulate a certain amount of the impurity in the bombarded layer, in others the concentration profiles have to be known. This work reports on
a) retentivity measurements concerning high dose implants in silicon and germanium in order to show the possible range of application;
b) concentration profiles of germanium implants in silicon the implanted dose varying from the low to the high concentration level.

II. Experimental / II.1 General

All implantations were carried out with our low energy implantation system - $1 \leq U \leq 70$ kV - consisting of a Veeco - Thomson CSF - mass separator; the beam handling system and a UHV-target chamber were constructed by ourselves.

The implantation was performed by repeatedly moving the substrates through the ion beam which was monitored throughout the run. Usually, the implantation energy was adjusted as to give a projected range of approximately 200 Å in silicon. The amount of ions retained in the substrate was measured by Rutherford backscattering (RBS) using a silicon particle detector. These measurements were carried out in the laboratory of Prof.Dr. J.W. Mayer, Caltech, USA.

Two methods were employed to measure the implantation profiles: Secondary ion mass spectrometry (SIMS) with a Cameca system at Tübingen, described in detail elsewhere (3), and low energy RBS with a high resolution electrostatic analyzer (ESA).

II.2 SIMS

To increase the depth resolution and the sensitivity of the apparatus the primary ion beam was scanned over an area of about 1 by 1 mm. Since ions coming from the edges of the sputtered area are likely to produce tails in the concentration profiles, only the central part of the total area ($\sim 1/3$) was taken for the analysis.

In studying the binary system Si-Ge we obtained the best secondary ion yields by using an oxygen-ion beam and an oxygen pressure of 10E(-4) torr in the sputtering chamber enhancing the formation of Ge^+-ions. Ge^+ was the strongest component in the total secondary ion yield in which also molecule ions, such as GeO^+, Ge_2^+ etc. showed up. Unfortunately the formation of Si_2O^+ and Si_2N^+ ions interfered with Ge^+-ions on the masses 70-76 but fortunately the share of Si_2O^+ (2x30+16) is negligible small. Therefore, the 7.8% abundant Ge isotope on mass 76 had to be measured. The implantation conditions were modified to meet this problem.

In order to convert the time scale of a SIMS measurement into a depth scale one has to know how the sputtering rate and/or the secondary ion yield varies with the composition of the layer. The sputtering rate changed relatively slowly by about 30% from pure Si to Ge-rich Si. This was found from samples exposed for different times. The depth of the removed layer was measured by optical interference techniques. Since we did not succeed in obtaining homogeneous Si-Ge standards with varying fractional composition, the secondary ion yields were assumed to be constant, the ratio of Si-ions to Ge-ions per sputtered atom being ~ 60. From these informations and the amount of Ge in the layer, determined from RBS, the interesting concentration profile was derived.

II.3 RBS measurements with an ESA

For measuring thin layers of thicknesses in the order of 100 Å and below, RBS at energies of about 100 keV offers advantages over working at energies of 1 MeV. In particular, the 100-fold increase in cross section allows the use of detectors with high dispersion. Our ESA - angle of deflection 90°, bending radius 30 cm - allows to analyze particle energies with a precision of about 0.5%. Whereas depth resolutions of a few 10 Å have been obtained at the target surface, the straggling in energy loss causes broadening of the spectra on the low energy side of the spectrum with increasing depth (fig.1) in agreement with previous work (4). It appears possible, however, to reduce this effect by mathematical analysis and to obtain depth resolutions of better quality.

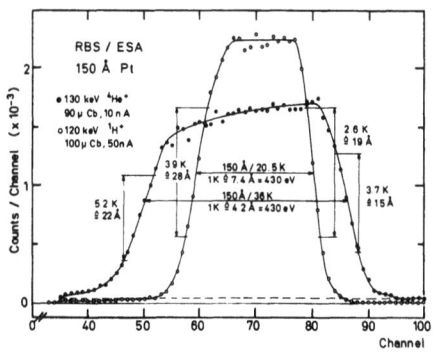

Fig.1:
High resolution backscattering analysis of a homogeneous Pt layer by using an electrostatic analyser.

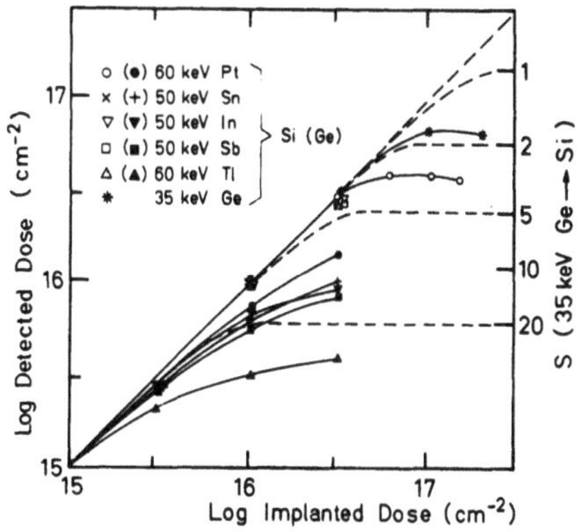

Fig.2:
Retention of implanted ions in Si and Ge as a function of dose. The stipled lines are calculated and correspond to sputtering coefficients $S(Si) = S(Ge)$ given on the right side. Note the substantially higher saturation level for Si.

To transform the energy spectrum into a depth scale one has to know the specific energy loss of the projectiles in the target material. The events per channel can be used to calculate volume concentrations of the target atoms by applying Braggs rule provided that the atomic density is known. Since the data on stopping powers in this energy range is little, we have measured the energy loss of light ions in silicon and germanium by two different methods:
i) evaporating thin films onto the surface of a silicon particle detector with a known thin dead layer and measuring the corresponding reduction of the signal; ii) evaporating thin films onto a gold backing and measuring the energy distribution of the backscattered particles.
The thickness of the evaporated layers was controlled with a Tally step device and with a standard vibrating quartz set-up. It turned out that the films contained substantial amounts of oxygen. This was seen in the RBS spectra where a carbon backing had been used.

Measurements were carried out in the laboratory of Dr. Siffert at CRN Strasbourg-Cronenbourg. From this data the number of atoms per unit area was derived, and with Braggs rule and the density of the crystalline material, dE/dx values for the pure elements were obtained. At 130 keV energy the figures are: 10.5 ± 1 and 17 ± 1 eV/Å for H-, D-, and He-ions, respectively, in both silicon and germanium.

III. Model calculations

As mentioned previously (2), a mathematical simulation of the implantation process at high doses has been carried out to better understand the relative importance of the different basic processes controlling the formation of binary systems. The present system under study, Si-Ge, is characterized by changes of the sputtering coefficient and of the range parameters with increasing Ge concentration, whereas the atomic densities of Si and Ge, and hence of Si-Ge, are practically the same. Since both constituents are miscible in any fraction, precipitation effects can be ruled out which would complicate the mathematical simulation considerably.

IV. Results and discussion

Fig.2 gives a survey on the possibilities and limitations of the implantation process concerning the formation of binary layers in the two most important elemental semiconductors silicon and germanium. Sputtering is seen to be a more controlling parameter for germanium than for silicon, a factor of about 3 being characteristic. The maximum amount and concentration of implanted ions are given by L/s and 1/s, respectively, s being the sputtering coefficient and L an effective depth related closely to the range parameters. The heavy ions are seen to fall into the bracket between about 5-20% for germanium; metallization, occurring at x-values in the vicinity of about 0.5, seems to be restricted to the lighter metal ions, such as Ti etc., in marked contrast to the situation for silicon. Similar problems will be encountered, if compound formation in germanium is attempted by implantation procedures.

The results of SIMS, RBS/ESA and computer simulations are shown together in figs. 3a - 3d, the samples being Si(Ge)-implants with increasing Ge dose. Whereas the peak positions of the low dose implants agree quite well with LSS-data by Gibbons, the corresponding range straggling is larger by about 20%. The tails on the high range side, comprising only a few percent of the total Ge content, resemble the profile patterns found in most implants in crystalline material, although a relative reduction of their magnitude would be expected as soon as the layers become amorphous. The other possibility is seen in a systematic error in the SIMS measurements, possibly due to redistribution effects from the periphery of the sputtered area. With increasing implantation dose the peaks shift towards the limiting position set by the range parameters in pure Ge. In addition, the highest dose implant shows a tendency towards a flat-topped, asymmetric profile, observed previously for Si(Al) implants (2), which is expected to be more pronounced at a dose of 3x10E17 Ge/cmxcm. The RBS/ESA spectra (fig.3c) have been obtained by using a dE/dx-value of 11,5 eV/Å which is on the high side of the experimental data of 10,5 ± 1 eV. Peak positions and FWHM's are in reasonable agreement with the SIMS data. On average, discrepancies are within an error of ± 10%. In fig.3d the results of computation, SIMS, and RES/ESA are compared for two different doses of Ge. The SIMS and the RBS/ESA profiles agree quite well over the whole depth scale. It appears that better agreement with the theoretical profile would be obtained if the range straggling values would be increased by about 10%.

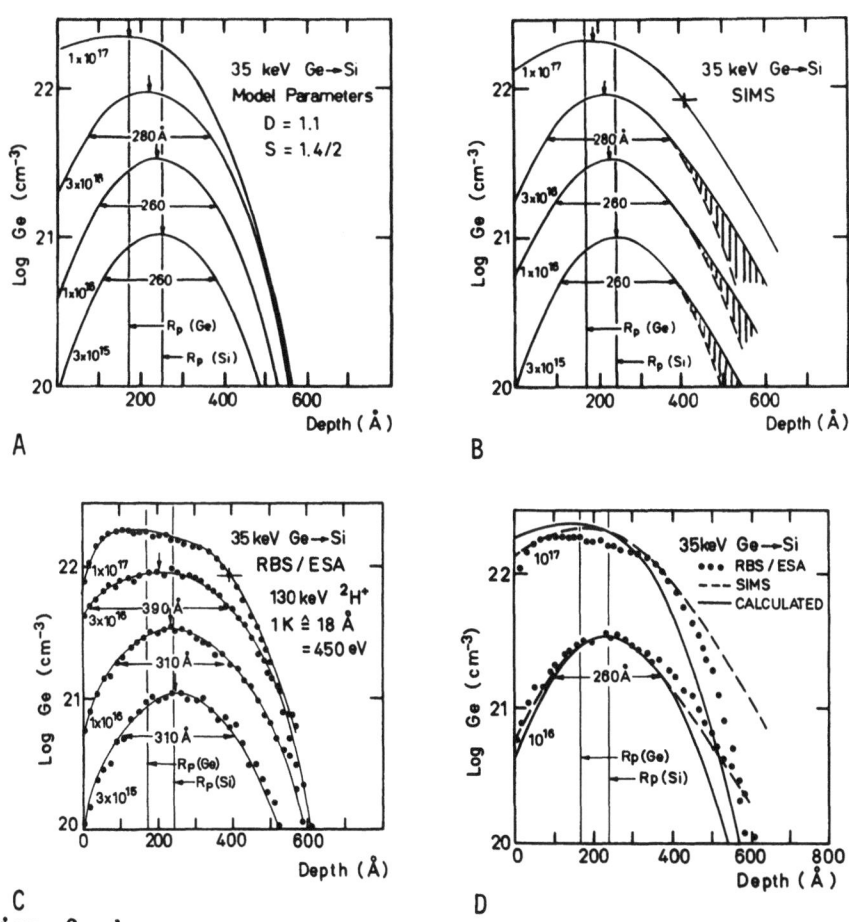

Figs. 3a-d:
a) computed profiles of Ge in Si(Ge); b) Ge-profiles measured by SIMS; c) Ge-profiles measured by RBS/ESA; d) comparison of a,b,c.

Conclusions

The available range of implanted concentrations of ions in germanium is reduced by about a factor of three as compared to silicon. Concentration profiles of high dose implants can be computed with reasonable accuracy in simple cases.
The RBS/ESA technique allows to measure concentration profiles with comparable precision as it is obtained by SIMS measurements.
We believe that the accuracy could be considerably imporved, since the systematical error of the measurements largely stem from stopping power and sputtering data which are not sufficiently well known at present time.

Acknowledgements

One of the authors (H.K.) wants to thank Prof. Dr. J.W. Mayer, Caltech, for helpful discussions and Dr. W.K. Chu, Caltech, and Dr. Siffert's group, CNR Strasbourg, for their help concerning RBS measurements.

References

1) See e.g.: Ion implantation in semiconductors and other materials, Proc. III. Int. Conf. at Yorktown Heights 1972, Plenum Press 1973, B.L. Crowder (Editor)
2) H. Kräutle and S. Kalbitzer, in ref. 1,pp.585-594
3) K.H. Gaukler, Quantitative analysis with electron microprobes and secondary ion mass spectrometry, Conference at Jülich (Germany) 1972, pp. 279-304
4) A. van Wijngaarden, B. Miremadi and W.E. Baylis, Can.J.Phys. $\underline{49}$ (1971) 2440

AN EXPERIMENTAL EQUIPMENT FOR ION IMPLANTATION

M. Setvak, J. Kral, Z. Hulek, L. Pina, A. Cako

Czech Tech. University, Faculty of Nuclear Science

Husova 5, 110 00 Prague, Czechoslovakia

ABSTRACT

A 150 keV ion implanter has been completed and put into operation in the Department of Electronics of the Faculty of Nuclear Science and Physical Engineering in Prague. The equipment consists of an duoplasmatron ion source with post-ionisation, sometimes called triplasmatron, a one-gap accelerator, a magnet mass separator and a target chamber in which the ion beam is scanned and deflected electrostatically to avoid neutral beam to reach the target. The pumping system consists of an oil diffusion pump, a turbomolecular pump and an ion--sputter pump at the target chamber. All parts of the vacuum system are made of stainless steel, copper gaskets are used everywhere except of ion source, target chamber can be heated to 400°C. Differential pumping is used and the vacuum in the target chamber is in the 10^{-8} Torr range. The ion beam will also be used for surface composition or depth profile measurements with a SIMS system which is being built.

THE EQUIPMENT

The machine is shown in fig. 1, the whole layout is shown schematically in fig. 2. The general idea was to build an ion implanter with the primary emphasis on clean target chamber, on the flexibility of the whole equipment and on low cost. The whole vacuum system is made of stainless steel standardised parts sealed with copper gaskets which enable the heating of the major part of the vacuum system up to 400°C. The variety of vacuum components makes it possible to change the equipment very quickly for different experiments. We have three independent vacuum systems that pump different sections of the machine.

Fig. 1 Assembled Ion Implanter.

The ion source and the accelerator are pumped down by a 2000 l/sec oil diffusion pump with a zeolite or a nitrogen trap. After disconnecting the accelerator and some other parts, the diffusion pump with zeolite trap can pump down the big volume of several tens of liters to the 10^{-10} Torr range. We are using differential pumping on two places where the beam has a cross-over. The vacuum system can be divided in two parts by a gate valve, so the ion source and the accelerator are all the time under vacuum. On fig. 3 are the spectra of residual gases in target chamber. Fig. 3a represents the spectrum of the target chamber alone after heating it to $200°C$. The chamber was pumped by an ion-sputter and a turbomolecular pump. The H_2 peak is present only when ion-sputter pump is on. The total pressure was $p_t = 5.10^{-8}$ Torr, the analysing quadrupole ion current $i_+ = 10^{-11}$ A. On fig. 3b is the spectrum of residual gases in the chamber during boron implantation, $p_t = 2.10^{-7}$ Torr, $i_+ = 10^{-10}$ A. The He peak originates from the duoplasmatron, where the working pressure of He is about 0.1 - 0.2 Torr and the plasma emerges through a hole of 0.7mm diameter. Before heating the chamber the main component of the pressure was H_2O.

A replaceable target chamber is being built with a SIMS, which will enable the measurement of the surface composition or a in-profile with argon beam from the same ion source as for implantation. Schematic representation of the system is on fig. 4. After implantation the sample is turned for better secondary emission and bombarded with argon beam focused to a diameter of several mm. A part of secondary ions enter a hole about 2 mm in a screen on the potential of the target, so there is no focussing between the target and the

EXPERIMENTAL EQUIPMENT FOR ION IMPLANTATION

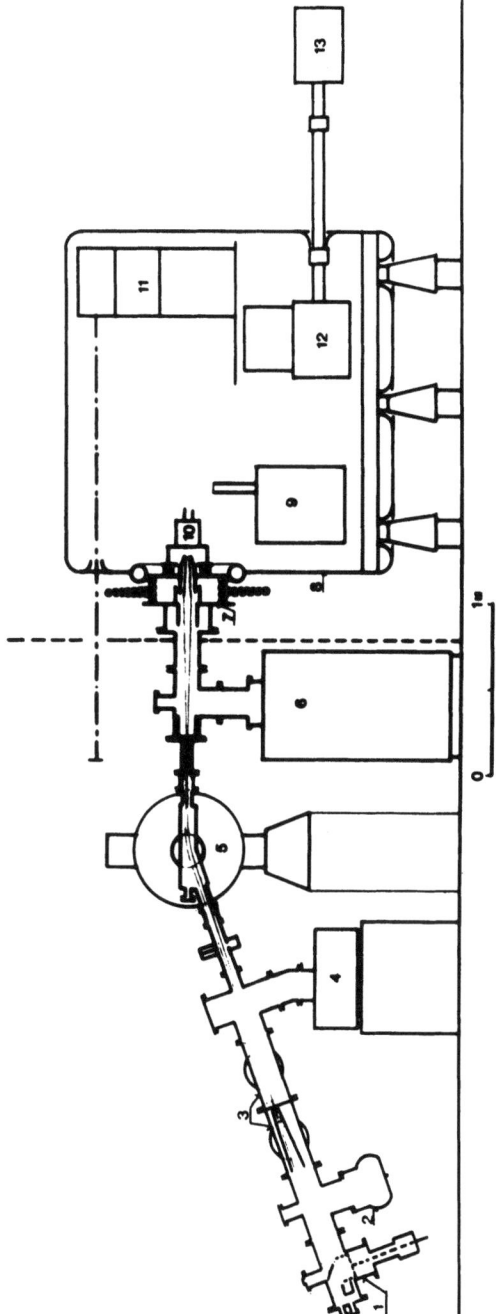

Fig.2 - General Layout of the Implantation Machine

1. Target chamber - 2. Ion sputter pump - 3. Beam scanning and switching system - 4. Turbomolecular pump - 5. Analysing magnet - 6. Diffusion pump - 7. One-gap accelerator - 8. High voltage terminal - 9. Extractor power supply - 10. Ion source - 11. Power supplies for ion source - 12. 380/220 V - 7.5 kVA generator - 13. Electromotor with isulating shaft.

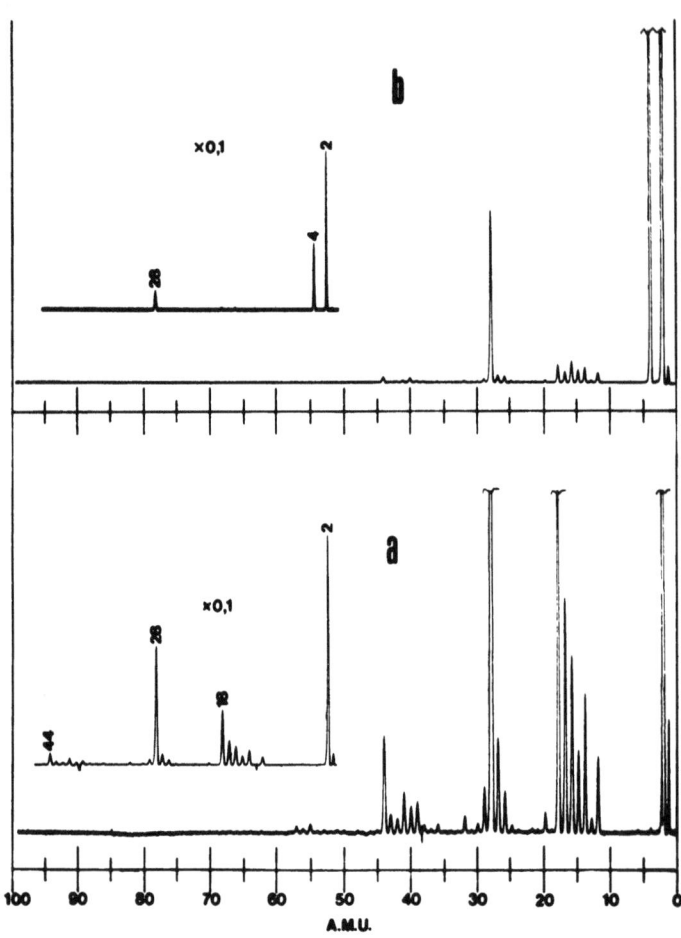

Fig. 3 - Spectra of Residual Gases in the Target Chamber.

screen. Secondary ions which passed the hole in the screen enter an electrostatic energy analyser. The ions are accelerated by a potential difference of 150 Volts between the screen and the analyser. This arrangement makes it possible that energetic ions and neutrals cannot enter the quadrupole mass analyser which follows after the energy analyser. This arrangement decreases the background of the quadrupole mass analyser. Some preliminary calculations and measurements were performed which showed that the resolution might be in the p.p.m. range.

The ion source consists of a duoplasmatron and a post-ionisation chamber with variable potential, sometime called a "triplasmatron" [1] - fig. 5. The duoplasmatron is working with helium, the wanted elements in the post-ionisation chamber being created by interaction

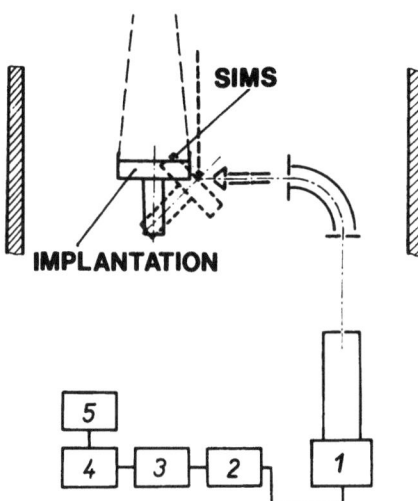

Fig. 4 - Schematic Representation of the Implantation/SIMS System. 1. Quadrupole mass spectrometer with electron multiplier - 2. Preamplifier - 3. Counter - 4. Ratemeter - 5. Recorder.

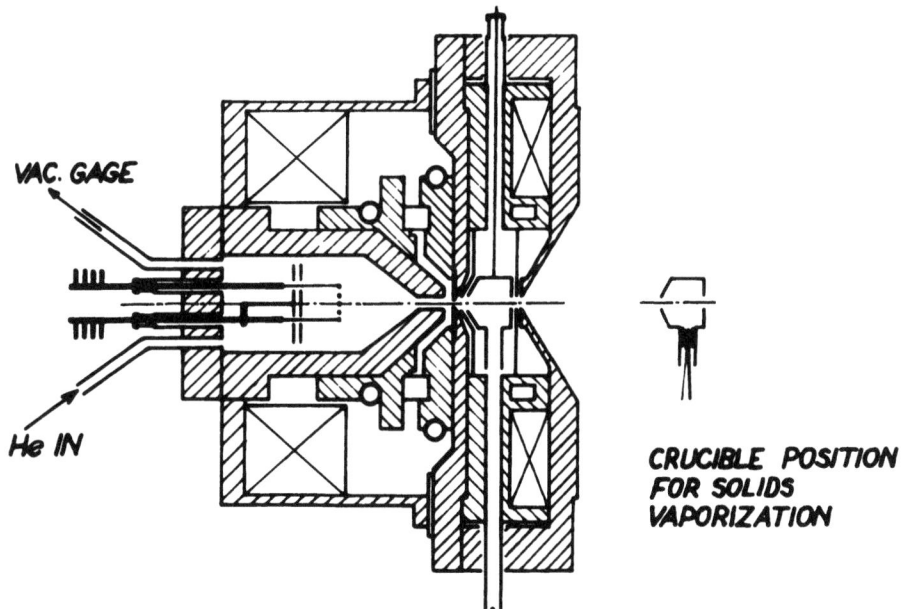

Fig. 5 - Schematic Representation of the Duoplasmatron Ion Source with a Post-Ionisation Chamber.

of helium plasma with gases or vapors of solid-state elements which are heated to high temperature in a micro-oven, which is inside the post-ionisation chamber. At the present time we can reach temperatures over $1000°C$ by heating the chamber and e.g. for boron implantation we use B_2O_3. A system has been built to enable the additional bombarding of the chamber and the expected temperatures should be over $2000°C$. The composition of the extracted ion beam depends on the pressure of helium in the duoplasmatron and on several parameters of the post-ionisation region such as added material pressure, magnetic field and on the potential of the ionisation chamber. Other parameters of the duoplasmatron almost do not have any influence on the ion beam composition, but the total beam current depends almost linearly on the duoplasmatron magnetic field and discharge current which is stabilised by a feed-back system controlling the cathode heating power in dependence on the signal proportional to the discharge current. The ion source is stabil throughout the whole life of the cathode which is about 100 hours. The total beam current extracted from the ion source is from 0.1 to several mA. The extraction voltage can be changed from several kV to about 30 kV. The extracted beam is accelerated in a two-cylinder accelerating lens with a potential difference 0-150 kV. The ion beam is focused using proper extraction / acceleration potential ratio. The useable ratio U_{ex}/U_{ac} is around 0.2, that means the beam energy of about 50 to 150 keV. [2].

REFERENCES

1. J. Aubert, G. Gautherin, C. Lejeune, Proc. of 2nd Int. Conf. on Ion Sources, Vienna, Sept. 11 - 15, 1972.

2. M. Setvak, J. Kral, Z. Hulek, L. Pina, Proc. of International Working Meeting on Ion Implantation Equipment, Swierk, Poland, Sept. 24 - 26, 1973.

DEVICES I

Ion Implantation into Polycrystalline Silicon

T. Hirao, T. Ohzone, S. Takayanagi and H. Hozumi

Central Research Lab. Matsushita Electric Ind. Co. Ltd.
Kadoma, Osaka, Japan

ABSTRACT

Some aspects of cencentration profiles and annealing behaviors of boron implanted in the polycrystalline silicon films have been studied in comparison with that implanted in single crystalline silicon.
The implantation of boron into the polycrystalline silicon definitely exhibited more skew in the concentration profiles than that in single crystalline silicon substrates. The location of peak concentration (R_p) was in good agreement for both single and polycrystalline silicon.
Annealing behaviors and apparent diffusion coefficients of boron were markedly affected by the crystalline states of the polycrystalline silicon films.

INTRODUCTION

Polycrystalline silicon films are beginning to have a wide spread impact on semiconductor device technology.
It is believed that ion implantation may be used for the fabrication of new devices in conjunction with polycrystalline silicon technology and offer extended freedom in device design.
There have been few reports on the behaviors of impurities ion-implanted into polycrystalline silicon films. (1-3)
In this paper, some aspects of concentration profiles and annealing behaviors of boron-implanted into polycrystalline silicon films deposited in N_2 and H_2 atmosphere, were studied in comparison with that in single crystalline silicon.

EXPERIMENTAL PROCEDURE

Polycrystalline silicon films in the thickness range of 0.5μ to 4μ were deposited on thermally grown SiO_2 via SiH_4 pyrolysis in N_2 atmosphere at $650°C$ and $700°C$, and H_2 atmosphere at $650°C$. The crystalline states of these films were examined by reflection electron-diffraction, surface carbon replicas, and scanning electron microscopy.

Ion Implantations were conducted at room temperature into (111) Si wafers offset $7°$ to the beam direction in order to minimize channeling effects and into polycrystalline silicon films set perpendicularly to the beam direction.

The concentration profiles and the peak location of boron atoms in single and polycrystalline silicon implanted with an energy in the range of 50keV to 150keV and a dose of $10^{15} ions/cm^2$, were measured using a Cameca secondary ion micro-analyzer (CSIMA).

Isochronal changes in electrical properties were studied in the temperature range of $360°C$ to $1000°C$ in N_2 atmosphere. The effective surface carrier density and Hall mobility were measured using Van der Pauw configuration, assuming that Hall mobility is equal to the drift mobility. The values of resistivity of as deposited polycrystalline silicon films were very high. In order to facilitate good electrical contact, contact regions were pre-implanted to a dose of $10^{15} ions/cm^2$ with 50keV B^+ and annealed at $650°C$ for 30 minutes in N_2 atmosphere.

Carrier concentration profiles were determined by combining Hall effect measurements and layer removal techniques by anodic oxidation. The electrolyte used for anodic oxidation was 0.04 KNO_3 + N-methylacetoamide.

RESULTS AND DISCUSSION

Polycrystalline Silicon Films

The films deposited in H_2 atmosphere consist of fairly uniform crystallites of 500Å in crystallite size, and the films deposited in N_2 atmosphere at $650°C$ and $700°C$ consist of slightly preferred (110) oriented crystallites of 2000Å in crystallite size, and (400) oriented crystallites of 2800Å, respectively. The crystallite size decreases with increasing depth from the surface.

Concentration Profiles and the Location of the Peak Concentration of Implanted Boron

Concentration profiles of boron implanted with an energy of 50keV and a dose of $10^{15} ions/cm^2$ were measured using CSIMA.

Experimentally it was found that the errosion rate of the polycrystalline silicon by primary ion beam was nearly linear with time within experimental errors, while the secondary ion yield such as Si^+ changed with increasing depth, as can be seen from Fig. 1. The abrupt change in Si^+ yield may be attributed

ION IMPLANTATION INTO POLYCRYSTALLINE SILICON

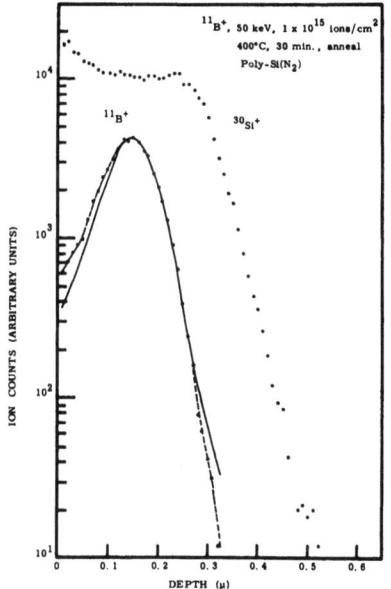

Fig. 1 Depth dependence of secondary ion yields and corrected boron yield (solid line) from polycrystalline Si.

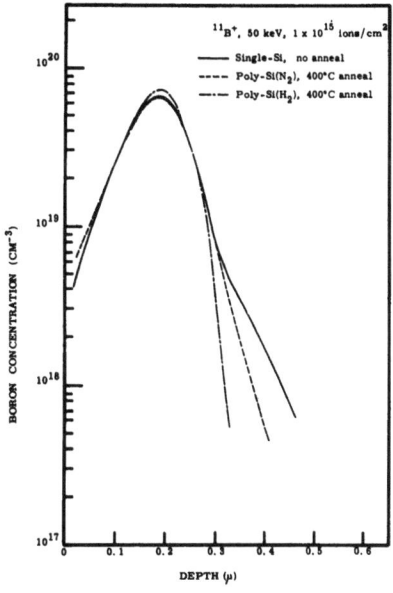

Fig. 2 Depth distribution of boron concentration in the polycrystalline Si deposited at 650°C.

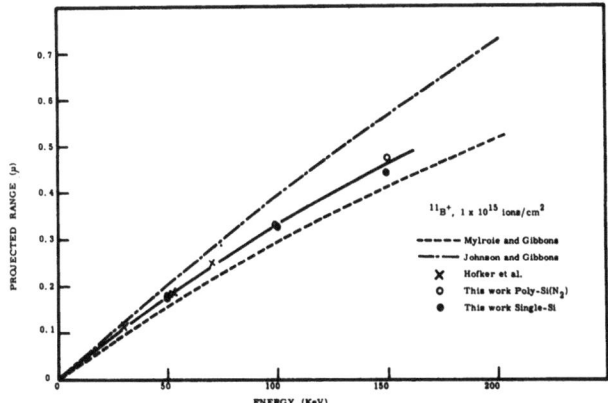

Fig. 3 Projected ranges versus energy for B^+-implanted single crystalline Si and polycrystalline Si deposited at 650°C.

to the charge up effect of the polycrystalline silicon films by the primary ion beam, because the resistivity in the region without boron atoms is very high. Provided that the secondary boron yields might change in the same manner as that of Si^+, the corrected boron concentration may be estimated from the secondary boron yield measured with CSIMA.

The concentration profiles of boron atoms in the polycrystalline silicon films deposited in N_2 and H_2 atmosphere and single crystalline silicon are shown in Fig. 2. The implantation of boron into polycrystalline silicon films definitely exhibited more skew in the profiles than that in single crystalline silicon.

Figure 3 shows a comparison of the measured location of peak boron concentration (R_p) in both single and polycrystalline silicon to those in single crystalline silicon reported so far by others. (4-6) The measured R_p were nearly the same with those in single crystalline silicon within experimental errors.

Annealing Behaviors of Implanted Boron

The isochronal annealing characteristics of the effective surface carrier density (N_S) and Hall mobility for the single and polycrystalline silicon boron-implanted are shown in Figs. 4 and 5. The typical reverse annealing stage observed in single crystalline silicon implanted with boron was not observed in polycrystalline silicon films deposited in N_2 atmosphere. N_S monotonously increases with increasing temperatures to $800°C$ above

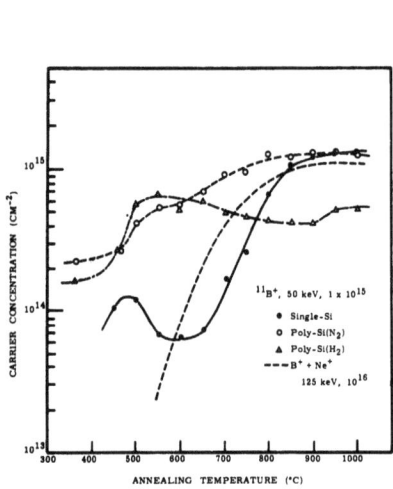

Fig. 4 The effective surface carrier concentration versus annealing temperature in the single and polycrystalline Si.

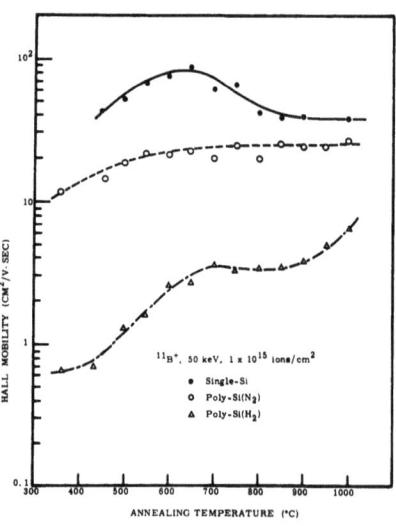

Fig. 5 Hall mobility versus annealing temperature in the polycrystalline Si.

which almost all of the boron atoms become electrically active. Similar tendency is observed in the annealing behaviors of the layer turned amorphous by ion implantation (dotted lines in Fig. 4). (7-8) In the films deposited in H_2 atmosphere, behaviors appear to be considerably different. Gradual decrease of N_S observed in the temperature range of 560°C to 900°C may be attributed to the captures of electrically active borons to the trapping centers in the crystallite boundaries. The fraction of electrically active borons after annealing at 800°C is approximately 40%.

The increase of Hall mobility with temperature in polycrystalline silicon can be considered to be due to the crystallite size growth.

The Comparison between the Total Boron Concentration Profiles and the Carrier Concentration Profiles and Diffusion Coefficients

Figures 6 and 7 show the comparison between the boron concentration profiles and the carrier concentration profiles in the single and polycrystalline silicon after annealing at 800°C and 1000°C for 30 minutes in N_2 atmosphere. The depth distributions of Hall mobility are also shown. For the films deposited

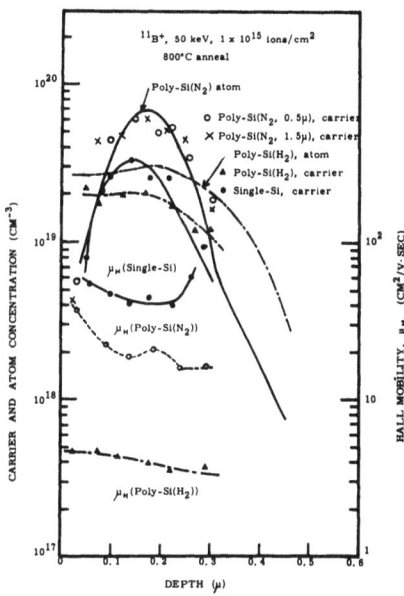

Fig. 6 Depth distributions of carrier and atom concentration, and Hall mobility profiles in the single and the polycrystalline Si.

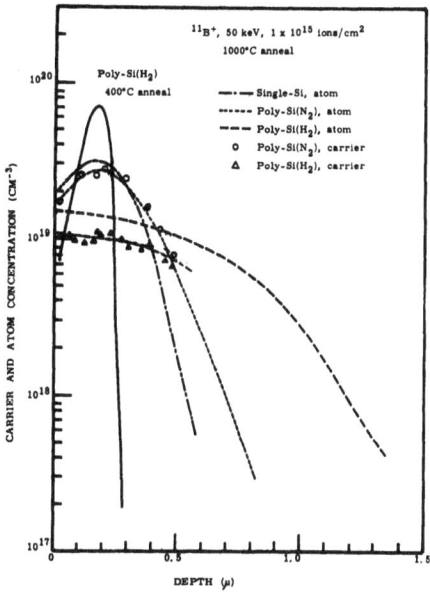

Fig. 7 Depth distribution of boron concentration and carrier concentration in the single and the polycrystalline Si.

Table I Apparent diffusion coefficients in the single and polycrystalline Si.

ANNEALING TEMPERATURE	Single-Si	Poly-Si(N_2) 650°C	Poly-Si(N_2) 700°C	Poly-Si(H_2) 650°C
800°C	-	-	-	5×10^{-14}
1000°C	5.5×10^{-14}	1.3×10^{-13}	$\sim 1.3 \times 10^{-13}$	8×10^{-13}

UNIT: $CM^2 \cdot SEC^{-1}$

in N_2 atmosphere, the profiles of boron concentration and carrier concentration are in good agreement after annealing.

A pronounced broadening in the depth distribution of boron atoms observed in the films deposited in H_2 atmosphere suggests that the diffusion along the crystallite boundaries occurs even at such a low temperature. Apparent diffusion coefficients can be obtained using these profiles before and after annealing. (8) The results are given in Table I.

ACKNOWLEDGEMENTS

The authors would like to express their appreciation to Mr. Y. Yaegashi for CSIMA measurements, Mr. H. Okazaki and Mr. A. Nishikawa for the preparations of the polycrystalline silicon films. They would also like to thank Miss M. Aoki for help in the electrical measurements, Miss Y. Yamamoto for computing the diffusion coefficients, Miss T. Kurahashi for help in making this manuscript.

REFERENCES

1. T. Hirao, T. Ohzone and H. Hozumi, Inst. Elect. Comm. Eng. Japan, EDD-73-100 (1973) (in Japanese)
2. T. Tsuchimoto, I. Yudasaka and T. Shirasu, 5th Symposium on "Ion Implantation in Semiconductor" Feb. (1974) at the institute Physical and Chemical Research. (in Japanese)
3. T. Yoshihara, R. Kado, H. Shibata, and A. Yasuoka, ibid.
4. W. K. Hofker, H. W. Werner, D. P. Oosthoek and H. A. M. de grefte, Proc. 3rd Intern. Conf. on Ion Implantation in Semiconductors, New York, 1972, P.133.
5. S. Mylroie and J. F. Gibbons, Proc. 3rd Intern. Conf. on Ion Implantation in Semiconductors, New York, 1972, P.243.
6. W. S. Johnson and J. F. Gibbons, "LSS Projected Range Statistics in Semiconductors" (Standard University Bookstore, Stanford, California, 1970)
7. N. Yoshihiro, T. Ikeda, N. Tamura, and T. Tokuyama, Proc. U.S.-Japan Seminar on Ion Implantation, Kyoto, 1971, P.33.
8. H. Ryssel, H. Muller, K. Schmid, and I. Ruge, Proc. 3rd Intern. Conf. on Ion Implantation in Semiconductors, New York, 1972, P.215

ION IMPLANTATION OF IMPURITIES INTO POLYCRYSTALLINE SILICON

T.TSUCHIMOTO, I.YUDASAKA & T.SHIRASU

Semiconductor & Integrated Circuits Division, Hitachi Ltd.

1450, Josuihoncho, Kodaira, Tokyo, JAPAN

ABSTRACT

The doping characteristics of polycrystalline silicon were observed by ion implantation of boron and phosphorus. The complicated conditions such as polycrystalline silicon film formation, doping and heat treatment were separated by the ion implantation for doping method. Obtained results were; (1) The sheet resistivity was decreased abruptly and was changed minimum to maximum values with annealing temperature. (2) The sheet resistivity was changed exponentially with implant dose and the gradient was -4.7. (3) The carrier mobility and the carrier activation ratio were increased with implant dose. (4) The activation energies were 0.23 to 0.06 eV corresponding to the implant dose. These results are explained that the trapping or the ejection of impurities from grain boundary and the obtained activation energies are interpreted as the potential barrier height at the grain boundary.

INTRODUCTION

The polycrystalline silicon (poly-Si) is now, currently used in silicon integrated devices. The use of poly-Si in silicon gate MOS structures is especially important.[1] In order to understand the wide range properties of poly-Si, Kamins [2] observed Hall mobility in chemically deposited poly-Si and explained most of the trends observed in his experiments by considering an inhomogenious model in which substantial number of traps existed at the grain bounderies between crystallites of silicon. Kamins, Manoliu and Tucker [3] studied the diffusion of electrically active impurities into poly-Si deposited by the thermal decomposition of silane, and Eversteyn and Put [4] studied in paticular the temperature dependence of the growth rate. Cowher and Sedgwick [5] measured Hall mobility and studied the irreproducible doping during growth.

In spite of these many works, the electrical characteristics of poly-Si are not fully understood yet because physical properties and therefore electrical properties are much influenced by formation and doping conditions. The ion implantation for poly-Si as a doping method can separate these complicated relations such as poly-Si film formation, doping and heat treatment. This is a very important advantage for the study of the characteristics.

EXPERIMENTAL

The poly-Si film was measured at the temperature of 620 °C by chemical vapour deposition of silane with thickness of 5000 Å on the silicon dioxide film, then doping impurities such as boron and phosphorus were implanted with the energy of 25 to 100 keV, doses of 5×10^{12} to $5 \times 10^{15}/cm^2$, and the annealings at various temperature were carried out in nitrogen atomosphere.

Sheet resistivity was measured by four point probe method below 10^8 ohm/□ , and the current-voltage characteristics of the poly-Si resistance were used for higher resistivity. The temperature characteristics were also measured by this resistance. The mobility and the number of carriers in poly-Si film were measured from Hall coefficient that was obtained by Van der Pauw's pattern, and the magnetic field intensity used for the measurement was 5000 gauss. The distribution of sheet resistivity in the film was measured by the the method of step etch using poly-Si etchant ($HF:HNO_3:CH_3COOH=$ 1:50:25).

RESULTS AND DISCUSSIONS

Fig.1 shows the relation between annealing temperature and sheet resistivity taking the implant dose as parameters. The sheet resistivity of boron implanted case was decreased abruptly at the temperature of 675 °C and was changed minimum to maximum values around 850 °C and 1000 °C, respectively, in 10 min. isochronal annealing. The sheet resistivity of phosphorus implanted case was more abrupt than Fig.1, and had same tendencies with boron implanted case.

The abrupt change is interpreted that the character of deposited amorphous silicon at 620 °C is changed to poly-Si by heat treatment of 675 °C and the implanted impurity atoms are changed to carriers as the substrate temperature increased to 850 °C. The increase of sheet resistivity between 900 to 1000 °C, that is so called "reverse annealing", is also observed at the case of boron implantation to single crystal substrate, but in this case, the

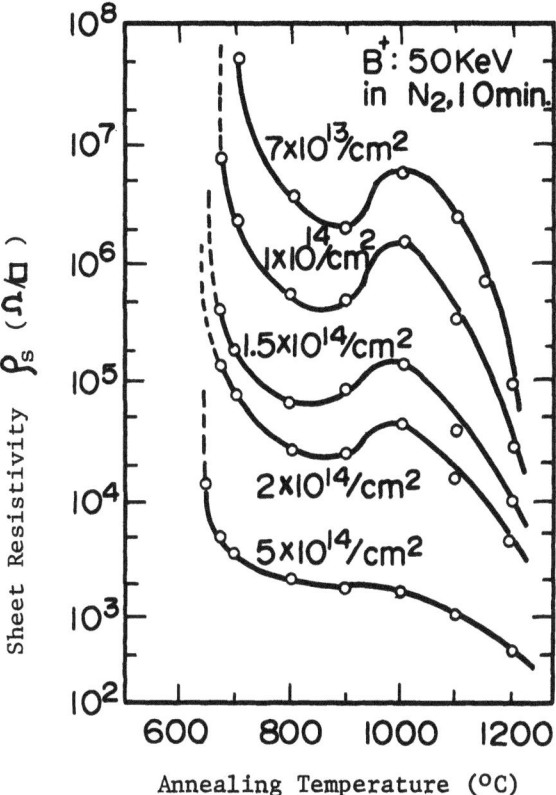

Fig.1 Relation between annealing temperature and sheet resistivity in boron implanted case.

reverse annealing occurred at the temperature between 550 to 650 °C, so it is hard to think the same phenomenon.

Though the mechanisms of carrier conduction and the change of poly-Si film with temperature are not yet perfectly understood, the phenomenon is explained as follows. The impurities begin to diffuse around the temperature of 850 °C and the diffusion causes the decrease of carrier mobility in poly-Si film (Fig.3) and therefore causes the increase of sheet resistivity. The grain size of poly-Si is considered begin to increase with the temperature around 1000 °C and the increase of grain size causes the decrease of total grain boundary area, and then the trapped impurity atoms at the grain boundary are emitted and are redistributed to the crystal domain. These cause the increase of carriers again and make a decrease of sheet resistivity again, that is, make a maximum curve at 1000 °C.

Fig.2 shows the change of sheet resistivity with implant dose, here the annealing was carried out at 1100 °C. It was noticed that

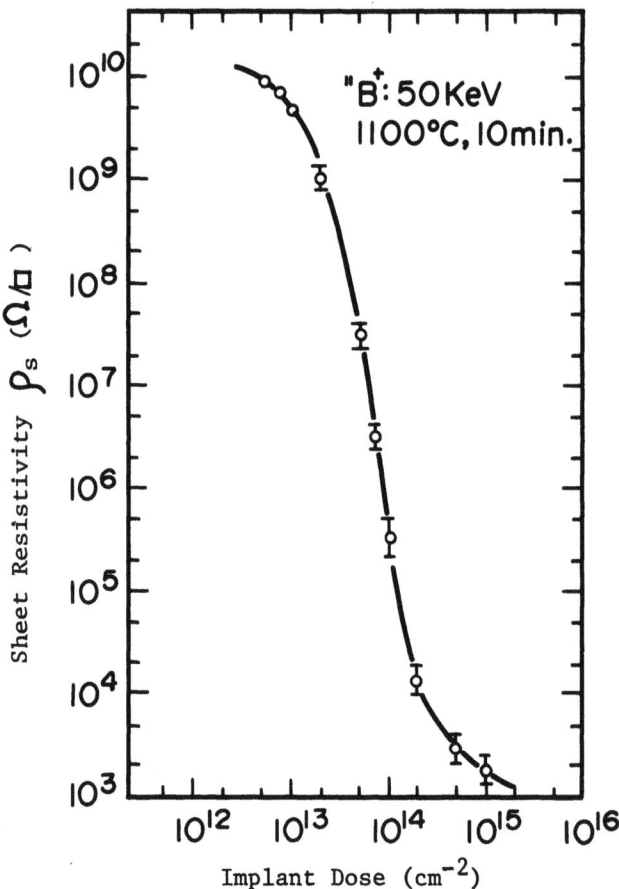

Fig.2 Change of sheet resistivity with implant dose.

the wide and steep change of sheet resistivity. The change was exponentially with implant dose and the obtained gradient was -4.7. This steep change is a big difference compared with the single crystal case which the gradient is about 1 or the less. This is attributed that the carrier mobility and the activation ratio of implanted impurities are both increased with increasing implant dose. (Fig.3).

The sheet resistivity of unimplanted poly-Si that only heat treated with 1100 °C was around 3×10^{10} ohm/□ , and the poly-Si below $10^{13}/cm^2$ implanted was around 10^{10} ohm/□ , so there was scarcely affected by the low dose implantation to the sheet resistivity. This is attributed that the low dose implantation, the implanted impurity atoms are distributed at first to the grain boundary by heat treatment, therefore the implanted atoms do not become carriers in this low dose implantation.

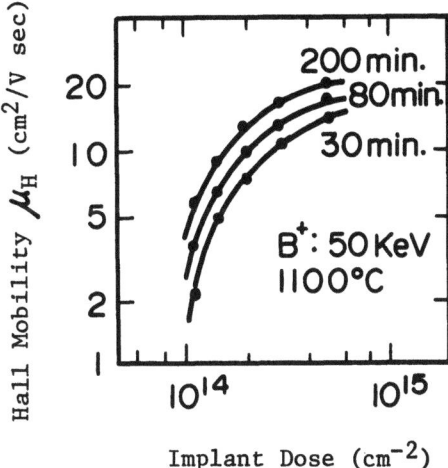

Fig.3 Relation of Hall mobility with implant dose.

In Fig.2, the gradient of sheet resistivity became rather mild at the dose of $2 \times 10^{14}/cm^2$. This will be due to the next reason. The capacity of grain boundary that traps implant atoms is satulated around $2 \times 10^{14}/cm^2$ dose implant, and when the implant dose exceeds this dose, most of the implanted atoms are distributed in crystal domain, so the increase of carriers and the change of mobility reach near to the single crystal, that is, the gradient becomes near 1. The trapped capacity of boron at grain boundary was calculated around $4 \times 10^{18}/cm^3$ at 5000 Å film thickness.

Fig.3 shows the relation of Hall mobility and implant dose. The mobility was increased with implant dose, and the mobility also was increased with annealing time. This is just the opposit to the case of single crystal. The increase of mobility with implant dose is explained that the height of potential barrier which exist at the grain boundary is decreased with increasing impurity concentration, that is, increase of implant dose. The mobility increase with annealing time is also explained that the each grain size become larger than before by heat treatment and the increase of grain size causes the decrease of total area of grain boundary, therefore the mobility increases.

The activation ratio, that is the ratio of measured carriers and implanted atoms, were also measured. The value was 60 to 70 % at 5×10^{14} to $1 \times 10^{15}/cm^2$ dose, while it was around 100 % in single crystal case on this implant dose, therefore the difference of 30 % is considered that the carriers are trapped at grain boundary.

Impurity distribution in poly-Si film are not generally well known, and the resistivity distribution in the film was measured instead of carrier distribution, and this may have a reciprocal

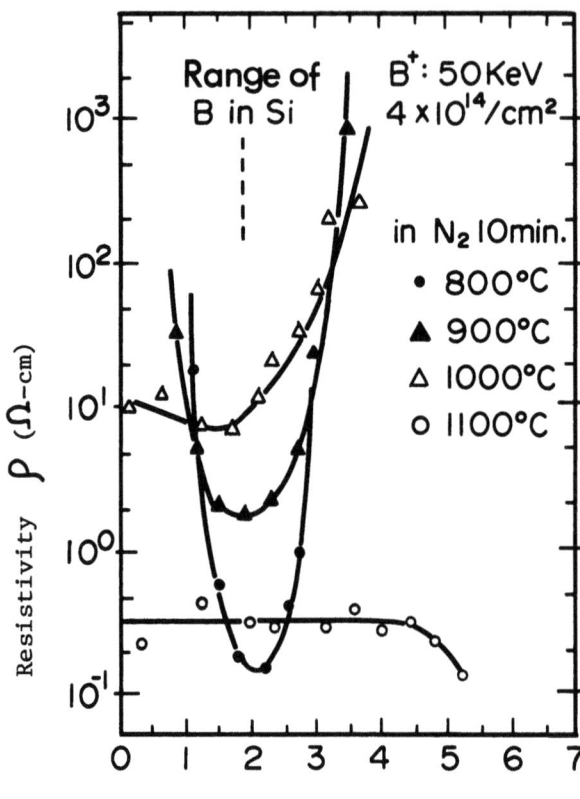

Fig.4 Resistivity distribution in poly-Si film.

relation to the carrier distribution. Fig.4 shows the resistivity distribution in poly-Si film that was formed at 600 °C in nitrogen atomosphere and was implanted $4\times10^{14}/cm^2$. Various annealing temperatures were chosen and then resistivity was measured with step etch, and the resistivity values were differentiated with the depth. There was rather concentrated distribution at 800 °C annealing. The minimum resistivity point at 800 °C annealing can be considered to correspond to the distribution peak of implant atoms. The range of boron in the single crystal silicon is also shown for the comparison. When the implant energies were increased, the minimum resistivity points were increased in depth at 800 °C annealing case. These points are also considered to correspond to the distribution peak of implant atoms at these energies.

Fig.5 shows the temperature dependence of sheet resistivity with Arrhenius plot. It was noticed that the temperature coefficient was negative and the logarithmic plots of sheet resistivity became streight line to the reciprocal of the temperature, and the

ION IMPLANTATION OF IMPURITIES

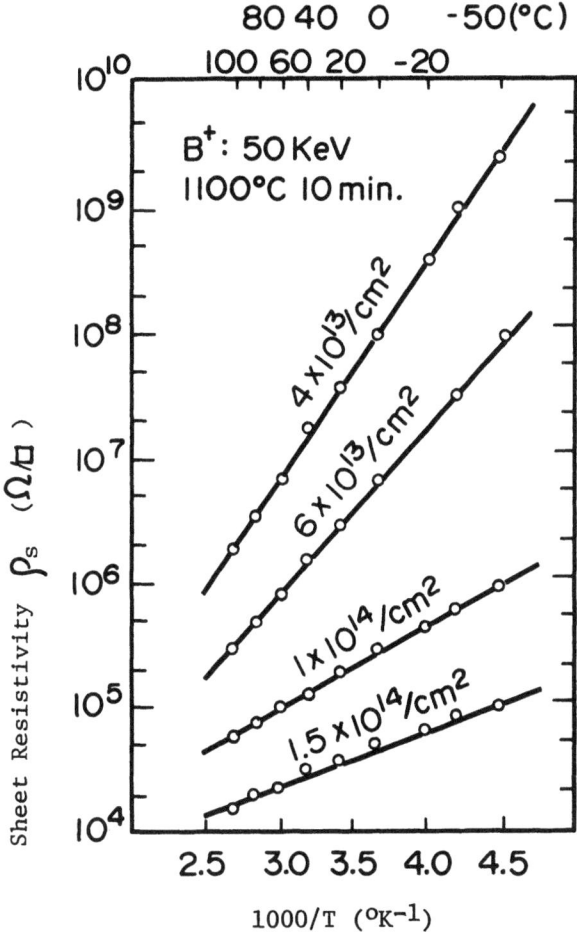

Fig.5 Temperature dependence of sheet resistivity.

tendency of temperature dependence was increased with sheet resistivity. These are interpreted as follows. The carriers are trapped at the grain boundary and the trapped carriers cause the formation of potential barrier and the increase of potential height. These trapped carriers are emitted to the conduction level as the temperature rises, and the height of potential barrier is decreased corresponding to the number of emitted carriers. That is, the number of carriers and carrier mobility become larger as the temperature rises. These cause the negative temperature coefficient of poly-Si resistivity. The large temperature dependence with higher sheet resistivity is that the ratio of carrier generation from these traps becomes larger compared with the total carriers. The activation energy and the temperature coefficient obtained from Fig.5 are shown in Table 1.

Table 1 Activation energies and temperature coefficients obtained from Fig.5.

Implant Dose (cm^{-2})	Activation Energy (eV)	*Temperature Coefficient (%/°C)
4×10^{13}	0.23	-7.8
6×10^{13}	0.18	-7.2
8×10^{13}	0.09	-2.8
1×10^{14}	0.08	-1.6
1.5×10^{14}	0.06	-1.6

B$^+$, 50 keV, in N$_2$, 1100 °C, 10 min., *Measured at 30 °C.

CONCLUSIONS

Obtained results and conclusions are as follows;
(1) In isochronal annealing, the sheet resistivity was decreased abruptly at the temperature of 675 °C, and was changed minimum to maximum values both in boron and phosphorus implanted case.
(2) The sheet resistivity varied exponentially with implant dose, and the obtained gradient was -4.7.
(3) The carrier mobility and the carrier activation ratio were increased with the implantation dose between 10^{14} to 10^{15}/cm^2.
(4) The activation energies, obtained from temperature-resistivity characteristics, were 0.23 to 0.06 eV corresponding to the doses of 4×10^{13} to 1.5×10^{14}/cm^2, and the temperature coefficient of resistance was -7.8 to -1.6 %/°C at 30 °C.

These results are explained qualitatively that the effect of potential barrier surrounding the grain boundary and the potential barrier is influenced by the trapping or the ejection of impurities from grain boundary as explained above, and the obtained activation energies are also interpreted as the potential barrier height formed at the grain boundary.

ACKNOWELDGEMENT

The authors express their sincere thanks to Drs. M.Ono and T.Agatsuma for their many of the valuable discussions.

REFERENCES

[1] L.L.Vadasz, A.S.Grove, T.A.Rowe and G.E.Moore; IEEE Spectrum, 6, 28 (1969)
[2] T.I.Kamins; J.Appl.Phys., 42, 4357 (1971)
[3] T.I.Kamins, J.Manoliu and R.N.Tucker; J.Appl.Phys.,43,83 (1972)
[4] F.C.Eversteyn and B.H.Put; J.Electrochem.Soc.,120,106 (1973)
[5] M.E.Cower and T.O.Sedgwick; J.Electrochem.Soc.,119,1565 (1972)

THE EFFECT OF PROTON BOMBARDMENT ON POROUS SILICON FORMATION

T. Yashiro, K. Saito and T. Suzuki

Musashino Electrical Communication Laboratory
Nippon Telegraph and Telephone Public Corporation
Musashino-shi, Tokyo, Japan

ABSTRACT

Influence of proton implantation for anodization of silicon in hydrofluoric acid has been investigated. This porous silicon can be used to form isolation layers for bipolar integrated circuits. Both p- and n-type silicon were used for the evaluation of proton implantation effects. Most implantations were carried out with 60-150 KeV beam of protons, at a total dose of 10^{13}-10^{16}/cm^2 and at room temperature. In the case of implantation in n-type silicon, enhancement in anodization was observed at proton doses ranging from 10^{15} to 10^{16}/cm^2, i.e., the implanted region was converted into porous silicon, while the unimplanted region remained unchanged. On the other hand, in p-type samples, anodization was observed in the implanted region. The mechanism of this phenomenon was investigated by heat treatment after proton implantation.

INTRODUCTION

In order to realize high performance and high degree of integration in silicon bipolar integrated circuits, there are continuous demands for the improvement of an isolation technique. Selective oxidation methods such as LOCOS (1) or ISOPLANAR (2) techniques are typical example of the improved isolation structure. However, necessity of a long oxidation time has created some serious problems.

Recently a novel isolation technique called IPOS has been proposed (3). The IPOS technique is constituted by selective porous silicon formation in silicon wafers by anodization in hydrofluoric acid, and subsequent oxidation for converting to isolating layers. Then, devices are formed on the islands. The advantage of the technique over other isolation techniques such as the ISOPLANAR method, is its extremely short oxidation time necessary to form thick insulating layers.

Selective formation of porous silicon by anodization in n-type layers such as n on p epitaxial wafers is made in two ways. In one method, called the n-type method, anodization is carried out using silicon nitride masks. In this case, illumination is necessary to supply holes at the silicon surface to attain the anodizing current (4). The other method, called the p-type method, is to diffuse p-type impurity prior to anodization. However, in these methods sideways spreading of the resultant isolating area beyond the original mask patterns was observed.

The present paper concerns the improvement of this IPOS technique using proton implantation. It was found that proton implantation is also a useful tool in the selective formation of porous silicon. Dependence of formation of selective anodized films on dose, implantation energy and anneal temperature will be discussed.

EXPERIMENTAL METHOD

Both n- and p-type wafers and n on p epitaxial wafers were used for these experiments. The sheet resistivities of n- and p-type wafers were 3-5 ohm-cm. The sheet resistivity of n-type epitaxial layers and p-type substrates was 0.3-0.6 ohm-cm and 2-10 ohm-cm, respectively. Protons were accelerated to 60-150 KeV and implanted in silicon at room temperature. Chemical vapor deposited silicon dioxide masks of 2 μm thick and plated gold masks of 1 μm thick were used for implantation of 60-70 KeV protons and for implantation of 150 KeV protons, respectively. A total dose of protons is $10^{13} - 10^{16}/cm^2$.

After implantation, masks were removed and aluminum was evaporated onto the opposite silicon surfaces as an electrode. This opposite electrode was covered with a wax. Anodization was carried out as shown in Fig.1 at a current density of between 10 and 400 mA/cm^2. After anodization, the samples were bevelled and stained to observe anodization progress. The staining was performed by an etchant containing 1 part of hydrofluoric acid and 20 parts of nitric acid.

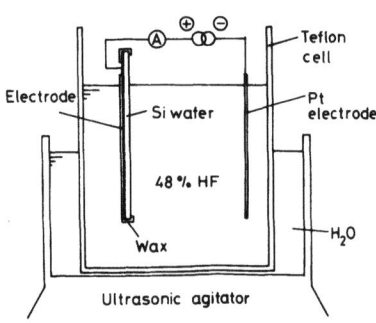

Fig. 1. Anode chamber cross-section view

POROUS SILICON FORMATION

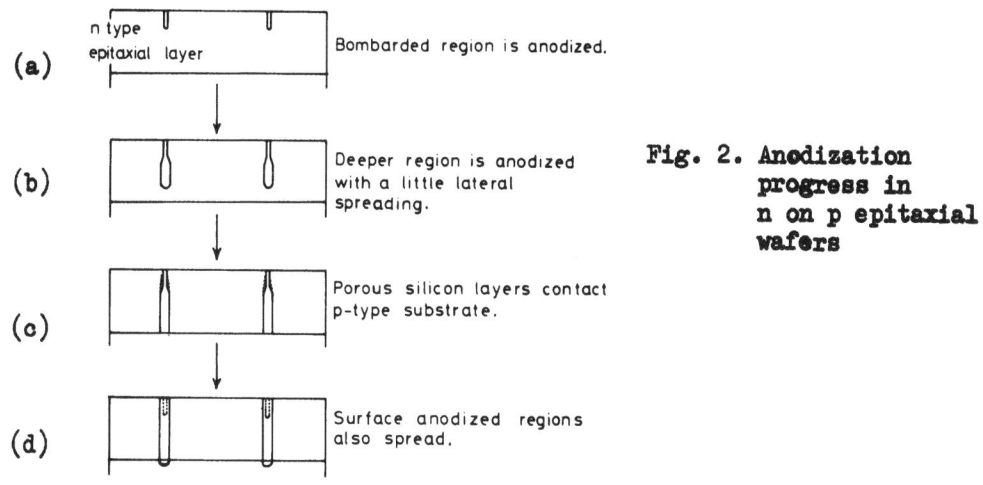

Fig. 2. Anodization progress in n on p epitaxial wafers

EXPERIMENTAL RESULTS AND DISSCUTION

SELECTIVE ANODIZATION ON N-TYPE SILICON

Anodization progress in n on p epitaxial wafers is investigated by varing proton dose, implantation energy, anodizing current density and anodization time. One example of anodization progress is shown schematically in Fig. 2. It is observed that anodization progress is restricted within the implanted regions at the initial stage as shown in Fig. 2(a). Thereafter, anodization proceeds deeper and slight lateral spreading is observed (Fig. 2(b)). On further treatment, porous layers contact the p-type substrate (Fig. 2(c)), and surface anodized regions also spread, as shown in Fig. 2(d).

In Fig. 3, a beveled sample corresponding to Fig. 2(c) is shown. The part shown by an arrow marked A is a beveled region and the width shown by an arrow B is the n-type epitaxial layer. The thickness of the layer is 3.2 μm. It can be seen that the anodized layers near the surface are narrow. On the other hand, the layers become wider at a depth more than 0.4 μm from the surface and reach the p-type substrate.

Table 1. Effect of dose for selective anodization of n-type or n on p epitaxial wafers

Dose (1/cm^2)	10^{13}	10^{14}	10^{15}	10^{16}
Selective Anodization	Not Observed	Partially Observed	Observed	Observed

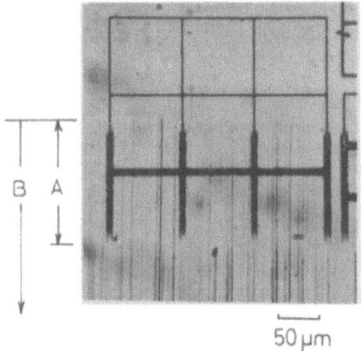

Fig. 3 Photomicrograph of beveled and stained sample after anodization.
A: N-type epitaxial layer
B: Beveled Region
Anodization: 190 mA/cm^2, 30 sec

Dose dependence of selective anodization is summarized in Table 1. In the case of n-type silicon, selective anodization was scarcely observed below a total dose of $10^{14}/cm^2$, regardless of implantation energy and silicon resistivity, while, it distinctly apeared at total doses of 10^{15} and $10^{16}/cm^2$ without light illumination.

A series of experiments on annealing has been performed, in which the implanted samples were heat treated in nitrogen atmosphere for 20 minutes prior to anodization. The results for 10^{15} and 10^{16} doses are summarized in Table 2. For the sample after 300°C annealing, selective anodization was observed. While, after 500°C and 800°C annealing, no selective anodization can be observed.

It has been reported that proton implantation at room temperature produces deep traps(5). Deep traps are mostly annealed out after 300°C treatment and, at the same time, shallow donors are produced. These results are also shown in Table 2. The donor concentration has

Table 2. Effect of annealing in N$_2$ for selective anodization of n-type or n on p epitaxial wafers.

Annealing	300°C	500°C	700°C	800°C
Selective Anodization	Observed		Not observed	
Shallow Donor	Produced	Have a peak	Disappear	
Deep Traps	Annealed out			

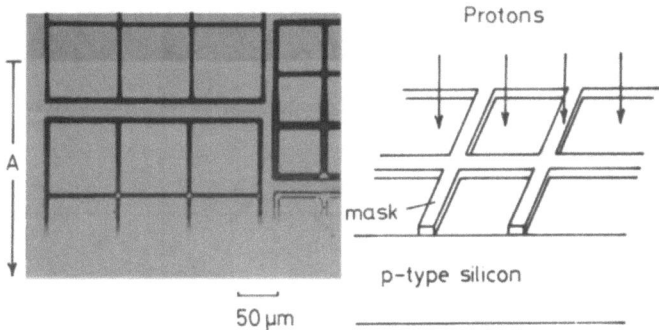

Fig. 4. Photomicrograph of beveled p-type sample after anodization

A: Beveled region
Proton dose: $1 \cdot 10^{15}/cm^2$
Energy: 60 KeV
Anodization: 25 mA/cm^2, 120 sec

a peak at about 500°C and disappears at above 700°C annealing. From the correlation of deep trap generation and selective anodization, it is suggested that selective anodization is caused by the increase in the number of deep traps brought about by the proton implantation. These deep traps act as generation - recombination centers and supply necessary holes for the anodization at the silicon - electrode interface. The same selective anodization was observed in the sample where gold was diffused partially to increase deep traps.

SELECTIVE ANODIZATION ON P-TYPE SILICON

In the p-type samples, on the other hand, no anodization was observed in the implanted region. The unimplanted region alone is anodized, as shown in Fig.4. Other difference between p-type samples and n-type samples is an aspect of anodized layer, i.e., in the p-type case, porous silicon layers can clearly be observed without staining.

Table 3. Effect of dose for selective anodization of p-type wafers

Dose (1/cm^2)	10^{13}	10^{14}	10^{15}	10^{16}
Selective Anodization	Not Observed	Partially Observed	Observed	Observed

Table 4. Effect of annealing in N_2 for selective anodization of p-type wafers

Annealing	300°C	500°C	700°C	800°C
Selective Anodization	Partially Observed	Observed		Not Observed

Table 3 shows the effect of dose at room temperature and Table 4 shows the effect of annealing in nitrogen for 20 minutes for $10^{15}/cm^2$ dose. In the case of a total dose of more than $10^{15}/cm^2$, selective anodization can clearly be observed. At 500°C annealing, where donor concentration has a peak, as shown in Table 2, selective anodization is most distinctly observed. After 800°C anealing, no selective anodization can be observed. This effect may be explained by the change of the specific resistivity due to the proton implantation.

CONCLUSION

Selective porous silicon formation after proton implantation in both n-type and p-type silicon was demonstrated. This effect may be applied to form thick insulating layers inlaid in silicon wafers, which serve as an isolation for silicon integrated circuits.

ACKNOWLEDGEMENTS

We express our thanks Dr.H.Harada and Mr.T.Honda for the use of implanter and the many valuable discussion of the ion implantation. We are indebted to Messrs. S.Nakajima, Y.Arita and T.Unagami for discussions and information about anodization and Messrs C.Tadachi and J.Hayashi for sample preparation.

REFERENCES

1) J.A.Appels and M.M.Paffen, Philips Res. Repts 26, 157, (1971)
2) W.D.Baker et.al., Electronics 29, 65, (1973)
3) Y.Watanabe et.al., Electrochemical Society Meeting, October 1973, p. 296.
4) D.R.Turner, J.Electrochem. Soc., 105, 402, (1958)
5) Y Ohmura et.al., Solid State Communications 11, 263, (1972)

P-TYPE DOPING OBSERVED IN SILICON IMPLANTED

WITH HIGH ENERGY CARBON IONS

J. Stephen, B. J. Smith and P. J. Hammersley

Electronics and Applied Physics Division

AERE, Harwell, Didcot, Oxon.

ABSTRACT

Some MOSTs in which ion implantation had been used to control the threshold voltage showed unexpectedly high p-type activity if the devices were annealed to only 500°C. This had been attributed to the p-type nature of the unannealed damage. In order to examine the phenomenon carbon ions, which do not produce doping centres in silicon, were implanted at energies high enough to permit capacitance-voltage profiles of the region between the surface of the silicon and the ion range to be observed.

This paper reports on some doping concentration profiles observed following annealing in the temperature range 175-800°C. The silicon used was 20Ω cm p-type oriented a few degrees from the <111> direction. Carbon ions were implanted with energies in the range 1.5 to 4.3 MeV using the Harwell van de Graaff accelerator.

Capacitance-voltage profiles show that in unannealed specimens two p-type doping peaks are present. Following annealing at about 400°C the deeper peak can no longer be observed, and the shallower peak increases in size. Finally after annealing at about 700°C no doping centres are observed.

INTRODUCTION

Some MOSTs in which ion implantation had been used to control the threshold voltage have shown unexpectedly high p-type activity if the devices have been annealed to only 500°C. In order to examine the effect without the ambiguity of impurity doping carbon ions have been implanted into silicon. Carbon does not produce impurity doping in silicon.

Whilst the phenomenon is particularly relevant to implantations at low energies, high energies have been used in this study because it is difficult to examine such effects close to the semiconductor surface. As the energy levels associated with the p-type centres are likely to be well away from the valence band the effect on the resistivity at room temperature is small. This is due to the low equilibrium concentration of ionised doping centres. This precludes the use of resistance techniques in the study. Therefore the capacitance voltage profiling method was used. This method is not normally satisfactory for measurements very near to the surface. Another reason for using the high energy implantation is to remove the region of interest from spurious surface effects which are not related to the implantation.

EXPERIMENTAL

Lopex <111> silicon wafers were used for the substrates. These were p-type of about 20Ω cm resistivity. A p-n junction was formed by implanting 40 keV phosphorus ions at a dose of 5×10^{14} cm^{-2} over the front surface and annealing at 800°C for 30 minutes. The back surface of wafer was made p$^+$ by implanting 1×10^{15} boron ions before annealing.

The high energy carbon ions were implanted on the Harwell van de Graaff accelerator at energies of 1.5, 2.5, 3.5 and 4.3 MeV. In each case the measured ion dose was 8×10^{11} cm^{-2}. The beam current was approximately 10^{-8} amps. During implantation the wafers were tilted 7° from the normal to the beam. The direction of the 7° misalignment was not known as the wafers had no orientation flat. The dose of the implanted ions was measured by integrating the collected charge and in order to minimise the charge lost by secondary electrons the target was mounted at the end of a tube 15 cm long with a 2 cm aperture at the other end. However some secondary electrons will have escaped back down the ion path giving a slightly higher measurement of dose than was actually implanted.

Following implantation the wafers were photolithographically etched to produce an array of 500 micron diameter mesa diodes. Each mesa was about 6 microns high and contained a p-n junction about .2 micron below the surface and the carbon implantation further down.

The mesa diodes were contacted by a metal probe. The capacitance voltage measurement equipment used has been described previously (1) and is a semi-automated system based on a Boonton 75D capacitance-resistance bridge. At each measurement point the diode voltage is automatically adjusted to balance the bridge and the capacitance and voltage data recorded automatically on punched tape. The tape is then processed by the Harwell computer to produce the doping profile by the normal method (2), and where necessary it modifies the calculated profile by applying the correction described by Kennedy and Klienfelder to allow for regions of quasi charge neutrality in the semiconductor.

Measurements were made following successive isochronal annealing at 175, 300, 400, 500, 600, 700 and 800°C for 30 minutes, the minimum temperature being that required to harden the photo-resist in the fabrication of the mesa diodes. The wafers were annealed in vacuum at 10^{-3} torr for 30 minutes at each temperature above 175°C.

RESULTS

Figure 1a,b,c & d shows the doping profiles observed at various stages in the annealing of the implanted wafers. It is seen that at the lowest temperatures two doping peaks occur. At intermediate temperatures the total doping increases and in some specimens the doping is too large to permit depletion of the whole profile. Following higher temperature annealing only the shallow peak remains and this decreases until it finally disappears at about 700°C. In the case of the 1.5 MeV profile some disturbance of the doping level remains even after annealing at 800°C. This is probably due to redistribution of the initial substrate doping caused by radiation enhance diffusion.

Figure 2 shows the total area under the profile, of the 2.5 MeV implanted specimen, plotted as a function of annealing temperature. The background doping has been subtracted. The accuracy of each point on the curve is about 20%, this being the variation of implantation dose across the wafer. At each annealing condition a suitable diode was selected for measurement. It was usually a different diode each time.

DISCUSSION

It is apparent that the effects produced by the carbon ions are deeper at the higher energies, and therefore bear some relationship to the range of the ions. The most obvious explanation is that they are associated in some way with the damage produced by the ion beam. Theoretical calculations of the position of the damage using

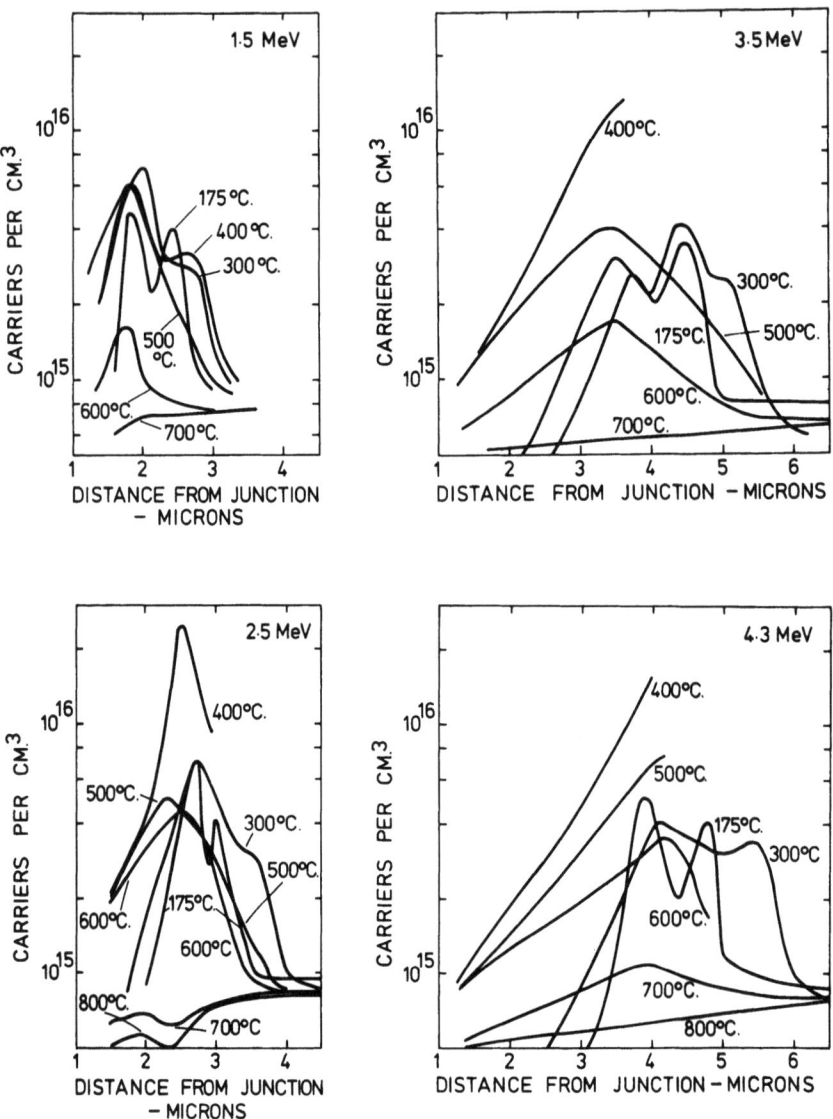

Figure 1. Capacitance-voltage doping profiles following carbon ion implantation at 1.5, 2.5, 3.5 and 4.5 MeV into silicon and annealing at various temperatures.

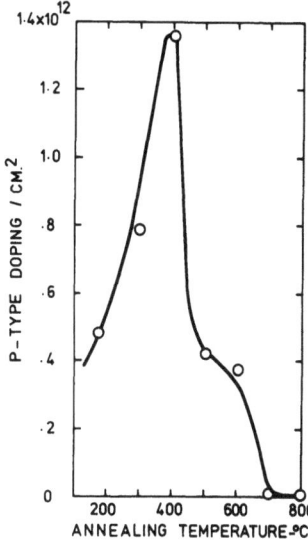
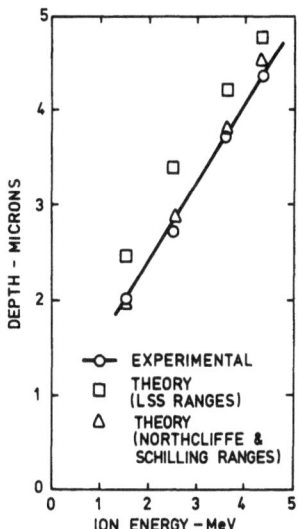

Figure 2. Variation of p-type doping with annealing temperature following implantation of 2.5 MeV carbon ions into silicon.

Figure 3. Variation of doping peak with ion energy.

the method of Manning and Mueller (3) gives the position of the damage peak shown in figure 3 for each energy used. This calculation is based on the projected range and straggling calculated by the computer program of Johnson and Gibbons. If, instead of using this value for the range the semi-empirical values of Northcliffe and Schilling (4) are used the damage peaks are predicted much closer to the surface and figure 3 shows these to be in good agreement with the observed position of the most shallow p-type doping peak. In making this comparison it is important to remember that the Northcliffe and Schilling only predict the total path length of the ion trajectory and .25µ was subtracted from this value to give an estimate of the projected range. Also the observed doping peak was measured from the p-n junction and the depth of the junction, .2µ, has to be added to give the depth from the surface.

The deeper doping peak observed after low temperature annealing is attributed to ion channelling during the implantation. Although the wafer was misoriented by 7° from the <111> direction this was not in a defined direction. Consequently some channelling could be expected. Variations in the degree of channelling also makes comparison between different wafers difficult. In general, changing the orientation of the wafer had the effect of changing the heights of the doping peaks but having no observed effect on the position. However the net doping after the 175°C anneal was approximately equal to half the number of ions implanted.

In order to obtain a satisfactory explanation of the annealing behaviour of the doping centres it would be necessary to understand the nature of the individual damage centres. This cannot be determined from the present measurements. However it is apparent that at least two types of doping centres are produced, one being a donor and the other an acceptor. Thus the net doping density in any region is the difference between the two concentrations. Hence the net doping effect is much smaller than the number of centres produced. It is also evident that whilst the number of acceptors always exceeded the number of donors the ratio of the two was not constant. This ratio may have been influenced by the degree of ion channelling during implantation. When the wafers were annealed the net doping concentration increased. It is unlikely that the total number of centres increased. Therefore it must be assumed that at about $400°C$ there was a more rapid annealing of donor type defects than acceptor defects. Finally all defects centres were annealed out at about $700°C$. The annealing behaviour of the damage produced by channelled ions is different from that produced by non-channelled ions. It anneals out at about $500°-600°C$. The annealing temperatures are consistent with the observations made by Pickar and Dalton (5) and Davies and Roosild (6) in the recovery of the minority carrier lifetime of carbon and helium ions implanted into silicon.

The importance of these effects in device fabrication is demonstrated by considering the possibility of a similar effect being produced by boron implantation. Boron has similar mass to carbon and probably produces a similar damage profile after annealing an implanted wafer to temperatures below $600°C$ the total doping concentration might be 100% higher than expected. This would not manifest itself in resistance measurements but would certainly show up when the damaged region was in the depletion region of a p-n junction or in the channel of an MOS transistor. These results show that for MOS applications such as threshold voltage control and p well implantation it is necessary to anneal at temperatures above $700°C$ if the effects of the damage produced by the low dose implantation are to be removed. The use of ion implantation for threshold voltage control combined with a low annealing temperature ($< 700°C$) is not a practical approach as the effect is very strongly dependent on the degree of damage.

CONCLUSIONS

It has been shown that the damage introduced by carbon implantation results in p-type activity. On annealing to $400°C$ the amount of activity increases but eventually disappears altogether on annealing to $700°C$. The position of the p-type activity occurs close to the region where maximum damage has occurred during implantation.

ACKNOWLEDGEMENTS

The authors wish to acknowledge the help of Mr. E. Sparrow and Mr. G. A. Gard with the implantations, Dr. M. D. Mathews for computer calculations of damage profiles, and Mr. E. M. Wittam for fabricating the mesa diodes.

REFERENCES

1. B. J. Smith, Capacitance-voltage measurements for determining impurity profiles in semiconductors. AERE Harwell R-7037 (1972).

2. J. Hilbrand and R. D. Gold, R.C.A. Review 21, 245 (1960).

3. I. Manning and G. P. Mueller, Comp. Phys. Comm. 7, 85 (1974).

4. L. C. Northcliffe and R. F. Schilling, Nuclear Data Tables 7, 233 (1970).

5. K. Pickar and J. V. Dalton, Rad. Effects 6, 89 (1970).

6. D. E. Davies and S. A. Roosild, App. Phys. Letters 17, 107 (1970).

NEGATIVE AND ANISOTROPIC MAGNETORESISTANCE IN PHOSPHORUS IMPLANTED SILICON

T. Itoh, M. Higashiura, and H. Sato*

School of Science and Engineering, Waseda University

Shinjuku-ku, Tokyo, Japan

ABSTRACT

The galvanomagnetic effects in phosphorus implanted n-channel MOSFET's and p-type silicon substrates were investigated. The transport phenomena in ion implanted MOSFET show the same tendency as unimplanted MOSFET to be related to the two-dimensional conduction except for increase in Hall mobility. The angular dependences of the magnetoresistance in phosphorus implanted p-type silicon substrates reflect the band structure of silicon at high annealing temperatures and for the light implantations. For heavy phosphorus implantations, the negative magnetoresistance is observed at 77 K and the angular dependences of the magnetoresistance are similar to those of n-channel MOSFET. This anomalous anisotropy of the magnetoresistance is found to be due to the spatial nonuniform distribution of the relaxation time.

INTRODUCTION

Ion implantation has been performed for the control of the threshold voltage in MOSFET in recent years. In connection with this, a number of transport properties, mainly carrier mobilities, in implanted MOSFET's have been studied extensively. Various scattering mechanisms were proposed on the basis of two-dimensional electron system (1). Measurements of the magnetoresistance have been used ascertain the band structure and scattering mechanism in bulk semiconductors. These measurements gave considerable informations about two-dimensional conduction in MOSFET (2).

In this paper, we experimentally study the magnetoresistances in phosphorus implanted n-channel MOSFET's and p-type silicon substrates, and theoretically calculate the magnetoresistance taking into account the spatial nonuniform distribution of the relaxation time.

EXPERIMENTAL

MOS Hall transistor was fabricated on p-type silicon. Crystal surface of (111) was used, with the current flow in the $\langle 11\bar{2}\rangle$ direction. Phosphorus ions were implanted with 50 keV through the thermally grown oxide with thickness of 500 Å. After implantation, oxide thickness of about 500 Å was chemically deposited at 900 °C to prevent dielectric breakdown by the gate field. The surface state density in unimplanted samples was about $2 \times 10^{11}/cm^2$.

Phosphorus ions were also implanted with 50 keV into p-type silicon substrates with (111) and (100) surface. Implanted samples were subsequently annealed at 1100 °C for 15 h or at 800 °C for 1 h.

RESULTS AND DISCUSSION

Figure 1 shows the temperature dependences of the mobility for unimplanted and phosphorus implanted n-channel MOSFET with 1.2×10^{12} ions/cm^2 as a function of gate voltages. Hall mobility in unimplanted MOSFET with high density of surface states is lower than that of devices by a standard fabrication process. Hall mobility in implanted MOSFET increases with implantation doses for light implantations. As a result of studying the effect of temperature on the mobility, it is seen that Coulomb scattering in n-channel MOSFET is reduced by phosphorus implantation. Introducing ionized donors in n-channel surface layer, the distribution of electrons has a maximum at a distance from the oxide-silicon interface where Coulomb scattering centers exist in excess (3). The experimental temperature dependence of the mobility at high temperatures shows the same dependence as unimplanted samples where phonon scattering in a two-dimensional gas is dominant scattering mechanism.

Angular dependence of the magnetoresistance with 10 kG in phosphorus implanted MOSFET at 77 K is shown in Fig. 2. The magnetoresistance in implanted MOSFET increases with implantation dose because of increase in carrier mobility. The negative magnetoresistance is observed in the phosphorus implanted MOSFET, similarly to unimplanted samples, with

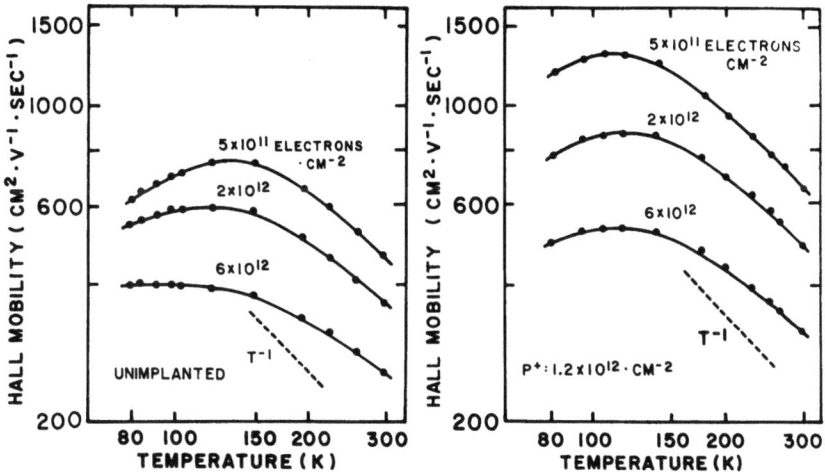

Fig. 1. Temperature dependences of the mobility in implanted MOSFET with 1.2×10^{12} ions/cm^2 and unimplanted one.

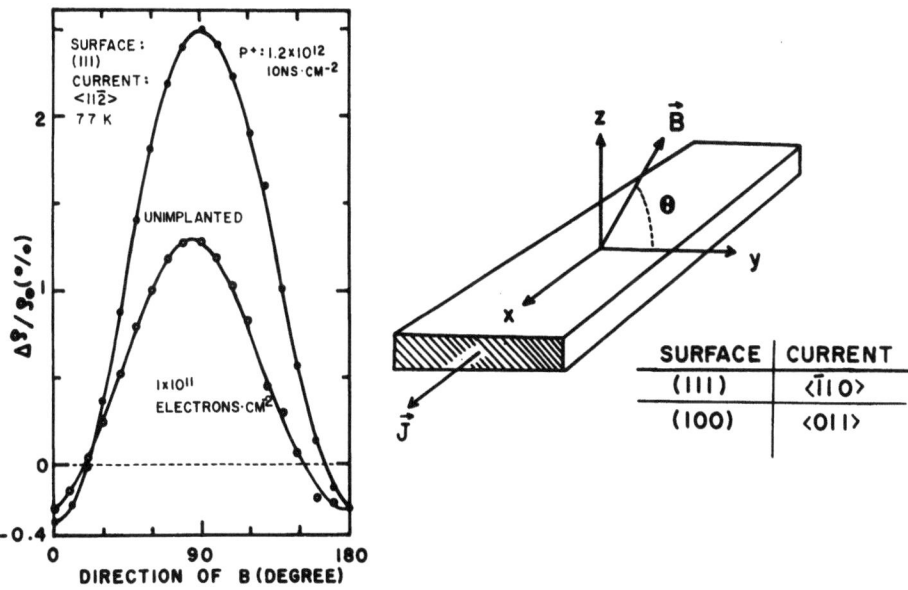

Fig. 2. (left) Angular dependence of the magnetoresistance in phosphorus implanted MOSFET at 77 K.

Fig. 3. (right) The direction of the magnetic field and the current flow in the p-type silicon.

magnetic field parallel to the surface at 77 K. The negative magnetoresistance in unimplanted MOSFET is related to the two-dimensional conduction of quantized electrons in the inversion layer along the semiconductor surface and depends only on the normal component of the magnetic field. However, considering potential profile near the silicon surface in implanted MOSFET, it does not seem that the observed effects are connected with two-dimensional gas.

The magnetoresistance in phosphorus implanted silicon was also studied. Ion implantation was performed at 50 keV with doses of 1×10^{13} to $1 \times 10^{15}/cm^2$. The influences of the annealing temperatures and implantation doses on the galvanomagnetic effects at 77 K are investigated. In contrast with MOSFET, the magnetoresistance decreases with implantation dose. Annealing at 1100 °C, the magnetoresistance decreases rapidly with implantation dose. This is due to the reduction of electron mobility by ionized impurity scattering. Even for the light implantations, dominant scattering mechanism is ionized impurity scattering in the samples annealed at 800 °C.

Angular dependences of the magnetoresistance with 10 kG in phosphorus implanted silicon are investigated at 77 K. The direction of the magnetic field and the current flow are shown in Fig. 3. The magnetic field is in a y-z plane and the angle of the magnetic field is measured from the y-axis. The magnetoresistance reflects the band structure in a bulk silicon and has a maximum in ⟨001⟩ direction of the magnetic field for the (111) crystal surface as indicated in Fig. 4. At room temperature phosphorus implanted silicon with 1×10^{13} ions/cm^2, annealed at 1100 °C for 15 h, has the same angular dependence as bulk silicon. At 77 K, however, this sample has a maximum at about 60 deg. and implanted silicon with 1×10^{15} ions/cm^2 at about 90 deg..

Figure 5 shows that annealing at 800 °C for 1 h, magnetoresistance curve shifts slightly in implanted silicon even with 1×10^{13} ions/cm^2 at room temperature and shifts quite largely at 77 K. The junction depth is 0.4 μ in the sample implanted with 1×10^{13} ions/cm^2 and annealed at 800 °C for 1 h. It is found from the junction depth that size effects as seen in thin films cannot be observed in implanted samples. For heavy implantations above 3×10^{14} ions/cm^2, negative magnetoresistance is observed with magnetic field parallel to the surface and angular dependence of the magnetoresistance is similar to that of n-channel MOSFET. This negative magnetoresistance is the same effect as observed in heavy doped n-type silicon and is different from that in MOSFET (4). All of the magnetoresistance, both positive and negative, have a square dependence of the magnetic field strength up to 10 kG.

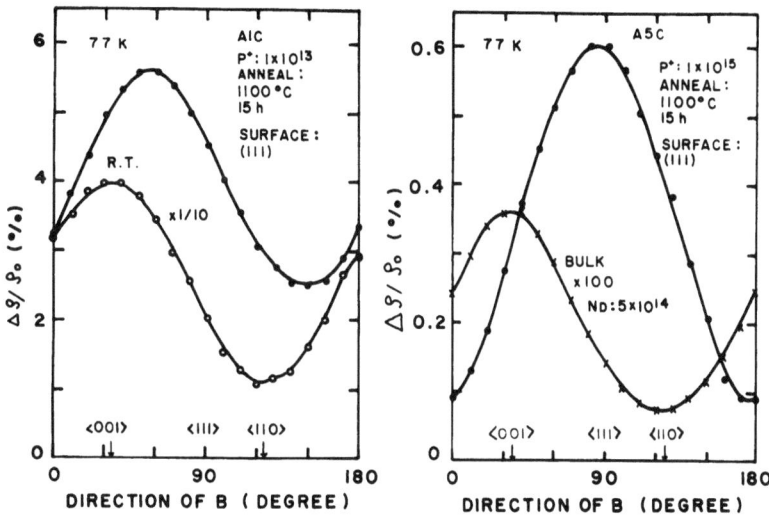

Fig. 4. Angular dependences of the magnetoresistance in phosphorus implanted silicon with (111) surface, annealed at 1100 °C for 15 h.

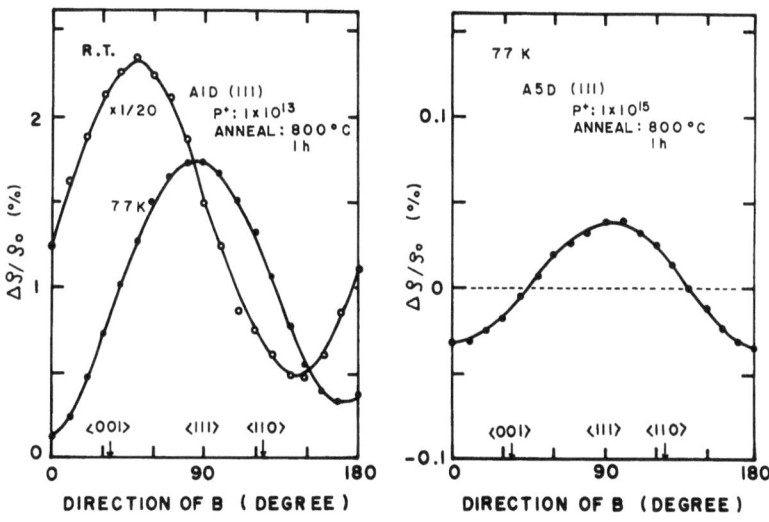

Fig. 5. Angular dependences of the magnetoresistance in phosphorus implanted silicon with (111) surface, annealed at 800 °C for 1 h.

Calculation of the magnetoresistance in implanted silicon with nonuniform distribution of the relaxation time was carried out. The calculated results are briefly mentioned. Both ionized impurity scattering and lattice scattering are considered. It is assumed that all donors with Gaussian distribution are ionized. Calculated effective magnetoresistance has a maximum at around 60 deg. in the case of silicon with total electrons of $1 \times 10^{14}/cm^2$. For heavy implantations, it is found that the position of a maximum is not drastically changed by the standard deviation in Gaussian distribution. This is in good agreement with the measured data.

It should be noticed that magnetoresistance measurements can be also used to analyze the epitaxial layer with nonuniform distribution of the impurities.

ACKNOWLEDGMENTS

The authors are indebted to Mr. H. Yoshida of Oki Research Lab. for preparing the samples.

REFERENCES

(1) F. Stern, Proc. 10th Int. Conf. Phys. Semiconductors, Cambridge, Mass., 1970, p. 451

(2) H. Sakaki and T. Sugano, Japan J. Appl. Phys. <u>10</u>, 1016 (1971)

(3) M. R. MacPherson, Solid-State Electron. <u>15</u>, 1319 (1972)

(4) W. Sasaki, J. Phys. Soc. Japan <u>12</u>, Suppl. 543 (1966)

* Present address : Central Research Lab., Mitsubishi Electric Corp., Amagasaki, Japan

ENHANCED RESIDUAL DISORDER IN SILICON FROM RECOIL IMPLANTATION OF OXYGEN AND NITROGEN BY ARSENIC IMPLANTS THROUGH DIELECTRIC LAYERS

T. W. Sigmon
Hewlett-Packard Laboratories, Palo Alto, CA. 94304

W. K. Chu, H. Müller* and J. W. Mayer
California Institute of Technology,† Pasadena, CA. 91109

ABSTRACT

Channeling effect measurements have been used to evaluate the enhanced residual disorder found in silicon after implantation of arsenic through thin oxide and nitride layers. The disorder is found to be a function of both the dielectric layer thickness and the arsenic dose. Measurements after anneal of the implants at 1000°C for 30 minutes indicate the amount of damage increases with As dose and dielectric layer thickness. Simulation of the "knock-on" implantation by oxygen or nitrogen implants allows the inference of the number of recoils as a function of As energy and layer thickness. For the case of a 480Å SiO_2 layer and 10^{16} As/cm² implanted at 200 keV approximately 4×10^{15}/cm² forward oxygen recoil atoms are produced. In general the residual damage for equivalent nitride layers was found to be approximately one-half that measured for oxide implants. Erosion of the films due to sputtering by the As implantation indicated that the number of atoms sputtered away is equivalent to 30-50Å of SiO_2 for a dose of 2×10^{16} As/cm².

INTRODUCTION

In As implantation through SiO_2 layers residual disorder has been found after annealing at 1000°C for 30 minutes.[1,2] In a recent paper we showed that this effect is due to oxygen recoils

*Permanent address: Lehrstuhl für Integrierte Schaltungen, Munich W. Germany

†Caltech work supported in part by O.N.R. (L. Cooper).

knocked into the silicon substrate by the As ions.[3] Since oxide masking is commonly used to define planar geometries for semiconductor devices and circuits degradation of performance by this residual damage at the periphery of the oxide cuts is expected. In this paper we have investigated the amount of residual disorder occuring as a function of oxide thickness and As ion dose by the use of backscattering techniques. We show that the residual disorder effect can be simulated by oxygen implants. Because nitrogen is a known dopant in silicon,[4] it was anticipated that damage created by knocked on nitrogen atoms would exhibit anneal behavior similar to that created by conventional dopants. Therefore, Si_3N_4 layers were also investigated as a possible alternative to the oxide layers for reducing the residual damage after implant.

EXPERIMENTAL TECHNIQUES

Silicon wafers of <111> orientation and resistivity of 2-4 ohm-cm were used throughout the experiments. Oxide layers of 200 to 800Å were grown by standard thermal techniques. Nitride layers were deposited pyrolytically and covered the thickness range of 200 to 700Å. Ellipsometry and step height measurements were used to determine the thickness of the films both before and after arsenic implantation. The arsenic implants were performed at 200 keV, with doses ranging between 2×10^{15} to 2×10^{16} atoms/cm^2. Arsenic implanted unoxidized wafers with oxygen and nitrogen pre- and post-implants were prepared for the simulation experiments.

Channeling analysis fo the arsenic implanted films and layers was carried out with 2.0 MeV He$^+$ ions on samples that were annealed at 1000°C for 30 minutes. The silicon portion of the aligned spectra was used as a measure of the residual disorder that remained in the samples after the anneal.

RESULTS AND DISCUSSION

A. Through Oxide Implants

In Figure 1 we have plotted aligned backscattering yields for annealed 200 keV 10^{16} As$^+$/cm^2 implants through SiO_2 films as a function of oxide thickness. The film thicknesses ranged from 196 to 776Å. For comparison the random and aligned yields from an As implanted unoxidized wafer are shown. Also illustrated in Figure 1 is the aligned and random yield of the arsenic distribution for the through oxide implants.

The data for the minimum yield clearly shows that the residual disorder increases with increasing oxide thickness. The disorder

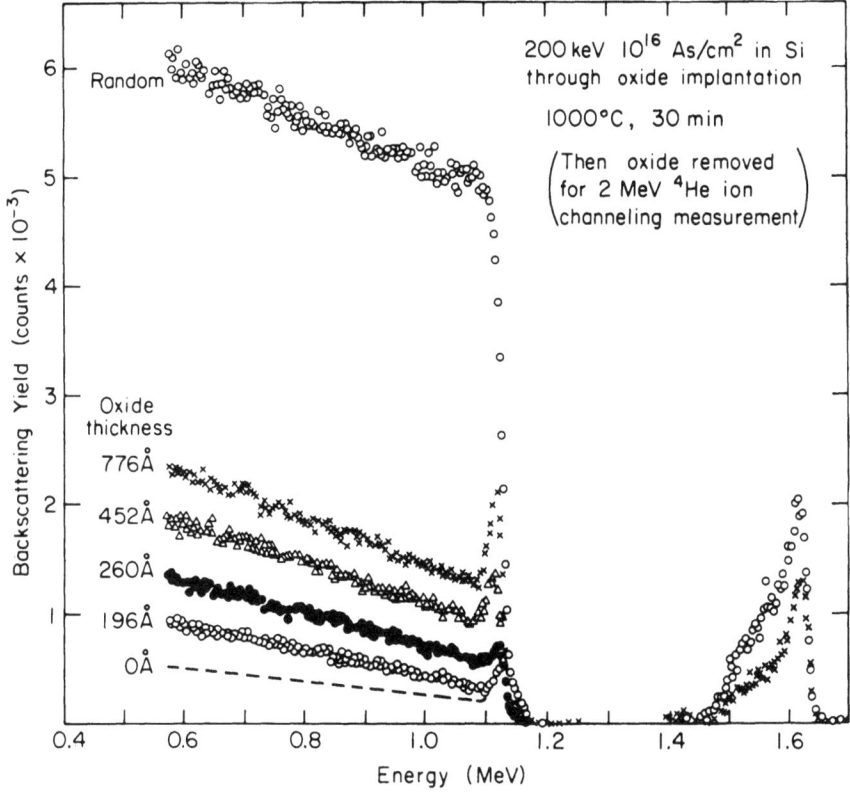

Figure 1

is observed close to the surface and extends about 900Å into the substrate for 200 keV implants through 776Å oxide layers. The number of scattering centers obtained for this sample by a linear background subtraction indicates that approximately 10 scattering centers are created per incident As ion.

With respect to the As distribution a surface peak is observed which exhibits a large fraction of As on nonsubstitutional sites. This part of the As distribution coincides with the damage distribution shown on the left. For depths beyond this peak the As has a higher substitutional fraction. Compared to the results on unoxidized samples the As distribution on through-oxide implants is skewed toward the surface. This indicates that the residual disorder caused by the oxygen affects the diffusion properties of the As.

The variation of residual disorder with As dose for an oxide thickness of 776Å is shown in Figure 2. From this data we see

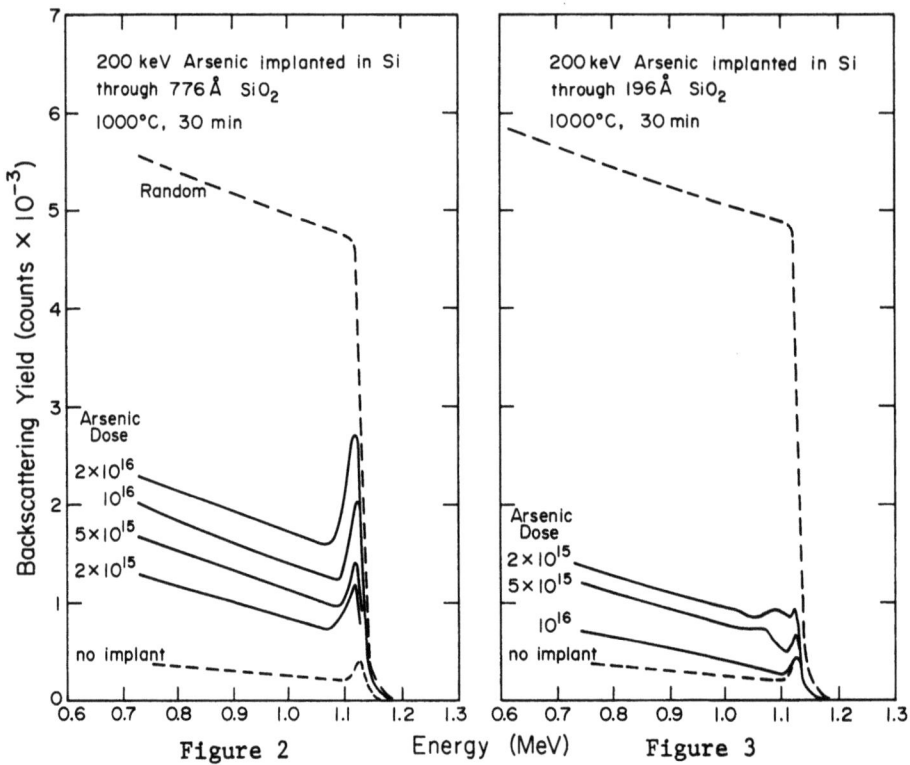

Figure 2

Figure 3

that the number of scattering centers increases with increasing As dose. Utilizing the variation in scattering centers from both Figures 1 and 2 allows determination of the number of scattering centers created per As atom per Å of oxide thickness. For the data shown this number is 1.3×10^{14} cm^{-2}/Å.

An intermediate case is shown in Figure 3 where the oxide thickness was 196Å and the observed damage is shown as a function of As dose. It can be clearly seen that two competitive effects are involved. As discussed in [5] doses of As at 200 keV in the range of 2×10^{14} to 2×10^{15} implanted into bare silicon do not anneal completely even at temperatures of 950 to 1050°C. Therefore for the thin oxide layers (such as shown in Figure 3) at doses of 0.2 to 2×10^{15}/cm^2 the disorder is dominated by incomplete recrystallization of the lattice. At higher doses the influence of the knock-on effect is beginning to increase and prevents proper annealing. Since the data indicates the residual damage is primarily created by knock-on oxygen for As doses greater than 5×10^{15} we conclude that the disorder picture presented in Figures 1 and 2 is due to oxygen recoils.

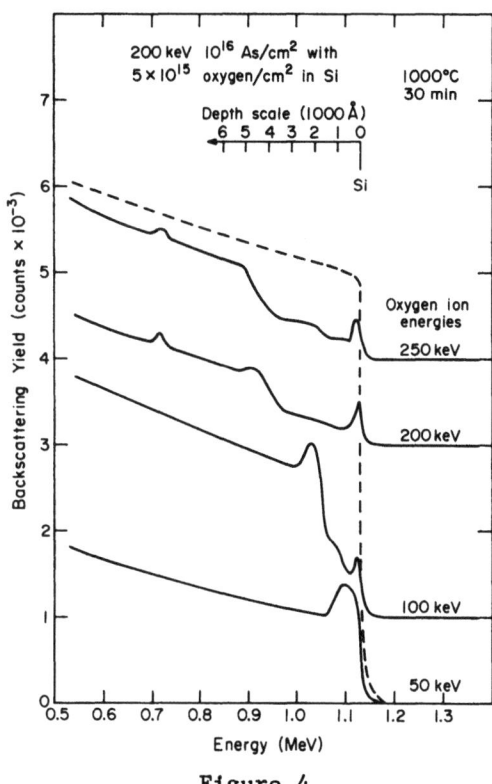

Figure 4

B. Simulation of Through Oxide Implantation

In order to determine that the location of the residual damage was associated with the implanted oxygen distribution implantation of oxygen at various energies into Si wafers that had been pre-implanted with 10^{16} As/cm^2 was performed. In Figure 4 we have plotted backscattering data for oxygen implanted at energies of 50 to 250 keV. All wafers were annealed for 1000°C for 30 minutes after the oxygen implantation. The LSS range for each oxygen implant is shown on the figure. It is seen that the maximum dechanneling occurs at depths comparable to the range of the oxygen. The nonsubstitutional fraction of the As also correlates with the oxygen range as indicated by the arrows.

Erosion of the oxide layers by the arsenic implantation has been measured. For a 2×10^{16} As/cm^2 implantation approximately 40Å (± 10Å) of oxide is sputtered away. Investigation of this effect as a function of oxide thickness indicates that there is little difference down to oxide layers of 100 to 200Å thickness.

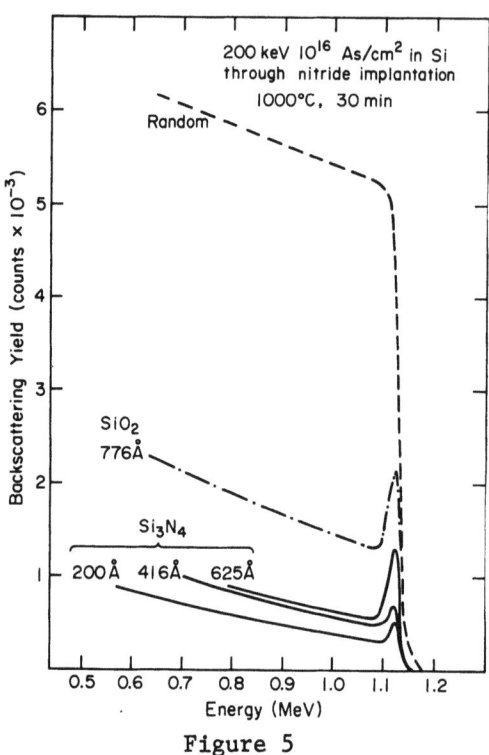

Figure 5

C. Through Nitride Implantation

The aligned spectra in Figure 5 illustrates the residual disorder for 200 keV 10^{16} As/cm^2 through various thicknesses of silicon nitride layers. The dot-dash line is the spectra for 10^{16} As through a 776Å oxide layer which contains approximately 10% more oxygen atoms/cm^2 than the 625Å nitride film contains nitrogen. Comparison of the spectra for the oxide and nitride layers indicates that twice as much disorder is produced in through oxide implants as in through nitride implants for equivalent layers. Aside from the factor of two quantitative difference, the through nitride implants exhibit the same behavior as through-oxide implants. The amount of disorder increases with nitride thickness and As dose. Also we find for thin nitride layers a strong influence from the residual disorder due to the As implantation for doses in the range of 0.2 to 2×10^{15} As/cm^2 (similar to that shown in Figure 3).

Nitrogen implants into samples previously implanted with As also produce disorder at depths comparable to the nitrogen range. Again the amount of disorder is about one-half that for an equivalent oxygen implant.

CONCLUSIONS

In summary we have measured the dependence of residual damage by As ion implantation through thin layers of SiO_2 and Si_3N_4 as a function of ion dose and layer thickness. The residual damage after anneal at 1000°C for 30 minutes increases with both dose and film thickness and is approximately twice as much for oxide films compared to nitride films. We attribute the residual damage to recoil implantation of oxygen or nitrogen atoms caused by penetration of As through the film. For the case of oxide films the number of scattering centers introduced by the As is 1.3×10^{14} cm^{-2} per Å of oxide. Transmission-electron microscopy studies are underway to clarify the nature of the defect centers.

REFERENCES

1. T. R. Cass and V. G. K. Reddi, Appl. Phys. Lett. 23, 268 (1973).

2. E. H. Bogardus and M. R. Poponiak, Appl. Phys. Lett. 23, 553 (1973).

3. W. K. Chu, H. Müller, J. W. Mayer and T. W. Sigmon, Appl. Phys. Lett. (to be published).

4. P. P. Pronko, J. B. Mitchell and J. Schewchun, Proceedings of the Second International Conference on Ion Implantation in Semiconductors, I. Ruge and J. Graul (Eds.), Springer-Verlag, Berlin.

5. W. K. Chu, H. Müller, J. W. Mayer and T. W. Sigmon. Paper III-5 This conference.

6. R. A. Moline, G. W. Reutlinger and J. C. North, Proceedings of the Fifth International Conference on Atomic Collisions in Solids (Gatlinburg, Tenn., No. 1973), Plenum Press, New York (to be published).

AN ANALYSIS OF ARSENIC ION IMPLANTATION FOR USE IN

SILICON BIPOLAR DEVICES

J. Chisholm[1], J. Stephen[2], J. Turner[2]
P. Dobson[3], R. Francis[3] and E. Williams[4]
(1) Ferranti Ltd., Wythenshaw, England
(2) A.E.R.E., Harwell, Oxon, England
(3) University of Birmingham, England
(4) Royal Radar Establishment, England

ABSTRACT

A study was made of arsenic ion-implantation in silicon for use in bipolar integrated circuits. Ion beam energies in the range 10 to 320 keV were used to implant 2×10^{15} to 2×10^{16} arsenic ions/cm^2 into the silicon. Annealing was carried out at temperatures in the range of 900 to 1250°C in the presence of oxygen. Electron channelling with a scanning electron microscope showed that the combination of the oxygen and the higher temperature anneals was adequate to produce a high quality single crystal surface.

Back-scattering experiments with 2 MeV helium ions confirmed the good quality single crystal nature of the surface. A careful analysis of the back-scattering random and aligned energy spectra showed that a high percentage of the arsenic atoms were substitutional in the crystal lattice. This high percentage was confirmed by surface resistivity and diffusion profile measurements. The surface resistivity variation in the implanted layers was better than 3%.

INTRODUCTION

A standard gaseous arsenic diffusion process is used at present for fabricating the n$^+$ buried layers for collector-diffusion-isolation (C.D.I.) bipolar structures (Ref. 1). A study has been made of an ion implanted arsenic diffusion source in order to assess the possibility of obtaining a sheet resistance comparable with that by gaseous diffusion but with an improved control over the uniformity of the sheet resistance. The implanted arsenic source must not produce any auto-doping effects in the p^- epitaxial layer which is grown on the implanted substrate. It is essential that the crystalline property of the silicon be maintained to ensure a high degree of perfection in the p^- epitaxial layer as low collector leakage currents and high current gains are required of the CDI transistors. An implanted source removes the problems associated with arsenosilicate glass and the contamination of the rear of the wafer.

EXPERIMENTAL

The sheet resistance of n^+ layers formed in p^- <100> 10 ohm-cm silicon wafers by ion implanting arsenic under various conditions of ion energy, dose, annealing temperature and time was measured using a modified four-point probe technique (Ref. 2). The arsenic implantations were done using the Lintott-Harwell separator equipped with mechanical scanning of the wafers. Those wafers implanted at a very high irradiance showed coloured striations due to ion beam heating. The wafers were annealed by inserting them in a conventional diffusion furnace operating at the indicated temperatures and passing dry oxygen for the first hour followed by dry nitrogen for the remaining time.

RESULTS

Table 1 shows the variation of sheet resistance (ohms/square) with ion dose and annealing temperature for an ion energy of 40 keV. Table 2 shows the variation of sheet resistance for a number of ion doses with annealing time and temperature. The sheet resistance exhibits the expected decrease with increasing ion dose, annealing temperature and time. The results agree with those of Drum (Ref. 3). The variation of sheet resistance with ion energy for a dose of 5×10^{15} ions/cm^2 is shown in Table 3. These results show that there is no need to use energies over 100 keV and satisfactory results were obtained between 40 and 80 keV. There is, in fact, a positive advantage in using a lower energy as this decreases the irradiance (power input/unit area) on the wafer and also makes masking easier.

Figure 1 compares a typical arsenic diffused profile with a typical ion implanted profile. The surface concentration for the implanted layer is lower (1.5×10^{19} As atoms/cm^3) than that for the diffused layer (2.3×10^{19} atoms/cm^3) while the

TABLE 1

Variation of sheet resistance with ion dose and annealing temperature

Ion Energy 40 keV	(Anneal Time 3 Hrs) Sheet Resistance ohms/square				
Dose	Annealing Temp (°C)				
	900	1000	1100	1200	1250
2×10^{15}	84	51	42	32	29
3×10^{15}	70	35	39		19
5×10^{15}	58	24	18	16	14
1×10^{16}	47	14	10.5		
2×10^{16}	32	9.4	5.6		

sheet resistance is lower for the implanted layer, 19 ohms/square compared with 25 ohms/square. The arsenic distribution profiles show that over 90% of the implanted arsenic becomes electrically active after diffusion over 1100°C and that little arsenic is lost by evaporation from the surface or by incorporation into the thin oxide layer grown during the first hour of the diffusion. The uniformity of the arsenic layers annealed at 1200° or 1250°C show uniformities of sheet resistance between 1% and 3%. There was no evidence in the sheet resistance measurement of the wafers implanted at a high irradiance that the ion beam heating effect had any influence on the uniformity of electrical activity.

TABLE 2

Variation of sheet resistance with ion dose and annealing time

Ion Energy 40 keV	Sheet Resistance ohms/square						
Anneal Temp. (°C)	1200	1200	1200	1200	1200	1200	1250
Anneal Time (Hrs)	2	4	7	8	16	21	3
Dose							
2×10^{15}	36	31		28	26		29
3×10^{15}						20	19
5×10^{15}	16	16	16		14	13	14
2×10^{16}					4.3		

TABLE 3

Variation of sheet resistance with ion energy

Ion dose 5×10^{15} ions/cm^2 Annealed 1200°C for 7 hours	
Energy keV	Sheet Resistance ohms/square
10	15.5
20	14.9
40	15.3
80	14.8
160	14.5
320	17.9*

*extrapolated from a dose of 3.85×10^{15} ions/cm^2

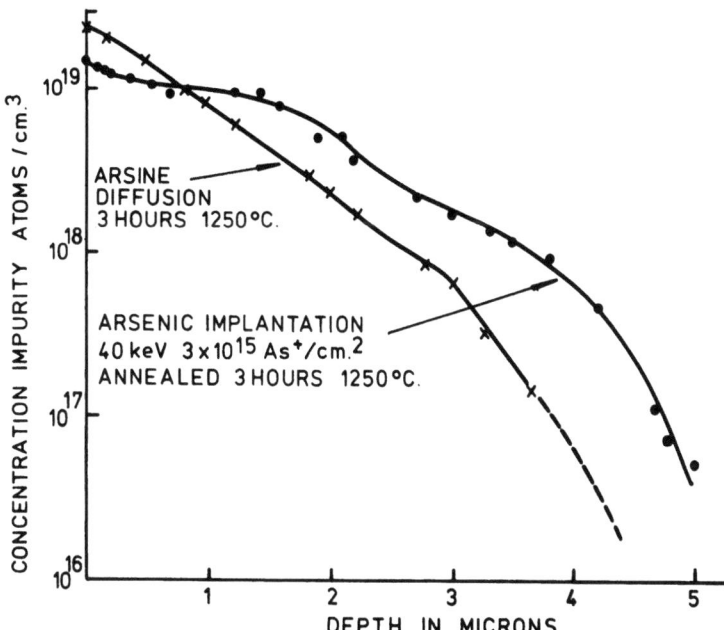

Figure 1. Typical Arsenic Diffused and Implanted Profiles.

Proton backscattering measurements have shown that a good quality single crystal was obtained when the diffusion temperatures exceeded 1100°C. The percentage of substitutional arsenic atoms also increased with annealing temperature which supports the results from the electrical profiles. Electron channelling on unannealed wafers showed that the surface was sufficiently damaged to prevent channelling and that the channelling recovered on annealing particularly at 1200°C.

Successful p type epitaxial layers were grown on the ion implanted and annealed wafers. To measure the autodoping effect, wafers with isolated implanted areas were produced. It was found that if the annealing temperature was less than 1100°C there was a significant leakage path between one n^+ region and the next due to the formation of a thin n type layer at the epitaxial interface. To obtain a very low leakage it was necessary to anneal at 1200°C for 8 hours or 1250°C for 3 hours.

From these results it has been established that the optimum conditions for the CDI buried n^+ layers are:

> Ion energy 40 keV
> Ion dose 2×10^{15} ions/cm^2
> Annealing temperature 1200°C
> Annealing time 1 hour in oxygen
>   6 hours in dry nitrogen.

The sheet resistance of the n^+ layer after this treatment is about 30 ohms/square.

CONCLUSIONS

The properties of annealed arsenic implanted layers in silicon have been studied and satisfactory epitaxial layers have been grown on the implanted surfaces. The sheet resistance of the layers showed the expected dependence on ion energy, dose and annealing temperature. The uniformity of the layers was between 1% and 3%. The optimum conditions of the CDI buried n^+ layers has been established.

ACKNOWLEDGEMENTS

The authors wish to thank Mr. G. Gard and Mr. D.C. Marshall of Harwell, for the implantation and the electron channelling observations respectively. Part of this work was carried out under a CVD Ministry of Defence Procurement Executive contract.

REFERENCES

(1) "C.D.I. and Competitive Techniques" by G. Bruchez, Microelectronics, 5, No. 4, p.45, 1974.

(2) This conference and "Measurement of doping uniformity in semi-conductor wafers" by B.J. Smith, J. Stephen and G.W. Hinder, A.E.R.E., Harwell Report R-7085, 1974.

(3) "Diffusion of ion implanted antimony and arsenic in silicon". C.M. Drum Abstract 84. Extended Abstracts Vol. 74-1 Electrochemical Society. May 1974.

ION-IMPLANTED PROFILES FOR HIGH-FREQUENCY (> 100 GHz)

IMPATT DIODES

D. H. Lee and R. S. Ying

Hughes Research Laboratories

Malibu, California 90265

ABSTRACT

Ion-implanted p(boron)- and n(arsenic)-type dopant profiles have been used in the fabrication of D-band double-drift-region (DDR) and single-drift-region (SDR) silicon IMPATT diodes for precise control of the impurity levels and dimensions of the space-charge and contact regions. Factors which influence the final diode designs are discussed. Continuous-wave output powers and conversion efficiencies obtained at 140 GHz are 80 mW and 2% for a DDR device and 140 mW with 2.8% for a complementary SDR didode.

INTRODUCTION

Silicon junction avalanche diodes, designed for operation above 100 GHz, require space-charge depletion widths several tenths of a micrometer. [1] In the fabrication of these structures it is important to precisely control the doping profiles over these dimensions for optimum microwave performance. Implantation is an attractive technique, because the charge and depth control of implanted profiles can be used to accurately adjust the doping levels which define the drift and contact regions of these devices. Examples of implanted boron and arsenic profiles are discussed for application to 140 GHz double-drift-region (DDR) and single-drift-region (SDR) IMPATT diodes.

RESULTS AND DISCUSSION

The predicted higher output power and conversion efficiency levels of a DDR diode over that of a SDR device have been confirmed experimentally at frequencies ≲100 GHz. [2-4] We will therefore consider the fabrication of the double-drift structure first.

One approach in the construction of DDR diodes is to overcompensate a uniformly doped n-type epitaxial layer with a multiple dose-energy implantation of boron ions to form identical back-to-back drift regions: one for electrons and one for holes. [2] After implantation the sample is annealed to remove the lattice disorder and bring the implanted profile to full electrical activity. A shallow boron diffusion is used for the p^+ contact after which the wafer is thinned, metallized, and mesa etched into individual diodes.

Design considerations show that the full depletion width of a 140 GHz DDR IMPATT diode is approximately 0.4 μm at a junction temperature of 200°C. To obtain the p-type drift region, three boron profiles were implanted into a 4×10^{17} cm^{-3}, 0.5 μm n-type epitaxial layer according to the dose-energy schedule given in Fig. 1. For all implanted profiles, the alignment between the major crystallographic axis of the substrate and the direction of the ion beam was adjusted to minimize channeling effects. The solid line represents the addition of Gaussians with projected ion range R_p and ion straggle ΔR_p values calculated from Lindhard, Scharff and Schiøtt (LSS) theory. [5] An estimate of the diffusional broadening is given by the dashed line for a 30 min heat treatment at 900°C and a boron diffusivity of $D = 3 \times 10^{-15}$ cm^2/sec.

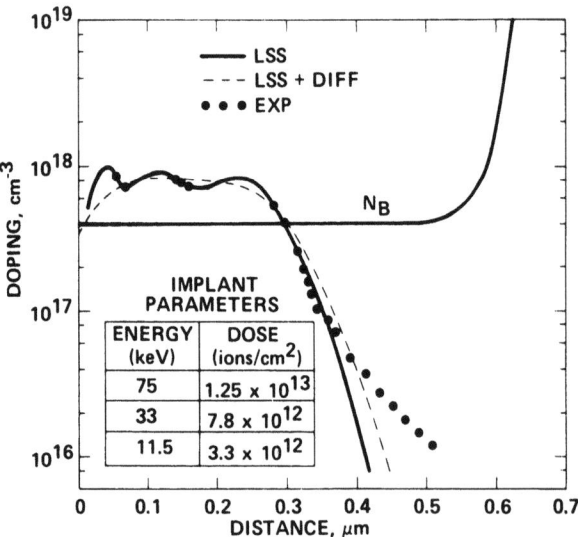

Fig. 1. Boron implanted profiles for overcompensation.

A measure of the final doping distribution (dots in Fig. 1) shows reasonable agreement with the predicted profile although the results are not sufficiently accurate to distinguish details of diffusional broadening. Moderate departures from the first order Gaussian approximation are evident in the leading edge of the highest energy implant at low doping; however, for this application these deviations have negligible effect on the location x_j of the p-n junction.

A composite profile of the completed device is given in Fig. 2. Clearly the junction abruptness and the desired uniformity of the p- and n-type profiles have not been achieved. Note that the doping concentration is plotted on a linear scale to further demonstrate the nearly linear impurity gradient behavior of the resultant p-n junction. This characteristic is a direct consequence of the ion straggling of the highest energy implant and represents a fundamental limitation on the formation of abrupt junctions by the method of overcompensation with implanted doping profiles. Another feature of the junction is the symmetry S which is defined as the ratio of the acceptor to donor impurities measured equidistant about x_j, as illustrated in the inset of Fig. 3. For linear graded and ideal abrupt junctions with p = n, S = 1. Overcompensation of a constant background N_B with a Gaussian distribution that has a peak concentration $2N_B$ of the opposite carrier type leads to a symmetry factor which varies with $x/\Delta R_p$ (Fig. 3). Maximum departure from S = 1 occurs when $x \simeq 0.85\ \Delta R_p$. Thus, to minimize possible dispersion effects in carrier transit times, the dimensions of the p- and n-type doping profiles should be much greater than this value. This condition, however, becomes increasingly more difficult to satisfy as the design center frequency increases above 100 GHz.

Because of the graded p-n junction, the width of the avalanche zone will extend over a greater portion of the depletion layer under reverse bias breakdown. This degrades the microwave performance, notably the conversion efficiency, compared to the ideal abrupt case. Nevertheless, output powers as high as 80 mW with 2% efficiencies have been obtained at 135 to 140 GHz with the implanted DDR IMPATT diodes profiled in Fig. 2.

Loss of microwave performance in the DDR diodes from unsymmetrical and graded junction effects suggests that comparable output characteristics may be obtained from a SDR device if more abrupt profiles are realized. Shallow epitaxial layer growth and implantation have been combined to form a complementary n^+-p-p^+ SDR IMPATT diode. The fabrication steps are schematically outlined in Fig. 4. Here, the approach is to overdope or tailor the background impurity level of a thin, high resistivity p^- epitaxial layer with different energy boron implantations to establish the p-type drift region. A single high-dose arsenic implant is chosen for the n^+ contact

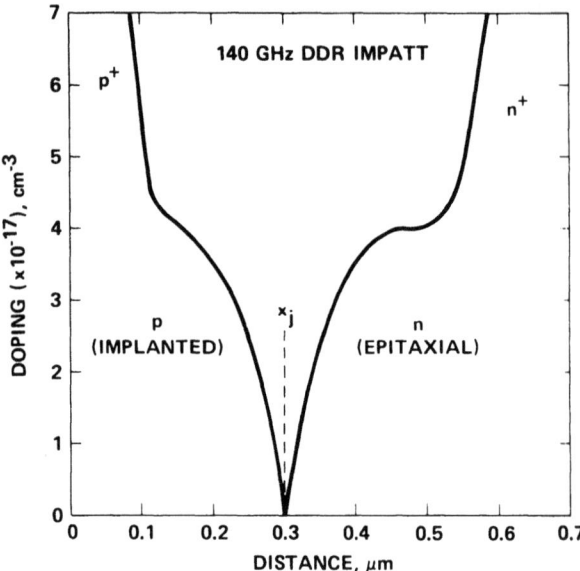

Fig. 2. Doping profile for a 140 GHz DDR IMPATT diode.

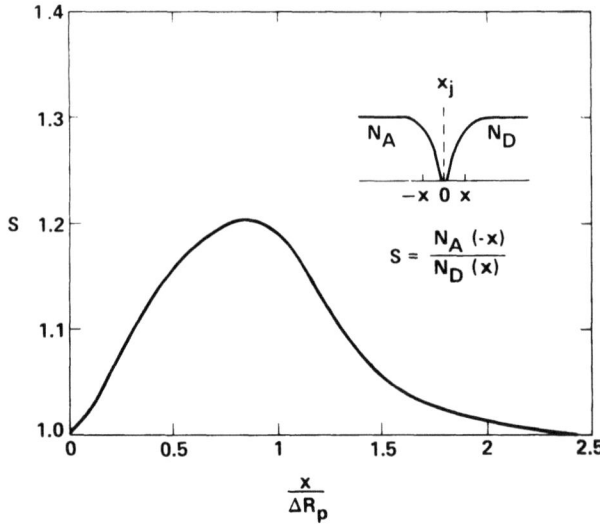

Fig. 3. Symmetry factor for a p-n junction formed by the method of overcompensation with Gaussian distributions.

HIGH-FREQUENCY IMPATT DIODES

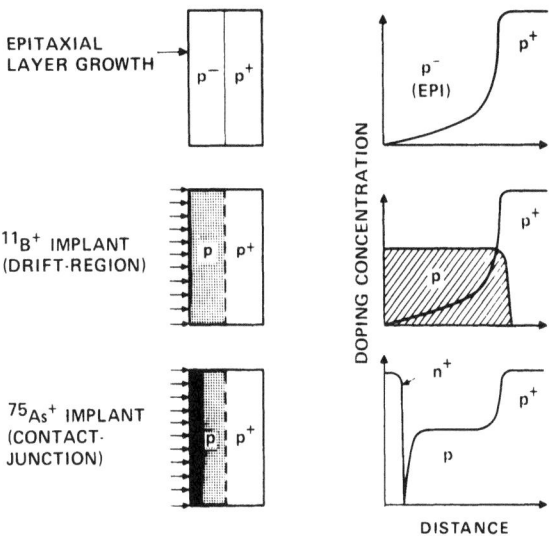

Fig. 4. Schematic representation of boron and arsenic implanted profiles for the fabrication of a n^+-p-p^+ IMPATT diode.

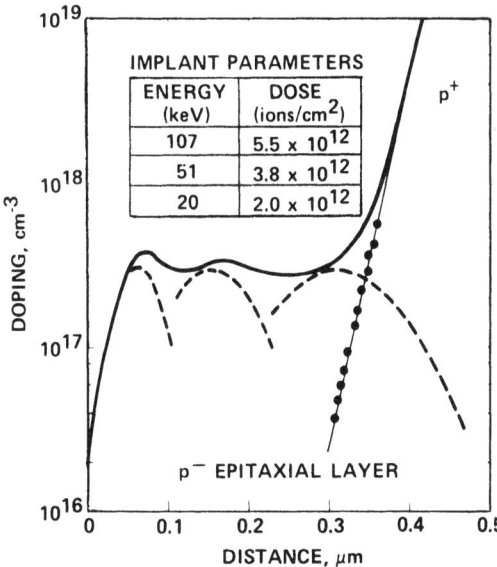

Fig. 5. High resistivity p-type epitaxial layer overdoped with implanted boron profiles.

because after the anneal it has moderate tail behavior at low doping and the concentration dependant diffusivity produces a very abrupt profile. [6]

Details of the boron implants for overdoping a 0.35 μm p^- epitaxial layer (solid dots) to a concentration 3×10^{17} cm^{-3} are given in Fig. 5. Arsenic ions were implanted at 70 keV to a dose of 1.9×10^{15} ions/cm^2 for the n^+ contact. Alpha-particle Rutherford backscattering measurements taken at 280 keV show that the implanted arsenic profile can be approximated by a Gaussian with R_p = 510 Å and ΔR_p = 190 Å. After a 900°C, 30 min anneal, there is some profile broadening and channeling effect measurements indicate that the amorphous layer formed by the arsenic implant has annealed to the preimplantation crystalline condition. Hall-effect measurements combined with layer removal by anodic oxide growth confirm the diffusional broadening and show a flat carrier concentration $1 - 2 \times 10^{20}$ cm^{-3} at the peak of the profile similar to previously reported results. [6] Figure 6 shows a composite doping profile of the completed n^+-p-p^+ structure. There is a reasonably abrupt doping transition at the p-p^+ interface and the doping values immediate to the n^+-p junction also exhibit abrupt characteristics even through there is some evidence, as determined from Hall-effect data, of an exponential tail in the annealed arsenic profile (dashed line in Fig. 6).

Several of these diodes were packaged into a quartz standoff and tested in both a reduced height waveguide cavity and in a full height hat-type circuit. Representative output power and efficiency measurements for one of the devices with the highest performance are given in Fig. 7 as a function of input power. In this case, the optimum frequency of oscillation was around 135 to 142 GHz with continuous-wave output power levels as high as 140 mW and 2 to 3% conversion efficiencies. These results represent one of the highest (power) X (frequency)^2 products ever achieved in this frequency range.

CONCLUSION

In the fabrication of high-frequency (>100 GHz) DDR IMPATT diodes by the method of overcompensation with implantation, charge and depth control for a p-type drift region are readily obtained with a series of different energy boron implantations. However, ion straggling effects produce an unsymmetrical junction with a nearly linear gradient impurity distribution. These doping characteristics degrade the output power and conversion efficiency because the avalanche zone occupies a larger fraction of the total space-charge depletion width.

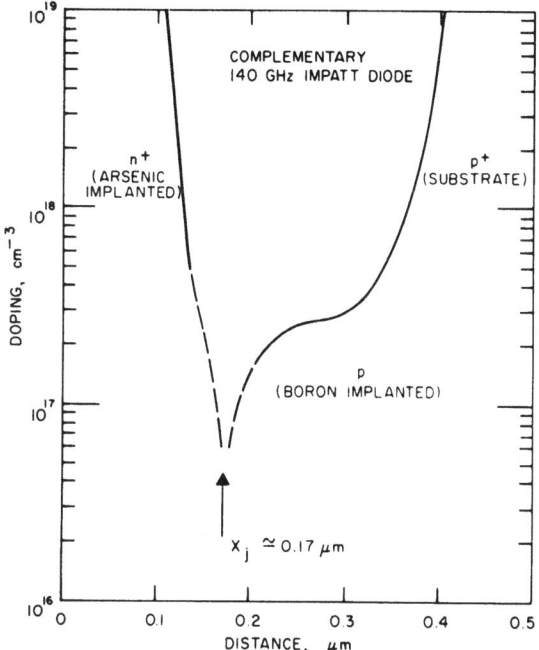

Fig. 6. Doping profile for a 140 GHz complementary SDR IMPATT diode.

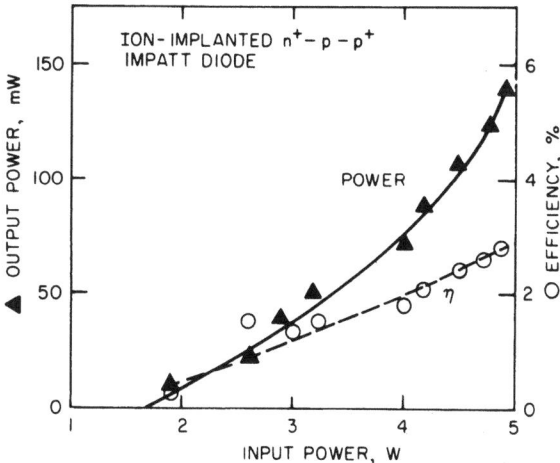

Fig. 7. Continuous-wave output power and conversion efficiency for the n^+-p-p^+ SDR IMPATT diode with the ion-implanted doping profile given in Fig. 6.

An alternative approach is to combine shallow epitaxial layer growth with boron and arsenic implantations to form complementary SDR IMPATT diodes. In this case, reasonably abrupt n^+-p junctions and controlled p-p^+ doping profiles allow n^+-p-p^+ devices to be constructed which have microwave results at 140 GHz that are comparable or better than those achieved with implanted p^+-p-n-n^+ diodes.

ACKNOWLEDGMENT

The authors wish to thank R. Hart (backscattering) and K. Weller (microwave measurements) for assistance, and M. Padilla and A. Willet for help with device fabrication.

REFERENCES

1. T. Misawa, Solid-State Electron. <u>15</u>, 457 (1972).

2. T.E. Seidel, R.E. Davis, and D.E. Iglesias, Proc. IEEE, <u>59</u>, 1222 (1971).

3. W.C. Niehaus, T.E. Seidel, and D.E. Iglesias, IEEE Trans. on Electron Devices, <u>ED-20</u>, 765 (1973).

4. D. H. Lee, R.S. Ying and D.M. Jamba, Proc. IEEE, <u>62</u>, 1025 (1974).

5. J. Lindhard, M. Scharff and H. Schiøtt, Kgl. Danske Vid. Selskab. Mat. Fys. Medd. <u>33</u>, 1 (1963).

6. F.N. Schwettmann, Appl. Phys. Lett., <u>22</u>, 570 (1973).

NOISE CHARACTERISTICS IN THE LOW FREQUENCY RANGE

OF ION-IMPLANTED-BASE-TRANSISTOR (NPN TYPE)

Tetsu KŌJI

Semiconductor Division, Nippon Electric Co., Ltd.

Shimonumabe, Kawasaki, Japan

ABSTRACT

As one of many examples associated with introduction of ion implantation technique to the production of transistors, there has been a formation of predeposition layer for the base, which has resulted in much superior uniformity for h_{FE} in a transistor wafer compared to that of the transistor wafer fabricated using conventional diffusion processes [1].

However, it has been well known that ion-implantation-induced defects can't be fully annealed out.

So, the low frequency noise performance which is extremely sensitive to defects was studied on transistors whose base region had been formed using the ion implantation deposition-diffusion process.

Measurements showed that the percentage of transistors having popcorn noise is dependent on the incident energy. X-ray examination of these transistor wafers showed that total number of defects increases with the incident energy.

From these experimental results, one can conclude that ion implantation-induced defects deteriorate the low frequency noise performance. Besides, it can be concluded that ion implantation-induced defects correspond to G-R (Generation-Recombination) centers, and G-R noise can be caused by the number of G-R centers less than that of G-R centers causing popcorn noise.

INTRODUCTION

Introduction of ion implantation technique to fabrication process for semiconductor devices has flourished remarkably in recent years, whose fruits have been reported.

As one among those, there has been a formation of a base region by the use of ion implantation deposition-diffusion rpocess, which has resulted in a superior uniformity of the surface concentration of the base region in a wafer compared to that by the conventional diffusion deposition-diffusion process. Accordingly, as for h_{FE}, transistors with the base region formed using the ion implantation deposition-diffusion process have had a superior uniformity in a wafer compared to that by the conventional process.

However, ion implantation technique as mentioned above has not only a good many advantages, but also some demerits. For example, as is well known, ion implantation-induced defects can't be fully annealed out. Accordingly, it is very doubtful if the ion-implanted-base-transistor as mentioned above can give the low frequency low noise characteristics, because of these characteristics sensitivity to the defects, especially.

In this paper, the low frequency noise characteristics of the ion-implanted-base transistor are reported in detail.

EXPERIMENT AND EXPERIMENTAL RESULT

1. FABRICATION PROCESS FOR ION-IMPLANTED-BASE-TRANSISTOR

Each sample was fabricated using the N type epitaxial wafer, whose base region was formed by the predeposition condition as shown in table 1. The emitter regions of all samples were formed using the conventional diffusion deposition-diffusion process.

The base regions of the ion-implanted-base-transistor were formed as the following.

First, $4 \times 10^{14}/cm^2$ $^{11}B^+$ ions were implanted at two incident energies of 50 keV & 150 keV. Subsequent annealing was carried out in a dry oxygen atmosphere at 900°C for 30 minutes. Secondly, by conventional diffusion process, the base-collector junction depth and the surface concentration of the base region for each sample were controlled equally.

Table 1. Predeposition condition for each sample. Sample A & B are ion-implanted sample, and sample C is diffused sample, respectively.

SAMPLE	DOSE (Si/B$^+$)	INCIDENT ENERGY
A	4×10^{14}/cm^2	50 keV
B	4×10^{14}/cm^2	150 keV
C	BCl3	

2. LOW FREQUENCY NOISE CHARACTERISTICS OF ION-IMPLANTED-BASE-TRANSISTOR

As an estimation of the low frequency noise characteristic of the ion-implanted-base-transistors, N.F. vs. f characteristics, N. L (Noise Level) characteristics and the generation rate of popcorn noise etc. were measured. In table 2, the distinctive features of the low frequency noise characteristics for the ion-implanted-base-transistors were inserted contrasting with that of the diffused-base-transistor by the conventional fabrication processes.

Table 2. Distinctive features of the low frequency noise characteristics for each sample. n value shows n in the following equation:
$$I_B = Is \exp\left(\frac{qV_{BE}}{nkT}\right)$$

SAMPLE	n VALUE	GENERATION RATE OF	
		POPCORN NOISE	G-R NOISE
A	1.73	50 %	100 %
B	1.89	100 %	100 %
C	1.39	1 %	1 %

As seen from table 2, the experimental results showed that n value showing indirectly the number of the G-R centers increases with the incident energy. Also, the generation rate of popcorn noise for the ion-implanted-base-transistors has the same tendency as n value.

In fig. 1, distribution of N. L for each sample is shown.

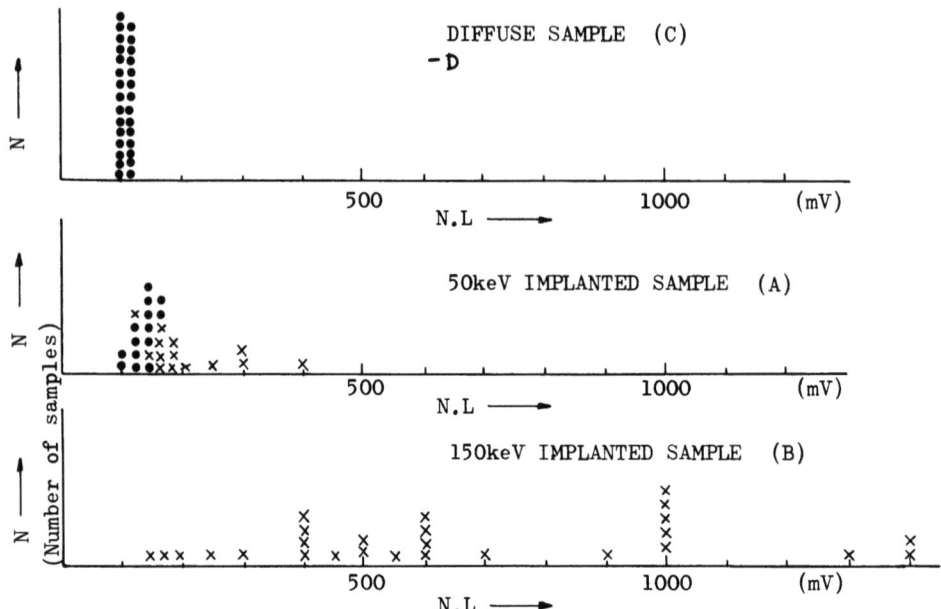

Fig.1. Distribution of N. L for each sample. One × mark shows one sample having popcorn noise, and one • mark does one sample having no popcorn noise (See the appendix for the explanation of N. L in the last page, if necessary.)

As shown in fig. 1, the dispersions of N. L for the ion implanted samples vary more widely than the diffused sample, and the dispersion increases with incident energy. These characteristics imply that only the ion implantation-induced defects don't make the noise sources, but many kind of contaminants (u, Fe, Cetc) precipitated in the defects do the noise sources. That is, the dispersions of N. L for the implanted samples might vary widely due to the dispersion of contaminant concentration in a wafer.

In fig. 2, N. F vs. f characteristics are shown.

As seen from fig. 2, the low frequency noise of the diffused-base-transistor usually consist of thermal noise, shot noise and 1/f noise, while those of the ion-implanted-

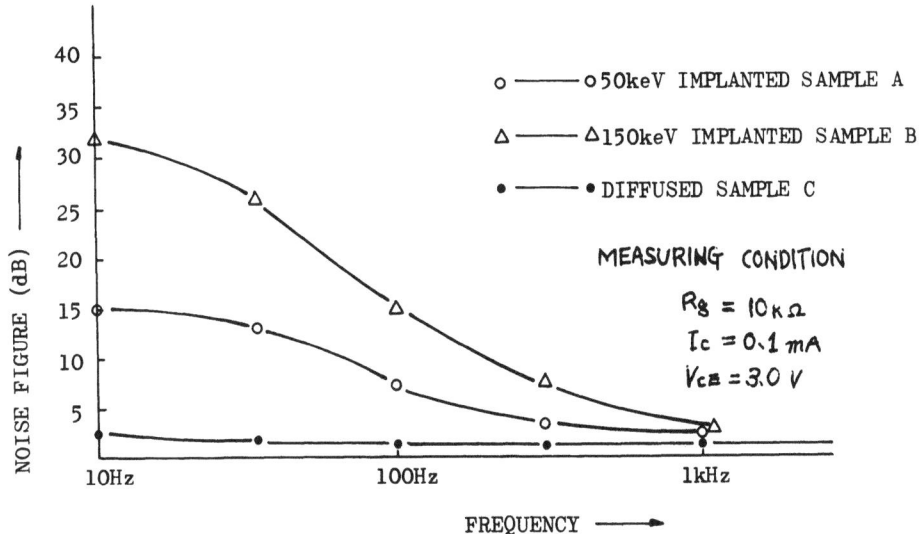

Fig.2. N.F vs f characteristics for each sample. The curves of the implanted samples have a shoulder at frequency of about 30 Hz.

base-transistors have G-R noise component besides the former components. As is well known, the low frequency noise of the diffused-base transistor seldom has G-R noise, which is caused by contaminants (Cu, Fe, C etc.) or diffusion-induced defects.

As for the generation mechanism of popcorn noise, many papers on popcorn noise have been reported [2], but its generation mechanism has not been clarified yet. It has been believed that it is caused by the heavy metals precipitated in defects.

The explanation of N.F vs. f characteristics is given by means of the following expression for noise factor [3]

$$F = 1 + \frac{1}{4kT\triangle F}\left(i_n^2 R_g + \frac{e_n^2}{R_g} + 2\gamma e_n i_n\right)$$

where k = Boltzmann's constant
       T = temperature in degrees Kelvin.
       $\triangle F$ = noise power bandwidth
       γ = correlation coefficient

The equivalent input noise current $\sqrt{\overline{in^2}}$ is proportional to the square root of the sum of shot noise, 1/f noise and G-R noise as the following equation:

$$\frac{\overline{in^2}}{\triangle f} \propto 2qI_B + \frac{Ko}{f^\alpha} Nts \cdot I_{BS}^n + \sum_{i=1}^{n} \frac{Ki\tau_i}{1+\omega^2 \tau_i^2} Ntdi I_{Bdi}^n$$

where
 I_B = base current
 q = charge on an electron
 α, n, Ko, Ki = arbitrary constant
 $\omega = 2\pi f$
 τi = relaxation time of minority carriers
 Nts = density of interface state
 Ntdi = density of G-R center at a certain energy level Ei of forbidden band gap in the bulk
 I_{BS}, I_{Bdi} = current passing through the interface state and G-R center in the bulk, respectively

In the equation shown above, the third term shows G-R noise component. As seen from fig. 2, N.F vs. f characteristics curves showing the low frequency noise for the implanted samples have a shoulder at a certain frequency (about 30 Hz). These phenomena show that the ion implantation-induced defects offer G-R centers which give a certain relaxation time of minority carriers.

As mentioned above, the ion-implanted-base-transistors have the inferior low frequency noise characteristics compared to the diffused-base-transistors by the conventional diffusion process.

As seen from table 3, X-ray examination of the predeposition layer for each sample showed that total number of defects increases with incident energy. These experimental results suppose the low frequency noise characteristics mentioned above.

Table 3. Measurement result of rocking curve for each predeposition layer. These measurements were done by means of the X-ray diffraction technique (a double crystal method) using a parrallel setting of (511)-(333) [5].

SAMPLE	PERCENT REFLECTION	θB
A	18.0 %	6.8"
B	6.9 %	16.6"
C	30.8 %	4.5"

Also these experimental results accord with the data by D.K. Brice etal [4].

CONCLUSION

Ion implantation-induced defects deteriorate remarkably the low frequency noise characteristics.

As the remarkable features, the low frequency noise deteriorated by the ion implantation-induced defects have G-R noise component having only a shoulder at a certain frequency and popcorn noise.

REFERENCES

[1] I. Sasaki, Seminar of Ion Implantation Technique (1973)
[2] for example S.T.Hsu etal, Solid-State Electron 13, 1055 (1970)
    T. Kōji, IEEE Electron Device (to be published)
[3] B. Crawford, Solid-State Communications McGRAW-HILL BOOK COMPANY 190 (1966)
[4] D.K Brice, ION IMPLANTATION, 107 Gordon and Breach Science Publishers (1971)
[5] T. Kōji, Trans. IECE 57-C No.1 (1974)

APPENDIX

N.L shows total noise summed up from the low frequency (about 0 Hz) component to high frequency (about 100 kHz) component.

N.L measurements were done using the circuit as shown below.

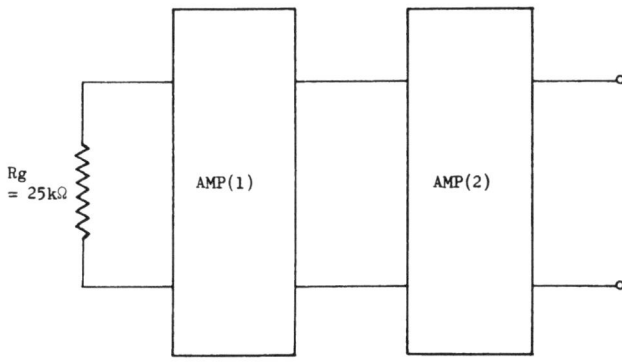

AMP(1) : RIAA - AMPLIFIER OF VOLTAGE GAIN 34dB AT 1kHz
AMP(2) : FLAT - AMPLIFIER OF VOLTAGE GAIN 60dB

DEVICES II

THE MEASUREMENT OF DOPING UNIFORMITY IN ION IMPLANTED WAFERS

J. Stephen, B. J. Smith and G. W. Hinder

Electronics and Applied Physics Division
A.E.R.E., Harwell, Didcot, Oxon.

ABSTRACT

This paper describes the measurement of the uniformity of sheet resistance of ion implanted layers in silicon wafers. A four point probe method is used with automatic data collection and analysis. Variations of 1% in sheet resistance have been mapped. The method is very flexible as it does not involve complex wafer preparation. It also allows further thermal annealing treatment to be given to the wafer.

INTRODUCTION

In recent years improved semiconductor doping techniques have permitted controlled doping over large areas. Uniformity of doping is important for circuits containing high value resistors (MacDougall et al. (1969)); MOSTs requiring controlled threshold voltage shifts (Aubuchion (1969)); and doped regions known as tubes or wells for complementary MOS circuitry (Dill et al. (1971)). Failure to control the doping concentration accurately and maintain a high degree of uniformity in such devices can seriously reduce the yield. Uniformity is also an essential requirement in the manufacture of large area position sensitive radiation detectors, semiconductor vidicon arrays and power diodes.

MEASUREMENT TECHNIQUE

This technique requires a minimum of wafer preparation and has been used for measurements of sheet resistance of doped layers up to about 5000Ω/square. The surface layer of the wafer is doped over its whole area by implantation to produce a layer of opposite doping type to that of the wafer. However it is possible to measure the uniformity of buried implanted layers by implanting into high resistivity wafers of the same doping type as the layer. The technique can also be used to measure the uniformity of unprocessed and diffused wafers.

An in-line four point probe as described in detail by Valdes (1954) has been used to measure the sheet resistance of the doped layers. The calculation of sheet resistance over the area of a finite wafer is not straightforward as the geometric position of the probes has to be taken into account. The sheet resistance of the conducting layer may be written as

$$R_s = K \cdot \frac{V}{I} \text{ ohms per square.}$$

The constant K tends to 4.53 for the in-line probe of an infinite conducting sheet of thickness less than 0.25 times the probe spacing. This relationship is valid for the conventional four point probe used in the centre of a 50 mm diameter wafer. As measurements are made over the whole implanted area it is necessary to correct for the distortion in the current flow pattern towards the edge of the wafer. Logan (1961) and Swartzendruber (1964a & b) have given the correction factors to be applied to the constant K. These correction factors assumed a uniform sheet resistance over the total area of the wafer and they should be modified for the non-uniform areas. The errors incurred by using the correction factors without modification are of a second order and can be neglected for measurements on nearly uniform wafers. If the conducting skin extends round the edges and over the back of the wafer as occurs in an unmasked diffused wafer, Logan (1961) has shown that the wafer can be treated as an infinite plane.

An important limitation on the four point probe method is imposed by the damage which the probe itself may produce on the surface. For example, implanting 40 keV phosphorus ions into p-type silicon produces an n-type layer approximately 800Å thick which cannot be probed with a four point probe exerting a force of 200 gms weight. Such a probe pierces the implanted layer and is found to make contact with the substrate. However, by using a probe force of 20 gms weight, the damage is less severe and it is possible to contact an 800Å thick layer without difficulty. Figure 1 is a photomicrograph taken in scanning electron microscope showing the damage caused to a silicon surface by a 200 gram probe. The damage produced by the 20 gm probe could not be seen in the microscope.

To permit a systematic probing the wafer is held on an insulated vacuum chuck free to rotate about a vertical axis. The vacuum chuck is moved in a perpendicular direction in the horizontal plane by a micrometer adjustment. The linear movement and the radial movement are calibrated. The use of a vacuum chuck to secure the wafer prevents it moving and the probes skidding as the load is applied. Measurements are made along a radius from the edge to the centre of the wafer before moving to the next radius.

The measurement current is derived from a Bradley type 132 calibrated current supply with a stability of $\pm$ .005%. The current is monitored by observing the voltage drop across a standard resistor. In order to ensure a good contact with all four probes on to high resistance material the current is first passed through the two inner probes, then switched to the outer pair of probes (Figure 2). The voltage across the four point probe is measured on a Solartron type LM1867 digital voltmeter connected to a data logging unit.

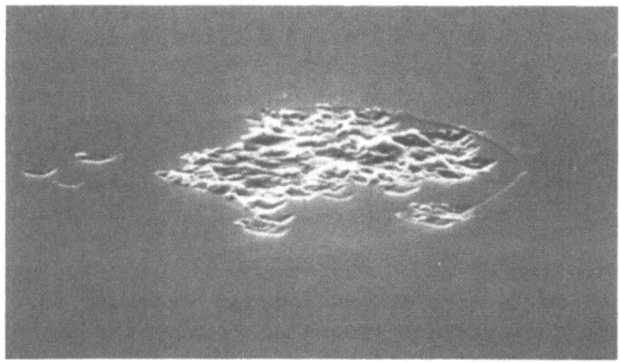

Fig. 1. Photomicrograph of damage to a silicon surface from a 200 gram 4 point probe (Magnification X2000)

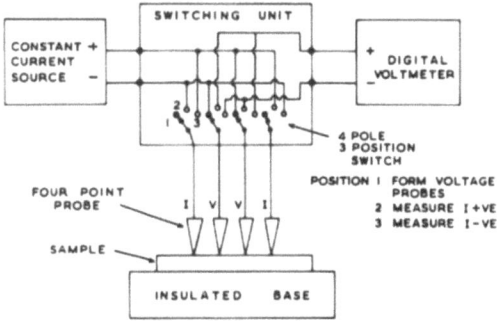

Fig. 2. Switching arrangements for 4 point probe measurements

A check is made that the absolute voltage value remains unchanged when the current flow is reversed. Common sources of discrepancy between voltages measured with opposite current flows are contamination of the wafer by an insulating film and uneven illumination of the wafer. By cleaning the wafers carefully and measuring the resistance in the dark the reproducibility of measurements in both current directions is better than 1%. This figure could probably be improved if the temperature of the wafer was kept constant. To minimise the temperature rise the current is usually less than 100 µA and for wafers with a sheet resistance greater than about 10^3 Ωcm 1 µA is used.

Other precautions that need to be taken have been reviewed more comprehensively by Dearnaley et al. (1973), and by Bullis (1967).

DATA ANALYSIS

A computer program has been written to analyse the recorded data. It first calculates the 4 point probe correction factors for each radial position. Then the sheet resistance for each measurement is calculated using the appropriate correction factor. The mean sheet resistance and the standard deviation of the distribution are calculated. Values are then sorted into groups according to their deviation from the mean value. The upper and lower limits of each group are defined in the input data as percentages from the mean. The populations of each group are displayed as a histogram.

The distribution of sheet resistance is also displayed as a contour map with an iso-resistance contour drawn for each histogram division. These contours are generated by dividing the area of the wafer into a large number of triangles with a know sheet resistance at each corner. Interpolating linearly between the values at the corners of each triangle determines the points at which any of the defined contours across the boundaries of the section. The contour is thus uniquely plotted across the section. Within a triangular section the contour is a straight line. In order to distinguish between the different contours the lines are chain dotted with the mark space ratio increasing for higher resistance contours. On the contour map the measurement positions are marked with crosses. The method is described in more detail by Smith et al. (1974).

USE OF THE TECHNIQUE

The technique described in the preceding sections has been used for the measurement of the uniformity of doping of silicon wafers by ion implantation and by diffusion. Measurements have also been made of the uniformity of undoped wafers and epitaxial layers. Non-uniformity of epitaxial doping or thickness shows up as a non-uniformity in sheet resistance. The technique has been used to study the effect of annealing ion implanted specimens and to observe changes in distribution of sheet resistance during the annealing process. As no metallic contacting techniques are required, it is a simple matter to follow the annealing behaviour of a wafer over a wide range of annealing temperature. It is usual to protect or cap the doped wafer surface with a layer of r.f. sputtered silicon dioxide which is readily removed in buffered hydrofluoric acid after annealing. The uniformity of oxidation processes have been studied by measurement of the uniformity of the silicon wafer after the oxide has been removed. For a wide range of conditions contours of sheet resistance separated by 1% of the mean value have been plotted across a wafer. This is consistent with the predictions of Bullis (1967) who calculated experimental errors of .68%.

For ion implanted specimens no difficulty has been experienced working in the range 10 to 1500 Ω/square. Above this range measurements were less reproducible but useful results have been obtained on specimens up to 8 kΩ/square. Using the Minirosion 20 gram large radius probe (type 4005.C) described previously, measurements have been made on very thin layers without breaking through to contact the underlying substrate. This type of layer has been made for example by implanting 40 keV arsenic ions into silicon giving an estimated thickness of approximately .03 microns.

A number of measurements have been made on diffused and epitaxial layers in the resistance range 3 to 1000 Ω/square. No difficulty was experienced with the measurements. Such layers were usually less uniform than implanted layers.

The two examples shown here are for typical applications. In each case all the output is generated by the Harwell central computer. The first, shown in figure 3, is of an ion implanted layer on a 2" diameter silicon wafer. The doping was 2.5×10^{15} boron ions per cm^2 implanted at 180 keV on the Harwell industrial ion implantation machine (Freeman et al. 1970). The measurement at the edge of wafer in the 6 o'clock position has been omitted as it was very close to the flat on the wafer. This position usually produces a high value.

In the second example the specimen was a 2 inch diameter silicon wafer doped with arsenic over its whole surface area by diffusion from an arsine source. (See Fig. 4). The circular contours are attributed to preferential doping at the edges of the wafer as the dopant gas passed along a stack of wafers. The gas flow was parallel to the line joining the centres of all the wafers in the stack.

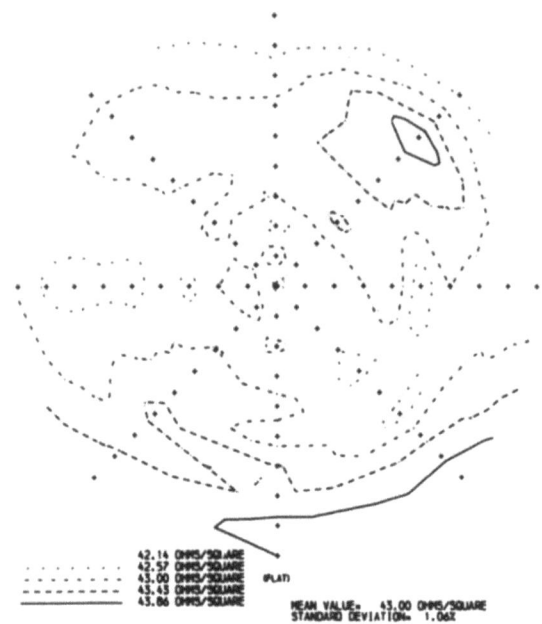

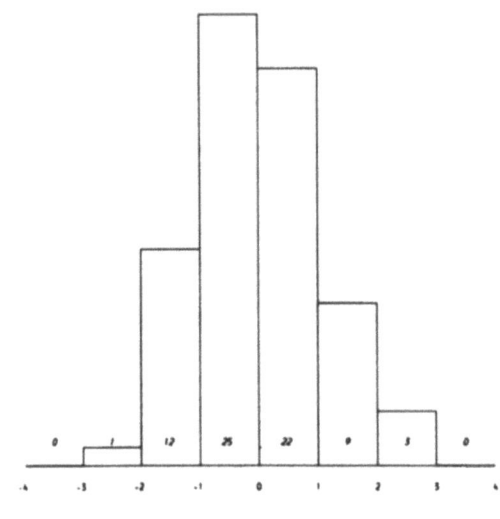

Fig. 3. Computer analysis of uniformity measurements: Boron implant

MEASUREMENT OF DOPING UNIFORMITY

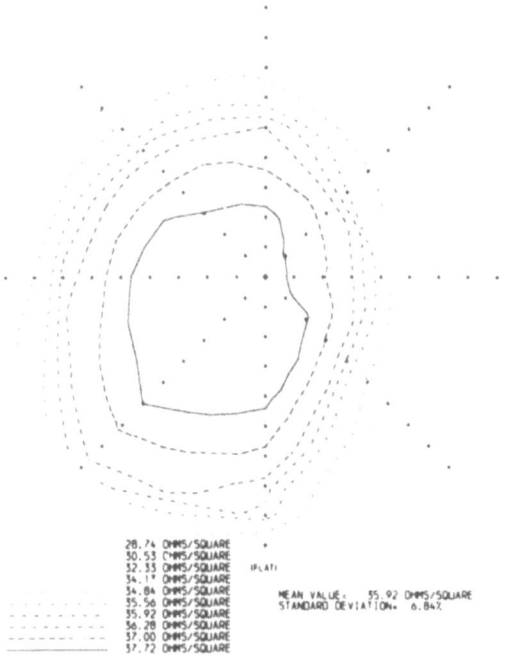

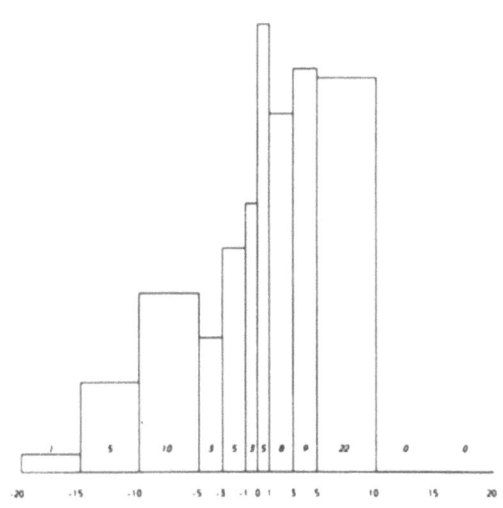

Fig. 4. Computer analysis of uniformity measurements: Arsenic diffusion

CONCLUSIONS

The measurements show that variations in sheet resistance of 1 percent may be plotted across a silicon wafer using the four point probe technique. The use of the four point method reduces the amount of wafer preparation involving photolithography, and, it allows further extensive thermal treatment to be given to the wafer without the risk of contamination from earlier contacting procedures.

ACKNOWLEDGEMENTS

The authors wish to acknowledge the assistance of E. M. Wittam in the preparation of the resistors, of G. Gard and D. Chivers for the ion implantation and of D. C. Marshall for the micrograph of the four point probe damage.

REFERENCES

Aubuchon, K. G., 1969. Proc. Int. Conf. Properties and uses of M.I.S. structures. Grenoble p.575.

Bullis, W. M., 1967. NBS report 9666.

Dearnaley, G., Freeman, J. H., Nelson, R. S. and Stephen, J. 1973. Ion Implantation, North Holland, Amsterdam, pp. 527-537.

Dill, N. G., Toombes, T. N., & Bauer, L. O., 1971. Proc. 2nd Int. Conf. Ion Implantation, Garmish (Springer-Verlag) p.315.

Freeman, J. N., Caldecourt, L. R., Done, K. C. W., and Francis, R.J., 1970 AERE R.6496.

Logan, M. A., 1961 Bell Syst. Tech. J. $\underline{40}$, 3, 885.

MacDougall, J. D., Manchester, K. E. & Roughan, P. E., 1969, Proc. I.E.E.E., $\underline{57}$, 9, 1538.

Smith, B. J., Stephen, J. & Hinder, G. W., 1974 AERE R.7085.

Swartzendruber, L. J., 1964a. Solid State Electronics, $\underline{7}$, 413.

Swartzendruber, L. J., 1964b. NBS Tech. Note 199.

Valdes, L. B., 1954. I.R.E. $\underline{42}$, 420.

STRESS ADJUSTMENT IN Si_3N_4 FILMS BY ION IMPLANTATION

G. W. Reutlinger and R. A. Moline

Bell Laboratories

Murray Hill, New Jersey 07974

ABSTRACT

Si_3N_4 films deposited at 730°C by current CVD techniques have an intrinsic tensile stress of $1-1.5 \times 10^{10}$ dynes/cm^2. Stress relief by cracking of the nitride film could impair the reliability of integrated circuits that incorporate these films as a mobile ion barrier. Cracks that originate at the surface may be prevented by adjusting the stress at the surface to be neutral or compressive.

We have adjusted the stress in a thin surface layer of Si_3N_4 film, pyrolytically grown at 730°C, from its intrinsic tensile value of $\sim 1.2 \times 10^{10}$ dynes/cm^2 to comparable compressive values by bombarding the surface with $1-6 \times 10^{14}$ Ar/cm^2 at 30 keV. The integrated stress (dynes/cm) in the film was measured by the optically-levered-laser technique. At a dose $\approx 4 \times 10^{14}$ Ar/cm^2, the integrated stress peaks at $\approx -4 \times 10^4$ dynes/cm (compression). Assuming the damaged region is 250Å thick, this corresponds to a (compressive) stress of -1.5×10^{10} dynes/cm^2. Very little change is seen at doses up to 6×10^{14} Ar/cm^2.

The influence of annealing on the stress was compared between implanted and unimplanted wafers. The stress (tensile) in the unimplanted wafer was virtually unchanged during 30 minute anneals as high as 800°C. Significant annealing was observed in the ion induced stress at temperatures $\geq 600°C$.

INTRODUCTION

A thin film of Si_3N_4 is often incorporated in integrated circuits to serve as a mobile ion barrier. Si_3N_4 films produced by current CVD techniques have been shown to have a high residual tensile stress. We have measured the stress in 1430Å thick films deposited at 730°C to be $1-1.5 \times 10^{10}$ dynes/cm². Grieco et al.[1] observed hairlike cracks in Si_3N_4 films greater than 4000Å thick, deposited at temperatures between 550° and 1225°C and found that the substrates were concave "toward the film" indicating tensile stress in the films. Bean et al.[2] have measured the breaking strength of Si rich films deposited from the $SiH_4 + NH_3$ reaction to be $.46 - .93 \times 10^{10}$ dynes/cm². The breaking strength of Si_3N_4 films used here is not known but may be comparable to the measured tensile stress of these films.

The cause of the residual tensile stress in CVD Si_3N_4 is not as yet understood. Tokuyama et al.[3] measured the thermal expansion coefficient of Si_3N_4, by X-ray techniques, to be 3.85×10^{-6}/°C which is a close match to that of Si for deposition temperatures of interest. They attribute the "thermal cracking" of Si_3N_4 films on Si to a relatively high Young's modulus which they measured to be 3.9×10^{13} dynes/cm².

EerNisse[4] has shown that the integrated stress (stress times the thickness of the film) in the surface layer of Si can be altered by ion bombardment and is compressive and approximately a linear function of energy deposited into atomic processes. Assuming the qualitative features of EerNisse's data apply to nonsingle-crystal material, then it should be possible to vary the integrated stress in a damaged surface layer and thereby change the radius of curvature of the slice. We have shown in this work that the stress in a thin (~250Å) layer at the surface of the Si_3N_4 film can be made compressive by bombardment with 30 keV argon ions. While the stress in the remainder of the Si_3N_4 film remains tensile, cracks that originate at the surface may be prevented.

Structure

The structure of our samples is shown in Fig. 1. Si_3N_4 was deposited onto a 5000Å thick layer of SiO_2 which was grown on 10.7 mil thick, 1.5" diameter, silicon slices. The oxidation was done in steam at 1050°C. The nitride was deposited pyrolytically at 730°C from the reaction of $N_2H_4 + SiH_4$ or at 875°C from the $NH_3 + SiH_4$ reaction. The N_2H_4 layers were measured by ellipsometry to be 1430Å thick and to have an index of refraction of 1.96 while the NH_3 layers were 1050Å thick.

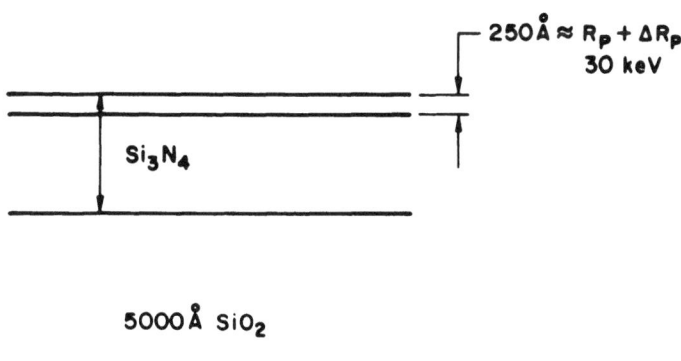
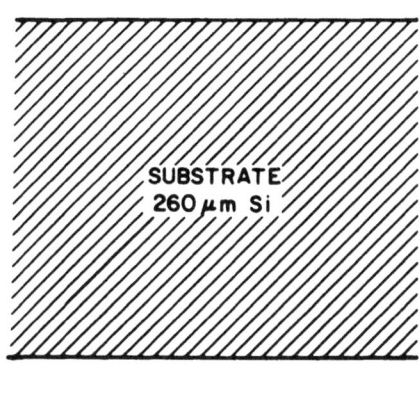

Fig. 1 Cross-section view of sample structure.

Procedure

The radius of curvature of the samples was measured before and after each incremental implantation dose using either the optically-levered-laser technique[5] or the Automatic Bragg Angle Control (ABAC) technique.[6] In the optically-levered-laser technique, the change in angle of reflection of a laser beam incident on the sample is measured as a function of position on the wafer. The ABAC technique is described elsewhere.[6] The average stress, σ, in the damaged layer was calculated from the measured change in stress induced curvature, $\Delta(R)$, using Eq. 1 of Ref. 6:

$$\sigma = \frac{Ed^2}{6(1-\gamma)t} \Delta(\frac{1}{R}), \qquad (1)$$

where E = Young's Modulus of silicon
       γ = Poisson's Ratio of silicon
       d = thickness of substrate
       t = thickness of the damaged layer which was estimated from LSS theory[7] to be 250Å.

Si_3N_4 Results

Figure 2 is a plot of the average stress in the 250Å damaged Si_3N_4 layer vs. argon dose for six of the samples studied. The stress in the damaged surface layer changes sign in the dose range of $0.5 - 2.5 \times 10^{14}/cm^2$, and the compressive stress saturates at about 1.5×10^{10} dynes/cm^2 at a 30 keV Ar dose of 4×10^{14} ions/cm^2. The scatter of data points in Fig. 2 is partly due to variations of stress in the as-deposited Si_3N_4 and partly due to limitations of the measurement apparatus to detect small changes of radius of curvatures. The smallest detectable laser beam deflection for the apparatus used here is 0.25 mm. The reproducibility of the measurement was determined by repeating the measurement five times on one sample. The sample was removed from the translation manipulator between each set of measurements. The average variance from the mean for the ten values was .16 mm. The total measurement uncertainty is thus ≈0.3 mm which translates to an uncertainty in the measured stress of 3×10^9 dynes/cm^2. Two of the data points in Figs. 2 and 3 carry error bars which are representative of the uncertainty of all points.

Two samples were given implantations and the stress was measured before and after each implantation. The implanted samples and an unimplanted control sample were then annealed in nitrogen at 600, 694, 798 and 898°C. They were held at each temperature for 30 minutes. Figure 3 is a combined plot of stress vs. argon dose and anneal temperatures.

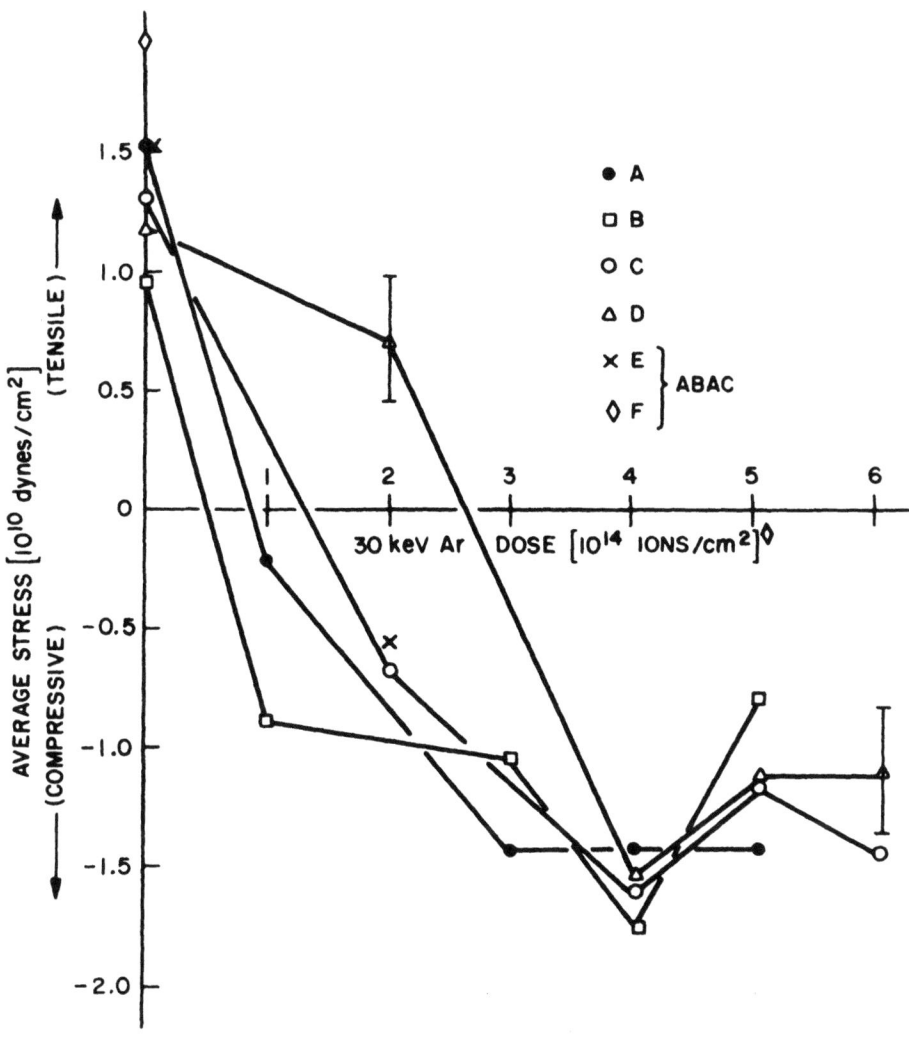

Fig. 2 Average stress in the 250Å damaged layer of Si_3N_4 vs. 30 keV argon dose. Representative error bars are shown. Samples A, B, C, and D were grown using N_2H_4 while E and F were grown using NH_3.

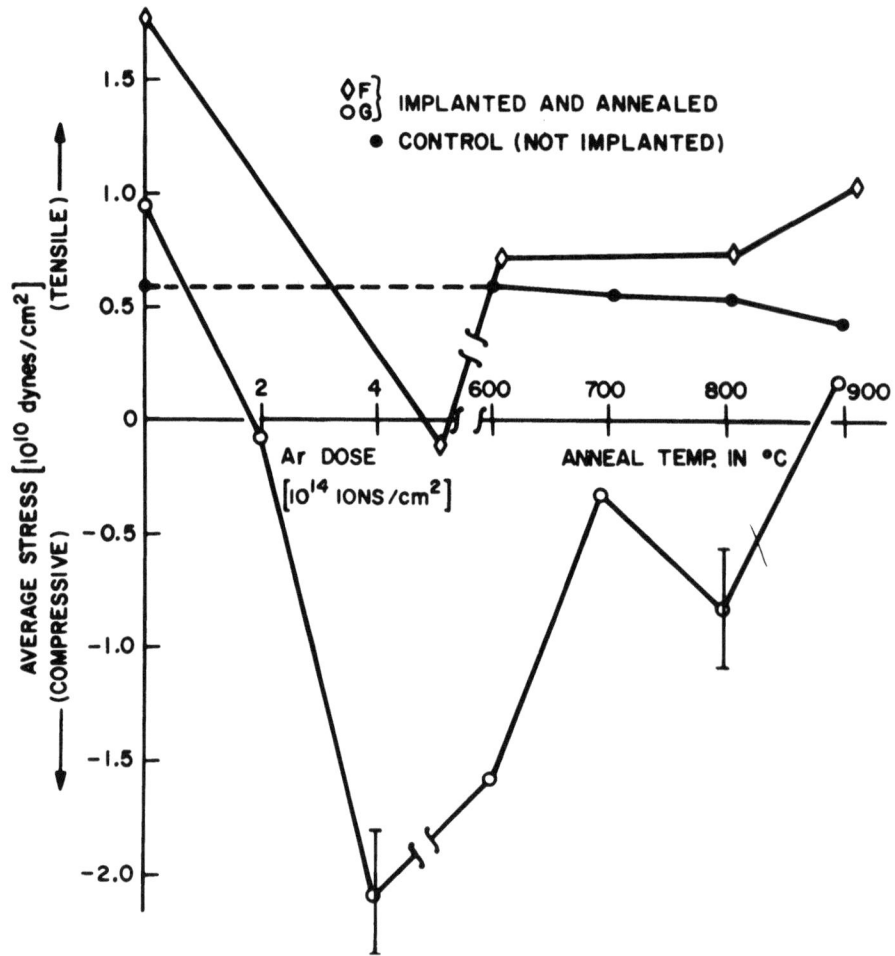

Fig. 3 Combined plot of stress vs. argon dose and annealing temperature. A control sample, which was annealed but not implanted is included in the plot. Sample G was grown using N_2H_4 while sample F was grown using NH_3.

Discussion

Recently, EerNisse[8] has shown that ion bombardment causes fused silica to compact.[4] Silicon dioxide films, thermally grown at 1050°C steam on polished silicon substrates, have shown a compressive stress of $2\text{-}4\times10^9$ dynes/cm². Figure 4 is a plot of average

STRESS ADJUSTMENT IN Si₃N₄ FILMS

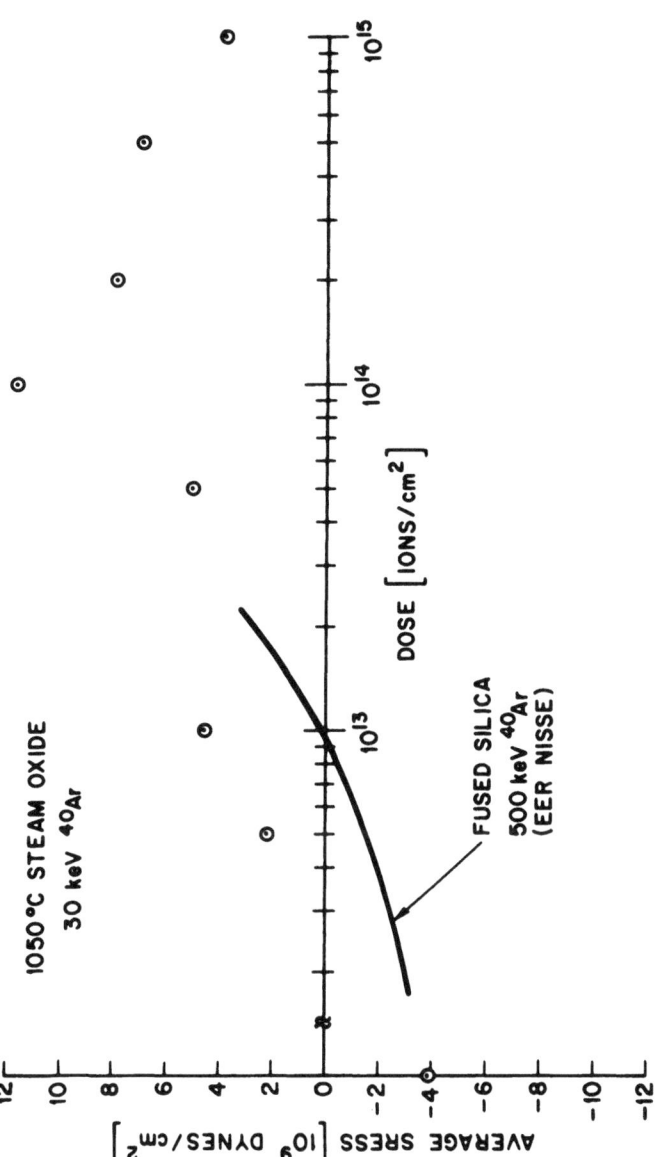

Fig. 4 Average stress in the 250Å damaged layer of thermal SiO_2 vs. 30 keV argon dose. The curve is taken from 500 keV argon damage data reported by EerNisse.[8]

stress in the 316Å (Rp + ΔRp) thick bombarded layer of steam grown SiO_2 vs. argon dose. The data points shown in the figure are for 30 keV ^{40}Ar ions and the curve represents EerNisse's[8] data for 500 keV ^{40}Ar ions in fused silica, assuming in his case the bombarded layer was 5000Å thick. These oxide stress measurements were made using the X-ray technique described by Rozgonyi and Ciesielka.[6]

Thus, although both fused silica and thermally grown SiO_2 compact as a result of ion damage, Si_3N_4 expands as a result of the same bombardment. In both cases studied here the sign of the stress changes during bombardment.

Conclusion

We have shown that the stress in a thin surface layer of Si_3N_4 can be made compressive by ion implantation and that significant annealing occurs during thermal cycles $\geq 600°C$. This is in contrast to the case of thermal SiO_2 where the material compacts during bombardment.

REFERENCES

1. M. J. Grieco, F. L. Worthing, B. Schwartz, "Silicon Nitride Thin Films from $SiCl_4$ Plus NH_3: Preparation and Properties," J. Electrochem. Soc. 11, No. 5, 525 (1968).
2. K. E. Bean, P. S. Glein, R. L. Yeakly, W. R. Runyan, "Some Properties of Vapor Deposited Silicon Nitride Films Using the SiH_4-NH_3-H_2 System," J. Electrochem. Soc. 114, No. 7, 733 (1967).
3. T. Tokuyama, Y. Fujii, Y. Sugita, S. Kishino, "Thermal Expansion Coefficient of a Pyrolitically Deposited Silicon Nitride Film," Japan J. Appl. Phys., 6, 1252-1253 (1967).
4. E. P. EerNisse, "Investigation of Ion Implantation Damage with Stress Measurements," Proc. 2nd International Conf. on Ion Implantation in Semiconductors, I. Ruge and J. Graul Ed., Springer-Verlag, Berlin-Heidelberg, New York, p. 17 (1971).
5. Private communication.
6. G. A. Rozgonyi, T. J. Ciesielka, "X-ray Determination of Stresses in Thin Films and Substrates by Automatic Bragg Angle Control," Rev. Sci. Instrum., 44, No. 8, August 1973.
7. W. S. Johnson, J. F. Gibbons, "Projected Range Statistics in Semiconductors," Stanford Univ. Book Store, Stanford, California (1970).
8. E. P. EerNisse, "Compaction of Ion-Implanted Fused Silica," J. Applied Phys., 45, 167-174 (1974).

ENHANCED OXIDATION OF SILICON BY ION IMPLANTATION AND ITS NOVEL APPLICATIONS

K.Nomura and Y.Hirose

Kitaitami Works, Mitsubishi Electric Corp.

4-1, Mizuhara, Itami, 664 Japan

Y.Akasaka, K.Horie and S.Kawazu

Central Reseach Lab., Mitsubishi Electric Corp.

Minamishimizu, Amagasaki 664 Japan

ABSTRACT

The enhancement of the oxidation of silicon due to ion implantation is observed to be significant (about 2.1 times at maximum) in the case of Sb^+, P^+, Sn^+ and Ar^+ implant with doses more than $5 \times 10^{14}/cm^2$. The annealing before oxidation does not affect the oxidation rate. The oxidation rate constant B, so called the parabolic constant, is not affected by ion implantation with any dose examined, but the constant A, which refers to linear rate constant, decreases at the dose about $5 \times 10^{14}/cm^2$. It is concluded that the residual damage is less effective, but the chemical effect caused by the implanted ions to the reaction between silicon and oxygen at the interface of Si-SiO_2 plays an important role in the enhancement of the oxidation rate. A novel application method of this effect to the fabrication of E/D MOS IC is also shown.

INTRODUCTION

Ion implantation has been extensively used as an impurity doping technique to semiconductors and some additional effects such as enhanced diffusion (1) of doped impurities or enhanced etching of dielectric layers (2,3) caused by ion implantation were reported. We found enhanced oxidation of silicon by ion implantation.

Dearnaley et al. (4) reported that in metals the oxidation rate of metals such as Ti and stainless steel was reduced or enhanced by ion implantation. They suggested that in the metal this effect was dominated by the electronegativity of implanted ions. Mayer et al.(5) reported the preliminary work on the enhanced oxidation of silicon. They observed by backscattering analysis that the content of oxygen at the sample surface after annealing in the Ar atmosphere are different with samples implanted with different ions. Very recently, it was reported by Fritzsche et al. (6) that N_2^+ ion implantation reduced the oxidation rate of silicon on account of the formation of silicon nitride compound.

We have investigated the effect of ion implantation to the oxidation rate of silicon systematically. The purpose of the present work are to understand the mechanism of the enhanced oxidation by ion implantation and also to apply the effect to the semiconductor device fabrication.

EXPERIMENT

Silicon wafers used were mostly ⟨111⟩ oriented, 10 — 20 ohm-cm, P-type crystals. Various kinds of ion species with different masses and electronegativities such as Sb^+, Ar^+, Sn^+ and P^+ were implanted at room temperature with the energy of 50 - 350 keV at the dose of 10^{13} - $2 \times 10^{15}/cm^2$. Silicon wafers selectively pre-implanted were oxidized at 900°C - 1100°C in the wet O_2 atmosphere. Some samples were annealed at 400°C - 1050°C in the dry N_2 atmosphere for 20 min. prior to the oxidation in order to study the damage effect. The thickness of the oxide layers was measured by ellipsometry. The location of the implanted ions after oxidation was studied by means of channeling analysis with 1.5 MeV He^+ ions. The etching rate of the oxide layer was measured in 10 seconds step with HF etchant (HF : H_2O = 1 : 10) at 25°C.

RESULTS AND DISCUSSION

Figure 1 shows the dependence of oxide thickness on implanted dose of 50 keV Sb^+ ion as a parameter of oxidation temperature. The enhancement of the oxidation is clearly seen in the cases of 900°C and 950°C, and also even at 1050°C, but it becomes less evident with the increase of the oxidation temperature probably due to the increase of the normal oxidation rate. In the dose range less than $2 \times 10^{14}/cm^2$, the enhancement of the oxidation is not observed. With heavier doses more than $5 \times 10^{14}/cm^2$, the oxidation rate is enhanced more and more as the dose increases. The dose dependence of the ratio of oxide thickness of Sb^+ and P^+ implanted region to that of unimplanted region for various oxidation times are shown Fig.2. The oxide thickness of Sb^+ implanted region is as large as 2.1 times of that of unimplanted region for 10 min. oxi-

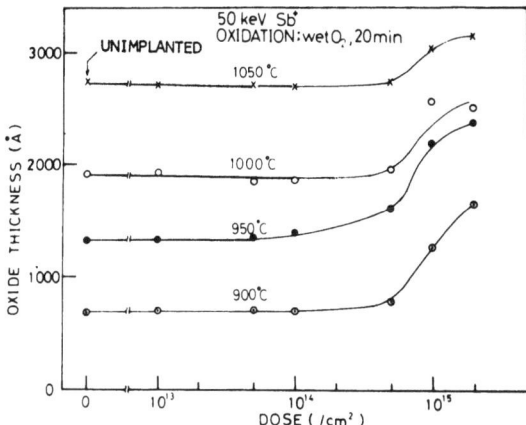

Fig.1

The dependence of oxide thickness as a function of Sb^+ ion dose for various oxidation temperature.

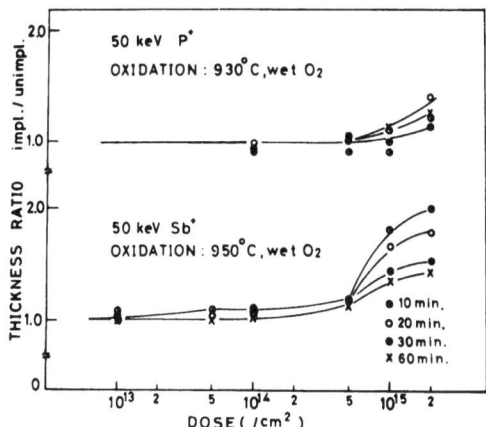

Fig.2

Ratio of oxide thickness of implanted region as a function of Sb^+ or P^+ ion doses for various oxidation time.

dation at 950°C. In the case of P^+ ion, it becomes 1.4 times of that of unimplanted region for 20 min. oxidation at 930°C. The ratio decreases with the increase of oxidation time but the absolute difference of the oxide thickness is almost constant in the oxidation time more than 30 min.

Figure 3 shows typical results concerning with ratio of oxide thickness of implanted region to that of unimplanted region as a function of oxidation time for 50 keV Sb^+ ion and for 200 keV Ar^+ ion. The enhancement in the case of Ar^+ implant is not so large as compared with that of Sb^+ implant.

One should consider at least two causes which affect the oxidation rate. One is damage effect and the other is chemical effect. We could not find Ar^+ ions by backscattering analysis in the sample

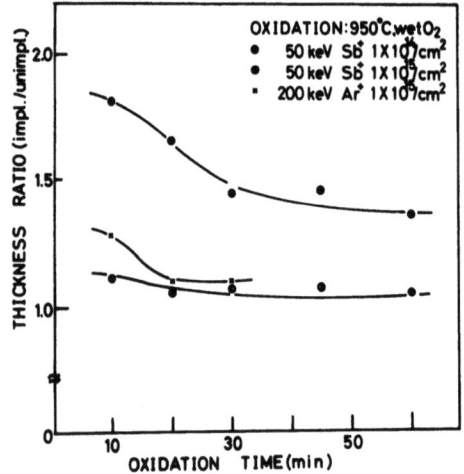

Fig. 3

Ratio of oxide thickness of implanted region to that of unimplanted region as a function of oxidation time for Sb^+ ion and Ar^+ ion implants.

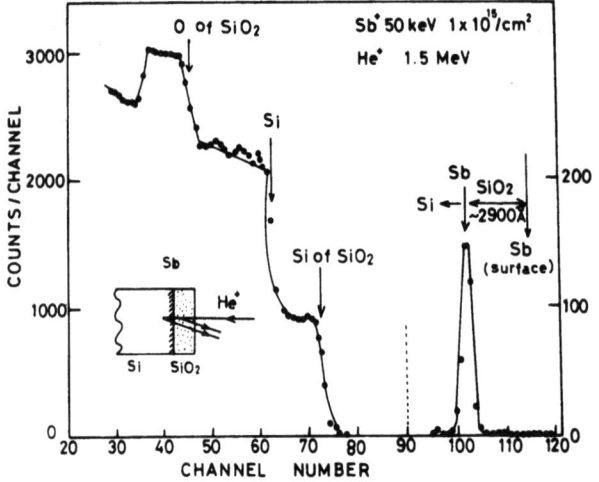

Fig. 4

Backscattering spectrum of Sb^+ implanted Si after 1100°C oxidation.

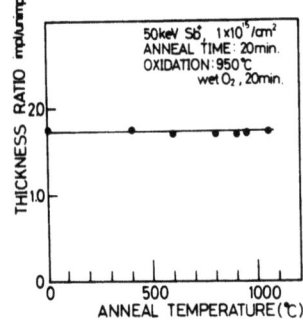

Fig. 5

Ratio of oxide thickness of implanted region to that of unimplanted region vs. pre-N_2 anneal temperature.

after 20 min. oxidation. On the contrary, it was found that almost all of Sb⁺ ions located in the silicon surface region just beneath the oxide layer as shown in Fig.4. This suggests that the damage effect does not seriously affect the oxidation rate, but the effect caused by the chemical characteristic of the implanted ions is dominant. This was confirmed by examining the pre-annealing effect to the oxidation rate shown in Fig.5. As clearly seen from the result, the oxidation rate does not change by annealing even at high temperatures. This result shows that the damage produced by ion implantation is not so effective to the enhancement of the oxidation.

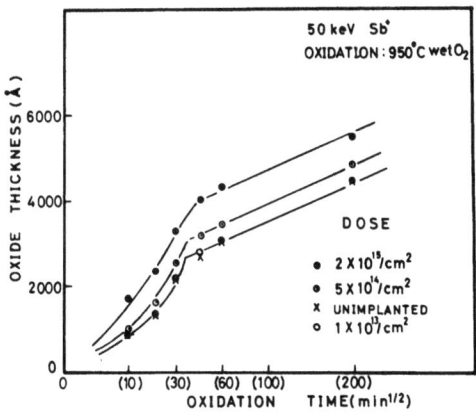

Fig.6

The dependence of oxide thickness as a function of the oxidation time for various doses of 50 keV Sb⁺ implants.

Figure 6 shows the dependence of oxide thickness as a function of root of oxidation time for various doses of 50 keV Sb⁺ implants. As well known, the dependence of grown oxide thickness on oxidation time is expressed approximately as

$$X^2 + AX = Bt$$

where, X is grown oxide thickness, t oxidation time, A and B are oxidation coefficients determined by experimental conditions, respectively. When the oxide thickness is thin, it has linear dependence on the oxidation time as $X \sim (B/A)t$. When it becomes thicker, it shows root dependence as $X \sim \sqrt{Bt}$, i.e. the parabolic dependence on the oxidation time. The linear relationship means that the oxidation rate is limited by the reaction between silicon and oxygen at the Si-SiO₂ interface and parabolic relationship means that the oxidation rate is limited by the diffusion of oxygen through a silicon oxide layer. B/A and B are refered to as the linear rate constant and parabolic rate constant, respectively. This figure shows this relationship clearly. No difference was observed between the unimplanted sample and the sample implanted to the dose of $1 \times 10^{13}/cm^2$. Obvious change was observed with heavily implanted samples in the linear relationship region. But in the parabolic region (more than about 30 min. oxidation), the oxidation curves go quite parallel irrespective of implanted doses. This suggests that the pre-implantation does not affect the parabolic term in the oxidation process, i.e. the pre-implantation does not change the diffusion parameters of oxygen through a oxide layer, but affects the interaction between oxygen and silicon at the Si-SiO$_2$ interface when the oxide thickness is thin. Oxidation coefficients A and B were calculated from these plots and are shown in Fig.7 as a function of Sb$^+$ dose.

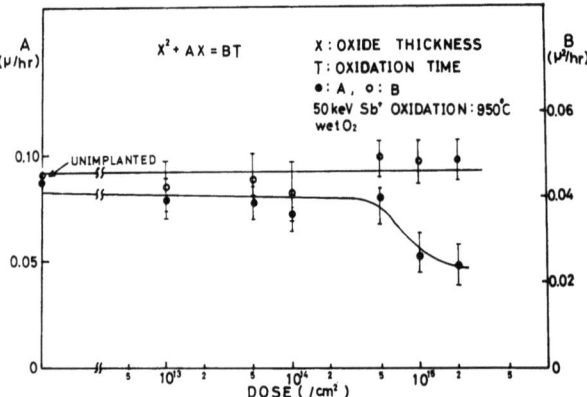

Fig.7 Variations of coefficients A and B in the above equation as a function of ion dose. B is refered to as the parabolic rate constant and A is derived from the linear rate constant B/A.

B is almost constant at any implanted dose. On the other hand, another constant A, which is derived from the linear rate constant B/A, begins to decrease at the dose of 5 x 10^{14}/cm^2. This dose coincides with the dose at which the enhancement of the oxidation becomes evident as shown in Figs.1 and 2. From these results mentioned so far, we can conclude that chemical effect is dominant in the enhancement of the oxidation induced by pre-implantation.

Figure 8 shows the chemical etching rate of the oxide grown by the enhanced oxidation. A conventional etchant(HF:H$_2$O = 1:10) was used. The etching rate of the oxide grown after the implantation was the same as that of oxide grown without implantation. This suggests that the quality of the oxide grown by the enhanced oxidation is as good as the oxide conventionaly grown. This property is very preferable, because the difference of the oxide thickness can be kept even after the chemical etching.

Figure 9 shows how to fabricate N-channel Si-gate E/D MOS by using the enhanced oxidation. After finishing the selective oxidation process for the field region, we implant Sn$^+$ ions to load MOS transistor region. Typical dose is $5 \times 10^{14}/cm^2$. Then gate oxide is grown. The gate oxide of the load MOS transistor becomes thicker than that of the driver MOS transistor on account of the enhanced oxidation caused by pre-implantation of Sb$^+$ ions. After that, boron ions are implanted without any mask. Boron ions are expected to stop in the gate oxide of the load MOS transistor, but to be implanted into the channel region of driver MOS transistor. Then, we proceed to the conventional Si-gate process. Merits derived by using this process are:
1) One photoresist process for channel doping is eliminated.
2) Photoresist process of the mask for Sn$^+$ implantation is very tolerable.

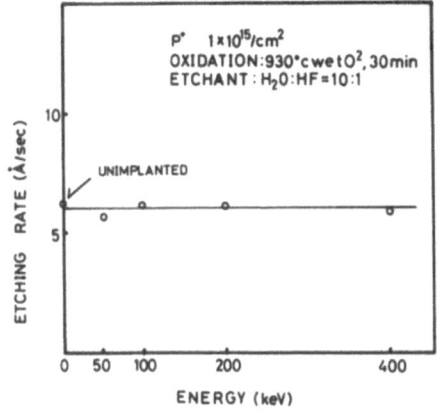

Fig.8

The dependence of the etching rate on the implanted energy of P$^+$ ions.

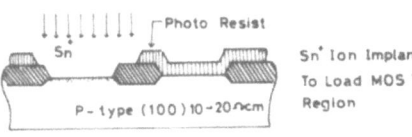

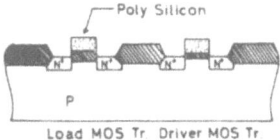

Fig.9

E-D MOS IC production process using enhanced oxidation effect.

Figure 10 shows the characteristics of both enhancement mode and depletion mode MOS transistors fabricated by the method mentioned above.

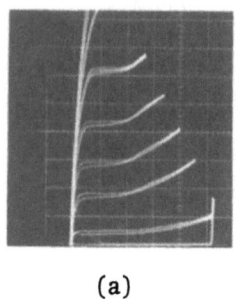
(a)

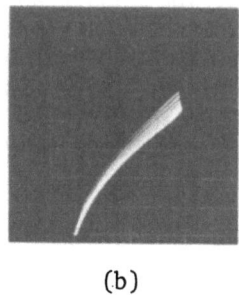
(b)

Fig.10

I-V characteristics of the N-channel MOS FET.
(a) E-MOS V_G=1V/step,V: 50 μA/div,H:5V/div.
(b) D-MOS V_G=1V/step,V: 20 μA/div,H:5V/div.

CONCLUSIONS

The conclusions obtained are: 1) A remarkable enhancement of the oxidation is induced by pre-ion implantation. 2) The chemical effect between oxygen and silicon at the interface of Si-SiO$_2$ induced by ion implantation is a main cause of the enhanced oxidation. 3) Damage produced by ion implantation affects the oxidation very little. In the case of Ar$^+$ implant, it affects at the early stage of the oxidation. 4) The application of this enhanced oxidation effect to E/D MOS devices is successful.

ACKNOWLEDGEMENT

We would like to thank Y.Yukimoto, Dr.H.Komiya and Y.Watari for helpful discussions. We also thank H.Sato, Y.Watakabe and K.Tsukamoto for technical assistance.

REFERENCES

1. G. J. Dienes et al. J. Appl. Phys. <u>29</u> 1713 (1958)

2. A. Monfret et al. Proc. 2nd Intern. Conf. on Ion Implantation in Semiconductors ed. I. Ruge and J. Graul (Springer-Verlag, 1971) p.389.

3. Y. Akasaka et al. Proc. 5th Conf. on Solid State Devices, Tokyo, 1973 p.493.

4. G. Dearnaley et al. Proc. III Inter. Conf. Ion Implantation in semiconductors and Other Materials, p.405.

5. J. W. Mayer private communication.

6. C. R. Fritzsche et al. J. Electrochem. Soc. <u>120</u> 1603 (1973)

ION IMPLANTED BURIED LAYERS APPLIED FOR NUCLEAR

DETECTOR TELESCOPES

A. Kostka
AEG-Telefunken
D-71 Heilbronn, Germany

S. Kalbitzer
Max-Planck-Institut für Kernphysik
D-69 Heidelberg, Germany

ABSTRACT

Boron doped ion implanted layers have been used to form buried contact layers within n-type silicon single crystals in order to fabricate solid-state ΔE-E detector telescope systems incorporating both their ΔE and E section within the same crystal. With ion energies of 3 and 8 MeV, thicknesses of the energy loss sections amounted to 3.5 and 10 μm, respectively. Spectra obtained with 5.5 MeV α-particles exhibit resolutions of approx. 100 keV (FWHM). Subtraction of noise and energy straggling indicates thickness variations of approx. $\pm$ 0.15 μm (Si) over an active area of 6 - 10 mm^2.

INTRODUCTION

One of the particular applications of the ion implantation technique is the formation of buried layers inside the bulk of bombarded specimens. The method allows for a precise control of doping, offers steep doping gradients, and results in a nearly complete complanarity of the buried layers and the surfaces they underlie. Using boron ions with energies from 1 to 10 MeV, the depth of the buried layers within silicon crystals extends from 1.5 to 15 μm, an order of magnitude well adapted to various applications. A particular problem encountered with these high-energy implants in semiconductor device technology concers the crystal quality of the penetrated zone between surface and buried layer. Whereas these requirements can considerably be lowered for crystal regions serving as contacts, the demands remain high where minority carrier transport has to take place with sufficiently low loss. This is in general the case for nuclear particle detectors, and in particular true for the system considered here, where the energy-loss (ΔE) section of an integrated nuclear particle telescope has been formed by a shallow phosphorus ion and a deep boron ion implant into n-type silicon /1,2/.

EXPERIMENTAL

Key parameters of such an integrated telescope structure are the sheet conductivity of the buried contact, and the residual radiation damage within the zone penetrated by the boron ions, since they control the RC time constant and the collection efficiency obtainable in the ΔE section. In order to determine the influence of boron ion dose and anneal temperature on these parameters, a series of v.-d.-Pauw sheet measurements /3/ and α-particle spectroscopy have been carried out using standard equipment. The boron ion dose was varied from 10^{12} to 3×10^{14} cm^{-2}, the anneal temperature from 550 to 1250 K.

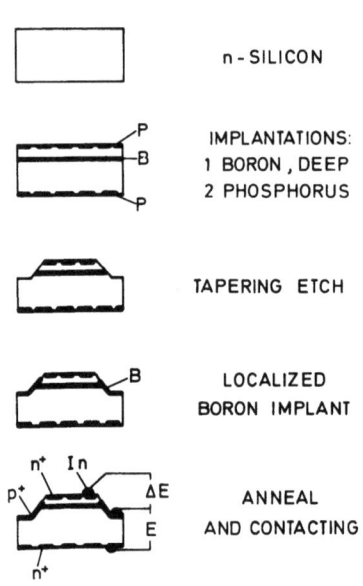

Fig.1: Scheme of telescope fabrication

As a quantitative measure of the influence of trapping on charge collection, the pulse height deficit detected from oscillographs of the preamplifier output after a certain time (1 μs) has been taken.

The various processing steps leading to the complete integrated n$^+$np$^+$nn$^+$ telescope structure are briefly illustrated in Fig. 1. As starting material, 400 Ωcm, n-type silicon wafers, (111)-oriented, of 20 mm diameter and 0.5 mm thickness were used. After the implantations defining the active structure (3×10^{13} boron/cm^2 of 3 or 8 MeV energy for the buried contact; 10^{14} phosphorus/cm^2 of 6 keV energy for the front and rear contact), a tapering etch was performed to make contact to the buried layer. An additional self-aligned, localized shallow boron implant (3×10^{14} cm^{-2}, 20 keV energy) outside the active detector area allows for the application of the electrical leads (indium pellets) after the anneal cycle, 30 minutes at 900 K. Final spectroscopic tests of the system were carried out using Am241 α-particles.

RESULTS AND DISCUSSION

Buried Layer and Penetrated Zone

The dose/anneal dependence of sheet resistance of the buried layer, and of trapping within the penetrated zone is shown in Fig. 2. Sheet resistances ϱ_s (Fig. 2a) from 10^5 down to approx. 10^2 $\Omega/\square$ have been obtained; the isochronal (30 min) annealing curves are similar to those known from lower energy implants, if not dose but concentration is used for comparison /3/. The accuracy in determining the pulse height deficit from photographs of oscilloscope traces (Fig. 2b) is limited to about $\pm$ 3 %; this restriction explains the field measurements have

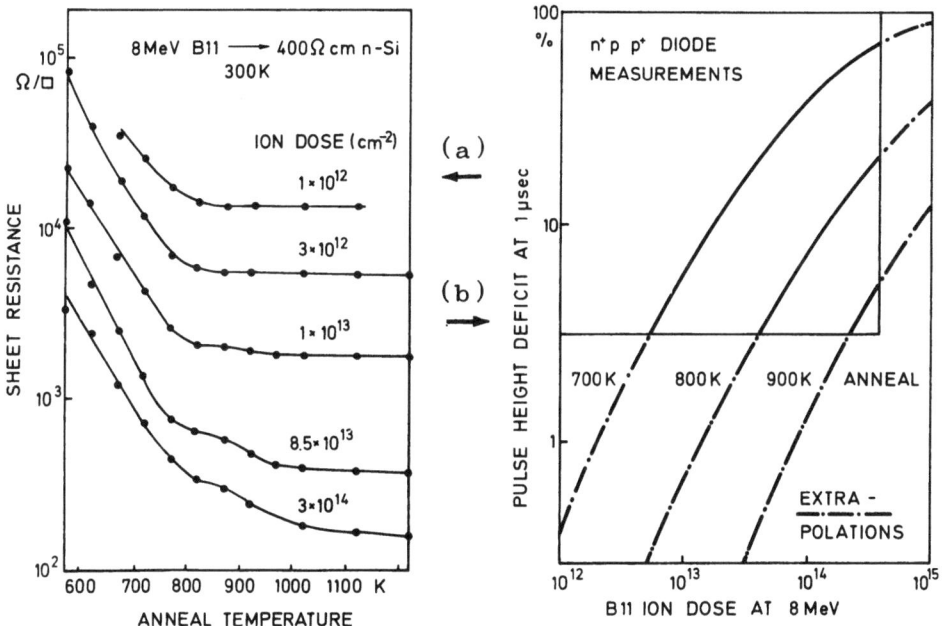

Fig. 2: Sheet resistance of buried layer (a) and pulse height deficit due to trapping within penetrated zone (b), for various ion doses and anneal conditions

been performed in, and the necessity of extrapolations especially for lower pulse height deficits. The results of Fig. 2a and Fig. 2b allow for the appropriate choice of dose/anneal temperature combination depending on requirements stated by the particular application. The combination used for the fabrication of the telescopes, 3×10^{13} cm^{-2} and 900 K, meets the following data: $\varrho_s \sim 10^3$ $\Omega/\square$, pulse-height deficit due to trapping $\ll$ 1 %, anneal temperature low enough to exclude diffusion and drastic bulk minority lifetime degradation.

Detector Telescopes

Am241 α-particle spectra obtained in two ΔE sections of different thickness are shown in Fig. 3 and 4. Pulser spectra are inserted for comparison. Additional data are compiled in Tab. 1. The diode characteristics of both samples are well suited for nuclear spectroscopy. The thicknesses, 3.5 and 10 µm, calculated from the energy deposited by the α-particles are in good agreement with ion range measurements performed previously /3/. The resolutions obtained, 95 and 100 keV (FWHM) exceed those reported up to now for integrated /1/ and discrete /4/ counters of comparable thickness. One of the reasons for this behavior is that integrated ΔE diodes exhibit appreciably less thickness variations compared with discrete counters of any origin. An upper limit of the thickness variations left may be calculated by quadratic subtraction of the pulser line width (noise) and the energy

Tab. 1: Characteristics of Ion Implanted ΔE Diodes

Boron Ion Energy	3 MeV	8 MeV
Effective Thickness	3.5 µm	10 µm
Active Area	6 mm^2	10 mm^2
Capacitance	370 pF	200 pF
RC Time Constant	$\sim$1.0 µs	$\sim$0.4 µs
Breakdown Voltage	12 V	85 V
Reverse Current at 5 V	1.5 µA	1.0 µA
Am241 α-Energy Deposited	470 keV	1.5 MeV
Line Width, L	95 keV	100 keV
Pulser Line Width (Noise), N	75 keV	45 keV
Energy Straggling, E	45 keV	75 keV
$(L^2 - N^2 - E^2)^{1/2}$ (FWHM)	40 keV	50 keV
Average Stopping Power	130 keV/µm	150 keV/µm
Thickness Variations ($\pm$HWHM)	$\pm$0.15 µm	$\pm$0.16 µm

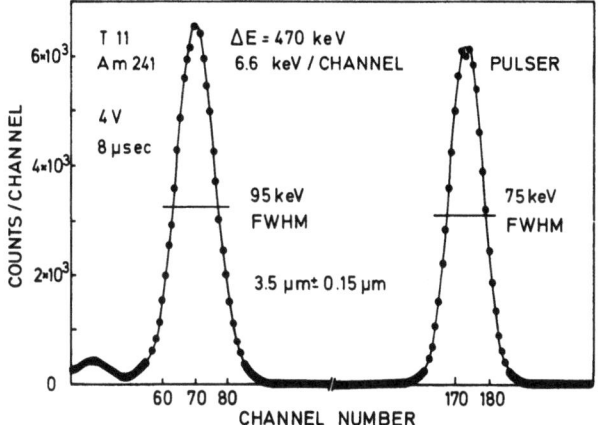

Fig. 3

α-particle spectrum obtained from integrated 3.5-μm-ΔE diode.

Pulser spectrum for comparison

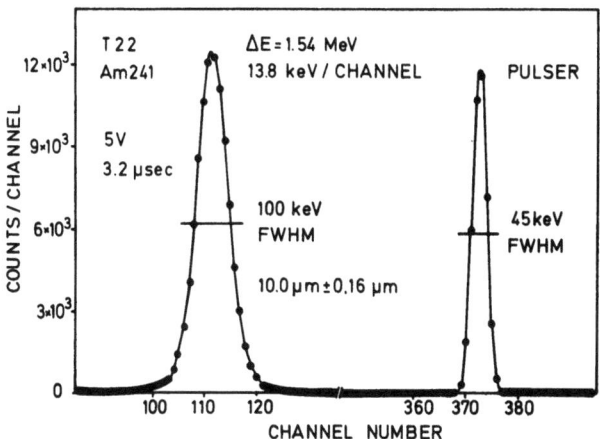

Fig. 4

α-particle spectrum obtained from integrated 10-μm-ΔE diode.

Pulser spectrum for comparison

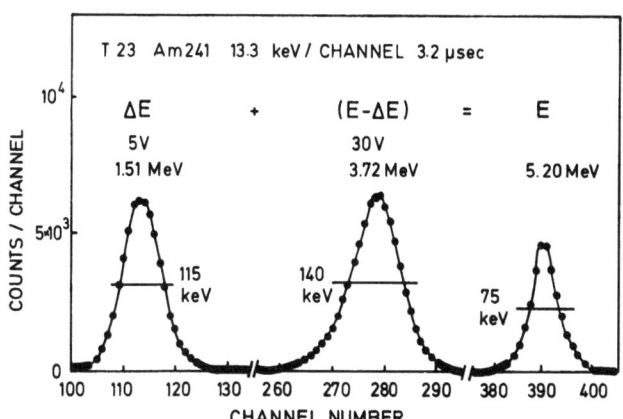

Fig. 5

α-particle spectra obtained from both sections of an integrated telescope.

Added spectrum for comparison

straggling within the ΔE diodes. This has been done in the Tab. 1, the results are approx. 50 keV corresponding to ± 0.15 μm Si (±HWHM). This compares well with ±0.3 μm Si for the best comparable discrete counters reported so far /4/, and to approx. ±0.5 μm Si of thin jet etched counters commercially available.

For the 3.5 μm diode, the high noise might be due to the high capacitance which cannot be compromised. In case of the 10 μm diode, a major part of the line width is contributed by the energy straggling and should be cancelled out by adding the signals of the ΔE and (E-ΔE) section of the telescope. This is illustrated in Fig. 5, where the added spectrum (E) exhibits a remarkably better resolution than the individual diodes. The difference between the sum of the energies deposited into the two sections, 5.23 MeV, and the initial particle energy, 5.47 MeV, corresponds to the total window thickness of the telescope, 1.5 μm Si. This is considerably less than would be expected from the thickness of the implanted p^+ layer remaining undepleted by the application of operation voltages, ~3 μm, (see /3/). In fact, many of the minority carriers created within the buried layer do not recombine, but are driven into the adjacent space-charge regions by the high built-in field existing inside the quasineutral zone as a result of the steep doping gradients of the implanted profile.

CONCLUSIONS

Implantation of high energy boron ions in silicon single crystals yields buried layers of μm-thickness with uniformities of ±0.15 μm. It is necessary to make a proper choice of implantation and anneal conditions in order to restore the crystal quality of the penetrated zone sufficiently. In this way, the requirements for the fabrication of nuclear particle detector telescopes could be fulfilled. The limitations in active area, 10 mm^2, are due to the RC-product of the resistive layer and the space-charge region capacitance.

REFERENCES

1. F.W.Martin, Nucl.Instr.Meth.72, 223 (1969)
2. A.Kostka, S.Kalbitzer, Appl.Phys.Lett.23, 704 (1973)
3. A.Kostka, S.Kalbitzer, Radiat.Eff.19, 77 (1973)
4. J.P.Ponpon, P.Siffert, F. Vazeille
    Nucl. Instr. Meth. 112, 465 (1973)

LIMITATIONS OF THE C-V TECHNIQUE FOR ION-IMPLANTED PROFILES*

C.P. Wu, E.C. Douglas and C.W. Mueller

RCA LABORATORIES

Princeton, New Jersey 08540, U.S.A.

A commonly used method for profiling the active impurity concentration in a semiconductor is the differential capacitance technique which relies on the application of the depletion approximation. The present investigation examines in detail the application and limitations of the depletion approximation when the impurity profile varies rapidly as is the case with ion-implanted profiles.

The analytical procedure consists of 1) assuming an impurity profile, 2) using Poisson's equation and the continuity equations to solve for the field and potential within the semiconductor for various applied voltages (across a Schottky diode), 3) calculating the capacitance of the system from the energy integral, and 4) using the standard C-V reduction procedures [1-3] to obtain the impurity profile for comparison with the initial profile. When solving Poisson's equation, the differential equation is converted into a set of finite difference equations with variable mesh sizes which are automatically chosen to insure the desired accuracy.

In order to test the mathematical model it is first applied to semiconductors with various values of constant impurity doping. In each case, the correct values of doping are obtained from the C-V

---

*This abstract summarizes the work presented at the Conference. A detailed paper will be published in the IEEE Transactions on Electron Devices.

reduction procedure indicating the applicability of the method to the constant doping case and illustrating the accuracy of the computations.

A high-low junction is next examined. Calculations of the space charge distribution and the majority carrier distribution are shown in relation to the impurity atom distribution. Deviations between the concentration profile as obtained from the C-V data and the initial assumed profile are shown to be due to the inadequacy of the depletion approximation in this case.

For typical ion-implanted Gaussian profiles, the edge of the depletion layer extends over a distance which is large compared with the half-width of the profile. This can introduce an "exponential tail" on the resulting calculated profile which was not present in the assumed distribution. The simulated C-V curve can also exhibit a non-monotonic behavior which causes the data to show a large concentration increase toward the surface. These effects have been reported in experimental measurements [4] and a comparison with our computations is given.

Finally, it is shown that as the original input profile becomes broader, the results yielded by the C-V technique approach true profile. It is concluded that the C-V technique should be applied with great caution to rapidly varying profiles.

REFERENCES

1. D.P. Kennedy, P.C. Murley and W. Kleinfelder, "On the Measurement of Impurity Atom Distributions in Silicon by the Differential Capacitance Technique, "IBM J. Res. Develop. Vol. 12, p.399, 1968.

2. D.P. Kennedy and R.R. O'Brien, "On the Measurement of Impurity Atom Distributions by the Differential Capacitance Technique," IBM J. Res. Develop. Vol. 13, p.212, 1969.

3. R.A. Moline, "Ion-Implanted Phosphorus in Silicon: Profiles Using C-V Analysis, "J. Appl. Phys. Vol. 42, p.3553, 1971.

4. Y. Zohta, "Rapid Determination of Semiconductor Doping Profiles in MOS Structures,"Solid State Electronics, Vol. 16, p.124, 1973.

GOLD IMPLANTATION IN SILICON: MOS C-V CHARACTERIZATION

F. N. Schwettmann, J. M. Herman, III, and T. M. Mosman

Texas Instruments, Inc.

Dallas, Texas 75222

ABSTRACT

Gold-doping by ion implantation offers an attractive alternative to conventional diffusion processes that are currently used to control minority carrier lifetime. This study reports on the effect of gold implantation on the properties of silicon as determined by a C-V analysis of MOS capacitors. The capacitor dielectric was formed by oxidation of the implanted wafer at 900°C. The three effects typical of gold-doping, i.e. substrate compensation, negative interface charge and reduced minority carrier lifetime are all observed. The redistribution of the electrically active gold between the interface and the bulk was studied as a function of dose, anneal temperature and time and substrate orientation and type. In each case, the results were typical of those previously reported for gold-doped capacitors formed by conventional diffusion processes. Gold doping by ion implantation appears to be viable process.

INTRODUCTION

In the fabrication of bipolar devices and intergrated circuits requiring rapid switching times, gold is used as a recombination center to control the minority carrier lifetime. The conventional process of doping a silicon slice with gold has in many cases been difficult to control. Therefore, ion implantation, which accurately controls the amount and location of gold incorporated in the slice, offers an attractive alternative.

The effect of gold doping by conventional diffusion processes on the C-V characteristics of an MOS capacitor has been reported by a number of groups [1-5]. Three major effects were observed. These include: introduction of a significant negative charge at the oxide-silicon interface, substrate compensation and a degradation of the minority carrier lifetime. Schultz et. al. [6] have modeled the redistribution of implanted gold and compared their predictions with neutron activation data. They show that the final gold distribution after a high temperature diffusion consists of a uniform concentration in the bulk with a significantly higher concentration at both interfaces.

The purpose of this paper is to report on one aspect of a study aimed at determining the feasability of gold-doping by ion implantation. It involves MOS C-V and 4-point probe resistivity measurements to determine the effect of gold implantation on the properties of silicon over a wide range of material and anneal conditions.

EXPERIMENTAL PROCEDURE

Silicon slices of 5-10 Ω-cm resistivity, both p- and n-type of (111) and (100) orientation were used. The wafers were chemmechanically polished on either one or both sides. Implantation was carried out using a Lintott implanter with a sputtering source. The implant energy was 30 keV with a beam current of 50µA. The dose was varied from 5×10^{12} to 5×10^{14} ions/cm^2. Under these conditions an amorphous layer was always formed as evidenced by a distinct "milky" appearance. Half of each slice was masked with aluminum foil so that a direct comparison could be made with an unimplanted sample processed through the same environment.

After implant, the slices were cleaned and oxidized at 900°C in steam for 25 minutes. A 5 minute heat-up period in oxygen preceeded the steam and a 5 minute cycle in nitrogen followed the steam cycle. The resulting oxide thickness varied from 700 to 1400 Å depending on the implant conditions and slice orientation. Post-oxidation anneals were performed in dry nitrogen at temperatures of 800 to 1200°C. Aluminum dots were then patterned on either the front or back surface.

Conventional high-frequency C-V measurements were made on the MOS structures at 1kHz and 1MHz [7]. Since the purpose of this study is to demonstrate the feasibility of gold-doping by ion implantation, the low frequency analysis of the C-V curves taking into account the effect of the deep gold levels was not attempted. Instead, the simple depletion approximation of Grove et. al. [7]

was used to evaluate Q_{ss} and the carrier concentration. This is clearly an approximation and is merely intended to show trends due to implant and anneal parameters and not provide absolute values. A more detailed analysis will be pursued in a later study.

RESULTS AND DISCUSSION

The tabulated projected range and projected range straggling for 30 keV gold implanted in silicon are 175Å and 20Å, respectively, [8]. These values give an as-implanted distribution that is completely contained within 300Å of the surface. Since 400 to 600Å of silicon are consumed in forming the oxide for the capacitor, the region in which the gold is initially located is completely converted to silicon dioxide. Preliminary backscattering and ion beam induced x-ray measurements [9] indicated that there is little segregation of the gold into the oxide and that a significant fraction remains in the silicon near the interface.

Typical 1MHz C-V curves indicating the effects of the gold implant are shown in Fig. 1. The three characteristics typical of gold doping are clearly evident. First, the curve for the implanted half of the slice is shifted to more positive voltages due to the negative gold charge at the interface.

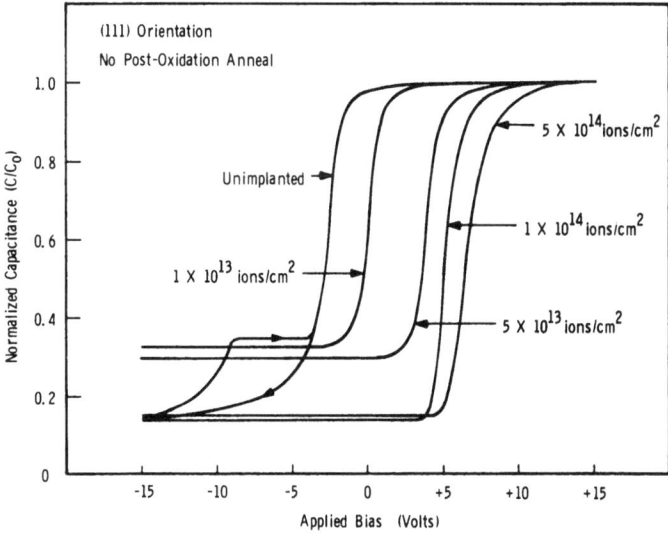

Fig. 1. Capacitance-voltage characteristics of unimplanted and 30 keV gold implanted n-type MOS capacitors with no post-oxidation anneal for doses of 1×10^{13}, 5×10^{13}, 1×10^{14} and 5×10^{14} ions/cm^2.

Second, the lifetime on the implanted side is degraded as shown by the lack of hysteresis in the inversion region due to the increase of minority carrier generation centers. Last, the minimum capacitance on the implanted side is lower than on the unimplanted half indicating a lower effective carrier concentration. Four-point probe readings on the back of the slice also demonstrate the compensation in the gold-implanted region.

As indicated in Fig. 1, an increase in the gold dose shifts the C-V curve to more positive values of applied bias. Also there is a decrease in the minimum capacitance indicating an increase in compensation as the dose increases. The curve shown for the unimplanted region is typical of that observed on each slice. The slightly higher Cmin for the highest dose is due to a somewhat higher carrier concentration on that slice prior to implantation. The calculated density of the electrically active gold centers appears to saturate at $1.2 \times 10^{12}/cm^2$ as the dose approaches 1×10^{15} ions/cm^2.

An interesting result is obtained when the MOS capacitor on the back surface of the lowest dose shown in Fig. 1 was measured. C-V curves nearly identical to those obtained on the front surface were generated. The sole difference was slightly lower gold interface charge ($\sim 10\%$) on the back of the wafer. The calculated density of the gold interface state for this sample is $5 \times 10^{11}/cm^2$, or a total of nearly $1 \times 10^{12}/cm^2$ when both surfaces are considered. This value is 10% of the implanted dose. If this state is located within the first 100Å of the surface, a surface concentration in excess of $5 \times 10^{17}/cm^3$ would be obtained. Using the resistivity as determining by 4-pt probe and spreading resistance measurements, in conjunction with the theoretical gold concentration-resistivity correlations reported by Bullis [10], the gold concentration in the bulk is approximately $2 \times 10^{14}/cm^3$. This is basically the picture shown by Schultz et. al. [6] for the total equilibrium gold concentration which is flat in the bulk and considerably higher at both surfaces. They indicate that this distribution should be achieved in several minutes at 900°C.

The thickness of the oxide grown in this study strongly suggests that a significant fraction of the gold redistribution occurs in the 5 minute cycle prior to the steam oxidation. On both the front and back surface of the slice, the half which was implanted with gold showed a thicker oxide. The increase in the thickness varied from about 1% for the lowest dose to about 7% for the highest dose. This was confirmed by both visual inspection and ellipsometry, in addition to the oxide capacitance measurements.

As the post-oxidation anneal temperature is increased, the gold interface charge decreases. At an anneal temperature of 1100°C only a small, barely measureable, Q_{Au} remains. At the same time that Q_{Au} is decreasing, the degree of compensation in the bulk is increasing. At doses above 1×10^{14} ions/cm^2 and anneal temperatures greater than 1000°C, near intrinsic silicon is formed. When the anneal time is increased from 30 minutes, the gold interface charge increases and then decreases for time over 60 minutes. It is interesting to note that the Q_{ss} on the unimplanted side follows a similar pattern. At each time, the Q_{Au} is greater than Q_{ss} by almost a factor of 2. Similar behavior has been observed by Brotherton [4] for gold-doped capacitors formed by conventional diffusion.

The results with p-type samples were similar to those obtained with the n-type wafers. A typical example of the gold concentration in the bulk is shown in Fig. 2. These data are again based on resistivity data and the theoretical calculations reported by Bullis [10]. The gold concentration increases with an increase in anneal temperature. The peak gold concentration is about a factor of 25% below C_{max}, the value calculated from the dose assuming a uniform distribution throughout the wafer. This difference could be due to inactive gold in the wafer, an inaccuracy in the theoretical calculations, or out-diffusion from the silicon.

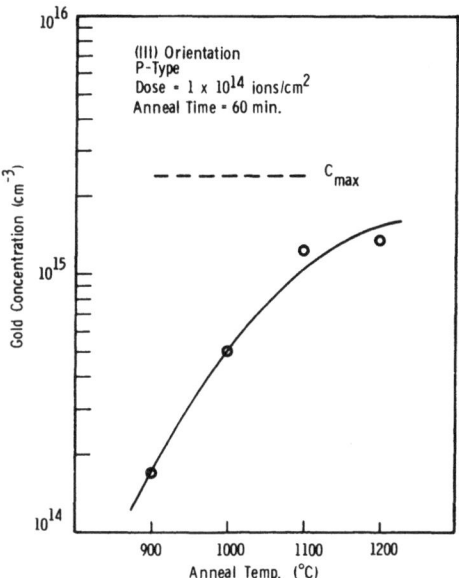

Fig. 2. Bulk gold concentration versus anneal temperature for p-type silicon implanted with gold at 30 keV and a dose of 1×10^{14} ions/cm^2. The anneal time is 60 minutes.

A number of experiments were also carried out with both n- and p-type (100) silicon. In the n-type samples a small positive shift in applied bias was observed. The amount of the Q_{Au} was nearly an order of magnitude lower than similarly implanted and annealed (111) wafers. Brotherton [4] has reported no detectable Q_{Au} in (100) silicon while Cagnina and Snow [2] made an extensive study of the variation of Q_{Au} with anneal parameters in (100) oriented silicon. The magnitude of the Q_{Au} observed in this study is in reasonable agreement with the results of Cagnina and Snow [2]. The Q_{SS} on the unimplanted standard is also almost an order of magnitude lower than that of the (111) samples. On (100) p-type samples, we detect no evidence of a gold interface charge for doses up to 1×10^{13} ions/cm^2.

In summary, gold-doping by ion implantation gives MOS capacitor characteristics very similar to those observed when gold is introduced by conventional diffusion processes. The decription [6] of the total gold distribution as nearly flat in the center with increasing concentration at each interface agrees well with the electrically active gold distribution inferred here from electrical measurements. As the gold dose is increased, the interface charge increases as does the bulk compensation. As the anneal temperature is increased, the interface charge decreases and the compensation in the bulk increases. The magnitude of the Q_{Au} varies in a similar way with anneal time as the Q_{SS} on the unimplanted side. Similar effects are observed for n- and p-type silicon with a considerably smaller gold interface charge in (100) material.

REFERENCES

[1] A. G. Nassibian, Solid-State Electron. 10, 879 (1967)
[2] S. F. Cagnina and E. H. Snow, J. Electrochem. Soc. 114 1165 (1967).
[3] D. R. Collins, J. Appl. Phys. 39, 4133 (1968).
[4] S. D. Brotherton, J. Appl. Phys. 42, 2085 (1971).
[5] P. Sixou and G. Nuzillat, Solid-State Electron. 15, 945 (1972).
[6] M. Schulz, A. Goetzberger, I. Franz and W. Langheinrich, Appl. Phys. 3, 275 (1974).
[7] A. S. Grove, B. E. Deal, E. H. Snow and C. T. Sah, Solid-State Electron. 8, 145 (1965).
[8] W. S. Johnson and J. F. Gibbons, "LSS Projected Range Statistics in Semiconductors" (Stanford University Bookstore, Stanford, California, 1970).
[9] T. J. Gray, private communication.
[10] W. M. Bullis, Solid-State electron. 9, 143 (1966).

THRESHOLD VOLTAGE SHIFT OF MOS-TRANSISTORS BY ION IMPLANTATION OF B, Al, Ga, P AND As

Hartmut Runge

Forschungslaboratorien der Siemens AG

D 8000 Munich 80, Germany

Introduction

Shifting the threshold voltage of p-channel MOS transistors by implanting B^+ ions has become a standard process for many semiconductor device manufacturers. Nevertheless problems still exist, as will be shown in this paper.

In the usual, simple model it is assumed, that the implanted ions act as an additional charge in the silicon near the SiO_2-Si interface. The threshold voltage shift ΔU_T can be calculated from:

$$\Delta U_T = (q/C_{Ox}) \cdot N_{Si} \qquad (1)$$

N_{Si} is the number of ions implanted in the silicon, that have become electrically active after annealing, q the elementary charge and C_{Ox} the oxide capacitance. The sign of ΔU_T corresponds to the donor or acceptor character of the implanted ion. N_{Si} can be calculated from the total number of ions N_I, that are implanted into the MOS-structure by considering that only the fraction $\beta \cdot N_I$ reaches the silicon. For a given oxide thickness β depends mainly upon the implantation energy. A second factor, the annealing rate α, takes into account, that only a part of the ions in the silicon act as donors or acceptors after the anneal. α depends strongly upon the annealing temperature. For ΔU_T therefore holds:

$$U_T = \alpha \cdot \beta \cdot (q/C_{Ox}) \cdot N_I \qquad (2)$$

In most threshold voltage shift implantations the annealing process is combined with the forming of the Al-contacts of the transistor at approximately 500°C. Utilizing Hall measurements the value of α for this temperature was determined by several authors / 1 /. For B^+-implantations with doses comparable to those in threshold shift experiments values α between 0,5 and 0,75 are found.

A great number of authors have reported on experiments to shift the threshold voltage with a B^+-implantation (for references see e.g. / 2 /). The experimental dependence of U_T on N_I is found to be:

$$\Delta U_T = \beta \cdot (q/C_{Ox}) \cdot N_I \qquad (3)$$

Compared with the theoretical result of equ. (2) the activation rate α seems to equal 1. The difference to the value of α taken from Hall measurements ($\alpha \lesssim 0,75$) does not seem to be too serious and until now has always been neglected.

If we try to shift the threshold voltage by implanting heavy ions such as Al, Ga, P or As, for the activation rate α after a 500°C anneal holds in all cases: $\alpha \ll 1$. If there is an error in the simple model of the threshold voltage shift, we can hope, that it will show up clearly with the implantation of heavy dopant ions.

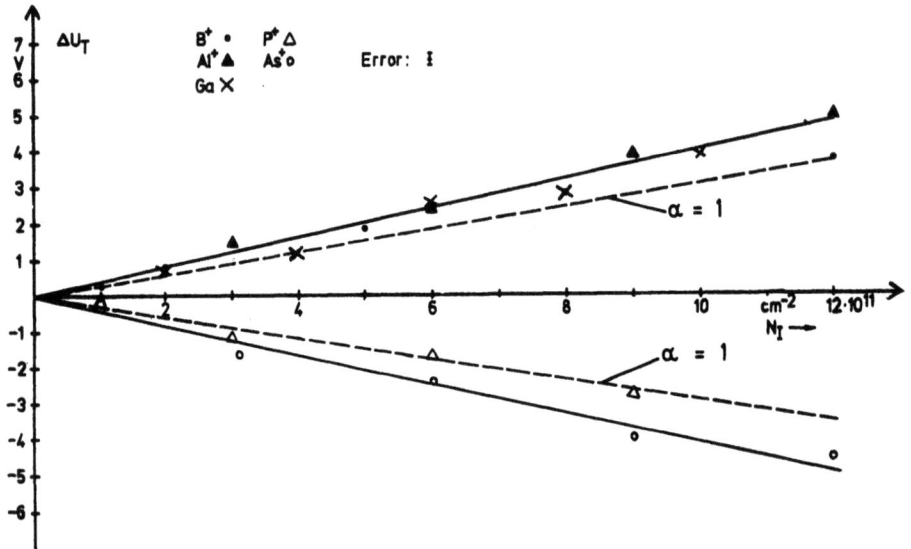

Fig.1: Threshold voltage shift by implantation of heavy ions. Implantation Energies: B^+ 35 keV, Al^+ 100 keV Ga^+ 207 keV, P^+ 110 keV, As^+ 270 keV.

Experiments

In figure 1 the experimental results of implantations into p-channel MOS-transistors are shown. The implantation energies were always chosen in such a way, that the maximum of the distribution of the ions approximately coincides with the SiO_2-Si interface. The dashed lines correspond to full electrical activation of the implanted ions, i.e. $\alpha = 1$. The B^+-implantation follows very closely this curve, as described above. More startling however is the fact, that the Al^+ and Ga^+ as well as the P^+, As^+ implantations seem to correspond and even to exceed the maximum activation curve.

Discussion

The small difference between theory and experiment that we observed in the B^+-implantations for the heavy dopant ions comes to almost one order of magnitude. If we calculate the dependence of ΔU_T on N_I with the correct α taken from Hall measurements the resulting curve for e.g. Ga^+ would almost coincide with the abscissa in figure 1. The simple model as it was hitherto used for calculating the threshold voltage shift therefore must be modified.
Not only the doping effect of the ion implantation has to be considered but also the eventual creation of deep levels and of interface states. In a p-channel transistor additionally implanted donors lead to a negative shift of the threshold voltage. If deep levels or interface states are created by the implantation, the threshold voltage is also shifted towards negative values. The total threshold voltage shift can be written as a sum of a term ΔU_T (DOP), depending on the doping effect and a second term ΔU_T (LEV), depending on the number of deep levels and interface states. The total threshold voltage shift therefore can be much larger than could be expected from the doping effect alone. The seemingly enlarged threshold voltage shift of the donor implantation in figure 1 can thus be explained.

The case of acceptor implantation is not so obvious. Additional acceptors in the silicon surface lead to a positive shift of the threshold voltage. Interface states and deep levels however would shift the threshold voltage in a negative direction. If again the sum ΔU_T (LEV) + ΔU_T (DOP) is formed, the total threshold voltage shift therefore would be smaller than expected from the doping effect alone. If many interface states or deep levels were created, the total threshold voltage shift might even be negative. This result completely contradicts the experimental facts shown in figure 1. For acceptor implantations there-

fore a more detailed analysis is necessary.
It was realized very early by different authors (/ 2 /, / 3 /, / 4 /) that implantation of acceptors in p-channel transistors leads to a p-type surface layer even at very low doses. In 1973 Sigmon and Swanson / 2 / discussed the influence of the surface layer on high frequency capacitance-voltage measurements. In figure 2 the band diagram of a MOS diode is shown under the assumption, that an homogeneous, ion implanted p-type surface layer extends to the depth w into the silicon. Between the surface layer and the n-type bulk material a pn - junction is formed with a space charge region of the width $l_p + l_n$. At negative gate voltages the surface layer is accumulated.
At more positive gate voltages depletion of the surface layer sets in and a depletion region of the width l_s is formed. When the gate voltage is further increased, the depletion region l_s extends to the space charge region $l_n + l_p$ of the pn junction. Now the entire surface layer is fully depleted of mobile carriers. In a corresponding transistor at this gate voltage no current will flow between source and drain. The threshold voltage of a transistor therefore is reached when the surface depletion region touches the space charge region of the pn-junction.

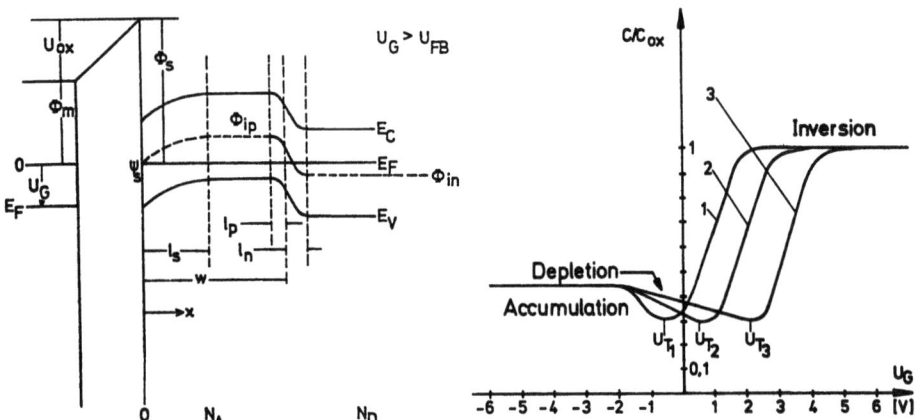

Fig.2: Band diagram   Fig.3: Expected C(U)-curves

Characteristics of a p-channel MOS diode with implanted p-type surface layer.
Curve 1: no deep levels or interface states.
Curve 2: some deep levels or interface states.
Curve 3: many deep levels or interface states.

In the C(U) curve (curve 1 of figure 3) at negative voltages the capacitance is given by the oxide capacitance and the capacitance of the pn - junction. At higher voltages the surface depletion region acts as a additional capacitance and the total capacitance drops, as the depletion region extends. A capacitance minimum is reached when the two depletion regions touch. The capacitance minimum therefore corresponds to the threshold voltage. With a further increase of the gate voltage the voltage drop over the pn - junction decreases and the total capacitance of the MOS diode increases. Finally, when the surface region is inverted, carriers from the bulk material can flow sideways under the gate region and the capacitance is determined by the oxide capacitance.

Extended Model

In view of the experimental results the model of Sigmon and Swanson has to be modified in such a manner, that the effects of surface states and deep levels can be included. For clearness sake in the following we will only consider deep levels, the discussion of interface states is analogous. In curve 2 of figure 3 we assume some deep levels and in curve 3 a great number of them. If the gate voltage is increased and depletion begins, the deep levels will change their charge state, when they cross the Fermi level. The development of the depletion region therefore in curve 2 and 3 occurs much slower than without deep levels (curves 1 of figure 3). The voltage at which the surface depletion region touches the pn - junction space charge region, i.e. the voltage, at which the capacitance minimum and the threshold voltage are reached, therefore is shifted towards positive values. In the presence of a surface layer deep levels actually shift the threshold voltage of a p-channel transistor in a positive direction. Our previous assumption, that the threshold voltage shift is a sum of a term ΔU_T (DOP) and a second term ΔU_T (LEV) therefore is also correct in case of acceptor implantations. The same discussion as for deep levels is also valid for interface states, that change their charge state when crossing the Fermi level.

The discussed model for the acceptor implantations is fully sustained by C(U) measurements. In figure 4 and 5 the results of Al- and Ga-implantations are shown. We see in curve b, that a surface layer is formed and a capacitance minimum appears. From curve c to curve e (resp. to curve f in figure 5) the number of deep levels or surface states increases, so that the capacitance minimum and hence the threshold voltage are shifted towards positive values.

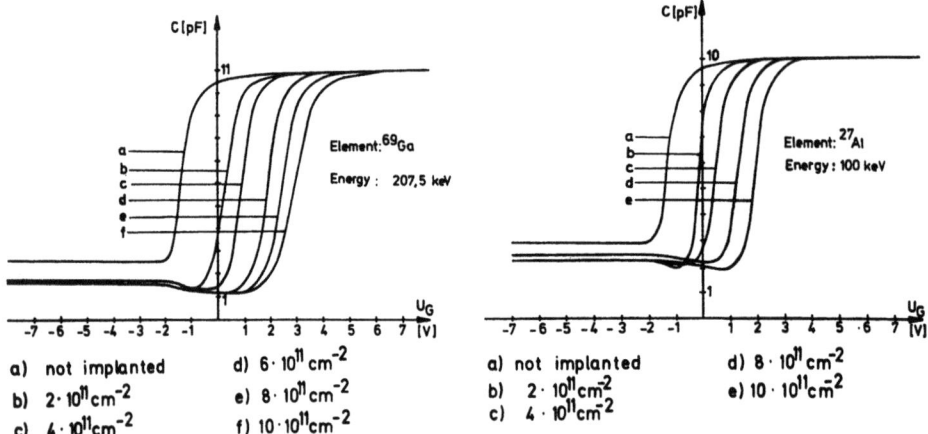

Fig.4: Ga$^+$-Implantation   Fig.5: Al$^+$-Implantation

Experimental C(U)-curves for acceptor-implanted p-channel MOS diodes.

C(U) measurements are a very useful tool in controlling the quality of MOS-processes. A steep descent from the oxide capacitance in a C(U)-curve is usually regarded as a sign of a low interface state density. If in acceptor implanted p-channel devices a p-type surface layer has been formed, the interface states however will not appear at the usual gate voltages, but at those negative gate voltages that correspond to the depletion region of fig.3.

Acknowledgement

I would like to thank Prof.Dr.H.Weiss for his steady interest and valuable discussions, Dr.E.F.Krimmel for generously endorsing my work and Miss H.Jarosch for the measurements.

Literature

/1/ J.W.Mayer, L.Erikson, J.A.Davies, Ion Impl.in Semicond., New York, (1970)
/2/ T.M.Sigmon, R.M.Swanson, Sol.St.Electr.,<u>16</u>,1217,(1973)
/3/ K.G. Aubuchon, Int.Conf.on Prop. and Use of MIS Struct., Grenoble, France (1969)
/4/ B.Höfflinger, K.D.Bigall, G.Zimmer, E.F.Krimmel, Siemens Forsch.u.Entwickl.Ber., <u>1</u>, 362, (1972)

NOISE CHARACTERISTICS OF ION-IMPLANTED MOS TRANSISTORS

K. Nakamura, O. Kudoh, M. Kamoshida and Y. Haneta

IC Division, Nippon Electric Co., Ltd.

1753 Shimonumabe, Nakahara-ku, Kawasaki, 211 Japan

ABSTRACT

The low frequency noise characteristics of p-channel MOS transistors implanted with $^{11}B^+$ or $^{31}P^+$ ions to doses of 10^{11}-10^{12}/cm^2 at acceleration energies of 20-150 keV and annealed above 1000°C were investigated. In the case of $^{11}B^+$-implantation, the noise spectra measured at lower drain current exhibited generation-recombination (G-R) noise components and increased with increasing acceleration energy or implant dose. Corresponding to this, the surface recombination velocity also increased. In the case of $^{31}P^+$ implantation, however, no G-R noise component was observed and surface recombination velocity was almost constant. At higher drain current, the G-R noise components disappeared and the noise voltages of $^{11}B^+$-implanted samples decreased, while those of $^{31}P^+$-implanted samples increased, with reference to the unimplanted sample value. This corresponds to the results of effective mobility measurements.

INTRODUCTION

Ion implantation technology is being used effectively for threshold voltage control of MOS transistors[1]. Although some electrical characteristics, such as threshold voltage, mobility or breakdown voltage of ion-implanted MOS transistors, have been investigated, there are only a few data reported on noise characteristics which can be considered to be more sensitive to the residual damage.

In this report, noise characteristics of $^{11}B^+$ or $^{31}P^+$-implanted p-channel MOS transistors will be reported in detail.

Acceleration energy, dose and some other implant condition dependences of the noise voltages will be shown and compared with the results of surface recombination velocity measurement.

EXPERIMENTAL PROCEDURE

Ring type p-channel MOS transistors (effective channel length L and width W were 11µ and 800µ, respectively) were fabricated on 5 Ω-cm n-type Si-wafers. Process steps were the same as conventionally used for fabrication of p-channel MOS transistors. Boron ions ($^{11}B^+$) or phosphorus ions ($^{31}P^+$) of 10^{11}-10^{12}/cm^2 were implanted at 20-150 keV through 1200-1500Å of gate oxides. Subsequent annealing was performed in nitrogen atmosphere at 1000-1200°C for 10-60 minutes. Some samples were annealed also in hydrogen atmosphere at 750°C for 20 minutes. Then, the wafers were metallized with pure aluminum by electron gun evaporator, annealed at 450°C in nitrogen atmosphere for 30 minutes for contact formation and encapsulated into TO-5 cases. Following the same procedures, gate-controlled diodes (gate electrode area is 1.63×10^{-3} cm^2) were fabricated on the same starting materials. The equivalent input noise voltages were measured with a Yokogawa Hewlett Packard 4470A transistor noise analyzer, at a fixed drain voltage V_D of 5V and drain currents of 1 µA - 10 mA within the frequency range of 10 to 10^6 Hz. The surface recombination velocities were obtained from gate-voltage-vs.-reverse-current characteristics measured at a drain voltage (V_D) of 0.1V.

RESULTS AND DISCUSSION

The frequency spectra of the equivalent input noise voltages of $^{11}B^+$-implanted transistors are shown in Fig. 1 with acceleration energies as parameters. In lower drain currents (Fig. 1A), the implanted samples exhibited excess noise different from normal 1/f noise. This is presumably due to the fact that generation-recombination (G-R) noise generated by ion-implantation-induced defects was superimposed on 1/f spectra[2]. Figure 2 shows the reverse current as a function of gate voltage measured with gate-controlled diodes. It can be found that the current peak, which corresponds to the surface generation current, increased with increasing acceleration energies. In unimplanted samples, surface generation current decreased at the onset of surface inversion. This current component gives the number of G-R centers at Si-SiO$_2$ interface. In implanted samples, generation current also decreased near the threshold voltage. In this case, however, the inversion occurs in the region away from the Si-SiO$_2$ interface. It can be considered that surface generation could contain contribution from ion-implanted subsurface regions. Figure 3 shows surface recombination velocity obtained from the data in Fig. 2 as a function of acceleration energy. The surface recombination velocity increased with increasing acceleration energy corresponding to the increae of G-R noise.

NOISE CHARACTERISTICS OF MOS TRANSISTORS 711

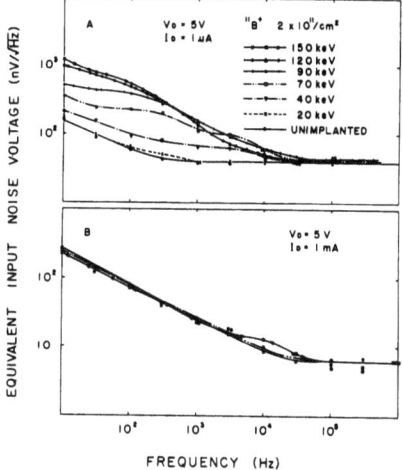

Fig. 1 Acceleration energy dependences of noise voltages of $^{11}B^+$-implanted transistors annealed at 1000°C for 10 minutes.

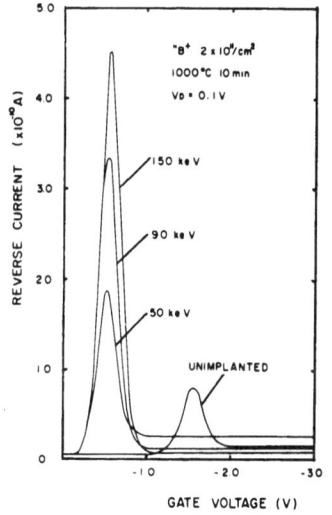

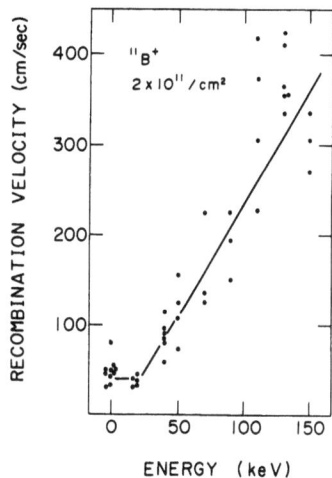

Fig. 2 Reverse current as a function of gate voltage of 11B+-implanted gate-controlled diode.

Fig. 3 Surface recombination velocity as a function of 11B+-ion acceleration energy.

The excess noise voltages of implanted samples decreased with increasing the drain voltage, as shown in Fig. 4. This could be due to the fact that trap levels are almost filled by carriers in higher drain current and generation-recombination process became ineffective. At higher drain current, the spectra showed 1/f spectra. There is only a little difference observed between

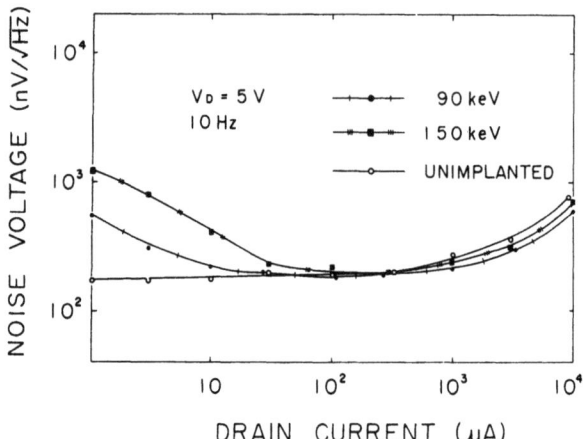

Fig. 4 Noise voltages of $^{11}B^+$-implanted transistors measured at 10 Hz as a function of drain current.

implanted and unimplanted samples, as shown in Fig. 1B.

Figure 5 shows dose dependences of noise voltages of $^{11}B^+$-implanted transistors. Again, here, noise voltages measured at lower drain current exhibited G-R noise component and increased with increasing the dose. The noise voltages of implanted samples decreased with increasing drain current, as shown in Fig. 6. Here, it can be found that higher drain currents are required for noise voltages of the samples implanted to higher doses, in order to decrease below the values for unimplanted samples.

At higher drain current, the noise voltages of implanted samples showed 1/f spectra, except the sample implanted to a dose of $10^{12}/cm^2$. The 1/f noise voltages of the implanted samples are somewhat smaller than that of unimplanted sample. This could be due to the decrease of surface scattering associated with the fact that carrier path is formed away from the $Si-SiO_2$ interface[3]. This reduction of 1/f noise voltages corresponds to increase of surface mobilities[4],[5].

Figure 7 shows frequency spectra of noise voltages of $^{31}P^+$-implanted transistors as a function of implant dose. In contrast to the case of $^{11}B^+$-implantation, no G-R noise component could be observed. The 1/f noise voltages increased with increasing the implant dose. This could be due to the increase of surface scattering associated with increase of surface electric field caused by donor implantation into n-type substrate. This also corresponds to the decrease of surface mobilities[5],[6],[7].

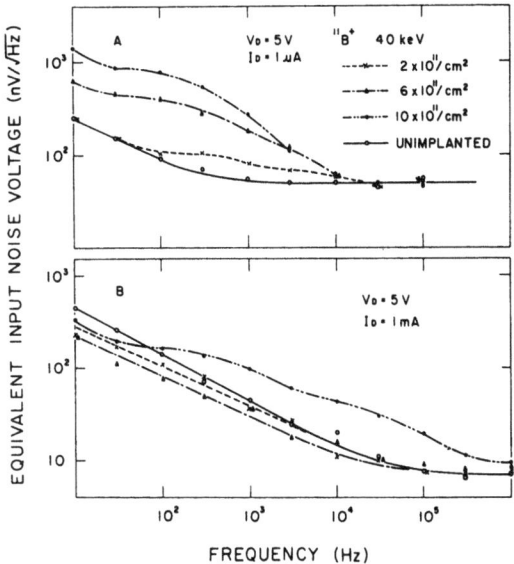

Fig. 5 Dose dependences of noise voltages of $^{11}B^+$-implanted transistors annealed at 1000°C for 10 minutes.

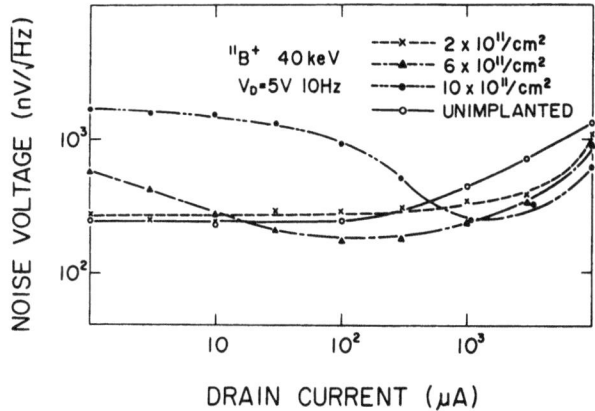

Fig. 6 Noise voltages of $^{11}B^+$-implanted transistors measured at 10 Hz as a function of drain current.

Figure 8 shows the surface recombination velocities as a function of $^{11}B^+$ or $^{31}P^+$ dose. The surface recombination velocities for $^{11}B^+$-implanted samples increased with increasing dose, while those for $^{31}P^+$-implanted samples were almost constant. One of the reasons of the difference between $^{11}B^+$ and $^{31}P^+$ implantation in measurements of noise voltages and surface recombination velocities might be the difference in mechanism of noise generation or surface generation due to the difference in surface band bending structures.

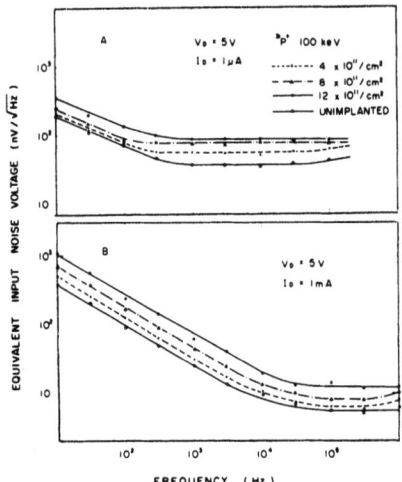

Fig. 7 Dose dependences of noise voltages of $^{31}P^+$-implanted transistors annealed at 1000°C for 10 minutes.

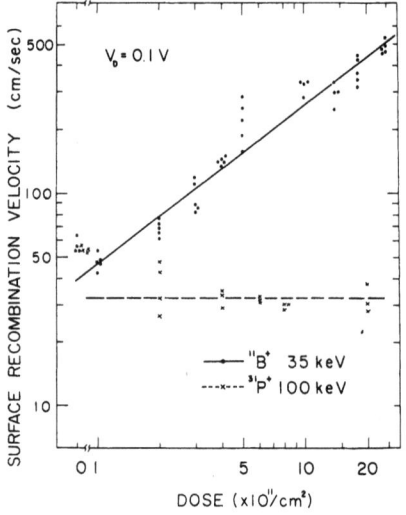

Fig. 8 Surface recombination velocity as a function of $^{11}B^+$ or $^{31}P^+$ implant dose.

Figure 9 shows the surface recombination velocity as a function of dose rate within the range of 0.1–10 nA/cm² at 2×10^{11}/cm². No dose rate dependences were observed in samples annealed at 480°C and 1000°C.

In the experiments described above, annealing temperature and time were fixed at 1000°C and 10 minutes. Annealing time depend-

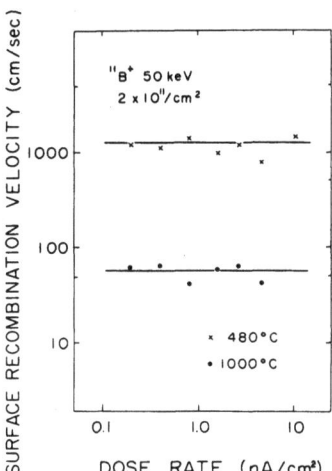

Fig. 9 Surface recombination velocity as a function of $^{11}B^+$-implant dose rate with annealing temperatures as parameters.

ences of noise voltages were investigated within the range of 10 to 60 minutes at 1000°C. It was found that longer annealing time does not give lower noise voltages than those obtained by 10 minutes annealing.

Effects of higher temperature annealing, up to 1200°C, and annealing in hydrogen were investigated. It was found that higher temperature annealings are not necessarily effective to reduce noise voltages. Annealing in hydrogen seems to reduce noise voltages to a certain extent. However, G-R noise did not disappear altogether.

ACKNOWLEDGEMENTS

The authors wish to thank Drs. S. Tsuneki, M. Yamashita and T. Okada for their continuous encouragement.

REFERENCES

1) for example M.R. MacPherson, Appl. Phys. Lett. 18, 502 (1971).
2) K. Nakamura and M. Kamoshida, J. Appl. Phys. 45, No. 9 (1974).
3) T.W. Sigmon and R. Swanson, Solid-State Electron, 16, 1217 (1973).
4) K. Nakamura and M. Kamoshida, J. Appl. Phys. 45, 334 (1974).
5) O. Kudoh, K. Nakamura and M. Kamoshida, J. Appl. Phys. 45, No. 10 (1974).
6) M. Kamoshida, Appl. Phys. Lett. 22, 404 (1973).
7) M. Kamoshida, Solid-State Electron, 17, 621 (1974).

IMPLANTATION PROFILE AND BURIED-CHANNEL DEPTH IN ION IMPLANTED MIS STRUCTURES

B. Höfflinger and L. Gabler

University Dortmund

46 Dortmund-Hombruch, Germany

1. INTRODUCTION

For many applications, the channel regions of MOS transistors are implanted with impurities causing counter-doping of the semiconductor. The result is the formation of a p-n junction and, under most operating conditions, a conducting channel buried at a certain depth in the semiconductor [1]. This phenomenon is reflected in the transistor characteristics by a threshold, a transconductance and a body effect differing from normal non-implanted transistors. It is the aim of this paper to describe the effects of ion implantation on the surface band bending (this alters the threshold and the body effect) and on the channel depth. They can be deduced from measurements of the body effect of implanted transistors, and they can be used to draw conclusions on the implanted impurity profile.

2. THEORY OF CHANNEL DEPTH AND IMPLANTATION PROFILE

In implanted MIS structures, at the onset of strong inversion the potential $\Phi(x)$ has a maximum

$$\Phi(x_p) = \Phi_{max} = 2\Phi_F + V_{SB} \qquad (1)$$

Here Φ_F is the bulk Fermi potential, V_{SB} the applied substrate bias. It is assumed that deep inside the semiconductor $\Phi=0$. Condition (1) was found to be a good analogon of the condition $\Phi(0) = 2\Phi_F + V_{SB}$ in normal non-implanted structures [2].

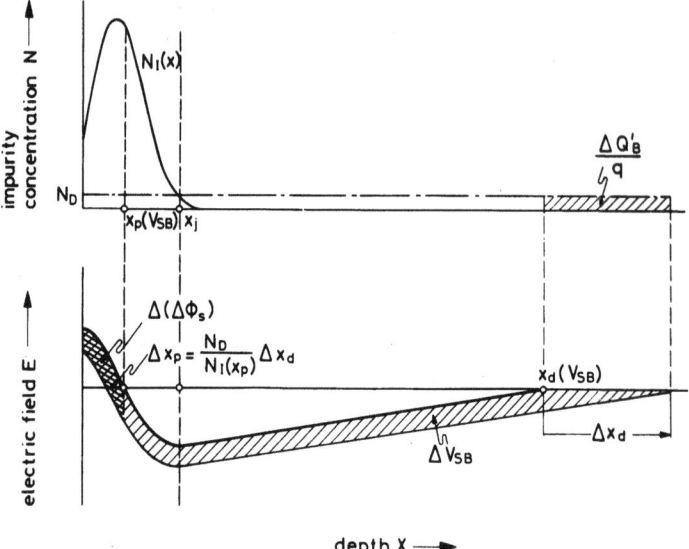

Fig. 1

Fig. 1 shows the impurity distribution and the associated electric field E for a given substrate bias V_{SB}. The total depletion depth is $x_d(V_{SB})$. At $x = x_p$, the field is zero. This is the location of the potential maximum Φ_{max} and the center of strong inversion, or in other words of the conducting channel. Therefore, x_p is defined as the channel depth. At the onset of strong inversion, the area under the field curve between x_p and x_d is $2\Phi_F + V_{SB}$. The area under E between $x=0$ and $x=x_p$ is the characteristic "surface band bending" $\Delta\Phi_S$. A change ΔV_{SB} in substrate bias has the following significant consequences [3]:

$$\Delta V_{SB} \approx (x_d - x_p) \Delta E \qquad (2)$$

$$\Delta(\Delta\Phi_S) \approx x_p \Delta E = \frac{x_p}{x_d - x_p} \Delta V_{SB} \qquad (3)$$

$$\frac{\Delta(\Delta\Phi_S)}{\Delta V_{SB}} = \frac{x_p}{x_d - x_p} \qquad (4)$$

ION IMPLANTED MIS STRUCTURES

Furthermore, the changes in channel depth Δx_d and depletion depth Δx_p are related through

$$\frac{\Delta x_d}{\Delta x_p} = -\frac{N_I(x_p)}{N_D(x_d)} \qquad (5)$$

The change ΔV_{SB} causes a change $\Delta(\Delta V_T)$ in the threshold voltage shift due to the uncovered bulk charge $\Delta Q_B'$ and the change $\Delta(\Delta \Phi_S)$ in surface band bending:

$$\Delta(\Delta V_T) = \frac{\Delta Q_B'}{C_o'} + \Delta(\Delta \Phi_S)$$

$$= \frac{t_{ox}}{\varepsilon_{ox}} q N_D \Delta x_d + \Delta(\Delta \Phi_S) \qquad (6)$$

Δx_d can be expressed in terms of ΔV_{SB}, using (2) and

$$\varepsilon_{Si} \Delta E = q N_D \Delta x_d$$

so that

$$\Delta(\Delta V_T) = \frac{t_{ox} \varepsilon_{Si}}{\varepsilon_{ox}} \frac{\Delta V_{SB}}{x_d - x_p} + \Delta(\Delta \Phi_S) \qquad (7)$$

If one increases V_{SB} in Fig. 1, a region is recognized, where $x_p = 0$, $\Delta \Phi_S = 0$ and $\Delta(\Delta \Phi_S) = 0$. In this region of large substrate bias, the differential threshold shift $dV_{SB}/d(\Delta V_T)$ is proportional to the depletion depth x_d according to eq. (7). Because of eq. (6), $d(\Delta V_T)/dx_d$ is proportional to the bulk doping density N_D. For decreasing V_{SB}, the channel or strong-inversion center x_p increases from zero and scans the implanted profile $N_I(x)$ according to Fig. 1 and eq. (5). Therefore, the measurement of the differential threshold shift due to substrate bias (in short "differential body effect") should reveal the surface band bending as well as the channel depth, and it should give supplemental information on the implanted profile very near to the surface (the range of x_p being from 0....2000 Å).

3. EVALUATION OF EXPERIMENTAL DATA

The body effect of ion-implanted p-channel MOS transistors due to an applied substrate bias was measured by plotting drain current I_D vs. gate voltage V_{GS} for small drain-source voltages of 50 mV. The $I_D - V_{GS}$ characteristics were extrapolated to zero drain current (or constant sub-threshold current in heavily implanted transistors [2]) and the resulting voltage was taken as the threshold voltage V_T. The threshold voltage shift due to V_{SB} is

$$\Delta V_T = V_T(0) - V_T(V_{SB})$$

The body effect characteristic V_{SB} vs. ΔV_T was then differentiated to obtain $dV_{SB}/d(\Delta V_T)$. This is shown in Fig. 2.

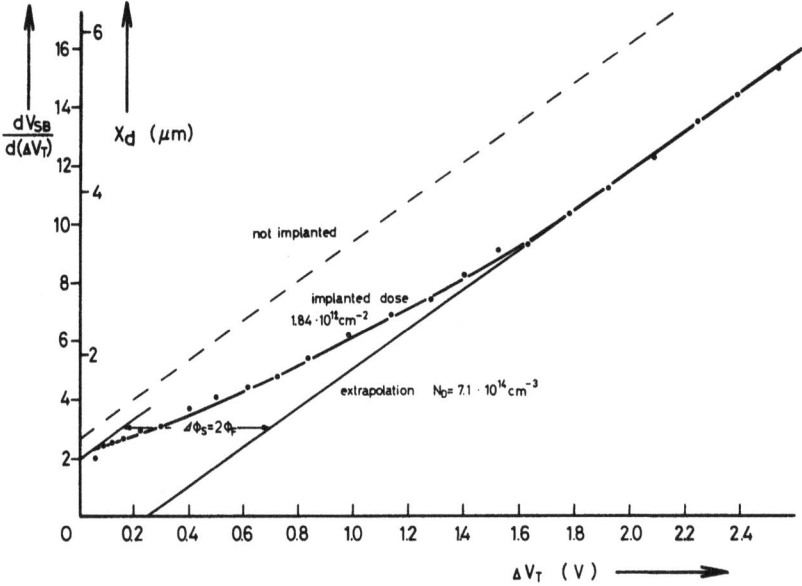

Fig. 2 Differential body effect

The implantation energy was 35 keV and the dose was $1,8 \cdot 10^{12} cm^{-2}$. For large ΔV_T or V_{SB} the resulting curve becomes a straight line. In this region, $\Delta(\Delta \Phi_S)$ approaches zero so that according to eq.(7) the ordinate of Fig. 2 is a direct measure of $x_d - x_p \approx x_d$, because x_d exceeds x_p by an order of magnitude or more. Also, the slope of the curve in this region is a direct measure of the bulk impurity concentration, the result being $N_D = 7,1 \cdot 10^{14} cm^{-3}$. Deviations of the curve in Fig. 2 from the linear extrapolation for smaller ΔV_T are the result of the surface band-bending term $\Delta \Phi_S$. Decreasing x_d by Δx_d, the threshold voltage variation is enhanced according to eq. (6) by a change $\Delta(\Delta \Phi_S)$ in surface band-bending. The resulting $\Delta \Phi_S$ is plotted in Fig. 3.

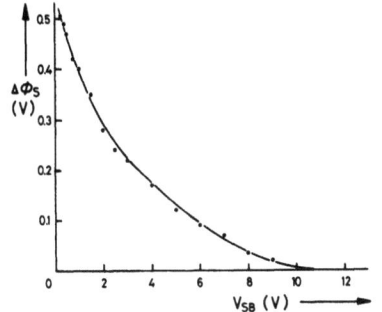

Fig. 3: Surface band bending $\Delta\Phi_S$ vs. substrate bias V_{SB}

The maximum $\Delta\Phi_S$ is 0,54 V, which is equal to $2\Phi_F$. For $\Delta\Phi_S > 2\Phi_F$, special considerations would be necessary [2]. The next processing step is to obtain $d(\Delta\Phi_S)/dV_{SB}$, which, in accordance with eq. (4), is the ratio of channel depth to depletion depth. This result is plotted in Fig. 4.

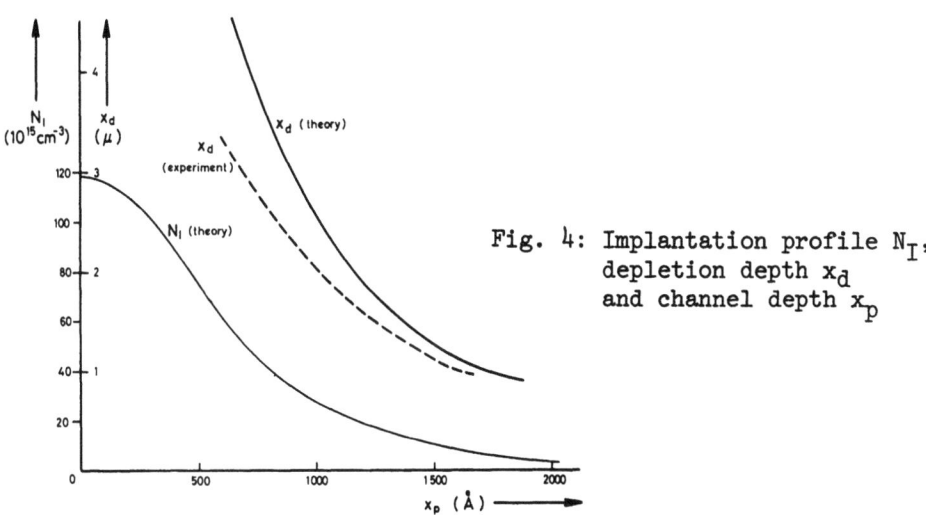

Fig. 4: Implantation profile N_I, depletion depth x_d and channel depth x_p

For minimum depletion depth of 1 μm at $V_{SB} = 0$, the channel depth has a maximum value of 1600 Å. The channel depth obtained from the experimental data is to be compared with a theoretical depth resulting from an implantation profile derived from data by Seidel [4]. The deviations point out that, following eq (5), a profile N_I with smaller values in the depth range 500 - 1600 Å would produce a better match of the channel depth obtained from the experimental data.

4. CONCLUSION

A set of significant relations between differential body effect, depletion depth, channel depth and implantation profile has been described and it has been shown, how these parameters can be obtained from measurements. This provides experimental evidence of the theory of ion-implanted buried-channel MIS structures and it gives a clue to the implantation profile. The information can be used in particular to taylor the implantation for a desired channel depth, for minimum variation of the channel depth under different operating conditions and for a reduced body effect of these structures.

5. ACKNOWLEDGMENT

The authors thank the Siemens AG, Munich, Germany, for the supply of ion-implanted MOS test samples.

6. REFERENCES

[1] B. Höfflinger, K.-D. Bigall, G. Zimmer, E.F. Krimmel : Ion - Implanted Low - Voltage MOS Circuits, Siemens Forsch. u. Entw. Ber. 1, 362 - 368, 1972

[2] W. Schemmert, L. Gabler and B. Höfflinger : Conductance of Ion-Implanted Buried - Channel MOS Transistors, to be published

[3] B. Höfflinger : Electronics of Ion - Implanted MIS Structures, Proc. Conf. Interfaces in Compound Semiconductors, Ft. Collins, Colorado, USA, February 1974

[4] T. Seidel : Distribution of Boron Implanted Silicon, Proc. II Int. Conf. Ion Implantation, 47 - 57, Berlin, Heidelberg, New York, Springer 1971

ELECTRICAL BEHAVIOR OF BORON-IMPLANTED MOS TRANSISTOR

S.KAWAZU and N.KOTANI

Central Reseach Lab., Mitsubishi Electric Corp.
Mizuhara, Itami 664, Japan

Y.WATAKABE

Kitaitami Works, Mitsubishi Electric Corp
Mizuhara, Itami 664, Japan

ABSTRACT

The modified impurity distribution equation which was obtained semi-empirically is proposed for B ion implanted SiO_2-Si structure. The experimental results of threshold voltage shift vs back gate bias characteristics were in good agreement with the theoretical values which were obtained with the modified impurity distribution equation.

The diffusion profiles of B which was implanted into SiO_2-Si structure are investigated by C-V measurments of MOS diodes and the measurments of threshold voltage vs back gate bias characteristics.

The samples having 1000-1100Å thick SiO_2 layer were implanted with B ion of 50 keV to dose of $1-32 \times 10^{11} cm^{-2}$ at room temperature. After B was implanted, the samples were annealed at $1085°C$ or $1100°C$ in nitrogen atomosphere.

INTRODUCTION

The threshold voltage control of MOS transistors by implanted B and P ions has become standard process for MOS-IC. However, there are still many problems to be solved. One of them is the redistribution of implanted ions due to the heat treatment after implantation. Threshold voltage is changed by the redistribution of implanted ions.

In this paper, the redistribution of implanted ions in SiO_2-Si structure after heat treatment was investigated by measuring MOS diode with C-V method and semi-empirical distribution equation. Using this equation, the threshold voltage shift of ion implanted MOS transistors was calculated.

EXPERIMENT

The wafers used in this experiment were $\langle 100 \rangle$ oriented, 10 ohm-cm, p-type Si crystals. After growing the gate oxide of 1000-1100Å thick, B ions were implanted at room temperature with energy of 30-150 keV at the dose of $1\text{-}32 \times 10^{11} cm^{-2}$. MOS transistors and MOS diode were fabricated using poly crystalline Si gate MOS process. The highest temperature used in poly crystalline Si gate process was 1085°C or 1100°C.

RESULTS AND DISCUSSION

Impurity Profiles

Figure 1 and 2 show the experimental results of the impurity distribution profiles of B ion implanted MOS diode. The theoretical curves was given by the account of equation (1).

$$N(x) = \frac{N_i}{\sqrt{\pi}\sqrt{4Dt + 2\Delta R_P^2}} \left[\exp\left\{-\frac{(x-R_P')^2}{4Dt + 2\Delta R_P^2}\right\} + \alpha_{ref} \cdot erf\left(\frac{R_P'}{\sqrt{2}\Delta R_P}\right) \right.$$

$$\left. \cdot \exp\left\{-\frac{(x+R_P')^2}{4Dt + 2\Delta R_P^2}\right\} \right] + N_a . \qquad \text{----------(1)}$$

Where α_{ref} is the reflectivity of B at the SiO_2-Si interface, N_i is ion dose, D is diffusion coefficient, x is distance from SiO_2-Si interface, and Na is impurity concentration of Si substrate, R_P' and ΔR_P are given as,[1]

$$R_P' = \frac{\Delta R_{PSi}}{\Delta R_{PSiO_2}} (R_{PSiO_2} - T_{ox}) , \qquad \text{-----------(2)}$$

$$\Delta R_P = \Delta R_{PSi} , \qquad \text{----------(3)}$$

where Tox is oxide thickness.

The experimental results are in good agreement with the theory.

Figure 3 shows the results of the calculated ion distribution from the equation (1) and the exact theory.[5]

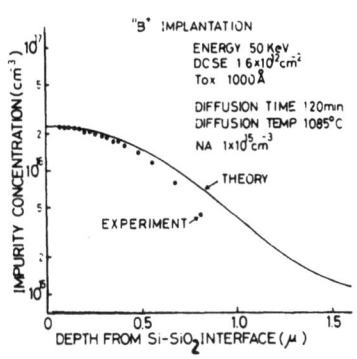
Fig.1

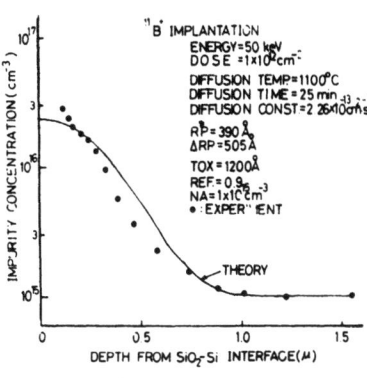
Fig.2

Impurity profiles of boron implanted MOS diode measured by C-V methode.

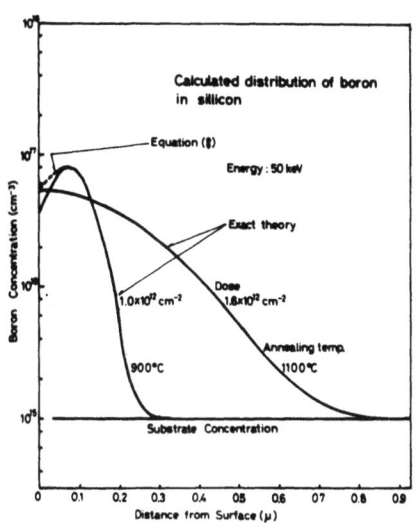

Fig.3

The calculating results of equation (1) comparison with exact theory.

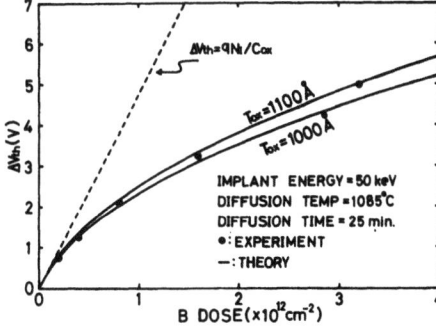

Fig.4

The threshold voltage shift by boron ion implantation.

At low temperature heat treatment, this equation are slightly different from exact theory near the surface. While at the high temperature heat treatment, we can give no attention to the difference of equation (1) and the exact theory. We made use of D after T.C.Chan et al[2], Rp_{Si} and ΔRp_{Si} after D.K. Brice[3] and Rp_{SiO_2} and ΔRp_{SiO_2} after S.Furukawa and H.Ishihara[4] for the calculation of equation (1).

Threshold Voltage Shift

Figure 4 shows the threshold voltage shift (ΔVth) vs ion dose characteristics. Vth is given by the drain current - gate voltage curve at triode region.

The threshold voltage of MOS transistor was calculated on the condition that the band bending at the Si surface is (2ϕf+Vbg).q. Where ϕf is difference between Fermi potential and mid gap of forbidden band, and Vbg is back gate bias voltage. Then, the theoretical threshold voltage shift is given by

$$\Delta Vth = \frac{Q_{BI}}{C_{ox}} - \frac{Q_B}{C_{ox}} + (2\phi f' - 2\phi b), \quad \text{-----(4)}$$

$$Q_{BI} = q \int_0^W N(x)\,dx, \quad Q_B = \sqrt{2\epsilon q N a \cdot 2\phi fo}, \quad \text{-----(5)}$$

where w is given by

$$2\phi f' + V_{bg} = \frac{q}{\epsilon} \int_0^W \int_0^x N(\alpha)\,d\alpha\,dx. \quad \text{-----(6)}$$

Figure 5 shows the threshold voltage vs back gate bias voltage characteristics. The experimental results are in good agreement with the theory.

Figure 6 shows the threshold voltage shift vs B ion implant energy characteristics. In the case of negative R$\dot{p}$ (low implant energy), the accuracy of equation (1) near the surface region is not so good, and there are slight differences between experiment and the theory. However, in the cace of positive R$\dot{p}$, this theory agrees well with experiment for wide energy range.

Effective Mobility

Figur 7 shows the effective mobility of triode region (μ_{tri}) and saturation region (μ_{sat}). The decrease of the effective mobility with ion dose is caused by impurity scattering. The ratio of μ_{tri}/μ_{sat} can be calculated from

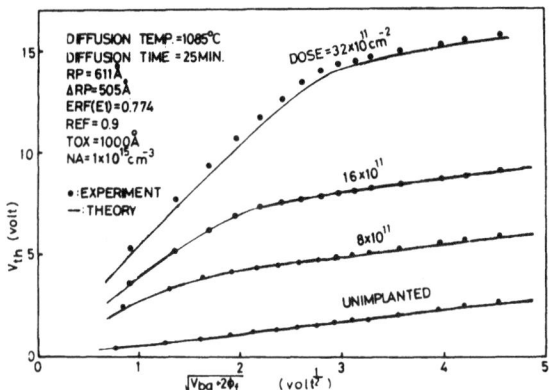

Fig.5

The characteristics of Vth vs back gate bias caused by boron ion implantation.

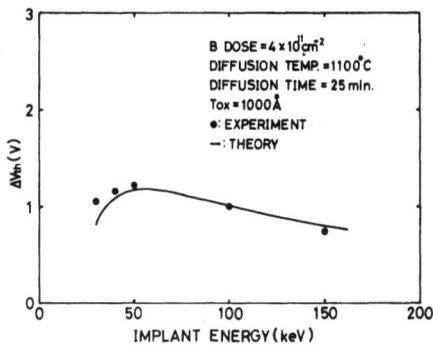

Fig.6

Threshold voltage shift vs implant energy characteristics.

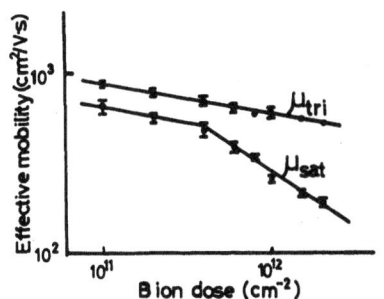

Fig.7

The effective mobility vs boron ion dose characteristics.

$$\mu_{sat} = \mu_{tri} \frac{2}{(V_g - V_{th})^2} \left[(V_g - V_{th}) \cdot V_{dj} - \frac{1}{2} V_{dj}^2 \right.$$

$$\left. - \frac{2}{3} K \left\{ (V_{dj} + 2\phi_f)^{3/2} - (2\phi_f)^{3/2} \right\} \right] , \quad \text{------- (7)}$$

where

$$V_{dj} = \frac{1}{2} \left[K^2 + 2(V_g - \Delta V) - K \left\{ K^2 + 4(V_g - \Delta V + 2\phi_f) \right\}^{1/2} \right] . \quad \text{------- (8)}$$

K is body effect constant.

Below the dose of 4×10^{11} cm^{-2}, the experimental data of the ratio μ_{tri}/μ_{sat} is 1.30±0.1 and the calculated value using the impurity concentration of substrate is 1.25.

In this dose region, the distribution of implanted ion has no influence on the mobility ratio. Above the dose of 4×10^{11} cm^{-2}, the mobility ratio is affected by implanted ion distribution.

CONCLUSION

The conclusions obtained are; 1) The diffusion profiles of B which are implanted into SiO_2-Si structure are in good agreement with equation (1) for wide annealing range. 2) The threshold voltage characteristics are calculated by using equation (1), and the caluculated values agree well with experimental data. 3) On the effective mobility ratio of the triode region to the saturation region; In the dose range where the threshold voltage shift is in propotion to the ion dose, the effective mobilty ratio is determined with impurity concentration of Si substrate. At the higher dose, the effective mobility ratio depends on the concentration of implanted B.

ACKNOWLEDGEMENT

We would like to thank Dr.H.Komiya for helpful discussions.

REFERENCES

(1) S.Furukawa and H.Ishihara: V1-5, this issue.
(2) T.C.Chan etal: Proc. IEEE., 58,(4),588, ('70)
(3) D.K.Brice: Sandia Laboratories Research Report ('71)
(4) H.Ishihara and S.Furukawa: Private communication.
(5) T.Tokuyama et al: Pro.4th Conf.Solid State Devices('73)

LIST OF AUTHORS

Adachi, S. 261
Akasaka, Y. 681
Ashburn, P. 463
Ashworth, V. 367
Augustyniak, W.M. 361

Baruch, P. 189
Baxter, D. 367
Bean, J.C. 229
Beserman, R. 457
Biersack, J.P. 211
Blanchard, B. 189
Borders, J.A. 361
Bourgoin, J.C. 385, 457
Brice, D.K. 399
Brosious, P.R. 457, 519

Cako, A. 591
Campbell, A.B. 291
Castaing, C. 189
Cembali, F. 219
Cheng, L.J. 457, 519
Chernow, F. 267
Chivers, D.J. 555
Chisholm, J. 641
Chu, W.K. 177, 633
Chung, C.H. 253
Corbett, J.W. 457, 519

Davies, J.A. 291, 493
DeBonte, W.J. 361
Demars, D. 235
Dobson, P. 641
Douglas, E.C. 695
Downing, D.L. 245

EerNisse, E.P. 437
Eisen, F.H. 3, 19
Engel, P.F. 267
Evans, B.D. 511

Favennec, P.N. 65
Feuerstein, A. 585
Fink, D. 211
Foti, G. 493
Francis, R. 641
Freeman, J.H. 555
Fröschen, W. 317
Fujimoto, M. 73
Furukawa, S. 125, 143, 423
Furuya, T. 347

Gabler, L. 717
Galloni, R. 219
Gamo, K. 35, 163, 533
Gard, G.A. 555
Gibbons, J.F. 57, 229, 413
Goh, K.H. 325
Gonda, S. 89
Grahmann, H. 585
Grant, W.A. 367
Grob, J.J. 547
Guernet, G. 235
Gyulai, J. 19

Hammersley, P.J. 619
Haneta, Y. 709
Harada, H. 73
Hasegawa, S. 525
Hasselbach, F. 585
Hemment, P.L.F. 27
Henoc, P. 439
Herman, III, J.M. 697

Hietel, B. 193
Higashiura, M. 627
Hinder, G.W. 555, 665
Hirao, T. 599
Hirose, Y. 577, 681
Höfflinger, B. 717
Hofker, W.K. 201
Horie, K. 681
Howe, L.M. 493
Howes, M.J. 463
Hozumi, H. 599
Hulek, Z. 591

Inada, T. 107
Ishihara, S. 163
Ishiwara, H. 143, 423
Itoh, N. 485
Itoh, T. 83, 627
Iwaki, M. 163
Izumi, T. 539

Kalbitzer, S. 547, 585, 689
Kamoshida, M. 709
Kang, H.S. 253
Karimoto, H. 525
Kasahara, J. 83
Kastl, R. 501
Kawamura, M. 423
Kawazu, S. 681, 723
Kimura, H. 335
Kimura, I. 163
Kobayashi, H. 183
Kobayashi, T. 47
Koeman, N.J. 201
Koike, K. 183
Kōji, T. 655
Komatsubara, K.F. 101
Kostka, A. 689
Kotani, N. 723
Kral, J. 591
Kranz, H. 169
Kraütle, H. 547, 585
Kudoh, O. 709
Kushiro, Y. 47

Lecrosnier, D.P. 439
Lee, D.H. 647
Lee, Y.H. 457, 519, 647
Lin, M.S. 35
Linker, G. 309

Look, D.C. 245
Lotti, R. 219
Ludvik, S. 155

Machi, Y. 261
Maekawa, S. 89
Magee, T.J. 229
Makita, Y. 89
Marine, J. 235
Masters, B.J. 501
Masuda, K. 35, 163, 473, 533
Matsumori, T. 133, 539
Matsumoto, T. 83
Matsumura, H. 125
Matsunami, N. 485
Mayer, J.W. 19, 177, 633
Meyer, O. 301, 309
Mitchell, J.B. 291, 493
Miyazaki, K. 133
Miyazaki, T. 41
Moline, R.A. 673
Monnier, J. 189
Moore, J.A. 493
Morabito, J.M. 115
Murakami, K. 183, 533
Morgan, D.V. 463
Morhange, J.F. 457
Mosman, T.M. 697
Mueller, C.W. 695
Müller, H. 19, 177, 633
Mylroie, S.W. 413

Nakamura, G. 577
Nakamura, K. 709
Namba, S. 35, 163, 533
Nishi, H. 347
Nomura, K. 681
Norris, C.B. 437

Ohmura, Y. 183
Ohnuki, Y. 107
Ohzone, T. 599
Okabayashi, H. 95
Oosthoek, D.P. 201

Palanki, H.R. 405
Park, Y.S. 245
Pelous, G.P. 439
Peng, J. 229
Picraux, S.T. 355

LIST OF AUTHORS

Pina, L. 591
Poate, J.M. 361
Prager, M. 585
Prissilinger, R. 547
Proctor, R.P.M. 367
Prussin, S. 449

Quillec, M. 235

Rao, E.V.K. 65
Ravetto, M. 235
Reutlinger, G.W. 673
Robinson, E. 375
Ruge, I. 169
Runge, H. 703
Rupprecht, H.S. 169
Ryssel, H. 169

Sahm, U. 317
Saito, K. 613
Sakurai, T. 347
Sasaki, A. 275
Sato, H. 627
Satya, A.V.S. 405
Scharpen, L. 155
Schewchun, J. 291
Schmid, K. 169
Schulz, F. 193
Schwettmann, F.N. 697
Sealy, B.J. 27
Setvak, M. 591
Shigetomi, S. 133
Shimada, T. 101
Shimizu, T. 525
Shin, B.K. 245
Shiraki, Y. 101
Shirasu, T. 605
Siffert. P. 547
Sigmon, T.W. 177, 633
Smith, B.J. 555, 619, 665
Stephen, J. 555, 619, 641, 665
Stephens, G.A. 375
Stephens, K.G. 27, 325

Stoneham, E.B. 57
Suzuki, T. 613

Tai, C.H. 253
Takagi, T. 275, 335, 341
Takai, M. 35
Takayanagi, S. 599
Tamura, M. 41, 571
Tanoue, H. 89, 285, 429
Thompson, D.A. 291
Tokuyama, T. 571
Tsai, J.C.C. 115
Tsuchimoto, T. 605
Tsurushima, T. 89, 285, 429
Turner, J. 641

Vook. F.L. 355

Watakabe, Y. 723
Weaver, H.E. 155
Welch, B. 19
Wellington, T.C. 367
Werner, H.W. 201
Williams, E. 641
Williams, J.S. 375
Wilson, I.H. 325
Wittmaack, K. 193
Wolf, G.K. 317
Wu, C.P. 695

Yamada, I. 275, 335, 341
Yamada, J. 423
Yanagawa, K. 341
Yashiro, T. 613
Yen, E.T. 501
Ying, R.S. 647
Yokota, K. 163
Yoon, H.W. 253
Yoshihiro, N. 571
Yudasaka, I. 605
Yukimoto, Y. 577

Zignani, F. 219

INDEX

Aluminum,
  superconducting transition temperature of ion-implanted thin films, 301-308
Aluminum-implanted silicon,
  threshold voltage shift of MOS transistors, 703-708
Aluminum-implanted zinc seledine,
  properties of, 245-252
Aluminum-implanted zinc telluride,
  luminescence properties of, 235-244
Amphoteric impurity in gallium phosphide,
  electrical properties of, 125-131
Annealing behavior in boron-implanted silicon, 599-604
Anodic oxidation of ion-implanted gallium arsenide, 19-25
Anodic stripping, 439-448
Anodization of silicon in hydrofluoric acid,
  influence of proton implantation on, 613-618
Antimony-implanted silicon,
  electrical behavior of heavily doped layers, 555-569
  oxidation of, 681-688
Argon bombardment of tantalum, 325-333
Argon implantation on thin molybdenum film on silicon, 347-354
Argon-implanted iron,
  corrosion and passivation behavior of, 367-373
Argon-implanted silicon
  oxidation of, 681-688
Arsenic implantation,
  in thin chromium film on silicon, 347-354
  in zinc-implanted $GaAs_{1-x}P_x$, 57-63
Arsenic-implanted cadmium telluride,
  electrical characteristics and radiation damage, 229-234
Arsenic-implanted silicon,
  anomalous annealing behavior of, 177-182
  atom and carrier profiles of, 163-168
  bipolar integrated circuits, use in, 641-646
  damage profiles of, 501-509
  electrical behavior in, 169-176
  electrical behavior of heavily doped layers, 555-569
  implantation profiles of, 155-162
  threshold voltage shift of MOS transistors, 703-708
  through thin oxide and nitride layers, enhanced residual disorder in, 633-639
Atom concentration profile in arsenic-implanted silicon, 163-168
Atom location in dysprosium-implanted nickel, 375-381
Atomic recoil process, 347-354
Auger electron spectroscopy (AES), 115-124, 155-162

Backscattering technique (see also Rutherford backscattering), 101-106, 125-131, 291-298, 347-354, 361-366, 473-483, 485-491, 641-646
Bipolar devices,
  use of arsenic-implanted silicon for, 641-646
Body effect of implanted transistors, measurements of, 717-722
Boron,
  damage and range distributions, 211-218
  redistribution through high-temperature proton irradiation, 189-192
Boron-difluoride-implanted silicon, ternary defects in, 449-456
Boron-fluroide-implanted silicon, ternary defects in, 449-456
Boron-implanted iron, magnetic and mechanical properties, 335-340
Boron-implanted MOS transistors,
  electrical behavior of, 723-728
  noise characteristics of, 709-715
Boron-implanted silicon,
  concentration profiles and annealing behavior of, 599-604
  deviated Gaussian profiles of, 183-188
  doping characteristics of, 605-612
  electrical behavior of heavily doped layers, 555-569
  nuclear detector telescopes, use in, 689-694
  secondary defects in, 439-448
  ternary defects in, 449-456
  threshold voltage shift of MOS transistors, 703-708
  tin diffusion, effect on secondary defects in, 577-584
Boron-implanted silicon dioxide, range distribution of, 193-200
Boron-implanted zinc telluride, luminescence properties of, 235-244
Buried channel depth in ion-implanted MIS structures, 717-722

Cadmium-implanted gallium arsenide, 35-40
Cadmium sulfide,
  nitrogen-implanted, energy level analysis of, 261-266
  polonium-implanted, deep penetration of, 267-274
Cadmium telluride, arsenic-implanted, electrical characteristics and radiation damage, 229-234
Capacitance/voltage measurements, 73-81, 155-162, 183-188, 463-469, 619-625, 695-696, 697-702
Carbon-implanted silicon,
  lattice disorder in, 519-524
  p-type doping in, 619-625
Carbon-12-implanted gallium phosphide,
  behavior of gallium and phosphorus damages in, 125-131
Carrier concentration profile in arsenic-implanted silicon, 163-168
Carriers, free, depth distribution of, 73-81
Cascade, average energy density within collision, 493-500
Cathodoluminescence, of ion-implanted zinc telluride, 235-244
Channeling effect, measurements of,
  dysprosium-implanted nickel, 375-381
  enhanced residual disorder in silicon, 633-639
  hydrogen in metals, 355-360
Chemical synthesis of compounds by implantation, 317-324
Chromium,
  deuterium-implanted lattice location of, 355-360
  thin film on silicon, arsenic bombardment on, 347-354
Chromium-implanted iron, corrosion and passivation behavior of, 367-373
Cluster ions, 341-346

INDEX

Compensating layers in gallium arsenide, 65–71
Concentration profiles, in boron-implanted silicon, 599–604
Conductivity modulation effect, in characterization of implanted layers, 463–469
Copper, gold-implanted, lattice location of, 361–366
Corrosion behavior of iron, effect of implantation on, 367–373
C-V measurements (see Capacitance/voltage)

Damage profiles
  arsenic-implanted silicon, 501–509
  extracted from forward bias of C-V curves, 463–469
Defect production in semiconductors, 385–398
Depletion region of silicon p-n junctions, introduction of radiation damage centers into, 463–469
Depth of p-n junction, measurement of, 107–114
Depth profile,
  of defects in silicon, 485–491
  of secondary defects in boron-implanted silicon, 439–448
Deuterium-implanted chromium, lattice location of, 355–360
Deuterium-implanted tungsten, lattice location of, 355–360
Devices, 597ff
Diamond, ion implantation in, 457–461
Diffusion models, used to obtain random fraction and depth profile of defects, 485–491
Dislocation networks, suppression by phosphorus-germanium double implantation, 571–576
Displacement damage profiles of neon-implanted magnesium oxide, 511–517
Displacement threshold energy, effect on number of knock-on atoms, 429–436

Doping characteristics of ion-implanted polycrystalline silicon, 605–612
Doping concentration profiles in carbon-implanted silicon, 619–625
Doping uniformity, in ion-implanted silicon wafers, 665–672
Double-drift-region (DDR) silicon IMPATT diodes, 647–654
Double-layer substrate, 423–428
Duoplasmatron ion source with post ionization (triplasmatron), 591–596
Dysprosium-implanted nickel, lattice disorder, atom location, and implant migration of, 375–381

Effective mobility ratio, 723–728
Electrical carrier profiles in phosphorus-implanted silicon, 219–225
Electrical properties,
  of amphoteric impurity in gallium phosphide, 125–131
  of arsenic-implanted cadmium telluride, 229–234
  of arsenic-implanted silicon, 169–176
  of proton-bombarded n-type gallium arsenide, 73–81
  of zinc-implanted GaAsP, 95–100
Electron paramagnetic resonance (EPR),
  lattice disorder in carbon- and nitrogen-ion-implanted silicon, 519–524
  measurements in ion-implanted diamonds, 457, 461
Electron spectroscopy, 27–34, 155–162
Electron spin resonance (ESR), 473–483, 525–531, 533–538, 539–545
Electrostatic analyzer (ESA), backscattering measurements of implantation profiles with, 585–590

Emission intensity of nitrogen-implanted GaAsP, 89-94
Encapsulation of ion-implanted gallium arsenide, 27-34
Energy conservation method (ECM) 429-436
Energy deposition distribution, 399-404
Energy level analysis of nitrogen-implanted cadmium sulfide, 261-266
Energy partition relations of solids bombarded with energetic ions, 429-436
EPR (see Electron paramagnetic resonance)
Equipment, experimental, for ion implantation, 591-596
ESR, (see Electron spin resonance)

fcc metals, formation of substitutional alloys in, 361-366
Fluorine-implanted silicon,
  distribution of fluorine in, 115-124
  ternary defects in, 449-456
Foreign-atom location, studies in nitrogen-implanted silicon carbide, 291-298
Forward bias C-V curves, damage profiles extracted from, 463-469

Gallium arsenide,
  cadmium-implanted, 35-40
  ion implantation in, 3-17
  ion-implanted, encapsulation of, 27-34
  oxygen-implanted, 65-71
  proton-bombarded n-type, 73-81
  silicon-doped n-type, 83-88
  silicon-implanted, 19-25, 41-46
  sulfur-implanted, 19-25
  tellurium-implanted, 19-25, 35-40
  zinc-implanted n-type, 83-88
$GaAs_{1-x}P_x$,
  nitrogen-implanted, 89-94
  zinc-implanted, 57-63, 95-100

Gallium damages in gallium phosphide, annealing behavior of, 125-131
Gallium, implantation in zinc-implanted $GaAs_{1-x}P_x$, 57-63
Gallium-implanted silicon, thresh-threshold voltage shift of MOS transistors, 703-708
Ga_2O_3, as a protective layer for gallium arsenide in implantation, 27-34
Gallium phosphide,
  carbon-12-implanted, 125-131
  magnesium-implanted, 107-114
  nitrogen-implanted, 101-106, 115-124, 133-139
  zinc-implanted, 107-114
Gaussian model, truncated, 405-411
Generation-recombination (G-R) centers,
  correspondence of implantation defects and, 655-661
  noise components in boron-implanted MOS transistors, 709-715
Germanium,
  chemical composition of high-dose implants in, 585-590
  ion-implanted, lattice disorder of, 493-500
GGG ($Gd_3Ga_5O_{12}$), surface-expanding phenomena in, 285-290
Gold-implanted copper, lattice location of, 361-366
Gold-implanted nickel, lattice location of, 361-366
Gold-implanted palladium, lattice location of, 361-366
Gold-implanted silicon, MOS C-V characterization, 697-702
Gold-implanted silver, lattice location of, 361-366

Hall effect measurements, 73-81, 107-114, 155-162
Helium-4 ion channeling, to determine lattice location of high-dose gold implantation, 361-366
High-dose implantation, 553ff

INDEX

IMPATT diodes, ion-implanted profiles for fabrication of DDR and SDR silicon, 647-654
Implanted ion species and concentration, influence on superconducting transition temperature, 301-308
Implantation profile in ion-implanted MIS structures, 717-722
In-depth profile detection limits, 115-124
Infrared spectroscopy, 155-162
"Inverse annealing" stage, 291-298
Ion bombardment, effects on
  evaporation speed, 317-324
  lattice disorder, 317-324
  superconducting transition temperature of thin films, 309-316
Ion-deposited energy, transport by recoiling target atoms, 399-404
Ion distribution, deformation of, 143-154
Ion implantation
  chemical aspects, 317-324
  dose and energy dependence, 177-182
  hydrogen in metals, 355-360
  in III-V compound semiconductors, 3-17
Ionization enhanced diffusion mechanisms (IED), athermal, 392-395
Ionized-cluster beam deposition method, 275-283, 341-346
Iron,
  argon-implanted, corrosion and passivation behavior of, 367-373
  boron-implanted, mechanical and magnetic properties, 335-340
  chromium-implanted, corrosion and passivation behavior of, 367-373
  iron-implanted, corrosion and passivation behavior of, 367-373

Iron (cont.)
  tantalum-implanted, corrosion and passivation behavior of, 367-373
Irradiance, as a measure of ion beam power density, 555-569
Irradiations, effect on superconductors, 309-316
Isochronal annealing, 519-524
  influence of paramagnetic defect centers, 533-538

Junction staining, 155-162

Kinchin-Pease formula, 429-436
Knock-on atoms, generation of, 429-436

Lattice disorder,
  dysprosium-implanted nickel, 375-381
  ESR to investigate, use of, 525-531
  in ion-bombarded elements, 317-324
  in ion-implanted silicon and germanium, 493-500
  in nitrogen-implanted silicon carbide, 291-298
  in silicon- and xenon-implanted silicon, 547-552
Lattice location of deuterium-implanted tungsten and chromium, 355-360
Linhard-Scharff-Schiott (LSS) theory, 115-124, 193-200, 405-411, 511-517
Liquid-phase epitaxy (LPE) method, 101-106
Lithium, implanted, damage and range distributions and diffusion profiles of, 211-218
LSS theory (see Linhard-Scharff-Schiott theory)
Luminescence measurements in ion-implanted diamonds, 457-461

Magnesium-implanted gallium phosphide, 107-114
Magnesium oxide, displacement damage in neon-implanted, 511-517

Magnet mass separator, 591-596
Magnetoresistance, negative and anisotropic in phosphorus-implanted silicon, 627-632
Metals, 299ff
Metal-silicon interface, effect of ion bombardment on, 47-354
MIS structures, implantation profile and buried-channel depth in, 717-722
Molybdenum, thin film on silicon, argon ion bombardment on, 347-354
MOS capacitors, C-V analysis of, 697-703
MOS transistors,
  boron-implanted
    electric behavior of, 723-728
    noise characteristics of, 709-715
  phosphorus-implanted, noise characteristics of, 709-715
  threshold voltage shift by ion implantation, 703-708
MOSFET,
  gate threshold voltage of, 423-428
  phosphorus-implanted n-type, galvanometric effects in, 627-632
Multilayer structures, ion implantation in, 405-411
Multiple scattering used to obtain random fraction and depth profile of defects, 485-491
Multistream diffusion (MSD), 267-274 (see also Plural-stream diffusion)

Neon-implanted magnesium oxide, displacement damage in, 511-517
Neon-implanted silicon-silicon dioxide, ESR spectra of radiation damage centers in, 539-545

Nickel,
  dysprosium-implanted, lattice disorder, atom location, and implant migration of, 375-381
  gold-implanted, lattice location of, 361-366
Nitrogen bombardment of tantalum, 325-333
Nitrogen-implanted cadmium sulfide, energy level analysis of, 261-266
Nitrogen-implanted $GaAs_{1-x}P_x$, 89-94
Nitrogen-implanted gallium phosphide,
  annealing behavior of, 133-139
  backscattering yields and photo-photoluminescence in, 101-106
  distribution of nitrogen in, 115-124
Nitrogen-implanted silicon,
  distribution of nitrogen in, 115-124
  lattice disorder in, 519-524
Nitrogen-implanted silicon carbide, lattice disorder and foreign-atom location studies of, 291-298
Nitrogen-implanted zinc selenide, converted from n- to p-type, 253-259
Noise characteristics,
  of ion-implanted base-transistor, 655-661
  of ion-implanted MOS transistors, 709-715
Nuclear reaction yield measurements, to study residual damage in nitrogen-implanted silicon carbide, 291-298

Optical absorption, measurements in ion-implanted diamonds, 457-461
Optically levered laser technique, 673-680
Optical reflectivity technique, 501-509

INDEX

Optical waveguides, 65–71
Oxidation of ion-implanted silicon, enhancement of, 681–688
Oxygen bombardment of tantalum, 325–333
Oxygen-implanted gallium arsenide, compensation in, 65–71
Oxygen-implanted silicon, distribution of oxygen in, 115–124
Oxygen-implanted silicon-silicon dioxide, ESR spectra of radiation damage centers in, 539, 545

Palladium, gold-implanted, lattice location of, 361–366
Paramagnetic defect centers, influence of isochronal annealing on, 533–538
Passivation, critical current density for, 367–373
Phosphorus damages in gallium phosphide, annealing behaviors of, 125–131
Phosphorus implantation in zinc-implanted $GaAs_{1-x}P_x$, 57–63
Phosphorus-germanium high-dose double implantation in silicon, 571–576
Phosphorus-implanted silicon,
  doping characteristics of, 605–612
  electrical behavior of heavily doped layers, 555–569
  electrical carrier profiles in, 219–225
  ESR line width of conduction electrons in, 525–531
  influence of isochronal annealing process on paramagnetic defect centers in, 533–538
  oxidation of, 681–688
  threshold voltage shift of MOS transistors, 703–708
Phosphorus-implanted MOS transistors, noise characteristics of, 709–715

Phosphorus-implanted n-channel MOSFETs and p-type silicon, galvanomagnetic effects in, 627–632
Phosphorus-implanted zinc selenide, properties of, 245–252
Photoluminescence spectra,
  of ion-implanted gallium phosphide, 107–114
  of ion-implanted zinc telluride, to measure impurity and damage distribution, 235–244
  of nitrogen-implanted gallium phosphide, 101–106
  of nitrogen-implanted GaAsP, 89–94
  of zinc-implanted n-type gallium arsenide, 83–88
Photoluminescent properties of zinc-implanted GaAsP
Plural scattering, to obtain random fraction and depth profile of defects, 485–491
Plural-stream diffusion (PSD), 267–274 (see also Multistream diffusion)
Polonium-implanted cadmium sulfide, deep penetration of, 267–274
Postprocessing, influence on electrical behavior of arsenic-implanted silicon, 169–176
Potentiostatic polarization measurements, 367–373
Profiles, 141ff
Profile construction, 417
Profile determination of ion-implanted gallium arsenide, 19–25
Proton implantation in silicon, effects of, 613–618

Quasi-Monte Carlo stimulation, 405–411

Radiation damage, 383ff
  in arsenic-implanted cadmium telluride, 229–234

Radiation damage centers,
  in ion-implanted silicon-silicon dioxide, 539-545
  introduction into depletion region of silicon p-n junctions, 463-469
Radiation-enhanced diffusion, 169-176
Radiotracer sectioning, 155-162
Raman scattering, measurements in implanted diamonds, 457-461
Random fraction caused by defects, derivation methods, 485-491
Range statistics, non-Gaussian, effect on energy deposition profiles, 413-421
Recoiling target atoms, transport of ion-deposited energy by, 399-404
Redistribution of impurities in silicon, 201-210
Residual disorder in silicon from arsenic implantation, 633-639
Rutherford backscattering (RBS), 27-34, 155-162, 375-381, 585-590 (see also backscattering)

Secondary defects in boron-implanted silicon, 439-448, 577-584
Secondary ion mass spectrometry (SIMS), 115-124, 585-590
Selenium-implanted gallium arsenide, 47-53
Semiconductors,
  II-VI compounds, 227ff
  III-V compounds, 1ff
Sheet resistivity, measurement of, 107-114, 555-569, 605-612, 641-646, 665-672
Silicon,
  aluminum-implanted, 703-708
  antimony-implanted, 555-569, 681-688
  argon-implanted, 347-354, 681-688
  arsenic-implanted, 169-176, 177-182, 347-354, 501-509, 555-569, 633-639, 641-646, 703-708
  boron-implanted, 183-188, 89-192, 193-200, 439-448, 449-456, 555-569, 577-584, 599-604, 605-612, 689-694, 703-708
  BF- and BF_2-implanted, 449-456
  carbon-implanted, 519-524, 619-625
  chemical composition of high-dose implants in, 585-590
  defects in ion-implanted layers on, 437-438
  depth profile defects in, 485-491
  fluorine-implanted, 115-124, 449-456
  gallium-implanted, 703-708
  gold-implanted, 697-702
  ion-implanted, lattice disorder in, 493-500
  ion-implanted wafers, doping uniformity in, 665-672
  nitrogen-implanted, 115-124, 519-524
  oxygen-implanted, 115-124
  p-n junctions, introduction of radiation damage into depletion region, 463-469
  p- and n-type implanted dopant profiles to fabricate IMPATT diodes, 647-654
  phosphorus-implanted, 219-225, 525-531, 533-538, 555-569, 605-612, 627-632, 681-688, 703-708
  phosphorus-germanium double implanted, 571-576
  proton bombardment of, 613-618
  redistribution of impurities in, 201-210
  silicon-implanted, 547-552
  tin-implanted, 681-688
  xenon-implanted, 547-552
Silicon carbide, nitrogen-implanted, 291-298
Silicon dioxide,
  boron-implanted, 193-200

Silicon dioxide (cont.)
  protective layer for gallium arsenide in implantation, 27-34
Silicon-implanted gallium arsenide, 19-25, 41-46, 83-88
Si_3N_4,
  protective layer for gallium arsenide implantation, 27-34
  stress adjustment in, by ion implantation, 673-680
Silicon-silicon dioxide double layer substrate, 423-428
  ESR spectra of radiation damage centers, 539-545
Silicon-silicon dioxide structure, boron-implanted diffusion profiles of, 723-728
Silicon dioxide layers, ion implanted on silicon, defects in, 437-438
Silver, gold-implanted, 361-366
Single-drift-region (SDR) silicon IMPATT diodes, 647-654
Single scattering, to obtain random fraction and depth profiles of defects, 485-491
Skewing, 413-421
Spin system, inhomogeneous character, 473-483
Stress adjustment in Si_3N_4 films by ion implantation, 673-680
Substitutional alloys in fcc metals, formation of, 361-366
Sulfur-implanted gallium arsenide, 19-25
Superconducting transition temperature of thin films,
  enhancement of, 301-308
  influence of heavy ion bombardment on, 309-316
Surface expansion of solids, ion bombardment induced, 285-290
Surface layer, ion implantation through, 405-411

Tantalum, ion bombarded, 325-333
Tantulum-implanted iron, corrosion and passivation behavior of, 367-373
Telescope, solid state E-E detector, 689-694
  nuclear detector, ion-implanted buried layers applied to, 689-694
Tellurium-implanted gallium arsenide, 19-25, 35-40
Temperature rise measurements of implanted silicon, 555-569
Ternary defects in implanted silicon, 449-456
Threshold voltage shift in MOS transistor, 703-708, 723-728
Thru-oxide implantation, excess damage layer with, 501-509
Tin diffusion, control of secondary defects in boron-implanted silicon, 577-584
Tin-implanted silicon, oxidation of, 681-688
Transistor, ion-implanted base, noise characteristics of, 655-661
Transmission electron microscopy (TEM), 27-34, 155-162, 439-448
Tungsten, deuterium-implanted, 355-360

"Vaporized Metal Cluster Ion Source," 275-283

Xenon-implanted silicon, lattice disorder in, 547-552

Zinc-implanted $GaAs_{1-x}P_x$,
  electrical and photoluminescent properties of, 95-100
  implantation of gallium arsenic and phosphorus in, 57-63
Zinc-implanted gallium phosphide, 107-114

Zinc-implanted n-type gallium
    arsenide, photoluminescence
    of, 83-88
Zinc selenide,
  aluminum-implanted, 245-252
  nitrogen-implanted, 253-259
  phosphorus-implanted, 245-252
ZnS:Mn dc electroluminescence
    cells, produced by low-energy
    ion implantation, 275-283
Zinc telluride, luminescence
    properties of boron- and
    aluminum-implanted, 235-244

If you have any concerns about our products,
you can contact us on
ProductSafety@springernature.com

In case Publisher is established outside the EU,
the EU authorized representative is:
**Springer Nature Customer Service Center GmbH
Europaplatz 3, 69115 Heidelberg, Germany**

Printed by Libri Plureos GmbH
in Hamburg, Germany